ESSENTIALS OF GEOCHEMISTRY

Second Edition

John V. Walther
Southern Methodist University

JONES AND BARTLETT PUBLISHERS

Sudbury, Massachusetts

BOSTON   TORONTO   LONDON   SINGAPORE

World Headquarters
Jones and Bartlett Publishers
40 Tall Pine Drive
Sudbury, MA 01776
978-443-5000
info@jbpub.com
www.jbpub.com

Jones and Bartlett Publishers
Canada
6339 Ormindale Way
Mississauga, Ontario
L5V 1J2
Canada

Jones and Bartlett Publishers
International
Barb House, Barb Mews
London W6 7PA
United Kingdom

Jones and Bartlett's books and products are available through most bookstores and online booksellers. To contact Jones and Bartlett Publishers directly, call 800-832-0034, fax 978-443-8000, or visit our website www.jbpub.com.

Substantial discounts on bulk quantities of Jones and Bartlett's publications are available to corporations, professional associations, and other qualified organizations. For details and specific discount information, contact the special sales department at Jones and Bartlett via the above contact information or send an email to specialsales@jbpub.com.

PRODUCTION CREDITS
Chief Executive Officer: Clayton Jones
Chief Operating Officer: Don W. Jones, Jr.
President, Higher Education and Professional Publishing: Robert W. Holland, Jr.
V.P., Sales and Marketing: William J. Kane
V.P., Design and Production: Anne Spencer
V.P., Manufacturing and Inventory Control: Therese Connell
Publisher, Higher Education: Cathleen Sether
Acquisitions Editor: Molly Steinbach
Managing Editor: Dean W. DeChambeau
Editorial Assistant: Caroline Perry
Production Manager: Louis C. Bruno, Jr.
Associate Production Editor: Leah Corrigan
Senior Marketing Manager: Andrea DeFronzo
Text Design: Anne Spencer
Photo Research Manager and Photographer: Kimberly Potvin
Assistant Photo Researcher: Jessica Elias
Cover Design: Kristin E. Ohlin
Composition: Achorn International
Cover Image: © Kateleigh/Dreamstime.com
Printing and Binding: Malloy, Inc.
Cover Printing: Malloy, Inc.

About the cover: A photo of Canary Springs Terraces, Yellowstone National Park, Wyoming. The springs emit pH 7.5–8.0, H_2S-rich water of temperatures to 80°C. The white color is from travertine ($CaCO_3$) being deposited by the CO_2 outgassing reaction $Ca^{2+} + 2HCO_3^- \rightarrow CaCO_3 + H_2O + CO_2 \uparrow$. The yellow color, for which the springs are named, stems from elemental sulfur produced by thermophiles (a type of Archaea prokaryote) in the waters. They use the abundant CO_2 present to produce organic matter (CH_2O) in the reaction $CO_2 + 2H_2S \rightarrow CH_2O + 2S + H_2O$. The orange color is due to carotenoids, pigments related to vitamin A, in thermophile microbial mats. Carotenoids protect these thermophiles from intense damaging sunlight.

Library of Congress Cataloging-in-Publication Data
Walther, John Victor.
    Essentials of geochemistry/John V. Walther — 2nd ed.
        p. cm.
    ISBN 978-0-7637-5922-3 (alk. paper)
    1. Geochemistry. I. Title
QE515.W35 2009
551.9—dc22   2008028564

6048

Printed in the United States of America
12 11 10 09 08   10 9 8 7 6 5 4 3 2 1

Brief Contents

Appendices

Contents

Appendices

Preface

Knowledge in the field of geochemistry is expanding rapidly. Choices are made when a textbook is revised. I have included some new topics of current research interest, such as boron isotopes, the global carbon cycle, and a geochemical perspective on the origin of life, but kept the order of the first edition of *Essentials of Geochemistry* and most of the material the same. As in the first edition, I have tried to introduce students to the important chemical approaches used to study the Earth. The study of geology and Earth sciences has become more environmentally oriented in recent years and, as a general trend, has also become more quantitative. In turn, mathematics and chemistry are an increasingly important part of an environmental science curriculum. In fact, most of the approaches to solving "environmental problems" are chemical in nature. Recognizing both of these trends, *Essentials of Geochemistry* brings an environmental and quantitative focus to the study of geochemistry.

The basis of modern quantitative geochemistry is thermodynamics. While geology/earth science students learn the basics of thermodynamics in courses taught in undergraduate chemistry and physics curricula, the applications of interest are different in geochemistry. To understand thermodynamics in modern geochemistry analysis, the student must consider magmas and minerals at high pressures and temperatures and, at the same time, take into account environmentally important processes on the Earth's surface and in its atmosphere. Consequently, thermodynamic principles are introduced and used throughout the book.

In my view, there are two basic ways to introduce geochemistry to students. The "formation of elements" approach descriptively goes through the formation of the Earth and the elements in it. The other, the "concepts of chemical equilibrium" approach, considers the chemical reactions that occur. In *Essentials of Geochemistry, Second Edition*, I use the "concepts of chemical equilibrium" approach; I set up the Earth in terms of changes in heat production and pressure driving chemical reactions and use the constructs of thermodynamics to teach the relations involved. This text is for those who want a quantitative treatment that integrates the principles of thermodynamics, solution chemistry, and kinetics into the study of Earth processes with an environmental approach. It is also for those who prefer not to spend a large amount of time at the beginning of the course on "formation of elements," that is, descriptive cosmochemistry.

The second edition includes an expanded number of problems at the end of each chapter with an answer key available to instructors from Jones and Bartlett Publishers. Electronic files of all the figures from the book (for which Jones and Bartlett Publishers has permission to reproduce digitally) are also available for download from the publisher's website, http://www.jbpub.com/science.

Essentials of Geochemistry, Second Edition, makes use of a computer program, SUPCRT92, to calculate thermodynamic properties so that once the procedure is known, student effort can focus on the implications of the relations rather than the rigors of long arithmetic calculations. The software package, SUPCRT92 (from Johnson, J. W., Oelkers, E. H. and Helgeson, H. C. 1992. SUPCRT: A software package for calculating the standard molal thermodynamic properties of minerals, gases, aqueous species, and reactions from 1 to 5000 bar and 0 to 1000°C. *Comput. Geosci.* v. 18, pp. 899–947) is available as a downloadable file from the book's Student Resources page on Jones and Bartlett Publishers' website (http://www.jbpub.com/science). An appendix in the book includes a brief explanation of the construction of SUPCRT92 and a set of instructions for its use to calculate properties of reactions that include minerals, gases, and/or ions. There is also a brief discussion of the theoretical basis for determining standard state properties of ions (e.g., Born equation) in the code. At the introductory level this is all that is needed. Interested students can read the original articles from which the code was developed.

Acknowledgments

For the second edition I would like to thank Pat Brady (Sandia National Laboratories), Mark Reed (University of Oregon), Jeremy Fein (University of Notre Dame), and my colleagues Kurt Ferguson and Bob Gregory for their chapter reviews.

John V. Walther
Southern Methodist University

About the Author

John V. Walther's graduate education culminated in a Ph.D. in geology from the University of California at Berkeley in 1978. While at Berkeley, he worked under the direction of the theoretical high-temperature/pressure solution geochemist, Prof. Harold C. Helgeson. He then spent 2 years at Yale University as a Gibbs Instructor in the Geology and Geophysics Department where he learned hydrothermal experimental techniques from Prof. Philip Orville. In 1985, he accepted an appointment in Northwestern University's Geological Sciences Department in Evanston, IL. At Northwestern, he rose to the rank of Professor, served as chair of the Department of Geological Sciences, and was the founding director of Northwestern's Environmental Science Program.

In 1994, Dr. Walther was appointed to the Matthews Chair of Geochemistry in the Huffington Department of Earth Sciences at Southern Methodist University. At S.M.U., he runs hydrothermal and wet-chemical laboratories researching fluid-mineral reactions in geological processes. His teaching responsibilities involve educating undergraduate and graduate students about the uses of geochemical principles to understand chemical interactions in the natural world around them and the impact of human beings on the environment.

1 | Introduction

Geochemistry is concerned with the chemical processes that are responsible for the distribution of *elements* in the solid earth, its oceans, and the atmosphere and how they have changed as a function of time. These changes can be natural, as in the precipitation of carbonates in the ocean, or anthropogenic, as with the introduction of CO_2 into the atmosphere from the burning of fossil fuels. Geochemistry is also concerned with an understanding of the earth's composition relative to the rest of the universe where it interfaces with the discipline of cosmochemistry.

An important way to view the chemical interactions that occur on the earth is to consider the earth as a closed system with a set of reservoirs and fluxes between the reservoirs as shown in **Figure 1-1**. This emphasizes the fact that each reservoir is controlled by interactions between other reservoirs. For instance, human influence on the environment, shown as the dashed lines in Figure 1-1, must be considered in tandem with natural processes that occur in the geochemical cycle of an element. These relationships are considered in more detail in later chapters.

■ Models in Geochemistry

This book does not cover all aspects of geochemistry but is devoted to outlining some of the approaches used to understand the chemical interactions in and on the earth. Particular emphasis is placed on those processes that involve water. It is water that allows life to exist on the earth, transports the material that turns sediments into rocks, and carries much of the material present in many ore deposits. The book assumes some familiarity with basic geologic terminology, the material in a high school chemistry class, and the mathematics of elementary calculus. In most programs, it fits into an upper-division geologic science undergraduate major's or a beginning graduate student's curriculum. Definitions of words given in italics when first used can be found at the end of the book in a glossary section.

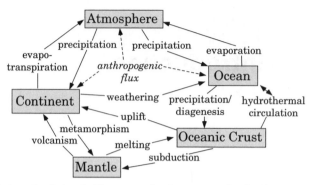

FIGURE 1-1 Global geochemical cycle. The arrows give the processes involved between the main earth reservoirs shown as boxes.

An understanding of chemical processes that are presently occurring and have occurred in the past is built on models/relations. These help to determine the factors that control the process and, therefore, the composition of the environments observed. The constituents of or variables used to describe the models may not actually be present. For instance, the amount of oxygen gas is often used to describe environments under highly reducing conditions in which it does not exist. Consider oxygen in equilibrium with Fe metal and magnetite that can be expressed by the reaction:

$$3Fe_{metal} + 2O_2 = Fe_3O_4 \tag{1.1}$$

The partial pressure of oxygen gas, O_2, in equilibrium with Fe metal and magnetite (Fe_3O_4) at 25°C and 1 bar (earth surface conditions) in the earth's atmosphere is calculated to be one part in 8×10^{88} parts of atmosphere. Because a *mole* of gas has *Avogadro's number* (6.0221×10^{23}) of gas molecules, this means that one molecule of O_2 is in 1.3×10^{65} moles of the earth's atmosphere. As a first approximation, the earth's atmosphere can be modeled as an ideal gas where $PV = nRT$. P, V, n, R, and T in this equation stand for the pressure (bar), the volume (cm^3), the number of moles of gas, the gas constant 83.1424 cm^3 bar mol^{-1} K^{-1}, and temperature (K), respectively. At 1 bar and 298 K, the amount of O_2 in equilibrium with Fe metal and magnetite is 1 molecule in a volume, V, of earth atmosphere of

$$V = \frac{nRT}{P} = \frac{1.3 \times 10^{65} \text{mol} \times 83 \text{ cm}^3\text{bar mol}^{-1} \text{ K}^{-1} \times 298\text{K}}{1 \text{ bar}} = 3.2 \times 10^{69} \text{ cm}^3 \tag{1.2}$$

The volume of the solar system is about 8.7×10^{44} cm^3; thus, there is calculated to be 1 molecule of O_2 per 3.7×10^{24} solar system volumes. Despite these problems, a model in which the partial pressure of O_2 is given as $10^{-88.9}$ bars is helpful in understanding the relationship between variables being considered.

Some of these models are simple, whereas others are more complex. They are all, however, simplifications of reality. These models/relations are then helpful only in their ability to predict and understand the observed relationships. If the constructed model/relation is not contradicted by observation, it is referred to as a law; however, laws can have their limits as well, even if they are not explicitly stated. For instance, there are limits on Newton's second law of motion. This law relates the *force* put on an object to its *acceleration*:

$$F = ma \qquad\qquad [1.3]$$

where F is a variable representing the force applied, m denotes the mass, and a stands for the acceleration of the mass. This law, however, is limited to classical mechanics and is not valid near the speed of light. Models/relations that are acceptable in a certain region of experience may not hold outside that region.

Two types of variables can be distinguished in a model: extensive and intensive. An *extensive variable* such as volume or mass depends on the amount that is present. Extensive variables are additive. For instance, with a volume of quartz of 10 cm^3 and another volume of quartz of 5 cm^3, the total volume of quartz is 15 cm^3. In contrast, *intensive variables* such as temperature or pressure are not additive. If a volume of quartz of 20 cm^3 at 25°C is cut in half, each half of the volume has the same temperature as the original volume, although its volume is now 10 cm^3. To make extensive variables such as volume or mass intensive variables, they can be normalized per mole of the substance. A mole is Avogadro's number (6.0221×10^{23}) of particles of the substance. For instance, the molar volume of quartz (SiO_2) at earth's surface pressure and temperature conditions is 22.688 cm^3 mol^{-1} no matter how much is present and is, therefore, an intensive variable.

As stated in the first law of thermodynamics, which is discussed in the following chapters, "Energy cannot be created or destroyed, only transferred from one state to another." This law merely states an observation. In every situation in which it has been considered, the law has not been violated. Consider an object of mass, m, falling in the earth's atmosphere because of gravity. It accelerates as the *potential energy* from being high in the gravitational field is converted to *kinetic energy* of $1/2\ m\ v^2$, where v is its *velocity*, as it accelerates. With increased time, the acceleration of the object decreases to zero, and the mass reaches a constant velocity in the atmosphere. At this point, the mass is still losing potential energy but is gaining no more kinetic energy. The potential energy that is lost is now converted to *heat energy* as the mass heats because of friction with the atmosphere. The conversion of one type of energy to another also occurs in chemical processes. In this case, the understanding of energy transfer is in the field of *thermodynamics*. To understand what is happening chemically, therefore, requires an understanding of thermodynamics. A strong underlying theme in this book is that geochemical processes can be understood as processes that transfer energy in thermodynamic models.

■ Processes in Science

The drive of science is to increase the understanding of the complexities of the natural world. Generally, this involves developing simple deterministic laws that can be used to understand the phenomena observed. In a deterministic or *phenomenological approach*, knowledge is created by relating observations of events and experiments to each other in a consistent manner and developing relationships or laws typically given by simple mathematical expressions. For instance, consider the relationship between mass, m, acceleration, a, and force F of $F = ma$. Small fluctuations in repeated observations/measurements are considered errors of measurement. In contrast in a *stochastic approach*, understanding of the dynamics of the system is gained by use of a random function, defined on a specific probability space. Such an approach uses a *Markov process*, a random process in which future probabilities are determined only by its most recent values. For instance, a stochastic approach can be used to understand rates of mineral surface dissolution by considering the probability of surface detachment of species under different constraints given by Markov processes. The rate of dissolution then predicts the amount of dissolution that has occurred but not the actual molecular sites of dissolution. In general, the deterministic model displays regular, smooth behavior at all scales, whereas stochastic models produce unpredicted behavior at some scales. With mineral surface dissolution, what occurs at the molecular scale is unpredictable.

Some deterministic systems, however, can produce highly erratic behavior. In most deterministic models, differential equations are used to describe changes in controlling variable with time. The erratic behavior can occur in *nonlinear systems*. Even in linear systems, that is systems in which the governing set of differential equations are composed of linear functions with constant rates of change, multiple roots can appear. In the case of nonlinear systems of dynamic governing differential equations, a quite complex behavior can be produced even though the system is completely deterministic. Such a system can possess *deterministic chaos*, or often simply called chaos. Chaos theory is not about disorder. Chaos occurs when minor changes can cause huge fluctuations in the values of a function. Although it is impossible to determine the state of a chaotic system exactly, it is generally easy to model the overall behavior of the system. These systems have *strange attractors*, a kind of regular pattern that never exactly recurs.

As a classic example, consider the Rössler attractor, in which the governing differential equations of a system in three dimensions (x,y,z) with three constants, a, b, and c, as a function of time (t) are given by the following:

$$\frac{dx(t)}{dt} = -(y(t) + z(t)) \qquad\qquad [1.4]$$

$$\frac{dy(t)}{dt} = x(t) + ay(t) \qquad\qquad [1.5]$$

and

$$\frac{dz(t)}{dt} = b + x(t)z(t) - cz(t) \qquad\qquad [1.6]$$

Only Equation 1.6 is nonlinear, containing the term $x(t)z(t)$. With a = b = 0.2 and c = 5.7, consider starting at time = 0 with $x(0) = -1$, $y(0) = 0$, and $z(0) = 0$. Given these equations, the position, that is, values of $x(t)$, $y(t)$, and $z(t)$ as a function of time, can be determined exactly. This is shown in **Figure 1-2**. The x, y,

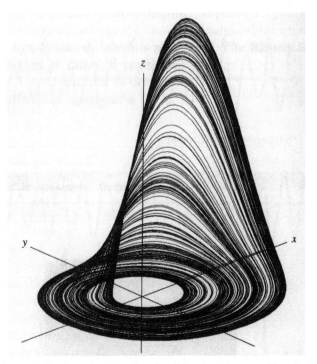

FIGURE 1-2 Projection of the Rössler attractor as a function of time in *x-y-z* space with a = b = 0.2 and c = 5.7. (From H.-O. Peitgen, *et al., Chaos and Fractals: New Frontiers of Science*, New York: Springer-Verlag, 1992.)

and z spend most of their time spiraling out from the origin in the x-y plane at $z = 0$ (stretch mode). At some distance from the origin, the equations begin to increase in z to a critical value and then reinsert themselves into the x-y plan at $z = 0$ (fold mode). Note that z does not approach a steady state nor any limit cycles with time.

In systems that display deterministic chaos, "errors" produced by round off from computer solving of the equations accumulate exponentially rather than canceling. Such systems are extremely sensitive to initial conditions so that small initial changes cause great differences long term. These describe open systems operating far from thermodynamic equilibrium, which exchange energy, matter, or entropy with its surroundings. These are discussed further in Chapter 13.

■ Standard Units of Measure

Before a discussion of important geochemical relationships can start, a certain amount of background information must be known. Because of the importance in understanding the basics in calculus and chemistry before the chemistry of the earth can be considered, a review of some of the important fundamentals is presented. Appendix A gives the abbreviations used for symbols and units as well as the physical constants required. Other conventions in notation used are also given there. Although many units are used to describe an object or mass, it is particularly helpful to know the Système International d'Unités (SI units), based on the meter, kilogram, second, kelvin, and mole, as given in **Table 1-1**.

Table 1-1   S.I. units of measurement

Unit	Name	Abbreviation	Value
Length	Meter	m	
Mass	Kilogram	kg	
Time	Second	s	
Temperature	Kelvin	K	
Volume	Liter	l	$(= 10^{-3}\ m^3)$
Force	Newton	N	$(= kg\ m\ s^{-2})$
Energy	Joule	J	$(= kg\ m^2\ s^{-2} = N\ m)$
Pressure	Pascal	Pa	$(= kg\ m^{-1}\ s^{-2} = N\ m^{-2})$
Electric charge	Coulomb	C	$(= 6.24151 \times 10^{18}\ electron\ charges)$
Power	Watt	W	$(= J\ s^{-1})$
Dynamic viscosity	Poiseuille	Pl	$(= kg\ m^{-1}\ s^{-1})$

This system of units is also referred to as MKS, indicating the units of length, mass, and time used. Sometimes CGS units (cm-gram-second) are more convenient and are used occasionally. Conventions on the multipliers of units are given at the end of Appendix A. Appendix B provides the conversion factors needed to convert between MKS and CGS units along with conversions to some other commonly used units of measure.

Often degrees Celsius, °C, rather than Kelvin, K, are used. This is because degrees Celsius are more commonly used in everyday life. The conversion is K = °C + 273.15. Because a unit of temperature is the same on both scales, the effects of a given change in temperature are identical, that is, $\Delta(K) = \Delta(°C)$. Also, a nonstandard SI unit of pressure, the bar, is used. The bar is defined as exactly 10^5 pascals. The bar is important because it is close to an atmosphere of pressure: There are 1.0133 bars in a standard atmosphere, the standard atmosphere being the average pressure at sea level because of its atmosphere. Clearly, with the onset of high- and low-pressure air masses, the real atmospheric pressure at a location can vary somewhat with time. If experiments in a laboratory at atmospheric pressure are done, in most instances, pressure differences can be ignored, and the results are considered valid for a pressure of 1 bar. If changes in pressure are important, then a small correction from the laboratory pressure to 1 bar can be made, as would also be needed to convert to the SI pressure unit pascals. The advantages of the bar are that it can easily be related to SI units, and its value under normal laboratory conditions can generally be taken as unity (unlike the pascal). This becomes important in thermodynamic models of systems. It stems from the fact that 1 bar is a good reference pressure because the logarithm of 1 is 0. As a result, bars will be the primary pressure unit used in this book and not pascals. Also, for most individuals, the effects of pressure in bars are easier to conceive of than in pascals because the bar's difference from atmospheric pressure is only about 1%.

The SI unit of energy is the joule, a newton-meter. Often, in the literature, however, values of energy are reported in calories. Values in calories are not necessarily equivalent because there is a gram-calorie, which is the amount of heat needed to raise the temperature of 1 g of water from 14.5°C to 15.5°C at 1 atmosphere of pressure. It is equivalent to 4.1855 J. An International Table calorie is different and has the value of exactly 4.1868 J. Thermochemical calories have values equal to exactly 4.184 J. When the calorie is used, it is the thermochemical calorie. Calories are employed more often than joules in this book because this is how thermodynamic variables determined by the computer program SUPCRT92 are reported. This computer program, which is available at the publisher's website (http://www.jbpub.com/catalog/9780763726423/), is used to make some of the calculations reported in the book. Again, if values in joules are required, the conversion is J = 0.23901 cal.

■ Mathematical Review

To look at the dynamics of a geochemical model, differential calculus needs to be understood, in particular the *differential*. Consider the function y, of the variable x, written as $y = y(x)$. The derivative of $y(x)$, sometimes written as $y'(x)$ and sometimes as dy/dx, is the slope of y or gradient of y at x, that is, the infinitesimal change in y with an infinitesimal change in x. This can be represented as

$$y'(x) = \lim_{\Delta x \to 0} \left(\frac{y(x + \Delta x) - y(x)}{\Delta x} \right) \qquad [1.7]$$

where $\lim_{\Delta x \to 0}$ is the value at the limit when the change in x, Δx, becomes infinitesimal and approaches zero. This "derivative" is important because it indicates how much the function, $y(x)$, is changing in response to the variable, x. Often, the variable of interest is time.

As an example, consider a car traveling down a road. If the distance it has traveled as a function of time is plotted, it could look like the plot on the left side of **Figure 1-3**. If the magnitude of the car's velocity on the speedometer as a function of time is plotted, it would look like the plot on the right side of Figure 1-3, being the derivative of distance with respect to time. Clearly, the velocity is how fast the distance is changing with time. If the velocity is zero, then the position is not changing. To determine the velocity in the absence of a speedometer at any point, the distance traveled over a small section of road, dy, that includes the point of interest (e.g., A) can be measured, and this number is divided by the small increment of time, dt, that has passed. If the distance becomes infinitesimally small, this becomes the slope of the line at the point (i.e., velocity).

Suppose at some time, t°, a car is a distance c down the road (given by point A in Figure 1-3). At this time, the relationship between distance, y, and time, t, is given by

$$y = at^2 + c \qquad [1.8]$$

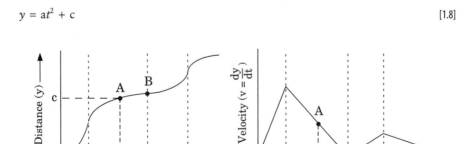

I FIGURE 1-3 Distance (left) and velocity (right) with time for a car traveling on a road.

where a and c are constants. That is, at $t = t°$, $y = c$ and the change in y with t is given by at^2. In the case of Figure 1-3, the constant a is a negative number because the car is slowing. Considering this function for the curve on the left between the two dashed lines that include A results in

$$(y + dy) = a\,(t + dt)^2 + c = at^2 + 2atdt + a\,(dt)^2 + c \qquad [1.9]$$

Making the substitution for y given in equation (1.8) gives

$$dy = 2atdt + a(dt)^2 \qquad [1.10]$$

Dividing both sides by dt yields

$$\frac{dy}{dt} = 2at + adt \qquad [1.11]$$

As dt goes to zero,

$$\frac{dy}{dt} = 2at \qquad [1.12]$$

or the velocity is a linear function of time, as shown on the right of Figure 1-3. Clearly, if the derivative of $y = f(x)$, that is, the gradient at x, is equal to zero, then the function at x reaches an inflection point as given by point B at time t^B or the function is at a maximum or minimum, as shown in **Figure 1-4**.

In most descriptions of the real world, functions depend on more than one variable, for instance, the function y might depend on the variables x, u, and z, that is, $y(x, u, z)$. A *partial differential* of the function is the value of the

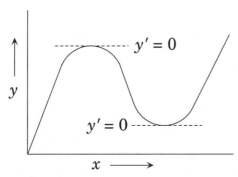

FIGURE 1-4 A single-valued function y of dependent variable x indicating that when the gradient of y, that is, y', is 0, the function is at a local maximum or minimum.

differential of the function with respect to one variable when holding all of the other variables constant. This can be expressed as

$$\left(\frac{\partial y}{\partial x}\right)_{u,z} = \lim_{\substack{\Delta x \to 0 \\ u \text{ and } z \text{ constant}}} \left(\frac{y(x + \Delta x) - y(x)}{\Delta x}\right) \qquad [1.13]$$

where ∂ is the partial derivative operator so that $\partial y/\partial x$ is the partial derivative of function y with respect to x. The u and z outside of the differential in parentheses indicate that these variables are to be held constant. Again, this indicates what the gradient is with respect to x, or how fast the function is changing at fixed values of u and z in response to changes in x. For instance, with the function

$$y = xz + u^2x^2 + u \qquad [1.14]$$

the partial differentials are

$$\left(\frac{\partial y}{\partial x}\right)_{u,z} = z + 2u^2x \qquad [1.15]$$

with u and z constant and

$$\left(\frac{\partial y}{\partial u}\right)_{x,z} = 2x^2u + 1 \qquad [1.16]$$

with x and z constant, whereas it is

$$\left(\frac{\partial y}{\partial z}\right)_{x,u} = x \qquad [1.17]$$

when x and u are constant. Equations 1.15, 1.16, and 1.17 then give the derivatives or "slopes" of the variables, that is, how fast y changes with x, u, and z, respectively, under the constraint that the other variables are held constant. Rules for differentiating a particular function can be found in an elementary calculus textbook.

As a real-life example of partial derivatives, imagine that you are standing on a hilly terrain with a compass. The elevation (think y) of the ground you are standing on is then a function of your position. Two variables, say distance north–south, N-S (think x), and distance east–west, E-W (think z), describe the changes in elevation as a function of position; therefore, the elevation (y) at any point can be described at a position in the N-S direction (x) and position in the E-W direction (z) or elevation $= f$(N-S, E-W). If the movements are limited to the N-S direction (i.e., E-W position is constant), then the partial derivative of elevation with respect to N-S movement is

$$\left(\frac{\partial(\text{elevation})}{\partial(\text{N-S})} \right)_{\text{E-W}}$$

This would be the slope of the land surface at the position considered in the N-S direction. If this were zero, the location would be along the bottom of a valley or along the top of a ridge where the slope changes sign with distance or would be at a point where the slope does not change sign but increases in one direction and decreases in the other for the N and S directions. When displaying a differential of x with respect to y, the symbol dx/dy is used, indicating that no values are held constant, whereas a partial differential is specified with the constant variables indicated outside of the parenthesis on the variable.

Another important concept is the *total differential* of y, dy. This gives the total change of y with infinitesimal changes in all of the variables that affect y. If three variables, x, u, and z dictate the change of y and they are linearly independent (each cannot be made up by a combination of the other variables), then for $y = y(x, u, z)$, the total differential of y is

$$dy = \left(\frac{\partial y}{\partial x} \right)_{u,z} dx + \left(\frac{\partial y}{\partial u} \right)_{y,z} du + \left(\frac{\partial y}{\partial z} \right)_{x,u} dz \qquad [1.18]$$

This is made up of the gradient in each variable times the change in each variable that affects the function. For the function given in Equation 1.14, the total differential is

$$dy = (z + 2u^2x)\, dx + (2x^2u + 1)\, du + x\, dz \qquad [1.19]$$

Going back to the idea of the hilly terrain, the total differential of elevation, d(elevation), would give the total change in elevation in any direction at a particular point. This could be written as

$$d(\text{elevation}) = \left(\frac{\partial(\text{elevation})}{\partial(\text{N-S})} \right)_{\text{E-W}} d(\text{N-S}) + \left(\frac{\partial(\text{elevation})}{\partial(\text{E-W})} \right)_{\text{N-S}} d(\text{E-W}) \qquad [1.20]$$

The first term on the right is the slope in the N-S direction times the distance traveled in the N-S direction, and the second term is the slope in the E-W direction times the distance traveled in the E-W direction. Clearly, by knowing this information, the change in elevation, d(elevation), could be determined in any direction.

As a chemical example of the use of total differentials, consider 1 mole of an *ideal gas* where the relation between pressure, volume, and temperature is described by the ideal gas law given by this equation:

$$P\overline{V} = RT \tag{1.21}$$

As stated earlier P, R, and T stand for pressure, the universal gas constant, and temperature in K, respectively. For a pure gas, $\overline{V}$ gives the molar volume of the gas; this is the volume of 1 mole of the gas. $\overline{V}$ is used rather than V with the number of moles of gas, n, present as in Equation 1.2 because it is an intensive variable that does not change with the amount of gas considered. $\overline{V}$ can be written as a function of P and T because R is constant as

$$\overline{V} = \overline{V}(P,T) = RT/P \tag{1.22}$$

The total differential of $\overline{V}$ is

$$d\overline{V} = \left(\frac{\partial \overline{V}}{\partial P}\right)_T dP + \left(\frac{\partial \overline{V}}{\partial T}\right)_P dT = -\frac{RT}{P^2}dP + \frac{R}{P}dT \tag{1.23}$$

If the total differential is exact, the order of differentiation does not matter so that

$$\left[\frac{\partial(\frac{\partial y}{\partial x})_z}{\partial z}\right]_x = \left[\frac{\partial(\frac{\partial y}{\partial z})_x}{\partial x}\right]_z \tag{1.24}$$

Therefore, the cross-partial derivatives in an expression for an exact total differential must be equal. For an ideal gas from Equation 1.23

$$\left[\frac{\partial(-\frac{RT}{P^2})}{\partial T}\right]_P = \left[\frac{\partial(\frac{R}{P})}{\partial P}\right]_T = -\frac{R}{P^2} \tag{1.25}$$

The derivatives of some variables are not exact differentials. For instance, consider changes in work for a mole of a substance, $\overline{W}$ with changes in pressure and temperature. Pressure-volume work (considered in Chapter 3 where thermodynamic models are introduced) is given by

$$d\overline{W} = -Pd\overline{V} \tag{1.26}$$

This expresses the fact that the work per mole done on a substance is equal to the pressure times the change in molar volume. Consider the work done on an ideal gas when changing pressure and temperature. From Equation 1.18, this can be written as

$$d\overline{W} = -P\left(-\frac{RT}{P^2}dP + \frac{R}{P}dT\right) = \left(\frac{RT}{P}\right)dP - RdT \qquad [1.27]$$

However, the cross-partial derivatives of $d\overline{W}$ are not equal as

$$\left(\frac{\partial(RT/P)}{\partial T}\right)_P \neq \left(\frac{\partial R}{\partial P}\right)_T \qquad [1.28]$$

That is, the order of differentiation matters. This implies that $d\overline{W}$ is not an exact differential and is dependent on the way changes in pressure and temperature are done in computing $\overline{W}$. Work is, therefore, not a *state variable* because its value depends on the pressure-temperature path of the calculation, as discussed in Chapter 3. With these mathematical constructs in mind, one should be able to understand the derivation of the necessary equations for the models constructed in basic geochemistry.

■ Portraying Changes in Variables

Most geochemical systems are complex and need many variables to describe them. Geochemists often want to show the relationship between a number of important variables under consideration while keeping the others constant. Obviously, on a flat sheet of paper with its two-dimensional surface, an x–y plot can be constructed to display the relationship between two variables. If three independent variables are to be shown, then a projection of the variables on paper can be constructed to give the illusion of depth so that the third variable can be viewed along a "depth axis."

In many cases, however, there are three variables to be considered, but only two are independent, for instance, the relationship between the amount of Al_2O_3, CaO, and $(FeO + MgO)$ in a mineral or suite of rocks. These relationships can be displayed by projecting the mineral composition into Al_2O_3 + CaO + $(FeO + MgO)$ space by stipulating that the amount of Al_2O_3 + CaO + $(FeO + MgO)$ = 100%. If the amount of two of the variables is known, the amount of the third can be determined because they add to 100%. That is, although there are three variables, only two are independent, and relationships can be plotted accurately on a two-dimensional piece of paper. This is done with what is called a triangular diagram for the three variables. With variables denoted as $A = Al_2O_3$, $C = CaO$, and $F = FeO + MgO$, triangular diagrams such as shown in **Figure 1-5** can be produced.

In this type of diagram, the closer a composition is to an apex, the more of that component is in the sample. The position of a composition on the diagram is determined by the *lever rule*, much like the force produced by a mass on a

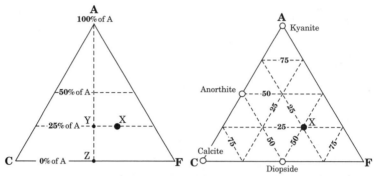

FIGURE 1-5 Triangular diagrams showing the relations between A = Al$_2$O$_3$, C = CaO, and F = (FeO + MgO) where Al$_2$O$_3$ + CaO +(FeO + MgO) = 100%. The left diagram shows for position X the lines needed to determine the % of A in the sample. Y and Z have equal amounts of C and F. The diagram on the right gives the position of some phases in the system as well as lines for 25%, 50%, and 75% of the A, C, and F components in the mixture.

lever. The position of a mass along a lever away from the fulcrum toward a particular end increases the force developed for that end. This concept can be used to determine the percentage of A in a sample that plots at point X in Figure 1-5. A line is first drawn through the point parallel to the C-F line to the edges of the triangle, as shown by a dashed line labeled 25% on the left diagram of the figure. A line is then drawn perpendicular to the line just drawn that goes from A to the C-F line at point Z, as given by the vertical dashed line. The percentage of A in sample X is equal to 100 times the ratio of the length of the line segment from Z to Y divided by the total length of the line segment from A to Z. A sample that plots anywhere along the line connecting C and F would have 0% of A. For sample X, the amount of A is 25%. The percentage of C in sample X can be found the same way. To do this, a line is drawn through the point of interest that is parallel to the A-F line. Then a line perpendicular to this line through the point that goes through C is drawn. The percentage of C is 100 times the ratio of the line segment between the line through the point and the A-F line to that of the length of the line segment from C to the A-F line. The percentage of F is then simply 100% – percentage of A – percentage of C. These relationships are shown on the right diagram of Figure 1-5. Lines denoting 25%, 50%, and 75% of each of the three components are shown. Sample X then has 50% F and 25% C, as well as 25% of A.

To plot a mineral or rock composition on a triangular A-C-F diagram, all oxide components except Al$_2$O$_3$, CaO, FeO, and MgO are ignored. Kyanite (Al$_2$SiO$_5$), therefore, plots at the pure Al$_2$O$_3$ or (A) position. Anorthite (CaAl$_2$Si$_2$O$_8$) with Al$_2$O$_3$/(Al$_2$O$_3$+ CaO) = 0.5 plots along the 50% A line. With 50% C and no F, anorthite plots at the halfway point along the A-C boundary line. Calcite (CaCO$_3$) would plot at the C position. Diopside (CaMgSi$_2$O$_6$)

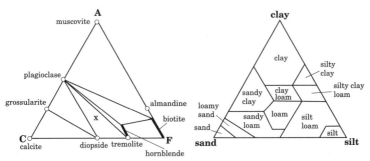

FIGURE 1-6 The left diagram plots the composition of minerals with their tie lines, giving the mineral assemblages found in amphibolite zone rocks of Orijärvi, Finland with quartz and microcline present. (Adapted from Turner, F. J., 1981.) The right diagram shows the nomenclature for soils based on the amount of clay, sand, and silt present. (From the U.S. Department of Agriculture.)

has no A but equal amounts of C and F, and thus, it plots at the midpoint of the line connecting C and F. Clearly, at any point on the diagram, A + C + F equals 100%.

Figure 1-6 displays the triangular diagrams showing the stable mineral assemblages for amphibolite facies metamorphic rocks and another that classifies soils according to grain size. On the left triangular diagram, minerals such as hornblende and biotite that exhibit a range of compositions plot as a field, as shown by the thick lines. If the composition of a rock is plotted on this diagram, the minerals that it contains are given by the minerals that bound the lines that enclose its composition on the diagram. For instance, a quartz plus microcline-bearing rock from Orijärvi, Finland in the amphibolite facies at composition x as shown on the left triangular diagram in Figure 1-6 would also contain the minerals anorthite, diopside, and tremolite. Examination of the soil classification diagram on the right side of Figure 1-6 indicates that a true loam has significant amounts of sand and silt, with somewhat lesser amounts of clay. A clay loam, as to be anticipated, plots in an area toward the clay apex from the true loam compositions with approximately equal concentrations of sand, silt, and clay.

■ Phases, Solutions, and Concentration Scales

A model can be constructed to help understand the controlling variables in a natural system. A few conventions from basic chemistry are needed to describe the compositions in the model. *Phases* are defined as distinct solids, liquids, or gases with uniform physical and chemical characteristics that can be mechanically separated from any other mass in the system. Because they can be mechanically separated, a rock composed of many crystals of quartz, plagioclase, and biotite would consist of three distinct solid phases. A solution of NaCl in water would be a single fluid phase because the NaCl cannot be physically separated from the water. A *solution* is a mixture of two or more chemical species, that is, entities, in one phase. The term *species* is used to denote a specific chemical entity such as a molecule or ion in a phase. If NaCl is dissolved in water, an NaCl solution is produced. Solutions also exist in solids and gases. An NaCl solution can be characterized in terms of the species present, which include H_2O molecules together with the ions Na^+ and Cl^-. The air one breathes is a single-phase solution made up of a mixture of gas species, the two most dominant being N_2 and O_2. The phase seawater is a liquid solution mostly made up of H_2O, but with significant concentrations of the species Na^+, Cl^-, Mg^{2+}, and SO_4^{2-}. Solid solutions like the Mg-Fe olivines are common. These can be considered to be a mixture of the species in endmember forsterite (Mg_2SiO_4) and fayalite (Fe_2SiO_4). An olivine of composition $Mg_{1.5}Fe_{0.5}SiO_4$ would then be a mixture of 75 mole % forsterite and 25 mole % fayalite. *Components* are chemical endmember formulas used to describe the composition of all of the phases in a system. Because most systems involve oxide mineral phases (e.g., Fe_2SiO_4), these are often given as simple single oxides such as FeO, Al_2O_3, and SiO_2. In a thermodynamic model of a system, the number of these endmember component formulas must be a minimum.

When considering solutions, one is typically interested in the relative amounts of entities rather than their absolute amount; therefore, solutions are often characterized by the *mole fractions* of the constituent or endmember species. This is the ratio of the number of moles of the species relative to the total number of moles of all of the species in the solution. The mole fraction of species A, X_A, is then equal to

$$X_A = \frac{\text{Moles of A}}{\text{Sum of moles of all species}} = \frac{\text{Mole \% of A}}{\text{Sum of mole \% of all species}} \qquad [1.29]$$

Consider dry air with its dominant species given in **Table 1-2**. These species in air behave nearly ideally so that the total volume of the mixture can be considered

Table 1-2 **Amount and molecular weight of the four main constituents of dry air**

Species	Volume (%)	Mole fraction	Molecular weight (g/mol)
N_2	78.09	0.7809	28.01
O_2	20.94	0.2094	32.00
Ar	0.94	0.0094	39.95
CO_2	0.0375	0.000375	44.01

to equal the sum of the individual volumes of the same amount of each gas species in the mixture at the same temperature and pressure; therefore, these volume percentages can be taken as the mole percentages of the species. Using relation (1.29) and the volume percentages in Table 1-2, the mole fractions of N_2, O_2, Ar, and CO_2 in air can be calculated as given.

When species are present in small concentrations, parts per million, *ppm*, are often used. These are either ppm by volume or ppm by weight. For CO_2 gas, ppm by volume in dry air is equal to its volume percentage times $10^4 = 375$ ppm (by volume). To determine ppm by weight, that is, grams of the species per million grams of the mixture, the molecular weights of N_2, O_2, Ar, and CO_2 as given in Table 1-2 need to be considered. The number of grams of each species in 1 mole of air is then its molecular weight times its mole fraction or 21.87, 6.70, 0.38, 0.0165 g, respectively. The total number of grams in 1 mole of air is, therefore, 28.97 g(air)/mole. The ppm of CO_2 in air by weight is

$$\frac{0.0156 \text{ g}(CO_2)/\text{mole}}{28.97 \text{ g(air)/mole}} 10^6 = 538 \text{ } CO_2 \text{(ppm weight)} \qquad [1.30]$$

This is greater than the ppm by volume because CO_2 has a greater molecular weight than the average molecular weight of the other gas species in air. If an entity is present in even smaller concentrations, parts per billion, *ppb*, are sometimes used.

In aqueous solutions, species that are present in small concentrations relative to the *solvent* species, H_2O, are termed *solutes*. Solute concentrations are often expressed on the *molarity* or *molality* concentration scales rather than with mole fractions. Molarity, specified with the symbol M, is defined as

$$\text{M} \equiv \frac{\text{moles of solute}}{1000 \text{ cm}^3 \text{ of solution}} \qquad [1.31]$$

and molality, specified with the symbol m, is given by

$$m \equiv \frac{\text{moles of solute}}{1000 \text{ g of solvent}} \qquad [1.32]$$

Molarity is a unit based on the volume of solution. This unit is convenient in a chemical laboratory where measurements can be done with a graduated cylinder or volumetric flask. Because the volume of a solution changes with temperature and pressure, the molarity of a given solution also changes with temperature and pressure; however, the molality, based on mass units, does not. Therefore, when changes of pressure and temperature are considered, molality or mole fraction is generally the preferred concentration unit. Molarity is defined on

| Table 1-3 | Concentration units |

Unit	Definition
Mole	Avogadro's number of "items" = 6.02214×10^{23}
Atomic weight	Number of grams in a mole of the substance
Molarity	1 mole of "items" dissolved in solution to give a total volume of 1 liter
Molality	1 mole of "items" dissolved in 1 kg of solvent
ppm (weight)	1 part in 10^6 parts by weight
ppm (volume)	1 part in 10^6 parts by volume

a total amount of solution and the molality only on the amount of solvent. The solvent is the chemical species that is present in the largest concentration, generally H_2O. M can be converted to m by making use of the following expression:

$$m = \frac{1000 \, M}{1000 \, \rho - (M \times S)}$$ [1.33]

where ρ is the density of the solution and S is the molecular weight of the solute. The different units of concentration considered are outlined in **Table 1-3**, and conversions between units are given in Appendix J.

A chemical reaction or equilibrium equation needs to be constructed so that it is balanced. That is, both the sum of the amount of each element and sum of charges on the species on one side of a reaction must equal those on the other side; therefore, the constructed relationship given by the reaction has the same meaning whether one is talking about a single species or a mole of species. This is consistent with the law of conservation of mass that states there must be exactly the same mass on both sides of a chemical reaction. For instance, the solubility reaction for calcite, $CaCO_3$, dissolving in pure water can be written as

$$CaCO_3 + H_2O = Ca^{2+} + HCO_3^- + OH^-$$ [1.34]

where Ca^{2+}, HCO_3^-, and OH^- are species in the aqueous solution. Both the number of each element (Ca, C, H, and O) and the sum of the charges (here summing to zero) are the same on both sides of the reaction. By convention, this reaction is read from left to right. One can, therefore, state that 1 mole of calcite dissolves to produce 1 mole of Ca^{2+} in solution. One refers to the species or entities on the left as *reactants* and those on the right as *products*. A reaction property indicating how much a quantity has changed during a reaction, for example, volume of reaction, is computed by subtracting the volume of the reactants from the volume of the products. This is the standard convention: Coefficients of products are taken as positive, whereas reactant coefficients are considered to be negative in determining reaction properties.

It is equally valid to write the calcite formula as $Ca_2(CO_3)_2$ so that the calcite dissolution reaction would be

$$Ca_2(CO_3)_2 + 2H_2O = 2Ca^{2+} + 2HCO_3^- + 2OH^- \qquad [1.35]$$

In this case, 1 mole of calcite ($Ca_2(CO_3)_2$) produces 2 moles of Ca^{2+}. It needs to be clear how a mole of a phase or species is written when one indicates that a mole of it is reacting. This becomes important when considering magmas (see Chapter 8) where the "quartz-like" melt species is considered to be Si_4O_8; therefore, it takes 4 moles of $SiO_{2(quartz)}$ to produce 1 mole of the Si_4O_8 species in the magma.

Rather than a molar property, such as the molar volume of a solution, one is often interested in describing a change in a property such as volume with changing composition of a single component in the solution at a particular molar concentration. This is termed a *partial molar quantity*. For the Zth molar property, this is typically signified with an overbar and is defined as

$$\bar{Z}_i \equiv \left(\frac{\partial Z}{\partial n_i} \right)_{P,T,n_j} \qquad [1.36]$$

where the partial derivative with respect to the moles of component i, n_i, is evaluated at constant P, T, and constant composition of all of the other n_j components except the ith one. The partial molar property is then the change in that property when 1 mole of i is added to an infinite amount of the solution in question.

Consider an infinitely large volume of H_2O at standard earth surface conditions (25°C and 1 bar). If 1 mole of water is added to this volume, it increases by about 18 cm³. This change in volume is then the molar volume of pure H_2O. Suppose instead that 1 mole of water is added to an infinitely large volume of the alcohol, *ethanol*. The volume increases by about 14 cm³. This is the partial molar volume of H_2O in pure ethanol. There is so much ethanol that each H_2O molecule is surrounded by an infinite sea of ethanol molecules, and this destroys the open tetrahedral structure of H_2O so that the volume increase on adding the H_2O is less. The partial molar volume of water in any composition of solution is defined as the increase in volume when 1 mole of H_2O is added to an infinite volume of the solution composition considered. This is shown in **Figure 1-7** for H_2O–ethanol mixtures. Often, other partial molar properties are used to describe the nature of solutions. Of particular interest is the partial molar Gibbs energy of a component in a solution (i.e., how the Gibbs energy of a solution changes when a mole of the component is added to an infinite amount of the composition considered). Knowledge of the nature of Gibbs energy awaits

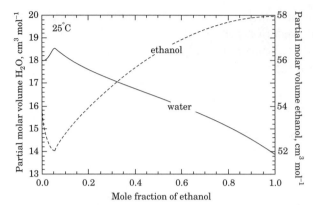

FIGURE 1-7 Partial molar volumes of water and ethanol in water–ethanol mixtures at 25°C and 1 bar as a function of the mole fraction of ethanol in the mixture.

the presentation of the laws of thermodynamics as applied to solutions given in Chapter 4.

The total volume change on adding a mixture of water and ethanol to a water–ethanol solution can be determined from a knowledge of the partial molar volume of both water and ethanol at the concentration of interest. The total differential of volume is then

$$dV = \left(\frac{\partial V}{\partial n_{\text{water}}} \right)_{P,T,n_{\text{ethanol}}} dn_{\text{water}} + \left(\frac{\partial V}{\partial n_{\text{ethanol}}} \right)_{P,T,n_{\text{water}}} dn_{\text{ethanol}} \qquad [1.37]$$

where the first term on the right is the partial molar volume of water times the change in the moles of water and the second term is the partial molar volume of ethanol times the change in the number of moles of ethanol.

The ideas developed here for treating chemical species in geochemical systems form the basis for the study of geochemistry. Before chemical reactions are considered, however, a basic understanding of the gross chemical composition of the earth and a development of sources of heat energy that drive reactions is helpful and is introduced in the next chapter.

Summary

Geochemistry is built on models of reality. Models can be either deterministic or probabilistic (stochastic); however, some deterministic models can lead to a chaotic system. Models are made of extensive and intensive variables. The values of these variables are generally given in SI units based on the meter, kilogram, second, and kelvin; however, °C is often used in place of K for temperature. The pressure unit most often used in geochemical calculations is the bar because it is nearly equal to atmospheric pressure. There are 10^5 pascals, the SI pressure unit, in 1 bar so that the conversion is straightforward. Although the SI energy unit is the joule, the older unit of calories is still encountered in many contributions in the literature.

In mathematical analysis of a system, differentials are often used. A differential is the infinitesimal change in, or slope of, one variable with respect to another. Partial differentials give this change while holding all other variables constant. The total differential gives the total change of a function with respect to all of the variables that affect the function. It is constructed by summing a series of partial derivative times the change in the variable.

A large number of variables can be important in the construction of a model to understand a geochemical process. If the three most important compositional variables can be summed to give most of the change, a triangular diagram can be constructed. These diagrams show relationships between compositions where the more of a compositional variable present the closer to its apex is its location.

Geochemical models are generally described in terms of phases, species, and components. Solutions encountered can be in gases, liquids, and solids. The composition of the system can be characterized in terms of mole fractions; however, parts per million and parts per billion are also used for elements at small concentrations. In aqueous solutions, the solutes, species that are present in small quantities relative to the H_2O present, are characterized either in terms of molarity or molality. The molarity of a given solution changes as pressure and temperature change; molality does not. The partial molar property of component i is the partial derivative of the property with respect to the number of moles of i evaluated at constant P and T as well as constant composition of all of the other components in solution.

Key Terms Introduced

acceleration	intensive variable
Avagadro's number	kinetic energy
component	lever rule
deterministic chaos	Markov process
differential	molality
element	molarity
ethanol	mole
extensive variable	mole fraction
force	nonlinear systems
heat energy	partial differential
ideal gas	partial molar quantity

phase   solvent
phenomenological approach   species
potential energy   state variable
ppb   stochastic approach
ppm   strange attractor
products   thermodynamics
reactants   total differential
solutes   velocity
solution

Questions

1. What is a derivative? A partial derivative?
2. List four different kinds of energy. Give a process for turning each to another kind of energy.
3. What are SI units? What are the basic SI units from which all of the others can be built?
4. Give the difference between an extensive and intensive variable.
5. What is the difference between ppm by volume and ppm by weight? Between molal and molar properties?
6. Explain the construction of a triangular composition diagram.
7. What are partial molar properties?

Problems

1. State the property the units are referring to and find the conversion factors need for the conversion of
   a. calories to joules   b. bars to pascals
   c. μm to km   d. cm^3 bars to joules
   e. m^3 pascals to joules
2. In the following list of variables, which are extensive and which are intensive?
   a. heat capacity   b. temperature   c. viscosity   d. energy
   e. number of moles   f. molar volume   g. mass
3. In thermodynamic models, the chemical components needed to describe a system must be a minimum. List a minimum set of components needed to describe the following systems:
   a. A rock consisting of Mg-Fe olivine $(Mg,Fe)_2SiO_4$ and Mg-Fe pyroxene $(Mg,Fe)SiO_3$ both of which display variable concentrations of Mg and Fe.
   b. A rock consisting of variable amounts of quartz (SiO_2), enstatite $(MgSiO_3)$, and forsterite (Mg_2SiO_4).
   c. Liquid water containing cubes of ice.

4. For reaction 1-1, if a kg of Fe_{metal} reacts, what mass of Fe_3O_4 is produced?

5. What is the molarity of a 2.5 m NaCl aqueous solution of density 1.05 kg liter^{-1}?

6. What is the molality of an aqueous solution of 114 ppm by weight of SiO_2?

7. A 1000 g aqueous solution contains 10 g of NaCl. The density of the solution is 1.068 g/ml. What is the molality, molarity, and mole fraction of NaCl in the solution?

8. Differentiate the functions, $y(x)$, where a and b are constants.

a. $y(x) = a + b$

b. $y(x) = \dfrac{1 - ax}{1 + x}$

c. $y(x) = e^{ax}$

d. $y(x) = 10^x$

9. Consider function $f(x,y)$ where the differential of f: $df(x,y) = (y + 3A) dx + (A + x)dy$. If A is a constant, compute the cross-differentials and determine whether f could be a state function.

10. Find the total differential of $u(x,y,z)$ where

$$u = \frac{xyz}{(x+y+z)}$$

11. The area, a, of a rectangle can be considered to be a function of its width, x, and length, y; therefore, $a = a(x,y)$ with $a = xy$. If x and y are considered to be independent variables and a is the dependent one, then other dependent variables are the perimeter, $p = 2x + 2y$, and the diagonal, $d = (x^2 + y^2)^{1/2}$.

Find the values of the following partial derivatives in terms of x, y, a, or a numerical value.

a. $\left(\dfrac{\partial a}{\partial x}\right)_y$ b. $\left(\dfrac{\partial x}{\partial y}\right)_d$ c. $\left(\dfrac{\partial p}{\partial x}\right)_y$

d. $\left(\dfrac{\partial x}{\partial y}\right)_p$ e. $\left(\dfrac{\partial d}{\partial y}\right)_x$ f. $\left(\dfrac{\partial p}{\partial y}\right)_x$

12. Given that $dG = -SdT + VdP$, derive the expression

$$dG = V\left(\frac{\partial G}{\partial V}\right)_T dV + \left[V\left(\frac{\partial P}{\partial T}\right)_V - S\right]dT$$

13. Given that $dE = TdS - PdV$ and $dA = -SdT - PdV$, derive the expression

$$\left(\frac{\partial E}{\partial V}\right)_A = T\left(\frac{\partial S}{\partial V}\right)_A + S\left(\frac{\partial T}{\partial V}\right)_A$$

14. A mass of iron oxide of 1.0 kg is heated in hydrogen gas until it is completely converted to 0.7 kg of metallic iron. What is the formula of the iron oxide?

15. Draw an equilateral triangle and label the apexes A, C, and F as in Figure 1-5. Mark the compositions of A = 70%, C = 20%, F = 10%, and A = 30%, C = 50%, F = 20% on the diagram as x and y, respectively.

References

Anderson, G. M. and Crerar, D. A., 1993, Mathematical background. In *Thermodynamics in Geochemistry: The Equilibrium Model*. Oxford University Press, New York, pp. 7–36.

Lide, D. R., editor, 2003, *Handbook of Chemistry and Physics*, 84th edition. CRC Press, Cleveland, 2616 pp.

Peitgen, H.-O., Jürgens, H. and Saupe, D., 1992, *Chaos and Fractals New Frontiers of Science*. Springer-Verlag, New York, 984 pp.

Turner, F. J., 1981, *Metamorphic Petrology, Mineralogical, Field and Tectonic Aspects*, 2nd edition, Hemisphere Publishing, Washington, D.C., 524 pp.

2 The Earth's Aggregate Physical and Chemical State

The universe was produced with what is called the Big Bang 13.7 billion years ago according to recent measurements of the *cosmic microwave background*. Its mass is thought to have started primarily as neutrons. The neutrons then began to decay producing protons and electrons. As a result, early on, the mass of the universe consisted mainly of H_2 gas with some He gas. The expanding gas dispersed to form galactic nebulae. Within the nebulae gravitational collapse led to the formation of stars. At the centers of stars, the energy released on collapse led to fusion reactions producing elements of greater *atomic number*. If the star was massive enough, elements up to the atomic number of Fe could be produced by these fusion reactions. If the star was 8 solar masses or greater at the end of its life, after it has produced a core of Fe, it imploded leading to the formation of a *supernovae*. This process produces a flux of free neutrons. Neutron capture by the previously formed elements, followed by the decay of the neutron to a proton and electron, led to the formation of the elements with atomic number greater than iron.

Our solar system was likely born from the death of a *red giant* star ejecting a volume of H_2 plus He created in the Big Bang along with 2% dust of heavier atoms created in previous stars and supernovae into empty space. Gravitational attraction in the volume began collapsing it. Centrifugal forces in the collapsing volume produced a rotating disk referred to as the *solar nebula*. Some 99.86% of the material in the solar nebula accreted to form the sun, with 0.14% present at greater distances from the center of rotation that conserved the angular momentum of the nebula. A large outward temperature gradient was produced as the gravitational energy, lost from contraction, was transformed to heat energy vaporizing the material present. The earth is thought to have formed 4.566 ± 0.002 billion years ago (Allègre et al., 1995a) along with the other planetary bodies in the solar system by gravitational attraction within eddies of the solar nebula.

The abundances of the elements relative to Si in the solar nebula are shown in **Figure 2-1**, and the more abundant elements are given in **Table 2-1**. Note the 12

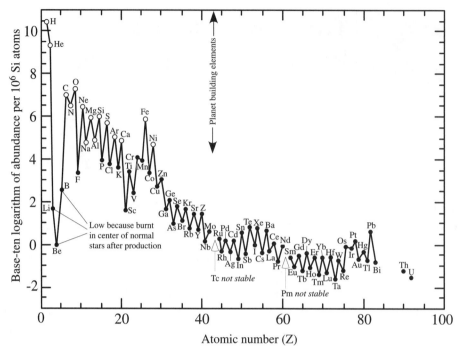

FIGURE 2-1 Relative cosmic abundance of the elements as a function of the number of protons in their nucleus, the atomic number. (Data from Anders, E. and Grevesse, N., 1989.)

orders of magnitude differences in their abundances. The general trend of decreasing abundance with increasing atomic number occurs because elements with a large number of protons are produced by fusion or neutron decay reactions with elements having lower numbers of protons. The exception is with Li, Be, and B, which are burnt in stars of normal mass.

The helium nucleus configuration (two protons + two neutrons) has increased stability over other configurations in a nucleus. This preference for even numbers is due to its increased symmetry. Neutrons and protons have half-integer angular momentum. A nucleus with an odd number of protons plus neutrons will have half-integer angular momentum and need to rotate by 720° rather than the 360° necessary for a nucleus with an even number of protons plus neutrons to remain in the same stable state. Nuclei that only rotate 360° rather than 720° are subject to less centrifugal force and are therefore more stable. This produces a saw tooth pattern of abundances in which the even atomic number elements are more abundant than their odd atomic number neighbors. Because no stable Be exists that has an even number of neutrons in its nucleus (see Figure 10-3), its abundance is particularly low. Figure 2-1 also identifies the elements with sufficient abundance to form the major constituents of planets (see Broecker, 1985, for a good discussion).

Table 2-1   Estimated atomic abundance relative to Si of the more abundant elements in the solar nebula

Element	Relative Abundance
H	27,900
He	2790
C	10.1
N	3.13
O	23.8
Ne	3.44
Na	0.054
Mg	1.074
Al	0.085
Si	1.0
P	0.0104
S	0.515
Ar	0.101
Ca	0.0611
Cr	0.0135
Fe	0.90
Ni	0.0493

Data from Anders and Grevesse, 1989.

Solid phases condensed from the elemental gas as temperatures cooled, producing *planetesimals*, solid objects between 10 m and 100 km in diameter. Planetesimals formed near the proto-sun were composed of Fe and Ni metals and oxides of Si, Al, Mg, and Fe, whereas planetesimals farther from the proto-sun formed ices composed of H_2O, NH_4, and CH_4 in a volume of H_2 and He. Accretion of the planetesimals produced the inner terrestrial planets, Mercury, Venus, Earth, and Mars. Ices, together with condensing H_2 and He, accreted to form the outer gaseous planets, Jupiter, Saturn, Uranus, and Neptune. During the formation of the terrestrial planets, the Fe and Ni metal-rich planetesimals accreted first with later accretion of silicate planetesimals.

The heat produced from the loss of gravitational attraction energy during the accretion process to form the earth allowed much of any remaining more dense Fe and Ni metals in the silicate material to melt and sink, aiding the formation of an Fe and Ni metal core. This left the silicate-rich material, referred to as the *mantle*, at greater distances from the earth's center. The formation of the earth's moon appears to require the impact of a Mars-sized body with the young earth. The atmosphere on the earth would have been lost in the collision;

therefore, rather than capturing primordial gases from the solar nebula, most of the volatile elements found in the earth's early atmosphere are thought to have accreted to the earth's surface after some cooling from capture of volatile-rich ice planetesimals and/or comets from the outer solar system. CO_2 and H_2O were then added to the atmosphere during thermal outgassing of the earth as *continental crust* was formed. If the bulk of the continental crust was formed before moon formation, then the later earth's atmosphere is likely dominated from capture of ice planetesimals and/or comets. Further cooling of the earth condensed the H_2O vapor from the atmosphere to form the oceans approximately 4.1 billion years ago. This left a CO_2-dominant, N_2-rich atmosphere before the CO_2 plus H_2O reacted with the silicate minerals present (see Chapter 16).

■ Meteorites

Much of our understanding of the formation and composition of planets in our solar system comes from the study of meteorites. Meteorites are the solid extraterrestrial material that strikes the surface of the earth. These are divided into three classes depending on composition: *irons*, which are principally a metallic Fe–Ni alloy; *stones*, which are composed of silicate minerals; and *stony-irons*, which have a subequal mixture of the compositions of irons and stones. Stones are further divided on whether they contain *chondrules*, nearly spherical silicate inclusions between 0.1 and 3 mm in diameter. Stones containing chondrules are termed *chondrites*. Those that do not contain chondrules are termed *achondrites*.

Of meteorites, a class of chondrites that contains carbon, called *carbonaceous chondrites*, is considered to be the most primitive chondrites because they have retained at least some of their volatile elements. They are thought by most to be good samples of the composition of the solar nebula except for its most volatile elements (i.e., H, C, N, O, and the noble gases) and a good model for the average composition of the nonvolatile elements in the earth. The assumption that there are no significant compositional gradients in the solar nebula before the planets began to differentiate is confirmed by measurements of the elemental spectra of the sun's photosphere. Although these are less precise than measurements of the concentration of elements in carbonaceous chondrites, they appear to give similar relative ratios of the nonvolatile elements.

Meteorites can also be divided into falls and finds. Falls are meteorites located from the trajectory of their fireball through the atmosphere. Finds are meteorites that have been found on the surface of the earth but were not observed to fall. Most falls are ordinary chondrites (**Figure 2-2**), whereas most finds

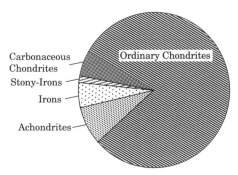

FIGURE 2-2 Relative abundance of meteorite types observed in falls. (Adapted from Sears, D. W. G. and Dodd, R. T., 1988.)

are irons. The meteorite falls give a better indication of the percentages of meteors that hit the earth because irons are more easily found than stones because they stand out from the silicate minerals present on the surface of the earth. Most meteorite collections, therefore, have a predominance of irons even though more stony meteorites have hit the earth.

■ Density of Earth Material

Figure 2-3 shows the density structure of the earth determined mainly from seismic wave velocities using a model of how the elasticity of the rocks affects the velocity. That is, the velocity of seismic waves that travel through the earth depends on the density and elasticity of the material through which they pass. Seismic wave travel times can be inverted to obtain the velocity structure of a seismic wave with depth in the earth. From the velocity structure, the solid earth is divided into a core, mantle, and crust. The ocean and atmosphere are then held to the solid earth by the force of *gravity*. The core has about one third of the mass of the earth, whereas the mantle has about two thirds of it. The uppermost layer, the crust, is less than 0.2% of the mass of the earth. Based on its location and composition, the crust is divided into continental crust and *oceanic crust*. As the names imply, generally, continental crust is under the continental land surface, and oceanic crust is under the ocean floor. The average density of rocks increases with depth in the earth. For the model shown in Figure 2-3, the average depth of the oceans would be near 3 km if they covered the entire earth, and below this, the average continental + oceanic crust would be about 24 km thick. This average oceanic + continental crust for a general earth model is divided into an upper crust of a density of 2.6 g cm^{-3} and a lower crust of 2.8 g cm^{-3}. The model then averages the thickness and density of oceanic and continental crust that are present.

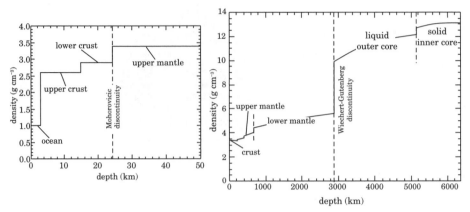

FIGURE 2-3 Average density versus depth in the earth determined from the Preliminary Reference Earth Model (PREM). (Adapted from Dziewonski, A. M. and Anderson, D. L., 1981.)

Oceanic and Continental Crust

Figure 2-4 shows a cross-section through average oceanic crust. It consists of 0.3 km of sediments and sedimentary rocks of both biogenic and terrigenous sources. Their proportion depends on the amount of time the crust has spent below zones of productivity of the surface ocean as opposed to the windward side of continental margins and volcanic centers. Below the sedimentary rocks are a layer of pillow basalts and then a layer of sheeted dikes for a combined thickness of 1.4 km. The lowermost layer of oceanic crust is a thickness of 4.7 km of layered gabbro (Shor and Raitt, 1969). Actual oceanic crust thickness varies from about 4 km at some ocean ridges to more than 10 km where

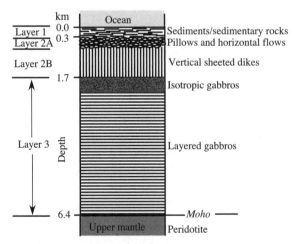

FIGURE 2-4 The characteristics with depth of average oceanic crust indicating the seismically determined layer designations.

volcanic plateaus produced from large outpourings of basaltic lava or a great thickness of continentally derived detrital sediments occur.

Continental crust is much thicker than oceanic crust, averaging 36 km in thickness but varying from about 20 to 80 km. The thickness is greatest under mountains. Continental crust's composition and density change with depth. The upper continental crust is *felsic*, a rock type with a large abundance of feldspars, an *alkali ± alkaline earth* and Al-rich mineral, and has a density of about 2.6 g cm^{-3}. With increasing depth, the continental crust becomes denser as the rocks become *mafic*. Mafic rocks are those that contain a majority of more dense Mg + Fe silicate minerals (olivine and proxene). The lower continental crust has a density that reaches about 3.0 g cm^{-3}.

Figure 2-5 displays a cross-section through typical continental crust of 40 km thickness showing how its composition changes with depth. Most of the crust is igneous and metamorphic, with about 7.5% of the total sedimentary rocks. Continental crust is not as effectively recycled back into the mantle as oceanic crust, and therefore, it is generally much older. As a result, it has often had a complex history of metamorphism and melting. This has produced *granulites* in the lower crust. Granulites are rocks that have undergone metamorphism at very

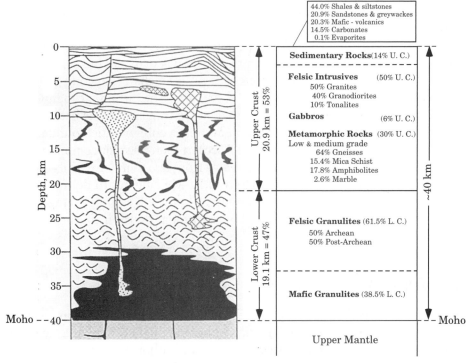

FIGURE 2-5 Diagrammatic standard profile of the continental crust together with its composition. (Adapted from Wedepohl, K. H., 1995.) U.C. = upper crust, L.C. = lower crust.

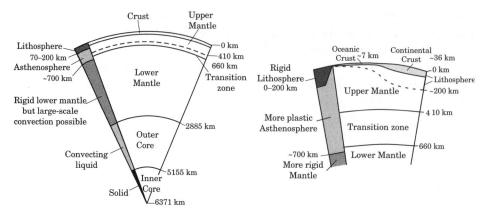

FIGURE 2-6 Major compositional divisions in the earth with depth. The diagram on the right is a blowup of the near surface region of the diagram on the left. The rheologic divisions are also shown on the left-hand sides of each diagram.

high temperatures with the destruction of any OH bearing minerals (e.g., micas and hornblende). This loss of water generally occurred by its incorporation into a lower density partial melt. The melt phase then rises as a diapir through the more dense rocks because of *buoyancy forces*, as shown diagrammatically on the left side of Figure 2-5.

The crust is distinguished from the mantle by the Mohorovicic seismic discontinuity (*Moho*) that marks the depth where rock densities increase dramatically. This increase in density is thought to reflect a change in composition to rocks with less than 45% SiO_2, termed *ultramafic*. The ultramafic mantle is divided into an upper mantle to a 400-km depth, a transition zone between 410 and 650 km, and a lower mantle from 650 to the core–mantle boundary at a 2885-km depth. Deep in the earth, at the Wiechert-Gutenberg discontinuity of the core–mantle boundary, the density increases from less than 6 to over 10 g cm^{-3} because of the compositional change from Fe-Mg silicates to dominantly Fe metal. The density structure, together with the knowledge of the average chemistry of the earth, leads to the gross compositional divisions in the earth, as summarized in **Figure 2-6**.

■ Rheologic Divisions of the Earth

The *rheologic divisions* that indicate the relative rigidity of layers in the earth are also shown in Figure 2-6. The *lithosphere* is the outer rigid layer of the earth that is produced at mid-oceanic ridges and subducted into the less rigid mantle below at subduction zones. The lithosphere has nearly a zero thickness at the ridge where it is produced by partial melts from *asthenosphere* upwelling. Its thickness increases as it is transported further from the ridge and cools, with heat dissipated in the ocean water above. The lithosphere, under old oceanic crust, can reach a thickness of 125 km.

Under continents, the lithosphere is generally thought to be between 60 and 200 km thick. Its thickness depends to a large extent on its tectonic and, therefore, thermal history. It is thickest under stable *shield* areas that have cooled since the Precambrian and thins with greater heat flux under active tectonic areas such as in mountain belts and back-arc basins as with the Basin and Range area of the western United States. Some investigators consider the continental lithosphere to reach 400 km in depth at some locations. A problem in determining its thickness is that in many places there does not appear to be a strong contrast in rigidity with depth, but the change occurs over a significant depth interval. **Figure 2-7** shows a cross-section through a stable ocean–continental crust boundary that demonstrates the relationship between layers thought to occur. The lithosphere–asthenosphere boundary is generally considered to coincide with the 1300°C isotherm. The lithosphere contains all of the crustal rocks and some of the ultramafic mantle as well.

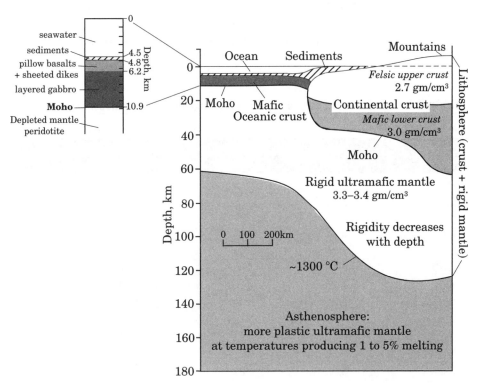

FIGURE 2-7 Cross-section of the surface layers of the earth showing compositional changes and the relation of oceanic and continental crust to the underlying mantle.

■ Composition of the Earth's Layers

Estimates of the core, mantle, and crust compositions are constrained by the entire earth composition derived from the composition of the solar nebula and by knowledge of the densities of the layers in the earth as given in Figure 2-3. The core must be metallic to generate a magnetic field, and this metal must be in a liquid state at very high pressures, as the outer core does not transmit seismic shear waves. The core, therefore, must be dominantly Fe and is thought to have 4% Ni and a little over 0.2% Co. From its density, there must also be lighter elements to make up approximately 10% of its mass. These light elements could be Si, S, O, C, or H. Most investigators rule out C and H because their high volatility at high temperatures would have made their condensation in the core unlikely. Some investigators believe the light element is Si because of its great abundance in the solar system, its low volatility, and its ability to alloy with Fe at very high pressures. Others argue for some mixture of Si, S, and O.

From estimates of the whole earth composition and subtracting out the core's composition, an estimate of the primitive silicate mantle composition, that is, the composition before the crust was fractionated out, can be obtained. Various researchers' estimates of the composition of this primitive mantle are given in **Table 2-2**. This is then the present average mantle + crust composition.

|Table 2-2   Estimates of the mantle + crust composition (= primitive mantle)

Oxide (wt %)	Ringwood (1991)	McDonough and Sun (1995)	Allègre et al. (1995)
SiO_2	44.76	45.0	46.12
Al_2O_3	4.46	4.45	4.09
FeO	8.43	8.05	7.49
MgO	37.23	37.8	37.77
CaO	3.60	3.55	3.23
Na_2O	0.61	0.36	0.36
K_2O	0.029	0.029	0.034
Cr_2O_3	0.43	0.384	0.38
MnO	0.14	0.135	0.149
TiO_2	0.21	0.20	0.18
NiO	0.241	0.25	0.25
CoO	0.013	0.013	0.07
P_2O_5	0.015	0.021	

Given its composition, primitive mantle (and, therefore, the mantle today) is dominated by FeO, MgO, and SiO_2 or ferromagnesian silicates.

As mentioned previously, the mantle and crust have significantly different compositions. Most igneous petrologists believe that both the oceanic and continental crust have been differentiated by a small partial melting of the mantle. The melt extractions partitioned incompatible elements (e.g., Na and K) from the mantle into the oceanic and continental crust. The oceanic crust is produced by a single partial melting event in the asthenospheric mantle, producing mid-oceanic ridge basalt (*MORB*), whereas the continental crust has undergone further differentiation by internal partial melt events producing a mafic lower crust and felsic upper crust. Estimates of the composition of the igneous part of oceanic crust as well as lower and upper continental crust are given in **Table 2-3**. Models based on studies of trace elements, that is, those found in low abundance, as well as isotope ratios of mantle rocks, indicate that only the upper one third of the mantle became depleted by crust formation. Compositional differences are, therefore, thought to exist between the upper and lower mantle.

SiO_2 and Al_2O_3 make up about three fourths of the mass of the continental crust and about two thirds of the mass of MORB; therefore, most bonding between atoms found in minerals in the crust is the bonding of O with Si and Al, which promotes the formation of *feldspars*. CaO is significantly more concentrated in MORB than in continental crust. MORB feldspars, therefore, are

Table 2-3	Average composition of mid-oceanic ridge basalt (MORB) and upper, lower, and total continental crust

Oxide	MORB*	Upper continental crust[†]	Lower continental crust[†]	Total continental crust[†]
Si	50.45	64.9	58.05	61.6
TiO_2	1.615	0.52	0.84	0.68
Al_2O_3	15.255	14.6	15.5	15.1
ΣFeO[‡]	10.426	4.4	7.34	6.28
MnO	—	0.07	0.12	0.10
MgO	7.576	2.2	5.2	3.7
CaO	11.303	4.1	6.8	5.5
Na_2O	2.679	3.5	2.86	3.2
K_2O	0.088	3.1	1.85	2.4
P_2O_5	—	0.15	0.20	0.18

*Data from Hofmann (1988).
[†]Data from Wedepohl (1995).
[‡]$\Sigma FeO = FeO + Fe_2O_3$.

more anorthite rich. Although the lower crust is more mafic than the upper crust, it is not as mafic as basalt. The upper mantle continues this compositional trend with an ultramafic composition of lower SiO_2 content being made up of Mg and Fe^{2+} silicates consisting mainly of Mg–Fe *olivine* but with some Mg–Fe *orthopyroxene* and *clinopyroxene*. The nature of the minerals olivine, orthopyroxene, and clinopyroxene are discussed in Chapter 5. The lower mantle is considered by most to be made up of three phases, Mg-*perovskite* ($MgSiO_3$ with approximately 4% Al_2O_3), Ca-perovskite ($CaSiO_3$), and Mg-*wüstite* ((Fe,Mg)O).

The change in mineralogy with depth in the mantle proposed by Ringwood (1991) is shown in **Figure 2-8**. Plagioclase, the Al-rich phase in ultramafic mantle rocks near the earth's surface, undergoes a phase transition to spinel at about 30 km depth, and a transformation from spinel to garnet occurs at 60 to 90 km below the surface depending on its composition. At increasing depth of about 300 km, the pyroxene components begin to form a solid solution with the garnet. Near the 400-km depth, the α-olivine $(Mg,Fe)_2SiO_4$ phase is transformed to the more dense β-olivine phase. At a depth of approximately 450 km, pyroxene of any composition is no longer stable, and the garnet solid solution, termed *majorite*, is the only Al phase present. At 500 km in depth, β-olivine begins to be transformed to spinel (γ-olivine) and is completely trans-

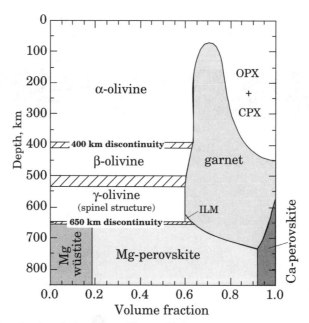

FIGURE 2-8 Volume fraction of mineral assemblages with depth in the mantle. OPX, orthopyroxene; CPX, clinopyroxene; ILM, ilmenite. (Adapted from Ringwood, A. E., 1991.)

formed at a depth of about 530 km. At greater depths, Ca and Mg in majorite become unstable, producing Ca-perovskite and Mg-ilmenite. The Mg-ilmenite is transformed at the 650-km depth boundary of the lower mantle where spinel is transformed to Mg-perovskite with the iron going into a Mg-wüstite structure.

The composition of the upper mantle under oceanic and continental crust is thought by many to be somewhat different, as shown in **Figure 2-9**. The mantle is *peridotite* in composition, a general term for olivine-rich rock that has some orthopyroxene and clinopyroxene. Mantle peridotite can be divided into primitive mantle peridotite, also termed "*pyrolite*" or fertile mantle, and depleted peridotite from which basalts have been extracted. Within the depleted peridotite under continents are thought to be segregations of *eclogite*, a rock containing garnet and pyroxene. It has been argued that under oceanic crust the depleted peridotite can be divided into a strongly depleted *harzburgite*, which contains almost exclusively olivine plus orthopyroxene, and a less depleted *lherzolite* layer, which contains a small amount of clinopyroxene as well.

MORB is thought by most investigators to be produced at mid-oceanic ridges by decompression melting of the asthenosphere at depths of 150 to 200 km (Klein and Langmuir, 1987), as indicated in **Figure 2-10**. Partial melts are formed that feed an axial magma chamber. It is thought by many that the degree of partial melting is between 8% and 20%. Within the magma chamber, the primary magmas produced are modified in composition by low pressure partial crystallization before MORB is erupted. Fracturing of the rocks above the chamber causes escape of magma into seawater to produce pillow basalts and sheeted dikes in the fractures. With cooling of the magma chamber, layered gabbros are formed, as shown in Figures 2-4 and 2-10. Because a magma chamber has been difficult

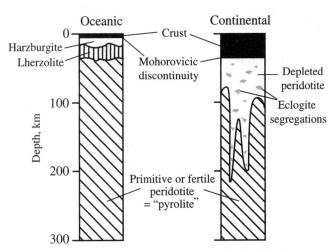

FIGURE 2-9 Chemical zonation with depth of the upper mantle under oceanic and continental crust. (Adapted from Ringwood, A. E., 1991.)

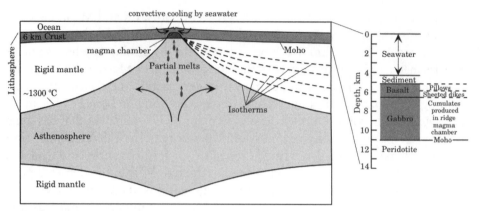

FIGURE 2-10 Structure and composition of the mid-oceanic ridge where basaltic oceanic crust is produced from partial melt diapirs separating and rising from upwelling asthenosphere.

to image by seismic refraction, it is thought by many investigators to be a transient feature. The cooling of the crust is predominantly by convection of seawater at the ridge crest giving way to conductive cooling at greater ages as distances from the ridge increase. Sediments are then deposited in a top layer as a function of time.

The elevation of oceanic crust at the ridge is about 2 km below sea level and sinks to 6 km below sea level in old oceanic crust. The newly formed ridge volume displaces seawater from ocean basins relative to older oceanic crust. Most investigators believe that large-scale seawater flooding of the continents documented in the rock record was due to increased displacement of seawater by increased oceanic ridge activity. For instance, in the Cretaceous, the rate of seafloor spreading is determined to be approximately 30% greater than today. This is calculated to have caused a 200-m rise in sea level that flooded a substantial portion of the continental interiors. A sedimentologic record of the flooding was left in the shallow continental seas produced.

The average age of the continental crust appears to be about 2.1 billion years from Pb isotopic considerations. Whether there was steady or rapid initial growth is still a matter of debate. The crust could be older than this average age, reflecting ages of recycled crust. **Figure 2-11** shows one model of how continental crust may have grown through time. This is an S-type growth model with both slower initial and recent growth. Mechanisms of crustal growth in the modern island arc regime could be from mantle plumes such as the one currently under Yellowstone in the western United States (see below) or magmatism related to continental rifting as presently occurring in the African rift valley. Also, magmas produced in ocean plate subduction as in the Andes mountains and tectonic addition of crustal material during ocean plate subduction in accretionary prisms and crustal underplating, as is presently occurring

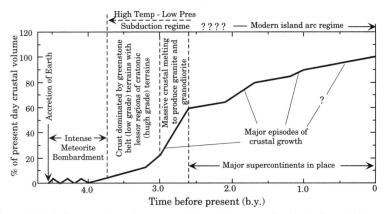

FIGURE 2-11 Model of growth of continental crust through time as outlined by Taylor and McLennan (1995).

in Alaska, may be important. Pervasive arguments have also been presented, however, that argue for early formation of all of the continental crust that has undergone recycling through the mantle with time but with no later growth (Armstrong, 1991). The growth processes listed above may be balanced by subduction of continental sediments and delamination of continental basement to keep the mass of the continents constant. If one argues for continental growth through time, then why, if the earth soon after accretion rapidly differentiated to form a core, did it not also differentiate at this time with respect to silicates to form the continental crust?

■ Pressure in the Earth

Pressure is force per unit area and is generally given in pascals, Pa ($= N\ m^{-2}$), bars ($= 10^{5}\ Pa$), or atmospheres ($= 9.87 \times 10^{4}\ Pa$). Force, F, is mass, m, times acceleration, a ($F = ma$). The force exerted at depth in the crust is equal to the mass, m, above the unit area of interest, A (m^{2}), times the *acceleration due to gravity*, g ($m\ s^{-2}$). g depends somewhat on location on and in the earth. It stays reasonably constant, however, and can be taken as $9.81\ m\ s^{-2}$ for calculation in the crust and upper mantle. This occurs because with increasing depth in the earth the pull of gravity of the rocks above decreases g but is balanced by increases from below due to being closer to the greater mass of higher density rocks in the core.

To determine pressure at a depth z, consider the diagram given in **Figure 2-12**. The volume of material above the unit area of interest is equal to zA, where z is the distance from the earth's surface to the unit area of interest. The mass, m, in this volume is the density of the rock, ρ_{R}, times the volume so that $m = \rho_{R}zA$.

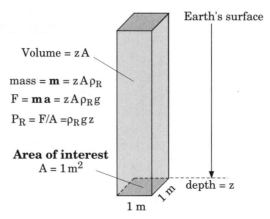

FIGURE 2-12 Control volume indicating that pressure, that is, force per unit area, at a depth in the earth is given by the density of the rocks, ρ_R, times the acceleration of gravity, g, times the depth, z.

The force exerted at depth z is then equal to $\rho_R g z A$. Because the pressure produced by the overlying rocks, P_R, is force per unit area at depth, z, in the crust

$$P_R = \frac{mg}{A} = \rho_R g z \qquad [2.1]$$

To evaluate Equation 2.1, knowledge of the density of rocks is needed. Mineral densities at earth surface conditions vary from quartz (2.65 g cm^{-3}), calcite (2.7 g cm^{-3}), and feldspars ($2.6–2.7 \text{ g cm}^{-3}$) to that of diopside ($3.3 \text{ g cm}^{-3}$), olivine ($3.2–4.4 \text{ g cm}^{-3}$), and garnet ($3.5–4.3 \text{ g cm}^{-3}$). The density of granite at earth surface conditions is $2.67 \pm 0.14 \text{ g cm}^{-3}$. Although the density of a rock increases somewhat with increasing pressure, mineral expansion with increased temperature with depth in the crust tends to offset the effect. Because rocks have less quartz and more olivine (become more mafic), however, rock density increases with depth. This is approximated in Figure 2-3 by having an upper crust of 2.6 g cm^{-3} and a lower crust of 2.8 g cm^{-3}. With this in mind, the density of rocks, ρ_R, when considering the entire earth's continental crust can be taken as approximately 2.7 g cm^{-3} ($= 2.7 \times 10^3 \text{ kg m}^{-3}$ in SI units).

From Equation 2.1, the pressure in pascals with depth, z, in meters is given by

$$P_R \text{ (pascals)} = 2.7 \times 10^3 \text{kg m}^{-3} \text{ (rock density)}$$
$$\times\ 9.81 \text{ m s}^{-2} \text{ (acceleration of gravity)} \times z \text{ (depth)} \qquad [2.2]$$

$$= 2.65 \times 10^4 \text{ N m}^{-3} \times z \text{ (meters)} \qquad [2.3]$$

or in bars as

$$P_R(\text{bars}) = P_R(\text{pascals}) \times 10^{-5} = 0.265 \text{ bar m}^{-1} \times z(\text{meters}) \qquad [2.4]$$

With a maximum thickness of about 80 km, the maximum pressure at the bottom of thickened continental crust would be about 2100 MPa (= 21 kbar). The pressure at the bottom of the average thickness of continental crust of 36 km is 954 MPa (= 9.54 kbar). Because of increases in the density of rocks with increasing depth, the assumption of a constant density of rocks (~2.7 g cm^{-3}), which is reasonable for most calculations in the upper crust, is not a reasonable assumption for the greater densities in the earth's lower crust or mantle; however, in general, it can be stated that in the crust the *lithostatic* (= rock) *pressure* is related to depth by

$$36 \text{ km}/9.54 \text{ kbar} \approx 3.8 \text{ km per kbar} \qquad [2.5]$$

At the bottom of 6 km of oceanic crust, the pressure from the overlying column of rocks would be 159 MPa (= 1.59 kbar). To this the pressure generated from the overlying column of seawater needs to be added. The additional fluid pressure, P_F, due to the ocean with an average density of approximately 1.03 g cm^{-3} (1.03 $\times$ 10^3 kg m^{-3}) is

$$\begin{aligned} P_F(\text{pascals}) &= 1.03 \times 10^3 \text{kg m}^{-3} \text{ (seawater density)} \\ &\quad \times 9.81 \text{ m s}^{-2} \text{ (acceleration of gravity)} \times h \text{ (depth)} \\ &= 1.01 \times 10^4 \text{N m}^{-3} \times h \text{ (m)} \end{aligned} \qquad [2.6]$$

or

$$P_F(\text{bars}) = P_F(\text{pascals}) \times 10^{-5} = 0.10 \text{ bar m}^{-1} \times h(\text{m}) \qquad [2.7]$$

With an average depth of the ocean of about 4.5 km, the added pressure would be 45 MPa (= 450 bars). Equations 2.6 and 2.7 can also be used to determine fluid pressure in the continental crust as well. If fluid at depth exists continuously to the surface, these equations indicate that for each kilometer below the surface of the continental crust the fluid pressure at depth increases approximately 10 kPa (= 100 bars). This difference between fluid pressure and rock (lithostatic) pressure is maintained by the strength of the rocks. At middle or greater depth in the earth's continental crust, the decreased strength of rocks due to increased temperature does not allow this pressure difference to exist. At these depths, fluid pressure approaches rock pressure as the mass of the overlying rocks is loaded on the fluid, as discussed in Chapter 9, where diagenesis and metamorphism of rocks are considered.

■ Heat Flow in the Earth

The *temperature* at the earth's surface varies with latitude and time of year because it is heated from the sun and the earth's axis of rotation changes its tilt with the seasons. This heat energy penetrates 10 m or less into the earth as evidenced by the constant temperature in caves and nearly constant temperature in many wine cellars; however, over periods of thousands of years, climatic changes could conceivably affect temperatures to depths of tens of meters. Below this depth, the temperature is not affected by solar radiation and does not vary with time scales on the order of hundreds of thousands of years or less. At greater time scales, tectonic activity such as mountain building can perturb the thermal structure of the crust.

The temperature in the earth as a function of increasing depth has been measured at the earth's surface below the influence of the solar heating and increases downward. This is consistent with the observation that magmas come to the earth's surface from depth. The *geothermal gradient*, the rate of change of temperature, T, with depth, that is dT/dz, generally ranges from approximately 15°C to 40°C km^{-1} at the earth's surface, depending on location. Over short time scales in localized areas, for instance, after a magma reaches the earth's surface and before it cools significantly, the geothermal gradient can be much higher. Because in general, however, the temperature increases downward, heat is being transported out of the earth. This heat is thought to be produced from the conversion of gravitational energy to heat energy with the sinking of denser material to the core during the early history of the earth (>4.4 Ga). It has also been argued that the earth was hit at this time by a Mars-sized body. If this is the case, then some of the heat that presently escapes from the earth's interior was produced by this impact. To this heat is added the heat produced from the decay of radioactive elements (especially K, U, and Th) residing predominantly in the continental crust.

Heat is a form of energy that can be thought of as "energy in motion" and is generally given in joules, J (N m), ergs (dyne cm), or calories (4.184 joules). A calorie is a unit based on the amount of heat necessary to raise the temperature of 1 g of water by 1°C at standard earth surface conditions. Energy units are a force × distance, which is an amount of *work*. Because work and energy have the same units, work is also a form of energy. Energy can also be specified as volume × pressure. Pressure is force per unit area so the dimensions of volume × pressure are also a force multiplied by a distance. The amount of heat that flows across a unit surface area per unit time is termed the *heat flux*, q (J s^{-1} m^{-2}), and is given for the conduction of heat by

$$q = -K \frac{dT}{dz} \tag{2.8}$$

where K is the *thermal conductivity* constant ($J\ K^{-1}\ m^{-1}\ s^{-1}$) of the material. Because a watt (W), the SI unit of power, is a $J\ s^{-1}$, K can also be given as $W\ K^{-1}\ m^{-1}$. The thermal conductivity constant gives the ability of the material to conduct heat.

From field measurements of dT/dz and with a laboratory determination of K, q can be determined. For silicate rock material found in the crust, values of K are between 1.7 and 3.3 $J\ K^{-1}\ m^{-1}\ s^{-1}$. As mentioned previously, most temperature gradients measured near the earth's surface vary between approximately 15°C and 40°C km^{-1}. Taking a median value of the geothermal gradient in this range of 25°C km^{-1} and a median thermal conductivity of 2.5 $J\ K^{-1}\ m^{-1}\ s^{-1}$, the average *conducive heat flux*, q, out of the earth's crust would be

$$q = 2.5\ J\ K^{-1}m^{-1}s^{-1}\ \text{(thermal conductivity)} \times 2.5 \times 10^2\ K m^{-1}\ \text{(temperature gradient)}$$
$$= 62.5\ mJ\ m^{-2}\ s^{-1}\ \text{(heat flux gradient)} = 62.5\ mW\ m^{-2} \qquad [2.9]$$

Sometimes the heat flux is given in heat flow units $= \mu cal\ cm^{-2}\ s^{-1}$. Converting joules to calories (see Appendix B), 62.5 $mW\ m^{-2} = 1.5$ heat flow units. With the radius of the earth, r, equal to 6371 km, the surface area of the earth ($4\pi r^2$) is $5.1 \times 10^{14}\ m^2$. Using this number along with knowing that there are 3.15×10^7 seconds in a year, the total conductive heat output of the earth, q_{cond} per year, that is, the loss of conductive heat through the earth's surface, is

$$q_{cond} = 62.5 \times 10^{-2}\ J\ m^{-2}s^{-1}\ \text{(heat flux)} \times 5.1 \times 10^{14}\ m^2 \text{(surface area)}$$
$$\times\ 3.15 \times 10^7\ s\ y^{-1}\ \text{(seconds in year)}$$
$$= 1.0 \times 10^{21}\ J\ y^{-1} \qquad [2.10]$$

or 32 TW (tera-watts). How does this quantity of conductive heat compare with anomalously heated areas on the earth where other heat transfer processes are operating?

Because of plate tectonics, approximately 23 km^3 of new molten material are added to the ocean ridges each year. This is determined by knowing that there are about 65,000 km of ocean ridge on the earth (for reference, the earth's circumference = 40,000 km). The spreading rate of the plates varies somewhat depending on location, with slow-spreading ridges spreading at a rate between 1 and 4 cm y^{-1} and fast-spreading ridges between 8 and 16 cm y^{-1}. The average spreading rate of all the plates is about 6.0 cm y^{-1} (3 cm y^{-1} in opposite directions). This gives an area of new crust production a year, A_c, of

$$A_c = 6.0 \times 10^{-5}\ km\ y^{-1}\ \text{(spreading rate)} \times 6.5 \times 10^4\ km\ \text{(ridge length)}$$
$$= 3.9\ km^2\ y^{-1}\ \text{(area produced)} \qquad [2.11]$$

This rate of production of new ocean crust area can be checked because there are $3.1 \times 10^8\ km^2$ of ocean crust on the surface of the earth. The average age of crust with a constant average spreading rate would then be

$$\frac{3.1 \times 10^8 \text{ km}^2}{3.9 \text{ km}^2 \text{y}^{-1}} = 79 \times 10^6 \text{ y} \qquad [2.12]$$

This 79 million year average age is consistent with the observed age distribution of ocean crust and suggests that the rate of movement of crustal plates has stayed, on average, near 6.0 cm/y for at least the last 200 million years.

With a thickness of oceanic crust of 6 km, the volume of new crust produced is then 23 km^3 y^{-1} = 23 $\times$ 10^9 m^3 y^{-1}. With a density of the crust, ρ_R = 2.7 $\times$ 10^3 kg m^{-3} (2.7 g cm^{-3}), there are 6.2 $\times$ 10^{13} kg of new rock added on average each year. How much heat does this newly produced oceanic crust release on cooling? To make this calculation, the amount of heat released on crystallization of magma, ΔH_{xtal}, and how much can be stored in rocks needs to be known. The amount of heat that can be stored in rocks is given at constant pressure as the *heat capacity*, C_P. The heat capacity is the amount of heat necessary to raise the temperature of an object by 1°C. Consider 1200°C magma injected at a ridge (a reasonable temperature for molten basalt) and cooling to 100°C (an average temperature in old 6-km thick crust). Using a heat of crystallization, ΔH_{xtal}, of silicate rocks of 400 kJ kg^{-1} and a C_P of rocks of 1.7 $\times$ 10^3 J kg^{-1} °C^{-1}, this magma produces a heat flux, q_{ridge}, of

$$
\begin{aligned}
q_{\text{ridge}} = &(4.0 \times 10^5 \text{ J kg}^{-1} \; (\Delta H_{\text{xtal}}) + (1200 - 100)\,°\text{C (temperature difference)} \\
&\times 1.7 \times 10^3 \text{ J kg}^{-1}°\text{C}^{-1}) \text{ (heat capacity)} \times 2.7 \times 10^3 \text{ kg m}^{-3}\text{(density)} \\
&\times 23 \times 10^9 \text{ m}^3 \text{ y}^{-1}\text{(volume produced)} = 1.4 \times 10^{20} \text{ J y}^{-1}
\end{aligned}
\qquad [2.13]
$$

A significant amount of this heat is lost rapidly at the oceanic ridges by hydrothermal convection of seawater through the hot rocks when magma is first injected into the oceanic crust (Figure 2-10). Much of it, however, is also lost over time frames on the order of millions of years as the crust slowly cools toward the steady-state conductive heat loss values in old oceanic crust as computed above. Relative to the average conductive heat loss of the entire crust, the heat released from magma injection at the ridges is approximately

$$\frac{1.4 \times 10^{20} \text{ J y}^{-1}}{1.0 \times 10^{21} \text{ J y}^{-1}} = 14\% \text{ of } q_{\text{cond}} \qquad [2.14]$$

What about other high heat flow areas? These can be identified as due to magma injected beneath *hot spots* and the heat produced at *volcanic arcs* and *back-arc basins* associated with ocean crust subduction zones. Hot spots are areas where magma is intruded into the crust from rising plumes of hot mantle material. Their locations are shown in **Figure 2-13**. In times past, these hot spots have supplied the heat necessary to produce large igneous provinces such as the continental basaltic lava traps of India and Siberia as well as oceanic

FIGURE 2-13 Present location of known hot spots with recent tracks across the lithosphere shown as lines. Iceland, Yellowstone, and Hawaii are significantly larger than the other hot spots shown. (Reprinted, with permission, from the *Annual Review of Earth and Planetary Sciences* Volume II © 1983 by *Annual Reviews* www.annualreviews.org.)

basaltic plateaus. The three largest of these hot spots are Hawaii, Yellowstone, and Iceland. Each of these three brings about 0.2 km³ y⁻¹ of magma on average into the shallow crust. All of the other 40 hot spots are significantly smaller compared with these three because their size distribution is near *log normal*, a distribution in which the logarithm of size has a normal distribution.

With about 40 hot spots, the total magma input into the crust each year would be approximately twice the sum of the three largest for a total of about 1.2 km³ y⁻¹. This amount of magma is then approximately 5% of the oceanic ridge input of magma. Except locally, therefore, hot spots do not add significant additional heat to the general transport of conductive heat out of the crust.

Volcanic arcs and back-arc basins are areas of abnormally high surface heat flow in subduction zone complexes, whereas heat flow over oceanic trenches is abnormally low. The low values over oceanic trenches are not surprising because cold material is being thrust under them. The high heat flow at the volcanic arc is also not surprising because magmas are being brought to the surface from interactions in the mantle due to the sinking oceanic lithosphere. The amount of subduction zone magma that is produced worldwide is approximately 30% of the ocean ridge production (Fisher and Schmincke, 1984). Volcanic arcs, therefore, add a significant amount of magma worldwide but less than 4% of the conductive heat budget of the earth.

What is the heat contribution of back-arc basins? **Figure 2-14** gives the surface heat flow across the southern Canadian Cordillera. It indicates that the young Juan de Fuca plate still has a high heat flux that has yet to decrease to old oceanic crust values as given by the Pacific plate. Why is high heat flow produced in back-arc basins if colder material is sinking in the mantle below them? The values are nearly twice that of older stable continental crust and reflect the thinning of the continental crust by a factor of 2. The heat energy necessary for this higher temperature gradient and heat flow is obtained from decreasing gravitational energy by *viscous heating* with the heat transported across the *mantle wedge* by asthenospheric upwelling. The mantle wedge is the wedge-shaped area in the mantle between the subducted ocean lithosphere and the continental lithosphere above. It is a similar situation to that of the initial heating of the earth from the sinking of the denser Fe core material to the center of the earth. The energy conversion from loss of gravitational energy in the sinking slab, ΔE_{grav}, is equal to the heat energy produced. This energy is then

$$\Delta E_{grav} = \Delta \rho g h \qquad [2.15]$$

where h is the distance material sinks in the gravitational field and $\Delta \rho$ is the density difference between the slab and the surrounding mantle before it comes to gravitational equilibrium.

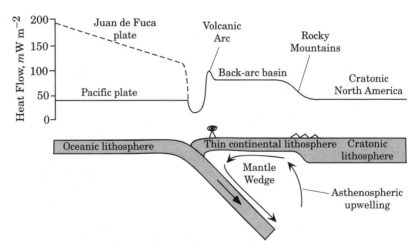

FIGURE 2-14 Heat flow across the Pacific–North American subduction zone complex in southern Canada. Note the low heat flow at the oceanic trench and high heat flow in the volcanic arc and back-arc basin relative to cratonic North America and the Pacific plate. (Adapted from Davis, E. E. and Lewis, T. J., 1984.)

The amount of this energy can be approximated by considering an oceanic lithosphere of 50 km thickness, a length of lithosphere sinking into the earth below the back-arc basin of 400 km (before it comes to gravitational equilibrium), and a length of 50,000 km of subduction zones on the earth's surface. If the slabs are subducting at a 45° angle with an average subduction rate of 3.0 cm y^{-1} and with a 0.1 g cm^{-3} difference in density between the lithosphere and the mantle material through which it sinks,

$$\Delta E_{grav} = 50 \text{ km (slab thickness)} \times 400 \text{ km (length subducted)}$$
$$\times 50,000 \text{ km (surface extent)} \times 0.1 \times 10^3 \text{ kg m}^{-3} \text{ (density difference)}$$
$$\times 9.81 \text{ m s}^{-2} (g) \times 3.0 \times 10^{-2} \text{ m y}^{-1} \text{ (velocity)} \times \sin 45 \text{ (angle)}$$
$$= 2.1 \times 10^{19} \text{ J y}^{-1}$$

[2.16]

This is 2.1×10^{19} J y$^{-1}/1.2 \times 10^{20}$ J y^{-1} = 17.5% of oceanic ridge heat production. The heat produced is distributed over a large area of the mantle below the crust contributing to q_{cond} but with a significant amount conducting out of the earth over a 100 km width along the back-arc basins. Here the surface heat flow is about 25 mW m^{-2} greater than the normal background heat flow away from tectonically active regions. This additional back-arc basin heat, ΔH_{bab}, is then

$$\Delta H_{bab} = 25 \times 10^{-3} \text{ J m}^{-2} \text{ s}^{-1} \text{ (heat flow)} \times 100 \times 10^3 \text{ m (basin width)} \times 50,000$$
$$\times 10^3 \text{ m (length)} \times 3.15 \times 10^7 \text{s y}^{-1} \text{ (seconds per year)}$$
$$= 3.9 \times 10^{18} \text{ J y}^{-1}$$

[2.17]

The relative amount of this heat flow in the back-arc basin to that produced by subduction is

$$\frac{3.9 \times 10^{18} \text{ J y}^{-1}}{2.1 \times 10^{19} \text{ J y}^{-1}} \text{ or } \sim 20\%$$

[2.18]

indicating a significant amount, but not all, of the heat from oceanic slab sinking is manifest in the back-arc basin. These numbers could be off by a factor of 2 given the simplicity of the calculations, but one now has a handle on the energy transfer processes that can cause chemical reactions to occur in the earth's crust.

■ Temperature in the Earth

Heat flow exerts a strong control on the temperature distribution in the crust. During mid-oceanic ridge volcanism and at volcanic islands like Hawaii and Iceland, basalt magma is brought to the earth's surface. Temperature of this material is approximately 1200°C. How does temperature change with depth in the crust away from these magma injections? Consider a thermal gradient of 25°C km^{-1} (a number used above and a reasonable number measured at the surface of old oceanic crust away from the ridges where the heat flow has become nearly steady state) and an ocean bottom temperature near 0°C. The temperature at the bottom of the 6-km thick oceanic crust, T_{oc}, would be

$$T_{oc} = 0°C + 25°C \, km^{-1} \times 6 \, km = 150°C \qquad [2.19]$$

Obviously, in newer oceanic crust the temperature gradient and, therefore, the bottom temperature would be greater because of the larger heat flux as indicated in Figure 2-14 for the Juan de Fuca plate.

A surface geothermal gradient of 25°C per km is also a reasonable number for average continental crust. With a surface temperature of 20°C, the temperature at the bottom of 35-km thick continental crust, T_{cc}, would be

$$T_{cc} = 25°C \, km^{-1} \times 35 \, km + 20°C = 895°C \qquad [2.20]$$

At 35 km (~10 kbar), laboratory experiments indicate that continental rocks would begin to melt at approximately 650°C. Nevertheless, seismic wave studies indicate the continental crust is not partially molten at this depth. This implies the gradient in temperature, dT/dz, measured at the surface must decrease with depth in the crust or the lower crust is more mafic than the upper crust and has lost its volatiles, which increases its melting temperature.

A lower gradient occurs because continental crust rocks contain significant concentrations of *isotopes* that undergo spontaneous radioactive decay and produce heat. The important ones for heat production are ^{238}U, ^{235}U, ^{232}Th, and ^{40}K, which produce heat of 3.0, 19.7, 0.88, and 0.88 J g^{-1} y^{-1}, respectively. The superscript number indicates the *atomic weight* of these particular isotopes. This is the number of protons plus neutrons in the nucleus of an atom of the isotope. Radioactive isotopes are discussed in Chapter 10. Given the abundance of radioactive isotopes the production of heat from a typical granite is approximately 2.5×10^{-6} Wm^{-3} and that for the upper crust as a whole is estimated to be 1×10^{-6} Wm^{-3}.

To determine how the geothermal gradient changes with depth, an infinitesimally small volume of rock can be considered, as shown diagrammatically in **Figure 2-15**. The cross-sectional area through which the heat fluxes is shown in gray and denoted as A. The heat flux per unit area, q in J m^{-2} s^{-1}, that passes through A leads to a total heat flux of $q(z)A$ into the volume at height z. The heat that accumulates in the volume per unit time is the heat that fluxes in at z minus the heat that fluxes out at $z + dz$. This difference is given by the gradient in $-Aq$ from z to $z + dz$ times the length, dz, so that

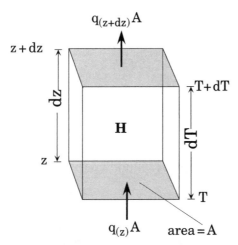

FIGURE 2-15 A control volume of rock given by $A\,dz$ with a flux of heat entering at z given by $q(z)\,A$ and a flux of heat exiting at $z + dz$ given by $q(z + dz)\,A$, where the geothermal gradient is dT/dz.

$$\text{Diffusive heat increase in the volume per unit time} = -dz\frac{d(Aq)}{dz} \tag{2.21}$$

Now consider that the volume contains radioactive elements that generate heat. If this heat production per unit volume per unit time is H ($\mathrm{J\,m^{-3}\,s^{-1}}$), then the total heat produced by radioactive decay per unit time is $A\,dz\,H$. The total increase in heat in the volume is

$$\text{Total heat increase in the volume per unit time} = -dz\frac{d(Aq)}{dz} + Adz\,H \tag{2.22}$$

As a function of time, the amount of heat in the volume is the heat produced or absorbed by any chemical reactions taking place per mole of reaction in the unit mass, ΔH_R, times the mass of the volume, times the molar reaction rate dR/dt. To this must be added the heat stored in the volume that is equal to the heat capacity, C_P, per unit mass times the mass of the volume times the rate that temperature increases in the volume, dT/dt. With the amount of mass in the volume considered of $A\,dz\,\rho_R$, where ρ_R is the density of the rock in the volume, the heat increase is given by

$$\text{Total heat increase in the volume per unit time} = A\,dz\,\rho_R\left(\Delta H_R\frac{dR}{dt} + C_P\frac{dT}{dt}\right) \tag{2.23}$$

In progressive metamorphic reactions (see Chapter 9), the heat absorbed during devolatilization reactions can be similar to that stored through the C_P of the rock. Ignoring these reactions would then lead to a calculation that was in error by a factor of about 2 (Walther and Orville, 1982).

Equating the expression in Equations (2.22) and (2.23) results in

$$A \, dz \, \rho_R \left(\Delta H_R \frac{dR}{dt} + C_P \frac{dT}{dt} \right) = -dz \frac{d(Aq)}{dz} + A \, dz \, H \qquad [2.24]$$

Because A does not depend on z,

$$\frac{d(Aq)}{dz} = A \frac{dq}{dz} \qquad [2.25]$$

Equation 2.24, therefore, can be simplified to

$$\rho_R \left(\Delta H_R \frac{dR}{dt} + C_P \frac{dT}{dt} \right) = -\frac{dq}{dz} + H \qquad [2.26]$$

From Equation 2.8, $q = -K(dT/dz)$. Equation 2.26 can, therefore, be expressed as

$$\rho_R \left(\Delta H_R \frac{dR}{dt} + C_P \frac{dT}{dt} \right) = K \frac{d^2T}{dz^2} + H \qquad [2.27]$$

Solving for dT/dt

$$\frac{dT}{dt} = \underset{\substack{\text{heat} \\ \text{flux}}}{\frac{K}{\rho_R C_P} \frac{d^2T}{dz^2}} + \underset{\substack{\text{heat} \\ \text{production}}}{\frac{H}{\rho_R C_P}} - \underset{\substack{\text{heat of} \\ \text{reaction}}}{\frac{\Delta H_R}{C_P} \frac{dR}{dt}} \qquad [2.28]$$

This is the one-dimensional heat conduction equation with chemical reactions occurring. The thermal diffusivity, κ of rocks is defined as

$$\kappa \equiv \frac{K}{\rho_R C_P} \qquad [2.29]$$

with units of length^2/time (m^2 s^{-1}). Equation (2.28), therefore, can also be written as

$$\frac{dT}{dt} = \kappa \frac{d^2T}{dz^2} + \frac{H}{\rho_R C_P} - \frac{\Delta H_R}{C_P} \frac{dR}{dt} \qquad [2.30]$$

Equations 2.29 and 2.30 indicate a number of variables that are needed to obtain the change in temperature with time; however, a *dimensional analysis* can be done on Equation 2.30. If a temperature change, dT, occurs in a characteristic time change, dt, the change propagates a distance, d, on the order of $d \sim (\kappa dT)^{1/2}$. Also, an expression with κ that gives a characteristic time required for a temperature change to propagate a distance, d, is $\sim d^2/\kappa$. These relationships are helpful to obtain estimates of thermal effects. For instance, the time necessary for heat from the formation of the earth's core to be propagated by conduction to the surface ($d = 2885$ km) with κ for the mantle of 1.5×10^{-6} m^2 s^{-1} is 5.5×10^{18} s or 1.8×10^{11} y. This implies that if heat from the core is reaching the earth's surface, a different heat transfer mechanism must be operating (i.e., convection).

At steady state, with no change in temperature and where chemical reactions are at equilibrium, $dT/dt = dR/dt = 0$ and

$$0 = \frac{K}{\rho_R C_P}\frac{d^2 T}{dz^2} + \frac{H}{\rho_R C_P} \qquad [2.31]$$

Solving for $d^2 T/dz^2$ gives

$$\frac{d^2 T}{dz^2} = -\frac{H}{K} \qquad [2.32]$$

With H and K constant, taking the indefinite integral of Equation 2.32 with respect to depth z gives

$$\frac{dT}{dz} = -\frac{H}{K}z + c_1 \qquad [2.33]$$

where c_1 is a constant of integration. If at the earth's surface $z = 0$ and $q_o = -K$ (dT/dz), then $c_1 = q_o/K$, where q_o is the heat flux per unit volume at the surface reflecting both the heat produced in the column and entering from below the column. Making this substitution,

$$\frac{dT}{dz} = -\frac{H}{K}z + \frac{q_o}{K} \qquad [2.34]$$

Integrating Equation 2.34 to obtain the temperature as a function of depth results in

$$T = -\frac{H}{2K}z^2 + \frac{q_o}{K}z + c_2 \qquad [2.35]$$

where c_2 is a constant of integration. Taking the average temperature at the surface of the earth ($z = 0$) from conductive heat loss as 10°C, Equation 2.35 can be written as

$$T = -\frac{H}{2K}z^2 + \frac{q_o}{K}z + 10°C \qquad [2.36]$$

where T is in °C. With $H = 0$, $K = 2.5$ J °C^{-1} m^{-1} s^{-1}, and $q_0 = 62.5$ mJ m^{-2} s^{-1}, Equation 2.36 becomes

$$T = \frac{62.5 \; m\text{J m}^{-2}\text{s}^{-1}}{2.5 \text{ J °C}^{-1}\text{m}^{-1}} z + 10\text{°C} \qquad [2.37]$$

Therefore, if H, the radiogenic heat input in the column, is zero, the temperature is a linear function of distance and all the surface heat flux must be accounted for by heat entering the column from below. The temperature as a function of depth from Equation 2.37 for 35-km thick continental crust with $H = 0$ is shown in **Figure 2-16**.

Given the concentration of heat-producing elements in the upper crust, the amount of heat produced in average upper continental crust, that contributes to the observed surface heat flow, is about 1×10^{-6} J m^{-3} s^{-1}. Taking this value for the crust as a whole, the temperature structure of the crust at steady state is given by

$$T = -\frac{1 \times 10^{-6} \text{ J kg}^{-1} \text{ s}^{-1}}{2 \times 2.5 \text{ J °C}^{-1} \text{ m}^{-1}} z^2 + \frac{62.5 \times 10^{-3} \text{ J m}^{-2} \text{ s}^{-1}}{2.5 \text{ J °C}^{-1} \text{ m}^{-1}} z + 10 \qquad [2.38]$$

where T is in °C. This geothermal gradient is also shown in Figure 2-16. It is thought by most investigators that the radioactive heat-producing elements are heavily concentrated in the upper part of the crust, with the lower crust being more mafic and, therefore, containing less radioactive heat-producing el-

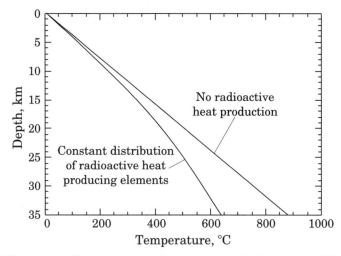

FIGURE 2-16 Temperature with depth in the earth's continental crust indicating that a 25°C per km gradient observed at the surface would give a temperature of 885°C at 35 km, whereas the actual temperature is near 600°C because of the addition of radiogenic heat in the crust.

ements. This decrease in radioactive elements with depth would cause the temperature at the base of the crust to be lower.

The general zone of partial melting in the earth does not begin to occur until the depth of the asthenosphere is encountered. As discussed above, this depth is variable depending on location but is generally between 65 and 220 km below continents. Below this depth, heat is believed to be transferred dominantly by convection of material except in the solid part of the core. Transporting hot material upward by a *convective heat flux* is a more efficient way to move heat, and temperature gradients are, therefore, much less than for conductive heat transfer. If the convection is vigorous, the temperature gradient becomes adiabatic. An adiabatic system is one that does not exchange heat with the surroundings.

To determine the *adiabatic gradient*, the constructs of thermodynamics are needed, which is not introduced until Chapter 3. It can be stated here, however, that for the upper mantle this calculation gives an adiabatic temperature gradient of approximately $0.5 \, °C \, km^{-1}$ and approximately $0.3 \, °C \, km^{-1}$ for the lower mantle. With this in mind, consider a mass of rock in the mantle that is convecting upward along an adiabatic gradient, as shown schematically in **Figure 2-17**. Depending on its composition, at some point, the mass of rock can start to melt as its melting curve is encountered, as given by point A in Figure 2-17. As melting occurs, the presence of melt plus solid buffers the temperature along the melting curve. Because the melt is generally less dense than the solid rock, at some point (point B in Figure 2-17), the melt separates from the rock and rise as a *diapir*. This melt expands to a greater degree than the original rock, and its adiabatic gradient is, therefore, greater, being approximately

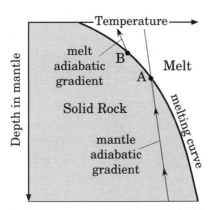

FIGURE 2-17 The temperature profile in convecting mantle rocks is given by an adiabatic gradient as shown. If the melting curve for a rock of a particular composition crosses the adiabatic gradient curve, a separate magma phase can be produced at point A. The temperature for a further rise of the rock is buffered along the melting curve until the magma phase can separate along its own adiabatic gradient given as point B.

$1.0\,°C\,km^{-1}$. This is then how more dense mantle material can be added to the earth's crust.

Summary

The earth is thought to have formed from planetesimals produced on cooling of the solar nebula. The physical and chemical nature of the layers in the earth is deduced by knowing its gross density from astronomic calculations and composition from meteorites. Seismic wave refractions and travel times are then used to obtain the thickness and density of each of the layers.

The earth consists of a core made of Fe metal with approximately 4% Ni metal and likely 10% of Si, S, and/or O. It is nearly 3500 km in thickness. The outer part of the core of approximately 2300 km thickness is molten and convects, producing the earth's magnetic field. Outward from the core is the mantle of the earth. It is dominantly made up of minerals of the oxides of SiO_2 and MgO with lesser amounts of FeO, Al_2O_3, and CaO. It is about 2860 km thick and makes up most of the mass of the earth. The lower mantle, from the core at a depth of 2885 km to the 650 km seismic discontinuity, is considered to be made up dominantly of Mg-perovskite, $MgSiO_3$, with approximately 18 volume % Mg-wüstite, $(Fe,Mg)O$, and approximately 8% Ca-perovskite, $CaSiO_3$. Above this lower mantle layer, from 650 km to approximately 400 km in depth, the mantle is considered to be dominantly composed of an olivine polymorph and garnet. From 400 km to the top of the mantle, olivine, garnet, orthopyroxene, and clinopyroxene dominate the mineralogy.

The top compositional layer of the earth resting on the mantle is the crust. The crust under the oceans is different from that in the continents. Oceanic crust is crystallized at mid-ocean ridges and accumulates sediments with age, having an average thickness of 6.4 km made up of 0.5 km of sediments on 1.4 km of pillow basalts that is underlain by 4.7 km of layered gabbro. Continental crust is significantly thicker, varying from about 20 to 80 km in thickness. The upper continental crust is felsic, and the lower continental crust is mafic.

Pressure in the earth is determined from $\rho_R\,g\,z$, where ρ_R is the average density of the material above the depth of interest, z, and g is the acceleration of gravity. If fluid exists in interconnected pores to the earth's surface, its pressure can be calculated in the same manner but using the density of the fluid rather than ρ_R.

Heat escapes from the earth primarily because of conduction of heat through the crust. Heat is also lost through the crust by intruding hot rocks at mid-oceanic ridges, hot spots, and volcanic arcs. These are, however, second-order relative to crustal conduction. In the mantle, the heat transport mechanism is convection. The temperature in the earth is given by the geothermal gradient dT/dz. At the earth's surface, it is determined to generally be between $15°C\,km^{-1}$

and 40°C km^{-1} and decreases with depth in the crust due to heat production from radioactive decay.

Key Terms Introduced

acceleration due to gravity
achondrites
adiabatic gradient
alkali
alkaline earth
asthenosphere
atomic number
atomic weight
back-arc basins
buoyancy forces
carbonaceous chondrites
chondrites
chondrules
clinopyroxene
conductive heat flux
continental crust
convective heat flux
cosmic microwave background
diapir
dimensional analysis
eclogite
feldspar
felsic
geothermal gradient
granulites
gravity
harzburgite
heat capacity
heat flux
hot spots
irons
isotopes

lherzolite
lithosphere
lithostatic pressure
log normal
mafic
majorite
mantle
mantle wedge
Moho
MORB
oceanic crust
olivine
orthopyroxene
peridotite
perovskite
planetesimals
pressure
pyrolite
red giant
rheologic divisions
shield
solar nebula
stones
stony-irons
supernovae
temperature
thermal conductivity
ultramafic
viscous heating
volcanic arcs
work
wüstite

Questions

1. What are chondrules?
2. Describe the density structure of the earth and its rheologic divisions.
3. What is the difference between crust and lithosphere?

4. How thick is the earth's crust?

5. What is the most common mineral in the earth? In the earth's crust?

6. Why is oceanic crust more silica-rich than the mantle and continental crust more silica-rich than oceanic crust?

7. Give a definition of heat.

8. Heat is fluxing toward the earth's surface. Where does it come from?

9. Give a definition of pressure.

10. How does a sinking lithospheric plate create heat energy? If energy is neither created nor destroyed, where does it come from?

11. Where does the heat produced from radioactive decay come from if energy is neither created nor destroyed?

12. Why is g, the acceleration of gravity on earth, not constant? What does it depend on?

Problems

1. Given in the table are measured velocities of galaxies (from magnitude of their red shifts) and their distance from an expansion center (from their decrease in star brightness). Plot the velocities against distances. The slope of the line through (0,0) is the Hubble constant (= 1/time). What is its value? The age since expansion is the inverse of the Hubble constant. What is the age in years of the universe from this calculation?

Velocity (km s^{-1})	Distance (10^{21} km)
1,900	0.9
3,600	1.9
6,100	2.8
6,100	3.4
7,300	3.7
10,000	4.6
16,000	7.3
20,000	9.3
30,000	12

2. Calculate the difference between lithostatic and fluid pressure at 5 km below the surface.

3. Calculate the pressure that develops below 10 km of mafic lower crust of 2.8 g cm^{-3} resting below 15 km of granite upper crust of density 2.6 g cm^{-3}.

4. With an average thermal conductivity of 2.5 J °C m^{-1} s^{-1} and a range of temperature gradients on the earth's surface between about 8 and

40°C km^{-1} compute the range of heat fluxes out of 1 m^2 of average crust. Give the answer in mW.

5. If the Hawaiian hot spot injects 0.5 km^3 of magma per year into the shallow crust, what is its average annual heat flux contribution?

6. How much viscous heat would be produced in a year by sinking a 1000-km length of lithospheric slab of 80-km thickness into the mantle for a length of 400 km before it comes to gravitational equilibrium if the average density difference between the slab and mantle is 0.08 g cm^{-3}? The angle of subduction is 30° with a rate of 3 cm per year.

7. Consider a model of the upper 10 km of continental crust with average granite composition that produces heat from radioactive decay of 2.8×10^{-2} J kg^{-1} y^{-1}. With the density of these rocks of 2.8 g cm^{-3}, what would be their contribution to the surface heat flux at steady state? Would this contribute the majority of average conductive heat output at the surface? Show your calculations.

8. Give the temperature at the bottom and calculate a steady-state isotherm through a 40-km crust given a surface temperature of 20°C, rocks with a thermal conductivity of 2.5 W m^{-1} °C^{-1}, and where every m^3 of rock produces 1.3 μW of heat with a heat flux per unit volume at the surface of 60 mJ m^{-2} s^{-1}.

9. Calculate the volume of the (a) crust and (b) mantle and (c) the crustal volume as a percentage of the mantle volume. Assume a crust of 24 km in thickness. (Note: the volume of a sphere or radius r is given by $4/3\pi r^3$.)

10. Use the values computed in Problem 8. Assume heat is produced in the crust and mantle with a uniform distribution of heat producing elements generating 1.5×10^{-6} W m^{-3} and 1.0×10^{-8} W m^{-3}, respectively. What is the heat output of the (a) crust and (b) mantle per second? If the heat from the mantle is added to the crust, what percentage of total heat flow at the earth's surface at steady-state is from the mantle?

11. With a surface temperature = 10°C and heat flux = 65 mJ m^{-2} s^{-1} above a 35 km crust, calculate temperature profiles at steady state through the crust. Assume no heat production and a constant heat production = 1.2×10^{-6} J m^{-3} s^{-1} with a thermal conductivity $K = 2.3$ J °C^{-1} m^{-1} s^{-1}.

12. With a surface temperature = 10°C and surface heat flux = 60 mJ m^{-2} s^{-1}, calculate temperature profiles at steady state through the crust with no heat production in the crustal material and also a constant heat production = 0.5×10^{-6} mJ kg^{-1} s^{-1}. Assume a 35-km thick crust, rock density, $\rho = 2.7$ g cm^{-3} and a thermal conductivity, $K = 2.5$ J °C^{-1} m^{-1} s^{-1}.

References

Allègre, C. J., Manhes, G. and Gopel, C., 1995a, The age of the earth, *Geochim. Cosmochim. Acta,* v. 59, pp. 1445–1456.

Allègre, C. J., Poirier, J. P., Humler, E. and Hofmann, A. W., 1995b, The chemical composition of the Earth. *Earth Planet. Sci. Lett.,* v. 134, pp. 515–526.

Anders, E. and Grevesse, N., 1989, Abundances of the elements: Meteoritic and solar. *Geochim. Cosmochim. Acta,* v. 53, pp. 197–214.

Armstrong, R. L., 1991, The persistent myth of crustal growth. *Australian Jour. Earth Sci.,* v. 38, pp. 613–630.

Broecker, W. S., 1985, *How to Build a Habitable Planet,* Eldigio Press, Palisades, New York, 291 pp.

Carslaw, H. S. and Jaeger, J. C., 1959, *Conduction of Heat in Solids,* 2nd edition. Oxford, New York, 520 pp.

Crough, S. T., 1983, Hotspot swells. *Annul. Rev. Earth Planet. Sci.,* v. 11, pp. 165–193.

Davis, E. E. and Lewis, T. J., 1984, Heat flow in a back-arc environment: Intermontane and Omineca crystalline belts, southern Canadian Cordillera. *Can. J. Earth Sci.,* v. 21, pp. 715–726.

Dziewonski, A. M. and Anderson, D. L., 1981, Preliminary reference Earth model. *Phys. Earth Planet. Int.,* v. 25, pp. 297–356.

Fisher, R. V. and Schmincke, H. U., 1984, *Pyroclastic Rocks.* Springer-Verlag, New York, 409 pp.

Hofmann, A. W., 1988, Chemical differentiation of the Earth: The relationship between mantle, continental crust, and oceanic crust. *Earth Planet. Sci. Lett.,* v. 90, pp. 297–314.

Hutchison, R., 1974, The formation of the earth. *Nature,* v. 250, pp. 556–568.

Klein, E. M. and Langmuir, C. H., 1987, Global correlations of ocean ridge basalt chemistry with axial depth and crustal thickness. *J. Geophysical Res.,* v. 92, pp. 8089–8115.

Mason, B., 1966, *Principles of Geochemistry,* 3rd edition. Wiley, New York, 329 pp.

McDonough, W. F. and Sun, S.-S., 1995, The composition of the Earth. *Chem. Geol.,* v. 120, pp. 223–253.

Ringwood, A. E., 1991, Phase transformations and their bearing on the constitution and dynamics of the mantle. *Geochim. Cosmochim. Acta,* v. 55, pp. 2083–2100.

Ronov, A. B. and Yaroshevsky, A. A., 1976, A new model for the chemical structure of the Earth's crust. *Geochem. Intern.,* v. 13, pp. 89–121.

Rudnick, R. L. and Fountain, D. M., 1995, Nature and composition of the continental crust: A lower crustal perspective. *Rev. Geophys.*, v. 33, pp. 267–310.

Sears, D. W. G. and Dodd, R. T., 1988, Overview and classification of meteorites. In Kerridge, J. F. and Matthews, M. S. (eds.), *Meteorites and the Early Solar System*. University of Arizona Press, Tucson, pp. 3–31.

Shor, G. S., Jr. and Raitt, R. W., 1969, Explosion seismic refraction studies of the crust and upper mantle in the Pacific and Indian oceans. In Hart, P. J. (ed.), *The Earth's Crust and Upper Mantle*. Geophysical Monograph 13. American Geophysical Union, Washington, DC, pp. 225–230.

Taylor, S. R. and McLennan, S. M., 1995, The geochemical evolution of the continental crust. *Rev. Geophys.*, v. 33, pp. 241–265.

Verhoogen, J., Turner, F. J., Weiss, L. E., and Wahrhaftig, C., 1970, Heat sources and the thermal evolution of the earth. In *The Earth, An Introduction to Physical Geology*. Holt, Rinehart and Winston, New York, pp. 637–661.

Walther, J. V. and Orville, P. M., 1982, Volatile production and transport in regional metamorphism. *Contrib. Mineral Petrol.*, v. 79, pp. 252–257.

Wedepohl, K. H., 1995, The composition of the continental crust. *Geochim. Cosmochim. Acta*, v. 59, pp. 1217–1232.

3

Introduction to Thermodynamics

Most earth processes involve the transfer of *energy* into or out of a mass or volume of material. Many forms of energy can be transferred, including chemical, heat, kinetic, gravitational, and mass. Thermodynamics deals with this flow of energy into and out of a *system* through its *boundaries* as it changes from one state to another. The *state of a system* is defined by its composition, the physical state of the phases it contains, and the *forces* operating across its boundaries, as shown schematically in **Figure 3-1**. In a geologic context, the most important forces are temperature, pressure, and *chemical potential*. Temperature is the force that causes heat to flow. Pressure is the force that causes pressure-volume work to be done, and chemical potential is the force that allows mass to be chemically transported into or out of the system. Other forces such as gravitational, electrical, and magnetic need to be considered when these become important.

The state of the system is given by measurable properties called state variables. *State variables* include temperature, pressure, volume, chemical composition of phases in a system, and the system's position in a gravitational field. *Phases* characterize the mass in the system and are homogeneous parts of the system with distinct boundaries, as outlined in Chapter 1. **Figure 3-2** depicts a section through an amphibolite, a high-grade metamorphic rock. It can be con-

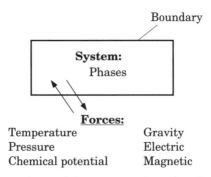

I FIGURE 3-1 The nature of constraints needed to construct a thermodynamic model of a system.

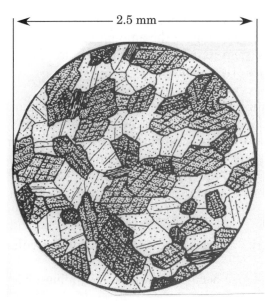

FIGURE 3-2 Drawing of an amphibolite from the Grand Canyon (modified from Williams et al., 1954) showing the four phase system *idioblastic* hornblende and *xenoblastic* plagioclase, biotite, and titanite.

sidered a four-phase system dominantly made up of hornblende and plagioclase but that also includes the phases biotite and titanite.

Consider a system made up of two phases, 1 mole of water plus 1 mole of ice, at −2°C and 1 bar where no forces other than pressure and temperature are important, as shown in **Figure 3-3**. The system ice + water at −2°C and 1 bar would not be an equilibrium system because with time it would undergo a process in which the water would freeze to form ice. If the system were 1 mole of liquid H_2O plus 1 mole of ice at 0°C and 1 bar, this would be an equilibrium system because it would not change in character with time. This equilibrium can be tested by raising the pressure slightly and then returning the pressure to its original value. During this process, a small amount of ice would melt during the time the pressure was increased, and an equal amount of water would freeze after the pressure was returned to 1 bar, bringing the ice and water back to their original amounts; therefore, a system is at *equilibrium*

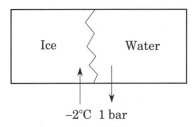

I FIGURE 3-3 A system of two phases: ice + water at −2°C and 1 bar.

1. When there is no change in its character with time
2. When altered by changing the amount of a force that operates on the system, the system returns to its original state if the changed force is returned to its original value

Given certain information about the state of a system, thermodynamics predicts the direction toward which the nonequilibrium system will change (i.e., the direction of the process that occurs as the system changes toward equilibrium) and determines when a system is at equilibrium. Consider the nonequilibrium system, water plus ice initially at −2°C and 1 bar. If it were involved in an *isolated system* process, the system would change under the following constraints:

1. The boundaries are rigid walls not allowing the volume of the system to change.
2. No heat exchange occurs with the surroundings. The system is thermally isolated.
3. No external forces act on the boundaries of the system.

As this system approached the equilibrium state, the temperature of the system would increase. This occurs because during the spontaneous process toward equilibrium heat is given off by the water when some of it is converted to ice and this heat raises the temperature of the system. The pressure of the system would also increase during the process because a mass of ice near −2°C and 1 bar takes up more space than the same mass of water and the volume is constrained to be constant.

State variables depend only on the state of the system and not on how the system got to the state it is in. State variables necessary to describe the two-phase system ice + water would be the forces operating on the system, for example, the temperature and pressure, plus the characterization of how the mass of the system is distributed, in this case, the moles of the water phase, n_w, and the moles of the ice phase, n_i. For this isolated system, during the spontaneous process of freezing some of the water to ice, the equilibrium state achieved would be characterized by the new n_i and n_w, as well as a new pressure and temperature.

Processes are the way a system changes from one state to another. Real processes do not occur in isolation but often can be modeled as a *closed-system* process. During a closed-system process

1. Heat exchange with the surroundings and volume changes can occur.
2. No forces operate to change the system except pressure and temperature.
3. Also, no mass is transferred in or out of the system.

Depending on constraints imposed during the process, different types of processes are defined:

Adiabatic processes occur with no heat exchange across the boundaries of the system.

Isobaric processes occur at constant pressure, but the boundaries of the system can move so that the system volume can change.

Constant volume processes occur with no movement of the boundaries of the system, but the pressure of the system may change.

When a system change involves the transfer of mass across its boundaries, it is said to be an *open system* rather than closed. **Table 3-1** gives the definition of some of the terms used in constructing a thermodynamic model of a system.

Thermodynamics considers the *Internal Energy*, *U*, of a system. This can be considered to be the total energy of the system made up of energy from three components: mass, motion, and potential. The largest is the energy due

|**Table 3-1**   **Thermodynamic terms**

System is any region of the universe large or small that is considered for analysis. It is defined by its chemical composition, the physical state of the phases in it, and the forces operating across its boundaries, for example, temperature, pressure, gravity. The boundaries distinguish the system from its surroundings. A system may be conceptual rather than real. Often, systems are constructed to be simplifications of reality.

State of a system is given by a set of state variables that characterize the mass and state of all the phases in and all the forces operating on the system.

State variable is an independent variable that defines the state of the system, for example, temperature and volume. The value of this variable depends on the state of the system and not on how the state was achieved.

Phase is a homogeneous mass with distinct boundaries. It can, in principle, be mechanically separated from other phases in the system.

Process is the way a system changes between two states. Three types of system processes that are often considered are as follows:

Isolated system process—no energy, including heat, work, matter, and potential energy, is exchanged with the system's surroundings through its boundaries.

Closed system process—energy can be exchanged due to pressure and temperature changes, but matter is not exchanged with the system's surroundings nor are any other forces operating on the system.

Open system process—all types of energy transfer; heat, work, mass, and positions in potential fields are allowed to be exchanged with the system's surroundings.

Equilibrium is a state of a system when none of its properties change with time, and if one of the state variables that describe it is disturbed slightly and returned to its original value, the system changes but returns to its original state.

to its mass, E_m, to be exact, its rest mass. The mass, m, is equated to energy by Einstein's special law of relativity:

$$E_m = mc^2 \qquad\qquad [3.1]$$

where c denotes the speed of light (3×10^8 m s^{-1}). For 1 mole (18 g) of H_2O

$$E_m^{H_2O} = \underset{\text{mass of } H_2O}{18 \times 10^{-3} \text{ kg mol}^{-1}} \times \underset{\text{speed of light}}{(3.0 \times 10^8 \text{m s}^{-1})^2} = 1.6 \times 10^{15} \text{J mol}^{-1} \qquad [3.2]$$

A system also has energy because of the kinetic motion of the atoms and other particles it contains. This kinetic energy, E_k, can be expressed by

$$E_k = \frac{1}{2} \sum_i m_i v_i^2 \qquad\qquad [3.3]$$

where the summation is taken over each atom, i, of mass, m_i, and velocity, v_i, as it vibrates in a solid or has dominantly translational and rotational motion in a gas or liquid. The vibrational energy is really a combination of kinetic and potential energy, the sum of which is equal to the maximum kinetic energy of the atom during a vibration. For the discussion here, no kinetic energy caused by changing the location of the system as a whole is considered.

E_k is temperature dependent, being 0 at absolute zero and increasing with increasing temperature because the velocity of the particles increases from 0 K. For H_2O, approximating it as an *ideal gas* so that it has only translational kinetic energy, the mean velocity of a molecule is 6.15×10^2 m s^{-1} at 1 bar and 0°C; therefore, the ideal gas kinetic energy of H_2O, $E_k^{H_2O}$, at these conditions would be

$$E_k^{H_2O} = 0.5 \times 18 \times 10^{-3} \text{ kg mol}^{-1} \times (6.15 \times 10^2 \text{ m s}^{-1})^2 = 3.4 \times 10^3 \text{ J mol}^{-1} \qquad [3.4]$$

Additionally, potential energy, E_p, can exist for a system because of its position in potential fields. These potential energy fields can be chemical, electrical, magnetic, or gravitational. For instance, H_2O on the top of a hill in a gravitational field (as exists on earth) has greater energy than the H_2O at the bottom of the hill. If the H_2O flows down the hill, it increases its kinetic energy as it loses its gravitational potential energy. A similar situation occurs with vibrational energy, as mentioned above. The maximum kinetic energy of vibration occurs at the center of vibration and decreases to zero as the particle approaches the point where it changes direction. At this point, the particle's potential energy is a maximum. The kinetic and potential energy sum to a constant during vibration. In both examples, energy gets transferred but not created or destroyed. This is what the first law of thermodynamics states, which is introduced below.

Note the 12 orders of magnitude difference between $E_m^{H_2O}$ and $E_k^{H_2O}$. This means the total U of H_2O cannot be measured to the precision of a value of $E_k^{H_2O}$ because mass changes can typically only be characterized to six or seven significant figures. That is, the total energy of H_2O caused by mass changes cannot be characterized to the accuracy of changes in its kinetic energy. This is generally not a problem, however, because in thermodynamics one is interested in characterizing the changes in U, that is, ΔU of a system during a process rather than its absolute value in any state. The changes can be measured accurately, whereas the total U is known with less absolute accuracy.

■ Changes in *U* Due to Changes in Heat, Work, and Mass

Changes in U, ΔU, are positive if the energy of the system increases. This can be done by adding *heat*, q; doing *work* on the system, w; or increasing its mass. Heat is the energy that flows across a system boundary in response to a temperature gradient. It can be thought of as a disorganized transfer of energy. Work is all other types of energy transfer across the system boundaries except for mass transfer in open systems. Work can be thought of as an organized transfer of energy because it does not increase the *entropy*, that is, randomness of the system (see below). To understand how work adds energy to a system, consider the work necessary to compress a gas. The expansion of the compressed gas back to its original volume can be used to have the system expend the energy to do work on its surroundings.

Work, w, is force, F, acting through a distance. For mechanical systems in one direction the relationship can be written as

$$\text{work} = \text{force required to move body} \times \text{distance body is moved} \qquad [3.5]$$

or joules = newtons × meters. Aligning the direction of the force *vector*, F, along the x direction from x_o to x, the amount of work can be computed from

$$w = \int_{x_o}^{x} F dx \qquad [3.6]$$

If the force continues to act on the body through time, the body is accelerated. With a constant force, the acceleration, a, is constant ($F = ma$). With F and a constant, the velocity varies as a function of time, t. A velocity v at x and v_o at $x_o = 0$ gives

$$a = \left(\frac{v - v_o}{t}\right) \text{ and } x = \left(\frac{v + v_o}{2}\right)t \qquad [3.7]$$

With $F = ma$ from Newton's second law of motion, work is then given by

$$w = Fx = max = m\left(\frac{v - v_0}{t}\right)\left(\frac{v + v_0}{2}\right)t = \frac{1}{2}mv^2 \qquad [3.8]$$

Because $1/2\ mv^2$ is kinetic energy, the work done by this constant force acting through a distance increases the system's kinetic energy and, therefore, its Internal Energy, U.

A pressure, P, exerted on a system is force, F, per unit area, A, on the boundaries of the system, as shown in **Figure 3-4**. From Equation 3.6 with $F = PA$

$$w = \int_{x_0}^{x} Fdx = -(PA)(x - x_0) = -P\Delta V \qquad [3.9]$$

where ΔV is the volume change. Because the work is done by changing the volume of the system because of pressure, the work is referred to as pressure-volume work. The minus sign indicates that if ΔV is negative so that a system decreases in volume, work is done on the system, and its energy U has increased.

Again, energy transfer can be by mass, heat, or work changes to a system. For work, the transfer is equal to force times distance (Nm), which is equivalent to a volume times pressure (m^3 Pa). The energy transferred by heat or work cannot be distinguished after they are put into a system because they both contribute to a system's kinetic energy. As demonstrated above, for H$_2$O, a system's Internal Energy, U, is a very large number relative to the small energy changes caused by processes involving transfers of work or heat that are considered in a thermodynamic model. These small changes, however, can be determined relative to some arbitrary scale along with any mass changes that may occur. Again, it is the change in U, ΔU, and not the total U of a system, that is needed to apply the laws of thermodynamics.

Because of the ability to change U independently by either heat, work, or mass changes, at a given instance in time, U is a function of the mass of the system plus any two independent variables of state that characterize the heat and work changes. If these are chosen to be T and V, then U is a function of the

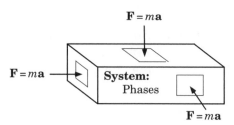

FIGURE 3-4 Pressure on the boundaries of a system as force per unit area. Any decrease in the volume of the system means pressure-volume work has been done on the system.

variables T, V along with the mass of the species $1, 2, \ldots k$ given as $n_1, n_2, \ldots$ n_k in the system. Again, species are a characterization of the mass of the system in terms of distinct molecular units of different types.

■ Extent of and Reversible Paths of Reaction

The changes in U for an isolated system process is always toward equilibrium and can be described by an advancement variable, ξ, called the *extent of reaction variable*. Consider a closed-system process in which the state of the system is changing because a reaction is taking place within the system with initial species of mass $n_1^o, n_2^o \ldots n_k^o$ at time $t = 0$. The change in U of the system as a function of time can be described by T, V, and ξ. ξ monitors the change in mass of the initial species and production of new species from the reaction that is taking place in the system. Because the reaction is taking place with a fixed ratio of the reactant and product species, only one variable, ξ, is needed to describe the relative changes in mass of the k species in the system. For instance, consider a closed-system process of forsterite, Mg_2SiO_4, dissolving in H_2O involving three solute and one mineral species:

$$4H^+ + Mg_2SiO_4 \rightarrow 2Mg^{2+} + H_4SiO_4 \qquad [3.10]$$
(aqueous) mineral   (aqueous) (aqueous)

With the stoichiometric reaction coefficient of species i in the reaction denoted by v_i and remembering that products are positive and reactants are negative, for Reaction 3.10 $v_{Mg^{2+}} = 2$, $v_{H_4SiO_4} = 1$, $v_{H^+} = -4$, and $v_{Mg_2SiO_4} = -1$; therefore, $d\xi$ can be written as

$$d\xi = \frac{dn_{Mg^{2+}}}{2} = \frac{dn_{H_4SiO_4}}{1} = \frac{dn_{H^+}}{-4} = \frac{dn_{Mg_2SiO_4}}{-1} \qquad [3.11]$$

Total differentials were introduced in Chapter 1. The total differential of U for a closed system undergoing a reaction is given by

$$dU = \left(\frac{\partial U}{\partial T}\right)_{V,\xi} dT + \left(\frac{\partial U}{\partial V}\right)_{T,\xi} dV + \left(\frac{\partial U}{\partial \xi}\right)_{T,V} d\xi \qquad [3.12]$$

Now consider a closed system at equilibrium, that is, where $d\xi = 0$, undergoing a transformation from state A to state B to state C, as shown in **Figure 3-5**. X and Y stand for any two independent state variables such as V and T that characterize the changes in heat and work in the system during the process between the states as given in Equation 3.12. The process is said to be *reversible* if there exists a change from state C to state B to state A such that the system returns through the variables X and Y in the opposite order. Any heat exchange with the surroundings and work done on the system takes place with the reverse

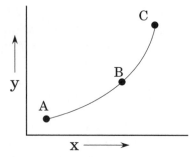

FIGURE 3-5 If a reaction is reversible, then the changes in variables of state x and y in going from A to B to C must be exactly reversed in going from C to B to A. (Adapted from Prigogine and Defay, 1954.)

sign in the reverse order; therefore, if a process is reversible, there exists a continuous succession of stable equilibrium states that form a path between one state and another.

In nature, there are no reversible processes because of things such as friction. Natural processes can, however, be very close to reversible. Consider the u-tube filled with water, as shown in **Figure 3-6**. As a function of time, the water in the left side of the tube would flow past its equilibrium position to an increasing height in the right arm and then flow past the equilibrium in the opposite direction, reversing its path. On its return trip, the water in the left side would increase to its original position if the fluid had zero viscosity and zero friction with the walls of the tube. Although the system is not originally at equilibrium, a change in the state of water in the tube and a reversible path to restore the original state of the system would have occurred. Between any two energy states of a system a reversible path can be described (e.g., zero viscosity and friction), even if it cannot be achieved in nature.

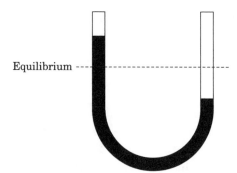

Equilibrium

FIGURE 3-6 U-tube filled with water. In a reversible process, the decrease in the height of water in the left-hand side with time would be matched by the height in the right-hand side and vice versa. (Adapted from Prigogine and Defay, 1954.)

■ First Law of Thermodynamics

J. Willard Gibbs (1873), building on the work of others, considered heat and work transfers in and out of a system. He outlined a first law of thermodynamics. It states that for a process, energy of a system is transferred, but it cannot be created or destroyed. For a closed system, this can be expressed for infinitesimal changes of state as

$$dU = \delta w + \delta q \tag{3.13}$$

where δq is the heat added to the system and δw the work done *on* the system. The notation δ is used to indicate that the change in the quantities of work and heat are path dependent and, therefore, are inexact differentials (see Chapter 1 for a discussion of differences between exact and inexact differentials). In other words, a different amount of work and heat can be added to the system to achieve the same change in Internal Energy. In the case of an open system where mass transfer is involved, energy conservation can be expressed as follows:

$$dU = \delta w + \delta q + c^2 \sum_i dm_i \tag{3.14}$$

where the summation is taken over all the components of mass, i, in the system. Both dU and dm_i depend on only the change in the state of the system and not on its path and are, therefore, exact differentials unlike δw and δq. Although δw can be any kind of work, the most important to consider is pressure-volume work. Generally, this will be the only kind of work considered; however, in Chapter 14, electrical work is considered when energy transfer in oxidation-reduction reactions is described.

Consider a closed-system adiabatic process so that $\delta q = 0$. Also restrict the system so that the only work is pressure-volume work and the system pressure, P, is equal to the external pressure, P_{ex}. As the system undergoes a finite change from state A to state B from the first law of thermodynamics written for a finite change

$$(\Delta U)_{\delta q = 0} = U_B - U_A = - \int_{V_A}^{V_B} P dV = \Delta w \tag{3.15}$$

where ΔU is the finite change in Internal Energy from state A to B, and therefore, Δw is the finite amount of work done. ΔU for this process can be determined by measuring the volume change and knowing the pressure, P, as a function of volume for the change in the system.

Alternatively, consider a closed-system isovolumetric process, that is, where $\Delta V = 0$, and again, assume the only work is pressure-volume work. In this case, $\Delta w = 0$. The first law of thermodynamics, therefore, can be written as follows:

$$(\Delta U)_{dV=0} = U_B - U_A = \Delta q \qquad [3.16]$$

ΔU for this process can be obtained by measuring the heat transferred into the system at constant V. Thus, changes in U can be determined for these restricted processes in which either work is done or heat is transferred. There is great difficulty, however, in obtaining ΔU for a process with both a heat and volume change.

Most reactions or processes observed on and in the earth occur at constant P not V and, therefore, involve the transfer of both work and heat. Writing the first law of thermodynamics for a closed system in differential form with only pressure-volume work gives

$$dU = \delta q - PdV \quad \text{or} \quad \delta q = dU + PdV. \qquad [3.17]$$

A new state function H, *enthalpy*, is created to describe heat changes at constant P. H is defined as

$$H \equiv U + PV \qquad [3.18]$$

It is clear that H is a state function because it is made up of only state variables. The total differential of H is

$$dH = dU + PdV + VdP \qquad [3.19]$$

From Equations 3.17 and 3.19,

$$\delta q - dH = VdP \qquad [3.20]$$

At constant P, $\delta q = dH$ so that dH is indeed the heat change, that is, the heat absorbed at constant P when only pressure-volume work is considered.

■ Second Law of Thermodynamics

Certain processes or reactions occur spontaneously. A gas opened to a larger volume always expands to occupy the additional space. Heat always flows from an object at higher temperature to one at lower temperature. A mole of halite (i.e., $NaCl$) put in a liter of water at 25°C and 1 bar dissolves completely; a mole of quartz does not. It would be helpful to have a variable to indicate under which conditions a spontaneous reaction proceeds and in which direction.

What is needed from a variable to indicate the observed behavior? This variable needs to be a state variable and, therefore, independent of path of the reaction. It should change in a characteristic manner, giving the direction of a reaction that proceeds spontaneously. This variable is entropy, S, which has a statistical basis that depends on the probability of finding the molecules in the system distributed in their most likely energy states. S then measures the variability of the distribution of energy among the individual atoms of the species in the system. For instance, the entropy of a substance increases as it changes from a solid to a liquid and finally to a gas with increasing temperature because the number of possible energy states open to the species of the substance increases. Entropy also increases when a substance such as a gas expands or when a substance is mixed with another in the same phase because, again, the number of energy states available to species of the substance increases.

With the second law of thermodynamics, the change in entropy for a closed system undergoing a reversible process is defined as the heat transfer into the system from the surroundings, dq_{rev}, per T. That is, the heat absorbed from the surroundings in going from state A to state B by a reversible process is not unique and, therefore, is not an exact differential. By dividing by its force, T, it becomes an exact thermodynamic function of the state of the system; therefore, the second law of thermodynamics defines the entropy change in terms of a reversible process between two states as

$$dS \equiv \frac{dq_{rev}}{T} \qquad [3.21]$$

where dq_{rev} is the heat absorbed from the surroundings during the reversible process. dS is postive if heat is absorbed and negative if heat is lost from the system. For a spontaneous process rather than a reversible process, additional entropy is produced so that in general

$$dS = \delta S_{heat} + \delta S_{irr} \qquad [3.22]$$

where δS_{heat} is the entropy added to the system by transfer of heat, and δS_{irr} is the irreversible entropy produced in the system. This irreversible entropy change produces no work nor transfers any heat into the system but increases the disorder of the system. δS_{irr} is then defined as follows:

$$\delta S_{irr} \equiv dS - \delta S_{heat} \qquad [3.23]$$

δS_{irr} does not change the Internal Energy of the system but increases its entropy. This is always positive if an irreversible reaction is taking place and is zero for a reversible reaction. To state the second law of thermodynamics,

$$\delta S_{irr} \geq 0 \qquad [3.24]$$

for all reactions and

$$\delta S_{irr} = 0 \qquad [3.25]$$

for any reversible reactions or along a reversible path between two states.

An isolated system process does not allow energy (heat or work) to be transported across its boundary. Consider an isolated system composed of two pure idea gases separated by a partition. If the partition is removed, the gases will spontaneously mix but the Internal Energy of the system stays the same. The increased randomness in this spontaneous process increases the system's entropy and is always positive. It is this entropy, δS_{irr}, that always increases in a spontaneous reaction.

Now consider a closed system process involving an irreversible reaction where only pressure-volume work is done. The heat adsorbed, δq, is therefore not equal to TdS but $T(dS - \delta S_{irr})$ as δS_{irr} leads to no heat production. The pressure-volume work done on the system is $-PdV$. From the first and second laws of thermodynamics, the Internal Energy change is

$$dU = \delta q + \delta w = T(dS - \delta S_{irr}) - PdV \qquad [3.26]$$

Equation 3.26 can also be written as

$$dU = TdS - PdV - T\delta S_{irr} \qquad [3.27]$$

Because $T\delta S_{irr}$ is always positive in a spontaneous reaction and zero in a reversible process:

$$dU \le TdS - PdV \qquad [3.28]$$

for any process. Therefore, in a process taking place at constant entropy and volume, $dU \le 0$. If $\delta S_{irr} = 0$ in this process, then $dU = 0$, and equilibrium exits. If $\delta S_{irr} > 0$, then $dU < 0$ when a spontaneous reaction occurs.

Although most reactions occur because of changing pressure and temperature conditions, these are forces that are imposed from the outside on a local system. The reactions themselves occur at constant T and P rather than at constant S and V; therefore, a function that indicates equilibrium or spontaneity at constant P and T is needed. This function is the *Gibbs energy*, G, defined as:

$$G \equiv U + PV - TS = H - TS \qquad [3.29]$$

G is the energy that is "free" to run processes at constant P and T and is zero at equilibrium. It is, therefore, often referred to as Gibbs "free" energy. The total differential of G can be written as

$$dG = dU + PdV + VdP - TdS - SdT \qquad [3.30]$$

Making the substitution for dU from Equation 3.27 for a spontaneous process gives

$$dG = -SdT + VdP - T\delta S_{irr} \qquad [3.31]$$

For a spontaneous process occurring at constant T and P (i.e., $dT = dP = 0$)

$$dG = -T\delta S_{irr} \qquad [3.32]$$

Because δS_{irr} is always positive for a spontaneous process from Equation 3.32, dG must be negative. When $T\delta S_{irr} = 0$, dG is 0, and there exists no "free" energy to do work; the system is at equilibrium. **Table 3-2** outlines the differentials considered in the thermodynamic model.

At this point, it is helpful to consider changes in enthalpy, dH. From the definition of H in Equation 3.18 and dH in Equation 3.19, it has been determined that

$$dH = d(U + PV) = dU + PdV + VdP. \qquad [3.33]$$

Making the substitution for dU from Equation 3.28 into Equation 3.33 gives

$$dH \leq TdS + VdP \qquad [3.34]$$

For a process occurring at constant S and P, therefore, $dH = 0$ would indicate equilibrium. **Table 3-3** gives the energy functions and their dependent variables.

Table 3-2   Differentials considered

dU = exact differential giving the change in Internal Energy during the process of going from one infinitesimal state to another; for example, from state A with U_A at time t to state B with U_B at time $t + dt$.

dH = exact differential giving the change in enthalpy during the process.

dG = exact differential giving the change in Gibbs energy during the process.

dS = exact differential giving the change in entropy during the process made up of entropy change, δS_{heat} due to heat exchange plus irreversible entropy change, δS_{irr}, due to mixing during the process.

δw = inexact differential giving the work done on the system, that is, the work put into the system minus the work taken out ($w_{in} - w_{out}$) during the process or reaction from time t to $t + dt$. δw can be any kind of work, e.g., pressure–volume work or work due to changes in electrical, gravitational, or magnetic fields.

δq = inexact differential giving the heat absorbed from the surroundings in the time interval from t to $t + dt$. If heat is lost to the surroundings, then δq is negative.

dm$_i$ = the exact differential giving the change in the mass of the system due to changing the number of moles of component i in the system.

| Table 3-3 | Thermodynamic potentials |

Type of energy	Dependent variables
Internal Energy, U	$U = U(S,V)$
Enthalpy, H	$H = H(S,P)$
Helmholtz energy, A	$A = A(T,V)$
Gibbs energy, G	$G = G(T,P)$

Although the analysis presented is helpful in determining whether a reaction is at equilibrium and if not which way a reaction proceeds, it is also important to characterize the rate of reactions that occur in nature in some way.

■ Rates of Reaction

Consider a chemical reaction with reactants, R_i, going to products, P_i, as a function of time, with v_i being the stoichiometric coefficients of the ith species in the reaction:

$$v_1 R_1 + v_2 R_2 + v_3 R_3 \rightarrow v_4 P_4 + v_5 P_5 + v_6 P_6 \qquad [3.35]$$

For each reactant or product, the change in mass from an initial mass, m_i^o, to a final mass, m_i, expressed in moles for the ith species is given by

$$m_i - m_i^o = v_i \xi \qquad [3.36]$$

ξ stands for the extent of reaction in a particular time interval with $m_i = m_i^o$ at $\xi = 0$. For instance, the dissolution of quartz in H_2O at 25°C and 1 bar at near neutral pH where H_4SiO_4 is the dominant Si aqueous species in solution can be written as

$$SiO_2 + 2H_2O \rightarrow H_4SiO_4 \qquad [3.37]$$

ξ (final) $= m_{H_4SiO_4}$ (final) $= 10^{-4}$ m because the solubility of quartz in pure H_2O at 25°C and 1 bar is 10^{-4} m; therefore, ξ would have changed from 0.0 to 10^{-4} during the course of the dissolution of quartz reaction to equilibrium.

Equation 3.36 can be written in differential form as

$$d_{in} n_i = v_i d\xi \qquad [3.38]$$

where $d_{in} n_i$ is the differential change in the number of moles of species i due to an internal reaction in the system. The rate of the reaction, r, is then given by

$$r = \frac{d\xi}{dt} \qquad\qquad [3.39]$$

To describe a system as a function of time, therefore, the state of the system as a function of ξ can be considered. Remembering from Equation 3.12, the change in the Internal Energy of a closed system is a function of two additional independent state variables, say V, T, as well as ξ. The total differential of U where time is considered can, therefore, be written as

$$dU = \left(\frac{\partial U}{\partial T}\right)_{V,\xi} dT + \left(\frac{\partial U}{\partial V}\right)_{T,\xi} dV + \left(\frac{\partial U}{\partial \xi}\right)_{V,T} d\xi \qquad\qquad [3.40]$$

If, during the reaction, V is constant, no work is done on the system. Further stipulating a constant T, then for this closed system

$$dU = \left(\frac{\partial U}{\partial \xi}\right)_{V,T} d\xi = TdS_{\text{irr}} \qquad\qquad [3.41]$$

The only entropy that can be produced from the change in the state of the closed system at constant V and T is that associated with an internal change in order given by dS_{irr}. A *partial equilibrium* exists as the system is in equilibrium with respect to the variables P and T but not with respect to the redistribution of matter between species in the system, that is, not with ξ. Defining the *chemical affinity*, A, as

$$A \equiv TdS_{\text{irr}} \qquad\qquad [3.42]$$

Because T and dS_{irr} are positive, A is always positive if a reaction is proceeding, whereas at equilibrium, $A = 0$.

The rate of reaction, r, can be considered in terms of change in the advancement variable, $d\xi$, with time driven by the chemical affinity, A, of the reaction. The link between $d\xi$ and time awaits the developments of chemical kinetics outlined in Chapter 13. Whether determining the size of A or the conditions for chemical equilibrium of a reaction, the values of G of species at the temperature and pressure of interest are needed.

■ Determining Values of Gibbs Energy

Changes in G along an equilibrium path where $\delta S_{\text{irr}} = 0$ as a function of temperature and pressure, say from P_1 and T_1 to P_2 and T_2, can be determined from Equation 3.31 by integration:

$$\Delta G = G_{P_2,T_2} - G_{P_1,T_1} = \int\limits_{T_1,P_1}^{T_2,P_2} dG = - \int\limits_{T_1,P_1}^{T_2,P_2} S dT + \int\limits_{T_1,P_1}^{T_2,P_2} V dP \qquad [3.43]$$

This leaves three problems in evaluating the values of G for a substance:

1. How is the integration done to evaluate the change in G with changes in temperature and pressure? That is, how are S and V determined as a function of temperature and pressure to do the integration in Equation 3.43?
2. This equation only considers changes in G and not absolute values. How is a value of G of a substance determined to compare across a reaction to evaluate whether $dG = 0$ and equilibrium has been reached?
3. How are compositional changes from pure phases dealt with? For instance, how is the Gibbs energy of O_2 in air calculated to determine the Gibbs energy change for the reaction $4Fe_3O_4 + O_2 = 6Fe_2O_3$ to know whether magnetite or hematite is stable in air?

This last question about changes in G with composition is addressed in Chapter 4. Only the Gibbs energy of pure phases is considered in this chapter. Besides volume changes, Equation 3.43 indicates that S must be known as a function of temperature. S cannot be measured directly, but it can be related to a measurable property of the phase, the *heat capacity*.

■ Heat Capacity

Heat capacity is the amount of heat needed to raise the temperature of a substance by 1°C. This heat can be determined at constant volume, C_V, or at constant pressure, C_P. Equation 3.17 states that $dU = \delta q - P dV$; therefore, with V constant, this becomes

$$(dU)_V = \delta q = (dT)C_V \qquad [3.44]$$

The change in Internal Energy of the substance under these conditions is then C_V times the temperature change. This increased energy as temperature increases is stored in various *modes* of motion in the substance. A mode is a particular independent manifestation of the phenomena. The maximum energy of $1/2RT$ for each mode per mole of atoms is reached at elevated temperatures. For solids, this energy is stored primarily in modes of vibration of the atoms in the substance. There are three possible directions of vibration with two modes per direction, one mode being the stored kinetic energy and the other a potential energy mode, giving a maximum vibrational energy of $3RT$. This implies a $C_{V(max)}$ per *mole-atom* of approximately 3R for minerals at high temperature from their vibration. A mole-atom is the number of atoms in moles in a mole of the species. For example, in 1 mole of wollastonite, $CaSiO_3$, there are 5 mole-atoms because there are five atoms in the formula. With R = 8.31424 J mol^{-1} K^{-1} = 1.98726 cal mole^{-1} K^{-1}, $C_{V(max)}$ is then about 25 J mole-atom^{-1} K^{-1} or 6.0 cal mole-atom^{-1} K^{-1} for wollastonite. (The value in calories is included because the program SUPCRT92 determines values in calories; see note below.)

Heat capacity is usually determined at constant pressure, given as C_P. The relationship between C_V and C_P is

$$C_P - C_V = \frac{V\alpha^2}{\beta}T$$

[3.45]

where α and β are the coefficients of isobaric thermal expansion and isothermal compression, respectively, as discussed later in this chapter. At elevated temperatures in the region where C_V reaches its maximum for solids of 3R per mole-atom,

$$C_P = 3R + \frac{V\alpha^2}{\beta}T$$

[3.46]

Because for solids α and β as well as V are nearly independent of pressure and temperature at conditions in the earth's crust, it is clear from Equation 3.46 that C_P will be nearly a linear function of temperature when C_V reaches its maximum value. As shown in **Figure 3-7**, this occurs at temperatures above approximately 700°C. As temperature decreases, some of the vibrational modes are inhibited so that C_V along with C_P decreases. At $T = 0$ K, no vibrations occur, and thus, C_P and C_V approach 0. Note the similarity in shape of the C_P curve of the three minerals given in Figure 3-7.

C_P is usually given in units of J mol^{-1} K^{-1} (or cal mole^{-1} K^{-1}) rather than J atom-mole^{-1} K^{-1} so that for wollastonite the atom amount needs to be multiplied by 5 to convert to the molar formula amount. Taking note of the behavior of C_P shown in Figure 3-7, the molar C_P at 1 bar of minerals was described by Maier and Kelley (1932) with the equation

$$C_P = a + bT - \frac{c}{T^2}$$

[3.47]

where a, b, and c are constant value fit parameters and T is in K. The first two terms, a and bT, give the linear behavior at high temperature, whereas c gives the decrease of C_P at lower temperatures as the vibrational modes become inhibited. As C_P measurements became more precise, the fit equation was modified to an "extended" Maier-Kelley C_P power function:

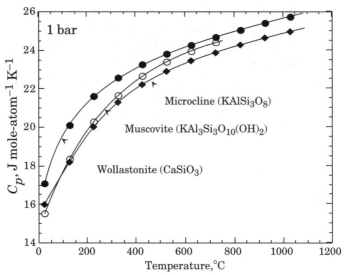

FIGURE 3-7 Joule per kelvin per mole-atom heat capacity of the indicated minerals as a function of temperature at 1 bar (calculated from SUPCRT92).

$$C_P = a + bT - \frac{c}{T^2} + dT^2 + e \ln T \tag{3.48}$$

with the additional fit coefficients d and e. Alternatively, precision in C_P can be increased by using the original Maier-Kelley C_P power function but dividing the C_P calculation into temperature intervals with the three term function fit to each interval. This is the way the C_P of minerals is calculated by SUPCRT92 and the Maier-Kelley coefficients are tabulated in Appendix F3.

Gases have three independent translational modes that together contribute a maximum of 3/2R or 12.5 J mole-atom^{-1} K^{-1} (3.0 cal mole-atom^{-1} K^{-1}) to C_V. For an ideal gas with $P\overline{V} = RT$, Equation 3.45 gives $C_P - C_V = R$. The C_P of an ideal gas, therefore, is 5/2R or 20.8 J mol^{-1} K^{-1} = 4.97 cal mol^{-1} K^{-1}. For real polyatomic gases, the effects of rotational modes and any vibrational modes that are present need to be added. Again, as with all real phases, C_P and C_V decrease to zero with decreasing temperature as 0 K is approached. The C_P of five representative gases as a function of temperature at 1 bar in units of J mol^{-1} K^{-1} are given in **Figure 3-8**. The larger the number of atoms per formula, the greater the C_P per mole are at elevated temperature. Also, Ar gas has a constant $C_P = 20.8$ J mol^{-1} K^{-1} and behaves as a monatomic ideal gas, whereas the other gases increase their C_P with increasing temperature in a manner similar to the minerals because of the additional modes of motion that are activated with increasing temperature. The same Maier-Kelley C_P function is used to characterize the C_P of gases as a function of temperature as used for

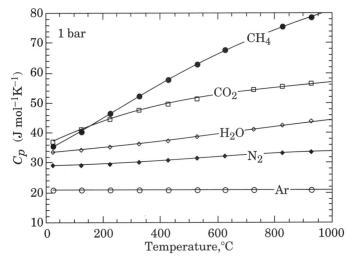

FIGURE 3-8 Molar heat capacity of the indicated gases in units of J mol⁻¹ K⁻¹ as a function of temperature at 1 bar (computed from SUPCRT92).

the minerals. Maier-Kelley C_P function coefficients for a number of gases, along with other thermodynamic parameters for gases, are given in Appendix F4. The C_P of H_2O shown in Figure 3-8 at temperatures below 100°C is for H_2O as a steam phase, that is, at pressures below 1 bar where H_2O is gaseous. H_2O as a liquid has distinctly different C_P, as well as other thermodynamic properties.

■ Properties of H₂O Liquid

Depending on pressure and temperature, H_2O can exist as a liquid, solid, vapor, or *supercritical fluid* phase, as shown in **Figure 3-9**. Supercritical fluid denotes a phase that is above the critical pressure and temperature of the substance (see below). Ice is somewhat anomalous as a solid because it is less dense than liquid water at the same pressure and temperature; therefore, the slope of the ice–liquid H_2O phase boundary is negative on the diagram, indicating that increasing pressure in the ice stability field causes ice to melt. The C_P of ice and gaseous H_2O can be calculated by the procedure outlined above. How is the C_P of H_2O as a liquid or supercritical fluid characterized? Consider the stability of H_2O liquid relative to vapor shown in Figure 3-9. The locus of pressures and temperatures where both a vapor and liquid coexist in equilibrium is termed the *saturation curve* (labeled as "vapor saturation" in Figure 3-9). At these conditions, the vapor is at its maximum concentration, that is, saturated per unit volume. Increasing the gas concentration any more by compressing it would cause the gas to revert to a liquid. The position of the saturation curve

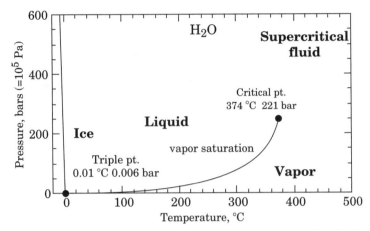

FIGURE 3-9 Pressure and temperature phase diagram for H_2O. Note the regions of ice, liquid, vapor, and supercritical fluid as well as the negative slope on the ice-liquid H_2O boundary.

is consistent with the well-known observation that at atmospheric pressure (1.013 bar) water boils at 100°C.

The saturation curve for H_2O indicates that if the temperature is increased above 100°C and H_2O is to remain as a liquid, the pressure must be increased above 1.013 bars. At 300°C, the pressure required is 85.8 bars. With increasing temperature and pressure along the saturation curve, the liquid becomes more gas like and the gas more liquid like. At high enough temperatures, there is no difference between the two and only one phase exists. The temperature and pressure along the saturation curve at which there is no difference between the gas and liquid phase is termed the *critical point* for the substance. The pressure and temperature at this point are then the critical pressure and temperature. For pure H_2O, this is 221.2 bars and 647.3 K (374.2°C). Values of the critical pressure and temperature of other substances that can be gases on and in the earth's crust are given in Appendix F4 where the thermodynamic data for gases are tabulated.

Because water is a polar molecule and has the potential for hydrogen bonding at the higher densities of a liquid as opposed to a gas, the C_P of supercritical water is a complex function of P and T, as shown in **Figure 3-10** (see Chapter 6 on aqueous solutions for a further discussion of the polar nature of H_2O). At the critical point where the liquid and gas lose their identity, the C_P of H_2O approaches infinity. Along an isobar at pressures above the critical pressure, C_P maximizes as a function of temperature, with the maximum decreasing with increasing pressure as shown in the figure. This greater C_P for liquid H_2O as opposed to gaseous H_2O (Figure 3-8) is due to the additional energy stored in the bonding between H_2O molecules in a liquid-like structure as opposed

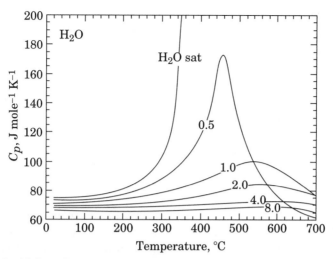

FIGURE 3-10 C_P of H_2O as a function of temperature in joules per mole per kelvin at the indicated pressures labeled in kbar. H_2O sat denotes the pressure of the liquid when saturated with vapor. (Data from Helgeson and Kirkham, 1974.)

to the lower energy stored between molecules in a vapor. A function in two independent thermodynamic variables with one being temperature is typically fit to the C_P values of H_2O in the region of interest for calculation purposes. The other variable is typically pressure, volume, or density (Johnson and Norton, 1991).

In summary, the heat capacities of solids and gases are expressed with a Maier-Kelley C_P type function, whereas liquids like water are expressed as more complex functions of two variables with one being temperature. These equations allow one to compute the C_P of minerals and gases at 1 bar and the C_P of liquid water as a function of both pressure and temperature.

■ Volume Changes

The thermodynamic model of a system also requires knowledge of volume changes (normally at the temperature of interest). The volume as a function of pressure and temperature of a substance is termed its *equation of state* (EOS) and along with the heat capacity allows the Gibbs energy changes to be calculated. The relationship of Gibbs energy to pressure at constant temperature along an equilibrium path from Equation 3.31 is

$$\left(\frac{\partial G}{\partial P}\right)_T = V \qquad\qquad [3.49]$$

If the pressure or temperature of a phase changes, its volume also changes. Consider first the volume changes of minerals. The effect of temperature at constant pressure and that of pressure at constant temperature are considered independently because this is the way the changes in volumes of most minerals are determined.

Measurements have been made of how much a mineral expands when heated at constant pressure. This is given by the *coefficient of isobaric thermal expansion*, α:

$$\alpha \equiv \frac{1}{V}\left(\frac{\partial V}{\partial T}\right)_P \tag{3.50}$$

Figure 3-11 shows for four representative minerals the percentage expansion at 1 bar as a function of temperature relative to 20°C. α is then the slope of the lines shown. α is relatively constant except for quartz, which undergoes a phase transition around 573°C. α for minerals varies from about 2 to 5×10^{-5} °C^{-1} except for quartz near the transition from its low to high temperature structure. Interestingly, quartz expands on heating significantly more than the other minerals. As temperatures increase above 400°C, α increases for the minerals shown because the curves are somewhat concave.

The volume change with increased pressure at constant temperature is termed the *coefficient of isothermal compression*, given by

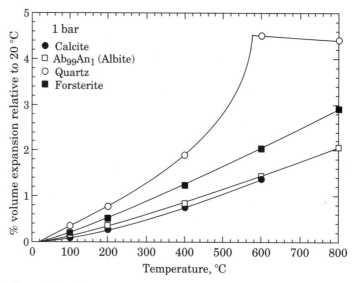

FIGURE 3-11 Expandability of four representative minerals as a function of temperature at 1 bar relative to 20°C. (Data from Skinner, 1966.)

$$\beta \equiv -\frac{1}{V}\left(\frac{\partial V}{\partial P}\right)_T \qquad\qquad [3.51]$$

This is the slope of the solid lines in **Figure 3-12** for the four representative minerals. As with thermal expansion, the compressibility of quartz is significantly greater than for the other minerals. The other minerals compress by between 0.8% and 2.5% as pressure is increased from 1 bar to 10 kbar at 20°C.

Is the density of forsterite greater or less at the base of the continental crust than at the earth's surface? Reasonable lower continental crust conditions are 700°C and 10 kbar. Forsterite expands 2.5% from 20 to 700°C at 1 bar. With the volume of forsterite at 20°C and 1 bar of 43.79 cm^3 mol^{-1}, the volume change of forsterite in increasing temperature to 700°C at 1 bar would be

$$0.025 \times 43.79 \text{ cm}^3 \text{ mol}^{-1} = 1.09 \text{ cm}^3 \text{ mol}^{-1} \qquad\qquad [3.52]$$

At 25°C, olivine (forsterite) compresses about 0.8% from 1 bar to 10 kbar for a volume change of

$$0.008 \times 43.79 \text{ cm}^3 \text{ mol}^{-1} = 0.35 \text{ cm}^3 \text{ mol}^{-1} \qquad\qquad [3.53]$$

It might be argued from Equations 3.52 and 3.53 that forsterite is somewhat less dense at the base of the continental crust. The opposite is true. This is because compressibility as a function of pressure to 10 kb at 700°C is greater than at 25°C. The decrease in volume is, however, quite small, generally less than 1% for most minerals. With minerals behaving similarly, the effects tend

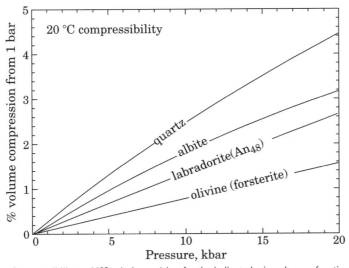

FIGURE 3-12 Compressibility at 20°C relative to 1 bar for the indicated minerals as a function of pressure. (Data from Birch, 1966.)

to cancel across reactions. For most purposes, therefore, V of minerals can be taken as a constant, independent of P and T, for thermodynamic calculations involving crustal processes. The exception may be for quartz stability involved in solid–solid reactions deep in the crust.

■ EOS for Fluids

Unlike minerals, liquids and gases undergo significant volume changes as a function of temperature and pressure. In the earth, liquids become more gas like as they are heated and gases more liquid like as pressure increases. As a result, the general term fluid is used to identify these dense gas–liquid phases. To quantify a fluid's change in volume, a number of different equations of state (EOS) have been developed. These include regression polynomial fits, van der Waals models, virial equations, and perturbation models.

Consider first the fluid H_2O that can exist as steam, liquid, or a supercritical fluid depending on the pressure and temperature conditions. Its molar volume, V_{H_2O}, changes significantly in the crust. At 8 kbar V_{H_2O} increases by 50% as temperature is increased from 25°C to 900°C. At lower pressures, isobars of V_{H_2O} increase even more dramatically with increasing temperature, as shown in **Figure 3-13**. These changes have dramatic effects on G of H_2O as a function of temperature and pressure. Because these volume changes are a complex function of the interactions between H_2O molecules for values to 10 kbar and 1000°C considered here, no simple model has predicted molar volume to the

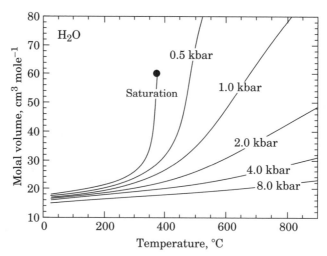

FIGURE 3-13 Molar volume of H_2O as a function of temperature along the indicated isobars. Saturation denotes the liquid-vapor saturation pressure. (Data from Helgeson and Kirkham, 1974.)

accuracy of the measurements; therefore, a regression polynomial fit is typically used. Used in this book and in SUPCRT92 to pressures of 1 kbar is the equation of Keenan and others (1969) for the *Helmholtz free energy*, *A*, of H_2O. With Helmholtz free energy the temperature and volume are minimized, rather than with Gibbs energy, which minimizes pressure and temperature. Because

$$P = - \left(\frac{\partial A}{\partial V} \right)_T \qquad [3.54]$$

the volume of H_2O as a function of pressure and temperature can be determined by an iterative method. At pressure to 1 kbar and 800°C, the computed values have 0.1% or better accuracy, whereas at temperatures below 220°C, the uncertainty is 0.01%. From 1 to 10 kbar, regression polynomials in three separate pressure and temperature regions from Burnham et al. (1969) are used to determine V_{H_2O}. These values are within the 0.6% error of the experimental measurements (Helgeson and Kirkham, 1974). Values of V_{H_2O} at integral values of pressure and temperature as calculated from these equations are given in Appendix E.

Except for regression polynomials fits, the other types of EOS equations typically start with the ideal gas law where

$$\overline{V} = \left(\frac{R}{P} \right) T \qquad [3.55]$$

The overbar on *V* indicates a molar quantity. For gases, the interactions between molecules are weak, and in the limit of 0 pressure, the interactions become negligible. Under these conditions, the ideal gas law is obeyed for all gases. The molar volume of an ideal gas is a linear function of *T* at constant pressure with a slope of R/*P*.

Figure 3-14 shows $\overline{V}$ measured for CO_2 at 0.1, 0.5, and 1 kbar along with the calculated $\overline{V}$ from the ideal gas law. It is clear that significant departure from ideal gas behavior occurs at pressures as low as 100 bars for temperatures below 400°C. Above 400°C, the departure from ideal gas behavior is smaller but becomes more significant as pressure is increased. At elevated temperatures, the behavior of gases like CO_2 can be modeled with a van der Waals type equation of state where the relationship between, *P*, and *T* is given by

$$P = \frac{RT}{\overline{V} - b_w} - \frac{a_w}{\overline{V}^2} \qquad [3.56]$$

a_w and b_w represent van der Waals constants that are unique for each gas/fluid. The b_w term can be considered to account for the volume occupied by the gas

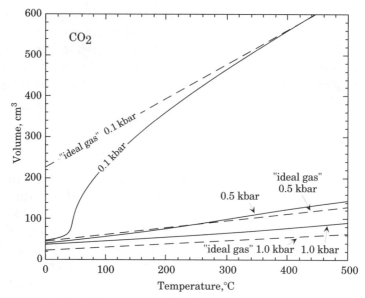

FIGURE 3-14 Molar volume of CO_2 at 0.1, 0.5, and 1.0 kbar as a function of temperature. Also shown as dashed lines are the ideal gas volumes at these pressures.

molecules that must be subtracted from the total volume of an ideal gas when gas density increases. The $a_w / \overline{V}^2$ term accounts for attraction between the gas molecules. As with molecular attraction, the term $a_w / \overline{V}^2$ increases as the concentration of molecules increase with increasing P. With $a_w = b_w = 0$, the equation reduces to the ideal gas law.

Geologists most often use the Redlich and Kwong (1949) modification of the van der Waals equation for gases/fluids because it gives more accurate results at higher densities than the original van der Waals equation. Other modifications of the van der Waals equation have also been considered (Anderko, 2000). The Redlich and Kwong equation has an attractive term that is temperature dependent:

$$P = \frac{RT}{\overline{V} - b_{RK}} - \frac{a_{RK}}{T^{1/2}\, \overline{V}(\overline{V} + b_{RK})} \qquad [3.57]$$

The Redlich–Kwong constants, a_{RK} and b_{RK}, for each gas/fluid can be calculated from their critical pressure and temperature. This is possible because all substances tend to behave similarly—that is, they have similar changes in volume with P and T near their critical points. a_{RK} and b_{RK} are given by

$$a_{RK} = \frac{0.4275 R^2 T_C^{2.5}}{P_C} \qquad [3.58]$$

and

$$b_{RK} = \frac{0.0866RT_C}{P_C} \qquad\qquad [3.59]$$

where P_C and T_C denote the critical pressure and temperature of the substance, respectively. Substituting a_{RK} and b_{RK} determined from Equations 3.58 and 3.59 into Equation 3.57 for the gas of interest allows the volume of a gas/fluid as a function of pressure and temperature to be approximated. This procedure gives a good fit to the experimental determinations of volume with pressure and temperature for CH_4, O_2, and N_2 for conditions in the earth's crust. When asymmetric gas molecules like CO_2 and H_2O are considered, modified parameters appear to give more accurate results. **Table 3-4** lists those parameters that were derived by Holloway (1976) that have a temperature-dependent a_{RK} and b_{RK}. For H_2O, the coefficients are valid for temperatures of 400°C and above. These are the parameters used in SUPCRT92 for calculating the thermodynamic properties of CO_2. For H_2O, the more accurate polynomial regressions outlined above are used.

Besides polynomial fits and modified van der Waals equations, virial and perturbation EOS are also used. A virial EOS for gases/fluids is constructed by starting with a reference system, typically the ideal gas equation as the first term in a power series expansion. The expansion is usually in pressure or density such as

$$P\bar{V} = RT(1 + BP + CP^2 + \ldots) \qquad\qquad [3.60]$$

where B and C stand for the second and third virial coefficient, respectively. At low pressure, this equation reduces to the ideal gas law.

In perturbation theory, the EOS is a molecular model for the Helmholtz free energy. This is done by splitting the energy into an unperturbed or reference part and a perturbed part. The reference part is typically an ideal gas, and the perturbed part generally uses a *Lennard-Jones 6-12 potential*, Γ_{LJ}, for the energy where

Table 3-4	Redlich-Kwong fit parameters for pure CO_2 and H_2O

Species	a_{RK} (T in °C)	b_{RK}
CO_2	$119.03 \times 10^6 - 71{,}400T + 21.57T^2$	29.7
H_2O	$201.8 \times 10^6 - 193{,}080T + 186.4T^2 - 0.071288T^3$	14.6

The parameters for H_2O are valid at 400°C and above.
Adapted from Holloway, 1976.

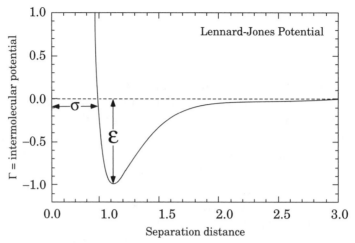

FIGURE 3-15 Energy of interaction between two particles given by the Lennard-Jones potential as a function of the distance of separation normalized to σ with $\sigma = \varepsilon = 1$.

$$\Gamma_{LJ} = 4\varepsilon \left[\left(\frac{\sigma}{r} \right)^{12} - \left(\frac{\sigma}{r} \right)^{6} \right]$$

$$\underset{\text{repulsive}}{} \quad \underset{\text{attractive}}{}$$

[3.61]

This potential models the interaction of nonpolar molecules as the sum of an attractive and repulsive energy term where σ gives the distance of separation at zero potential, ε denotes the negative of the energy at the equilibrium distance, and r is the distance the molecules are separated (**Figure 3-15**). When polar molecules are considered terms for *dipole–dipole*, $\Gamma_{\text{dip-dip}}$, and induced dipole–dipole, $\Gamma_{\text{ind-dip}}$, interactions are also included so that the total potential energy, Γ_{TOT}, is given by

$$\Gamma_{TOT} = \Gamma_{ideal} + \Gamma_{LJ} + \Gamma_{dip\text{-}dip} + \Gamma_{ind\text{-}dip}$$

[3.62]

Additional terms that account for the perturbation of the real system from the interactions given by Equation 3.62 can also be added (Churakov and Gottschalk, 2003).

■ Changes in Gibbs Energy with *P* and *T*

How is the Gibbs energy between two different pressures and temperatures, that is, from P_1 and T_1 to P_2 and T_2, calculated? The calculation can be done along any path in pressure-temperature space between the two pressures and temperatures because Gibbs energy as a state function is independent of the path taken to calculate it. The normal path used is to increase temperature at 1 bar pressure and then calculate the increase in pressure at the temperature of interest, as shown in **Figure 3-16**. This is done because the temperature change requires knowledge of the C_P of the substance and most measurements of C_P are done at atmospheric pressure (1 bar).

From the definition of Gibbs energy of a substance (Equation 3.29) at P_1 and T_2

$$G_{P_1,T_2} = H_{P_1,T_2} - T_2 \, S_{P_1,T_2} \tag{3.63}$$

where the subscripts indicate the pressure and temperature considered. H_{P_1,T_2} in Equation 3.63 is related to H_{P_1,T_1} by

$$H_{P_1,T_2} = H_{P_1,T_1} + \int_{T_1}^{T_2} C_{P_{(P_1)}} \, dT \tag{3.64}$$

$C_{P_{(P1)}}$ in Equation 3.64 denotes the heat capacity at pressure P_1.

S_{P_1,T_2} in Equation 3.63 can be determined from S_{P_1,T_1} by considering Equation 3.34 along an equilibrium path so that $dH = TdS + VdP$. At constant pressure, $dS = dH/T$ so that

$$S_{P_1,T_2} = S_{P_1,T_1} + \int_{T_1}^{T_2} \frac{C_{P_{(P_1)}}}{T} \, dT \tag{3.65}$$

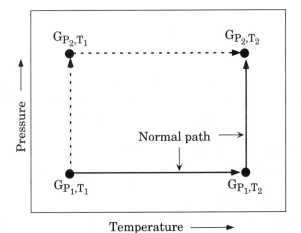

FIGURE 3-16 Pressure and temperature path used to calculate the Gibbs energy change from pressure P_1 and temperature T_1 to P_2 and T_2.

Substituting Equations 3.64 and 3.65 into the expression for G_{P_1,T_2} in Equation 3.63 gives

$$G_{P_1,T_2} = H_{P_1,T_1} - T_2 S_{P_1,T_1} + \int_{T_1}^{T_2} C_{P_{(P_1)}} dT - T_2 \int_{T_1}^{T_2} \frac{C_{P_{(P_1)}}}{T} dT \qquad [3.66]$$

Again, from Equation 3.29, G_{P_1,T_1} can be written as

$$G_{P_1,T_1} = H_{P_1,T_1} - T_1 S_{P_1,T_1} \qquad [3.67]$$

Combining Equations 3.66 and 3.67 results in

$$G_{P_1,T_2} = G_{P_1,T_1} - S_{P_1,T_1}(T_2 - T_1) + \int_{T_1}^{T_2} C_{P_{(P_1)}} dT - T_2 \int_{T_1}^{T_2} \frac{C_{P_{(P_1)}}}{T} dT \qquad [3.68]$$

Equation 3.68 gives the change in G of a substance as a function of temperature at pressure P_1. Also required is knowledge of G as a function of pressure at T_2. The expression is

$$G_{P_2,T_2} = G_{P_1,T_2} + \int_{P_1}^{P_2} V_{T_2} dP \qquad [3.69]$$

V_{T_2} denotes the volume of the substance at a temperature of T_2. This expression can be deduced by considering Equation 3.43 at a constant temperature of T_2.

Substituting the value of G_{P_1,T_2} from Equation 3.68 into Equation 3.69 results in

$$G_{P_2,T_2} = G_{P_1,T_1} - S_{P_1,T_1}(T_2 - T_1) + \int_{T_1}^{T_2} C_{P_{(P_1)}} dT - T_2 \int_{T_1}^{T_2} \frac{C_{P_{(P_1)}}}{T} dT + \int_{P_1}^{P_2} V_{T_2} dP \qquad [3.70]$$

Equation 3.70 indicates that to determine a value of G_{P_2,T_2}, the measurable properties of C_P as a function of temperature at P_1 as well as V as a function of pressure at T_2 are required. Also, some way to obtain values of S_{P_1,T_1} and G_{P_1,T_1} is needed.

Although V and C_P can be measured along the appropriate pressure and temperature paths, how are S_{P_1,T_1} and G_{P_1,T_1} determined? This is the second question from above to be considered. For if S_{P_1,T_1} and G_{P_1,T_1} can be determined, a unique value of G_{P_2,T_2} can be calculated from Equation 3.70. These values can be compared across a reaction to see whether phases are in equilibrium and if not which way the reaction will go. To determine S_{P_1,T_1} and G_{P_1,T_1} a *reference state* must be considered.

The third law of thermodynamics outlined by J. Willard Gibbs recognizes that the absolute entropy of any substance has a calorimetric part, S_{heat} (whose change is given by δS_{heat} above), because of its kinetic energy and a configurational part, S_{config}, because of mixing in the substance. At absolute zero temperature, the calorimetric entropy of any substance is zero because all lattice vibrations have stopped. Because any pure, crystalline, perfectly ordered substance has zero configurational entropy, its absolute entropy, $S_{abs} = S_{heat} + S_{config}$, is 0 at 0 K, and all other substances at any temperature or configurational change are positive. S_{abs} is, therefore, measurable and considering Equation 3.65 is given by

$$S_{abs} = \int_0^T \frac{C_P}{T} dT + S_{config} \qquad [3.71]$$

S_{P_1, T_1} of a substance can be determined by calculating its configurational entropy and measuring its C_P as a function of temperature from 0 K to the temperature T_1 at a pressure of P_1. A value of G_{P_1, T_1} is still needed to evaluate Equation 3.70. S_{P_1, T_1} could be determined because the third law of thermodynamics gave a starting point of 0 at 0 K for the calculation. How is a starting value of Gibbs energy at some conditions determined? This then becomes a problem of determining H, the enthalpy, in some state because once H is determined, knowing $G = H - TS$, G can also be calculated. Determining a unique value of H is not possible, but it is not required if a reference state is considered.

■ Reference States

Figure 3-17 gives a procedure for calculating the enthalpy of reaction, ΔH_R, for

$$CaCO_3 = CaCO_3 \qquad [3.72]$$
calcite   aragonite

at 25°C and 1 bar. First, the enthalpy of formation of aragonite from the elements, $\Delta H_{f, aragonite}$, can be measured by determining the heat absorbed during the reaction

$$Ca + C + O_2 \rightarrow CaCO_3 \qquad [3.73]$$

The heat absorbed by an identical reaction to form calcite can also be measured to give $\Delta H_{f,calcite}$ at 25°C and 1 bar. ΔH_R for Reaction 3.72 can then be computed from

$$\Delta H_{R,(Eq.\ 3.72)} = \Delta H_{f,aragonite} - \Delta H_{f,calcite} \qquad [3.74]$$

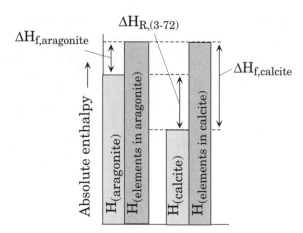

I FIGURE 3-17 Relationship between the enthalpy of formation of calcite and aragonite from their elements.

What has been done is to calculate the enthalpy of the reaction from a measurement of the enthalpy of formation of the substance from the elements on both sides of the reaction. Because there is exactly the same number of moles of each element on both sides of a balanced chemical reaction, the procedure works for any reaction; therefore, if $\Delta H_{f,sub}$ are available for all of the substances in a reaction, they can be subtracted across a reaction to determine ΔH_R. Because values of $\Delta H_{f,sub}$ are used to determine ΔH_R, the calculation is relative to the elements, and the thermodynamic values of the elements are exactly the same on each side; thus, one can arbitrarily fix their value. The typical convention is to assume the elements in their most stable state (e.g., O_2 and Si) at 25°C and 1 bar have zero enthalpy.

To calculate entropy changes, two different conventions are commonly used. One is to use third law (absolute) entropies as outlined above, and the other is to use entropies of formation from the elements. Using the third law entropy convention, the entropy of a substance is taken as the measured absolute (third law) entropy, S_{abs}, determined from Equation 3.71; therefore, because $G = H - TS$, ΔG of a substance using this convention is

$$\Delta G_{S_{abs}} = \Delta H_{f,sub} - TS_{abs} \qquad [3.75]$$

Therefore, $\Delta G_{S_{abs}}$ for the elements (e.g., O_2, Si) at 25°C and 1 bar in their most stable state under this absolute entropy convention are equal to $-298.15 \times S_{abs}$.

With the other convention, S as well as H of the element in its stable state at 25°C and 1 bar is assigned a value equal to zero. G of the elements for this convention, at 25°C and 1 bar, are then equal to zero. The change in S, H, and G of formation of the substance from the elements is computed, and this value is assigned to the substance to give $\Delta S_{f,sub}$, $\Delta H_{f,sub}$, and $\Delta G_{f,sub}$, respectively. That is, for a substance, its entropy is

$$\Delta S_{f,sub} = \left(S_{sub} - \sum S_{element} \right) \qquad [3.76]$$

similar to the convention for enthalpy of

$$\Delta H_{f,sub} = \left(H_{sub} - \sum H_{element} \right) \qquad [3.77]$$

The Gibbs energy of the substance with this convention is

$$\Delta G_{f,sub} = \left(G_{sub} - \sum G_{element} \right) = \Delta H_{f,sub} - T\Delta S_{f,sub} \qquad [3.78]$$

where the summation is taken over all the elements in the substance.

Although G_{sub} cannot be determined, the value of $\Delta G_{S_{abs},sub}$ or $\Delta G_{f,sub}$ for a balanced reaction is measurable. At equilibrium using the absolute entropy convention at any P and T,

$$\sum_{reactants} \Delta G_{S_{abs},species} = \sum_{products} \Delta G_{S_{abs},species} \qquad [3.79]$$

whereas with the Gibbs energy of formation convention,

$$\sum_{reactants} \Delta G_{f,species} = \sum_{products} \Delta G_{f,species} \qquad [3.80]$$

Again, for a reaction, G of the elements cancels across the reaction so their values do not matter. It needs to be clear, however, as to whether an absolute (i.e., third law) S or $S = 0$ at 25°C and 1 bar convention is used in computing the Gibbs energy. The relationship between the two is given by

$$\Delta G_{S_{abs},sub} = \Delta G_{f,sub} + \left(298.15 \sum S_{element} \right) \qquad [3.81]$$

where the summation is over the absolute entropies of all the elements.

A third reference state, which is often more convenient at high pressures and temperatures, is also used. G, H, and S of the elements are defined to be 0 at any temperature and pressure, not just 25°C and 1 bar as with the formation from the elements reference state. With this reference state a determination of the change in the thermodynamic properties of the elements with increasing P and T is not needed. Again, values of these elements cancel across a reaction, so to assume they have a value of 0 at the pressure and temperature of interest is a reasonable convention. When using this convention that G, H, and S for the elements are equal to 0 at any pressure and temperature, the values are termed *apparent values* of formation from the elements. The relationship between apparent G of formation of a substance from the elements at pressure and temperature, $app\ \Delta G_{f,\text{sub},P,T}$, and G of formation relative to 1 bar and 25°C, $\Delta G_{f,\text{sub},1,25°C}$, is

$$app\Delta G_{f,\text{sub},P,T} = \Delta G_{f,\text{sub},1,25°C} + (G_{\text{sub},P,T} - G_{\text{sub},1,25°C}) \tag{3.82}$$

The apparent H and S of formation at any P and T are similarly given by

$$app\Delta H_{f,\text{sub},P,T} = \Delta H_{f,\text{sub},1,25°C} + (H_{\text{sub},P,T} - H_{\text{sub},1,25°C}) \tag{3.83}$$

and

$$app\Delta S_{f,\text{sub},P,T} = \Delta S_{f,\text{sub},1,25°C} + (S_{\text{sub},P,T} - S_{\text{sub},1,25°C}) \tag{3.84}$$

respectively. Using the values of G, H, and S determined from absolute entropies or from formation or apparent formation from the elements allows numerical values of G, H, and S to be computed for a substance, and these values give the correct change in the thermodynamic properties when summed across a balanced reaction. It should be clear by context which type of values are being used. The notation of $\Delta G_{S_{abs},\text{sub}}$, $\Delta G_{f,\text{sub},P,T}$, or app $\Delta G_{f,\text{sub},P,T}$, therefore, will be dropped, and just ΔG will be used in the rest of the book; however, it is important to know which type of value is reported when values are obtained from multiple sources. The three different conventions are outlined in **Table 3-5**.

Using the reference states outlined, with a knowledge of C_P and V, the Gibbs energy of a reaction can be determined. These calculations can, however, be complex and tedious. A number of computer programs have been written to perform the task. These include THERMOCALC (Powell and Holland, 1988), GEO-CALC (Berman et al., 1987), and SUPCRT92 (Johnson et al.,

|Table 3-5   Conventions regarding the determination of Gibbs energy

Third law entropy	Formation from elements	Apparent formation from elements
$\Delta G_{S_{abs}} = \Delta H_f - TS_{abs}$	$\Delta G_f = \Delta H_f - T\Delta S_f$	app $\Delta G_f = \Delta H_f - T\Delta S_f - (\Sigma\ G_{element}$ from 298.15 K to temperature of interest)

*Many computer programs have been written to do the thermodynamic calculations of interest. Used in this book is the software package SUPCRT92 (Johnson, Oelkers, and Helgeson, 1992). It is available at http://www.jbpub.com/catalog/9780763726423/ on the Student Resources page. Instructions for loading and use of the software on a PC or Macintosh compatible computer for a sample reaction are given in Appendix F1 and appear on the aforementioned website. Some problems at the end of this chapter and others in the book make use of SUPCRT92.

1992). THERMOCALC and GEO-CALC use the absolute entropy reference state convention, whereas SUPCRT92 uses apparent formation from the elements values. SUPCRT92 is used in this book because of its ability to calculate the thermodynamic properties of aqueous species as well as gases and minerals.* Apparent ΔG of formation values with $G_{elements} = 0$ at all pressures and temperatures, therefore, is used as a reference state unless another is specified.

■ Simple Phase Relations

Because of the requirement that at equilibrium reactants and products must have the same Gibbs energy, the pressure and temperature stability of phases at equilibrium can be considered. For each species A in a reaction, $dG^A = -S^A dT + V^A dP$ so that for the reaction as a whole

$$d(\Delta G_R) = -\Delta S_R dT + \Delta V_R dP \qquad [3.85]$$

where ΔG_R, ΔS_R, and ΔV_R give the Gibbs energy, entropy, and volume of reaction, respectively. If the reaction is at equilibrium, $\Delta G_R = 0$. For the reaction to remain at equilibrium while the system is changing its pressure and temperature, ΔG_R must remain 0. Putting this constraint on Equation 3.85 and rearranging, the *Clapeyron equation* is obtained:

$$\left(\frac{\partial P}{\partial T}\right)_{\Delta G_R = 0} = \frac{\Delta S_R}{\Delta V_R} \qquad [3.86]$$

where the $\Delta G_R = 0$ outside the brackets indicates equilibrium is to be maintained. For a reaction, therefore, to maintain equilibrium, the slope of the equilibrium boundary on a pressure-temperature diagram is given by the ratio of ΔS_R to ΔV_R.

Figure 3-18 shows the equilibrium boundary for the solid–solid reaction:

$$SiO_2 + Ca_3Al_2Si_3O_{12} = CaAl_2Si_2O_8 + 2\ CaSiO_3 \qquad [3.87]$$

quartz   grossular   anorthite   wollastonite

computed from SUPCRT92. The phase boundary is linear with a positive slope; therefore, $\Delta S_R/\Delta V_R$ must have a positive constant value independent of pressure and temperature along the equilibrium boundary. As was outlined above, the volume of minerals can be considered to be independent of pressure and temperature in the crust. With this constraint, ΔV_R is a constant. This implies that ΔS_R must also be nearly constant to keep $\Delta S_R/\Delta V_R$ constant.

As a state variable at equilibrium, the cross-partial differentials of the Gibbs energy of reaction in pressure and temperature must be equal. That is, whether

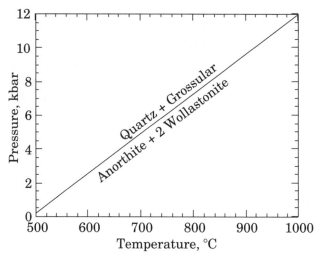

FIGURE 3-18 Pressure-temperature phase diagram for the indicated solid–solid reaction computed from SUPCRT92.

the differential of Gibbs energy with respect to temperature is taken first and then the pressure differential is evaluated or differentiation with respect to pressure is done first and then the temperature differential determined, the results are equal. At equilibrium, Equation 3.85 implies

$$- \left(\frac{\partial \Delta S_R}{\partial P} \right)_T = \left(\frac{\partial \Delta V_R}{\partial T} \right)_P \qquad \text{[3.88]}$$

with ΔV_R constant, that is, independent of T, it is clear from Equation 3.88 that ΔS_R is also constant and independent of P.

What about the temperature dependence of ΔS_R? It can be shown that the C_P of a complex mineral can be approximated by adding the C_P of its constituent oxides as given by what is known as *Kopp's rule*. This is consistent with the behavior shown in Figure 3-7 at elevated temperature where the C_P of minerals when plotted on a mole-atom basis are similar. Because there are equal numbers of mineral oxide components on both sides of a solid–solid reaction, ΔC_{P_R} is near 0 for a reaction involving only solids. Recall that

$$\left(\frac{\partial \Delta S_R}{\partial T} \right)_P = \frac{\Delta C_{P_R}}{T} \qquad \text{[3.89]}$$

This implies that if $\Delta C_{P_R} = 0$, then ΔS_R is also independent of temperature. The reason that reactions involving only solids generally have positive slopes is that volume and entropy typically change in the same direction. For instance, if the temperature is increased, both the volume and entropy of a solid phase generally increase.

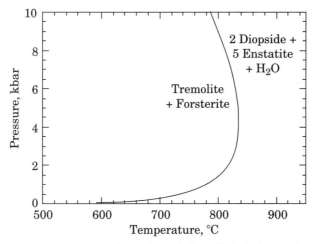

Figure 3-19 Pressure-temperature phase diagram for the indicated dehydration reaction computed from SUPCRT92.

Figure 3-19 shows the phase boundary for the reaction

$$Ca_2Mg_5Si_8O_{22}(OH)_2 + Mg_2SiO_4 = 2CaMgSi_2O_6 + 5MgSiO_3 + H_2O \qquad [3.90]$$

$$\quad\text{tremolite}\qquad\qquad\text{forsterite}\qquad\text{diopside}\qquad\text{enstatite}$$

computed from SUPCRT92. At low pressures, the volume of H_2O is large. This makes ΔV_R large, which causes the equilibrium boundary slope as a function of pressure and temperature to be relatively small as surmised by considering Equation 3.86. As pressure increases, the volume of H_2O decreases and therefore ΔV_R decreases. This causes the equilibrium slope as shown in Figure 3-19 to increase with increasing pressure. In fact, at pressures above 5 kbar, ΔV_R changes signs and becomes negative as H_2O is compressed to such an extent that it occupies less space in a free H_2O phase than it does caged in the tremolite structure. These negative slopes occur for most reactions involving hydrous minerals at high enough pressure unless they melt first.

Summary

Thermodynamic models of systems have great applicability to the real world. The first law of thermodynamics states the observation that energy is conserved when a system changes state. It defines the Internal Energy, U, of a system that can be thought of as the energy of mass + kinetic energy + potential energy. During a process occurring for a system from one state to another, dU is given by

$$dU = \delta w + \delta q + c^2 \sum_i dm_i$$

where δw is the work done on the system, δq is the heat added to the system, dm_i is the change in mass of the i components in the system, and c is the speed of light. The work done on a system along with mass changes can be considered organized energy transfer, whereas the flux of heat causes disorganized energy transfer.

The second law of thermodynamics determines the direction of reactions. It defines the entropy produced, dS, from a reversible process as

$$dS = \frac{dq_{rev}}{T}$$

where dq_{rev} is the heat adsorbed from the surroundings during the reversible process. For nonreversible, that is, spontaneous, processes, an additional irreversible entropy, δS_{irr}, is also produced. $\delta S_{irr} = 0$ when a reaction is at equilibrium or along an equilibrium path. δS_{irr} is positive and increases in a spontaneous reaction so that

$$dS = \delta S_{heat} + \delta S_{irr}$$

where δS_{heat} denotes the entropy obtained from the flux of heat into the system. The Internal Energy change for a spontaneous reaction is therefore given by

$$dU = \delta q + \delta w = T(dS - \delta S_{irr}) - PdV$$

or

$$dU < TdS - PdV$$

At constant temperature and pressure, systems tend toward the lowest Gibbs energy, G, so that for a closed system that therefore does not involve the transfer of mass

$$dG < -SdT + VdP$$

where $G \equiv U + PV - TS$. The change in Gibbs energy along an equilibrium path of a closed system is given by

$$dG = -SdT + VdP$$

To evaluate dG and compute the Gibbs energy of a phase or species at a particular pressure and temperature, the measurable properties of heat capacity, C_P, and volume as a function of temperature and pressure are needed. Also, the Gibbs energy of formation from the elements at a reference pressure and temperature is required. This allows the Gibbs energy across a reaction to be computed because there are the same quantities of elements on both sides of a balanced reaction. In Chapter 4, Gibbs energies of solutions that contain mixtures of components rather than the pure component phases considered here are analyzed.

Key Terms Introduced

adiabatic

apparent values

boundaries

chemical affinity

chemical potential

Clapeyron equation

closed system

coefficient of isobaric thermal
    expansion

coefficient of isothermal
    compression

critical point

dipole

energy

enthalpy

entropy

equation of state

equilibrium

extent of reaction variable

forces

Gibbs energy

heat

heat capacity

Helmholtz free energy

ideal gas

idioblastic

Internal Energy

isobaric

isolated system

Kopp's rule

Lennard-Jones 6-12 potential

mode

mole-atom

open system

partial equilibrium

phases

reference state

reversible

saturation curve

state of a system

state variables

supercritical fluid

system

vector

work

xenoblastic

Questions

1. What is work? Heat? Entropy? How are they defined?
2. What is a phase?
3. What does thermodynamic reversibility mean?
4. What is the difference between an isolated, closed, and open system?
5. What are the three types of energy a system can have?
6. What measurable values are needed to determine the Gibbs energy of a phase?
7. What is chemical affinity? A partial equilibrium?
8. How does the heat capacity at constant pressure change for a solid? A gas? H_2O?
9. In terms of a thermodynamic model, what determines the slope of a univariant curve on a pressure–temperature phase diagram?

Problems

1. If a force of 1.0 N acts on a mass of 1.0 kg, what is the acceleration after 1 minute? What is its velocity? How much work has been done on the mass?

2. Near the earth's surface, what is the pressure exerted by a cube of rock of density 2.7 g cm^{-3} that is 10 m on each edge? Why couldn't one overcome this pressure and lift the cube with one's bare hands?

3. If a newton of force is operating on each cm^2 of the boundaries of a system, what pressure is the system under? If at this pressure the volume decreased by 10 cm^3, how many joules of energy were added to the system?

4. Determine the C_P of 1 mole of muscovite at 25°C and at 500°C at 1 bar using the Maier-Kelley equation.

5. Calculate the difference in the non-mass energy of 1 mole of H_2O as an ideal gas at 1 atm and 0°C where all the oxygen is mass 18 g mol^{-1}, O^{18}, from that where all the oxygen is mass 16 g mol^{-1}, O^{16}. In natural H_2O, the average mass of O^{18} is 0.200%, $O^{17} = 0.038\%$ and that of $O^{16} = 99.762\%$. What is the difference between the total energy of natural H_2O and pure O^{16} H_2O?

6. Consider a mineral where the decrease in volume from 1 bar to 1 kbar is greater at 600°C than at 25°C. Will the C_P of the mineral between 25 and 600°C be different if it was measured at 1 kbar rather than at 1 bar? Why?

7. What is the maximum work that can be obtained by expanding 25 g of CO_2, behaving as an ideal gas from 1 to 10 liters at 25°C?

8. Give expressions for α and β for an ideal gas.

9. Starting with the total differential of S show that the adiabatic temperature gradient is given by

$$\left(\frac{\partial T}{\partial P}\right)_S = \frac{\alpha V T}{C_P}.$$

10. Using the relationship in Problem 9, calculate the temperature at a depth of 35 km for basaltic magma that has reached the surface with a temperature of 1200°C (1473 K) under the constraint that the process was so rapid that no heat was lost. (For basaltic magma, $\rho = 2.61$ g/cm^3 $\alpha = 1 \times 10^{-4}$°C^{-1}, $C_P = 0.2$ cal g^{-1}°C^{-1}). Assume that the volume of the magma is independent of pressure and use Equation 2.4 to calculate the pressure at depth. Conversion between cal and cm^3 bar is given in Appendix B.

11. What is the expression for dP for a gas that obeys the van der Waals equation?

12. Calculate the volume change for forsterite from 1 bar and 25°C to 20 kbar and 800°C. (Assume α and β are independent of pressure and temperature. $\alpha = 44 \times 10^{-6}$ K^{-1}, $\beta = 8 \times 10^{-7}$ bar^{-1}, $V_{25°C,1\ bar} = 43.79$ cm^3/mole.)

13. Calculate the enthalpy change for 100 g of $CaCO_3$ when the temperature is increased from 25°C to 900°C. Assume the molar heat capacity at constant pressure (calories/mole K) over the temperature range, T, is approximated by the Maier-Kelley heat capacity power function.

14. Kyanite, andalusite, and sillimanite (Al_2SiO_5 polymorphs) are common minerals in metapelites.

    a. Using the data given below, calculate the stability relations and construct an Al_2SiO_5 phase diagram as a function of pressure and temperature. (Assume $\Delta C_P = 0$ and that ΔV and ΔH are independent of temperature and pressure.)

	Data at 1 bar and 298 K		
	ΔH_R	ΔS_R	ΔV_R
	kJ/mole	J/mole K	cm^3/mole
Kyanite = Sillimanite	7.52	12.97	5.81
Andalusite = Sillimanite	3.21	3.77	–1.63

    b. At what temperature and pressure is the mineral assemblage andalusite + kyanite + sillimanite stable? Compare the results with that calculated with SUPCRT92 or Figure 49 of Helgeson et al. (1978).

15. The origin of red-bed sandstones, in which the grains are coated with minute amounts of hematite, Fe_2O_3, has long been controversial. A key question in the controversy is whether hematite is stable in water at low temperatures. Calculate whether hematite or goethite, $FeO(OH)$, is stable in the presence of water at 25°C and 1 bar (G_f(hematite) = –178,155 cal/mol, G_f(goethite) = –116,766 cal/mol, $G_f(H_2O)$ = –56,688 cal/mol).

16. Suppose you woke up one winter morning to a bucket of water at –2°C, 1 atm on your back porch. You kick it, and it spontaneously freezes to ice at –2°C (this reaction gives off 80 cal/g of heat). Entropy, S, is a measure of disorder of a system; therefore, $S_{water} > S_{ice}$. How is this process possible? Doesn't this imply S has decreased for a spontaneous process? Explain.

17. Describe any two processes for the same system involving the same change in U but where heat put into and work done on the system are different.

Problems Making Use of SUPCRT92

18. By building a new file of pressures and temperatures along the liquid–vapor saturation curve of H_2O for the phase H_2O:

a. Construct a diagram giving the density of water as a function of temperature at vapor saturation pressure from 0°C to 370°C.

b. Construct a pressure versus temperature diagram between 100°C and 370°C to indicate where the liquid phase as opposed to the vapor phase is stable.

19. By building a new file of pressures at 1 bar and temperatures from 0°C to 1000°C at 100°C increments, calculate the heat capacity of wollastonite in cal/K. Convert these numbers to joules per mole-atom per K and compare with Figure 3-7.

20. Determine the equilibrium pressure-temperature univariant phase boundary for the following reactions at the specified conditions. (Suggest using the "univariant curve option" in SUPCRT92. Note that with pure phases the log **K** of reaction = 0.0 and the temperatures should be somewhere between 0°C and 1000°C.)

a. Forsterite + talc = 5 enstatite + H_2O (between 200 and 5000 bars)

Compare results with Chernosky (1976).

b. Muscovite = corundum + K-feldspar + H_2O (between 500 and 9000 bars)

Compare results with Chatterjee and Johannes (1974).

c. Brucite = periclase + H_2O (between 200 and 2500 bars)

Compare results with Barnes and Ernst (1963).

21. Determine the Gibbs energy of H_2O as a steam phase (H_2O, g) at 25°C and 1 bar, 100 bars, and 1.0 kbar. Why is it the same value in SUPCRT92?

22. Calculate the molar volume of H_2O at 1 kbar and 100°C, 200°C, 300°C, 400°C, 500°C, 600°C, and 700°C. Compare with Figure 3-13.

References

Anderko, A., 2000, Cubic and generalized van der Waals equations. In Sengers, J., Kayser, R., Peters, C., and White, H., Jr. (eds.), *Equations of State for Fluids and Fluid Mixtures*, Part I. Elsevier Science, Amsterdam, pp. 75–126.

Barnes, H. L. and Ernst, W. G., 1963, Ideality and ionization in hydrothermal fluids: The system MgO-H_2O-NaOH. *Amer. Jour. Sci.*, v. 261, pp. 129–150.

Berman, R. G., Brown, T. H. and Perkins, E. H., 1987, GEO-CALC: Software for calculation and display of P-T-X phase diagrams. *Amer. Mineral.*, v. 72, pp. 861–862.

Birch, F., 1966, Compressibility: Elastic constants. In Clark, S. P. (ed.), *Handbook of Physical Constants*. Memoir 97. Geol Soc. Amer. Pub., New York, pp. 97–173.

Burnham, C. W., Holloway, J. R. and Davis, N. F., 1969, Thermodynamic properties of water to 1000 °C and 10,000 bars. *Geol. Soc. America Spec. Paper*, 132, 96 pp.

Chatterjee, N. D. and Johannes, W., 1974, Thermal stability and standard thermodynamic properties of synthetic 2M1-muscovite, KAl$_2$[AlSi$_3$O$_{10}$(OH)$_2$]. *Contrib. Mineral. Petrol.*, v. 48, pp. 89–114.

Chernosky, J. V., 1976, The stability of anthophyllite: A reevaluation based on new experimental data. *Amer. Mineral.*, v. 61, pp. 1145–1155.

Churakov, S. V. and Gottschalk, M., 2003, Perturbation theory based equation of state for polar molecular fluids. I. Pure fluids. *Geochim. Cosmochim. Acta*, v. 67, pp. 2397–2414.

Denbigh, K., 1981, *The Principles of Chemical Equilibrium*. 4th ed., Cambridge University Press, Cambridge, 494 pp.

Gibbs, J. W., 1873, Graphical methods in the thermodynamics of fluids, *Trans. Connecticut Academy*, vol. ii, pp. 309–342.

Helgeson, H. C. and Kirkham, D. H., 1974, Theoretical prediction of the thermodynamic behavior of aqueous electrolytes at high pressures and temperatures. I. Summary of the thermodynamic/electrostatic properties of the solvent, *Amer. Jour. Sci.*, v. 274, pp. 1089–1198.

Helgeson, H. C., Delany, J. M., Nesbitt, H. W. and Bird, D. K., 1978, Summary and critique of the thermodynamic properties of rock-forming minerals. *Amer. Jour. Sci.*, v. 278-A, pp. 1–229.

Holloway, J. R., 1976, Fugacity and activity of molecular species in supercritical fluids. In Fraser, D. G. (ed.), *Thermodynamics in Geology*. D. Reidel Publishing Co., Dordrecht-Holland, pp. 161–181.

Johnson, J. W. and Norton, D., 1991, Critical phenomena in hydrothermal system: State, thermodynamic, electrostatic, and transport properties of H$_2$O in the critical region, *Amer. Jour. Sci.*, v. 291, pp. 541–648.

Johnson, J. W., Oelkers, E. H. and Helgeson, H. C., 1992, SUPCRT92: A software package for calculating the standard molal thermodynamic properties of minerals, gases, aqueous species, and reactions from 1 to 5000 bar and 0 to 1000°C. *Comput. Geosci.*, v. 18, pp. 899–947.

Keenan, J. H., Keyes, F. G., Hill, P. G. and Moore, J. G., 1969, *Steam Tables*, John Wiley & Sons, 162 pp.

Klotz, I. M., 1964, *Chemical Thermodynamics Basic Theory and Methods*. W. A. Benjamin, New York, 468 pp.

Maier, C. G. and Kelley, K. K., 1932, An equation for the representation of high temperature heat content data, *Jour. Amer. Chem. Soc.*, v. 54, pp. 3243–3246.

Powell, R. and Holland, T. J. B., 1988, An internally consistent thermodynamic dataset with uncertainties and correlations. 3. Applications to geobarometery, worked examples and a computer program. *J. Metamorph. Geol.*, v. 6, pp. 173–204.

Prigogine, I. and Defay, R., 1954, *Chemical Thermodynamics*. Longman Group Ltd., London, 543 pp.

Redlich, O. and Kwong, J. N. S., 1949, An equation of state: Fugacities of gaseous solutions, *Chem. Rev.*, v. 44, pp. 233–244.

Skinner, B. J., 1966, Thermal expansion. In Clark, S. P. (ed.), *Handbook of Physical Constants*. Memoir 97. Geol. Soc. Amer. Pub., New York, pp. 75–96.

Williams, H., Turner, F. J., and Gilbert, C. M., 1954, *Petrography: An Introduction to the Study of Rocks in Thin Section*, W. H. Freeman & Co., San Francisco, 406 pp.

4 Solutions and Simple Phase Relations

Most of the mass in phases found in and on the earth resides in solutions of two or more constituents. Thermodynamic models need to consider the energetics of these solutions. These can be gaseous, aqueous, magmas, or solid solutions in minerals. These solutions are characterized in terms of endmembers. For instance, a Mg-Fe olivine of intermediate composition is considered a solution of forsterite (Mg_2SiO_4) and fayalite (Fe_2SiO_4) endmembers. Air can be considered a solution of various pure gas endmembers. These endmembers can be taken as components in a thermodynamic analysis. As presented in Chapter 3, the Gibbs energy of these pure components can be calculated. How is the Gibbs energy of the component in a mixture determined, however?

Because of the importance of changes in Gibbs energy of a component with composition in a solution, a new thermodynamic variable is defined. The change in Gibbs energy as a function of the change in the number of moles of the ith component in a solution is termed the chemical potential of component i and is represented by the symbol μ_i so that

$$\mu_i \equiv \left(\frac{\partial G}{\partial n_i} \right)_{P,T,n_j}$$

[4.1]

where the P, T, and n_j outside the partial differential indicate pressure, temperature, and the moles of all of components except the ith are to be held constant. From the definition of partial molal quantities outlined in Chapter 1, the chemical potential of i is the partial molal Gibbs energy of component i. This is the change in Gibbs energy with a change in a mole of the component while keeping constant all the other variables that affect Gibbs energy in the solution.

For example, an olivine with a composition of $Mg_{1.4}Fe_{0.6}SiO_4$ can be characterized as a mixture of the components Fo (Mg_2SiO_4) and Fa (Fe_2SiO_4). The mole fraction of the Fo component in this olivine, $X_{Fo} = 1.4/2 = 0.7$, and the mole fraction of Fa component, $X_{Fa} = 0.6/2 = 0.3$. Knowing the amount of the component in the mixture, how is the component's Gibbs energy in the solution

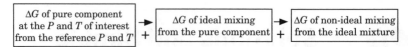

I FIGURE 4-1 Calculations needed to be summed to determine the ΔG of a component in a solution.

determined? The calculation of the thermodynamic properties of a component in a solution is separated into three steps, as given in **Figure 4-1**. First, the Gibbs energy of a mole of the pure endmember component of the solution at the temperature and pressure of interest is calculated. For component i, this can be denoted as μ_i^o. The discussion given in Chapter 3 outlines how this is done, and the thermodynamic data to do the calculation are generally available (e.g., SUPCRT92). To this value, changes in Gibbs energy due to changes in composition from the endmember component composition to the composition of interest are added. This later determination is separated into two parts. The changes due to what is referred to as *ideal mixing* of the endmember component with other components in the solution are calculated. To this is added the change caused by nonideal mixing. What is ideal as opposed to nonideal mixing and how does this affect the Gibbs energy of a component in a solution?

■ Ideal Mixing

The Gibbs energy of a mole of a solution, s, of components A and B, μ_s, is given by

$$\mu_s = X_A \mu_A + X_B \mu_B \qquad [4.2]$$

where the molar Gibbs energy of components A and B as they exist in the solution are designated as μ_A and μ_B, respectively. For instance,

$$\mu_{(Mg_{1.4}Fe_{0.6}SiO_4)} = 0.7\mu_{Fo} + 0.3\mu_{Fa} \qquad [4.3]$$

where μ_{Fo} and μ_{Fa} stand, respectively, for the Gibbs energy of forsterite and fayalite in the solid solution rather than the Gibbs energy of a mole of pure forsterite, μ_{Fo}^o, and pure fayalite, μ_{Fa}^o.

Now consider a *mechanical mixture* of the composition of the solution of interest. A mechanical mixture has the correct amount of each pure component in the solution but no chemical interaction between the components has actually taken place. How a mechanical mixture of A and B is constructed is shown in **Figure 4-2**. The system has the correct mole fraction of A and B for the solution; however, A in the mixture interacts only with other A and B only with other B. The Gibbs energy of a mole of mechanical mixture of A and B would be

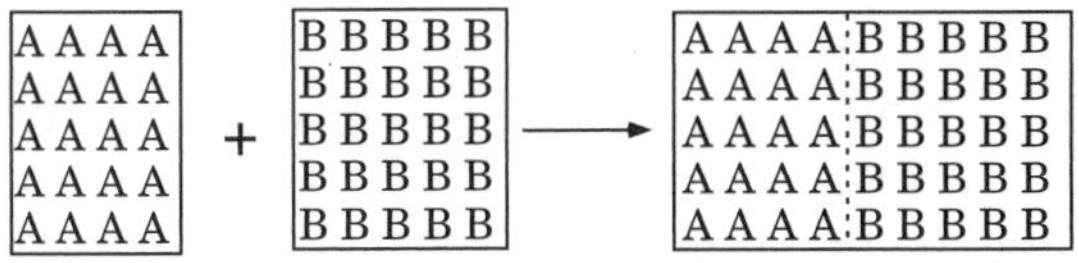

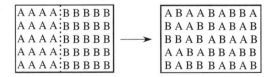

FIGURE 4-2 The combining of pure A and B in the correct proportions to produce what is meant by a mechanical mixture of A and B whose composition is characterized by X_A and X_B but no mixing has occurred.

$$\mu_{\text{mech mix}} = X_A \mu_A^o + X_B \mu_B^o \qquad [4.4]$$

where the superscript o on μ_A^o and μ_B^o indicates these are the Gibbs energy of pure endmembers A and B, respectively. In the case of the olivine $Mg_{1.4}Fe_{0.6}SiO_4$, this is

$$\mu_{\text{mech mix}}^{\text{olivine}} = 0.7\mu_{\text{Fo}}^o + 0.3\mu_{\text{Fa}}^o \qquad [4.5]$$

If A and B consist of gas molecules or in general most species able to form a solution, it is clear that after the mechanical mixture is formed it would spontaneously mix to produce a random mixture (i.e., solution) of uniform composition throughout, as shown in **Figure 4-3**. Because mixing is spontaneous, the Gibbs energy of the solution, μ_s, must be less than the Gibbs energy of the mechanical mixture of the components, $\mu_{\text{mech mix}}$. The difference in molar Gibbs energy between the solution and the mechanical mixture, can be specified as

$$\Delta\mu_{\text{mix}} = \mu_s - \mu_{\text{mech mix}} \qquad [4.6]$$

Again, if mixing is spontaneous, $\Delta\mu_{\text{mix}} < 0$. From the definition of Gibbs energy, this difference on mixing can be written as

$$\Delta\mu_{\text{mix}} = \Delta\bar{H}_{\text{mix}} - T\Delta\bar{S}_{\text{mix}} \qquad [4.7]$$

where again the overbars on H and S indicate these are molar values. $\Delta\bar{H}_{\text{mix}}$ and $\Delta\bar{S}_{\text{mix}}$ are then the change in enthalpy and entropy on mixing the pure components to produce a mole of the mixture, respectively.

Mixing is ideal if all the forces of interaction between A and B, that is, A–A, B–B, A–B, are the same so that the enthalpy, $\Delta\bar{H}_{\text{mix}}$, and the volume, $\Delta\bar{V}_{\text{mix}}$, of

FIGURE 4-3 Mixing of A and B from a mechanical mixture to a random mixture. The Gibbs energy associated with this spontaneous process, termed $\Delta\mu_{\text{mix}}$, is negative.

mixing are zero. Clearly, with $\Delta\mu_{mix} < 0$ and with $\Delta\overline{H}_{mix} = 0$ for an ideal mixture, $\Delta\overline{S}_{mix}$ must be greater than zero. In fact, the only contribution to $\Delta\mu_{mix}$ for an ideal mixture is the increase in S because of increased randomness on mixing. Ideal gases mix ideally because there is neither interaction between molecules in the gas nor changes in volume on mixing. The entropy of ideal mixing can be derived from *statistical mechanics* by calculating each component's configurational entropy and for each *component i* in the mixture this is given by

$$\overline{S}_i = -RX_i \ln X_i \qquad [4.8]$$

For the solution as a whole, the entropy change on mixing is then

$$\Delta\overline{S}_{id\ mix} = -R\sum_{i=1}^{c} X_i \ln X_i \qquad [4.9]$$

where R, the molar gas constant, is equal to the Boltzmann's constant times Avogadro's number, and c is the number of independent endmember components in the mixture.

Consideration of Equations 4.6, 4.7, and 4.9 for ideal mixing allows the molar Gibbs energy of an ideal solution, μ_s^{ideal}, to be written as

$$\mu_s^{ideal} = \mu_{mech\ mix} - T\Delta\overline{S}_{id\ mix} = \mu_{mech\ mix} + RT\sum_{i=1}^{c} X_i \ln X_i \qquad [4.10]$$

Because $X_B = 1 - X_A$, an ideal solution of components A and B in a solution with one mole of mixing sites per mole of the phase is

$$\mu_s^{ideal} = \underbrace{X_A\mu_A^o + (1 - X_A)\ \mu_B^o}_{\text{mechanical mixture}} + \underbrace{X_A RT \ln X_A + (1 - X_A)RT \ln(1 - X_A)}_{\text{ideal mixing}} \qquad [4.11]$$

where μ_i^o is the molar Gibbs energy of pure component i at the pressure and temperature of interest. In **Figure 4-4**, μ_s^{ideal} for a two-component mixture A–B is plotted against the mole fraction of A in the solution; μ_s^{ideal} is always lower than the Gibbs energy of the mechanical mixture of the pure components or for that matter of the Gibbs energy of any two bounding compositions of the mixture. Ideal solutions are, therefore, stable and do not unmix for any compositions from pure B to pure A.

Also shown in Figure 4-4 is a tangent to the line giving μ_s^{ideal} at $X_A = 0.6$. This line can be used to obtain μ_A and μ_B for $X_A = 0.6$. The slope of the tangent of any binary Gibbs energy versus X_A line from Equation 4.2 is given by

$$\left(\frac{d\mu_s}{dX_A}\right)_{P,T} = \mu_A - \mu_B \qquad [4.12]$$

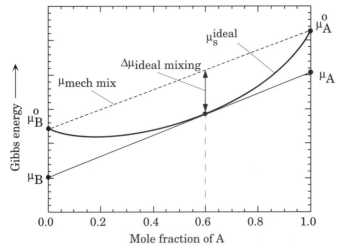

FIGURE 4-4 Gibbs energy of the mechanical mixture of components A and B together with the negative contribution from ideal mixing to produce the curve labeled μ_s^{ideal}, which gives the Gibbs energy of an ideal mixture of A and B as a function of the mole fraction of A. The tangent to this curve projected to the end-member compositions give the chemical potential of the endmembers in the solution as shown for $X_A = 0.6$.

because $X_B = (1 - X_A)$. The equation of the tangent line is then

$$y = (\mu_A - \mu_B) \, X_A + \mu_B \qquad [4.13]$$

and the value on the y-axis at $X_A = 0.0$ equals μ_B and at $X_A = 1.0$ equals μ_A for the $X_A = 0.6$ tangent.

When a complete solution from pure B to pure A does *not* exist, there must be another energy term that is important. Often the energetic interactions between A and B are significant so that mixing is not ideal. This interaction is frequently modeled with what is called a *regular solution*. A regular solution has an additional Gibbs energy term in the mixture accounting for interactions of unlike components of the form

$$\Delta\mu_{reg} = \omega \, X_A \, X_B \qquad [4.14]$$

where ω represents the regular solution constant that is independent of composition reflecting the extent of interaction of A and B. The regular solution constant can be either positive or negative. If ω is positive, this term increases the Gibbs energy of the solution. When the terms for ideal mixing and regular solution effects are added to the mechanical mixture term, the Gibbs energy of a regular solution

$$\mu_s^{reg} = \mu_{mech \; mix} + \Delta\mu_{id \; mix} + \Delta\mu_{reg} \qquad [4.15]$$

is obtained. For a phase with the two components, A and B, Equation 4.15 can be written as

$$\mu_s^{reg} = X_A\mu_A^o + (1 - X_A)\mu_B^o + X_ART\ln X_A + (1 - X_A)RT\ln(1 - X_A) + \omega X_A(1 - X_A)$$

$$\underbrace{\qquad\qquad\qquad\quad}_{\text{mechanical mixture}} \qquad \underbrace{\qquad\qquad\qquad\quad}_{\text{ideal mixing}} \qquad \underbrace{\qquad}_{\text{regular solution [4.16]}}$$

μ_s^{reg} is plotted at T_1 as a function of the mole fraction of A with ω positive in **Figure 4-5**. μ_s^{reg} is greater than a mechanical mixture of the compositions labeled x and y for all mixtures between X_A of 0.2 and 0.8. At the temperature T_1, when $0.2 < X_A < 0.8$, the lowest Gibbs energy is given by a mechanical mixture of compositions x and y, and all mixtures of compositions between x and y are thermodynamically unstable. At lower and higher X_A, the Gibbs energy of the regular mixture is lower so that a variable composition solid solution would exist in these regions. Again, with $0.2 < X_A < 0.8$, different proportions of the x and y phases are stable. Outside of this region there would only be a single stable solution phase of variable composition.

From Equation 4.16, the ideal mixing terms contribute a larger negative value to μ_s^{reg} as temperature is increased, while the regular mixing term is constant if ω is not a function of temperature. As temperature is increased, therefore, the ability to mix is increased. This is shown in **Figure 4-6**, where the composition and temperature phase diagram indicates that the miscibility gap, termed the *solvus* between A and B, becomes smaller as temperature is increased. At temperatures greater than those along the solvus, a complete solid

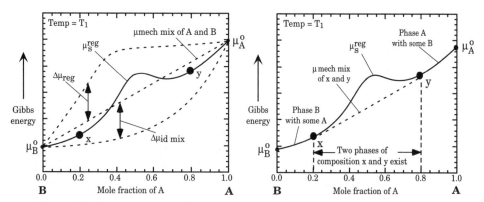

FIGURE 4-5 The Gibbs energy of a regular solution as a function of composition. In the left diagram, the mechanical mix, ideal solution, and regular solution contributions to μ_s of the regular solution are shown. In the right diagram, the region where the miscibility gap occurs (between compositions x and y) is indicated. In this composition region, the phases x and y are stable together. Outside of this region, solutions of variable composition can exist.

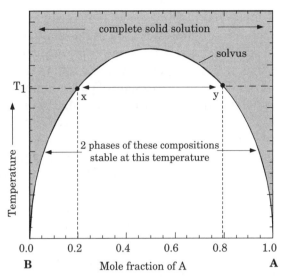

FIGURE 4-6 Phase diagram for A–B mixtures at constant pressure as a function of temperature produced from a regular solution. The miscibility gap given by the solvus indicates the boundary between a single solid solution (shown in gray) and where two phases are stable (shown in white). For the temperature, T_1, the points x and y give the boundaries of the miscibility gap consistent with Figure 4-5.

solution is possible for all compositions from A to B. Minerals that show immiscibility include the feldspars and carbonates and are discussed in Chapter 5.

■ Gibbs Energy of a Species with Composition in a Mixture

Given the brief introduction to the Gibbs energy changes of a solution on mixing of components, consider the Gibbs energy of a particular species in the solution. Representing the species O_2 in air or in fact any species in a solution as A, the chemical potential of A in the solution, μ_A, is calculated by splitting the change in G from the reference state into two separate determinations:

$$\mu_A = \mu_A^o + \mu_A^* \qquad [4.17]$$

where μ_A^o stands for the molar Gibbs energy in going from the reference state (elements in their most stable state at 25°C and 1 bar with Gibbs energies of elements taken as zero) to what is called the *standard state* of A. μ_A^* then gives the change of molar G from this standard state to the state of interest. This splitting of the μ_A calculation into two terms allows the use of pure A or endmember properties for part of the calculation, similar to the procedure in Chapter 3. This split can be made at any state of the system that is convenient because μ_A is independent of the path taken to calculate it between states. For aqueous species, a state of interaction of the species with only H_2O molecules at the temperature and pressure of interest is used as the standard state. What is required to specify a standard state is the same as for any calculation of G. This is the pressure, temperature, composition, and configuration of the matter in the state. For A in a mixture, a standard state of pure component species A at the temperature of interest and 1 bar is often chosen. This implies μ_A^o gives the change in G for 1 mole of pure A from the reference state to the T of interest at 1 bar, given by point x in **Figure 4-7**.

$\Delta \mu_A^*$ must then give the change in G (from point x) with increased P at the T of interest (point y) and also the departure of G with composition from pure A at the pressure and temperature of interest (point z). Although it is straightforward to calculate the standard state change of Gibbs energy, μ_A^o, knowing the heat capacity of pure A from what has been done in Chapter 3, what about the nonstandard state change, μ_A^*?

To determine μ_A^* with changes in pressure and composition, *partial pressure* must be understood. In 1811, Dalton investigated the relation between P and T for mixtures of pure gas, given as A, B, and C, in a constant volume, V, at low pressure where they act "ideally," that is, where they are at such a low concentration the molecules in them do not interact significantly and the gases, therefore, obey the ideal gas law. For each gas separately occupying the given volume, V, he found that

$$P_i = n_i \frac{RT}{V} \tag{4.18}$$

where P_i and n_i are the pressure and the number of moles of gas i, respectively; therefore, at constant temperature, if $n_A \neq n_B \neq n_C$, it is clear that $P_A \neq P_B \neq P_C$.

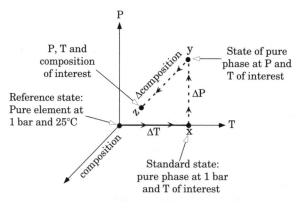

FIGURE 4-7 Schematic diagram representing the path taken to compute the Gibbs energy change from the reference state to the standard state, μ^o, and finally to the state of interest of $\mu^o + \mu^*$ that proceeds with temperature, then pressure, and finally composition of the substance.

When these gases of amounts n_A, n_B, and n_C are mixed together in the same volume, the pressure, P_{total}, was shown by Dalton to be given by

$$P_{total} = P_A + P_B + P_C \qquad [4.19]$$

P_A, P_B, and P_C are called partial pressures of gases A, B, and C in the mixture, respectively, as given by Equation 4.18 when only the subscripted gas occupied the volume. Combining Equations 4.18 and 4.19 results in

$$P_{total} = (n_A + n_B + n_C)\frac{RT}{V} \qquad [4.20]$$

or for gas A

$$\frac{P_A}{P_{total}} = \frac{n_A}{n_A + n_B + n_C} = X_A \qquad [4.21]$$

This gives Dalton's law of partial pressure of a gas in a mixture, which is generally written as

$$P_A = X_A P_{total} \qquad [4.22]$$

where P_A is the partial pressure of gas A in the gas mixture of total pressure, P_{total}.

Remember from the combined first and second law of thermodynamics discussed in Chapter 3 that the change in G as a function of pressure at constant temperature is the pressure-volume work done on the system. For a mole (or any fixed mass) of gas, its pressure-volume-temperature behavior can be characterized by

$$PV = k(T) \qquad [4.23]$$

where $k(T)$ is a temperature-dependent constant. In the case of 1 mole of an ideal gas, $k(T) = RT$. **Figure 4-8** shows this function, $PV = k(T)$, for $k(T)$ equal to 1 and 4. The work, that is, Gibbs energy, added to the system in going from one pressure to another is the area under the curve from the lower pressure to a higher pressure. In general, the change in the molar Gibbs energy with pressure is given by

$$d\mu_T = (\overline{V}dP)_T \qquad [4.24]$$

where the subscript T indicates the differential is to be evaluated at constant temperature. Both sides of Equation 4.24 can be integrated as a function of pressure for a gas in a gas mixture. For gas A in a mixture, the molar nonstandard state change in Gibbs energy, μ_A^*, from the 1 bar pure gas standard state pressure is then

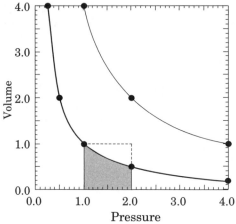

FIGURE 4-8 Volume versus pressure diagram at constant temperature for $PV = 1.0$ (heavy solid line) and $PV = 4.0$ (light solid line). PV work put into a gas with a change in pressure is the area under the curve. For $PV = 1.0$ and increasing pressure from 1 to 2, the work done on the system is $\ln 2 - \ln 1 = 0.693$ PV units, as shown by the gray shaded area.

$$\mu_A^* = \int_1^{P_A} \left(\frac{d\mu_A}{dP} \right)_T dP = \int_1^{P_A} \overline{V}_A dP \qquad [4.25]$$

where P_A is the partial pressure of A in the mixture and $\overline{V}_A$ is the partial molar volume of A in the mixture. Making the substitution for 1 mole of A in an ideal gas mixture from Equation 4.18 where $P_A \overline{V}_A = RT$ gives

$$\mu_A^* = \int_1^{P_A} \frac{RT}{P_A} dP = RT(\ln P_A - \ln 1) = RT \ln P_A \qquad [4.26]$$

$\ln P_A$ denotes the natural logarithm of P_A. If $P_A = 2$ and $RT = 1$, the work added to the system would be the shaded area under the curve in Figure 4-8.

For the ideal gas, A, in a gas mixture its Gibbs energy is, therefore, determined by calculating the change in Gibbs energy from the reference state to the standard state of pure gas at the temperature of interest and 1 bar, $\mu_{A,1bar,T}^\circ$. To this is added the Gibbs energy change from the standard state to a state where the gas is at the pressure and temperature of interest in a mixture at a partial pressure P_A, μ_A^*, given by Equation 4.26. The molar Gibbs energy of an ideal gas in a gas mixture is then given as

$$\mu_A = \mu_{A,1bar,T}^\circ + RT \ln P_A \qquad [4.27]$$

where $RT \ln P_A$ is the work done on 1 mole of the pure gas when compressing/expanding it from a state where $P = 1$ to partial pressure P_A in a gas mixture of final total pressure P. Making the substitution given by Dalton's law in Equation 4.22 results in

$$\mu_A = \mu^o_{A,1bar,T} + RT \ln X_A P_{total} \qquad [4.28]$$

Separating the terms in the logarithm yields

$$\mu_A = \mu^o_{A,1bar,T} + RT \ln X_A + RT \ln P_{total} \qquad [4.29]$$

The last term on the right side is the Gibbs energy change for (work done on) pure gas A in increasing its pressure from 1 bar to P_{total}. If this is combined with the first term on the right, the standard state is changed to that of pure A at the P and T of interest. Equation 4.29 therefore becomes

$$\mu_A = \mu^o_{A,P,T} + RT \ln X_A \qquad [4.30]$$

where the subscript P,T on the standard state term indicates that a different standard state has been used (point y in Figure 4-7).

Combining Equation 4.2 for an ideal solution with Equation 4.30 for A and B in a binary mixture gives

$$\mu^{ideal}_s = X_A(\mu^o_{A,P,T} + RT \ln X_A) + X_B(\mu^o_{B,P,T} + RT \ln X_B) \qquad [4.31]$$

so that

$$\mu^{ideal}_s = \underbrace{X_A \mu^o_{A,P,T} + X_B \mu^o_{B,P,T}}_{\text{mechanical mixture}} + \underbrace{X_A RT \ln X_A + X_B RT \ln X_B}_{\text{ideal mixing}} \qquad [4.32]$$

as given in Equation 4.11. The properties of an ideal solution are outlined in **Table 4-1.**

Table 4-1   **Properties of ideal mixtures with a pure phase at the P and T of interest standard state**

Each component: $\mu_i = \mu^o_i + RT \ln X_i$

Solution as a whole: $\mu^{ideal}_s = \sum_i X_i \mu^o_i + RT \sum_i X_i \ln X_i$

$\overline{V}^{ideal}_s = \overline{V}_{\text{mech mix}} = \sum_i X_i \overline{V}^o_i$

$\Delta \overline{H}^{ideal}_s = 0$

Real Mixtures

Real gases at normal pressures are not ideal so that μ_A^* in Equation 4.26 does not equal $RT \ln P_A$. A new variable, f, called *fugacity*, can be defined however and thought of as the thermodynamic partial pressure. The variable f then is equal to the value that gives the correct value of μ_A^* when using the expression $\mu_A^* = RT \ln f_A$. The relationship between f_A and P_A is

$$\chi_A = \frac{f_A}{P_A} \qquad [4.33]$$

where χ_A is termed the *fugacity coefficient* of gas A. For a real gas, A, the chemical potential becomes

$$\mu_A = \mu_{A,\,1\text{bar},T}^{o} + RT \ln f_A \qquad [4.34]$$

The concept of fugacity can be generalized for any substance, not just gases. Consider a solid that contains some A sealed in a vacuum container, as shown in **Figure 4-9**. The thermodynamic model stipulates a vapor of A of some very small partial pressure develops in contact with the solid. At equilibrium, μ_A (vapor) $= \mu_A$ (solid), and given the standard state outlined above for gases

$$\mu_A(\text{solid}) = \mu_A(\text{vapor}) = \mu_A^{o}(\text{vapor}) + RT \ln f_A^{\text{vapor}} \qquad [4.35]$$

μ_A in the solid, based on a gas standard state, is, therefore, equal to μ_A^{o} (vapor) $+ RT \ln f_A^{\text{vapor}}$. For minerals in the crust, V can generally be taken as independent of pressure and temperature. This makes the calculation of the Gibbs energy as a function of pressure straightforward. The change in μ_{solid}^{o} as a function of pressure at constant temperature, similar to a pure gas as given in Equation 4.25, can be written as

$$\int_1^P \left(\frac{d\mu_{\text{solid}}^{o}}{dP} \right)_T dP = \int_1^P \overline{V}_{\text{solid}}^{o}\, dP \qquad [4.36]$$

I FIGURE 4-9 Solid and its vapor in equilibrium in a closed system.

If $\overline{V}^o_{solid}$ is independent of pressure, the integration in Equation 4.36 is equal to $\overline{V}^o_{solid}(P-1)$. It is reasonable, therefore, to take the standard state for mineral solids as the pure endmember solid at the temperature and pressure of interest. In this case, rather than $\mu^o_{solid, 1\,bar,\,T}$, the standard chemical potential $\mu^o_{solid,\,P,T}$ is used that contains the change in G of the pure endmember mineral from the reference P and T to the P and T of interest as given in Equation 4.36. μ^*_{solid} then contains only the change in G as a function of composition from the pure endmember phase.

The same form of $\mu*$ can be retained, but to indicate that the pressure integration is done in the standard state term, a_i, termed the *activity* of i, is used instead of f_i, the fugacity. a_i is defined as

$$a_i \equiv \frac{f_i}{f^o_i} \qquad [4.37]$$

where f^o_i is the fugacity of pure endmember i at the pressure of interest. The chemical potential of a solid phase i, using a pure solid phase at the pressure and temperature of interest standard state, is

$$\mu_i = \mu^o_{i,P,T} + RT \ln \frac{f_i}{f^o_i} = \mu^o_{i,P,T} + RT \ln a_i \qquad [4.38]$$

For a pure endmember solid phase $f_i/f^o_i = 1$ so $a_i = 1$, and therefore, $RT \ln a_i = 0$. In this case, $\mu_i = \mu^o_{i,P,T}$. If a mineral is not pure but mixing on the sites is totally random, that is ideal, then like ideal gases

$$\mu^*_i = n_i RT \ln X_i \qquad [4.39]$$

where there is mixing on n_i moles of sites per mole of the mineral.

Consider Fe-Mg mixing in olivine. The nonstandard state Gibbs energy for forsterite, Mg_2SiO_4 ($Fo_{(solid\,sol)}$), component in this mixture with ideal mixing on two identical Mg sites would be given by

$$\mu^*_{Fo\,(solid\,sol)} = 2RT \ln X_{Fo} = RT \ln X^2_{Fo} \qquad [4.40]$$

Mixing on sites in minerals is considered in greater detail in Chapter 5, when mineral chemistry is discussed. When considering liquids (e.g., H_2O), a "gas" standard state (pure component at 1 bar and any T) is often used. It is sometimes more convenient, however, to use a "solid" (pure component at any P and T) standard state.

■ Law of Mass Action

Consider the general reaction between reactant, R_i, and product, P_i, species:

$$n_1 R_1 + n_2 R_2 + n_3 R_3 = n_4 P_4 + n_5 P_5 + n_6 P_6 \qquad \text{[4.41]}$$

With a standard state stipulated as the pure component species at the pressure and temperature of interest, the chemical potential of each species, i, in the reaction can be written as

$$\mu_i = \mu_i^o + RT \ln a_i \qquad \text{[4.42]}$$

With n_i moles of species i, the total Gibbs energy contributed by species, i, in Reaction 4.41 is

$$n_i \mu_i = n_i \mu_i^o + RT \ln a_i^{n_i} \qquad \text{[4.43]}$$

where the mole number of i in the second term on the right is brought inside the logarithmic term.

The total Gibbs energy change of Reaction 4.41 is

$$\Delta G_R = \sum_{P_i} n_i \mu_{P_i}^o - \sum_{R_i} n_i \mu_{R_i}^o + \sum_{P_i} RT \ln a_{P_i}^{n_i} - \sum_{R_i} RT \ln a_{R_i}^{n_i} \qquad \text{[4.44]}$$

where the summations over product, P_i, and reactant, R_i, species are separated into different terms. With the standard state Gibbs energy of reaction, ΔG_R^o, defined as

$$\Delta G_R^o \equiv \sum_{P_i} n_i \mu_{P_i}^o - \sum_{R_i} n_i \mu_{R_i}^o \qquad \text{[4.45]}$$

ΔG_R in Equation 4.44 becomes

$$\Delta G_R = \Delta G_R^o + RT \ln \left[\frac{a_{P_4}^{n_{P_4}} a_{P_5}^{n_{P_5}} a_{P_6}^{n_{P_6}}}{a_{R_1}^{n_{R_1}} a_{R_2}^{n_{R_2}} a_{R_3}^{n_{R_3}}} \right] \qquad \text{[4.46]}$$

The term in the logarithm in Equation 4.46 is called the *activity product* for the reaction. It is typically designated with the symbol Q. For Reaction 4.41, this is

$$Q \equiv \frac{a_{P_4}^{n_{P_4}} a_{P_5}^{n_{P_5}} a_{P_6}^{n_{P_6}}}{a_{R_1}^{n_{R_1}} a_{R_2}^{n_{R_2}} a_{R_3}^{n_{R_3}}} \qquad \text{[4.47]}$$

The powers on the activities, a_i, are the stoichiometric coefficients of the ith reactant (R_i) or product (P_i) species in the reaction. Combining Equations 4.46 and 4.47 gives

$$\Delta G_R = \Delta G_R^o + RT \ln Q \qquad [4.48]$$

At equilibrium, $\Delta G_R = 0$, and the equilibrium activity product is designated with a bold capital **K** and termed the *equilibrium constant* so that

$$-\frac{\Delta G_R^o}{RT} = \ln \mathbf{K} = 2.303 \log \mathbf{K} \qquad [4.49]$$

This is the *law of mass action*. The coefficient, 2.303, in the last term gives the conversion from the natural logarithm to the commonly used base-10 logarithm. Equation 4.49 indicates that the equilibrium constant depends on the standard states chosen for the species in the reaction. Different standard states for different species in a reaction can be combined in ΔG_R^o and used to compute **K**. Standard states, therefore, of

1. The pure phase at the pressure and temperature of interest for a mineral component in a solid solution;
2. The pure phase at 1 bar and the temperature of interest for a gas component in a fluid;
3. A hypothetical one molal aqueous solution in which the species behave as though they are at infinite dilution at the P and T of interest for an aqueous species can be combined to calculate **K** (see below).

The dependence of **K** on standard states needs to be kept in mind when equilibrium constants are considered.

■ Determination of the Gibbs Energy of a Real Gas in a Mixture

The Gibbs energy of a gas i in a mixture can be calculated from Equation 4.34 if its fugacity, f_i, can be determined. For a pure gas, its volume can be measured as a function of pressure at constant temperature, and the work done on the gas is the area under the curve similar to the plot shown in Figure 4-8. This work is then equal to $RT \ln f_i$. As given in Equation 4.33, f_i is related to pressure through the fugacity coefficient. In Chapter 3, the Redlich and Kwong model (Equation 3.56) was used to calculate the volume of a pure real gas as a function of pressure and temperature. The fugacity coefficient of pure gas i, χ_i, from this model is

$$\ln\chi_i = Z_i - 1 - \ln(Z_i - BP) - \frac{A^2}{B}\ln\left(1 + \frac{BP}{Z_i}\right) \tag{4.50}$$

where

$$B = \frac{b_{RW}}{RT}, \quad A^2 = \frac{a_{RW}}{R^2 T^{2.5}}, \quad \text{and} \tag{4.51}$$

Z_i, the compressibility factor of pure gas i, is given by

$$Z_i = \frac{PV_i}{RT} \tag{4.52}$$

To compute Z_i at a particular pressure and temperature, the volume of pure gas i must be known. This can be determined with Equation 3.72 from Chapter 3 through an iterative method. Knowing Z_i, Equation 4.50 can be used to obtain χ_i and therefore f_i from Equation 4.33.

Redlich and Kwong proposed that for mixtures of gases with like molecules, an expression similar to Equation 3.72 from Chapter 3 can be used:

$$P = \frac{RT}{\overline{V} - b_{RK}^{mix}} - \frac{a_{RK}^{mix}}{T^{1/2}\overline{V}(\overline{V} + b_{RK}^{mix})} \tag{4.53}$$

where $\overline{V}$ is the molar volume of the gas mixture. The constants a_{RK}^{mix} and b_{RK}^{mix} are determined from the pure gases by noting that

$$a_{RK}^{mix} = \sum_i \sum_j a_{RK}^{i-j} X_i X_j \tag{4.54}$$

and

$$b_{RK}^{mix} = \sum_i b_{RK}^i \tag{4.55}$$

where

$$a_{RK}^{i-j} = (a_{RK}^i a_{RK}^j)^{1/2} \tag{4.56}$$

The superscripts i and j denote the ith and jth gas in the mixture, respectively. The summations are then taken over all the gases in the mixture; therefore, the Redlich-Kwong model for gas mixtures requires only the a_{RK} and b_{RK} for the individual pure gases. It works well for gases of the same type (e.g., spherical) because the equations are symmetric. For mixtures of spherical and polar molecules such as mixtures of CO_2 and H_2O, other mixing models have been proposed. For supercritical mixtures of CO_2 and H_2O, a method outlined by Holloway (1977) as corrected in Flowers (1979) where a_{RK}^{mix} in Equation 4.54 is a function of temperature is often used (see also Kerrick and Jacobs, 1981). The equations that are produced are quite complex. A good discussion of calculation of fugacities of gas mixtures is given in Chapter 5 of Prausnitz (1969).

■ Gibbs Energy of Aqueous Species

Aqueous solutions have two kinds of species that mix with H_2O: charged and uncharged. The charged species are ions (e.g., Na^+, Cl^-, and Ca^{2+}). Uncharged aqueous species include dissolved gases (e.g., O_2 and CH_4), associated weak acids (e.g., H_4SiO_4 and H_2CO_3), and organic molecules (e.g., benzene and C_6H_6). Because of the small mole fractions of other species relative to H_2O, a concentration scale based on molality, m_i, rather than mole fraction, is typically used, that is, moles of solute i per 1000 g of H_2O. Because there are 55.51 moles of H_2O in 1000 g of water, the relationship between m_i and X_i is

$$\frac{m_i}{55.51 + m_i} = X_i \qquad [4.57]$$

At the low molalities typical of most solute concentrations ($<10^{-2}$), the denominator in Equation 4.57 can be taken as 55.51. Under these conditions

$$m_i = 55.51 X_i \qquad [4.58]$$

m_i and X_i are, therefore, directly proportional to each other at low concentrations. This is important when considering standard states for aqueous species (see Figure 4-11).

To understand how the Gibbs energy of aqueous species in a solution with H_2O is computed, the relationship between *activity coefficients*, activities, and standard states must be understood. As indicated in Equation 4.37, the activity of i, a_i, is defined as the ratio of two fugacities

$$a_i \equiv \frac{f_i}{f_i^o} \qquad [4.59]$$

where f_i^o is the fugacity of pure endmember i at the pressure and temperature of interest, based on a gas standard state. **Figure 4-10** shows an activity versus mole fraction diagram for a mixture of a neutral aqueous component A and H_2O based on a pure component phase at the pressure and temperature of interest standard state. With this standard state, ideal mixing is given by the long dashed lines in Figure 4-10 where $a_A = X_A$ and $a_{H_2O} = X_{H_2O}$. There is positive

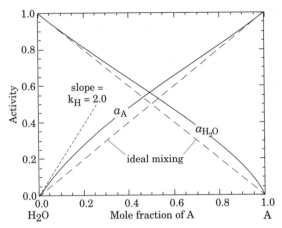

FIGURE 4-10 Activities of mixtures of a typical neutral species and H_2O as function of the mole fraction of *A* based on a pure component at the pressure and temperature of interest standard state. Note the positive departure from ideality, but activity approaches mole fraction as the mole fraction of either H_2O or *A* approaches 1.

departure from ideal mixing for both the neutral species and H_2O in the mixture for the example given by the solid lines in Figure 4-10. Other situations can occur that produce negative departure.

It must be true that as X_A approaches 1 and, therefore, the standard state for A is approached, a_A must approach 1. On the opposite end of the concentration scale at very low X_A, a_A is not given by $a_A = X_A$ but rather by

$$a_A = k_H X_A \qquad [4.60]$$

where k_H is a constant. Equation 4.60 is called *Henry's law*, and the composition region where Equation 4.60 is valid is termed the Henry's law region. In Figure 4-10, k_H equals 2 in the Henry's law region. This region is of most concern because neutral species in H_2O at earth surface conditions typically have concentrations in this region. The activity of H_2O in the mixture is given by X_{H_2O} when the neutral species is in the Henry's law region.

Figure 4-11 is a blowup of the low concentration-activity corner of Figure 4-10. Recall from Equation 4.42 that the molar Gibbs energy of species A in solution is given by

$$\mu_A = \mu_A^\circ + RT \ln a_A \qquad [4.61]$$

Consider a pure component phase at the pressure and temperature of interest standard state. This is also called a *Raoult's law* standard state. Using this standard state, the chemical potential of species A is specified as

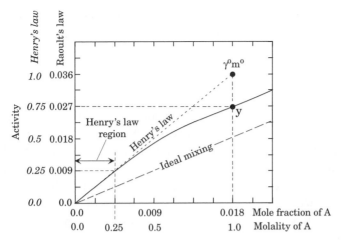

FIGURE 4-11 A blowup of the low concentration region of Figure 4-10 showing activity versus composition for a species that obeys Henry's law. With a 1 molal standard state referenced to infinite dilution, the activity = 1 in the standard state given by $\gamma^\circ = m^\circ = 1$.

$$\mu_A = \mu_A^\circ + RT \ln \Gamma_A X_A \qquad [4.62]$$

where Γ_A is the Raoult's law activity coefficient. Along the line labeled "ideal mixing" in Figure 4-11, $\Gamma_A = 1$ for the activity coefficient of the neutral species. The solid line gives the "real" activity versus composition relation for the mixing. At $X_A = 0.018$, the Raoult's law activity (point y) is 0.027, and therefore, $\Gamma_A = a_A / X_A = 0.027/0.018 = 1.5$ rather than the value of 1 for ideal mixing.

Now consider a change in standard state for A. Remember that μ_A must be independent of the standard state chosen. The typical standard state for aqueous species on which to define activity is a 1 molal solution where the species particles do not interact with each other but only with the H_2O molecules in solution, that is, the behavior of the species in what is called an *infinitely dilute* aqueous solution; however, because a 1-molal concentration of these species is required, the standard state is hypothetical. Activity coefficients based on a molality concentration scale are then defined from this definition of the standard state and activity is given by

$$a_i \equiv \frac{f_i}{f_i^\circ} = \frac{\gamma_i m_i}{\gamma_i^\circ m_i^\circ} \qquad [4.63]$$

where γ_i° and m° are the activity coefficient and molality in the standard state. γ_i and γ_i° can be taken as dimensionless. In this hypothetical solution, there are no interactions between the i species so the behavior is ideal—that is, $\gamma_i^\circ = 1$ at $m_i^\circ = 1$ molal.

γ_i must, therefore, account for the nonideal behavior as a function of the molality, m_i, from infinite dilution to the concentrated solution. Equation 4.61 becomes

$$\mu_A = \mu_A^o + RT \ln \gamma_A m_A \qquad [4.64]$$

Re-examine Figure 4-11. This time consider the molality concentration scale for A given below the mole fraction scale. Also consider the Henry's law region of concentration. If a Henry's law standard state is used in the Henry's law region, a_A must equal m_A so that $\gamma_A = 1$. For instance, at $0.25\ m$ the Henry's law activity must be 0.25. This is shown as the set of numbers labeled Henry's law on the figure. This Henry's law scale is extended to activity = 1.0. This must be the Henry's law standard state and is shown by the point at 1 molal labeled $\gamma^o m^o$.

For a "real" one molal solution ($X_A = 0.018$) in Equation 4.64, $\gamma_A = 0.75$ (point y). As indicated above, based on a Raoult's law standard state, Γ_i would be 1.5 at this concentration. In the Henry's law region for the "real" solution, $\gamma_i = 1$, whereas $\Gamma_i = 2$. So in dilute solutions in the Henry's law region with the hypothetical 1 molal standard state outlined, the chemical potential is given by

$$\mu_A = \mu_A^o + RT \ln m_A \qquad [4.65]$$

Stronger interactions typically occur even in the Henry's law composition region if in addition to neutral species there are also charged species in the solution. These interactions between neutral and charged species are accounted for by the activity coefficient term, γ_A, in Equation 4.64. When neutral species are involved, γ_A is often denoted as γ_n. γ_n is typically computed with a Setchénow (1892) type equation:

$$\log \gamma_n = k_S I \qquad [4.66]$$

where k_S is a constant independent of concentration but is different for each electrolyte interacting with the neutral species. k_S is typically about 0.12 ± 0.1 at earth surface conditions, as indicated by the values reported in **Table 4-2** for the indicated neutral species in NaCl solutions. I denotes the *ionic strength* of the solution and is computed from

$$I = \frac{1}{2} \sum_j m_j Z_j^2 \qquad [4.67]$$

where Z_j indicates the charge on the jth species of molality m_j. The summation is taken over all the j charged species in solution. I then gives the concentra-

Table 4-2 Values of the Setchénow coefficient for uncharged species in NaCl solutions at 25°C

Aqueous species	Setchénow coefficient, k_S
H_2S	0.020
H_4SiO_4	0.080
H_2	0.094
CH_4	0.129
O_2	0.132
$CO_{2(aq)}$	0.231

Values from Harned and Owen (1958), Millero and Schrieber (1982), Marshall and Chen (1982), and Millero (1983).

tion of positive or negative charge in the solution with the higher charged species having a greater effect. As a point of reference, average lake and river waters have $I \sim 0.01$ m, whereas seawater has $I \sim 0.5$ m. For an electrolyte like NaCl, its ionic strength is equal to its molality. The activity for neutral species n, a_n, with its molality being m_n can be computed from

$$a_n = \gamma_n m_n \tag{4.68}$$

In the Henry's law region at low ionic strengths of about 0.1 m or less, γ_n is often assumed to be unity. This is because using Equation 4.66 with $k_S = 0.12$ and $I = 0.1$, $\gamma_n = 1.028$. As required by the standard state, when the ionic strength is decreased toward zero, Equation 4.66 indicates that γ_n approaches 1.

Charged Species in Solution

Now consider the activities of charged species in solution. When partial pressures over solutions of *strong electrolytes* like HCl are measured, they exhibit the behavior shown in **Figure 4-12**. At these observed low pressures, the partial pressures given in Figure 4-12 show ideal behavior and can be taken as fugacities. The fugacity of the electrolyte as a function of molality has a zero slope at infinite dilution, and the slope increases continuously with increasing molality. No Henry's law region exists, therefore, to define an infinitely dilute standard state for HCl.

A strong electrolyte dissolves by dissociating almost completely to ions in H_2O. For instance, HCl in H_2O at 25°C exists almost entirely as H^+ and Cl^- ions. With two species in solution it makes sense to plot fugacity as a function of molality squared, as shown in Figure 4-12B. In this case, the slope near infinite dilution is finite. In this region, the relationship between fugacity and molality for a 1 to 1 strong electrolyte, i, like HCl is

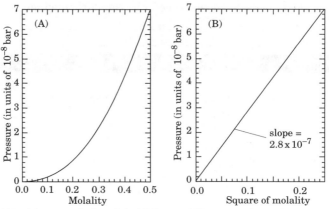

FIGURE 4-12 (A) Partial pressure (= fugacity) of HCl over a HCl solution of the indicated molality and (B) molality squared at 25°C and 1 bar. (Adapted from Pitzer, K. S. and Brewer, L., 1961.)

$$\left(\frac{f_i}{m_i^2}\right) = k_c \qquad [4.69]$$

where k_c is a constant as a function of molality (for HCl at 25°C, $k_c = 2.8 \times 10^{-7}$).

The behavior shown in Figure 4-12 is due to the fact that 1 mole of HCl produces 2 moles of species. Using the 1 molal infinite dilution standard state for an electrolyte i, which produces two aqueous species, requires that

$$\lim_{m_i \to 0} \left(\frac{a_i}{m_i^2}\right) = 1 \qquad [4.70]$$

The activity is defined from Equation 4.61, however, in terms of relative fugacity so

$$\left(\frac{a_i}{f_i/f_i^o}\right) = 1 \qquad [4.71]$$

Substituting Equation 4.69 into Equation 4.71 yields

$$\left(\frac{a_i}{k_c m_i^2 / f_i^o}\right) = 1 \qquad [4.72]$$

For both Equations 4.70 and 4.72 to be correct, $f_i^o = k_c$, where k_c is the Henry's law constant when fugacity is plotted against m_i^2. That is, a 1 molal solution where the standard state fugacity, f_i^o, is equal to k_c is the required standard state. Using this standard state, Equation 4.68 is satisfied.

Single Ion Properties

Although the necessary standard state has been outlined, the relationship between the activity of charged ions in solution to that of the electrolyte as a whole is important. The activity of a strong electrolyte, a_i, must be the product of the activity of all the species it produces in solution. To generalize the behavior in Equation 4.72 for an electrolyte, $A_{n+}B_{n-}$ requires that

$$a_i = (a_+)^{n^+}(a_-)^{n^-} \tag{4.73}$$

where n^+ and n^- are the stoichiometric number of positive and negative ions, respectively, produced when the electrolyte is put in H_2O. a_i can be measured; however, the individual ion activities, a_+ and a_-, cannot be independently determined because the solution must retain charge balance—that is, there is no way to separately measure the change in a single ion activity with concentration. To get around this problem, a *mean ionic activity*, $a_\pm$, is defined as

$$a_\pm = a_i^{1/n} = ((a_+)^{n^+}(a_-)^{n^-})^{1/n} \tag{4.74}$$

where n represents the total number of ions produced by an electrolyte so that

$$n = n^+ + n^- \tag{4.75}$$

$a_\pm$ then gives the "average" behavior of all the ions in solution. For NaCl, the mean ionic activity is

$$a_\pm = a_{NaCl}^{1/2} = (a_{Na^+} a_{Cl^-})^{1/2} \tag{4.76}$$

whereas for $CaCl_2$, it is

$$a_\pm = a_{CaCl_2}^{1/3} = ((a_{Ca^{2+}})(a_{Cl^-})^2)^{1/3} \tag{4.77}$$

When the infinitely dilute state is approached, the ions need to behave ideally. This implies that

$$\frac{a_+}{m_+} \to 1 \text{ as } m_+ \to 0 \tag{4.78}$$

and

$$\frac{a_-}{m_-} \to 1 \text{ as } m_- \to 0 \tag{4.79}$$

where a_+ and a_- are the activities and m_+ and m_- are the molalities of the cation and anion species, respectively. The individual cation and anion activity coefficients are defined as

$$\gamma_+ = \frac{a_+}{m_+} \tag{4.80}$$

and

$$\gamma_- = \frac{a_-}{m_-} \tag{4.81}$$

This requires the mean ionic activity coefficient, $\gamma_\pm$, to be equal to

$$\gamma_\pm = ((\gamma_+)^{n^+}(\gamma_-)^{n^-})^{1/n} = \frac{a_\pm}{m_\pm} \tag{4.82}$$

where $m_\pm$ is the mean molality of an electrolyte of molality, m_i. Because of Equation 4.74, $m_\pm$ in Equation 4.82 must be

$$m_\pm = ((n^+ m_+)^{n^+}(n^- m_-)^{n^-})^{1/n} \tag{4.83}$$

This means that for a 1 to 1 electrolyte like NaCl or CaSO$_4$

$$m_\pm = m_i = m_+ = m_- \tag{4.84}$$

where m_+ and m_- are the molalities of the cation and anion produced by the electrolyte, respectively. For a 1 to 2 electrolyte like CaCl$_2$ with $n^+ = 1$ and $n^- = 2$,

$$m_\pm = ((m_i)(2m_i)^2)^{1/3} = 4^{1/3}m_i \tag{4.85}$$

to retain consistency with Equation 4.83.

Nonideality of Charged Species

Even when the number of species in solution is accounted for with mean ionic values, electrolytes like HCl, NaCl, and CaCl$_2$ deviate from Henry's law behavior much more rapidly than uncharged species. This occurs because the charge on the species produces a long range energy of interaction with other charged species in solution because of the coulombic forces present. Considering the electrostatic energies involved, Peter Debye and Erich Hückel (1923) worked out a simple theoretical model of these interactions as a function of a solution's ionic strength from the ideal infinite dilution solution of H$_2$O standard state. They derived an expression for the mean ionic molal activity coefficient of an electrolyte. This Debye-Hückel activity coefficient expression can be written as

$$\log \gamma_\pm = -\frac{A|Z^+ Z^-|I^{1/2}}{1 + \mathring{a}BI^{1/2}} \tag{4.86}$$

where Z is the charge of the superscripted ion in the electrolyte and $\mathring{a}$ denotes the electrolyte's mean size or distance of closest approach of the ions in solution. A and B are constants independent of ionic strength that can be calculated from the volumetric and dielectric properties of H_2O, which are a function of temperature and pressure. Values for A and B as a function of temperature at H_2O *steam saturation pressure* are given in **Table 4-3**. A more extensive compilation is given in Appendix G.

A number of assumptions were made in deriving the Debye-Hückel activity coefficient expression. These include assuming the bulk value of the dielectric properties of H_2O could be used rather than considering H_2O to be made up of distinct polar molecules. The departure from ideality was also considered to be due solely to coulombic interactions. In the derivation, a spherically symmetrical *Poisson distribution* was combined with the *Boltzmann distribution* and a linear approximation made. Although theoretically correct in the limit of infinite dilution, the model's departure from real solution behavior increases as the concentration of the electrolyte increases.

| Table 4-3 | Debye-Hückel A and B parameters of liquid H_2O at vapor saturation pressures except at 100°C and below, which are at 1 bar |

Temperature (°C)	A ($kg^{0.5}$ $mol^{-0.5}$)	B × 10^{-8} ($kg^{0.5}$ $mol^{-0.5}$ cm^{-1})
0	0.4913	0.3247
5	0.4943	0.3254
10	0.4976	0.3261
15	0.5012	0.3268
20	0.5050	0.3275
25	0.5092	0.3283
30	0.5135	0.3291
35	0.5182	0.3299
40	0.5231	0.3307
50	0.5336	0.3325
70	0.5574	0.3362
100	0.5998	0.3422
150	0.6898	0.3533
200	0.8099	0.3655
250	0.9785	0.3792
300	1.2555	0.3965
350	1.9252	0.4256

A = $1.82484 \times 10^6 \rho^{1/2}/(\varepsilon T)^{3/2}$ and B = $50.29159 \times 10^8 \rho^{1/2}/(\varepsilon T)^{1/2}$, where ρ is the density and ε the dielectric constant of H_2O, respectively (values from Helgeson and Kirkham, 1974).

The Debye-Hückel equation was developed as an expression for the change in the activity of an electrolyte from infinite dilution as a function of its molality expressed as ionic strength. To use it for most earth processes, it needs to be modified for use with complex mixtures of electrolytes and nonelectrolytes. For instance, with a mixture of Na^+, Ca^{2+}, Cl^-, and SO_4^{2-} species in solution, which endmember electrolytes are used? Is the Na^+ associated with Cl^- or SO_4^{2-} to obtain properties of the total electrolyte? Also, what are the effects of the other electrolytes in solution on the electrolyte considered? What is needed is an expression for an individual ion in a complicated mixture containing many electrolytes.

To characterize the activity or activity coefficient of a single ion, either the MacInnes convention (MacInnes, 1919) or H^+ conventional properties are used. With the MacInnes convention, $\gamma_\pm$ of KCl, $\gamma_\pm^{KCl}$, is given by

$$\gamma_\pm^{KCl} = (\gamma_{K^+}\gamma_{Cl^-})^{1/2} = \gamma_{K^+} = \gamma_{Cl^-} \qquad [4.87]$$

That is, K^+ and Cl^- are considered to behave the same in solution. One of the arguments for doing this is that K^+ and Cl^- have the same magnitude of charge and are very similar in size; however, water molecules interacting with K^+ and Cl^- are not completely charge symmetric; therefore, the energetics with positive and negative ions are somewhat different.

In contrast, when using H^+ conventional properties, everything is referenced to the properties of H^+. That is, the Gibbs energy of formation of H^+ in its hypothetical 1 molal state, where the H^+ particles behave as though they are in an infinite sea of H_2O, is taken as zero. Using this convention

$$\gamma_\pm^{HCl} = (\gamma_{H^+}\gamma_{Cl^-})^{1/2} = \gamma_{Cl^-} \qquad [4.88]$$

With either convention, after the activity coefficient of a single ion is defined, all of the other single ion activity coefficients can be obtained from the definition of mean ionic activity coefficient given in Equation 4.84. For instance, γ_{Na^+} can be obtained from the properties of γ_{Cl^-} by noting that

$$\gamma_{Na^+} = \frac{(\gamma_\pm^{NaCl})^2}{\gamma_{Cl^-}} \qquad [4.89]$$

This means for the MacInnes convention

$$\gamma_{Na^+} = \frac{(\gamma_\pm^{NaCl})^2}{\gamma_\pm^{KCl}} \qquad [4.90]$$

whereas for H^+ conventional properties the value is

$$\gamma_{Na^+} = \frac{(\gamma_\pm^{NaCl})^2}{\gamma_\pm^{HCl}} \qquad [4.91]$$

For the two calculation procedures to give identical values requires $\gamma_\pm^{HCl}$ = $\gamma_\pm^{KCl}$. This is a good approximation at low ionic strengths but not at higher ionic strengths (**Figure 4-13**). Using either convention the Debye-Hückel equation of an individual charged species, c, is

$$\log \gamma_c = -\frac{A Z_c^2 I^{1/2}}{1 + \mathring{a}_c B I^{1/2}}$$ [4.92]

where Z_c and $\mathring{a}_c$ are the charge and "size" of the cth charged species, respectively. For two species of the same charge, γ_c varies only because $\mathring{a}_c$ can be different. The ion size parameter $\mathring{a}_c$ is then an experimentally determined parameter. The values shown in **Table 4-4** are for the MacInnes convention. $\mathring{a}_c$ of K^+ and Cl^- are, therefore, the same. Also, these "sizes" are not the diameters of the ions measured in crystals but are larger. For instance, the diameters of Na^+ and Al^{3+} obtained from their "size" in crystals are 1.9 Å and 1.0 Å, respectively, whereas their experimentally determined $\mathring{a}_c$ are 4 Å and 9 Å, respectively. Table 4-4 indicates $\mathring{a}_c$ is larger for ions with smaller crystallographic sizes and higher charge.

At low I, $\mathring{a}BI^{1/2}$ in Equation 4.92 will be small compared with 1 so the denominator of Equation 4.92 approaches unity. This implies at low I ($< \sim 10^{-3}\ m$) the activity coefficient can be computed from

$$\log \gamma_c = -A Z_c^2 I^{1/2}$$ [4.93]

This is referred to as the *Debye-Hückel limiting law*. In the limit, as I approaches 0, $\log \gamma_c$ approaches 0, and, therefore, γ_c becomes 1 as is necessary as infinitely dilute conditions are approached.

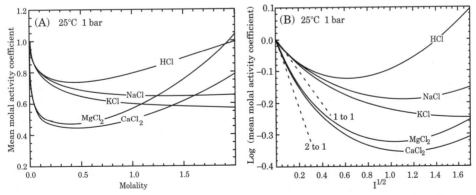

FIGURE 4-13 (A) Mean ionic activity coefficient of the indicated electrolyte at 25°C as a function of its molality and (B) logarithm of the activity coefficient as a function of the square root of its ionic strength. In (B), the dashed lines give Debye-Hückel limiting law behavior for electrolytes possessing species with a single positive and single negative charge (1 to 1) and those with double positive and single negative charge (2 to 1), as computed from Equation 4.93.

Table 4-4	Ion size parameter, $å_c$, for the Debye-Hückel equation using the MacInnes convention

Ion	$å_c \times 10^8$ (cm)
H^+, Al^{3+}, Fe^{3+}	9
Mg^{2+}	8
Li^+, Ca^{2+}, Fe^{2+}	6
Sr^{2+}, Ba^{2+}	5
CO_3^-, Pb^{2+}	4.5
Na^+, HCO_3^-	4–4.5
Hg^{2+}, SO_4^{2-}	4
OH^-, F^-	3.5
K^+, Cl^-, NO_3^-	3
Rb^+, Cs^+, NH^{4+}	2.5

Adapted from Klotz (1964).

For all singly charged species at 25°C and 1 bar, it follows from Equation 4.93 that

$$-\log \gamma_c = 0.509 \, I^{1/2} \tag{4.94}$$

at low ionic strengths. Measured values of $\gamma_\pm$ for some strong electrolytes as a function of their molalities are shown in Figure 4-13A. In Figure 4-13B log $\gamma_\pm$ as a function of $I^{1/2}$ is plotted. Activity coefficients of all 1 to 1 electrolytes are similar at low ionic strength as given by Equation 4.94.

At 25°C and 1 bar, $åB$ is typically close to 1 for NaCl-rich solutions. At these conditions, Equation 4.92 can be recast as

$$\log \gamma_c = -\frac{A Z_c^2 I^{1/2}}{1 + I^{1/2}} \tag{4.95}$$

This is the Güntelburg approximation and implies $å = 3.04$ Å at 25°C. The Debye-Hückel expression and Güntelburg equation appear to describe the behavior of activity coefficients of charged species to ionic strengths as high as 0.01 m at earth surface conditions (**Figure 4-14**). This includes most fresh waters encountered.

To account for the departure of the Güntelburg equation at greater ionic strengths such as with saline waters, Guggenheim made a modification of the equation by adding an additional term, bI, often referred to as the *Debye-Hückel extended term*, to produce the expression

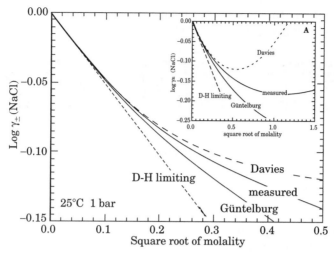

FIGURE 4-14 The logarithm of the measured mean ionic activity coefficient of NaCl at 25°C as a function of the square root of its molality together with values from equations used to describe its behavior as a function of its ionic strength. The inset indicates behavior at higher molality.

$$\log \gamma_c = -\frac{AZ_c^2 I^{1/2}}{1 + I^{1/2}} + b_c I \tag{4.96}$$

where b_c is the extended term constant that depends on the nature of ion c as well as other ions in solution. This equation appears to describe the behavior of activity coefficients of ions to ionic strengths up to 0.1 m (~0.32 $m^{1/2}$) at earth surface conditions.

To simplify the Guggenheim equation, Davies argued for fixing the value of b_c to 0.2 for all electrolytes at 25°C so that

$$\log \gamma_c = -\frac{AZ_c^2 I^{1/2}}{1 + I^{1/2}} + 0.2I \tag{4.97}$$

This equation gives a good description of the mean ionic activity coefficients in NaCl-rich solutions up to nearly 0.1 m and does not have an ion specific term.

The equations outlined can be used to describe the activity coefficients of charged species in solution as a function of the concentration of all the charged species through the solution's ionic strength. If there are significant concentrations of uncharged aqueous species in solution along with the charged species, the interactions of the uncharged species on the charged species become important. These interactions are not considered in the Debye-Hückel formulation and another term must be added (Walther, 2001). In this case, the Davies equation becomes

Table 4-5	Standard states and conventions for determining the Gibbs energy of components in the indicated phases		

Phase	Typical standard state	Term and symbol for nonstandard state value of i	Value of nonstandard state term for ideal mixing
Gas	Pure phase at 1 bar and the T of interest	Fugacity = $f_i = \chi_i P_i$	Partial pressure = P_i
Mineral	Pure endmember phase at the P and T of interest	Activity = $a_i = \Gamma_i X_i$	Mole fraction = X_i
Liquid	Pure endmember phase at the P and T of interest	Activity = $a_i = \Gamma_i x_i$	Mole fraction = X_i
Aqueous species	1 molal solution with species behaving as though they are in an infinite sea of H_2O at the P and T of interest	Activity = $a_i = \gamma_i m_i$	Molality = m_i

$$\log \gamma_c = -\frac{A Z_c^2 I^{1/2}}{1 + I^{1/2}} + 0.2I + 0.5 Z_c^2 \sum_j k_j m_j \qquad [4.98]$$

where the summation on j in the last term is over all the neutral species in the solution with k_j being the Setchénow coefficient of the jth neutral species of molality m_j. A summary of the standard states and the method to characterize nonideal interactions for gases, minerals, liquids, and aqueous species is given in **Table 4-5**.

◼ The Gibbs-Dühem Equation, Open Systems, and the Phase Rule

As outlined earlier, the Gibbs energy per mole of a system depends on the pressure, temperature, and the chemical composition of the system being considered, that is, $G = f(P,T,n_i)$. The total differential of Gibbs energy of a system due to these variables is given as

$$dG = \left(\frac{\partial G}{\partial T}\right)_{P,n_i} dT + \left(\frac{\partial G}{\partial P}\right)_{T,n_i} dP + \sum_{i=1}^{k} \mu_i dn_i \qquad [4.99]$$

where the subscripts P or T on the derivatives indicate temperature or pressure are to be held constant, whereas n_i indicates the moles of all the i components are to be held constant. The total number of components, k, must be the minimum number needed to describe the chemical composition of all the phases in the system to write a total differential. That is, they must completely describe the chemical composition of all of the phases in the system; however, none of the components specified can be described by a combination of the other components.

Consider a system, such as a rock, made up of the five mineral phases as listed in **Table 4-6**. The minimum number of components needed to describe all the phases in this system is four. These components can be chosen as CaO, MgO, CO_2, and SiO_2. Alternatively, the set of components, $CaSiO_3$, $MgSiO_3$, CO_2, and SiO_2 is equally valid. For the latter set, a mole of calcite would be specified as a mole of $CaSiO_3$ minus a mole of SiO_2 plus a mole of CO_2; therefore, either component set can describe the varying compositions in the phases in the system.

From the combined first and second law of thermodynamics, the Gibbs energy change for the system as a whole is

$$dG = -SdT + VdP + \sum_{i=1}^{k} \mu_i dn_i \qquad [4.100]$$

Comparing Equations 4.99 and 4.100, it is clear that

$$\left(\frac{\partial G}{\partial T}\right)_{P,n_i} = -S \qquad [4.101]$$

and

$$\left(\frac{\partial G}{\partial P}\right)_{T,n_i} = V \qquad [4.102]$$

which is consistent with what was developed in Chapter 3.

The total Gibbs energy of a system is equal to the change in Gibbs energy of a mole of each component, μ_i, times the number of moles of that component in the system

Table 4-6   **A five-phase mineral system found in a rock**

Mineral	Formula
Dolomite	$CaMg(CO_3)_2$
Calcite	$CaCO_3$
Wollastonite	$CaSiO_3$
Enstatite	$MgSiO_3$
Quartz	SiO_2

$$G = \sum_{i=1}^{k} n_i \mu_i \qquad \text{[4.103]}$$

Differentiating G in Equation 4.103 as the product of n_i and μ_i for each component gives the total change of G as

$$dG = \sum_{i=1}^{k} \mu_i dn_i + \sum_{i=1}^{k} n_i d\mu_i \qquad \text{[4.104]}$$

Subtracting Equation 4.100 from Equation 4.104 results in

$$0 = SdT - VdP + \sum_{i=1}^{k} n_i d\mu_i \qquad \text{[4.105]}$$

This is the *Gibbs-Dühem equation*. It indicates the way changes in temperature, pressure, and chemical potential must be related to maintain equilibrium for a system whether it consists of a single phase or a number of phases.

Consider a system made up of p different phases. For each phase, A, the Gibbs-Dühem expression is

$$0 = S_i^A dT - V_i^A dP + \sum_{i=1}^{k} n_i^A d\mu_i^A \qquad \text{[4.106]}$$

where S_i^A, V_i^A, n_i^A, and μ_i^A denote the entropy, volume, number of moles of component i, and chemical potential of component i in the Ath phase, respectively. If there are p phases there will be p such equations. By knowing the composition of phase A, n_i^A is known, whereas the entropy, S_i^A, and volume, V_i^A, can be determined for a particular pressure and temperature.

At equilibrium, the chemical potential of each component in each phase must be the same; therefore, the unknowns for the set of p equations are T, P, and the μ_i. This means there are $2 + k$ unknowns, where k is the number of components in the system. For the system to be completely determined, p (number of phases) must equal the $2 + k$ unknowns. The number of degrees of freedom, f, is the number of intensive variables that can be changed for a system of phases and still maintain equilibrium. This is given by the difference between $k + 2$ and p or

$$f = k + 2 - p \qquad \text{[4.107]}$$

This is the *phase rule*. It is helpful in evaluating experimental observations, constructing phase diagrams, and determining whether rock systems observed in the field are open to components such as H_2O or CO_2.

■ Simple Phase Relations Involving Mixtures

Figure 4-15 shows a schematic pressure and temperature phase diagram for the NaCl-H$_2$O system. As a two-component system, the *phase rule*, Equation 4.107, indicates four phases must be present at an invariant point. Point P is an *invariant point* in the system where the four phases, hydrohalite (NaCl•2H$_2$O), halite (NaCl), liquid, and vapor, coexist. Point E is also an invariant eutectic point where halite and ice coexist with the liquid and vapor phase.

To understand the relations shown in Figure 4-15, it is helpful to first consider the pure H$_2$O and pure NaCl phase diagram. In the single component system, H$_2$O, the invariant point (i.e., triple point) of H$_2$O is the pressure and temperature where ice, liquid, and vapor coexist. This occurs at 0.006 bars and −20.8°C and is labeled as TP(H$_2$O) in Figure 4-15. The triple point in the NaCl system where halite, liquid NaCl, and NaCl vapor coexist is at 0.0003 bars and 801°C and is labeled as TP(NaCl) in Figure 4-15. In these one-component

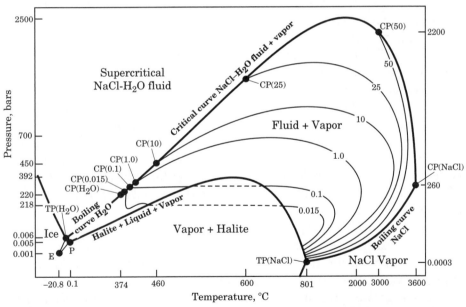

FIGURE 4-15 Schematic phase diagram for the NaCl–H$_2$O system. (Adapted from Bischoff, J. L. and Pitzer, K. S., 1989.) E denotes the NaCl-H$_2$O eutectic (hydrohalite-ice-liquid-vapor), and P is the invariant point of halite-hydrohalite-liquid-vapor. TP stands for the triple point in the indicated pure endmember systems. CP gives the critical point for the wt% NaCl indicated. The numbers on the lines also indicate the wt% of NaCl connecting the critical curve and the halite + liquid + vapor curve. Above the curve, a single supercritical phase exists. Below the curve, an NaCl-rich aqueous fluid phase is in equilibrium with an H$_2$O-rich NaCl vapor phase.

systems, increasing pressure and temperature along their saturation (i.e., boiling curves) retains equilibrium with both a vapor and fluid for each of the systems. At a pressure and temperature of 221.2 bars and 374.2°C along the pure H_2O boiling curve, a critical endpoint for H_2O (labeled CP(H_2O)) is reached. The critical endpoint occurs at about 260 bars and 3600°C for NaCl along the pure NaCl boiling curve (labeled CP(NaCl)). These critical points indicate the pressure and temperature above which only one fluid phase can exist in the one component system, whereas below the critical point either a liquid or vapor exists above or below the boiling curve, respectively (see the discussion about the pressure and temperature phase diagram for pure H_2O given for Figure 3-9).

Now consider the two-component NaCl–H_2O system. When either a small amount of NaCl is added to liquid H_2O or a small amount of H_2O is added to liquid NaCl, the pressure and temperature of the critical endpoint changes. The critical endpoints with 0.015, 0.1, 1.0, 10, and 50 wt% NaCl added to H_2O are shown. The set of all critical endpoints gives the critical curve. The critical curve in the system H_2O–NaCl extends from the critical point of pure H_2O (CP(H_2O)) to pure NaCl (CP(NaCl)). The pure H_2O and pure NaCl boiling curves can be considered to exist on two parallel planes at different depths. The critical curve is then a surface between the two planes.

The halite with liquid and vapor curve is also a projection from the triple point of NaCl (TP(NaCl)) and that of H_2O (TP(H_2O)) through points E and P. Halite with liquid and vapor curve does not intersect the critical curve. Halite only becomes stable with both liquid and vapor below 392 bars. These halite stability relations become important for mid-ocean ridge environments where magmas on the order of 1200°C but at low pressures are injected into shallow oceanic crust and interact with seawater (3.2 wt% NaCl). Also, the large fluid + vapor region above 392 bars disappears above 2500 bars in the NaCl–H_2O system where only a single supercritical NaCl fluid phase exists; therefore, above 2500 bars, a complete solution exists at elevated temperatures between pure H_2O and pure NaCl.

Figure 4-16 is a schematic phase diagram for the NaCl–H_2O system at 500 bars pressure consistent with Figure 4-15. At 500 bars, the vapor + fluid region is at higher temperatures than the solubility of halite in a fluid. At lower pressures, the vapor + liquid region would impinge on the halite + fluid boundary, whereas at greater pressures, the vapor + liquid region becomes smaller and would not exist above 2500 bars. Because natural fluids typically have low NaCl concentrations ($X_{NaCl} < 0.1$), the figure indicates a two-phase (liquid + vapor) fluid exists over a significant temperature range at 500 bars.

Figure 4-17 is a schematic phase diagram for the SiO_2–H_2O system. Because the solubility of SiO_2 in H_2O is much less than NaCl in H_2O, the quartz solubility curve with liquid and vapor intersects the critical composition curve very

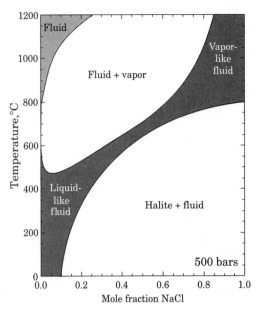

FIGURE 4-16 Schematic phase diagram for the NaCl–H_2O system at 500 bars as a function of temperature and the mole fraction of NaCl that is consistent with the relations given in Figure 4-15.

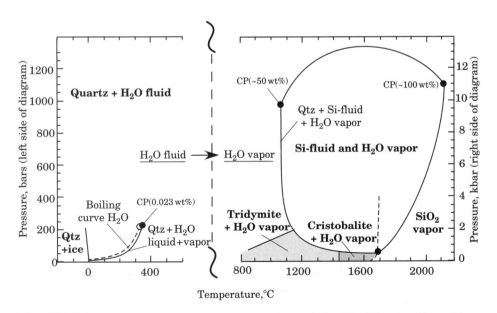

FIGURE 4-17 Schematic pressure and temperature phase diagram for the SiO_2–H_2O system. (Adapted from Krauskopf, K. B. and Bird, D. K., 1995.) The numbers indicate the wt% of SiO_2 in the system.

close to the pure H_2O critical point (at 0.023 wt% SiO_2). This means there is a loop made up of the boiling curve for H_2O (given schematically as a dashed line) and the quartz + liquid + vapor curve that is nearly on top of it. A fluid + vapor field, therefore, does not occur until temperatures are quite high (above 1000°C). For most pressure and temperature conditions in the crust, a single fluid phase in equilibrium with quartz exists. This diagram for SiO_2 is useful when considering any silicate mineral because all silicate minerals have low solubilities in H_2O at temperatures below 1000°C.

FIGURE 4-18 shows the phase diagram for CO_2–H_2O. The boiling curve for pure CO_2 extends to its critical point at 31°C and 72.9 bars. A small amount of H_2O can dissolve in CO_2 at this pressure, which produces a very small CO_2–H_2O critical composition curve to the lower critical endpoint of 31.5°C and 74 bars. Similar to the NaCl–H_2O system, this produces a small loop between the boiling curve where liquid CO_2 is stable and a critical composition curve of CO_2-rich vapor + H_2O-rich liquid at pressures below 74 bars. The two-fluid phase region of CO_2-rich vapor and H_2O-rich fluid extends to an upper critical composition curve that decreases in temperature as pressure is increased from

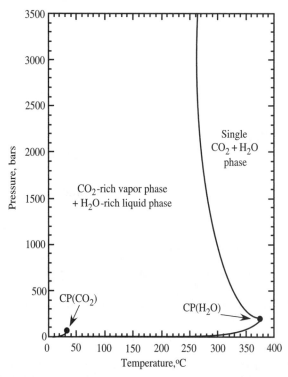

FIGURE 4-18 Schematic pressure and temperature phase diagram for the CO_2–H_2O system where CP denotes the critical point of the indicated phase. (Data from Tödheide, K. and Franck, E. U., 1963.)

the critical point of H_2O at 221.2 bars and 374.2°C. It reaches a temperature minimum at 266°C and 2450 bars. At greater pressure, its temperature increases very slightly with increasing pressure. It is clear from the diagram that at a depth in the crust where temperatures are greater than 300°C, a single fluid phase is produced from reactions of water-rich sediments and carbonate minerals; however, if the fluid cools on transport to the earth's surface two-fluid phases develop.

Summary

The thermodynamic analysis of solutions and simple phase diagrams outlined in this chapter should help place constraints on reactions that are observed and the conditions for stability of the nonpure phases that are encountered in earth processes. The change in the Gibbs energy of a component or species A with change in concentration at constant pressure and temperature is given by the chemical potential, μ_A. To calculate μ_A in a mixture, the Gibbs energy of the pure component is determined at the pressure and temperature of interest from the reference conditions as outlined in Chapter 3. To this is added the change due to ideal mixing from the pure component to the composition of interest. For component A, this is given by $X_A RT \ln X_A$. Nonideal contributions to the components are then added to obtain the Gibbs energy of the component in a mixture.

For a species i, μ_i is determined by splitting the calculation into a standard state term, μ_i°, to which a nonstandard state term, μ_i^*, is added. Both of these states are defined by their pressure, temperature, composition, and configuration of the matter present.

For a "gas phase" standard state, μ_i° gives the Gibbs energy per mole of the pure substance i at the temperature of interest and 1 bar. μ_i^* then gives the Gibbs energy with change in pressure and composition to the state of interest. By considering the form of the ideal gas law, $\mu_i^* = RT \ln f_i$, where f_i is termed the fugacity of i. When ideal gases are considered, f_i becomes a partial pressure, P_i, and $\mu_i = \mu_i^\circ + RT \ln P_i$.

For a "solid phase" standard state, μ° gives the Gibbs energy per mole of the pure substance at the temperature and pressure of interest. μ^* then accounts for the Gibbs energy from the pure phase substance state to the composition of interest. In this case, $\mu_i = \mu_i^\circ + RT \ln a_i$, where a_i, the activity of i, accounts for difference in molar Gibbs energy due to changes in composition.

For aqueous species, a hypothetical standard state of a mole of species at infinite dilution in H_2O is chosen; therefore, a mole of species interacts only with water molecules at the pressure and temperature of interest. The non-standard state term then accounts for changes in Gibbs energy with addition of other species to this infinitely dilute aqueous solution. On a molality concentration scale, $a_i = \gamma_i m_i$, where γ_i is the molal activity coefficient. The Debye and Hückel equation allows calculation of γ_i for interactions of charged species in solution with increased concentration. The Setchénow equation can be used to calculate the activity coefficient for uncharged aqueous species in solution.

Two-component phase diagrams for $NaCl–H_2O$, $SiO_2–H_2O$, and $CO_2–H_2O$ together with the phase rule can be used to understand the location of four-phase invariant points, critical points, critical curves, and two-phase regions as a function of pressure and temperature in these systems. This allows phase stability as a function of composition with changing pressure and temperature to be considered.

Key Terms Introduced

activity
activity coefficients
activity product
Boltzmann distribution
chemical potential
components
Debye-Hückel extended term
Debye-Hückel limiting law
equilibrium constant
fugacity
fugacity coefficient
Gibbs-Dühem equation
H_2O steam saturation pressure
Henry's law
ideal mixing
infinitely dilute

invariant point
ionic strength
law of mass action
mean ionic activity
mechanical mixture
partial pressure
phase rule
Poisson distribution
Raoult's law
reference state
regular solution
solvus
standard state
statistical mechanics
steam saturation pressure
strong electrolytes

Questions

1. Why does the equilibrium constant of a reaction depend on the standard state?
2. What is a solubility product? Solid solution?
3. What is the relation between the Gibbs-Dühem equation and the phase rule?
4. Describe a regular solution.
5. Why does adding a second component to a one-component system sometimes lower the melting temperature (albite added to anorthite) and sometime raise it (anorthite added to albite)?
6. What is an exothermic reaction? Give an example.
7. Give the phase rule and explain the terms.
8. Is the activity of H_4SiO_4 unity in an infinitely dilute aqueous solution? Why?

Problems

1. Construct a diagram at a temperature of 300 K for G of mixing of A with B from a mole fraction of 0.0 to 1.0 of A. Assume mixing is ideal and that $\mu_A^o = -100$ cal mol^{-1} and $m_B^o = -300$ cal mol^{-1}.

2. If the system in Problem 1 does not mix ideally but displays regular solution behavior with $\omega = 1500$ cal mol^{-1}, make a similar plot of G of mixing versus mole fraction of A. Indicate the region of mole fractions where A and B will unmix.

3. Calculate the standard enthalpy, Gibbs energy change, and the value of the equilibrium constant for the following reactions at 25°C and 1 bar. Which side is more stable if the phases are pure? Indicate whether the reaction as written is endothermic or exothermic.
   a. Brucite = periclase + H_2O (steam)
   b. Quartz + muscovite = K-feldspar + andalusite + H_2O (steam)

4. Determine the standard Gibbs energy and equilibrium constant at 25°C and 1 bar of the reaction:

   Albite = nepheline + 2 quartz

   With pure albite, nepheline, and quartz using the typical pure phase standard state, the activities of these phases and, therefore, the equilibrium constant will be unity. Why is the calculated equilibrium constant from values in the back of the book not equal to 1?

5. Which is more stable in air of 21% O_2 at standard conditions, tin or cassiterite? Show your calculation.

6. Is magnesite or nesquehonite more stable in water at standard conditions?

7. Consider the system $MgO-SiO_2-H_2O$. List three mineral species and three aqueous species in this system. With these six species, what is the number of phases present? At a fixed pressure, can temperature be varied and keep all the phases in equilibrium?

8. Calculate the ionic strength of the following solutions:

   a. 0.015 m NaCl   b. 0.015 m CaCl$_2$ c. 0.15 m BaSO$_4$

9. Calculate the activity coefficients of H^+, Cl^-, and Na^+ at 25°C and 1 bar as a function of ionic strength between 0.0 and 0.1 using the Debye-Hückel Equation 4.92.

10. Calculate the activity coefficient as a function of ionic strength between 0.0 and 0.1 for the ions Ca^{2+}, Mg^{2+}, and SO_4^{2-} using the Debye-Hückel Equation 4.92.

11. At 25°C, at what pressure will quartz be converted to coesite?

12. If the solubility of Ag$_2$SO$_4$ at 25°C is 0.9 g/100 g H_2O, calculate the solubility product. The solubility product is the product of the concentration

of the aqueous species in the solubility reaction raised to the power of the coefficient of the species in the reaction.

13. Assume that talc ($Mg_3Si_4O_{10}(OH)_2$), magnesite ($MgCO_3$), quartz (SiO_2), and enstatite ($MgSiO_3$) coexist at equilibrium.
   a. What is the minimum number of components in this system?
   b. Give two possible sets of components.
   c. How many degrees of freedom exist in this assemblage?
   d. At an arbitrary constant value of P and T, can μ_{H_2O} and μ_{CO_2} vary in this assemblage?

14. Consider the vein shown in the figure below found in a rock from the Hashknife ore deposit (adapted from H. C. Helgeson class presentation).

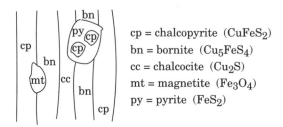

cp = chalcopyrite ($CuFeS_2$)
bn = bornite (Cu_5FeS_4)
cc = chalcocite (Cu_2S)
mt = magnetite (Fe_3O_4)
py = pyrite (FeS_2)

   a. How many components are there in this system (as represented by the vein)?
   b. List all equilibrium phase assemblages found in this vein (i.e., those phases that coexist in what appears to be equilibrium). What criteria are you using to establish equilibrium?
   c. Draw a triangular composition diagram for the subsystem Cu–Fe–S, and illustrate the relationships in b above by means of tie lines.
   d. Why are the minerals chalcopyrite and chalcocite separated? Reinforce your answer with a chemical reaction.

15. Write the equilibrium constant expression and compute the equilibrium constant for the following reactions at 25°C and 1 bar. Indicate which side is most stable.
   a. analcime + quartz = albite + H_2O
   b. kaolinite + H_2O = 2 gibbsite + 2 quartz
   c. diopside + calcite = akermanite + CO_2
   d. talc + 3 calcite + 3 CO_2 = 3 dolomite + 4 quartz + H_2O

16. Using both the Debye-Hückel law and limiting law, calculate the mean stoichiometric activity coefficient of NaCl in a
   a. 0.1 molal NaCl solution at 25°C and 1 bar
   b. 0.1 molal NaCl solution at 200°C and vapor sat pressure

Problems Making Use of SUPCRT92

17. Determine the vapor pressure (fugacity) of water at 1 bar and 25°C, 75°C, and 100°C using the reaction H_2O (liquid) = H_2O (gas). What happens at temperatures above 100°C?

18. At 15°C, at what pressure are magnesite and nesquehonite in equilibrium in water if the magnesite is pure but the activity of nesquehonite is only 0.002?

References

Bischoff, J. L. and Pitzer, K. S., 1989, Liquid-vapor relations for the system $NaCl$-H_2O: Summary of the P-T-x surface from 300° to 500°C. *Amer. Jour. Sci.*, v. 289, pp. 217–248.

Debye, P. and Hückel, E., 1923, The theory of electrolytes II: The limiting law of electrical conductivity. *Physik. Z. Leipzig*, v. 24, pp. 305–325.

Flowers, G. C., 1979, Correction of Holloway's (1977) adaptation of the modified Redlich-Kwong equation of state for calculation of the fugacities of molecular species in supercritical fluids of geological interest. *Contrib. Mineral. Petrol.*, v. 69, pp. 315–318.

Harned, H. S. and Owen, B. B., 1958, *The Physical Chemistry of Electrolyte Solutions*. Reinhold Book Corp., New York, 803 pp.

Helgeson, H. C. and Kirkham, D. H., 1974, Theoretical prediction of the thermodynamic properties of aqueous electrolytes at high pressures and temperatures. II. Debye-Hückel parameters for activity. *Amer. Jour. Sci.*, v. 274, pp. 1199–1261.

Holloway, J. R., 1977, Fugacity and activity of molecular species in supercritical fluids. In Fraser, D. G. (ed.) *Thermodynamics in Geology*. D. Reidel, Dordrecht-Holland, pp. 161–181.

Kerrick, D. K. and Jacobs, G. K., 1981, A modified Redlich-Kwong equation for H_2O, CO_2, and H_2O-CO_2 mixtures at elevated temperatures and pressures. *Amer. Jour. Sci.*, v. 281, pp. 735–767.

Klotz, I. M., 1964, *Chemical Thermodynamics Basic Theory and Methods*. W. A. Benjamin, New York 468 pp.

Krauskopf, K. B. and Bird, D. K., 1995, *Introduction to Geochemistry*, 3rd edition. McGraw-Hill, New York, 647 pp.

MacInnes, D. A., 1919, The activities of the ions of strong electrolytes. *J. Am. Chem. Soc.*, v. 41, pp. 1086–1092.

Marshall, W. L. and Chen, C. A., 1982, Amorphous silica solubilities. V. Prediction of solubility behavior in aqueous mixed electrolyte solutions to 300°C. *Geochem. Cosmochim. Acta*, v. 46, pp. 289–291.

Millero, F. J., 1983, The estimation of the pK*$_{HA}$ of acids in sea water using Pitzer equations. *Geochem. Cosmochim. Acta*, v. 48, pp. 571–581.

Millero, F. J. and Schrieber, D. R., 1982, Use of the ion pairing model to estimate activity coefficients of the ionic components of natural waters. *Amer. Jour. Sci.*, v. 282, pp. 1508–1540.

Pitzer, K. S. and Brewer, L., 1961, *Thermodynamics*, 2nd edition. McGraw Hill, New York, 723 pp.

Prausnitz, J. M., 1969, *Molecular Thermodynamics of Fluid-phase Equilibria*. Prentice-Hall, Englewood Cliffs, NJ, 523 pp.

Tödheide, K. and Franck, E. U., 1963, Das Zweiphasengebiet und die kritische Kurve im Kohlendioxid-Wasser bis zu Drucken von 3500 bar. *Zeit. Phys. Chem. N. F.*, v. 37, pp. 387–401.

Walther, J. V., 2001, Experimental determination and analysis of the solubility of corundum in 0.1 and 0.5 m NaCl solutions between 400 and 600°C from 0.5 to 2.0 kbar. *Geochem. Cosmochim. Acta*, v. 65, pp. 2843–2951.

Other Helpful References

Anderson, G. M., 2005, *Thermodynamics of Natural Systems*, 2nd edition. Cambridge University Press, Cambridge, 648 pp.

Denbigh, K., 1981, *The Principles of Chemical Equilibrium*, 4th edition. Cambridge University Press, Cambridge, 494 pp.

5 Mineral Chemistry

Minerals are naturally occurring homogeneous inorganic solids with definite chemical compositions and specific repeating arrangements of atoms. Glasses are disordered solids without a repeating arrangement of atoms and are therefore not minerals. Rocks are made up of minerals with extrusive igneous rocks also containing glass, some sedimentary rocks containing organic matter. The abundance of a particular mineral depends on the availability of its constituent elements and whether the bonding is stable for the physical conditions present and its ability to withstand alteration.

■ Elements

Figure 5-1 gives the Periodic Table of the elements. The elements are arranged into horizontal rows designated as "periods" with vertical columns termed "groups" labeled with Roman numerals. The subscripted number in front of the element symbol is its *atomic number*. This is the number of protons in the nucleus of an atom of the element (see Table 1 in Appendix C for a key to the element symbols). As implied by the name Periodic Table, many atomic, electrical, and chemical properties of elements are periodic in nature, with elements in each group behaving similarly. The table is laid out in three divisions: a set of main elements, a set of similar *transition elements*, and a set of very similar inner transition elements. The properties of elements are periodic because they depend on the arrangement of electrons in outer atomic *orbitals* that repeat at various distances from the nucleus. Properties depend less on the nature of the protons or neutrons present than on the number of electrons in the outer orbitals.

The failure of classical mechanics for small particles and the discovery that energy was quantized into units of hv, where h stands for Planck's constant (= 6.626×10^{-34} J s) and v gives the particle's velocity, led to the development of quantum mechanics. This outlines how atomic particles have the characteristics of waves, and waves have the characteristics of particles as given in the de Broglie relation:

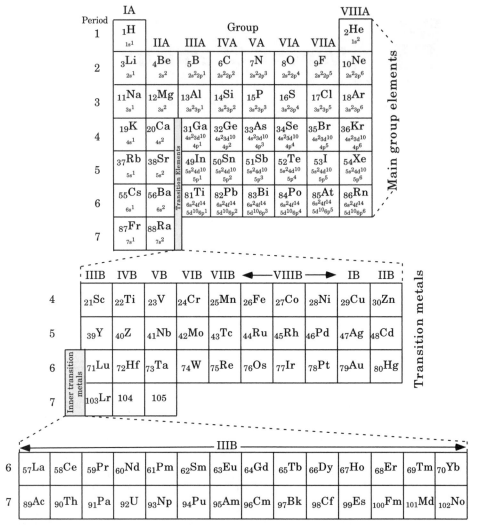

FIGURE 5-1 Periodic Table of the elements showing the natural division of elements into the variable elements in the main group, the more similar transition metals, and the very similar inner transition metals. The subscript before the symbol for each element is its atomic number. The small print in the boxes below the symbols for the main group elements indicates the electrons in the outer electron orbitals of the element (see text).

$$\lambda = \frac{h}{p} \tag{5.1}$$

where p denotes the momentum of a particle of mass, m, given by

$$p = mv \tag{5.2}$$

The greater the mass or velocity of a particle, therefore, the smaller is its wavelength.

Particles can, therefore, be described by wave functions as given by the Schrödinger equation. Wave functions can be thought of as a probability distribution for the particle's location. For instance, if one rolls two dice a large number of times and observes the outcome, the sum of the values on the dice will be 2 or 12 1/36th of the time and 7 1/6th of the time. The "wave function" for the values of the two dice would then have a maximum of 1/6th at 7 and a minimum of 1/36th at 2 and 12. This wave function can be considered a property of the two dice. Wave functions of position of electrons in atoms and molecules are termed orbitals. These give the probability of finding the electron at a particular location and are based on the quantification of their energy due to their distance from the nucleus, the magnitude of their angular momentum, and the orientation of their angular momentum vector.

Based on applying the *Schrödinger equation* to the hydrogen atom, there are four types of electron orbitals giving the probability for the position of electrons that orbit the atom's nucleus. They are denoted as s, p, d, and f orbital types. These orbital types contain 1, 3, 5, and 7 orbitals, respectively. Each orbital contains up to two electrons of opposite spin. The s, p, d, and f orbitals, therefore, give the probability distribution of the location of 2, 6, 10, and 14 electrons, respectively. Orbital types repeat at different average distances from the nucleus as given in **Figure 5-2**, and these are referred to as shells. There are up to seven different shells around a nucleus. Figure 5-2 is somewhat schematic in that each atom or ion has a unique set of energy levels determined by its nuclear charge and the number of electrons present, as given by the Schrödinger equation. The pattern of relative energies of orbitals, however, is correct for all atoms. A comparison of Figures 5-1 and 5-2 indicates that with the increasing number of a period in the Periodic Table, the orbitals are at increasing average distances from the nucleus. Electrons generally occupy the closest, least energetic, orbitals first. The 1s orbital is, therefore, the first filled. The neutrally charged elements H and He have one and two electrons, respectively, in a 1s orbital. Higher energy orbitals are filled in the order of increasing energy shown when elements with more electrons are present. Given for the main group elements in Figure 5-1 in smaller print under the element's symbol are the electrons filled in the last or outermost orbital shell for the neutrally charged element indicated. The superscript after the orbital type gives the number of electrons occupying that orbital type.

For instance, sulfur, $_{16}S$, with an atomic number of 16 has 16 electrons in the neutrally charged S atom. As given in Figure 5-2, the first shell orbital is filled (period 1). All of the second shell orbitals are filled, 2s and 2p (period 2), and in the third shell (period 3, the outermost shell), two electrons occupy the 3s orbital; four electrons occupy the 3p orbitals. This is designated as $3s^2 3p^4$

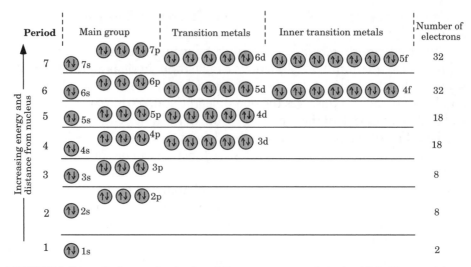

FIGURE 5-2 Schematic diagram showing the relative average energy of atomic orbitals. The up and down arrows in each orbital represent the opposite spins of the two electrons that can occupy the orbital. The number to the right of a row of orbitals indicates to which shell the orbital belongs, with the letter indicating the type of *orbital subshell*. The 4s orbitals are at a lower energy than the 3d orbitals, the 5s orbitals are at a lower energy than the 4d orbitals, the 6s orbitals are at a lower energy than 4f and 5d orbitals, and the 7s orbitals are at lower energy than 5f orbitals.

in Figure 5-1. A complete list of filling of the orbitals for neutrally charged atoms of all of the elements is given in Table 2 of Appendix C.

Now consider the neutrally charged element krypton, $_{36}$Kr. The electrons will distribute themselves with two present in 1s to fill shell 1, and then shell 2 will be filled with two occupying 2s and six in 2p. Similar to shell 2, in shell 3, two electrons will occupy 3s and six in 3p. In shell 4, however, not only are the 4s and 4p orbitals filled, but all of the 3d orbitals (first transition metal group) are as well. This is designated as $4s^23d^{10}4p^6$ in Figure 5-1. With all of the orbitals in period 4 filled, Kr is very stable and in general cannot lower its energy by forming bonds that share electrons with another element. That is, it cannot increase its stability by obtaining a bonding electron from another element, nor is it stabilized by giving up an electron. Kr, like the other inert noble gases given in group VIIIA of the Periodic Table, has all of the orbitals in its outer shell filled. The electrons required to fill each shell increase from 2 in the first shell to 8, 8, 18, 18, and finally 32 in shells 2 to 6, respectively. The complete filling of these shells produces the noble gases $_2$He, $_{10}$Ne, $_{18}$Ar, $_{36}$Kr, $_{54}$Xe, and $_{86}$Rn. Because atoms of these elements do not increase their stability by bonding with another atom of another element, they remain as neutral atoms with a *valence* of 0 as given in Table 1 in Appendix C.

With the lowest energy electron orbital, 1s, filled first, the first period of the periodic table contains only the elements H and He. In the second and third periods, in addition to the s orbital, the three p orbitals must be filled, resulting in eight additional elements in each of these periods. The 3d orbitals are not filled in the third period because they are at a greater energy than the 4s orbital. This means the 3d orbitals are filled after the 4s orbital in the fourth period. A similar situation occurs in the fifth period where the 4d orbitals are filled after the 5s orbital. These additional 5d orbitals mean the fourth and fifth periods each have an additional 10 elements so that there are 18 elements in these periods. This is handled in the Periodic Table by inserting what are termed *transition metals* into the table (see below). In the sixth period, one also encounters f orbitals so that the 6s, 4f, 5d, and then 6p all must be filled, which requires 32 electrons. Similarly, in period 7, one fills the 7s and then 5f orbitals before filling the 6d orbitals. In periods 6 and 7, in addition to the extra 10 transition elements in d orbitals, the elements for the f orbitals are inserted in the Periodic Table as what are called inner transition metals.

On the left side of the Periodic Table, in group IA are the reactive soft metals, which are also called the alkali elements of Li, Na, K, Rb, Cs, and Fr as well as hydrogen. As indicated in Table 2 in Appendix C as well as on the Periodic Table, each of these elements has a single electron in an outermost s orbital shell. These elements tend to bond or to be stabilized by giving up this electron to form an ion, giving it a valence charge of +1 with a noble gas configuration of orbitals. Alkali elements as well as hydrogen, therefore, give up one electron in *electrostatic* bonds and exist as the +1 charged ions—H^+, Li^+, Na^+, K^+, Rb^+, Cs^+, and Fr^+—in aqueous solutions.

The elements in group IIA are referred to as the *alkaline earth elements* and have two electrons in an outer s orbital in addition to the noble gas core of electrons. They, therefore, develop a valence of +2 by giving up these electrons. These elements will exist as Be^{+2}, Mg^{+2}, Ca^{+2}, Sr^{+2}, Ba^{+2}, and Ra^{+2} in aqueous solutions or will give up the two electrons to other elements in electrostatic bonds. The elements in group VIIA need to gain an electron to attain a stable noble gas electron orbital configuration. These elements, therefore, have a valence of −1; exist with a single negative charge as F^-, Cl^-, Br^-, I^-, and At^- ions in aqueous solutions; and attract an electron from another element in electrostatic bonds to form a noble gas electron orbital configuration. They are referred to as the *halogens*.

Complications arise in bonding electrons and, therefore, valence of the elements between the alkaline earths and halogens. Transition metals are elements with incompletely filled d orbitals. The energy differences between the d orbitals correspond to the energies for visible light. Many transition metals, therefore, have a marked color (e.g., Fe and Cu). Transition metals are often used by humankind because they are chemically stable. Most have more than one oxidation (i.e., valence) state. The precious metals (group IB), sometimes referred to as the coinage metals, Cu, Ag, and Au, should have a valence of +1 because of a single electron in an outer s orbital in the neutral element, but Cu can also have a valence of +2 and Au a valence of +3 because of the small energy differences between d orbital electrons. Group IIB elements, containing Zn, Cd, and Hg, have valences of +2 because they contain two outer shell s electrons, but Hg can also exhibit a valence of +1. The possible valences for all the elements are given in Table 1 of Appendix C.

The *actinides* are a series of 15 chemical elements of the Periodic Table with atomic numbers 89 to 103. The first four of the group are the naturally occurring elements actinium, thorium, protactinium and uranium. The others have been made artificially by radioactive bombardment. All isotopes of all of the actinides are radioactive. Ionic radii of the typical +3 oxidation state of actinide ions show a contraction with increasing atomic number. Th, however, is most commonly in a +4 oxidation state and Pa in the +5 oxidation state.

Rare earth elements (REEs) are a group of 17 chemical elements made up of scandium, yttrium, and the *lanthanides*. The lanthanides are the 15 consecutive elements in the first row of the inner transition elements of the Periodic Table starting with lanthanum and including lutetium (atomic numbers 57–71). They are very similar in chemical properties and vary only in the number of 4f orbital electrons. These are divided into the light REEs (LREEs) from La to Sm plus Sc and Y and the heavy REEs (HREEs) from Gd to Lu. Because more of the 4f orbitals are filled, the heavy REEs have somewhat smaller ionic radii than the light rare earths. With their smaller size, heavy REEs are more easily accommodated in most crystal structures, particularly amphibole and garnet.

The REEs provide an array of trace elements in minerals with similar chemical properties but with different selectivity for minerals. This makes them good tracers for some geologic processes. For instance, Ce is oxidized to Ce^{4+} under the oxidizing conditions present in today's oceans. In this state, it precipitates from seawater in manganese nodules on the sea floor. Seawater, therefore, exhibits a distinctive negative Ce anomaly relative to other REEs. This anomaly measured in *carbonates* can be used to help understand palaeo-redox conditions and the extent of seawater-rock interactions. Unique among REEs is europium, Eu, because it can have a valence of +2 as well as the typical valence of REEs of +3. Because the Eu^{2+} radius is similar to Ca^{2+} (**Table 5-1**), it readily substitutes for Ca^{2+} in minerals. Any melt that has crystallized a Ca-rich phase such as plagioclase is, therefore, depleted in Eu as the Eu is incorporated into the Ca-rich mineral. A "negative Eu anomaly" relative to the other REE is then present in the residual melt. This is helpful for understanding the crystallization of Ca-rich phases from melts.

Table 5-1 Cation radii of some common elements for the indicated coordination with O^{2-} plus a number of anion radii

Cation	Radii (Å)	Cation	Radii (Å)	Cation	Radii (Å)
Rb^+	1.68 (8)	Fe^{2+}	0.69 (6) L		
	1.81 (12)		0.86 (6) H	Eu^{2+}	1.17 (8)
K^+	1.59 (8)	Mg^{2+}	0.80 (6)	Eu^{3+}	0.947 (6)
	1.68 (12)		0.97 (8)		
Ba^{2+}	1.50 (8)	Fe^{3+}	0.63 (6) L		
	1.68 (12)		0.73 (6) H		
				Anion	**Radii (Å)**
Na^+	1.10 (6)				
	1.24 (8)	Ti^{4+}	0.69 (6)	Cl^-	1.72 (6)
	1.40 (9)				
Ca^{2+}	1.08 (6)				1.30 (4)
	1.20 (8)	Al^{3+}	0.47 (4)	O^{2-}	1.32 (6)
	1.26 (9)		0.61 (6)		1.34 (8)
Mn^{2+}	0.75 (6) L				
	0.91 (6) H	Si^{4+}	0.34 (4)	F^-	1.25 (6)

Coordination given in parentheses. H and L after a radius indicate bonding electrons in high and low spin state, respectively.

Data from Whittaker and Muntus, 1970.

■ Mineral Stability

Minerals form to lower the internal energy of the mass in the system. This is accomplished by building a three-dimensional repeating arrangement of atoms using chemical bonds of the shortest possible length. Because of the abundance of oxygen, these are generally bonds with oxygen atoms. In bonding, oxygen can accommodate two additional electrons in its outer electron shell, giving it a valence of −2. O^{2-} then bonds with elements that donate electrons, giving them a positive valence. These are termed *cations*.

The arrangement of these oxygen atoms to obtain the shortest distance between oxygen and cations is dictated to a large degree by closest packing rules of solid geometry. Portraying atoms as spheres, the number of spheres of one size (gray) that can be packed around a sphere of a different size (black) is shown in **Figure 5-3**. This is termed the coordination number of the atom. This packing depends on the radius ratio of the two types of atoms, that is, the radius of black sphere/radius of coordinating spheres. The ranges of radius ratios for each type of coordination are given in Figure 5-3. The greater the

Cubo–octahedral Coordination

Radius ratio > 1.0
Coordination number = **12**
Black sphere located at the center of a cube
with coordinating spheres at the midpoints
of the edges of the cube.

Cubic Coordination

0.732 ≤ Radius ratio ≤ 1.0
Coordination number = **8**
Black sphere at the center of a cube with the
coordinating spheres at each corner.

Octahedral Coordination

0.414 < Radius ratio ≤ 0.732
Coordination number = **6**
Black sphere at the center of an
octagon with the coordinating spheres
at each corner.

Tetrahedral Coordination

0.225 < Radius ratio ≤ 0.414
Coordination number = **4**
Black sphere at the center of a tetrahedron with the coordinating spheres
at each corner.

Trigonal Coordination

0.155 < Radius ratio ≤ 0.225
Coordination number = **3**
Black sphere between three coordinating
spheres.

Linear Coordination

Radius ratio ≤ 0.155
Coordination number = **2**
Black sphere between two coordinating spheres.

FIGURE 5-3 Coordination polyhedron for a black sphere surrounded by spheres of different coordination numbers indicating the coordination and possible radius ratios. Types of diagrams used to represent each coordination are also shown.

difference between the radius of the two types of atoms, the lower is its coordination number.

What are the sizes of cations that are coordinated around the O^{2-}? Table 5-1 gives the "effective radii" of some cations in coordination with O^{2-} as well as the effective radii of a number of *anions* besides O^{2-}. These effective radii are obtained by producing a scheme by which summing cation and anion radii gives the distances between cations and anions measured in a crystal. These cation radii depend on the nature of the anion, the cation's valence, the number of anions in its coordination environment, whether the bonding electrons are in a high or low spin state, the extent of covalent bonding, the extent of distortion of the polyhedral site, and the degree of electron delocalization.

Whether the bonding electrons are in a high or low spin state depends on the separation of 5d-orbital electrons into two high and three low energy orbitals in transition metals because the orbitals are not spherically symmetric. If electron–electron repulsion energy is large due to the position of O^{2-} in the ligand (structure) relative to the energy of occupying a high-energy d-orbital, then high spin states occur as electrons fill high-energy orbitals after the low-energy d-orbitals are occupied by only one electron each. If the electron–electron repulsion energy is small relative to the split in d-orbital energies, then all of the low-energy orbitals are filled first in what is termed the low spin state. In molecular orbital theory, the phenomenon is referred to as ligand field stabilization. In minerals, it is called *crystal field stabilization.*

Although the number of O^{2-} coordinated with the cation and the spin state of the bonding electrons, where appropriate, are considered, other factors are ignored because they are second order. The effective radii, therefore, are only approximate but help in our understanding of cation–O^{2-} bonding in minerals and possible atomic substitutions that can occur. The values given are for crustal temperatures and pressures. At greater depths, radii can contract somewhat. As an example of the use of atomic radii, consider fourfold coordination of O^{2-} around Si^{+4}. With the radius of $Si^{4+} = 0.34$ Å and that of $O^{2-} = 1.30$ Å, the radius ratio of $Si^{4+}/O^{2-} = 0.34$ Å/1.30 Å $= 0.262$, consistent with tetrahedral closest packing coordination. Al^{3+} can coordinate with 4 or 6 O^{2-}, having radius ratios of Al^{3+} to O^{2-} of 0.47Å/1.30Å $= 0.362$ and 0.61Å/1.32Å $= 0.462$, respectively.

■ Bonding Between Atoms

If anions are packed around a cation, how do the atoms bond, that is, interact to lower their energy? Bonding between atoms can be of four different types: ionic, covalent, metallic, and Van der Waals bonds (**Table 5-2**). All bonding is based on electrostatic interactions, the interactions between charges in the electron orbitals around the atoms. Most bonds in solids are a mixture of ionic and covalent types. Consider the bonding between O and Si. Si is a nonmetal of intermediate *electronegativity* with an oxidation state of +4 and coordination of 4 with oxygen. Electronegativity is a measure of the ability of an

| Table 5-2   Bonding types between atoms

Ionic bonds are produced when the outermost electrons are removed from one atom, the electron donor, and reside in the electron orbitals of another atom, the electron acceptor. Because of this charge transfer, strong coulombic bonds of 3 to 8 eV (electron volts) per bond are produced. The electron donors reside on the left-hand side of the Periodic Table of elements, whereas the electron acceptors are on the right. The electron donors then produce cations, whereas the electron acceptors are referred to as anions. The bonding in the mineral halite (NaCl) is the classic example. There is almost complete removal of the outermost electron from Na to produce the cation Na^+ and its incorporation into the outer electron orbital of Cl to give the anion Cl^-.

Covalent bonds form where a pair of electrons of equivalent energy and opposite spins are shared between atoms. These bonds are quite stable because the sharing allows the stable electronic configuration for both elements to occur, yielding bond energies of 3 to 8 eV per bond. Covalent bonds exist in the gases H_2, N_2, O_2, and Cl_2.

Metallic bonds occur when valence electrons are released by a metal atom. These electrons move around the array of cations, binding them together in the structure. The freely moving electrons form bonds with energies between 1 and 3 eV per bond. Metallic bonds occur in sulfide minerals that do not have sufficient electrons to fill all eight orbitals in their outermost shell. These electrons give sulfides their metallic luster.

Van der Waals bonds arise in a number of ways, all due to dipole–dipole interactions. That is, many molecules, although electrically neutral, have time-averaged concentrations of electrons on one side of the molecule, creating a distortion of the charge distribution. This creates a positive and negative end to the molecule. In some instances these permanent dipoles induce oppositely oriented instantaneous dipole moments in other molecules. Another type of Van der Waal bond is the hydrogen bond. Hydrogen bonds are formed with a hydrogen atom that is already part of a molecule. These positively charged hydrogens bond with a pair of unshared electrons on an electronegative atom in the same molecule or that of an adjacent molecule. Because these bonds only have energies between 0.02 and 0.5 eV, they are weak. Van der Waals bonds are important in understanding the nature of liquids and gases as well as the bonding between the electrically neutral sheets in phyllosilicates.

atom to attract an electron. With O-Si bonding as well as bonding of oxygen with other cations, the bond is mostly covalent in character, but the sharing of electrons is generally uneven, which causes the covalent bond to have some ionic character. One identifies these mixed type bonds as *polar bonds* or intermediate bonds. The fundamental unit in silicate mineral bonding is, therefore, the polar bonded SiO_4^{4-} *tetrahedron.*

■ Structures of Minerals

The knowledge of structure of a mineral is important for understanding a mineral's density, hardness, cleavage, and x-ray diffraction pattern. It is also helpful for understanding the degree of solid solution, its melting properties, and exsolution phenomena. Minerals can be classified according to their structure into one of the following groups:

native elements
sulfides
sulfates
sulfosalts
oxides
hydroxides
halides
carbonates
nitrates
borates
phosphates
tungstates
silicates

Of these divisions, native elements, sulfides, sulfates, oxides, hydroxides, carbonates, and silicates are the most common in nature.

When elements are inert enough they can exist in their native state. These include the coinage metals in group IB, Au, Ag, and Cu, as well as the metals Pt and Fe. The nonmetals S and C are also commonly found in their native state, with C crystallizing in the graphite or diamond structure depending on pressure and temperature.

The class of minerals known as the sulfides contains reduced sulfur with a −1 or −2 oxidation (valence) state (see Chapter 14). Sulfides include many important ore minerals. Two common sulfides are pyrite (FeS_2) and chalcopyrite ($CuFeS_2$). The sulfate class of minerals, on the other hand, contains sulfur in the oxidized +6 state with S^{+6} coordinated to 4 O^{2-} producing an SO_4^{2-} group in the mineral. The sulfates include the common *evaporite mineral*, gypsum ($CaSO_4 \bullet 2H_2O$).

The oxide class of minerals can be divided into simple and multiple oxides. The simple oxides consist of a single type of atom bonded with oxygen. Ice is a simple oxide mineral. Others include corundum (Al_2O_3), hematite (Fe_2O_3), rutile (TiO_2), and magnetite (Fe_3O_4). The minerals spinel ($MgAl_2O_4$) and chromite ($FeCr_2O_4$) are common multiple oxide minerals. The lower mantle is made up of the oxide phases silicon substituted Mg-perovskite (($Mg,Fe)SiO_3$), Mg-wüstite (($Mg,Fe)O$), and silicon substituted Ca-perovskite (($Ca,Mg)SiO_3$). Because of the large volume of the mantle, this makes perovskite the most abundant mineral in the earth.

Hydroxide minerals have OH^- groups bound to cations in their structure. The presence of OH^- causes the bond strength to cations to be weaker than in oxides so that they tend to have lower mineral hardness. Brucite ($Mg(OH)_2$), goethite ($FeO \bullet OH$), and gibbsite ($Al(OH)_3$) are common hydroxides.

Carbonate minerals are important because they are abundant on the earth's surface, being formed by chemical or biologic precipitation from seawater, for example, calcite ($CaCO_3$) and aragonite ($CaCO_3$). Dolomite ($CaMg(CO_3)_2$) is often formed in reactions in which modified seawater reacts with $CaCO_3$. The carbonates are built on isolated anionic complexes of CO_3^{2-} units. Bond strengths within these units are quite strong with no sharing of electrons between CO_3^{2-} units.

The silicate class of minerals is built on the silicon–oxygen tetrahedron. It makes up the preponderance of minerals in the earth's upper mantle and crust. Because silicates are the most common type of minerals encountered on the earth's surface, their structures are of particular interest.

■ Structures of Silicate Minerals

How silicon–oxygen tetrahedra are bonded together determines the type of silicate structure produced. The six possibilities are shown in **Figure 5-4**. In a silicon–oxygen tetrahedron, each O^{2-} gives one half of its bonding energy to Si. The rest of the energy is free to bond with another cation. If this is a Si in another silicon–oxygen tetrahedron, then one refers to the oxygen as a *bridging oxygen*. The bridging oxygens for sorosilicate and cyclosilicate structures are indicated with vertical arrows in Figure 5-4. For other structures in which tetrahedra are shown without explicitly showing the oxygens, the bridging oxygens are at the juncture of the corners of two attached silicon–oxygen tetrahedra.

In nesosilicates, which are also referred to as orthosilicates, the silicon–oxygen tetrahedra have no shared corners with other tetrahedra, that is, they have no bridging oxygens. Nesosilicates are bonded together through connecting cations. They tend to be closely packed, giving these minerals high densities. Common nesosilicates include the *olivines* (e.g., forsterite, Mg_2SiO_4), garnets (e.g., almandine, $Fe_3Al_2Si_3O_{12}$), and zircon ($ZrSiO_4$). The tetrahedral cation-to-O ratio in nesosilicates is 1:4.

With the sharing of one corner oxygen between silicon–oxygen tetrahedra, the class of silicates termed sorosilicates is produced. The tetrahedral cation-to-O ratio, therefore, becomes 2:7 in these isolated double silicon–oxygen tetrahedral groups. Although not extremely common, important sorosilicates include the epidote minerals such as clinozoisite ($Ca_2Al_3O(SiO_4)(Si_2O_7)(OH)$). Clinozoisite is formed during metamorphism of basaltic rocks. This mineral has both a single silicon–oxygen tetrahedron as well as a double silicon–oxygen tetrahedral group in its structure.

As the name implies, cyclosilicates are made up of rings. They can have three, four, or six linked silicon–oxygen tetrahedra in each ring, which results in a tetrahedral cation-to-O ratio of 1:3. Common cyclosilicates include beryl

$(Be_3Al_2Si_6O_{18})$ and cordierite $((Mg,Fe)_2Al_3(AlSi_5)O_{18} \bullet nH_2O)$. As indicated in its structural formula, in cordierite, one of the six tetrahedral sites is occupied by Al rather than Si. Tourmaline is a complex cyclosilicate built on six member silicon–oxygen rings in which the structure typically contains B, Al, Fe, Mg, and Li atoms.

Inosilicates or chain silicates consist of extended chains of silicon–oxygen tetrahedra. These can be either single or double chains. In a single-chain inosilicate

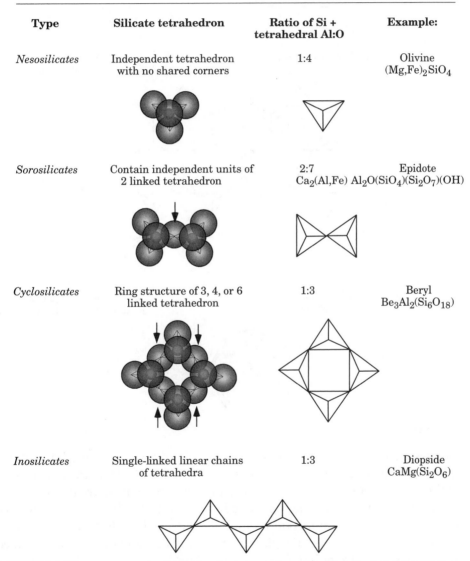

Type	Silicate tetrahedron	Ratio of Si + tetrahedral Al:O	Example:
Nesosilicates	Independent tetrahedron with no shared corners	1:4	Olivine $(Mg,Fe)_2SiO_4$
Sorosilicates	Contain independent units of 2 linked tetrahedron	2:7	Epidote $Ca_2(Al,Fe)\ Al_2O(SiO_4)(Si_2O_7)(OH)$
Cyclosilicates	Ring structure of 3, 4, or 6 linked tetrahedron	1:3	Beryl $Be_3Al_2(Si_6O_{18})$
Inosilicates	Single-linked linear chains of tetrahedra	1:3	Diopside $CaMg(Si_2O_6)$

FIGURE 5-4 Silicate structures showing the bonding between the oxygen tetrahedra that distinguish the difference between nesosilicates, sorosilicates, cyclosilicates, inosilicates, phyllosilicates, and tectosilicates.

(Continues)

Type	Silicate tetrahedron	Ratio of Si + tetrahedral Al:O	Example:
	Double cross-linked chains of tetrahedra	4:11	Tremolite $Ca_2Mg_5(Si_8O_{22})(OH)_2$
Phyllosilicates	Sheets of tetrahedra	2:5	Muscovite $KAl_2(AlSi_3O_{10})(OH)_2$
Tectosilicates	3-dimensional networks of linked tetrahedra	1:2	Microcline $K(AlSi_3)O_8$ Quartz SiO_2

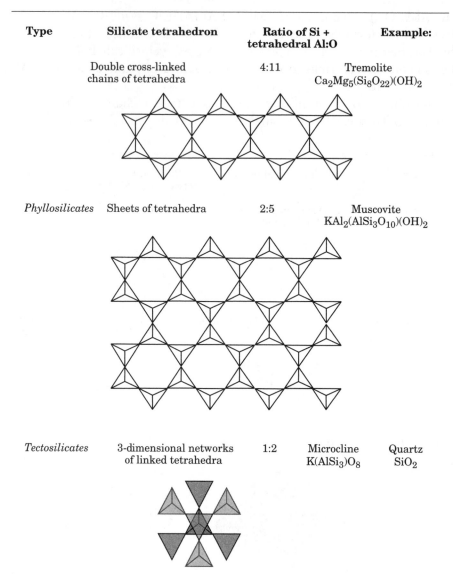

I FIGURE 5-4 *Continued*

each tetrahedron shares two of its oxygens with adjacent tetrahedra. This results in tetrahedral cation-to-O ratio of 1:3. In a double-chain inosilicate, half of the tetrahedra share three oxygens, and the other half share two oxygens, producing a structure having a tetrahedral cation-to-O ratio of 4:11. The inosilicates include two important common groups of Fe, Mg, and Ca silicates: the single-chained pyroxenes (e.g., enstatite, $MgSiO_3$) and the doubled-chained amphiboles (e.g., tremolite, $Ca_2Mg_5Si_8O_{22}(OH)_2$).

In phyllosilicates or sheet silicates, the silicon–oxygen tetrahedra are polymerized into extended sheets. Three of the four oxygens in the tetrahedron are shared with other tetrahedra, leading to a tetrahedral cation-to-O ratio of 2:5. Bonding between two silicon–oxygen tetrahedral sheets is by way of cations that reside in octahedral sites between the oxygens of the two tetrahedral sheets. Bonding between these tetrahedral-octahedral-tetrahedral composite sheets is through 12-fold sites. These can be vacant or contain Na or K. Because bonding in these 12-fold sites is weak, the structure is easily cleaved along the sheets, giving these minerals their flaky crystal habit. Sheet silicates include the commonly observed micas and clay minerals.

Tectosilicates or framework silicates such as quartz (SiO_2) and feldspars (e.g., albite, $NaAlSi_3O_8$) are bonded at all four corners of the silicon–oxygen tetrahedron to another tetrahedron in a three-dimensional structure. They make up nearly three fourths of the volume of the earth's crust. The tetrahedral cation-to-O ratio is 1:2. In the feldspar albite ($NaAlSi_3O_8$), one fourth of the tetrahedral sites are occupied by Al, whereas in anorthite ($CaAl_2Si_2O_8$), it is one half of the sites. In general, the greater the linking of the tetrahedra, the greater the Si content relative to other cations in a mineral structure; therefore, high-silica-containing rocks typically have large amounts of tectosilicates.

■ Clay Minerals

Clays are perhaps the largest and most compositionally diverse group of minerals. There are two uses of the term clay. One refers to the size of a mineral— that is, a mineral grain that is less than 5 mm in diameter is termed a clay-sized mineral. The other usage is that for a group of structurally complex hydrous phyllosilicates, as outlined in **Figure 5-5**.

Clay minerals are made up of two types of layers of atoms, tetrahedral, and octahedral layers. The tetrahedral layer consists of a layer of interconnected

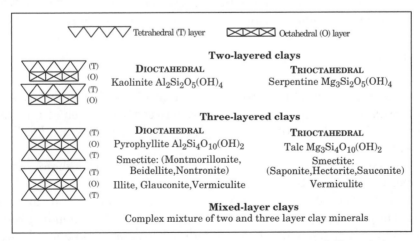

Tetrahedral (T) layer   Octahedral (O) layer

Two-layered clays

DIOCTAHEDRAL	**TRIOCTAHEDRAL**
Kaolinite $Al_2Si_2O_5(OH)_4$	Serpentine $Mg_3Si_2O_5(OH)_4$

Three-layered clays

DIOCTAHEDRAL	**TRIOCTAHEDRAL**
Pyrophyllite $Al_2Si_4O_{10}(OH)_2$	Talc $Mg_3Si_4O_{10}(OH)_2$
Smectite: (Montmorillonite, Beidellite, Nontronite)	Smectite: (Saponite, Hectorite, Sauconite)
Illite, Glauconite, Vermiculite	Vermiculite

Mixed-layer clays
Complex mixture of two and three layer clay minerals

FIGURE 5-5 Classification of clay minerals. Strictly speaking, serpentine, pyrophyllite, and talc are not clay minerals but are shown to illustrate compositions of the simplest two- and three-layer phyllosilicates.

Si^{4+} and Al^{3+} with O^{2-} in tetrahedra in a phyllosilicate type layer. The octahedral layer is a layer of Al^{3+}, Mg^{2+}, or Fe^{2+} bonded to 6 O^{2-} or OH^- in an octahedron. The octahedral layer is termed dioctahedral if it contains two Al^{3+} in sixfold (octahedral) coordination with O^{2-} or OH^-, with one of every three octahedral sites vacant, or tri-octahedral if there are three (Mg^{2+} + Fe^{2+}) coordinated with 6 O^{2-} or OH^- so that all the octahedral sites are occupied. The tetrahedral and octahedral layers are joined to form the two-layered clays (T-O), three-layered clays (T-O-T), or mixed-layer clays consisting of mixtures of two- and three-layered clays.

■ Polymorphism

Dimorphism occurs when a mineral with a given composition can exist in two structural forms, such as with graphite and diamond or α- and β-quartz. When three or more structures exist such as the clays kaolinite, dickite, and nacrite, they are termed *polymorphs*. These changes in structure may involve changes in bonding, site arrangements, and/or cation coordination. What determines conditions under which a particular polymorph will be stable? Consider the phase rule for dimorphic phases. With a single compositional variable and two phases, the phase rule states the equilibrium between them will be univariant in pressure-temperature space. The Gibbs energy difference, ΔG_R, between the two phases along this univariant boundary must be zero. That is, for the reaction between phases α and β:

$$\Delta G_R = G_{\alpha\text{-phase}} - G_{\beta\text{-phase}} = 0.0 \qquad [5.3]$$

Other thermodynamic variables, however, can be different for the two dimorphs along the equilibrium boundary. As discussed in Chapter 3, dP/dT of the equilibrium boundary for two minerals can be obtained from

$$\frac{dP}{dT} = \frac{\Delta S_R}{\Delta V_R} = k \qquad [5.4]$$

where ΔS_R and ΔV_R are the entropy and volume of reaction, respectively, and k is approximately constant for reactions between solids such as mineral dimorphs. If ΔS_R and ΔV_R of the dimorph transition are known, dP/dT of the dimorph reaction can be calculated.

In a *first-order transformation*, there are differences in enthalpy, entropy, volume, and heat capacity of the two phases at the transformation. This type of transformation requires a breaking of at least one bond to form the new phase. The graphite/diamond transition is first order because bonding between C in graphite is hexagonal, whereas bonding between C in diamond is cubic.

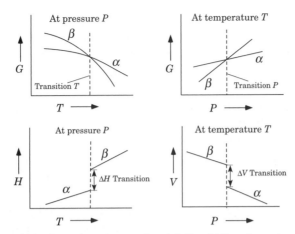

FIGURE 5-6 Changes in the thermodynamic properties of G, H, and V for a first-order phase transformation as a function of temperature and pressure.

Figure 5-6 shows how the thermodynamic properties change with pressure and temperature for a first-order transition between polymorph α stable at lower temperature and higher pressure and β stable at higher temperature and lower pressure. At constant pressure, P, with increasing temperature the α-phase has a lower Gibbs energy and is stable until the transition temperature is reached. Above this temperature, the β-phase has the lower Gibbs energy. At a fixed temperature, T, the β-phase is stable at pressures below the transition pressure, and the α-phase is stable at higher pressures. Unlike the Gibbs energy, there is an enthalpy, ΔH_{tran}, and volume, ΔV_{tran}, of transition. This implies there is also an entropy of transition, ΔS_{tran}. The derivative properties of ΔS_{tran} and ΔV_{tran} of heat capacity and the *coefficient of isothermal compressibility*, respectively, must therefore be infinite at a transformation T and P.

In a *lambda transformation*, ΔH_{tran}, ΔS_{tran}, $\Delta C_{P,tran}$, and ΔV_{tran} are continuous functions because the transition occurs over a range of temperatures and pressures, as given in **Figure 5-7**. The α/β-quartz transition is a displacive lambda

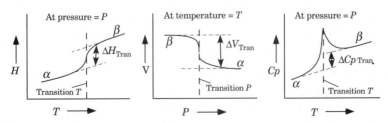

FIGURE 5-7 Changes in the thermodynamic properties for a lambda type transformation as a function of pressure and temperature.

transformation involving minor atomic adjustments with no breaking of Si–O bonds. C_P across the transition at constant pressure reaches a maximum value at the transition T. The temperature region over which the lambda transition occurs can vary widely because of substitutional order/disorder of elements in minerals, as discussed later in the chapter.

Thermodynamic properties of minerals that undergo lambda transformations can be approximated by extrapolating the values of the property on either side of the transformation to a transition T, as shown by the short dashed lines in Figure 5-7. A pseudo-value of transition of the property is then added at the transition T or P. For instance, the C_P properties of phases that exhibit lambda transitions can be approximated by using a Maier-Kelley type equation for the C_P of both the low and high temperature phase as given by the dashed lines in Figure 5-7. To this, one adds a pseudo-ΔC_P of transition as indicated. Using this characterization, the derivative thermodynamic properties close to the transformation will incorporate some error. For minerals that display lambda transitions, however, this is how the thermodynamic functions are characterized by SUPCRT92.

■ Important Polymorphic Transformations

Important polymorphic transformations occur between minerals with compositions of SiO_2, Al_2SiO_5, $CaCO_3$, and $KAlSi_3O_8$. The first two compositions show little solid solution so that the stability of each stable phase defines a unique field in pressure-temperature space. The latter two can have significant solid solution that influences their stability at a given pressure and temperature.

Figure 5-8 is the pressure-temperature phase diagram for SiO_2 together with a 6°C per km geotherm. Maximum temperatures in the crust are about 1200°C (basalt magma), and maximum pressures are about 22 kbar (bottom of thickened continental crust). Because geotherms in the earth's crust are of 6°C per km or higher, the area of possible pressure and temperature conditions for the earth's crust is shown in light gray in Figure 5-8. For crustal pressure and temperature conditions, the stable SiO_2 phase will be either low or high quartz, that is, α- or β-quartz, except for tridymite at low pressures above 867°C. The α- to β-quartz transition is not quenchable. β-quartz, therefore, formed at depth in the crust will revert to α-quartz when observed at the earth's surface. For this reason, quartz is not typically specified as either α or β.

As shown in Figure 5-8 the SiO_2 polymorph coesite is only stable at higher pressures than possible in crustal rocks. The presence of coesite in a rock indicates its formation in the mantle; however, mantle rocks typically do not have enough silica in them to stabilize an SiO_2 polymorph phase. Coesite-containing rocks observed at the surface are SiO_2-rich crustal rocks that have experienced

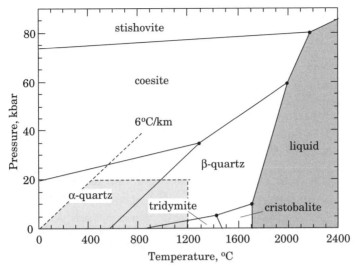

FIGURE 5-8 Pressure-temperature phase diagram for SiO_2, including a 6°C per km geotherm. Shown in light gray is the field of pressure and temperature conditions possible in the earth's crust. (Phase stability adapted from Zoltai, T. and Stout, J. H., 1984.)

mantle pressures by being rapidly subducted to great depths in subduction zones. In such an environment, high pressures at relatively low temperatures can be developed bringing the SiO_2-rich rocks into the coesite stability field.

Figure 5-8 indicates that stishovite is formed at even higher pressures than coesite. These pressures are not obtained by subduction of SiO_2-rich crustal rocks that are brought back to the surface. The only known natural environment of formation is by the extreme pressures produced by meteorite impact. Stishovite has been observed in meteorite impact craters. Noncrystalline forms of SiO_2 also exist that include the minerals chalcedony and opal. These two minerals are metastable at all pressures and temperatures and at the earth's surface will revert to quartz given enough time. (For a discussion of metastability, see Chapter 13.) The pressure-temperature boundaries of the transformation from one SiO_2 polymorph to another have positive slopes except for the tridymite-cristobalite transformation. Considering Equation 5.2, this indicates the higher temperature phase has both a higher entropy and higher volume as is typical of most solid–solid phase transformation reactions.

The Al_2SiO_5 phase diagram indicating the andalusite, kyanite, and silli-manite polymorph stability fields is given in **Figure 5-9a**. Depending on pressure and temperature conditions in the crust, any of these phases can be stable in Al-rich crustal rocks, termed *pelites*. Pelites are rocks that formed from material originally deposited as mud. The importance of this diagram stems from the fact that pelitic rocks are the most common rock type in most sedimentary

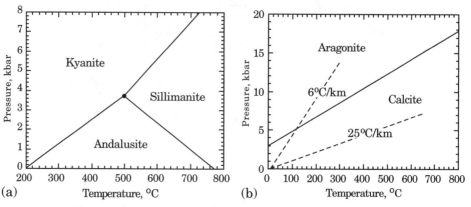

FIGURE 5-9 (a) Pressure-temperature phase diagram for the aluminosilicates. (Adapted from Holdaway, M. J. and Mukhopadhyay, B., 1993.) (b) Pressure-temperature phase diagram for calcium carbonate. (Adapted from Helgeson, H. C. et al., 1978.) Six and 25°C per km geotherms are shown on the $CaCO_3$ phase diagram.

basins. As these rocks get buried, depending on changes in pressure and temperature, geotherms can intersect any of the three-phase fields shown and often transect the equilibrium boundary between two of the minerals. Because they show little solid solution, pressures and temperatures of these phase boundaries are well constrained. The presence, therefore, of one of the Al_2SiO_5 phases in a rock identifies a region of pressure and temperature at which the mineral formed. Because of the unlikely possibility of obtaining rocks that formed at the exact temperature and pressure of the phase boundary, the presence of two or three different Al_2SiO_5 phases together in rocks from an area suggests polymetamorphism (more than one metamorphic event) or possibly outlines the pressure-temperature history of the rock. That is, it is likely that the phases formed at different times when the rock was subjected to different pressure and temperature regimes. See the discussion of metamorphism in Chapter 9 to help understand mineral formation at depth in the crust.

Figure 5-9b gives the pressure-temperature phase stability for $CaCO_3$ phases, calcite, and aragonite, along with 6°C per km and 25°C per km geotherms. The pressure scale in Figure 5-9b is over two times greater than that in Figure 5-9a. Aragonite is more dense than calcite and, therefore, is stable at higher pressure for a given temperature. Figure 5-9b indicates that aragonite is not stable at the earth's surface. Clearly, along a typical 25°C per km geotherm found in normally heated crust, aragonite will never become stable. It is only when the geotherm is lowered by rapid burial of cold surface rocks that aragonite becomes the stable $CaCO_3$ phase. This means the presence of metamorphic aragonite in crustal rocks indicates that surface rocks have been buried to mid-crustal depths before the temperature has increased to the more nearly steady-state geothermal gradients that are typical of tectonically inactive crust. If decreasing pressure-temperature conditions of a rock cross the calcite-aragonite equilibrium boundary at temperatures very much above 100°C, the reaction of aragonite back to calcite is so fast that aragonite is converted to calcite and not preserved.

Aragonite, however, is also formed at the earth's surface. Why has it formed if it is not stable relative to calcite? Aragonite is produced as shell material by some organisms. This disequilibrium situation is not completely understood but appears to be due in part to the inhibiting effects of Mg in seawater on calcite growth. This allows the solution to become supersaturated with calcite so that the solution can saturate with and precipitate aragonite. Maintenance of disequilibrium is considered in Chapter 13, where kinetics of reactions are considered.

■ Compositional Changes in Minerals

Most mineral phases show significant solid solution. That is, just like the spontaneous mixing of two endmember gases that are added to a volume, spontaneous mixing of endmember compositions of solids occurs by exchanging atoms on crystallographic sites. This lowers their Gibbs energy and makes them more stable. Minerals, therefore, often show isomorphic substitutions, that is, solid solutions that do not alter the crystalline structure but produce compositional changes. The substitution can take place on various atomic sites in the mineral as given for some minerals of importance in **Table 5-3**.

The olivine structure has a site formula of $(M1)(M2)(T)O_4$, where M1 and M2 indicate two different types of octahedral cation sites, and T stands for a tetrahedral cation site. The M2 site is slightly larger than the M1 site. The tetrahedral site in olivine is occupied by Si^{4+}, and the octahedral sites are dominantly occupied by a combination of Fe^{2+} and Mg^{2+}. Given the olivine structure, compositions (solid solution) between pure forsterite, Mg_2SiO_4, and pure fayalite, Fe_2SiO_4, with Fe^{2+} for Mg^{2+} substitution on the M1 and M2 sites and Si occupying the T site can occur; therefore, Mg-Fe olivine is sometimes written as $(Mg,Fe)_2SiO_4$ to indicate the Fe-Mg mixing. Fe-Mg exchange occurs because of the identical valence and similar radii of Mg^{2+} and Fe^{2+} ions (Table 5-1) with a ratio of radii of Mg^{2+}/Fe^{2+} (high spin) of 0.80Å/0.86Å = 0.93. This allows complete substitution of Mg^{2+} for Fe^{2+} in olivine. Less common in high-calcium environments is substitution of Ca^{2+} (radius = 1.08Å) into the larger M2 site producing olivines from monticellite, $CaMgSiO_4$, to kirschsteinite, $CaFeSiO_4$, with complete substitution of Fe^{2+} for Mg^{2+} on M1. Solid solution is extensive in olivines because they are stable at elevated temperatures and, as discussed in Chapter 4, the extent of solid solution in a mineral increases with increasing temperature.

Table 5-3 Site formulas for various minerals, with T, M, N, A, and OH indicating the tetrahedral, octahedral, feldspar, cubo-octahedral, and OH site location in the mineral, respectively

Mineral: (example) formula	Formula indicating crystallographic sites
Olivine: (forsterite) Mg_2SiO_4	$(M1)(M2)(T)O_4$
Pyroxene: (diopside) $CaMgSi_2O_6$	$(M1)(M2)(T)_2O_6$
Amphibole: (pargasite) $NaCa_2Mg_4AlAl_2Si_6O_{22}(OH)_2$	$(A)(M4)_2(M1)_2(M2)_2(M3)\,(T1)_4(T2)_4O_{22}(OH)_2$
Mica: (phlogopite) $KMg_3AlSi_3O_{10}(OH)_2$	$(A)(M1)(M2)_2(T)_4O_{10}(OH)_2$
Garnet: (grossular) $Ca_3Al_2Si_3O_{12}$	$(A)_3(M)_2(T)_3O_{12}$
Feldspar: (anorthite) $CaAl_2Si_2O_8$	$(N)(T1)_2(T2)_2O_8$
Carbonate: (dolomite) $CaMg(CO_3)_2$	$(M1)(M2)(CO_3)_2$

The N site in feldspar is not symmetric and can have a coordination between 9 and 12 oxygens depending on temperature. It can be thought of as a pseudo–cubo-octahedral site. Numbers following the site designations indicate significant structural differences between the same site type, and subscripts indicate how many of the sites there are.

Pyroxene cation site formulas can be designated as $(M2)(M1)(T)_2O_6$, where the size difference between the octahedral M2 and M1 sites is greater than in olivine. With pyroxene as with olivines, there is complete Mg-Fe exchange, producing pyroxenes from endmember enstatite ($Mg_2Si_2O_6$) to endmember ferrosilite ($Fe_2Si_2O_6$). The M2 site is larger than the M1 site, and Ca^{2+} can occupy it. With Ca^{2+} occupying the M2 site and with Fe-Mg substitution on the M1 site, the compositions from diopside ($CaMgSi_2O_6$) to hedenbergite ($CaFeSi_2O_6$) are produced. Mg^{2+} and Fe^{2+} can also occupy the M2 site along with Ca^{2+}, but a large miscibility gap occurs between Ca and Mg + Fe substitutions in M2. Ca^{2+} is too large to occupy the M1 site. To accommodate Ca^{2+} in all octahedral sites, wollastonite ($Ca_2Si_2O_6$) results from a transformation with displaced silicon–oxygen tetrahedra in its chains relative to pyroxenes. Augite, $(Ca,Na)(Mg,Fe,Al)(Si,Al)_2O_6$, is a common pyroxene solid solution that includes the coupled *jadeite exchange* of Na^+ for Ca^{2+} in the M2 site together with Al^{3+} exchange for Mg^{2+} or Fe^{2+} in the M1 site. Coupled exchanges occur to maintain charge balance in the structure. Complete jadeite exchange produces the endmember mineral phase, jadeite, $NaAlSiO_6$. Augite also has the coupled *plagioclase exchange* of Na^+ substituting for Ca^{2+} in the M2 site coupled with Si^{4+} for Al^{3+} in the tetrahedral T site (**Table 5-4**).

Amphiboles have site formulas of $(A)(M4)_2(M1)_2(M2)_2(M3)(T1)_4(T2)_4$ $O_{22}(OH)_2$. Because they contain eight tetrahedral sites of types T1 and T2, seven octahedral sites of types M1, M2, M3, and M4, a cubo-octahedral site, A, and an OH site, they have an extensive range of composition. All of these sites can accommodate a number of different ions. When all of the octahedral sites are occupied by Mg^{2+} and all of the tetrahedral sites by Si^{4+}, leaving the cubo-octahedral site empty for charge balance, magnesio-anthophyllite $(Mg_7Si_8O_{22}(OH)_2)$ is produced. By replacing Mg^{2+} with Fe^{2+}, all compositions to ferro-anthophyllite $(Fe_7Si_8O_{22}(OH)_2)$ exist. Because the two M4 sites are large enough to accommodate Ca^{2+}, tremolite $Ca_2Mg_5Si_8O_{22}(OH)_2$ to actino-lite $(Ca_2Fe_5Si_8O_{22}(OH)_2)$ compositions also can be produced. Also occurring is the coupled *tschermaks exchange*, where Fe^{2+} plus Mg^{2+} is replaced by Al^{3+} in the octahedral sites coupled with Al^{3+} for Si^{4+} exchange in the tetrahedral sites to retain charge balance. Amphiboles also have coupled *edenite exchange* consisting of an empty A-site plus Si^{4+} exchanged by Na^+ in the empty A-site and Al^{3+} for Si^{4+} in the tetrahedral site to produce amphiboles with an eden-ite, $NaCa_2Mg_5AlSi_7O_{22}(OH)_2$, component. Additionally, Cl^- and F^- can sub-stitute for OH^- in all amphibole compositions. When this is significant, the phases are referred to as a chloro-phase or fluoro-phase. Because of these com-plexities, solid solutions of amphiboles that contain some Na^+ and display multiple substitutions are given the general name hornblende. Hornblende then has a chemical formula of $(Ca,Na)_{2-3}(Mg,Fe,Al)_5 Si_4(Si,Al)_4O_{22}(OH,F,Cl)_2$.

Another important class of silicates in which solid solutions are important is micas. Micas are either dioctahedral or trioctahedral, depending on whether two or three of the available octahedral sites are occupied. They can be thought of as being built on the minerals pyrophyllite $(Al_2Si_4O_{10}(OH)_2)$ and talc

Table 5-4	Common coupled mineral substitutions where [ ] is a vacancy in the A-site

Name of substitution	Coupled Reaction
Jadeite	$Ca^{2+} + (Mg^{2+} + Fe^{2+}) \rightarrow Na^+ + Al^{3+}$
Plagioclase	$Ca^{2+} + Al^{3+} \rightarrow Na^+ + Si^{4+}$
Tschermaks	$(Fe^{2+} + Mg^{2+}) + Si^{4+} \rightarrow Al^{3+} + Al^{3+}$
Edenite	$[ ] + Si^{4+} \rightarrow Na^+ + Al^{3+}$

$(Mg_3Si_4O_{10}(OH)_2)$, respectively, with site formulas of $(A)(M1)(M2)_2(T)_4$ $O_{10}(OH)_2$. In pyrophyllite, the A site is vacant. Two of the three M sites are occupied by Al^{3+}, and all of the T sites are occupied by Si^{4+}. In talc, the A site is vacant. All of the M sites are occupied by Mg^{2+}, and all of the T sites are occupied by Si^{4+}. For either of these minerals to be micas, the A site must be occupied. With a coupled substitution of K^+ in the vacant A cubo-octahedral site and Al^{3+} for an Si^{4+} site, pyrophyllite becomes the dioctahedral mica muscovite $(KAl_2AlSi_3O_{10}(OH)_2)$. This mica, like all dioctahedral micas, is sometimes referred to as a white mica because the lack of Fe in its structure makes it colorless in thin section. Starting with talc, a coupled substitution of K^+ in the vacant A cubo-octahedral site and Al^{3+} for an Si^{4+} in the tetrahedral site produces the tri-octahedral mica phlogopite $(KMg_3AlSi_3O_{10}(OH)_2)$. Biotite, $K(Mg,Fe)_3AlSi_3O_{10}(OH)_2$, is a common K-rich trioctahedral mica along the Mg and Fe join. Muscovite and biotite can coexist in equilibrium because there is little solid solution between them.

Garnets have site formulas of $(A)_3(M)_2(T)_3O_{12}$. In grossular, $Ca_3Al_2Si_3O_{12}$, the A site contains Ca^{2+}, the M site is occupied by Al^{3+}, and Si^{4+} occupies the T site. Substitution of Fe^{3+} for Al^{3+} in the M site produces andradite, Ca_3Fe_2 Si_3O_{12}, whereas substitution of Mg^{2+} for Ca^{2+} in the A site produces pyrope, $Mg_3Al_2Si_3O_{12}$. With Fe^{2+} substitution in the A site almandine, $Fe_3Al_2Si_3O_{12}$, and with Mn^{2+}, spessartine, $Mn_3Al_2Si_3O_{12}$, occur. Complete solid solution between all of these endmembers is thought to exist at very high temperatures.

Feldspars have site formulas of $(N)(T1)_2(T2)_2O_8$. They show an array of substitutions in their N site together with various amounts of Si^{4+} and Al^{3+} in the tetrahedral sites to maintain charge balance. Particularly common is K^+ in the N site, with one fourth of the tetrahedral sites occupied by Al^{+3} and three fourths by Si^{4+} producing K-feldspar $(KAlSi_3O_8)$. K-feldspar also commonly has some Na^+ replacing K^+. There are three K-feldspar polymorphs: microcline, orthoclase, and sanidine. Their structures differ in the amount of ordering of Al^{3+} and Si^{4+} between the T1 and T2 sites. At low temperature, each of the tetrahedral sites becomes distinct, which can be designated as $T1_a$, $T1_b$, $T2_a$, and $T2_b$. In maximum microcline, Al^{3+} is confined to the $T1_a$ site, whereas in high sanidine, Al^{3+} is randomly distributed between all the T1 and T2 sites. Orthoclase is intermediate in character with nearly equal Al in $T1_a$ and $T1_b$ but less Al^{+3} in the T2 than in the T1 sites. Because this ordering does not change the volume measurably, there is little pressure dependence for polymorph stability, but there is a significant temperature dependence. Maximum microcline is stable at the lowest temperatures with orthoclase, sanidine becoming stable with increasing temperatures.

Na^+ can also occupy the N site to produce albite $(NaAlSi_3O_8)$. As with K-feldspar there is a change from completely ordered Al^{3+} in T1 in low-temperature albite to random distribution between T1 and T2 in high-temperature *mono-*

clinic albite (monalbite). Phase relations for Na^+ and K^+ alkali feldspar solid solutions at 1 kbar are shown in **Figure 5-10**. With both Na and K present, there is complete solid solution between $KAlSi_3O_8$ and $NaAlSi_3O_8$ at high temperatures but little solid solution at low temperatures. This occurs because the radius ratio of K^+ to Na^+ of $1.63Å/1.40Å = 1.16$ is significantly larger than unity.

Plagioclase feldspars are a solid solution from albite ($NaAlSi_3O_8$) to anorthite ($CaAl(AlSi_2)O_8$) with the coupled plagioclase exchange of Ca^{2+} plus Al^{3+} for Na^+ plus Si^{4+}. (Plagioclase stability is discussed relative to igneous melts in Chapter 8.) Rocks can crystallize with both a K-feldspar and plagioclase phase. **Figure 5-11** gives the ternary feldspar diagram at 1 kbar showing the limit of solid solution at 650°C, 750°C, and 900°C. Notice that Ca-rich plagioclase feldspars cannot contain much K, even at high temperature. Likewise, K-rich alkali feldspars cannot contain much Ca even at high temperature. Also, consistent with Figure 5-10, along the pure albite–K-feldspar join, no solid solution exists at 650°C between about 35 and 50 mole % of the K-feldspar endmember.

The presence of a solvus, as is the case of K-Na feldspars, which limits complete solid solution between end members, indicates a positive excess Gibbs energy of mixing must be important. In the last chapter, a regular solution model, which is also termed a one-parameter or two-suffix *Margules (1895) formulation* of

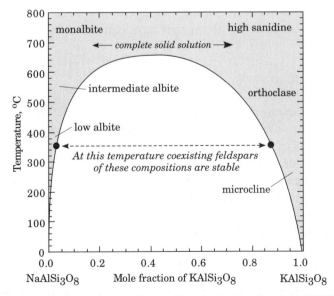

FIGURE 5-10 Alkali feldspar solvus at 1 kbar. (Adapted from Smith, P. and Parsons, I., 1974.) The values at any temperature along the solvus give the composition of the stable Na-rich and K-rich phases at the indicated temperature. Solid solution exists from these solvus compositions to those of the pure endmember phases as given by the shaded area.

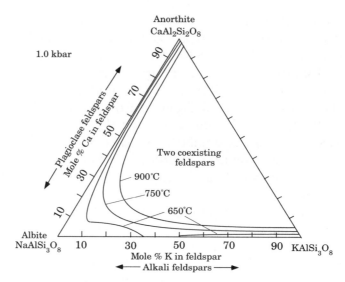

FIGURE 5-11 Ternary feldspar phase diagram at 1 kbar (adapted from Seck, H. A., 1971.) showing the extent of solid solution and two-phase region at 650°C, 750°C, and 900°C. At 650°C, there is a solvus for alkali feldspars as given in Figure 5-10, whereas at 750°C and 900°C, complete solid solution exists.

$$\Delta\mu_{reg} = \omega X_A X_B = \omega(1 - X_B)X_B \qquad [5.5]$$

was introduced to account for the excess Gibbs energy of mixing. This is a three-term *virial equation* with the first term equal to zero as Equation 5.5 can be rewritten as

$$\Delta\mu_{reg} = 0 + \omega X_B + (-\omega)X_B^2. \qquad [5.6]$$

This equation is symmetric about a mole fraction of 0.5. The inherent symmetry of the regular solution model cannot describe the alkali feldspar solvus or for that matter the asymmetry found in most other solvi that occur between endmember mineral compositions.

The model can be extended to asymmetric situations such as the alkali feldspars by considering that K-rich feldspar is a regular solution that mixes with Na-rich feldspar, which is also a regular solution. In this case, the total excess Gibbs energy of mixing of these two regular solutions is

$$\Delta\mu_{ex} = X_{Na}(\omega_K X_K X_{Na}) + X_K(\omega_{Na} X_{Na} X_K). \qquad [5.7]$$

This produces what is termed a subregular solution or a two parameter Margules equation. With $X_K + X_{Na} = 1$, Equation 5.7 can be rearranged to

$$\Delta\mu_{ex} = 0 + \omega_K X_K + (\omega_{Na} - 2\omega_K)X_K^2 + (\omega_K - \omega_{Na})X_K^3 \qquad [5.8]$$

indicating that this is a four-term virial equation. Equation 5.7 or 5.8 allows most nonsymmetric solvi to be described to reasonable accuracy.

Carbonates have site formulas of $(M1)(M2)(CO_3)_2$. They are abundant in many sedimentary rocks as well as their metamorphic equivalents. Many different cations can be accommodated in their M1 and M2 octahedral sites. The most common are Ca^{2+} and Mg^{2+} producing the minerals calcite ($CaCO_3$), magnesite ($MgCO_3$), and dolomite ($CaMg(CO_3)_2$). Because the radius ratio of octahedral Ca to Mg is 1.08Å/0.80Å = 1.35, Ca-Mg carbonates, similar to K-Na feldspars, form a complete solid solution at high temperatures but display little solid solution at low temperatures. Also, with decreasing temperature, ordering of Ca on M1 and Mg on M2 occurs in dolomite. **Figure 5-12** shows the stability of calcite relative to dolomite at 2 kbar as a function of temperature. Calcite can take more Mg into its structure than the dolomite structure can incorporate Ca at a particular temperature, but at temperatures above 1040°C, a complete solid solution exists.

Calcite presently precipitates as cements from seawater where the molecular ratio of Mg to Ca is 5:1. The cement typically contains about 12 mole % $MgCO_3$ (high Mg-calcite). Reef building and shallow marine organisms also produce skeletal high-Mg calcite. There is still some debate as to whether this high-Mg calcite is in equilibrium with present-day seawater (Mackenzie et al., 1983; Thorstenson and Plummer, 1978). Much of the concern stems from the biogenic origin of the calcite. Determinations of the composition of Mg-calcites,

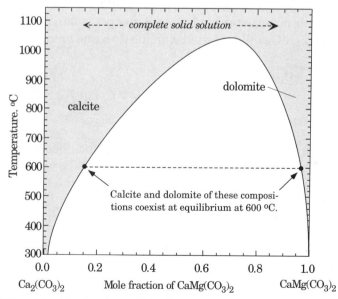

I FIGURE 5-12 The calcite-dolomite solvus at 2 kbar. (Adapted from Anovitz, L. M. and Essene, E. J., 1987.)

precipitated from seawater with varying Mg concentrations, suggest that it is (Mucci and Morse, 1984). For instance, evidence indicates low-Mg calcite precipitated in the Late Cretaceous seas when the Mg/Ca ratio of seawater was lower (see Figure 16-1). Temperature also appears to play a role in determining the Mg content of calcite. The Mg content of calcite decreases from the equator to high latitudes. This may be related to a pole-ward decrease in CO_3^{2-} concentrations of surface seawater with decreased temperature.

During diagenesis, the high Mg-calcite is often converted to low Mg-calcite (<5 mole % $MgCO_3$). This likely occurs by changes in the composition of the reacting pore solution from that of normal seawater as the high Mg-calcite dissolves. With Mg/Ca near 0.14 stoichiometric dissolution of the high Mg-calcite lowers the Mg/Ca of the pore solution. A low Mg-calcite of <5 mole % $MgCO_3$ in equilibrium with the new pore solution then precipitates. Because of the slow kinetics of calcite dissolution/precipitation near equilibrium, equilibrium between the calcite and solution is only approached but not normally reached.

■ Solid Solutions Involving Multiple Sites

The minerals outlined above display solid solution on more than one site. How does this affect their stability from a thermodynamic standpoint in a reaction? It can be deduced from the discussion in Chapter 4 on the thermodynamics of solutions that for a mineral like Mg-Fe olivine, $(Mg,Fe)_2SiO_4$, with two moles of Mg-Fe sites per mole of olivine, the ideal mixing Gibbs energy of this phase would be twice that of the phase with only 1 mole of sites per mole of mineral. The Gibbs energy of $(Mg,Fe)_2SiO_4$ solid solution with ideal mixing on the two sites is then

$$\Delta G_{\text{id mix}}^{\text{olivine}} = \Delta G_{\text{mech mix}}^{\text{olivine}} + \Delta G_{\text{id mixing}}^{\text{olivine}}$$

$$= X_{\text{Fo}} \Delta G_{\text{Fo}}^{\text{o}} + X_{\text{Fa}} \Delta G_{\text{Fa}}^{\text{o}} + 2(X_{\text{Fo}}RT \ln X_{\text{Fo}} + X_{\text{Fa}}RT \ln X_{\text{Fa}}) \quad \text{[5.9]}$$

$$\underbrace{\qquad\qquad\qquad}_{\text{mechanical mixture}} \qquad \underbrace{\qquad\qquad\qquad}_{\text{ideal mixing}}$$

where X is the mole fraction of the subscripted endmember olivine in the mineral. The last term on the right in Equation 5.9 can also be written with the two brought into the logarithm terms so that

$$\Delta G_{\text{id mixing}}^{\text{olivine}} = X_{\text{Fo}}RT \ln X_{\text{Fo}}^2 + X_{\text{Fa}}RT \ln X_{\text{Fa}}^2 \quad \text{[5.10]}$$

In general, the ideal mixing term for the Gibbs energy of a phase with mixing of A, B, and C on α equivalent sites is given by

$$\Delta G_{\text{id mixing}} = X_A RT \ln X_A^{\alpha} + X_B RT \ln X_B^{\alpha} + X_C RT \ln X_C^{\alpha} \quad \text{[5.11]}$$

where X_i^α is the mole fraction of the ith element substituted on α number of equivalent sites. The Gibbs energy of the A component in the mixed phase with ideal mixing on α sites of one kind is then

$$\Delta G_{id\ mix}^{A} = \Delta G_A^o + X_A RT \ln X_A^\alpha \tag{5.12}$$

where ΔG_A^o is the Gibbs energy of the pure A mineral with α sites at the pressure and temperature of interest. For instance, the Gibbs energy of pure forsterite (Mg_2SiO_4) in an olivine solid solution is

$$\Delta G_{id\ mix}^{Fo} = \Delta G_{Fo}^o + X_{Fo} RT \ln X_{Fo}^2 \tag{5.13}$$

Now consider garnet, with a site formula of $(A)_3(M)_2Si_3O_{12}$, having two distinct types of cation sites, A and M. Shown as a rectangle in **Figure 5-13** is the possible Fe^{2+} for Ca^{2+} exchange in the three A sites on the horizontal axis and Fe^{3+} for Al^{3+} exchange in the two M sites on the vertical axis. The amount of grossular in a garnet is given by the perpendicular bisector extending from the grossular corner. The location of compositions with 0, 50, and 100 mole % of the grossular endmember in both site types is shown on the diagram. The open circle marks the garnet composition (($Fe^{2+} = 0.2$, $Ca^{2+} = 0.8$)$_3$($Al^{3+} = 0.7$,

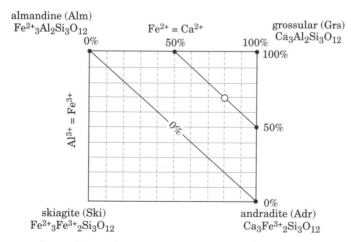

almandine (Alm)
$Fe^{2+}{}_3Al_2Si_3O_{12}$
0% $Fe^{2+} = Ca^{2+}$ grossular (Grs)
                    50% $Ca_3Al_2Si_3O_{12}$
                         100%
$Al^{3+} = Fe^{3+}$
                         50%
                         0%
skiagite (Ski)   andradite (Adr)
$Fe^{2+}{}_3Fe^{3+}{}_2Si_3O_{12}$ $Ca_3Fe^{3+}{}_2Si_3O_{12}$

FIGURE 5-13 The Fe^{2+} for Ca^{2+} and Fe^{3+} for Al^{3+} exchange rectangle for garnet compositions given by $(Fe^{2+},Ca^{2+})_3(Al^{3+},Fe^{3+})_2Si_3O_{12}$. Percent values indicate the molar percent of the grossular endmember in each site. The open circle marks the composition of a garnet with $X_{Grs} = 0.5$, $X_{Alm} = 0.2$, and $X_{Adr} = 0.3$.

$Fe^{3+} = 0.3)_2Si_3O_{12})$ that has 50% of the grossular endmember. Because the extent of compositional changes can be displayed on a plane surface, only three of the four endmembers are needed to describe the solid solution compositions in this phase. If, somewhat arbitrarily, grossular (Grs), almandine (Alm), and andradite (Adr) are chosen, then the composition of these garnets can be plotted on a triangular diagram as given in **Figure 5-14**. A garnet with composition given by the open circle in Figure 5-13 that has 50% of the grossular endmember will plot with its composition along the 50% grossular line two fifths of the way between the Grs–Adr and Grs–Alm bounding lines. Using the lever rule with this composition, the percent of Adr relative to Alm is $(1 - 2/5) \times 100 = 60\%$. Because Adr + Alm is 50% of the garnet, the mole fractions of the three endmember components are then $X_{Grs} = 0.5$, $X_{Adr} = 0.3$, and $X_{Alm} = 0.2$. Figure 5-14 indicates with an open circle where the garnet would plot on the triangular diagram.

The Gibbs energy of this garnet $((X_{Fe}^{2+} = 0.2, X_{Ca}^{2+} = 0.8)_3(X_{Al}^{3+} = 0.7, X_{Fe}^{3+} = 0.3)_2Si_3O_{12})$ can be calculated from

$$\Delta G_{garnet} = X_{Grs} \Delta G^{o}_{Grs} + X_{Alm}\Delta G^{o}_{Alm} + X_{Adr}\Delta G^{o}_{Adr}$$

$$+ RT(X_{Grs} \ln a_{Grs} + X_{Alm} \ln a_{Alm} + X_{Adr} \ln a_{Adr}) \qquad [5.14]$$

The ideal mixing on sites activity model of the endmember phases in the outlined compositional space for garnet based on a pure endmember phase at the pressure and temperature of interest would be

$$a_{Grs} = (X_{Ca^{2+}})^3 (X_{Al^{3+}})^2 \qquad [5.15]$$

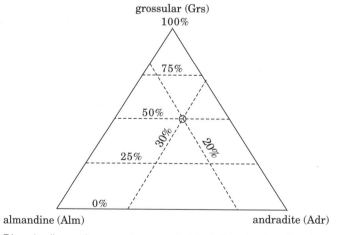

FIGURE 5-14 Triangular diagram for a garnet composed of the indicated endmembers. The composition marked with an open circle is given by $X_{Grs} = 0.5$, $X_{Alm} = 0.2$, and $X_{Adr} = 0.3$.

$$a_{Alm} = (X_{Fe^{2+}})^3 (X_{Al^{3+}})^2 \qquad\qquad [5.16]$$

and

$$a_{Adr} = (X_{Ca^{2+}})^3 (X_{Fe^{3+}})^2 \qquad\qquad [5.17]$$

Consider the reaction

$$3CaAl_2Si_2O_8 = SiO_2 + 2Al_2SiO_5 + Ca_3Al_2Si_3O_{12} \qquad [5.18]$$
$$\text{anorthite} \qquad \text{quartz} \qquad \text{kyanite} \qquad \text{grossular}$$

with an equilibrium constant expression of

$$K_{(5.18)} = \frac{a_{Qtz} a_{Ky}^2 a_{Grs}}{a_{An}^3}. \qquad\qquad [5.19]$$

The activity of the grossularite component in the garnet discussed above as given by Equation 5.15, would be

$$a_{Grs} = (X_{Ca^{2+}})^3 (X_{Al^{3+}})^2 = (0.8)^3 (0.7)^2 = 0.25 \qquad [5.20]$$

rather than a value of unity as would be the case for a pure grossular garnet. An activity of 0.25 would be used for the grossular component in Equation 5.19. The analysis given here can be extended if there are more than two different types of sites in a particular mineral.

In some situations, rather than ideal mixing on each site, there is an effect of substitution of one type of ion on a site affecting the occupancy of a nearby site. For instance, local charge balance appears to occur in the plagioclase (Na + Si $\leftrightarrow$ Ca + Al), tschermak (Mg + Si $\leftrightarrow$ Ca + Al), and edenite ([] + Si $\leftrightarrow$ Na + Al) substitutions. When such coupling occurs, the distribution of the mixing on one site is dependent on the mixing of the other. Because less randomness can occur when substitutions are coupled, the negative contribution to Gibbs energy on mixing will be less and the Gibbs energy of the mixture is greater than that given above for two independent sites. If the substitution on one site dictates that on the other, from an energy standpoint, mixing on only one of the two sites needs to be considered. For instance, in plagioclase ($NaAlSi_3O_8$–$CaAl_2Si_2O_8$) with Na + Si $\leftrightarrow$ Ca + Al exchange assuming ideal mixing and coupled substitution, the activity of the albite component is

$$a_{Ab} = X_{Na^+} \qquad\qquad [5.21]$$

and the anorthite component is

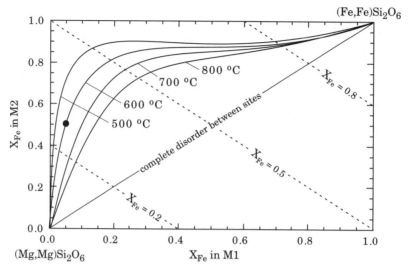

$$a_{An} = X_{Ca^{2+}} \tag{5.22}$$

despite the fact that there is also Si for Al exchange on the tetrahedral sites.

Consider cation exchange in pyroxenes, with a site formula of $(M1)(M2)(T)_2$ O_6. In pyroxenes, the M1 and M2 sites are significantly different in size but become more similar with increased temperature. Distribution of Mg and Fe between the two sites in pyroxene is, therefore, temperature dependent. Occupancy of each site must be determined by x-ray crystal structure determination. At lower temperature, more Fe is stable in the M2 site than in the M1 site, but at higher temperature, the concentration of Fe between sites becomes more evenly distributed. The distribution of Mg and Fe between the sites can be characterized by the *exchange reaction*:

$$Fe_{M2} + Mg_{M1} = Fe_{M1} + Mg_{M2}. \tag{5.23}$$

The distribution coefficient of the reaction is given by

$$K_D = \frac{X_{Fe_{M1}} X_{Mg_{M2}}}{X_{Fe_{M2}} X_{Mg_{M1}}}. \tag{5.24}$$

Although K_D can be a function of both pressure and temperature, given the small volume of reaction in Equation 5.23, K_D has very little pressure dependence. Because K_D is a strong function of temperature but not pressure, the K_D can be used as a geothermometer. This means a determination of K_D will give the temperature at which the pyroxene formed. **Figure 5-15** shows the site occupancy in Fe-Mg pyroxenes as a function of temperature. As temperature increases, the mixing becomes more disordered between sites as the M1 and M2 sites begin to act more and more similarly.

Consider the pyroxene with the distribution of Fe and Mg between M1 and M2 given by

$$(X_{Mg} = 0.95, X_{Fe} = 0.05)_{M1}(X_{Fe} = 0.5, X_{Mg} = 0.5)_{M2}Si_2O_6 \qquad [5.25]$$

This composition plots on the 600°C line given by the filled circle in Figure 5-15. This would then be its temperature of last equilibration. Because of the limit on disorder due to intercrystalline exchange, less randomness occurs—that is, a higher Gibbs energy is present than with complete disorder between sites but more than with a coupled substitution between sites.

■ Intercrystalline Exchange

Intercrystalline exchange between phases can also be present. For instance, Fe^{2+} and Mg^{2+} exchange between Fe–Mg garnet with endmember compositions $Fe_3Al_2Si_3O_{12}$ (almandine) and $Mg_3Al_2Si_3O_{12}$ (pyrope) together with Fe–Mg exchange in biotite with endmember compositions $KFe_3AlSi_3O_{10}(OH)_2$ (annite) and $KMg_3AlSi_3O_{12}(OH)_2$ (phlogopite) can occur. The exchange reaction is

$$\underset{\text{Alm}}{Fe_2Al_2Si_3O_{12}} + \underset{\text{Phl}}{KMg_3AlSi_3O_{12}(OH)_2} = \underset{\text{Prp}}{Mg_3Al_2Si_3O_{12}} + \underset{\text{Ann}}{KFe_3AlSi_3O_{10}(OH)_2} \qquad [5.26]$$

with an equilibrium constant expression of

$$K_D = \frac{X_{Prp}\, X_{Ann}}{X_{Alm}\, X_{Phl}}. \qquad [5.27]$$

If K_D is a strong function of temperature but not pressure, then this K_D can also be used as a geothermometer. Likewise, if K_D is a strong function of pressure but not temperature, then K_D can be used as a geobarometer to help determine the pressure of formation of a rock. In the case of Fe–Mg exchange between biotite and garnet, both temperature and pressure affect K_D, but the temperature effect is more significant. The temperature can be determined if the pressure can be independently calculated or estimated. Because of its low sensitivity, it would be difficult to determine pressure with this exchange even if the temperature could be independently determined. **Figure 5-16** shows the K_D at 2.07 kbar plotted as a function of inverse temperature. Consider a rock containing Fe-Mg

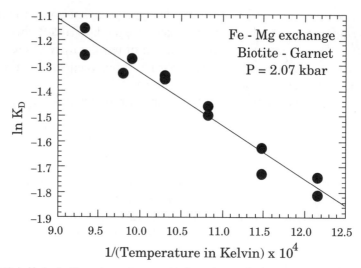

FIGURE 5-16 In K_D for Fe-Mg exchange between biotite and garnet in the presence of H_2O as a function of the inverse of temperature in K at 2.07 kbar. (Data from Ferry, J. M. and Spear, F. S., 1978.) This diagram is for low-Al biotites and is changed somewhat if Al in biotite is increased.

biotite and garnet. If the rock formed near 2 kbar, the plot could be used to determine the temperature of formation from a determination of the Fe-Mg composition of its biotite and garnet.

Summary

The Periodic Table of the elements is built on the similarities of electrons in outer electron orbitals denoted as s, p, d, and f. Elements bond by electron transfer to fill these orbitals. As a first approximation, atoms can be modeled as spheres of a fixed size. Bonding then occurs by satisfying valence charges and minimizing the atom to atom distances. Because of the dominance of oxygen and silicon in the earth's mantle and crust, the polar Si–O bond is particularly important. Four compacted oxygens of valence −2 have a center cavity that can accommodate a silicon atom of valence +4. The basic building block of silicate minerals is, therefore, the silicon–oxygen tetrahedron, SiO_4^{4-}. The four extra electrons in the tetrahedron create four bonds with other species. Long chains, sheets, and three-dimensional frameworks can be produced with cations occupying sites and bonding to other tetrahedrons.

Structural variation exists between mineral polymorphs. Across the phase transformation, the Gibbs energy is a continuous function of pressure and temperature. If bonds are broken in the formation of the new polymorph, the transformation will be first order, for example, graphite–diamond. First-order transformations have discontinuities in the derivative Gibbs energy properties as a function of pressure and temperature. Lambda transformations involve bond displacement, and the C_P reaches a maximum as a function of temperature across the transformation (e.g., α- and β-quartz).

Most minerals display compositional variations. These can be modeled as exchanges of atoms on sites in the mineral structure. Depending on their coordination environment, atomic sites in silicate minerals are termed tetrahedral, octahedral, N-feldspar, cubo-octahedral, and OH sites. Substitutions on the sites are governed by the ion's size and charge. To maintain charge balance, coupled substitutions can occur. These include plagioclase ($Na^+ + Si^{4+} \rightarrow Ca^{2+} + Al^{3+}$), jadeite ($Na^+ + Al^{3+} \rightarrow Ca^{2+} + (Mg^{2+}, Fe^{2+})$), tschermaks ($Al^{3+} + Al^{3+} \rightarrow (Fe^{2+}, Mg^{2+}) + Si^{4+}$), and edenite (empty cubo-octahedral + $Si^{4+} \rightarrow Na^+ + Al^{3+}$) exchange.

The Gibbs energy of ideal mixing on α equivalent A sites contributes $X_A RT \ln X_A^\alpha$ to an endmember mineral component, where X_A is the mole fraction of the endmember composition. With three different types of sites, A, B, and C, in a mineral with α, β, and χ of each type, respectively, the Gibbs energy is decreased by $X_A RT \ln X_A^\alpha + X_B RT \ln X_B^\beta + X_C RT \ln X_C^\chi$ when ideal site mixing occurs. X_i gives the mole fraction of the endmember composition of the ith site.

If, however, substitution on one site affects the mixing on another site, less randomness of mixing will occur. With completely coupled substitutions, mixing on only one type of site needs to be considered to determine its change in Gibbs energy with mixing. Often the coupling is not complete and is temperature dependent. In this case, an exchange reaction between sites in a mineral can be written. Exchange reactions of cations between minerals can also occur. The equilibrium constants of these reactions, K_D, can be strong functions of temperature and pressure. Determining the concentrations of cations on sites in minerals together with knowledge of the K_D can be used to determine at what temperature and pressure the minerals last equilibrated.

Key Terms Introduced

actinides
alkaline earth elements
anion
atomic number
bridging oxygen
carbonate
cations
coefficient of isothermal
    compression

crystal field stabilization
dimorphism
edenite exchange
electronegativity
electrostatic
evaporite mineral
exchange reaction
first-order transformation
halogens

jadeite exchange
lambda transformation
lanthanides
Margules formulation
minerals
monoclinic
olivine
orbital
orbital subshell
pelite

plagioclase exchange
polar bonds
polymorphs
rare earth elements (REEs)
Schrödinger equation
tetrahedron
transition elements/metals
tschermaks exchange
valence
virial equation

Questions

1. Give a definition of a mineral.
2. What is the Periodic Table of the elements? Why do elements in a column have similar properties?
3. What is a solid solution? An ideal solution?
4. What is a coordination number? Polymorphism?
5. What is a silicon–oxygen tetrahedron? Does it have a charge?
6. List the different types of silicate minerals.
7. How are 1:1, 2:1, and 2:2 clay minerals different?
8. Define polymorphs. Give two examples.
9. How does a lambda differ from a first-order transformation?
10. Give the site formula for olivine, pyroxene, amphibole, mica, garnet, feldspar, and carbonate.
11. How are site formulas used to determine the thermodynamic activities of minerals?

Problems

1. Complete the following chart using the Periodic Table and Appendix C1.

Element	Symbol	Group	Period	Atomic number	Atomic weight	Valance states
		IIB	5			
	Te					
				2		
Silver						
					26.98	
Iron						

2. Using Appendix C1, calculate the weight % of each element in the mineral anorthite.

3. What is the coordination number of an octahedron? What should be the coordination number of chorine atoms around a magnesium atom?

4. Give the oxidation number of N in the following species:
   a. N_2 b. HNO_3
   c. NH_4^+ d. NO_2

5. Using triangles to depict silicate tetrahedra, show the structure of
   a. pyroxene   b. mica

6. For each of the following silicate groups: (a) Name the structural group (e.g., framework and sorosilicate). (b) Give the general chemical formula. (c) Give the ratio of tetrahedral sites to oxygens in tetrahedral coordination.
   a. alkali feldspar   b. pyroxenes   c. dioctahedral micas
   d. garnets   e. amphiboles

7. Write a generalized compositional formula for Na-K-Ca feldspars (e.g., Mg-Fe olivines would be $Mg_XFe_{2-X}SiO_4$ with $0 < X < 2$).

8. At 25°C, calculate the molar Gibbs energy of the alkali feldspar solid solution at a mole fraction of albite of 0.0, 0.3, 0.5, 0.7, and 1.0 in an albite–max–microcline mixture. Assume ideal site mixing. Sketch a Gibbs energy versus X_{Ab} diagram with the computed values. Show the separate contributions from a mechanical mixture and ideal mixing.

9. Consider a garnet of composition:

$$(X_{Mg} = 0.61, X_{Fe} = 0.26, X_{Mn} = 0.08, X_{Ca} = 0.05)_3(X_{Al} = 0.87, X_{Fe^{3+}} = 0.013)_2Si_3O_{12}$$

   Calculate the ideal site mixing activities of the following garnet components
   a. pyrope ($Mg_3Al_2Si_3O_{12}$)
   b. grossular ($Ca_3Al_2Si_3O_{12}$)
   c. almandine ($Fe_3Al_2Si_3O_{12}$)
   d. andradite ($Ca_3Fe_2^{3+}Si_3O_{12}$)

10. Enstatite and diopside show limited solid solution with Margules parameters of $\omega_{En} = 26{,}125 - 0.0384P$ and $\omega_{Di} = 32{,}301 - 0.0067P$ in J atom^{-1} with P in bars (Perkins and Vielzeuf, 1992). Calculate the molar G of mixing as a function of the mole fraction of diopside in the solution at 1000°C and 10 kbar at 0.05 increments. Plot your results and indicate the region of immiscibility.

11. Iron formations that originally contained fayalite but have been oxidized and metamorphosed at high pressure and temperature can contain the olivine, laihunite, with iron that is fully oxidized. What is the exchange reaction that must occur and the composition of laihunite?

Problem Making Use of SUPCRT92

12. Calculate the univariant curve for the breakdown of tremolite to enstatite + diopside + quartz + H_2O between 1 and 12 kbar and 800°C to 1000°C. Calculate a similar breakdown curve for a fluoro-tremolite with F^- replacing 20% of the OH^- in the structure. Assume ideal mixing. Plot the results and explain the shapes and relative positions of the curves.

References

Anovitz, L. M. and Essene, E. J., 1987, Phase equilibria in the system $CaCO_3$-$MgCO_3$-$FeCO_3$. *J. Petrol.*, v. 28, pp. 389–414.

Ferry, J. M. and Spear, F. S., 1978, Experimental calibration of the partitioning of Fe and Mg between biotite and garnet. *Contrib. Mineral. Petrol.*, v. 66, pp. 113–117.

Helgeson, H. C., Delany, J. M., Nesbitt, H. W. and Bird, D. K., 1978, Summary and critique of the thermodynamic properties of rock-forming minerals. *Amer. Jour. Sci.*, v. 278-A, pp. 1–229.

Holdaway, M. J. and Mukhopadhyay, B., 1993, A reevaluation of the stability relations of andalusite: thermochemical data and phase diagram for the aluminum silicates. *Am. Mineralogist*, v. 78, pp. 298–315.

Hutchison, R., 1974, The formation of the earth. *Nature*, v. 250, pp. 556–568.

Mackenzie, F. T., Bischoff, W. D., Bishop, F. C., Loijens, M., Schoonmaker, J. and Wollast, R., 1983, Magnesian calcites: Low-temperature occurrence, solubility and solid-solution behavior. In *Carbonates: Mineral and Chemistry*, Reviews in Mineralogy v. 11, Mineralogy Society of America, Blacksburg, Virginia, pp. 97–144.

Margules, M.,1895, Über die Zusammensetzung der gesättigten Dämpfe von Mischungen, *Sitzungsberichte der Akademie der Wissenschaften Wien*, Math. Cl., 104. Abt. IIa, pp. 1243–1278.

Mason, B., 1966, *Principles of Geochemistry*, 3rd edition. Wiley, New York, p. 329.

Mucci, A. and Morse, J. W., 1984, The solubility of calcite in seawater solutions of various magnesium concentrations, $I_t = 0.697$ m at 25°C and one atmosphere total pressure, *Geochim. Cosmochim. Acta*, v. 48, pp. 815–822.

Perkins, D. and Vielzeuf, D., 1992, Experimental investigation of Fe-Mg distribution between olivine and clinopyroxene: Implications for mixing of Fe-Mg in clinopyroxene and garnet-clinopyroxene thermometry, *Am. Mineralogist*, v. 77, pp. 774–783.

Ringwood, A. E., 1991, Phase transformations and their bearing on the constitution and dynamics of the mantle. *Geochim. Cosmochim. Acta*, v. 55, pp. 2083–2110.

Saxena, S. and Ghose, G., 1971, Mg^{2+}-Fe^{2+} order-disorder and the thermodynamics of the orthopyroxene crystalline solution. *Am. Mineralogist*, v. 56, pp. 532–559.

Seck, H. A., 1971, Koexistierende Alkali-feldspar und Plagioklase in System $NaAlSi_3O_8$-$KAlSi_3O_8$-$CaAl_2Si_2O_8$-H_2O bei temperaturen von 650°C bis 900°C. *News Jahrb. Mineral. Abhandlungen*, v. 115, pp. 315–345.

Smith, P. and Parsons, I., 1974, The alkali-feldspar solvus at 1 kilobar water-vapour pressure. *Mineral. Mag.*, v. 39, pp. 747–767.

Thorstenson, D. C. and Plummer, L. N., 1978, Reply: Equilibrium criteria for two-component solids reacting with fixed composition in an aqueous phase—Example: The magnesium calcites, *Amer. Jour. Sci.*, v. 278, pp. 1478–1488.

Whittacker, E. J. W. and Muntus, R., 1970, Ionic radii for use in geochemistry. *Geochim. Cosmochim. Acta*, v. 34, pp. 945–956.

Zoltai, T. and Stout, J. H., 1984, *Mineralogy: Concepts and Principles.* Burgess Publishing Co., Minneapolis, Minnesota, 505 pp.

Other Helpful References

Mason, B. and Moore, C. B., 1982, Some thermodynamics and crystal chemistry. In *Principles of Geochemistry*, 4th edition. John Wiley & Sons, New York, pp. 65–91.

Spear, F. S., 1993a, Activity models for phases of petrologic importance. In *Metamorphic Phase Equilibrium and Pressure-Temperature-Time Paths.* Mineralogical Soc. Amer., Washington, DC, pp. 176–239.

Spear, F. S., 1993b, Heterogeneous phase equilibria. In *Metamorphic Phase Equilibrium and Pressure-Temperature-Time Paths.* Mineralogical Soc. Amer., Washington, DC, pp. 241–288.

6 | Aqueous Solutions

Because of its ubiquity and importance in sustaining life, the chemistry of water and aqueous solutions has been studied extensively. Water, H_2O, is prevalent in the atmosphere, exists in fractures and pores in rocks at depth, is bound in minerals like micas and clays, and is, of course, the predominate constituent of seawater, rivers, and lakes. With the valence of oxygen of -2 and that of hydrogen of $+1$, the structure of a water molecule can be depicted as shown in **Figure 6-1**.

The O–H bonding distance in a water molecule is 0.97 Å (1.0 Å = 10^{-8} cm) with an H–O–H bond angle of 105°, close to the tetrahedral angle of 109.5°. The O–H bonding electron orbitals produce, on average, a negative charge near the oxygen atom and a net positive charge in the vicinity of the two hydrogens. H_2O can, therefore, be modeled as an *electric dipole* (with a moment of 5.7×10^{-7} C m) having a separate positive and negative charge, although the entire molecule is neutrally charged as shown in **Figure 6-2a**. The *electrical moment* gives the charge times distance of separation. If considered in more detail, because the positive charge on the water molecule is contributed by two separate hydrogen atoms, H_2O has some *quadrupole* characteristics, as shown schematically in **Figure 6-2b**.

The nearly tetrahedral nature of the bonding orbitals causes solid H_2O (ice) to exist in an open, tetrahedral-like, framework structure. On heating to 0°C at 1 bar ice (density of 0.9168 g ml^{-1}) melts to liquid water of 9% greater density (0.999987 g ml^{-1}) as the open tetrahedral-like structure is replaced by a more densely packed coordination in the liquid. This is why ice floats on liq-

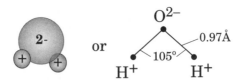

FIGURE 6-1 Two representations of the neutral water molecule showing the distribution of charge in the molecule.

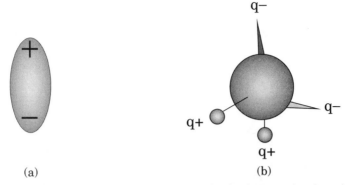

(a)   (b)

FIGURE 6-2 Water molecule represented with (a) a dipole charge showing the separation of + and −
changes in the molecule and (b) quadrupole charge where $q+$ denotes hydrogens and $q−$ represents the
free orbitals containing electron lone pairs in the oxygen atom.

uid water when typically the solid of a compound sinks in its liquid. The high
melting point of ice (0°C) relative to other hydrogen bonded compounds suggests that a kind of quasi-tetrahedral framework is retained in the liquid. The
density of liquid water increases as temperature is raised from 0°C to 4°C as
the open structure in the liquid is destroyed. Above this temperature, the density begins to decrease as the water molecules separate to greater distances because of increased vibrational energy as temperature is increased. Water has
its greatest density, therefore, at 4°C. The high boiling point of liquid H_2O,
100°C at 1 bar, relative to other hydrogen bonded liquids suggests strong intermolecular hydrogen bonding forces continue to exist in the liquid state.

The dipolar nature of liquid H_2O is important for stabilizing charged species
like Na^+ and Cl^- in solution. To help understand this, consider the force of
attraction, F_a, between two oppositely charged particles in a vacuum as given
by Coulomb's law,

$$F_a = \frac{1}{4\pi P_o} \frac{q^+ q^-}{d^2}$$ [6.1]

where q is the charge of the superscripted positive and negative species and d
is their distance of separation. The constant that relates the charge to its force
is typically written as $4\pi P_o$, where P_o is termed the *permittivity* constant (8.8542
$\times 10^{-12}$ C^2 N^{-1} m^{-2}), which gives the *polarizability* in a vacuum.

If a substance is put between the charges rather than having a vacuum,
F_a decreases. To account for this behavior, a relative polarizability, called the
dielectric constant, ε, is defined as

$$\varepsilon \equiv \frac{P}{P_o}$$ [6.2]

where P is the value of the polarizability constant with a substance, termed the dielectric, between the charges. As a ratio of polarizabilities, ε is a dimensionless quantity. With a dielectric present, Coulomb's law can be written as

$$F_a = \frac{1}{4\pi P_o} \frac{q^+ q^-}{\varepsilon d^2}$$

[6.3]

Table 6-1 presents the values of ε for some substances. The relatively high value of ε for H_2O decreases somewhat with increasing temperature. It is clear from an examination of Equation 6.3 that increasing ε would cause the force of attraction between two oppositely charged species in water to decrease; therefore, rather than combining to form a precipitate, a high ε in water stabilizes the separated charged species, keeping them dissolved in solution.

The ions Na^+ and Cl^- with attractive force, F_a, between them in an aqueous solution are shown schematically in **Figure 6-3**. The water dipoles orient

Table 6-1   Dielectric constant, ε, of the indicated substance

ε	Substance
1.0	Vacuum
1.00054	Dry air (25°C)
	Non-hydrogen-bonded
1.6	CO_2 (0°C)
2.2	CCl_4 (25°C)
2.3	Benzene, C_6H_6 (25°C)
	Hydrogen-bonded
16.9	NH_3 (25°C)
84	HF (0°C)
78.3	H_2O (25°C)
77.9	D_2O (25°C)
87.9	H_2O (0°C and 1 bar) *liquid*
80.4	H_2O (20°C and 1 bar)
55.5	H_2O (100°C and 1 bar)
25.5	H_2O (300°C and 1 kbar)
19.9	H_2O (400°C and 2 kbar)
17.6	H_2O (500°C and 4 kbar)
Solids	
5.4–7.0	Mica (25°C)
4.3	Quartz (25°C)
5.7	Diamond
6.1	NaCl (25°C)

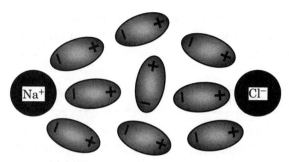

FIGURE 6-3 Schematic diagram showing orientated water dipoles between Na^+ and Cl^- that decrease the net force of attraction between these ions.

themselves so that on average the positively charged ends tend to face Cl^- and the negatively charged ends face toward the Na^+ because of the coulombic attractions between opposite charges. Thermal motion of the water molecules keeps the dipoles from becoming perfectly aligned. The dipole alignment, however, produces a force operating in the opposite direction of that between Na^+ and Cl^- because the charges on the dipoles are orientated oppositely to the charge direction between Na^+ and Cl^-. The net force between Na^+ and Cl^-, therefore, decreases. As a result, the dielectric, H_2O, lowers the energy of attraction between the charged species, stabilizing Na^+ and Cl^- in solution. In other words, they are kept apart by the dielectric H_2O so that they do not combine to form NaCl. With higher temperature, thermal agitation causes less orientation to occur and the dielectric constant of H_2O, ε_{H_2O} decreases.

Although many elements exist as single ions in aqueous solutions, certain combinations of elements have particular stability in a single species. These polyatomic ions are held together by covalent bonds. The names of some common polyatomic ions are given in **Table 6-2**. Sometimes hydrogen is also bonded

Table 6-2   **Some simple polyatomic ions common in aqueous solutions**

Name	Formula
Ammonium	NH_4^+
Bicarbonate	HCO_3^-
Carbonate	CO_3^{2-}
Chromate	CrO_4^{2-}
Dichromate	$Cr_2O_7^{2-}$
Phosphate	PO_4^{3-}
Sulfate	SO_4^{2-}
Sulfite	SO_3^{2-}

to the ion, and the name is modified by the prefix word hydrogen for a single hydrogen atom and dihydrogen for two hydrogens, in the name. One, therefore, has hydrogen phosphate, HPO_4^{2-}, and dihydrogen phosphate, $H_2PO_4^-$, ions.

■ Acids, Bases, and *pH*

An *acid* is a substance that dissociates in water giving free protons, that is H^+, to solution. Although often written as H^+, it is sometimes written as H_3O^+ because the free proton tends to form a cluster with H_2O molecules. When consideration of the clustering is important, H^+ is probably better characterized with tetrahedral coordination of H_2O as $H^+ \bullet (H_2O)_4$; however, in most situations, the representation as H^+ is sufficient.

H_2O undergoes some self-dissociation to produce H^+ and OH^- by the reaction

$$H_2O = H^+ + OH^-$$ [6.4]

Consider a standard state for H_2O of pure H_2O at the pressure and temperature of interest together with a standard state for H^+ and OH^- of a 1 molal solution but with the species behaving as though they are in an infinite sea of H_2O solution at the pressure and temperature of interest. The equilibrium constant for Reaction 6.4 can then be written as

$$K_{H_2O} = \frac{a_{H^+}\, a_{OH^-}}{a_{H_2O}}$$ [6.5]

Remember from what has been discussed in Chapter 4 on the law of mass action that the equilibrium constant for a reaction can be calculated from

$$\log K = \frac{\Delta G_R^o}{-2.3026\, RT}$$ [6.6]

where ΔG_R^o is the standard state Gibbs energy of reaction. The standard state Gibbs energies of the species found in SUPCRT92 (see Appendix F) at 25°C (298.15 K) and 1 bar allow the equilibrium constant of Reaction 6.4 to be calculated, giving

$$\log K_{H_2O} = \frac{(0.0\ \text{cal mol}^{-1}) + (-37{,}595\ \text{cal mol}^{-1}) - (-56{,}688\ \text{cal mol}^{-1})}{-2.3026 \times 1.9873\ \text{cal mol}^{-1}\text{K}^{-1} \times 298.15\ \text{K}} = -13.995$$ [6.7]

so that $K_{H_2O} = 10^{-13.995}$. This implies that very little self-dissociation occurs, and the solution remains predominately H_2O dipoles. With the standard state chosen for H_2O, a_{H_2O} can be taken as unity so that at 25°C and 1 bar

$$a_{H^+}\, a_{OH^-} = 10^{-13.995}$$ [6.8]

The molal activity coefficients of the charged species, H^+ and OH^-, can also be taken as unity because these species are at such low concentration in pure H_2O that they approximate the conditions of an infinitely dilute solution where $\gamma_i = 1$. The sum of charges in the entire solution must be zero. With H^+ and OH^- as the only charged species in pure H_2O, it is clear that $m_{H^+} = m_{OH^-}$ and with unit activity coefficients, $a_{H^+} = a_{OH^-}$. As a result, Equation 6.8 can be written as

$$(a_{H^+})^2 = 10^{-13.995} \qquad\qquad [6.9]$$

or

$$a_{H^+} = 10^{-6.997} \qquad\qquad [6.10]$$

at 25°C and 1 bar. pH is defined as the negative of the base-10 logarithm of the hydrogen ion activity of a solution. For pure H_2O at 1 bar and 25°C, this gives $pH = 6.997$. Because pHs are typically only determined to two decimal places, the pH of pure H_2O is taken as 7.00. A solution that has a greater activity of H^+ than pure H_2O ($pH < 7$) is termed an acid solution.

A base is a substance yielding an OH^- species on dissolving in water. The term basic solution is not used, but rather, solutions with OH^- concentrations greater than pure H_2O are termed *alkaline solutions*. The alkali elements—H, Li, Na, K, Rb, and Cs—readily form solid compounds such as $NaOH$ and KOH, which are bases when added to water. These are then alkaline solutions because they contain significant quantities of alkalis as well as OH^-. When OH^- is increased in solution, Reaction 6.4 proceeds to the left and H^+ is consumed until a new equilibrium is reached. Alkaline solutions therefore have a $pH > 7$ at 1 bar and 25°C.

When the acid HCl is added to an aqueous solution, the reaction

$$HCl = H^+ + Cl^- \qquad\qquad [6.11]$$

occurs with an equilibrium constant expression at 25°C and 1 bar, which can be written as

$$K_{HCl} = \frac{a_{H^+}\, a_{Cl^-}}{a_{HCl}} = 10^{6.5} \qquad\qquad [6.12]$$

In this solution, associated HCl is virtually undetectable as nearly all of the acid exists as H^+ and Cl^-. This can be demonstrated for a 1.0 m HCl solution. For the purposes of this calculation, assume the activity coefficients are near unity so that $a_{H^+} = m_{H^+}$. Before adding 1.0 mole of HCl to 1000 g of H_2O, the H^+ concentration is 10^{-7}. Ignoring this insignificant amount of H^+, after addition of the HCl

$$1.0 = m_{HCl} + m_{H^+} \quad \text{or} \quad m_{HCl} = 1.0 - m_{H^+} \tag{6.13}$$

Charge balance in the solution requires $m_{H^+} = m_{Cl^-} + m_{OH^-}$. With the m_{OH^-} small relative to m_{Cl^-} in this acid solution, $m_{Cl^-} = m_{H^+}$ is a reasonable approximation. The equilibrium constant in Equation 6.12 can, therefore, be written as

$$K_{HCl} = \frac{m_{H^+}^2}{1.0 - m_{H^+}} = 10^{6.5} \tag{6.14}$$

Solving for m_{H^+}, the value $m_{H^+} = 0.9999998$ is obtained. The 1 m HCl is, therefore, considered "completely" disassociated, although about 0.0000002 m of associated HCl is calculated to exist in the solution.

The strength of an acid or base can be specified in terms of its *normality*. Consider a 1.0 N (normal) acid solution. 1.0 N acid is the quantity of acid necessary to produce 1 mole of H^+ per liter of solution; therefore, 1.0 M HCl solutions are 1.0 N, whereas 1.0 M H_2SO_4 solutions are 2.0 N as 2 moles of H^+ are released into solution for each mole of H_2SO_4 added. Because a 1.0 N acid solution has $m_{H^+} = 1.0 = 10^{0.0}$, the *pH* of this solution is 0.0 if activity coefficients are near unity. A 10 N acid solution would have a *pH* = −1.0. A 1.0 N base has an OH^- concentration equal to 1.0 M (~1.0 m). Because $a_{H^+} \times a_{OH^-} = 10^{-14}$, a_{H^+} must equal about 10^{-14} or the solution has a *pH* of 14.0.

Acids such as HCl and H_2SO_4 are termed *strong acids* because they disassociate almost completely in H_2O. There are many acids where a significant amount of the acid remains as the associated neutrally charged species in solution. These are referred to as *weak acids*. Probably the most important weak acid in understanding earth processes is *carbonic acid*. It can be denoted as $CO_{2(aq)}$ but is often also written with an explicit water molecule included as the species H_2CO_3. In this latter formulation, the standard molal Gibbs energy of H_2CO_3 is equivalent to the sum of that of $CO_{2(aq)}$ plus an H_2O molecule.

In water, carbonic acid disassociates by the reaction

$$H_2O + CO_{2(aq)} = HCO_3^- + H^+ \tag{6.15}$$

with an equilibrium constant at 25°C and 1 bar that can be calculated from the Gibbs energies in SUPCRT92 for H_2O and the aqueous species by considering Equation 6.6. This gives

$$\log K_{CO_{2(aq)}} = \frac{(-140,282 + 0.0) - (-56,688 + -92,250) \text{ cal mol}^{-1}}{-2.3026 \times 1.9873 \text{ cal mol}^{-1} \text{K}^{-1} \times 298.15\text{K}} = -6.34 \tag{6.16}$$

so that

$$K_{CO_{2(aq)}} = \frac{a_{HCO_3^-} \, a_{H^+}}{a_{H_2O} \, a_{CO_{2(aq)}}} = 10^{-6.34} \qquad [6.17]$$

The HCO_3^- produced by Reaction 6.15, termed *bicarbonate*, can disassociate to produce acid and *carbonate*, CO_3^{2-}, by the reaction

$$HCO_3^- = CO_3^{2-} + H^+ \qquad [6.18]$$

The equilibrium constant of this bicarbonate disassociation reaction at 25°C and 1 bar is

$$\log K_{HCO_3^-} = \frac{(-126,191 + 0.0) - (-140,282) \text{ cal mol}^{-1}}{-2.3026 \times 1.9873 \text{ cal mol}^{-1} K^{-1} \times 298.15 \text{ K}} = -10.33 \qquad [6.19]$$

so that

$$K_{HCO_3^-} = \frac{a_{CO_3^{2-}} \, a_{H^+}}{a_{HCO_3^-}} = 10^{-10.33} \qquad [6.20]$$

The equilibrium constant expressions are related by the activity of H^+. Under what *pH* conditions are each of the three carbonate species, $CO_{2(aq)}$, HCO_3^-, and CO_3^{2-} dominant? Equation 6.17 with $a_{H_2O} = 1$ can be written as

$$\frac{\gamma_{HCO_3^-} \, m_{HCO_3^-}}{\gamma_{CO_{2(aq)}} \, m_{CO_{2(aq)}}} = \frac{10^{-6.34}}{a_{H^+}} \qquad [6.21]$$

and Equation 6.20 as

$$\frac{\gamma_{CO_3^{2-}} \, m_{CO_3^{2-}}}{\gamma_{HCO_3^-} \, m_{HCO_3^-}} = \frac{10^{-10.33}}{a_{H^+}} \qquad [6.22]$$

If the activity coefficient ratios in Equations 6.21 and 6.22 are taken as unity, these equations imply $CO_{2(aq)}$ dominates when $pH < 6.34$, whereas between a *pH* of 6.34 and 10.33, HCO_3^- dominates, and at $pH > 10.33$, CO_3^{2-} dominates. The calculated percentages of each species at a particular *pH* are shown in **Figure 6-4**. With changes in activity coefficients as the ionic strength of the solution changes, the concentration of carbonate in solution at a particular *pH* will also change so that the lines in Figure 6-4 can shift somewhat.

To determine the concentration of species in solution where activity coefficient effects are important, activity coefficient expressions need to be incorporated into the calculation. In the pure carbonate system, there are three possible carbon-containing species, $CO_{2(aq)}$, HCO_3^-, and CO_3^{2-}, as well as three hydrogen-containing species H^+, OH^-, and H_2O. Six equations are needed to solve for the distribution of species in this solution. These include the two

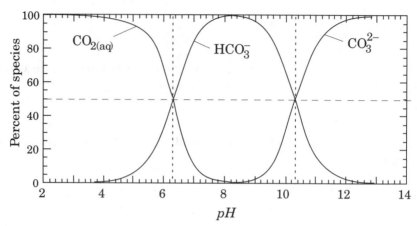

I FIGURE 6-4 Relative abundance of carbonate species in water at 25°C and 1 bar as a function of *pH*.

equilibrium constant expressions between carbon-containing species given in Equations 6.17 and 6.20 as well as the equation for water disassociation in Equation 6.5. A fourth equation comes from knowledge of the total inorganic carbon contributed by species to the solution, $CO_{2\,total}$, of

$$CO_{2total} = m_{CO_{2(aq)}} + m_{HCO_3^-} + m_{CO_3^{2-}} \qquad [6.23]$$

Because the solution must be charge balanced, a fifth relationship is

$$m_{H^+} = m_{HCO_3^-} + 2m_{CO_3^{2-}} + m_{OH^-} \qquad [6.24]$$

The sixth constraint is that the mole fractions of all the i species in the solution, X_i, sum to unity:

$$X_{CO_{2(aq)}} + X_{HCO_3^-} + X_{CO_3^{2-}} + X_{OH^-} + X_{H^+} + X_{H_2O} = 1.0 \qquad [6.25]$$

With these six equations together with expressions to relate the activities of species to their concentrations, such as the Debye-Hückel activity coefficient expression for charged species and the Setchénow equation for neutral species as discussed in Chapter 4, a more precise determination of the concentrations of all of the species in solution can be made.

Problems occur in solving the set of equations when the concentration of solutes in solution increases. Consider seawater, which to a first approximation is a 0.5 m NaCl + 0.05 m MgSO₄ solution. With complete disassociation of NaCl and MgSO₄ in the solution, there are 1.1 moles of solute species (i.e., Na⁺, Cl⁻, Mg²⁺, and SO₄²⁻) per 1000 g of water. With the molecular weight of H₂O of 18.015 g mol⁻¹, there are 55.51 moles of H₂O in 1000 g of water or about 50 H_2O molecules around each ion. Considering the three-dimensional nature of the interactions, it is probable that significant interaction between oppositely charged ions produces species of lower charge in solution. That is, in seawater at 25°C, there are significant concentrations of species such as $MgCl^+$, $NaSO_4^-$, and $MgSO_4^0$. These species are called aqueous *complexes* because they involve the complexing together of the simple cations, Mg^{2+} and Na^+, with the anions, Cl^- and SO_4^{2-}. From an energetic standpoint, this occurs because the coulombic attraction of the oppositely charged ions is stronger than the attraction of simple ions to water dipoles. Formation of complexes containing elements in a mineral structure leads to greater solubility of the mineral in solution.

Consider the solubility of halite in seawater, a common mineral in evaporite deposits:

$$NaCl = Na^+ + Cl^-$$
halite   [6.26]

Formation of species like $NaSO_4^-$ and $MgCl^+$ from SO_4^{2-} and Mg^{2+} in seawater will lower the concentration of the Na^+ and Cl^- species in solution. This causes Reaction 6.26 to proceed to the right, increasing the solubility of halite. To solve for the distribution of all of the species in such solutions, the equilibrium constant expressions for the formation of these aqueous complexes as given in the reactions

$$Na^+ + SO_4^{2-} = NaSO_4^- \qquad K_{(6.27)} = \frac{a_{NaSO_4^-}}{a_{Na^+}\, a_{SO_4^{2-}}}$$ [6.27]

and

$$Mg^{2+} + Cl^- = MgCl^+ \qquad K_{(6.28)} = \frac{a_{MgCl^+}}{a_{Mg^{2+}}\, a_{Cl^-}}$$ [6.28]

need to be considered. A calculation of the distribution of species in solution can often become very complex with the presence of aqueous complexes and the need to consider activity coefficient effects. Other than for the simplest cases, the number of equations gets large, and a numerical solution by computer calculation is typically done. Some of the codes used for this purpose are given in **Table 6-3**.

Shown in **Figure 6-5** is the distribution of carbonate species at 25°C and 1 bar as a function of *pH* where the dissolved inorganic carbon in solution totals 2×10^{-3} moles. The solid lines are values in pure water where activity coefficients are taken as unity. The dashed lines are for seawater with a salinity of

| **Table 6-3** | **Some available distribution of species calculation software** |

EQ3NR written and maintained by Tom Wolery (1992) at Lawrence Livermore Nat. Lab. See EQ3NR, A Computer Program for Geochemical Aqueous Speciation-Solubility Calculations: Theoretical Manual, User's Guide, and Related Documentation. Version 7.0. UCRL-MA-110662 PT III. http://geosciences.llnl.gov/esd/geochem/EQ36manuals/eq3nr.pdf.

MINEQL was developed by Westall et al. in 1976 after an earlier **REDEQLO** model of Morel and Morgan. A later modification is **MINTEQA2** (Allison, G. D., Brown, D. S. and Novo-Gradac, K. J., 1990, MINTEQA2/PRODEFA2, A Geochemical Assessment Model for Environmental Systems. Version 3. User's manual. Environmental Research Lab, U.S. Environmental Protection Agency, Athens, Georgia, USA). Scientific Software Group, P.O. Box 708188, Sandy, Utah 84070. **Visual MINTEQ** ver. 2.12, is an interface for Windows 95 or later operating systems at http://www.lwr.kth.se/english/OurSoftware/Vminteq/.

PHREEQE program and its derivatives were developed by the U.S. Geological Survey with an eye toward groundwater chemistry. The current version is **PHREEQC** (Version 2) by David Parkhurst of the U.S.G.S. http://water.usgs.gov/software/geochemical.html. A web-based version **WEB-PHREEQ** by Bernhardt Saini-Eidukat is also available at http://www.ndsu.nodak.edu/webphreeq/.

SOLVEQ written and maintained by Mark Reed (U. Oregon); see Reed, M. II. (1998). Calculation of simultaneous chemical equilibria in aqueous–mineral–gas systems and its application to modeling hydrothermal processes. In Richards J., Larson P. (eds.), Techniques in Hydrothermal Ore Deposits Geology. Reviews in Econ. Geol., vol. 10, pp. 109–124.

The Geochemist's Workbench® Release 4. A commercial product offered by C. Bethke. See Bethke (1996). Geochemical Reaction Modeling, Concepts and Applications. Oxford Univ. Press or http://www.geology.uiuc.edn/Hydrogeology/GWBUsersGuide.pdf.

WATEQ4F, first developed by Truesdell and Jones (1974); see Ball, J. W. and Nordstrom, D. K., 1991, User's Manual for WATEQ4F, With Revised Thermodynamic Database and Test Cases for Calculating Speciation of Major, Tracc, and Redox Elements in Natural Waters. U.S. Geological Survey Open-File Report 91-183,189 p. Available at http://water.usgs.gov/software/wateq4f.html.

35 g per 1000 g of solution where the effects of all of the species, including boric acid, are considered. $CO_{2(aq)}$ is less stable in salt solutions depressing its concentration over that of pure H_2O. Also, CO_3^{2-} has a higher concentration than the other carbonate species at $pH > 8.9$ in seawater at 25°C and 1 bar but only above a pH of 10.3 in the pure water system. This indicates that CO_3^{2-} is a significant carbonate species in surface seawater ($pH = 8.1$ to 8.35). This would not be the conclusion from considering the pure water system where, as indicated in Figure 6-5, an insignificant amount of CO_3^{2-} is present at these pHs. The reason that CO_3^{2-} increases its stability relative to HCO_3^- in seawater is that activity coefficients of charged species decrease with the square of charge of the species (see Figure 4-13). As a doubly charged species, CO_3^{2-} decreases its activity coefficient to a greater extent than HCO_3^-. Decreased activity coefficients then decrease the species activity, lowering its Gibbs energy and make it more stable.

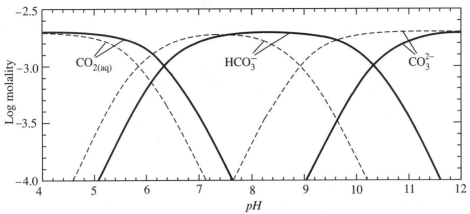

FIGURE 6-5 Concentrations of carbonate species where the total inorganic carbon in solution is 2 millimoles. Solid lines are for pure H_2O, and dashed lines for seawater at 25°C and 1 bar as a function of *pH*. (Adapted from Zeebe and Wolf-Gladrow, 2001.)

■ *pH* Buffers

Consider a small amount of acid, say 10^{-3} moles of HCl, added to 1000 g of pure H_2O. Using Equation 6.12, the *pH* is calculated to change from 7 to approximately 3. Now consider the input of water to the oceans that comes predominately from rain and river water. These waters are acidic solutions generally with a *pH* between 4.5 and 6.5. Ocean water, on the other hand, is alkaline with a *pH* between 7.7 and 8.35. Why does ocean water not become increasingly acidic with continued addition of this acidic water?

To answer this question, consider a simplified artificial "seawater" made by adding 0.01 moles of the salt Na_2CO_3 (sodium carbonate) and 0.01 moles of the salt $NaHCO_3$ (sodium bicarbonate) to 1000 g of H_2O. Both Na_2CO_3 and $NaHCO_3$ dissolve completely, producing a solution that contains 0.03 *m* of Na^+, 0.01 *m* of CO_3^{2-}, and 0.01 *m* of HCO_3^-. The *pH* of this solution with equal CO_3^{2-} and HCO_3^- is ~10.33, as indicated in Figure 6-4 or calculated from Equation 6.22 if one assumes $\gamma_{HCO_3^-} = \gamma_{CO_3^{2-}}$ at the low ionic strength of the solution. Now consider what happens if 10^{-3} moles of HCl are added to 1000 g of this solution. After addition of the acid, the reaction

$$H^+ + CO_3^{2-} = HCO_3^-$$ [6.29]

proceeds to the right. Most of the 0.001 moles of acid added are consumed by the *carbonate ion* to produce bicarbonate, increasing $m_{HCO_3^-}$ to about 0.011 and decreasing $m_{CO_3^{2-}}$ to nearly 0.009. From Equation 6.22

$$a_{H^+} = \frac{m_{HCO_3^-}}{m_{CO_3^{2-}}} K_{HCO_3^-} = \frac{0.011}{0.009} 10^{-10.33} = 10^{-10.24}$$ [6.30]

or a pH of 10.24 is produced. The addition of the acid, therefore, has only changed the pH of the solution from 10.33 to ~10.24 rather than to a $pH = 3$, as is the case with pure H_2O.

What if 10^{-3} moles of base is added to the solution instead? In this case, the reaction

$$OH^- + HCO_3^- = H_2O + CO_3^{2-}$$ [6.31]

proceeds to the right. Again the ratio of HCO_3^- to CO_3^{2-} remains nearly constant so that the pH changes little. A solution that keeps the pH nearly constant when acid or base is added is called a *pH buffer*. *pH* buffering occurs when a weak acid (e.g., HCO_3^-) and the salt of the acid (e.g., CO_3^{2-}) are present at significant concentrations in a solution.

The pH buffering of ocean water to keep the pH between 7.7 and 8.35 is more complicated than just the reactions given above and depends to some extent on reactions that are rate dependent. The range of values of pH occurs because of small differences in temperature, ionic strength, and total CO_2. Calcite plays a part in buffering pH where it is present. In Chapter 7, pH buffering in seawater is discussed in more detail.

■ Alkalinity

To determine the amount of $CO_{2(aq)}$, HCO_3^-, and CO_3^{2-} in a solution, *alkalinity* and carbonate titration curves need to be understood. As indicated above, depending on pH, CO_2 in an aqueous solution can exist as any of the three different species: carbonic acid, $CO_{2(aq)}$, bicarbonate, HCO_3^-, or carbonate ion, CO_3^{2-}. How much of each of these species exists in surface water in equilibrium with atmospheric CO_2?

With a concentration of CO_2 gas in the atmosphere, $CO_{2(g)}$, an equilibrium is set up with dissolved $CO_{2(aq)}$, carbonic acid, in surface water that can be written as

$$CO_{2(g)} = CO_{2(aq)}$$ [6.32]

The equilibrium constant for this reaction at 25°C and 1 bar as calculated from SUPCRT92 data is

$$\log K_{CO_{2(g)}} = \frac{(-92,250) - (-94,254) \text{ cal mol}^{-1}}{-2.3026 \times 1.9873 \text{ cal mol}^{-1} K^{-1} \times 298.15K} = -1.47$$ [6.33]

The equilibrium constant for this reaction is then

$$K_{CO_{2(g)}} = \frac{a_{CO_{2(aq)}}}{f_{CO_{2(g)}}} = 10^{-1.47} \qquad [6.34]$$

The CO_2 gas, at its low concentration in the atmosphere, behaves nearly ideally so that $f_{CO_2} = P_{CO_2}$. This allows Equation 6.34 to be written as

$$a_{CO_{2(aq)}} = K_{CO_{2(g)}} P_{CO_2} \qquad [6.35]$$

The relationship between carbonic acid and bicarbonate has already been considered in Reaction 6.15 for which the equilibrium constant expression is given in Equation 6.17. Combining Equations 6.17 and 6.35 results in

$$a_{H^+} a_{HCO_3^-} = K_{CO_{2(aq)}} K_{CO_{2(g)}} P_{CO_2} \qquad [6.36]$$

At constant P_{CO_2}, therefore, H^+ is inversely related to HCO_3^- concentration. Because a solution is always charge balanced

$$m_{Na^+} + m_{K^+} + 2m_{Ca^{2+}} \ldots = m_{Cl^-} + m_{HCO_3^-} + 2m_{CO_3^{2-}} \ldots \qquad [6.37]$$

where the left-hand side of the equation tallies the charges on all of the cations in solution and the right-hand side all the anions. Ions such as Na^+, K^+, Ca^{2+}, Cl^- are *conservative* in that their molalities are generally unaffected by changes in P, T, or pH. Clearly, at elevated P and T when aqueous complexes form or at extremes of pH where species such as $Ca(OH)_{2(aq)}$ become stable these aqueous species would not be conservative; however, in normal waters near the earth's surface where the concept is relevant, they are conservative. Equation 6.37 can therefore be rewritten as

$$\sum \text{conservative cations} - \sum \text{conservative anions} = \qquad [6.38]$$
$$m_{HCO_3^-} + 2m_{CO_3^{2-}} + m_{OH^-} - m_{H^+} + m_{H_3SiO_4^-} + m_{B(OH)_4^-} + m_{HS^-} + m_{\text{organic anions}}$$

where the summations are taken in charge equivalents. The right-hand side of Equation 6.38, which tallies the species affected by pH, is termed the *total alkalinity*. In natural water, often the only nonconservative species important in determining alkalinity are HCO_3^- and CO_3^{2-}. In seawater, however, the species $B(OH)_4^-$ is approximately 3% of total alkalinity.

The species OH^- and H^+ as well as other charged nonconservative species generally become significant in the summation only at very high or low pH. In somewhat acid to somewhat alkaline solutions, the total alkalinity is, therefore, often equal to the *carbonate alkalinity* given by $m_{HCO_3^-} + 2m_{CO_3^{2-}}$. Under these conditions, carbonate alkalinity is equal to the summation of the conservative species and is therefore also conservative. Carbonate alkalinity is independent of P_{CO_2} because P_{CO_2} is not directly involved in charge balance. For instance, increasing P_{CO_2} increases $m_{HCO_3^-}$. The reaction that occurs can be written as

$$CO_{2(g)} + H_2O + CO_3^{2-} = 2\ HCO_3^-$$ [6.39]

or

$$CO_{2(g)} + H_2O = H^+ + HCO_3^-$$ [6.40]

In Reaction 6.39, the alkalinity gained by $m_{HCO_3^-}$ is balanced by loss of $m_{CO_3^{2-}}$. In Reaction 6.40, alkalinity gained by $m_{HCO_3^-}$ is balanced by increase of m_{H^+}. Also, if a solution does not exchange with a CO_2 reservoir like the atmosphere, the total dissolved carbonate, ΣCO_2, can also be considered to be conserved because the sum of the concentrations of species $CO_{2(aq)}$, HCO_3^-, and CO_3^{2-} defines the total inorganic carbon in solution. The concentrations of bicarbonate and carbonate ion needed to calculate the alkalinity are determined by titrating the solution with acid and determining the amount of acid necessary to change pH of the solution to a pH near 4.3 (see below). This is reported as total alkalinity or as carbonate alkalinity if carbonate species are the only non-conservative species that are important.

■ Alkalinity Titration

Consider titrating 1 liter of $5 \times 10^{-3}\ m$ Na_2CO_3 with 1 molar HCl in a vessel where no gas phase is present. How does pH vary when adding increments of HCl to the solution? Assume for this exercise that activity coefficients as well as the activity of H_2O are unity. To start with

$$m_{Na^+} = 10^{-2}m \quad \text{and} \quad \sum CO_2 = 5 \times 10^{-3}m$$ [6.41]

in the solution where

$$\sum CO_2 = m_{CO_{2(aq)}} + m_{HCO_3^-} + m_{CO_3^{2-}}$$ [6.42]

Charge balance in the solution before the addition of acid implies

$$m_{Na^+} = m_{HCO_3^-} + 2m_{CO_3^{2-}} + m_{OH^-} - m_{H^+}$$ [6.43]

Also, before HCl is added m_{H^+} is negligible in this alkaline solution; therefore, Equation 6.43 simplifies to

$$m_{Na^+} = m_{HCO_3^-} + 2m_{CO_3^{2-}} + m_{OH^-} \qquad [6.44]$$

With an alkaline solution, $m_{CO_{2(aq)}}$ is also small so that Equation 6.42 can be written as

$$\sum CO_2 = m_{HCO_3^-} + m_{CO_3^{2-}} \qquad [6.45]$$

Multiplying Equation 6.45 by 2 and subtracting it from Equation 6.44 results in

$$m_{Na^+} - 2\sum CO_2 = -m_{HCO_3^-} + m_{OH^-} \qquad [6.46]$$

Remember that $m_{Na^+} = 2\sum CO_2$ because of the 1:2 ratio of Na and CO_2 in Na_2CO_3. Equation 6.46 can, therefore, be written as

$$m_{HCO_3^-} = m_{OH^-} \qquad [6.47]$$

From Equation 6.5 with $a_{H_2O} = 1$ and making the substitution for m_{OH^-} given in Equation 6.47 results in

$$m_{HCO_3^-} = \frac{K_{H_2O}}{m_{H^+}} \qquad [6.48]$$

The CO_3^{2-} can be determined by considering the disassociation of bicarbonate, Reaction 6.18, with an equilibrium constant expression given in Equation 6.20. Substituting Equation 6.48 into Equation 6.20 gives

$$m_{CO_3^{2-}} = \frac{K_{HCO_3^-} K_{H_2O}}{m_{H^+}^2} \qquad [6.49]$$

Substituting Equations 6.48, 6.49, and 6.5 into Equation 6.44 produces

$$m_{Na^+} = \frac{K_{H_2O}}{m_{H^+}} + 2\frac{K_{HCO_3^-} K_{H_2O}}{m_{H^+}^2} + \frac{K_{H_2O}}{m_{H^+}} \qquad [6.50]$$

Using the 25°C and 1 bar values of $K_{H_2O} = 10^{-14.00}$ and $K_{HCO_3^-} = 10^{-10.33}$ together with $m_{Na^+} = 10^{-2}$ m and multiplying both sides by $(m_{H^+})^2$, Equation 6.50 becomes

$$10^{-2}(m_{H^+})^2 = 2 \times 10^{-14} m_{H^+} + 2 \times 10^{-24.33} \qquad [6.51]$$

Solving for m_{H^+} by using the quadratic equation, $m_{H^+} = 10^{-10.97}$; therefore, the $pH = 10.97$ before the 1 molar HCl is added to the solution. With addition of HCl, the charge balance in solution becomes

$$m_{Na^+} - m_{Cl^-} = m_{HCO_3^-} + 2m_{CO_3^{2-}} + m_{OH^-} - m_{H^+}$$

[6.52]

With x ml of HCl added and with 10^{-2} m of the conservative ion Na^+ in solution, Equation 6.52 can be written as

$$10^{-2} - 10^{-3} x = m_{HCO_3^-} + 2m_{CO_3^{2-}} + m_{OH^-} - m_{H^+}$$

[6.53]

Combining Equations 6.7 and 6.20 gives

$$m_{CO_3^{2-}} = \frac{K_{HCO_3^-} K_{CO_{2(aq)}} m_{CO_{2(aq)}}}{m_{H^+}^2}$$

[6.54]

By substituting Equations 6.17, 6.54, and 6.5 into Equation 6.53,

$$10^{-2} - 10^{-3} x = \frac{K_{CO_{2(aq)}} m_{CO_{2(aq)}}}{m_{H^+}} + 2\frac{K_{HCO_3^-} K_{CO_{2(aq)}} m_{CO_{2(aq)}}}{m_{H^+}^2} + \frac{K_{H_2O}}{m_{H^+}} - m_{H^+}$$

[6.55]

Substituting values given in Equations 6.17 and 6.53 into the expression for total inorganic carbon in solution, Equation 6.42, becomes

$$5 \times 10^{-3} = m_{CO_{2(aq)}} + \frac{K_{CO_{2(aq)}} m_{CO_{2(aq)}}}{m_{H^+}} + 2\frac{K_{HCO_3^-} K_{CO_{2(aq)}} m_{CO_{2(aq)}}}{m_{H^+}^2}$$

[6.56]

Combining Equations 6.55 and 6.56 results in

$$10^{-3} x = 10^{-2} - \frac{5 \times 10^{-3} K_{CO_{2(aq)}} (1 + 2K_{HCO_3^-} / m_{H^+})}{m_{H^+} + K_{CO_{2(aq)}} + K_{CO_{2(aq)}} K_{HCO_3^-} / m_{H^+}} - \frac{10^{-14}}{m_{H^+}} + m_{H^+}$$

[6.57]

Values of x can be input into Equation 6.57, and the pH can be calculated. This produces the heavy solid curve shown in **Figure 6-6**. The curve contains two inflection points (endpoints) at 5 ml of HCl (point B), where $pH = 8.35$, and at 10 ml of HCl (point D), where $pH = 4.32$. With the alkaline solution starting at point A, acid is added to convert all CO_3^{2-} to HCO_3^-. This occurs at endpoint B where there are equivalent molalities of CO_3^{2-} and $CO_{2(aq)}$. At this point, these molalities are two orders of magnitude below $m_{HCO_3^-}$ so that virtually all of the CO_3^{2-} has been converted to HCO_3^-. For each mole of acid added, 1 mole of CO_3^{2-} is converted to HCO_3^- so that the moles of acid used to change the pH from point A to point B (5×10^{-3} moles) give the molality of CO_3^{2-} in the 1000 g solution. With continued titration of acid, all of the carbonate in solution is converted into $CO_{2(aq)}$ at point D. This equivalence point is when $m_{H^+} = m_{HCO_3^-}$, but again, this is two orders of magnitude below the concentration of $CO_{2(aq)}$ in solution. The amount of acid titrated to point D

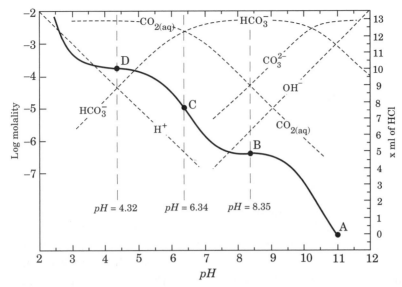

FIGURE 6-6 Acid titration of a 1000-ml alkaline carbonate solution of *pH* 11 shown as the heavy solid line. The short dash lines give the log concentration of the indicated species in solution as a function of *pH* as given on the left vertical axis. Beginning at point A, the volume of acid added is given on the right scale, and the resultant change in *pH* to a *pH* below 3 is shown. Note the change in the slope of the curve at points B and D where a small addition of acid changes the *pH* rapidly. At these points, only one carbon-containing species is significant, and thus, the solution does not buffer the *pH* with addition of acid. (Adapted from Drever, 1988.)

$(1 \times 10^{-2}$ moles) is then equal to the carbonate alkalinity $(m_{HCO_3^-} + 2m_{CO_3^{2-}})$ of 1×10^{-2} *m*; therefore, CO_3^{2-} is the only inorganic carbon containing species in the original $pH = 10.97$ solution, as indicated by the species concentrations given by the short dash lines in Figure 6-6. At the endpoints, *pH* changes so rapidly with addition of a small amount of acid that knowledge of the exact *pH* is not so important.

Alkalinity of water is the capacity of species in solution to act as a base by reacting with protons. An alkalinity titration is a useful analytical procedure that provides information on the distribution of negatively charged species in solution, their *pH* buffering ability, and therefore, the stability of the *pH* of the solution. For instance, such a titration done on lake water can be used to help predict the effect of a particular extent of acid rain on the lake's *pH*.

■ Mineral Solubility

With an understanding of the effects of changes in *pH* on the concentration of species in solution, mineral solubility controls on solution composition can be investigated. Consider first the mineral brucite (Brc) dissolving in H_2O:

$$Mg(OH)_2 = Mg^{2+} + 2OH^- \qquad [6.58]$$
$$\quad Brc$$

The equilibrium constant expression for this reaction is

$$K^{Brc} = \frac{a_{Mg^{2+}}\, a^2_{OH^-}}{a_{Brc}} \qquad [6.59]$$

This is related to the standard state Gibbs energy of reaction, ΔG^o_R, by

$$\log K^{Brc} = -\frac{\Delta G^o_R}{2.303RT} \qquad [6.60]$$

A standard state for brucite of the pure mineral at the pressure and temperature of interest and for the aqueous species, a 1 molal solution whose properties are those of infinite dilution in H_2O at the pressure and temperature of interest can be used. ΔG^o_R for the brucite solubility Reaction 6.58 can then be computed at 25°C and 1 bar from values of the Gibbs energies of the species with SUPCRT92 data. This gives

$$\log K^{Brc} = \frac{(-108,505 + 2(-37,595)) - (-199,646)\ \text{cal mol}^{-1})}{-2.3026 \times 1.9873\ \text{cal mol}^{-1}\ K^{-1} \times 298.15\ K} = -11.7 \qquad [6.61]$$

or

$$K^{Brc} = 2.0 \times 10^{-12} \qquad [6.62]$$

If brucite is pure, so that there is no nonstandard state contribution to its Gibbs energy, $a_{Brc} = 1$. The concentrations of Mg^{2+} and OH^- are low, being near the infinite dilution standard state for the aqueous species. Under these conditions, the approximation of $\gamma = 1$ for the aqueous species can be made so that K^{Brc} in Equation 6.59 is given by

$$K^{Brc}_{sp} = m_{Mg^{2+}}\, m^2_{OH^-} \qquad [6.63]$$

The subscript sp indicates that K^{Brc}_{sp} is the molal *solubility product* of brucite. A molal solubility product is equal to the product of the species molalities raised to their stoichiometric coefficients in the solubility reaction. K_{sp} of a solid is equal to the equilibrium constant, K, when the activity of the solid is unity and activity coefficients of the aqueous species can also be taken as 1. At 25°C and 1 bar, K^{Brc}_{sp} is then $K^{Brc} = 2.0 \times 10^{-12}$. Often in the literature, the two terms are equated. This is reasonable for dilute aqueous solutions; however, for concentrated solutions such as seawater, activity is significantly different from molal concentration.

For brucite solubility in pure H_2O, the only significant magnesium species in solution is Mg^{2+}. Each mole of brucite dissolved per 1000 g of water releases

1 mole of Mg^{2+} and 2 moles of OH^- to solution. The moles of Mg^{2+} added to 1000 g of H_2O, therefore, equal the solubility of brucite, s, in terms of molality. The moles of OH^- per 1000 g of H_2O equal $2s$ and

$$K_{sp}^{Brc} = m_{Mg^{2+}} \, m_{OH^-}^2 = s \times (2s)^2 = 2.0 \times 10^{-12} \tag{6.64}$$

so that

$$4s^3 = 2.0 \times 10^{-12} \tag{6.65}$$

or the amount of brucite dissolved per kg of H_2O is $7.9 \times 10^{-5} \, m$. Now consider brucite solubility in a $0.1 \, m$ $MgCl_2$ solution, and assume that activity coefficients are unity. Before addition of the brucite, the solution has 0.1 mole of Mg^{2+}. The solubility product of brucite in this solution can be written as

$$K_{sp}^{Brc} = m_{Mg^{2+}} \, m_{OH^-}^2 = (0.1 + s) \, (2s)^2 = 2.0 \times 10^{-12} \tag{6.66}$$

resulting in

$$4s^3 + 0.2s^2 = 2.0 \times 10^{-12} \tag{6.67}$$

Because s is a small number, $4s^3$ is small compared with $0.2s^2$. It can be ignored, and s is calculated to be $3.2 \times 10^{-6} \, m$. This indicates a decrease in the solubility of brucite in a $MgCl_2$ solution over that in pure water. This is what is called the *common ion effect*. That is, the solubility of a mineral decreases relative to its value in pure H_2O in a solution where any ions that occur in the ion solubility reaction for the mineral are also present in solution. Remember, however, that whenever aqueous complexes are present, they tend to mitigate this effect.

Often, it is important to know whether a solution is saturated, supersaturated, or undersaturated with respect to a particular mineral. To make this determination an *ion activity product*, K_{iap}, is calculated given by the product of the measured molalities of the species in solution raised to their stoichiometric coefficients in the solubility reaction. If $K_{iap}/K_{sp} = 1$, the solution is saturated; if greater than 1, it is supersaturated, and if less than 1, it is undersaturated relative to equilibrium with the mineral.

■ Solubility of Carbonate Minerals

As an example of the solubility of a carbonate mineral, consider calcite in pure water. Because surface waters are typically in equilibrium with CO_2 in the atmosphere, it is appropriate to calculate the solubility for a fixed partial pressure of CO_2. In this case, $f_{CO_2} \sim P_{CO_2} = 10^{-3.5}$. From Equation 6.34, this implies

$$a_{CO_{2(aq)}} = 10^{-1.47} \, 10^{-3.5} = 10^{-4.97} \tag{6.68}$$

Equation 6.17 becomes

$$a_{H^+}a_{HCO_3^-} = 10^{-6.34} \, 10^{-4.97} = 10^{-11.31} \tag{6.69}$$

or

$$a_{HCO_3^-} = \frac{10^{-11.31}}{a_{H^+}} \tag{6.70}$$

Equation 6.20 can be written as

$$a_{CO_3^{2-}} = 10^{-10.33} \frac{a_{HCO_3^-}}{a_{H^+}} \tag{6.71}$$

or considering Equation 6.70

$$a_{CO_3^{2-}} = \frac{10^{-21.64}}{(a_{H^+})^2} \tag{6.72}$$

Equation 6.8 can be solved for a_{OH^-} to give

$$a_{OH^-} = \frac{10^{-14.0}}{a_{H^+}} \tag{6.73}$$

The solubility reaction for calcite is

$$CaCO_3 = Ca^{2+} + CO_3^{2-} \tag{6.74}$$

The equilibrium constant for Reaction 6.74 with calcite as the $CaCO_3$ phase, K_{Cal}, at 25°C and 1 bar from the SUPCRT92 data is computed to be

$$\log K_{Cal} = \frac{(-132,120 + -126,191) - (-269,880) \text{ cal mol}^{-1}}{-2.3026 \times 1.9873 \text{ cal mol}^{-1}\text{K}^{-1} \times 298.15 \text{ K}} = -8.48 \tag{6.75}$$

The equilibrium constant expression for Reaction 6.74 becomes

$$a_{Ca^{2+}} = \frac{10^{-8.48}}{a_{CO_3^{2-}}} \tag{6.76}$$

and substituting from Equation 6.72 gives

$$a_{Ca^{2+}} = 10^{13.16}(a_{H^+})^2 \tag{6.77}$$

Because of charge balance in solution, with calcite dissolution in pure H_2O

$$2m_{Ca^{2+}} + m_{H^+} = 2m_{CO_3^{2-}} + m_{HCO_3^-} + m_{OH^-} \qquad [6.78]$$

Assuming activity coefficients are unity so that the values in Equations 6.70, 6.72, 6.73, and 6.77 can be substituted into Equation 6.78 gives

$$2 \times 10^{13.16}(m_{H^+})^2 + m_{H^+} = 2 \times \frac{10^{-21.64}}{(m_{H^+})^2} + \frac{10^{-11.31}}{m_{H^+}} + \frac{10^{-14.0}}{m_{H^+}} \qquad [6.79]$$

After multiplying both sides by $m_{H^+}^2$ this can be rewritten as

$$10^{13.46} m_{H^+}^4 + m_{H^+}^3 - 10^{-11.31} m_{H^+} = 10^{-21.34} \qquad [6.80]$$

Solving this equation for m_{H^+} by noting that the second term on the left is unimportant results in $m_{H^+} = 10^{-8.26}$ or the solution $pH = 8.26$. Using this value from Equation 6.77, $m_{Ca^{2+}} = 10^{-3.36}$ or 4.4×10^{-4} $m_{Ca^{2+}}$ and from Equation 6.70, $10^{-3.05} = 8.9 \times 10^{-4}$ $m_{HCO_3^-}$. Calcite solubility in equilibrium with CO_2 in the atmosphere results in a concentration of dissolved CO_2 greater than Ca in solution.

■ Solubility of Silicate Minerals

Consider the solubility of silica polymorphs at earth surface conditions. As pointed out in Chapter 5 on mineral chemistry, there are a number of common SiO_2 polymorphs. The most soluble form of SiO_2 is amorphous silica (AmSi). Its solubility reaction can be written as

$$SiO_{2(AmSi)} = SiO_{2(aq)} \qquad [6.81]$$

The equilibrium constant expression at 25°C and 1 bar for Reaction 6.81 from SUPCRT92 data is

$$\log K_{(AmSi)} = \frac{(-199,190 \text{ cal mol}^{-1}) - (-202,892 \text{ cal mol}^{-1})}{-2.3026 \times 1.9873 \text{ cal mol}^{-1}\text{K}^{-1} \times 298.15 \text{ K}} = -2.71 \qquad [6.82]$$

The least soluble SiO_2 polymorph and, therefore, the thermodynamically most stable Si mineral at 25°C and 1 bar is quartz (Qtz) with a solubility reaction of

$$SiO_{2(Qtz)} = SiO_{2(aq)} \qquad [6.83]$$

and an equilibrium constant at 25°C and 1 bar of

$$\log K_{(Qtz)} = \frac{(-199,190 \text{ cal mol}^{-1}) - (-204,646 \text{ cal mol}^{-1})}{-2.3026 \times 1.9873 \text{ cal mol}^{-1}\text{K}^{-1} \times 298.15 \text{ K}} = -4.00 \qquad [6.84]$$

The neutrally charged Si aqueous species in solution, $SiO_{2(aq)}$, is often also written as $H_4SiO_4^0$ and called silicic acid. From a thermodynamic standpoint, these species are equivalent in pure H_2O as the reaction

$$SiO_{2(aq)} + 2H_2O = H_4SiO_4^0 \qquad [6.85]$$

is considered to have a $\Delta G_R^0 = 0.0$ at all pressures and temperatures. The molar Gibbs energy of $SiO_{2(aq)}$ and $H_4SiO_4^0$ at any pressure and temperature, therefore, differs only by the Gibbs energy of 2 moles of water molecules.

$H_4SiO_4^0$ is a very weak acid that dissociates in alkaline solutions, giving up H^+. The reaction can be written as

$$H_4SiO_4^0 = H_3SiO_4^- + H^+ \qquad [6.86]$$

with a molal equilibrium constant at 25°C and 1 bar of $K_{(6.86)} = 10^{-9.82}$ (Baes and Mesmer, 1976). In other words, the concentration of $H_3SiO_4^-$ increases with pH. At a pH of 9.82, there are equal activities of $H_4SiO_4^0$ and $H_3SiO_4^-$, and at higher pH, the concentration of $H_3SiO_4^-$ is greater than $H_4SiO_4^0$. The relationships are outlined in **Figure 6-7**.

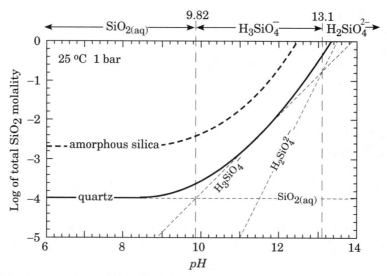

FIGURE 6-7 Solubility of quartz (thick solid line) and amorphous silica (thick dash line) as a function of pH at 25°C and 1 bar. The short dash lines show the contributions of the indicated species to the total quartz solubility. The long dash vertical lines divide the pH scale into regions where $SiO_{2(aq)}$, $H_3SiO_4^-$, and $H_2SiO_4^{2-}$ are the dominant Si aqueous species in the solution.

At even more alkaline conditions, $H_3SiO_4^-$ disassociates by the reaction

$$H_3SiO_4^- = H_2SiO_4^{2-} + H^+ \tag{6.87}$$

with a molal equilibrium constant at 25°C and 1 bar of $K_{(6.87)} = 10^{-13.10}$ (Baes and Mesmer, 1976). At a $pH = 13.1$, there is, therefore, an equal activity of $H_3SiO_4^-$ and $H_2SiO_4^{2-}$ in solution. Because most waters of interest have a $pH < 9$, where $H_4SiO_4^0$ (i.e., $SiO_{2(aq)}$) is dominant, this species is often considered to be the only Si species in solution; however, at very alkaline conditions, other silica species become stable, and these solutions in equilibrium with silicate minerals can have very high concentrations of dissolved Si.

■ Solubility of Al-Containing Minerals

Another important constituent of minerals in the earth's crust is Al. Al-rich phases are not very soluble in H_2O at 25°C. The stable Al phase at 25°C and 1 bar in aqueous solutions is gibbsite (Gbs), $Al(OH)_3$. The solubility of gibbsite is complicated by the fact that in equilibrium with H_2O there are five important aqueous Al species whose concentrations are pH dependent. These can be represented as Al^{3+}, $AlOH^{2+}$, $Al(OH)_2^+$, $Al(OH)_3^0$, and $Al(OH)_4^-$, although Al^{3+} likely exists in aqueous solutions dominantly as octahedral $Al(H_2O)_6^{3+}$.

To determine the solubility of gibbsite, the solubility reaction to each of these species can be summed. Generally, it is reasonable to assume gibbsite and H_2O are pure so their activities can be taken as unity. In this case, the solubility reactions and the logarithm of their equilibrium constants are

$$3H^+ + Al(OH)_{3(Gbs)} = Al^{3+} + 3H_2O \tag{6.88}$$

$$\log K_{(6.88)} = \log a_{Al^{3+}} - 3 \log a_{H^+}$$

$$= \frac{((-116,510) + (3 \times (-56,688)) - (-276,168) - (3 \times 0.0))\ \text{cal mol}^{-1}}{-2.3026 \times 1.9873\ \text{cal mol}^{-1}\text{K}^{-1} \times 298.15\ \text{K}} = 7.63 \tag{6.89}$$

$$2H^+ + Al(OH)_{3(Gbs)} = AlOH^{2+} + 2H_2O \tag{6.90}$$

$$\log K_{(6.90)} = \log a_{AlOH^{2+}} - 2 \log a_{H^+}$$

$$= \frac{((-166,425) + (2 \times (-56,688)) - (-276,168) - (2 \times 0.0))\ \text{cal mol}^{-1}}{-2.3026 \times 1.9873\ \text{cal mol}^{-1}\text{K}^{-1} \times 298.15\ \text{K}} = 2.66 \tag{6.91}$$

$$H^+ + Al(OH)_{3(Gbs)} = Al(OH)_2^+ + H_2O \qquad [6.92]$$

$$\log K_{(6.92)} = \log a_{Al(OH)_2^+} - \log a_{H^+}$$

$$= \frac{((-214{,}987) + (-56{,}688) - (-276{,}168) - (2 \times 0.0))\ \text{cal mol}^{-1}}{-2.3026 \times 1.9873\ \text{cal mol}^{-1}\text{K}^{-1} \times 298.15\ \text{K}} = -3.29 \qquad [6.93]$$

$$Al(OH)_{3(Gbs)} = Al(OH)_3^0 \qquad [6.94]$$

$$\log K_{(6.94)} = \log a_{Al(OH)_3^0}$$

$$= \frac{((-263{,}321) - (-276{,}168))\ \text{cal mol}^{-1}}{-2.3026 \times 1.9873\ \text{cal mol}^{-1}\text{K}^{-1} \times 298.15\ \text{K}} = -9.42 \qquad [6.95]$$

and

$$H_2O + Al(OH)_{3(Gbs)} = Al(OH)_4^- + H^+ \qquad [6.96]$$

$$\log K_{(6.96)} = \log a_{Al(OH)_4^-} + \log a_{H^+}$$

$$= \frac{((-312{,}087) + (0.0) - (-56{,}688) - (-276{,}168))\ \text{cal mol}^{-1}}{-2.3026 \times 1.9873\ \text{cal mol}^{-1}\text{K}^{-1} \times 298.15\ \text{K}} = -15.22 \qquad [6.97]$$

In **Figure 6-8**, plotted against *pH* are the log molalities of Al species in solution at 25°C and 1 bar calculated from Equations 6.88 to 6.97 assuming activity coefficients are unity. Unlike quartz solubility, gibbsite solubility increases both at

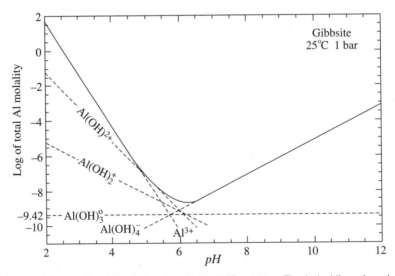

FIGURE 6-8 Solubility of gibbsite as a function of *pH* at 25°C and 1 bar. The dashed lines show the concentrations of the indicated Al aqueous species in solution. The solid line represents the sums of these contributions to produce the overall solubility of gibbsite.

low and high pH relative to near neutral pH. This occurs because positively and negatively charged Al species can be stable in solution. Gibbsite solubility is over two orders of magnitude lower than quartz solubility at near neutral pH but becomes greater than quartz solubility at pH below about 4 at 25°C and 1 bar.

The results for the Al_2O_3–H_2O and SiO_2–H_2O systems can be combined and mineral–fluid equilibria in the system SiO_2–Al_2O_3–H_2O considered. This system contains the possible mineral phases given in **Table 6-4** that are plotted on a triangular concentration diagram with apices of Al_2O_3, SiO_2, and H_2O in **Figure 6-9**. The phases that are not stable at 25°C and 1 bar are given in parentheses. The tie lines indicate that with H_2O present, equilibrium exists between gibbsite (Gbs), kaolinite (Kln), and H_2O or between quartz (Qtz), kaolinite (Kln), and H_2O at 25°C and 1 bar. Additionally, amorphous silica (AmSi) is sometimes present as a metastable phase with kaolinite and H_2O. The solubility reaction for kaolinite can be written as

$$Al_2Si_2O_5(OH)_4 + 6H^+ = 2Al^{3+} + 2SiO_{2(aq)} + 5H_2O \qquad [6.98]$$

with an equilibrium constant at 25°C and 1 bar from SUPCRT92 data of

$$\log K_{Kln} = 2 \log (a_{Al^{3+}} / a_{H^+}^3) + 2 \log a_{SiO_{2(aq)}} + 5 \log a_{H_2O} - a_{Kln}$$

$$= \frac{(2 \times (-116,510) + 2 \times (-199,190) + 5 \times (-56,688) - (-905,614) - 6 \times (0.0)) \, cal \, mol^{-1}}{-2.3026 \times 1.9873 \, cal \, mol^{-1}K^{-1} \times 298.15 \, K}$$

$$[6.99]$$

so that $\log K_{Kln} = 6.76$.

| Table 6-4 | **Names, abbreviations, and formulas for mineral phases in the system SiO_2–Al_2O_3–H_2O** |

Mineral	Abbreviation	Formula
Quartz	Qtz	SiO_2
Amorphous silica	AmSi	SiO_2
Corundum	Crn	Al_2O_3
Diaspore	Dsp	$AlO(OH)$
Boehmite	Bhm	$AlO(OH)$
Gibbsite	Gbs	$Al(OH)_3$
Andalusite	And	Al_2SiO_5
Kyanite	Ky	Al_2SiO_5
Sillimanite	Sil	Al_2SiO_5
Kaolinite	Kln	$Al_2Si_2O_5(OH)_4$
Pyrophyllite	Prl	$Al_2Si_4O_{10}(OH)_2$

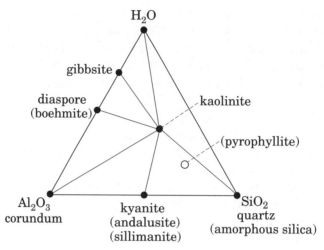

The solubility reactions for quartz, amorphous silica, and gibbsite as well as kaolinite can be used to consider the stability of these minerals in equilibrium with H_2O in the system Al_2O_3–SiO_2–H_2O. With the activities of the minerals and H_2O taken as unity so that a standard state of a pure phase at the pressure and temperature of interest is used for mineral and water, the set of equilibrium constant equations above can be cast in terms of the two variables, $\log (a_{Al^{3+}}/a_{H^+}^3)$ and $\log a_{SiO_{2(aq)}}$, as

(Gbs) $\log (a_{Al^{3+}}/(a_{H^+})^3) = 7.63$ [6.100]

(Kln) $\log (a_{Al^{3+}}/(a_{H^+})^3) + \log a_{SiO_{2(aq)}} = 6.76/2.0 = 3.38$ [6.101]

(Qtz) $\log a_{SiO_{2(aq)}} = -4.00$ [6.102]

and

(AmSi) $\log a_{SiO_{2(aq)}} = -2.71$ [6.103]

Figure 6-10 shows stability lines for equilibrium with the indicated phase in terms of $a_{SiO_{2(aq)}}$ and $a_{Al^{3+}}/(a_{H^+})^3$ as given by Equations 6.100 to 6.103. This type of diagram drawn at constant pressure and temperature is termed an *activity–activity diagram*. These lines then limit the range of values of the indicated variables for a particular phase. If the value for a solution plots at greater values than those given by the solution field, then it would be super-

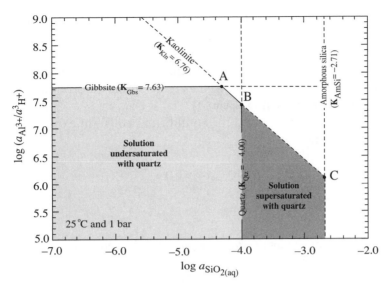

FIGURE 6-10 Log (activity of Al^{3+}/(activity of H^+)^3) versus the log (activity of $SiO_{2(aq)}$) at 25°C and 1 bar showing the solubility of gibbsite, kaolinite, and quartz as solid lines. The light dash lines are metastable extensions of the equilibrium. Also shown in dark gray is the solution undersaturated with metastable amorphous silica rather than quartz.

saturated relative to the indicated phase. As in Figure 6-9, the activity–activity diagram in Figure 6-10 shows that gibbsite and kaolinite can reach solubility equilibrium with the solution as given by point A or with kaolinite and quartz as given by point B. If a solution is supersaturated with quartz but maintains equilibrium with kaolinite and amorphous silica, it would have a composition given by point C.

Knowing that

$$\log (a_{Al^{3+}}/(a_{H^+})^3) = \log a_{Al^{3+}} + 3\, pH \qquad [6.104]$$

and because generally

$$\log a_{SiO_{2(aq)}} = \log m_{SiO_{2(aq)}} \qquad [6.105]$$

the compositions of solutions can be plotted on this diagram if the total molality of Al, Si, and pH can be measured and the pH is at or below 5 where Al^{3+} is the dominant species (Figure 6-8). If the pH is above 5, then the amount of Al in solution that is contributed by Al^{3+} must also be determined to plot its composition on Figure 6-10.

By adding the oxide component $KO_{0.5}$, the system expands to $AlO_{1.5}$–SiO_2–$KO_{0.5}$–H_2O. In this system, the phases K-feldspar (Kfs), $KAlSi_3O_8$, and muscovite

(Ms), $KAl_3Si_3O_{10}(OH)_2$, are present, as well as the phases already considered. The phase relations can be displayed in a triangular diagram if the value of one of the four components is fixed. If it is stipulated that all of the phases are in equilibrium with H_2O with $a_{H_2O} = 1$, then at 25°C and 1 bar, the phase relations that occur with quartz as well as meta-stable amorphous silica as the SiO_2 solid phase are shown in **Figure 6-11**.

The solubility reactions and the equilibrium constant expressions for the two additional minerals are

$$KAlSi_3O_8 + 4H^+ = Al^{3+} + K^+ + 3SiO_{2(aq)} + 2H_2O \qquad [6.106]$$

with

$$\log K_{(6.106)} = \log(a_{Al^{3+}}/a_{H^+}^3) + \log(a_{K^+}/a_{H^+}) + 3\log a_{SiO_{2(aq)}} + 2\log a_{H_2O} - \log a_{Kfs} \qquad [6.107]$$

and

$$KAl_3Si_3O_{10}(OH)_2 + 10H^+ = 3Al^{3+} + K^+ + 3SiO_{2(aq)} + 5H_2O \qquad [6.108]$$

with

$$\log K_{(6.106)} = 3\log(a_{Al^{3+}}/a_{H^+}^3) + \log(a_{K^+}/a_{H^+}) + 3\log a_{SiO_{2(aq)}} + 5\log a_{H_2O} - \log a_{Mu} \qquad [6.109]$$

Again, with the activity of the minerals a_{Kfs} and a_{Mu} as well as H_2O equal to 1, there are now three variables to consider: $\log(a_{K^+}/a_{H^+})$, $(\log(a_{Al^{3+}}/a_{H^+}^3))$, and log $a_{SiO_{2(aq)}}$. These can be portrayed in terms of two variables by "balancing" on the third. For instance, to consider the equilibrium between muscovite and K-feldspar, the solution can be considered to be saturated with both these phases so that both equilibrium constant expressions 6.107 and 6.109 hold. This re-

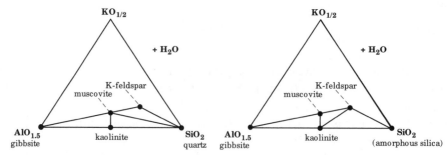

FIGURE 6-11 Triangular diagrams for the system $AlO_{1.5}$–SiO_2–$KO_{0.5}$–H_2O showing the relative stability of the phases with quartz (left side) and with metastable amorphous silica (right side) at 25°C and 1 bar.

sults in two equations and three unknowns. With $a_{H_2O} = a_{Kfs} = a_{Mu} = 1$, three times Equation 6.107 becomes

$$3\log K_{(6.106)} = 3\log(a_{Al^{3+}}/a_{H^+}^3) + 3\log(a_{K^+}/a_{H^+}) + 9\log a_{SiO_{2(aq)}} \qquad [6.110]$$

Subtracting Equation 6.109 from Equation 6.110 results in an expression that no longer contains the variable $\log(a_{Al^{3+}}/a_{H^+}^3)$ given as

$$3\log K_{(6.106)} - \log K_{(6.108)} = 2\log(a_{K^+}/a_{H^+}) + 6\log a_{SiO_{2(aq)}} \qquad [6.111]$$

This procedure has balanced on Al (i.e., $(a_{Al^{3+}}/a_{H^+}^3)$. Clearly, (a_{K^+}/a_{H^+}) or $a_{SiO_{2(aq)}}$ could be eliminated from the two equations rather than $(a_{Al^{3+}}/a_{H^+}^3)$ by balancing on K or SiO$_2$ instead. Typically, reactions between minerals are balanced on Al, which implies Al is conserved between the mineral phases. This is done because the typically low concentration of Al in most aqueous solutions indicates it is generally conserved in reactions between solid phases.

Equation 6.111 plots as a straight line of slope -3 as given by the line between point A and B in **Figure 6-12**. Boundaries with other phases can be determined in a similar manner. In this four-component system at constant

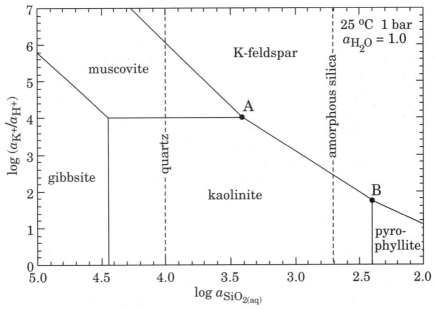

FIGURE 6-12 Stability of minerals in the system K$_2$O–Al$_2$O$_3$–SiO$_2$–H$_2$O at 25°C and 1 bar with $a_{H_2O} = 1.0$ in terms of the logarithm of the activity of K$^+$ over the activity of H$^+$ versus the logarithm of the activity of SiO$_{2(aq)}$ in solution.

temperature and pressure with the activity of H_2O fixed, a mineral is stable over a range of (a_{K^+}/a_{H^+}) and $a_{SiO_{2(aq)}}$; therefore, minerals that contain Al plot as fields on the diagram. The dashed lines labeled quartz and amorphous silica indicate where the solutions become saturated with these phases. They plot as vertical lines independent of $\log(a_{K^+}/a_{H^+})$ because these phases contain no K^+ or Al^{3+} and thus their stability does not depend on $\log(a_{Al^{3+}}/a_{H^+}^3)$ or $\log(a_{K^+}/a_{H^+})$. The presence of these solid SiO_2 phases, therefore, fixes the activity of $SiO2_{(aq)}$ in solution to the indicated value.

These activity–activity diagrams can be constructed for other systems and are useful in determining the extent of alteration in natural systems. These diagrams are a simplification of natural systems, and this needs to be kept in mind. For instance, a di-octrahedral illite is the stable phase in the natural system rather than muscovite. The properties of the illite are, however, similar to those of the muscovite phase shown. The importance of activity – activity diagrams stems from the fact that at earth surface conditions changes in temperature and pressure typically are small and therefore other thermodynamic variables control the stability of the minerals and fluids present. Their activities on the diagram can then be directly related to solution compositions measured in the field and the reaction path toward equilibrium determined.

■ Reaction Path Calculations

Besides determining the equilibrium distribution of species in a system that includes minerals, it is also important to understand how aqueous solution concentrations change with the extent of a reaction when minerals dissolve in or precipitate from a solution. This requires modeling a nonequilibrium system containing minerals as well as a fluid phase during the time frame it reacts before it comes to equilibrium. Many different kinds of reaction can occur in this system. These include reactions occurring solely in the fluid phase, reactions solely in the solid phase, and those that occur at the interface between the minerals and the fluid phase. Consider first reactions in the fluid phase. These are reactions between aqueous species such as

$$H_2O + Al^{3+} = Al(OH)^{2+} + H^+ \qquad [6.112]$$

They typically come to equilibrium in a matter of seconds or less, even at 25°C. Fluids can, therefore, be considered to be in local internal homogeneous equilibrium. This, however, does not preclude concentration gradients in the fluid with distance. As a function of time, however, these gradients are being reduced by diffusion and flow processes. As a reasonable approximation, in most cases, fluids in sediments and rocks can be considered to have uniform concentration between enclosing grains over cm distances on time frames of a year.

This is not true of concentration gradients, however, that occur in mineral grains. Diffusion in solid phases is typically very slow except at magmatic temperatures (>800°C), so gradients are preserved. For instance, compositional zoning in a plagioclase produced when it crystallizes from a magma can be preserved for millions of years at earth surface conditions. Reactions in solids are, therefore, typically very slow relative to the time frames of interest at earth surface conditions.

Reactions at the interface between minerals and aqueous solutions occupy a middle time frame. Minerals can dissolve or precipitate over time frames of years to thousands of years at 25°C; however, in deep crustal rocks where temperatures are significantly higher, surface reactions become more rapid than solute diffusion and flow. Under these conditions, solute diffusion and flow occur at the slowest rate. (This situation is discussed in Chapter 9, where metamorphic reactions are considered.) At the earth's surface, however, to help understand reactions among minerals and fluid, they can be modeled assuming diffusion in the solids is so slow that it is not important and assuming the fluid phase reactions are so fast that they maintain equilibrium no matter how rapidly material is added. The rate, therefore, of dissolution or precipitation of mineral phases at their surfaces is rate controlling. These reactions are most often of greatest interest when considering the approach toward equilibrium between minerals and solutions at the earth's surface.

The model system that is typically used is displayed in **Figure 6-13**. Consider a fixed concentration of components in an aqueous fluid. Given a sufficient number of constraints, the equilibrium aqueous species distribution in this fluid can be calculated. These constraints are the pressure, temperature, concentration of elements, buffered activities of aqueous species (e.g., pH and solubility constraints), and the equilibrium constants for reactions between aqueous species as given by reactions like Equation 6.112.

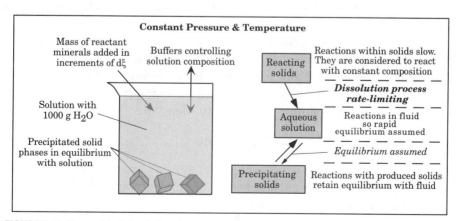

FIGURE 6-13 The model of dissolution of reactant minerals and precipitation of secondary minerals in a solution as a function of $d\xi$ that is discussed in the text.

Consider a solution of interest with 1000 g of H_2O. The first consideration is whether the solution itself is stable or will mineral phases precipitate out of it with time? Whether this initial solution is supersaturated with any mineral phases can be determined. If it is, there are two possible choices for the calculation. The precipitation of the minerals can be suppressed so that they are not allowed to precipitate out of solution. For instance, the precipitation of quartz can be suppressed to allow the solution to reach concentrations of Si as high as saturation with amorphous silica. The second possibility is to remove enough solutes from the solution so that it is no longer supersaturated with any minerals. That is, material is subtracted from the solution to precipitate enough of the supersaturated mineral to produce a solution that is saturated. In either case, after the problems with supersaturation are handled, a distribution of speciation calculation of the initial solution is done.

To this initial solution, a small amount, stipulated as $d\xi$, of reactant mineral is dissolved. That is, the mass of the reactant mineral components of the amount $d\xi$ is added to the solution, and with its new composition, a new distribution of species calculation is done. Any phases that saturate with this added material can be determined and their mass subtracted from the solution to bring it into equilibrium with the supersaturated phases. This procedure is repeated adding (titrating) small masses of reactant of $d\xi$ to the solution. The progress of the reaction is then characterized in terms of $d\xi$, with the amounts of precipitated minerals and changes in solution composition as a function of $d\xi$ determined. Eventually, as the concentration of species in solution increases with the addition of reactant mineral material, saturation with the reactant minerals occurs, and the entire system comes to equilibrium. When this occurs, the reactant material of $d\xi$ added to solution precipitates out the same amount of reactant mineral, leaving the solution composition the same. In some codes (e.g., PHREEQE), the opposite approach is taken, and you have to explicitly designate what phases should precipitate rather than suppressing those that should not appear.

As an example, consider K-feldspar dissolution at 25°C and 1 bar (**Figure 6-14**). What happens when increments of K-feldspar of $d\xi$ are added to pure H_2O? As determined by looking at the dashed lines in Figure 6-8 for the solubility of gibbsite, the dominant Al species produced at neutral *pH* would be $Al(OH)_4^-$. The initial feldspar dissolution reaction can, therefore, be written as

$$2H_2O + KAlSi_3O_8 \rightarrow K^+ + Al(OH)_4^- + 3SiO_{2(aq)} \qquad [6.113]$$
$$Ksp$$

The $Al(OH)_4^-$ produced would react with the small amount of H^+ in pure water according to the reaction

$$Al(OH)_4^- + H^+ \rightarrow Al(OH)_3^o + H_2O \qquad [6.114]$$

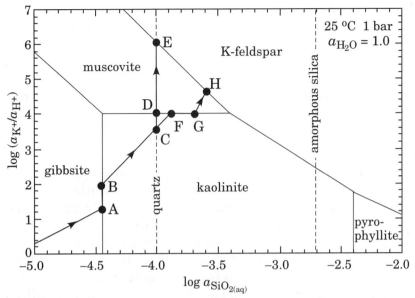

FIGURE 6-14 Log activity diagram at 25°C and 1 bar in the system $K_2O–Al_2O_3–SiO_2–H_2O$. Shown are reaction paths of K-feldspar dissolution in distilled water. One path precipitates quartz at its saturation, and in the other, the solution becomes supersaturated with quartz before the solution reaches equilibrium with K-feldspar.

This would make the solution slightly alkaline. As the feldspar dissolves increasing $d\xi$, it contributes Al to solution. Because of the low solubility of Al in solution at some $d\xi$, the solution would saturate with gibbsite. The dissolution reaction then becomes

$$KAlSi_3O_8 + 2H_2O \rightarrow Al(OH)_3 + K^+ + OH^- + 3SiO_{2(aq)} \qquad [6.115]$$
$$\text{Ksp} \qquad\qquad \text{Gbs}$$

The production of OH^-, K^+, and $SiO_{2(aq)}$ would cause the solution to move along the path shown with an arrow toward A in Figure 6-14. At point A, the $SiO_{2(aq)}$ released from the feldspar would react with the gibbsite to form kaolinite according to the reaction

$$2Al(OH)_3 + 2SiO_{2(aq)} \rightarrow Al_2Si_2O_5(OH)_4 + H_2O \qquad [6.116]$$
$$\text{Gbs} \qquad\qquad \text{Kln}$$

Reaction 6.116 buffers the concentration of $SiO_{2(aq)}$ in solution. With further dissolution, the K^+ added to solution moves the composition from A to B. At this point, with increased addition of $SiO_{2(aq)}$ to solution from the feldspar, all of the $Al(OH)_3$ produced by Reaction 6.115 will have reacted by Reaction 6.116 going to the right. The concentration of $SiO_{2(aq)}$ is no longer buffered in solution by this equilibrium.

The continued dissolution of feldspar increases $SiO_{2(aq)}$ in solution as well as K^+ as the solution changes composition along the path from B to C. At point C, the solution becomes saturated with quartz, and quartz starts to precipitate. Continued feldspar dissolution changes the solution composition along the line from C to D. At D, the solution has a high enough concentration of K^+ that kaolinite reacts with the solution and precipitates muscovite by the reaction

$$3Al_2Si_2O_5(OH)_4 + 2K^+ \rightarrow 2KAl_3Si_3O_{10}(OH)_2 + 2H^+ + 3H_2O \qquad [6.117]$$
$$\text{Kln} \qquad\qquad\qquad\qquad \text{Mus}$$

When the kaolinite is exhausted, the solution composition changes along the path from D to E where it reaches equilibrium with K-feldspar. At this point, no further reaction takes place, and the final assemblage is K-feldspar, muscovite, and quartz with the solution species activities given by point E.

If quartz did not precipitate (a common situation at 25°C), then the solution would continue past C until composition F is reached. At this point, Reaction 6.117 occurs because K^+ is supplied from the dissolving feldspar until the kaolinite is depleted at point G. The solution is no longer buffered to a constant $\log(a_{K^+}/a_{H^+})$ by the presence of both kaolinite and muscovite and changes composition to point H where it comes to equilibrium with K-feldspar. At point H, the final assemblage contains K-feldspar and muscovite along with a fluid with the indicated activities of species in solution.

Computer codes have been written to do the calculations described here and can be used to analyze more complicated situations. A number of these are presented in **Table 6-5**. Each of these has its own set of thermodynamic properties of minerals, gases, and solution species, as well as models for handling the changes in energetics of solid solutions and activity coefficients in aqueous solutions.

Table 6-5   **Mass transfer software**

EQ3/6 written and maintained by Tom Wolery (1992) at Lawrence Livermore Nat. Lab. A Software Package for Geochemical Modeling of Aqueous Systems: Package Overview and Installation Guide. Version 7.0. http://geosciences.llnl.gov/esd/geochem/EQ3manuals/eq36pkg.pdf.
SOLVEQ written and maintained by Mark Reed (Univ. Oregon); see Reed, M. H. (1998). Calculation of simultaneous chemical equilibria in aqueous-mineral-gas systems and its application to modeling hydrothermal processes. In: Richards, J. and Larson, P. (eds.) Techniques in Hydrothermal Ore Deposits Geology. Reviews in Econ. Geol., v. 10, pp. 109–124.
The Geochemist's Workbench®. Release 4. A commercial product offered by Craig Bethke. See Bethke, C. M. (1996). Geochemical Reaction Modeling, Concepts and Applications. Oxford Univ. Press, 397 pp.

Summary

Because of its structure with O–H bonds at 105°, the water molecule can be modeled as an electric dipole. The dipole properties of water stabilize charged species, that is, ions in solution. The concentration of ions depends on the solution's pH. Water undergoes some self-ionization, which at 25°C and 1 bar is given by $m_{H^+} m_{OH^-} = 10^{-14}$; therefore, because the activity coefficient of H^+ in pure H_2O is unity, $a_{H^+} = 10^{-7}$ or $pH = 7$. Solutions in which $a_{H^+} > 10^{-7}$ are termed acids, and when $a_{H^+} < 10^{-7}$, solutions are alkaline. Acids are substances that produce H^+ when added to water. They are considered strong if they disassociate almost completely in water and weak if a significant amount of the acid remains un-ionized in solution. Carbonic acid, $CO_{2(aq)}$, is the most important weak acid on the earth. It disassociates to produce bicarbonate, HCO_3^-, and carbonate, CO_3^{2-}.

To determine the distribution of species in a complicated solution, a large set of simultaneous equations need to be solved. These include equilibrium constant expressions for the aqueous species in solution, activity coefficient expression, and element mass and charge balance. One of the problems is that significant aqueous complexes can form in concentrated solutions, and their equilibrium constant expressions need to be evaluated.

Some solutions, including alkaline seawater, act as pH buffers to keep the pH nearly constant when acids or bases are added to the solution. pH buffering occurs when a weak acid and the salt of the weak acid are present. The total alkalinity of a solution gives the ability of a solution to neutralize H^+. It can be determined by titrating the solution with acid.

Solution compositions are often controlled by mineral solubility reactions. Solubility products of minerals are given by the product of molalities of all of the species in the solubility reaction raised to the power of their coefficients in the reaction. It can be shown that if a reaction species already exists in a solution, the solubility of the mineral is lower than if the mineral dissolves in pure water. This is called the common ion effect.

The last part of the chapter considers the solubility of minerals in aqueous solutions, demonstrating that SiO_2 polymorph solubility at 25°C and 1 bar is independent of pH at $pH < 8$. Gibbsite, $Al(OH)_3$, in contrast, has high solubility at both high and low pH relative to $pH \sim 6.5$. Activity–activity diagrams can be constructed to consider relationships between fluids and minerals in systems where temperature and pressure are less important. These diagrams can be used to map paths of dissolving reactant minerals in solution and calculate the extent of changes in solution composition and the precipitation of any secondary minerals as the solution comes to equilibrium with the minerals that were reacted.

Some basic concepts necessary to understand chemical equilibrium in aqueous solutions have been introduced in this chapter. With these concepts in mind, the controls on the chemistry of natural waters discussed in the next chapter can be understood.

Key Terms Introduced

acid
activity–activity diagram
alkaline solutions
alkalinity
bicarbonate
carbonate
carbonate alkalinity
carbonate ion
carbonic acid
common ion effect
complexes
conservative
dielectric constant

electric dipole
electrical moment
ion activity product
normality
permittivity
pH
pH buffer
polarizability
quadrupole
solubility product
strong acid
total alkalinity
weak acid

Questions

1. What is pH? A dipole? An alkaline solution?
2. Why are charged species typically more stable than uncharged species in water?
3. Why is neutral pH equal to 7 at earth surface conditions?
4. What makes an acid strong as opposed to weak? Give an example of each.
5. How does carbonate differ from bicarbonate?
6. What is an aqueous complex?
7. Why is the pH of ocean water alkaline when the water added from rivers is acidic?
8. What is a pH buffer? Carbonate alkalinity?
9. How is the solubility of a mineral added to a solution containing one of the constituents of the mineral different than the mineral's solubility in pure H_2O?
10. Why does solid amorphous silica have a greater solubility in solution than quartz?
11. Why does gibbsite solubility increase at both acid and alkaline pH relative to near neutral pH, whereas quartz solubility only increases at alkaline pH?

12. Explain how an activity-activity diagram at constant pressure and temperature is constructed.
13. Explain how the solution changes as a function of time when K-feldspar is dissolved in H_2O. Describe the equilibrium state.

Problems

1. What is the pH of a solution if the $OH^- = 1.59 \times 10^{-11}$ m?
2. (a) Calculate the pH of a 25°C and 1 bar solution that contains 3×10^{-5} m of hydroxyl ion. (b) If a solution has a $pH = -0.1$ what is the molality of hydrogen ion in solution?
3. As a strong base, what is the pH of a 0.03 m NaOH solution?
4. At standard conditions, calculate the pH of a solution of 0.00010 mol of Si in 1 kg of pure water.
5. Give the equilibrium constant expression and calculate the equilibrium constants at 25°C and 1 bar for the following reaction with both disordered and ordered dolomite from the values in the back of the book.

$$CaMg(CO_3)_2 + Ca^{2+} = 2CaCO_3 + Mg^{2+}$$
$$\quad\text{dolomite} \qquad\qquad\qquad \text{calcite}$$

   If most ground waters have greater concentrations of Ca^{2+} than Mg^{2+}, is disordered or ordered dolomite more stable in solution?
6. Using the values in Appendix F5, calculate the disassociation constant of acetic acid (CH_3COOH) at standard conditions. What is the pH of a $10^{-1.25}$ m solution of acetic acid?
7. The concentration of a saturated solution of $BaSO_4$ is 3.90×10^{-5} m. Calculate the solubility product (K_{sp}) for barium sulfate at 25°C.
8. Using the equilibrium constant in Equation 6.12 and the Gibbs energies of the ions in the back of the book, calculate the standard state Gibbs energy of HCl at 25°C.
9. Calculate the H^+ concentration and the pH of a solution of 0.0001 m total CO_2 at 25°C, and calculate also what fraction of $CO_{2(aq)}$ has dissociated. You can assume no appreciable CO_3^{2-} and activity coefficients as well as the activity of H_2O are unity.
10. What is the pH of pure water in equilibrium with atmospheric P_{CO_2} ($10^{-3.5}$ atm) at 25°C? Assume activity coefficients and the activity of H_2O are unity. Because the solution is acidic, OH^- and CO_3^{2-} concentrations are significantly below others in solution.
11. The following waters were sampled from dunites and peridotites (Barnes, I. and O'Neil, J. R., 1969.)

Sample	Burro Mtn Monterey Co, CA	Cazadero A Sonoma Co, CA	John Day Warm Spring Grant Co, OR	Adobe Canyon Stanislaus Co, CA	Blackbird Valley Santa Clara Co, CA
pH	11.54	11.77	11.25	11.78	12.01
$T\,°C$	20	18	31	15.6	10.6
Ca^{+2} (total) $mg\ 1^{-1}$	40	53	35	48	51
Mg^{+2} (total) $mg\ l^{-1}$	0.3	0.3	0.1	0.4	0.06

a. What is the saturation state of the water sampled at Burro Mtn relative to brucite? (At 20°C the log $(a_{Mg^{2+}}/a_{H^+}^2) = 16.63$ in equilibrium with brucite.) Assume activity coefficients are unity. If this water is in equilibrium with pure brucite, calculate the percent of Mg in solution that would be Mg^{2+}.

b. Calculate the log $(m_{Ca^{2+}}/a_{H^+}^2)$ for each of the waters above and plot them as a function of temperature to estimate a 25°C value of log $(m_{Ca^{2+}}/a_{H^+}^2)$.

c. Use the Gibbs energy values in the back of the book (or SUPCRT92) to calculate log $(a_{Ca^{2+}}/a_{H^+}^2)$ in equilibrium with pure diopside, antigorite, and tremolite.

d. If all of the phases are pure except diopside, what is its activity so that log $(m_{Ca^{2+}}/a_{H^+}^2)$ in the water sample is equal to the calculated log $(a_{Ca^{2+}}/a_{H^+}^2)$ for this assemblage.

12. The solubility of amorphous silica in pure water is about 120 ppm of SiO_2 at 25°C. The solution contains silicic acid, H_4SiO_4, whose first dissociation constant is $10^{-9.9}$. What is the *pH* of a solution saturated with amorphous silica? Assume that activity coefficients and the activity of H_2O are unity.

13. Calculate the *pH* of an aqueous solution at 25°C that contains 1.0 m acetic acid (CH_3COOH) if the acetic acid is 5% ionized. Assume that activity coefficients are unity.

14. It has been proposed to sequester carbon dioxide emitted from fossil-fuel power plants in the ocean. One method is to take the exhaust gas, which typically has partial pressures of CO_2 of about 0.15 bar and react it with seawater + calcite before adding it to the ocean. Using the procedure outlined in the book, calculate

a. *pH* of this solution

b. Amount of calcite dissolved

c. Total moles of CO_2 in solution per mole of $CaCO_3$ dissolved

For the calculations, assume standard conditions and that the solution is initially pure water with unit activity coefficients. Compare with values for seawater given in Caldeira, K. and Rau, G. H. (2000).

Problems Making Use of SUPCRT92

15. What is neutral pH at
   a. 5°C and 1 bar?
   b. 95°C and 1 bar?

16. Construct a pH versus temperature diagram between 0°C and 350°C at H_2O saturation showing the boundary where $CO_{2(aq)}$ HCO_3^- and CO_3^{2-} are the dominant species. Assume the activity coefficients of the species are unity. Also, plot neutral pH as a function of temperature on the diagram. Which species becomes dominant at neutral pH as temperature increases?

References

Baes, C. F. Jr. and Mesmer, R. E., 1976, *The Hydrolysis of Cations*. Wiley-Interscience, New York, 489 pp.

Barnes, I. and O'Neil, J. R., 1969, The relationship between fluids on some fresh alpine-type ultramafics and possible modern serpentinization, Western United States. *Geol. Soc. Amer. Bull.*, v. 80, pp. 1947–1960.

Caldeira, K. and Rau, G. H., 2000, Accelerating carbon dissolution to sequester carbon dioxide in the ocean: Geochemical implications. *Geophysics Research Letters*, v. 27, pp. 225–228.

Drever, J. I., 1988, *The Geochemistry of Natural Waters*, 2nd edition. Prentice Hall, Englewood Cliffs, 437 pp.

Stumm, W. and Morgan, J. J., 1981, *Aquatic Chemistry: An Introduction Emphasizing Chemical Equilibria in Natural Waters*. John Wiley & Sons, New York, 780 pp.

Zeebe, R. E. and Wolf-Gladrow, D., 2001, *CO₂ in Seawater: Equilibrium, Kinetics, Isotopes*. Elsevier Oceanography Series, v. 65, Elsevier Science B. V., 346 pp.

7 Chemistry of Natural Waters

What controls the chemistry of natural waters? It is important to realize that water is in a constant state of reaction with rocks as it moves from one environment to another. To characterize the change in composition with this movement of water between environments, a model of reservoirs and fluxes of water into and out of the reservoirs can be constructed. A *flux* is the amount of energy or mass that flows from one place to another in unit time (e.g., kg/yr).

Flux **x** → | Reservoir A | Flux **y** → | Reservoir B | Flux **z** →

■ Hydrologic Cycle

If the earth, as a whole, is considered, this transfer of water is referred to as the *hydrologic cycle*. Water is not lost or gained by the entire earth but merely cycles between the reservoirs. When the flux of water into a reservoir is equal to the flux out, the reservoir is in *steady state*, and the amount of water remains constant with time. This steady state should not be confused with chemical equilibrium because the amount in the reservoir changes if the fluxes in or out change. A *non–steady-state* condition occurs when the fluxes in and out of a reservoir are different. In this case, the size of the reservoir will change with time. The time water spends in a reservoir, whether a reservoir is in steady state or not, depends on the total fluxes in and out. When steady state is not achieved, the amount of component i in the reservoir, n_i, at any instant in time is given by

$$n_i = n_i^o + \int_0^t \left(\frac{dn_i^{in}}{dt} - \frac{dn_i^{out}}{dt} \right) dt \tag{7.1}$$

where n_i^o is the amount of component i in the reservoir at time $t = 0$, whereas dn_i^{in}/dt and dn_i^{out}/dt are the infinitesimal flux of i into and out of the reservoir, respectively.

Consider H_2O in the earth's ice reservoir. This is found mainly in ice caps at the earth's north and south poles and in alpine glaciers. Ice is produced each year as snow falls on the ice pack and melts in the warmer climates near its edges. Over the last 100,000 years it is clear that the amount of H_2O in the ice caps + glaciers reservoir has changed significantly as continental ice sheets have waxed and waned between the ice ages with glacial advances and retreats during interglacial periods. With increased melting of the ice sheets in the last 20,000 years, the flux of H_2O out of the ice reservoir has been greater than that fluxing in. This increases the flux of water into the ocean reservoir, leading to rising sea levels and, therefore, increasing the amount of water in the ocean reservoir. On a shorter time scale, however, for example, over a decade, the reservoirs can be considered to be in steady state. In this case, the flux of water falling as snow on ice caps and glaciers can be considered to be equal to the amount of ice that melts and *sublimates* back to the atmosphere. With a time scale shortened to a month or less, it is clear that the ice reservoir would not be in steady state as ice growth and melting changes with time because of yearly seasonal changes. These changes, however, balance each other out to some extent between the northern and southern hemispheres.

Whether a reservoir is in steady state or not, therefore, depends on the time frame. **Figure 7-1** is a simplified present-day steady-state hydrologic cycle for the earth in terms of yearly fluxes. Reservoirs of H_2O are displayed as boxes with values given in units of Eg, and the fluxes between reservoirs are shown with arrows and given in units of Eg yr^{-1}. Eg stands for exagram = 10^{18} g. An important aspect of the hydrologic cycle is the large size difference between the ocean and atmospheric reservoirs. Figure 7-1 also indicates that most fresh water is tied up in ice rather than groundwater. A net transfer of water from evaporation of the ocean to precipitation on land is returned to the ocean by rivers with a smaller flux contributed by groundwater.

The residence time, t_R, of a reservoir is the average amount of time a substance spends in the reservoir and at steady state is calculated from

$$t_R \text{(in years)} = \frac{\text{amount in reservoir (g)}}{\text{flux into or out of reservoir (g yr}^{-1})} \qquad [7.2]$$

The greater the residence time, the greater is the time frame over which a reservoir can be considered to be in steady state. t_R for H_2O in the atmosphere (atm) from the fluxes given in Figure 7-1 is

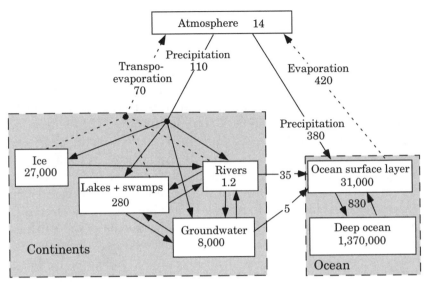

FIGURE 7-1 The yearly hydrologic cycle with boxes denoting reservoirs and arrows the fluxes between reservoirs. The dashed arrows give fluxes into the atmosphere. Reservoirs are in units of Eg and the fluxes shown are in Eg yr^{-1} of H$_2$O. (Modified after O'Neill, 1993.)

$$t_R(\text{in years}) = \frac{14 \text{ Eg}}{(70 + 420) \text{ Eg yr}^{-1}} = 0.029 \text{ yr} = 10.6 \text{ days} \qquad [7.3]$$

t_R for H$_2$O in the ocean is about 3300 years. The hydrologic cycle and similar geochemical cycles for other components are used to consider the time scales for changes in water chemistry, including contamination effects. If a contaminant is introduced with the flux of water into the atmosphere, its concentration could change rapidly because of its short residence time in the atmosphere. Introducing one of the major components of seawater into the oceans with residence times greater than 10^6 years would cause ocean concentrations to change much more slowly. To understand the changes in the composition of a reservoir with time, it is important to assess the controls on the changes in composition of components in each of the reservoirs. Many good references on geochemical cycles are available (e.g., Garrels et al., 1975; Mackenzie, 1998).

■ Rainwater

Consider the flux of water in rain out of the atmospheric reservoir. Rainwater is not pure H$_2$O because it reacts with the composition of air through which it falls (gasses, dust particles, and soot). The major gas species in dry air are given in **Table 7-1**. Except for CO_2, the gas species are unreactive with H_2O. With CO_2 in normal moist air, however, the following reaction occurs

$$CO_{2(gas)} = CO_{2(aq)} \tag{7.4}$$

where $CO_{2(aq)}$ denotes the carbonic acid species in water as introduced in Chapter 6.

Using the SUPCRT92 data and remembering that $\log K = -\Delta G_R^o/2.3026RT$, an equilibrium constant for Reaction 7.4 at 25°C and 1 bar is computed to be

$$\log K_{CO_{2(gas)}} = \frac{(-92,250) - (-94,254) \text{ cal mol}^{-1}}{-2.3026 \times 1.9873 \text{ cal mol}^{-1}\text{K}^{-1} \times 298.15 \text{ K}} = -1.47 \tag{7.5}$$

so that

$$K_{CO_{2(gas)}} = \frac{a_{CO_{2(aq)}}}{f_{CO_2}} = 10^{-1.47} \tag{7.6}$$

where the standard state for $CO_{2(aq)}$ is taken as a 1 molal solution with the properties of the species at infinite dilution at 25°C and 1 bar, and for the fugacity of CO_2, f_{CO_2}, the standard state is the pure ideal gas at 25°C and 1 bar. To analyze the effects of CO_2 gas adsorption in water, its partial pressure from Table 7-1 of 377 ppm/10^6 ppm per atm = $10^{-3.4}$ bar can be taken as its fugacity, f_{CO_2}. Additionally, the activity coefficient of neutrally charged $CO_{2(aq)}$ can be considered to be unity; therefore, from Equation 7.6

$$m_{CO_{2(aq)}} = 10^{-3.4}\, 10^{-1.47} = 10^{-4.87} \tag{7.7}$$

Some of the carbonic acid, $CO_{2(aq)}$, produced undergoes disassociation in water to HCO_3^- and H^+. Thus, the water in uncontaminated air has an acid pH because of its reaction with CO_2 in the atmosphere. For each H^+ produced, an

| Table 7-1 | Composition of dry air |

Species	Concentration (ppm by volume)
N	780,900
O_2	209,400
Ar	9,300
CO_2	377
Ne	18
He	5.2

HCO_3^- is also produced. The starting concentration of H^+ in pure H_2O ($pH = 7$) is only 10^{-7} m. For the CO_2-containing solution, because of charge balance, the approximation

$$m_{HCO_3^-} = m_{H^+} \qquad [7.8]$$

is, therefore, reasonable. Remember from Chapter 6 that at 25°C and 1 bar the equilibrium constant of the carbonic acid dissociation reaction is

$$K_{CO_{2(aq)}} = \frac{a_{HCO_3^-} \, a_{H^+}}{a_{CO_{2(aq)}} \, a_{H_2O}} = 10^{-6.34} \qquad [7.9]$$

Assuming unit activity coefficients of the aqueous species and substituting the value for $m_{CO_{2(aq)}}$ from Equation 7.7 and $m_{HCO_3^-}$ from Equation 7.8 into Equation 7.9 gives

$$(a_{H^+})^2 = 10^{-6.34} \, 10^{-4.87} \quad \text{or} \quad a_{H^+} = 10^{-5.6} \qquad [7.10]$$

Thus, pure rainwater in equilibrium with the atmosphere is computed to have a pH of 5.6 at 25°C and 1 bar.

Rainwater can also contain significant amounts of NaCl and $MgSO_4$ derived from the incorporation of oceanic sea salt spray as it falls through the atmosphere. The presence of Al indicates that wind-blown mineral dust particles have also been incorporated on the rain's descent through the atmosphere. The dissolution of these dust particles also contributes the cations K^+, Ca^{2+}, Na^+, and Mg^{2+} to rainwater. Typical concentrations of major ions in rainwater are given in **Table 7-2**. Note the higher Na^+ and Cl^- in marine and coastal rain

Table 7-2	Composition of rainwater uncontaminated by humans in ppm by weight

Ion	Continental rain	Marine and coastal rain
H^+	pH = 4–6	pH = 5–6
Na^+	0.2–1.0	1.0–5.0
Mg^{2+}	0.05–0.5	0.4–1.5
K^+	0.02–0.3	0.2–0.6
Ca^{2+}	0.02–3.0	0.2–1.5
NH_4^+	0.1–0.5	0.01–0.05
Cl^-	0.2–2.0	1.0–10
SO_4^{2-}	1.0–3.0	1.0–3.0
NO_3^-	0.4–1.3	0.1–0.5

From Berner and Berner (1996).

from the contribution of sea salt spray and increased K^+ in continental rain from mineral dust. The greater NH_4^+ and NO_3^- in continental rain are due to the presence of concentrations of organic compounds in air above vegetated continental land masses.

■ River Water

After the somewhat acidic rain reaches the land surface, it can react with mineral and organic matter that it encounters on its way through the soil and upper rock units before it is incorporated into a groundwater reservoir. Some of this groundwater can enter stream channels where it becomes river water. Other water derived from rain runs off the land surface directly into streams and rivers.

Water in rivers has a wide range of compositions because of the variety of reactions that occur with the minerals and organic matter on the land surface it contacts before it enters the river as well as reactions that occur during river-water transport. Rivers transport both dissolved and particulate material. The particulate material can be carried as a *suspended load* of fine particles of material or as a *bed load* of coarser material moving along the river bottom. Most rivers carry a greater amount of material in the suspended load than in either their bed load or dissolved load. The bed load can change with time and from place to place in a river and makes up to 50% of the total particulate load; however, if a stream is flowing in a channel of consolidated bedrock, as in deep gorges with steep gradients, the bed load is likely to be low.

During flooding, both suspended and bed loads typically increase dramatically, but with greater increases in the flux of particulate matter from the suspended load as larger particles are suspended. The amount of particulate load depends on the nature of rocks in the river basin and increases with mean continental elevation. **Table 7-3** lists estimates of the annual dissolved and particulate load for the world's 10 largest rivers. Note both the large dissolved and particulate load carried by the Amazon River relative to other rivers. Clearly, it dominates all other rivers in terms of its ability to transport material. The Zaire River has a substantial flow, but because of its low elevation and crystalline bedrock, it carries a relatively small amount of material.

Table 7-4 shows the average major elemental concentrations of particulate and dissolved loads in the world's rivers that do not occur in a gas species (e.g., O_2, H_2O, CO_2, and H_2S). The particulate load is where most of the less soluble elements of Al and Fe and most of the Si are transported. The more soluble Na and Ca are transported to a greater extent in the dissolved load. The dissolution of Na and Ca from minerals leads to their depletion in soils relative to surficial rocks, as indicated in Table 7-4. P is retained in the soil along with Si and is carried primarily in the particulate load.

Table 7-3	Annual discharge, dissolved, and particulate loads of the world's 10 largest rivers

River	Location	Water discharge (km^3/yr)	Dissolved load (10^{12}g/yr)	Particulate load (10^{12}g/yr)	Dissolved/ particulate (ratio)
Amazon	S. America	6300	275	1200	0.23
Zaire (Congo)	Africa	1250	41	43	0.95
Orinoco	S. America	1100	32	150	0.21
Yangtze	Asia	900	247	478	0.53
Brahmaputra	Asia	603	61	540	0.11
Mississippi	N. America	580	125	210	0.60
Yenisei	Asia	560	68	13	5.2
Lena	Asia	525	49	18	2.7
Mekong	Asia	470	57	160	0.36
Ganges	Asia	450	75	520	0.14

From Berner and Berner (1996).

The suspended load along with the bed load is deposited in deltas, estuaries, and along coastlines. Small particles in the suspended load typically *flocculate* and aggregate or are agglomerated by organisms in fecal material. The dissolved load fluxes into the ocean. The dissolved material, however, can be modified by dissolution/precipitation processes during mixing of fresh and saline waters in the estuary as well as reactions with bottom sediments and the overlying water. Typically, nutrients (organic C, N, P, Si) are more abundant in river water than seawater and are recycled in estuary environments. The average dissolved species chemical composition of river water from various continents is given in **Table 7-5**. The dissolved element concentrations differ somewhat in average river water from a particular continent; however, Ca^{2+} is the dominant cation and HCO_3^- the main anion in average river water from every continent, derived mainly from the dissolution of limestone. Dolomites are a major source of Mg^{2+}, whereas Cl^-, SO_4^{2-}, and some of the Na^+ are present from the weathering of evaporite deposits. K^+, $SiO_{2(aq)}$, and some of the Na^+ are present from the weathering of silicate minerals. Species of Al are not present in appreciable concentrations because of the precipitation of Al-rich secondary clay minerals.

Gibbs (1970) argued that the dissolved load composition of rivers is controlled mainly by atmospheric precipitation, rock weathering, and evaporation/mineral precipitation. Rivers that have low *total dissolved solids* (TDS) are considered by Gibbs to be precipitation dominant and to have a high

Table 7-4 Major element abundances in continental material and rivers

Element	Continent Surficial rock (mg/g)	Continent Soils (mg/g)	River Particulate load (mg/g)	River Dissolved load (mg/l)	River Particulate load (10^6 tons/yr)	River Dissolved load (10^6 tons/yr)	Wt% ratio River particulate/ rock	Wt% ratio Particulate/ (particulate + dissolved)
Al	69.3	71	94.0	0.05	1457	2	1.35	0.999
Ca	45.0	35	21.5	13.4	333	501	0.48	0.40
Fe	35.9	40	48.0	0.04	744	1.5	1.33	0.988
K	24.4	14	20.0	1.30	310	49	0.82	0.86
Mg	16.4	5	11.8	3.35	183	125	0.72	0.59
Na	14.2	5	7.1	5.15	110	193	0.50	0.36
Si	275	330	285	4.85	4418	181	1.04	0.96
P	0.61	0.8	1.15	0.025	18	1.0	1.89	0.96

From Berner and Berner (1996).

Table 7-5 Major species and total dissolved solids in rivers together with discharge and runoff ratios for continents

| | River water concentrations (mg/l) | | | | | | | | | Water | |
	Ca^{2+}	Mg^{2+}	Na^+	K^+	Cl^-	SO_4^{2-}	HCO_3^-	SiO_2	TDS	Discharge (10^3 km/yr)	Runoff Ratio
Africa											
Actual	5.7	2.2	4.4	1.4	4.1	4.2	26.9	12.0	60.5	3.41	0.28
Natural	5.3	2.2	3.8	1.4	3.4	3.2	26.7	12.0	57.8		
Asia											
Actual	17.8	4.6	8.7	1.7	10.0	13.3	67.1	11.0	134.6	12.47	0.54
Natural	16.6	4.3	6.6	1.6	7.6	9.7	66.2	11.0	123.5		
South America											
Actual	6.3	1.4	3.3	1.0	4.1	3.8	24.4	10.3	54.6	11.04	0.41
Natural	6.3	1.4	3.3	1.0	4.1	3.5	24.4	10.3	54.3		
North America											
Actual	21.2	4.9	8.4	1.5	9.2	18.0	72.3	7.2	142.6	5.53	0.38
Natural	20.1	4.9	6.5	1.5	7.0	14.9	71.4	7.2	133.5		
Europe											
Actual	31.7	6.7	16.5	1.8	20.0	35.5	86.0	6.8	212.8	2.56	0.42
Natural	24.2	5.2	3.2	1.1	4.7	15.1	80.1	6.8	140.3		
Oceania											
Actual	15.2	3.8	7.6	1.1	6.8	7.7	65.6	16.3	125.3	2.40	—
Natural	15.0	3.8	7.0	1.1	5.9	6.5	65.1	16.3	120.3		
World											
Actual	14.7	3.7	7.2	1.4	8.3	11.5	53.0	10.4	110.1	37.4	0.46
Natural	13.4	3.4	5.2	1.3	5.8	5.3	52.0	10.4	99.6	37.4	0.46
Pollution	1.3	0.3	2.0	0.1	2.5	6.2	1.0	0	10.5		
% Polluted	9	8	28	7	30	54	2	0			

Actual, measured values including pollution; Natural, values with anthropogenic influences removed. TDS = total dissolved solids. Runoff ratio is (runoff per unit area/average rainfall per unit area) from Berner and Berner (1996).

$Na^+/(Na^+ + Ca^{2+})$ ratio. Rock weathering increases the TDS and decreases the $Na^+/(Na^+ + Ca^{2+})$ ratio in river water. With evaporation of river water, the TDS increase further, and calcite precipitates, increasing the $Na^+/(Na^+ + Ca^{2+})$ ratio of the water. Stallard and Edmond (1983) disagreed with Gibbs, arguing that bedrock geology and extent of erosion are the primary factors controlling river solute compositions. Rivers with low TDS and high $Na^+/(Na^+ + Ca^{2+})$ result from weathering of relatively unreactive already weathered siliceous igneous and metamorphic rocks. The high TDS and $Na^+/(Na^+ + Ca^{2+})$ ratios result from weathering of evaporites (which have high Na/Ca ratios) even when these are exposed in a small percentage of the drainage area.

Figure 7-2 is a ternary diagram of compositions of major rivers plotted as a function of moles of Si, HCO_3^-, and charge (equivalents) from $Cl^- + SO_4^{2-}$ in a liter of river water as given by Berner and Berner (1996). Figure 7-2 is consistent with Si being contributed from dissolution of silicate rocks, the HCO_3^- coming mainly from carbonates plus atmospheric CO_2 and the $Cl^- + SO_4^{2-}$ from evaporites. For instance, the Negro River of the Amazon Basin flows dominantly through silica-rich sands. The Fraser River of British Columbia, Canada contacts large sections of calcite and magnesite-rich rocks. The Pecos River in

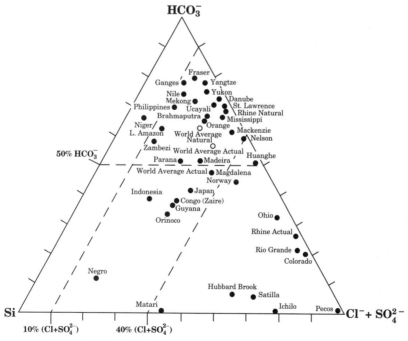

FIGURE 7-2 Percentage of Si (μmol/l), HCO_3^- (μmol/l), and $Cl^- + SO_4^{2-}$ (μEq/l) dissolved in major rivers. (Modified after Berner and Berner, 1996.)

west Texas transverses a watershed with significant amounts of Permian evaporites. Most rivers have greater than 50% HCO_3^- and between 10 and 40% $Cl^- + SO_4^{2-}$. Interestingly, silicates make up about 83%, carbonates about 16%, and evaporites 1.3% of the outcrop area of the world's river basins. The concentrations in Figure 7-2 are consistent with the fact that dissolution of evaporite minerals is significantly more rapid than calcite or dolomite and that carbonates dissolve significantly more rapidly than silicate minerals. Most of the world's rivers transport their dissolved load to the ocean, making them the major flux for most elements to seawater.

■ Seawater

The chemistry of the ocean is an important and complex subject. Average ocean depth is 3730 meters. Based on their temperature structure, oceans can be divided into surface and deep zones, with a transition zone in between. The surface zone of the ocean extends from the surface to between 50 and 300 meters, making up about 2% of the ocean's volume. It consists of water heated by the sun to an average temperature of about 18°C but is warmer near the equator and cooler near the poles. Small changes in its density can occur because of changes in salinity from atmospheric evaporation and precipitation, ice formation, and temperature changes.

At the north and south poles, surface water sinks because of its somewhat greater density, whereas deep water rises near the equator. Deep zone water has an average temperature of about 3.5°C and makes up 80% of the ocean. Because of changes in the density because of differences in salinity as well as temperature with depth, the transition zone between the surface and deep waters generally occurs over about 1 km of depth. It is often referred to as the *pycnocline* (density gradient) or the *thermocline* (temperature gradient) of the ocean.

Table 7-6 shows the concentrations of the main elements in seawater of a salinity of 35 ppt (parts per thousand by weight) with sulfur reported as SO_4^{2-} and dissolved carbon as HCO_3^-. Seawater, as a first approximation, is a 0.5 molal NaCl solution that includes an order of magnitude less $MgSO_4$ in solution. In Table 7-6, mg/kg (i.e., ppm by weight) can be converted into molality by noting that a salinity of 35 ppt is 35 g of solutes per 1000 g of solution; therefore, 1 kg of seawater has 1000 g − 35 g = 965 g of H_2O. Because molality is moles per 1.0 kg of H_2O, it is given by

$$\text{molality} = \frac{\text{mg per kg of solution (ppm)}}{\text{molecular weight}} \times \frac{1}{965} \qquad [7.11]$$

Table 7-7 shows the concentrations of the major solute species of seawater in molality determined from the compositions given in Table 7-6 at 25°C.

Table 7-6 Average major element concentrations of seawater with 35 ppt total dissolved solids excluding H and O

Element	mg/kg	Molality
Cl	19,350	0.566
Na	10,760	0.485
(SO_4^{2-})	2,710	0.0292
Mg	1,290	0.0550
Ca	411	0.0106
K	399	0.0106
(HCO_3^-)	142	0.0024

S and C given as SO_4^{2-} and HCO_3^-, respectively. From Drever (1988).

Table 7-7 indicates that although the total Ca in seawater is 0.01 m, Ca^{2+} makes up less than 60%, as the aqueous complex $CaCl^+$ is also significant. Other distributions of species models give somewhat different amounts of aqueous complexes. For instance, the amount of Ca^{2+} can be as high as 90% of the 0.01 m of total Ca in some calculation schemes. Concentrations of minor solutes of interest in seawater are given in **Table 7-8.**

Elements in seawater can be divided into *conservative elements*, *recycled elements*, and *scavenged elements* (**Figure 7-3**). Conservative elements are those in which the ratio of concentration relative to other conservative elements remains the same in the ocean so that their concentration profile with depth remains constant. For instance, when evaporation and precipitation processes occur in the surface layer, the ratios of the conservative components remain the same. The major elements in seawater are conservative because the rate of mixing of

Table 7-7 Concentrations of most abundant species in average ocean water at 25°C and 1 bar given by Drever (1988) as calculated by Bethke (1996)

Species	Molality	Species	Molality
Cl^-	0.5500	$NaSO_4^-$	0.0064
Na^+	0.4754	Ca^{2+}	0.0060
Mg^{2+}	0.0398	$MgSO_4$	0.0058
SO_4^{2-}	0.0161	$CaCl^+$	0.0038
K^+	0.0103	$NaCl$	0.0028
$MgCl^+$	0.0091	HCO_3^-	0.0015
$CaSO_4$	0.00083	$NaHCO_3$	0.00044
pH	8.3	Ionic strength	0.687

Table 7-8	CO_3^{2-}, $SiO_{2(aq)}$, and O_2 concentration in open ocean water at 25°C and 1 bar

Species	Molality
CO_3^{2-}	0.5×10^{-4} to 3.0×10^{-4}
$SiO_{2(aq)}$	8.3×10^{-6} to 1.7×10^{-4}
O_2	3.1×10^{-6} to 1.9×10^{-4}

seawater in the world's oceans is rapid relative to the rate of addition to or removal from the ocean of these components. They typically have long residence times in the ocean (greater than 10^6 years). In contrast, the recycled and scavenged elements vary significantly from place to place in the ocean. Recycled elements are used by organisms in the *photic zone*, and their availability can limit growth and production. This is particularly true of P, N, and Si, which are important nutrient elements in seawater. After death of an organism, during sinking toward the ocean floor, these elements are returned to seawater where they can be incorporated into new organisms. As a result, the depth profile of these elements tends to increase with depth in the ocean. Scavenged elements are those that are absorbed onto the surfaces of small particles brought to the ocean by rivers or by atmospheric precipitation. These elements equilibrate with seawater before being amalgamated into larger particles and sinking toward the ocean bottom. They typically have concentration profiles that decrease with increasing depth in the ocean.

The ocean surface layer is often slightly supersaturated with O_2 at 25°C and 1 bar. Water in equilibrium with the atmosphere contains 0.25 milli-mol kg^{-1} of O_2. In the surface layer, a balance exists between its production from pho-

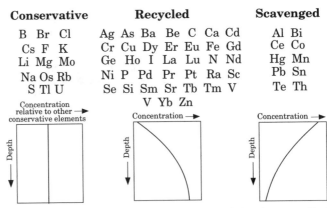

FIGURE 7-3 Elements divided into those that are conservative, recycled, and scavenged in the oceans. (Data from Brown et al., 1989.)

tosynthetic reactions together with incorporation of O_2 from air bubbles produced from wave action and its rate of diffusion into the atmosphere. Below this surface zone, an *oxygen minimum zone* generally develops, as shown for the North Pacific in **Figure 7-4**. This decrease in O_2 is produced by the consumption of oxygen during oxidation of organic matter, produced in the surface zone, as it sinks toward the ocean bottom. At greater depths, O_2 increases because of the supply of cold oxygenated water that sinks to the bottom of the oceans at the earth's north and south poles. In today's oceans, the amount of sinking organic matter is not enough to consume all of the O_2 in the oxygen minimum zone so the oceans remain *aerobic*. In the geologic past, events have occurred to produce an exceptionally high surface productivity of organic matter. It takes only about 3 mg of organic carbon to consume all of the O_2 in a kilogram of water containing 0.25 milli-mol kg^{-1} of O_2. At times in the geologic past, the high productivity caused the oceans to become *anaerobic* over significant depths.

Table 7-9 lists the major input and removal processes for some components in seawater over short, medium, and long time frames. As noted above, the major input mechanism of solutes to the ocean is rivers. Clearly, to stay at steady-state concentration, the removal processes must be of similar magnitude to input processes. This begs this question: Has the composition of seawater stayed constant through geologic time? The answer is not obvious. Clearly, the salinity of seawater changes somewhat with the waxing and waning of

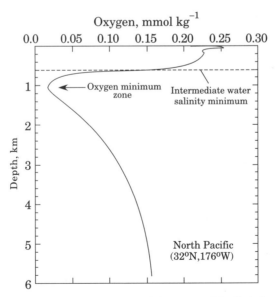

FIGURE 7-4 GEOSECS expedition profile of O_2 with depth from the north Pacific indicating the oxygen minimum zone and continued increase in O_2 with depth in deep water. (Modified after Broecker and Peng, 1982.)

| Table 7-9 | Major input and removal processes for components in seawater |

Component	Input	Removal (time frame)*
Chlorine	Rivers (including pollution)	Net sea–air transfer (short)
		Pore water burial (medium)
		Evaporative NaCl deposition (long)
Sodium	Rivers (including pollution)	Net sea–air transfer (short)
		Pore water burial (medium)
		Cation exchange (medium)
		Basalt–seawater reaction (medium)
		Evaporative NaCl deposition (long)
Sulfur	Rivers (including pollution)	Net sea air transfer (short)
		Biogenic pyrite formation (medium)
	Polluted rain + dry deposition	Evaporative $CaSO_4$ deposition (long)
Magnesium	Rivers	Net sea–air transfer (short)
		Volcanic–seawater reaction (medium)
		Biogenic Mg-calcite deposition (medium)
Potassium	Rivers	Net sea–air transfer (short)
	High T volcanic–seawater reaction	Fixation on clay near river mouths (medium)
		Low T volcanic–seawater reaction (medium)
Calcium	Rivers	Biogenic $CaCO_3$ deposition (medium)
	Volcanic–seawater reaction	Evaporitic $CaSO_4$ deposition (long)
	Cation exchange	
	Atmospheric transfer	Atmospheric transfer (short)
Carbon	Rivers	$CaCO_3$ deposition (medium)
	Biogenic pyrite formation	
Silica	Rivers	Biogenic silica deposition (medium)
	Basalt–seawater exchange	
Phosphorus	Rivers (including pollution)	Adsorption on volcanic ferric oxides (short)
	Rain and dry fallout	$CaCO_3$ deposition (medium)
		Burial of organic matter with P and in phosphate rock (long)
Nitrogen	Rivers (including pollution)	Denitrification (medium)
	N_2 fixation	Burial of organic N (long)
	Rain and dry deposition	

*Short, medium, and long time frames denote processes that are important over years, thousands of years, and millions of years, respectively.
From Berner and Berner (1996).

continental ice sheets and changes in the amount sequestered in evaporites. As discussed in Chapter 16, it has been argued that over the Phanerozoic the salinity of the ocean has decreased from more than 50% at the start of the Cambrian to the present day value of 34.7%. Global temperature change, and the advent of life on earth has changed interactions of the ocean with the atmosphere to a significant extent. Other changes could also have occurred. There are times in the past (Cambrian to Devonian) when mid-oceanic ridge (MOR) activity was likely greater than today. Arguments have been presented suggesting this led to increased loss of Mg from the flux of seawater through the mid-oceanic ridge basalt (MORB), discussed later in this chapter. Also, the higher stands of seawater on continents produced from the decreasing volume of the ocean basins due to more MOR can increase Mg loss from seawater through dolomitization of calcium carbonate in the shelf limestones encountered. These and other changes are a matter of great current debate. Besides global temperature, atmospheric composition, MORB hydrothermal interactions, eustacy, and midoceanic ridge spreading rate changes, the long-term composition of seawater also depends on changes in volcanism, tectonism, and the deposition and erosion of minerals.

The concentration of Cl in the ocean is presently increasing with time because there are no significant evaporite basins forming and significant Cl is added by humans from pollution sources. Because, however, the residence time of Cl in the oceans is 87 million years, the change in concentration of Cl in seawater on human time scales is insignificant. Sulfur, like Cl, is also presently increasing in the ocean because it is added from rivers and is not being removed in significant quantities as gypsum ($CaSO_4 \bullet 2H_2O$) in evaporites. Sulfur is also being added from rain that incorporates S from fossil fuel burning. Na is more closely in balance because of its removal by *cation exchange* on clays and by its incorporation into submarine basalt during its alteration by seawater. In these processes, Na and K are removed from seawater, whereas Ca is released from the clay or basalt to seawater. Ca is controlled to a large extent by the precipitation of calcium carbonate minerals. On long time scales, the supply of Ca to the oceans can increase with the dissolution of calcium silicates exposed during mountain uplift events. This increases the river flux of Ca to the oceans. The location of the *carbonate compensation depth* (CCD) then changes to allow more Ca to accumulate in sediments (see below). The flux of Si into the ocean by rivers obtained from the dissolution of silicate rocks on the continents is removed by planktonic organisms such as diatoms and radiolarians, which build their shells of SiO_2. This Si is then incorporated into ocean bottom sediments. Si can also be removed by reactions of seawater with ocean floor basalts during the production of clay minerals.

Carbonate Saturation

There is abundant calcite and aragonite in contact with surface seawater (e.g., coral reefs), particularly in the tropics. Is seawater saturated with these phases? In Chapter 6, the equilibrium constant of the reaction

$$CaCO_3 = Ca^{2+} + CO_3^{2-} \qquad\qquad [7.12]$$

for the mineral calcite was calculated to be $10^{-8.48}$ at 25°C and 1 bar from the SUPCRT92 data. Values of $K_{(7.12)}$ for calcite of between $10^{-8.5}$ and $10^{-8.3}$ have been obtained at 25°C and 1 bar using other thermodynamic data compilations. For aragonite $K_{(7.12)}$ values between $10^{-8.3}$ and $10^{-8.1}$ are reported. With the concentration of Ca^{2+} in seawater of about 6.0×10^{-3} m (Table 7-7) and with the activity coefficients of the doubly charged species of Ca^{2+} equal to about 0.2 and that of CO_3^{2-}, $\gamma_{CO_3^{2-}} = 0.039$ (Millero and Pierrot, 1998), the concentration of CO_3^{2-}, $m_{CO_3^{2-}}$, in equilibrium with pure calcite or aragonite in seawater can be calculated. The equilibrium constant for Reaction 7.12 is

$$K_{(7.12)} = \gamma_{Ca^{2+}} m_{Ca^{2+}} \gamma_{CO_3^{2-}} m_{CO_3^{2-}} = (0.2) \times (6.0 \times 10^{-3}) \times (0.039) \times m_{CO_3^{2-}} \qquad [7.13]$$

For equilibrium with calcite, 68 milli-mole $< m_{CO_3^{2-}} <$ 107 milli-mole, and for aragonite, 107 milli-mole $< m_{CO_3^{2-}} <$ 170 milli-mole. These values are within the 50 to 300 milli-mole given for $m_{CO_3^{2-}}$ in Table 7-8. Average seawater is near equilibrium with these phases. As temperature decreases, $K_{(7.12)}$ of calcite and aragonite increases somewhat. These carbonates, therefore, become less stable in colder waters. This is in large part why carbonate reefs are confined to the tropics and do not exist at higher latitudes.

Carbonate produced in the surface zone of the oceans tends to dissolve as it sinks to the ocean floor. To consider the effects of increased pressure on calcite solubility with greater depths in the ocean, the relationship between calcite solubility and $CO_{2(gas)}$ fugacity is needed. This can be obtained by summing the equilibrium constants for Reaction 7.12 with those given in Equations 7.6 and 7.9 and the inverse of the equilibrium constant of the bicarbonate Reaction 6.15 in Chapter 6. The following set of equations is obtained:

$CaCO_3 = Ca^{2+} + CO_3^{2-}$	$K_{Cal} = 10^{-8.5}$ to $10^{-8.3}$	[7.12]
plus: $CO_{2(gas)} = CO_{2(aq)}$	$K_{CO_2} = 10^{-1.47}$	[7.4]
plus: $CO_{2(aq)} + H_2O = H^+ + HCO_3^-$	$K_{CO_{2(aq)}} = 10^{-6.34}$	[7.9]
plus: $CO_3^{2-} + H^+ = HCO_3^-$	$K_{(7.14)} = 1/K_{HCO_3^-} = 10^{10.33}$	[7.14]

Summing these equations results in

$$CaCO_3 + CO_{2(gas)} + H_2O = Ca^{2+} + 2HCO_3^- \quad K_{(7.15)} = 10^{-5.99} \text{ to } 10^{-5.79} \qquad [7.15]$$

This equation indicates increased concentrations of $CO_{2(gas)}$, that is, increased fugacity of CO_2 gas in contact with calcite causes it to dissolve. Taking the activity of calcite and H_2O as unity, the equilibrium constant expression for Reaction 7.15 is

$$K_{(7.15)} = \frac{a_{Ca^{2+}} \, a^2_{HCO_3^-}}{f_{CO_{2(gas)}}} \qquad [7.16]$$

With increasing pressure at depth in the ocean, the fugacity of $CO_{2(gas)}$ in equilibrium with carbonate minerals increases. This causes $K_{(7.15)}$ to decrease and, therefore, the waters to become undersaturated with calcite and aragonite so that $CaCO_3$ dissolves at depth. This is compounded by the decreased temperature below the thermocline, which also contributes to the degree of undersaturation. The depth at which seawater reaches equilibrium with aragonite or calcite is the *lysocline* depth. Below this depth, $CaCO_3$ is undersaturated in the ocean. For the location shown in **Figure 7-5**, this would be about 0.5 km for aragonite and 3.3 km for calcite. The CCD is the depth in which all of the sinking carbonate has dissolved. Below this depth carbonates do not accumulate in sediments. This is somewhat below the lysocline and depends on the size of the flux of calcium carbonate downward. At the location given in Figure 7-5, carbonates are not accumulating in sediments.

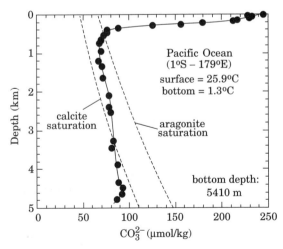

FIGURE 7-5 CO_3^{2-} measured by the GEOSECS expedition. The lysocline depth for aragonite and calcite is where their solubility lines cross the CO_3^{2-} concentration line as a function of depth. (Data from Takahashi et al., 1980.)

pH of Seawater

The pH of seawater is measured to be between 7.5 and 8.4 with surface waters generally being close to 8.2. In Chapter 6 the pH of water in equilibrium with calcite in pure H_2O and atmospheric CO_2 was determined to be 8.26 at 25°C and 1 bar. Calcite plays a roll in keeping seawater pH constant, that is, buffering it. The dissolution of calcite can absorb H^+ by the reaction

$$CaCO_3 + H^+ \rightarrow Ca^{2+} + HCO_3^- \qquad [7.17]$$

and OH^- is absorbed when calcite precipitates according to the reaction

$$OH^- + Ca^{2+} + HCO_3^- \rightarrow CaCO_3 + H_2O \qquad [7.18]$$

There is also a buffering effect of the significant bicarbonate concentration in seawater from the reaction

$$OH^- + HCO_3^- = CO_3^{2-} + H_2O \qquad [7.19]$$

The carbonate ion concentration is small, however, about an order and a half magnitude below HCO_3^- (Tables 7-7 and 7-8) so that the buffering capacity of

$$H^+ + CO_3^{2-} = HCO_3^- \qquad [7.20]$$

is limited. There is an even smaller amount of boron (B) in seawater (4.45 ppb) that also has a buffering effect because of the reactions

$$H^+ + H_2BO_3^- = H_3BO_3 \qquad [7.21]$$

and

$$OH^- + H_3BO_3 = H_2BO_3^- \qquad [7.22]$$

With the presence of the above buffer reactions and the significant mixing of surface waters, the pH of open ocean surface water is generally between 8.0 and 8.35.

Figure 7-6 shows the typical change of pH with depth in the ocean. Note the lower pH of deep waters relative to surface waters. This occurs because of the greater CO_2 solubility with decreasing temperature and greater pressures as depth increases (see Reaction 6.40). The minimum in pH above 1 km in depth occurs because the oxidation of organic matter

$$CH_2O + O_2 \rightarrow CO_{2(aq)} + H_2O \qquad [7.23]$$

produces carbonic acid, $CO_{2(aq)}$. This corresponds to the oxygen minimum zone, as shown in Figure 7-4.

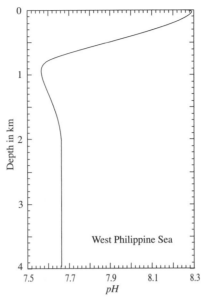

FIGURE 7-6 pH as a function of depth in the West Philippine Sea. (Modified after Chen and Mong-Hsiu, 1996).

Atmospheric CO_2 and *pH*

The partial pressure of CO_2, p_{CO_2}, in equilibrium with the surface ocean can be measured by allowing seawater to equilibrate with a small evacuated volume and measuring its CO_2 content. These determinations are shown as a function of latitude across the equator in **Figure 7-7**. When p_{CO_2} is greater than that in equilibrium with the atmosphere, there will be a flux of CO_2 out of seawater, and when p_{CO_2} is less than in equilibrium with the atmosphere, CO_2 will flux into seawater. The flux of CO_2 in and out of surface water changes seasonally. Decreasing temperature causes CO_2 to flux into seawater because of the greater solubility of CO_2 gas at lower temperatures. The maximum near the equator corresponds to upwelling of CO_2-rich deep water into the lower pressure surface environment with CO_2 degassing into the atmosphere. The minimum at 40° N occurs because of CO_2 uptake by planktonic blooms at this latitude; therefore, increased use of nutrients in the ocean, which leads to increased productivity, will cause a drawdown in atmospheric CO_2 into the ocean. Over a 100-year time scale or greater, the surface ocean can be considered to be in equilibrium with the atmosphere.

Because inorganic carbon in seawater is present as the three species $CO_{2(aq)}$, HCO_3^-, and CO_3^{2-} and the equilibrium constants between these species are known (Equations 6.17 and 6.20) and related through pH, knowledge of any two of the following variables: pH, total inorganic carbon (ΣCO_2), total alkalinity (T_{ALK}), activities of $CO_{2(aq)}$, HCO_3^-, and CO_3^{2-}, allows the others to be

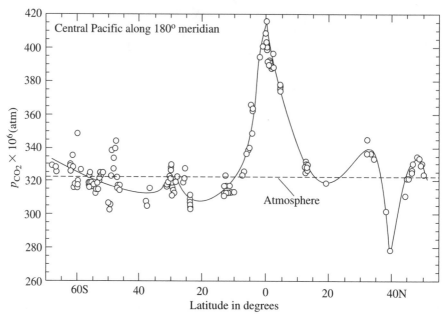

FIGURE 7-7 Partial pressure of CO_2 in equilibrium with ocean surface water as a function of latitude determined by *GEOSECS* between October 1973 and February 1974. (Modified after Broecker et al., 1979.)

calculated. T_{ALK} is the acid needed to react all of the carbonate species in solution to $CO_{2(aq)}$. This is then the number of moles of acid needed to lower seawater pH to 4.3. In seawater, because of species like $B(OH)_4^-$, this alkalinity is greater that that just due to the carbonate species and is, therefore, termed "total alkalinity" rather than carbonate alkalinity. ΣCO_2 can be determined by a *coulometric titration*. This technique measures the moles of electrons needed to reduce the inorganic carbon in solution.

If f_{CO_2} in equilibrium with $CO_{2(aq)}$ is known, the molality of CO_2, $m_{CO_{2(aq)}}$, in seawater can be obtained by rearranging Equation 7.6 to give

$$m_{CO_{2(aq)}} = \frac{K_{CO_{2(gas)}} f_{CO_2}}{\gamma_{CO_{2(aq)}}}$$

[7.24]

where $K_{CO_{2(gas)}} = 10^{-1.47}$ at 25°C. Using Equation 4.66 with an ionic strength of seawater = 0.687 m (Table 7-7) and a Stechénow coefficient of $CO_{2(aq)} = 0.231$ (Table 4-2), produces $\gamma_{CO_{2(aq)}} = 1.44$. Because of the low partial pressures of CO_2, f_{CO_2} can be taken as equal to p_{CO_2}. For instance, if $p_{CO_2} = 280$ m atm, the preindustrial value for the atmosphere determined from inclusions of air in ice, then $m_{CO_{2(aq)}} = 6.6 \times 10^{-6}$.

Knowing $m_{CO_{2(aq)}}$, the bicarbonate concentration in seawater, HCO_3^-, can be determined from Equation 6.17 at 25°C knowing pH. With $\gamma_{HCO_3^-} = 0.6$

(Figure 4-13) and a typical pH of surface water of 8.3 as the preindustrial pH value, this gives

$$K_{CO_{2(aq)}} = 4.57 \times 10^{-7} = \frac{\gamma_{HCO_3^-} m_{HCO_3^-} a_{H^+}}{\gamma_{CO_{2(aq)}} m_{CO_{2(aq)}}} = \frac{0.6 \; m_{HCO_3^-} \times 10^{-8.3}}{1.44 \; \times \; 6.6 \times 10^{-6}} \qquad [7.25]$$

or $m_{HCO_3^-} = 1,444$ μmol. A lower pH will decrease $m_{HCO_3^-}$ somewhat. From Equation 6.20, the carbonate concentration can be determined knowing $\gamma_{CO_3^{2-}}$ ($\gamma_{CO_3^{2-}} = 0.039$ in seawater) (Millero and Pierrot, 1998)

$$K_{HCO_3^-} = 4.68 \times 10^{-11} = \frac{\gamma_{CO_3^{2-}} m_{CO_3^{2-}} a_{H^+}}{\gamma_{HCO_3^{2-}} m_{HCO_3^{2-}}} = \frac{0.039 \; m_{CO_3^{2-}} \times 10^{-8.3}}{0.6 \times 1.44 \times 10^{-3}} \qquad [7.26]$$

or $m_{CO_3^{2-}} = 207$ μmol. Other calculation schemes, seawater pHs, and atmospheric CO_2 partial pressures give 50 μmol $< m_{CO_3^{2-}} < 350$ μmol for ocean surface waters.

Rather than using the two measurable properties CO_2 partial pressure (e.g., $m_{CO_{2(aq)}}$) and pH, another way to determine inorganic carbon speciation in seawater is from measurements of T_{ALK}, and ΣCO_2. T_{ALK} is given by

$$T_{ALK} = m_{HCO_3^-} + 2m_{CO_3^{2-}} + m_{B(OH)_4^-} + m_{OH^-} + \dots \qquad [7.27]$$

where $m_{OH^-} < 1.6$ μmol, and the concentration of the rest of the species in seawater affected by pH as its pH is lowered to 4.3, which needs to be included in Equation 7.27 (e.g., $m_{HPO_4^{2-}}$, $m_{PO_4^{3-}}$, $m_{H_3SiO_4^-}$, m_{NH_3}), is even smaller. A reasonable approximation, therefore, is

$$T_{ALK} = m_{HCO_3^-} + 2m_{CO_3^{2-}} + m_{B(OH)_4^-} \qquad [7.28]$$

Because ΣCO_2 in seawater is given by

$$\Sigma CO_2 = m_{HCO_3^-} + m_{CO_3^{2-}} + m_{CO_{2(aq)}} \qquad [7.29]$$

subtracting Equation 7.29 from Equation 7.28 gives

$$m_{CO_3^{2-}} = T_{ALK} - \Sigma CO_2 - m_{B(OH)_4^-} + m_{CO_{2(aq)}} \qquad [7.30]$$

Given in **Figure 7-8** are measurements of T_{ALK} and ΣCO_2 from the Pacific and Atlantic oceans (Murray, 2001). Surface T_{ALK} and ΣCO_2 vary somewhat with location and behave somewhat differently with depth at different locations. Consider $T_{ALK} = 2300$ μmol and $\Sigma CO_2 = 2000$ μmol so that $[B(OH)_4^-] \sim 86$ μmol and $[CO_{2(aq)}] \sim 7$ μmol. From Equation 7.30, $[CO_3^{2-}] = 221$ μmol.

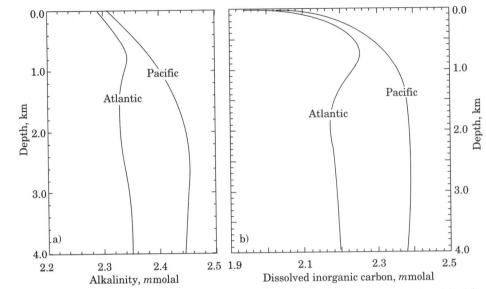

FIGURE 7-8 (a) Total alkalinity and (b) total inorganic carbon measured in seawater as a function of depth in the Atlantic and Pacific Oceans. (Modified after Murray, 2001.)

Shown in **Figure 7-9** is $[CO_3^{2-}]$ determined in ocean surface water and the reported range of calcite and aragonite solubilities at 25°C given the present Ca^{2+} concentration of seawater. Note the general inverse correlation between the data in Figures 7-7 and 7-9. This occurs because combining Equations 7-2 to 7-4 gives

$$\frac{f_{CO_2}}{a_{CO_3^{2-}}} = \frac{a_{H^+}^2}{K_{HCO_3^-}\ K_{CO_{2(aq)}}\ K_{CO_{2(gas)}}} \tag{7.31}$$

Changes in pH and the equilibrium constants in Equation 7.31 with temperature are small, therefore, f_{CO_2} is inversely proportional to $a_{CO_3^{2-}}$ in the surface ocean. Figure 7-9 indicates that, except possibly at high latitudes where temperatures are lower, most surface seawater is supersaturated with calcite and aragonite.

Past Atmospheric CO₂ Concentrations from Ocean Carbonate Chemistry

One way to determine past atmospheric CO_2 and ocean inorganic carbon concentrations is to first consider the concentration of Ca^{2+} and Mg^{2+} from fluid inclusions it evaporates as a function of their age, as shown in **Figure 7-10**. These can be considered a proxy for these concentrations in seawater. Note the decrease in Ca^{2+} and increase in Mg^{2+} as the present is approached. The average calcite CCD derived from observing deep-sea cores has changed from ~3.5 km 100 million years ago to ~4.8 km today (Van Andel, 1975). Assuming the

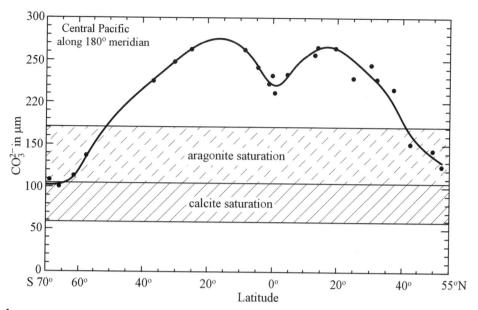

FIGURE 7-9 Concentration of CO_3^{2-} in surface ocean water as a function of latitude calculated by Broecker et al. (1979) from GEOSECS measurements of alkalinity and total inorganic carbon between October 1973 and February 1974. The shaded areas give the range of values for saturation with calcite and aragonite calculated from Equation 7.13 at 25°C. At the lower temperatures at the poles saturation CO_3^{2-} would be greater.

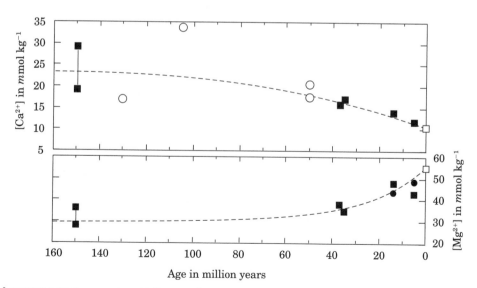

FIGURE 7-10 Concentration of Ca^{2+} and Mg^{2+} determined from fluid inclusions (■, Horita et al., 2002; ●, Zimmerman, 2000), and considering echinoderm Mg/Ca (○, Dickson, 2002) as a function of age, □ shows the present day value. (Modified after Tyrrell and Zeebe, 2004.)

degree of sea surface calcite supersaturation, W, tracks the CCD a model of changes in W through time can be constructed. W is given by

$$W = \frac{[Ca^{2+}] \times [CO_3^{2-}]}{K_{Cal}} \qquad [7.32]$$

where K_{Cal} is the solubility product of calcite. Knowing W, $[Ca^{2+}]$ and K_{Cal} values of $[CO_3^{2-}]$ can be computed as shown in **Figure 7-11** from the calculations of Tyrrell and Zeebe (2004). Note the increase in $[CO_3^{2-}]$ toward the present. This is consistent with the increase in surface pH as determined by Person and Palmer (2000) from boron isotope measurements (see the discussion on boron isotopes in Chapter 11); therefore, from the Cretaceous to the preindustrial present, the ocean appears to have become more alkaline.

Evaporation of Seawater

Surface seawater is near calcite saturation, and calcite precipitates rapidly with the initial loss of H_2O during evaporation at 25°C, as given by

$$Ca^{2+} + HCO_3^- \rightarrow CaCO_3 + H^+ \qquad [7.33]$$
$$\text{calcite}$$

Because the concentration of calcium species in solution is greater than the alkalinity, the precipitation of calcite removes all of the alkalinity. Further evaporation increases the Ca in solution until gypsum begins to precipitate when about 30% of the original volume of seawater remains (**Figure 7-12**).

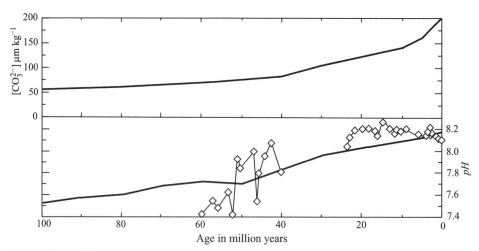

FIGURE 7-11 CO_3^{2-} and pH of surface water in equilibrium with the atmosphere as a function of age given as heavy solid lines modified after Tyrrell and Zeebe (2004). The independently determined pH data shown as diamonds are from the data of Pearson and Palmer (2000).

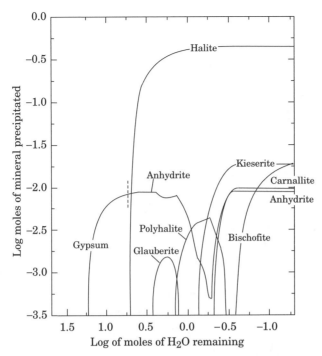

FIGURE 7-12 Calculated moles of minerals precipitated from evaporating seawater as a function of moles of H$_2$O remaining after initial calcite precipitation. Left side of diagram starts with 55.51 moles of H$_2$O in the seawater. (Modified after Harvie et al., 1980.)

$$Ca^{2+} + SO_4^{2-} + 2H_2O \rightarrow CaSO_4 \cdot 2H_2O \qquad\qquad [7.34]$$
$$\text{gypsum}$$

Because SO_4^{2-} is greater than Ca in solution at this point, the precipitation of $CaSO_4 \cdot 2H_2O$ removes the Ca with SO_4^{2-} concentrations increasing as evaporation continues. As the activity of water decreases to about 0.78, or 90% of the H$_2$O is evaporated, gypsum is converted into anhydrite, as shown by the vertical dashed line in Figure 7-12 (see also **Figure 7-13**).

$$CaSO_4 \cdot 2H_2O \rightarrow CaSO_4 + 2H_2O \qquad\qquad [7.35]$$
$$\text{gypsum} \qquad \text{anhydrite}$$

Halite begins to precipitate soon after this when seawater is evaporated to about 9.5% of its original volume by the reaction

$$Na^+ + Cl^- \rightarrow NaCl \qquad\qquad [7.36]$$
$$\text{halite}$$

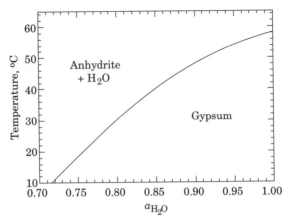

FIGURE 7-13 Stability of gypsum and anhydrite as a function of temperature and activity of H_2O at 1 bar. (Modified after Hardie, 1967.)

Because Cl is greater than Na in solution, the precipitation of halite removes most of the Na from solution, whereas the Cl concentration increases on evaporation. With about 5% of the solution remaining, glauberite forms by reaction of the remaining brine with anhydrite:

$$CaSO_4 + 2Na^+ + SO_4^{2-} \rightarrow Na_2Ca(SO_4)_2 \qquad [7.37]$$
anhydrite   glauberite

Further reaction produces a number of Mg and K salts. As shown in Figure 7-12, polyhalite forms first by the reaction of anhydrite with glauberite:

$$Na_2Ca(SO_4)_2 + CaSO_4 + 2K^+ + Mg^{2+} + SO_4^{2-} + 2H_2O \rightarrow$$
glauberite   anhydrite

$$K_2MgCa_2(SO_4)_4 \bullet 2H_2O + 2Na^+ \qquad [7.38]$$
polyhalite

The final equilibrium assemblage contains halite, kieserite ($MgSO_4 \bullet H_2O$), bischofite ($MgCl_2 \bullet 6H_2O$), carnallite ($KMgCl_3 \bullet 6H_2O$), and anhydrite. If, however, early formed minerals are not allowed to react with the brine produced or the brine is diluted with fresh seawater, the amount and character of the minerals change. Marine evaporites preserved in the geological record appear to be of two different types. One type is similar to what will precipitate out of modern evaporated seawater with $MgSO_4$ salts, polyhalite, kieserite, bischofite, and carnallite. The other type contains NaCl, KCl, and carnallite but is free of $MgSO_4$ salts. The implication of these differences is discussed in Chapter 16.

Gypsum-Anhydrite Stability

Very often, extensive gypsum beds are preserved in evaporites. The precipitated gypsum can be converted to anhydrite if the activity of water decreases enough. The stability of gypsum relative to anhydrite at 1 bar pressure as a function of temperature and H_2O activity is shown in Figure 7-13. The solid line indicates the equilibrium boundary for the reaction

$$CaSO_4 \bullet 2H_2O = CaSO_4 + 2H_2O \qquad [7.39]$$
$$\text{gypsum} \qquad \text{anhydrite}$$

At low temperatures, anhydrite is stable only at significantly reduced activities of H_2O, but as temperature increases, gypsum in a fluid with an activity of H_2O equal to unity is converted to anhydrite at approximately 58°C.

The pressure dependence of gypsum stability can be considered by assuming that at the shallow depths considered the fluid is under hydrostatic fluid pressure. That is, during burial, it experiences the pressure of an overlying column of fluid, P_F. The rock containing gypsum is under a greater lithostatic pressure, P_L, because of the rock's greater density. The pressure on gypsum in contact with the fluid, however, is at hydrostatic pressure. The mineral is then under some deviatoric stress that reduces its stability over that of a hydrostatically stressed state to some degree. Ignoring this effect, the change in the equilibrium for Reaction 7.39 as a function of pressure and temperature at 25°C and 1 bar is

$$\frac{dP}{dT} = \frac{\Delta S_R}{\Delta V_R} = \frac{(46.36 + 2 \times 16.712 - 25.50)}{(74.31 + 2 \times 18.1 - 45.94)}$$

$$= \frac{54.28 \text{ cal K}^{-1} \text{ mol}^{-1}}{64.6 \text{ cm}^3 \text{ mol}^{-1} \times 0.239 \text{ cal cm}^{-3} \text{bar}^{-1}} = 35.2 \text{ bar K}^{-1} \qquad [7.40]$$

The values at 25°C and 1 bar used to evaluate Equation 7.40 are from SUPCRT92 except for $S_{Gp}^o = 46.36$ cal K^{-1} mol^{-1} and $V_{Gp}^o = 74.31$ cm^3 from Berner (1971). The slope of the equilibrium boundary is greater than the slope of a hydrostatic pressure-temperature gradient for pure H_2O in normal crust. From Chapter 2, this would be about

$$\frac{100 \text{ bar km}^{-1}}{25°C \text{ km}^{-1}} = 4 \text{ bar } °C^{-1} \qquad [7.41]$$

Gypsum formed at the surface, therefore, should convert to anhydrite on burial at temperatures lower than 58°C, even in pure H_2O because of the pressure effect. For saline water with its lower activity of H_2O and greater density, the conversion of gypsum to anhydrite on burial in a sedimentary pile would occur even closer to the earth's surface.

■ Seafloor Hydrothermal Systems

Basaltic magma is intruded to a height of about 2 km below the ocean floor at mid-oceanic ridges (MOR). The heat supplied by this magma causes hydrothermal convection of seawater through the oceanic crust. Both deep and shallow convection cells are thought to be produced. Shallow convection is established through the very permeable pillow lavas in the top layer of the ocean crust. In these shallow convection cells, seawater is heated by the warm rocks it encounters and often mixes at depth with down-flowing colder seawater. It returns to the seafloor in diffuse up-flow zones where the fluid is generally only a few degrees above normal seawater bottom temperatures of 2°C to 4°C.

Deep convection down-flow occurs along through-going fractures and other permeable zones in the pillow basalts and sheeted dikes below. This allows some seawater to convect to the base of the vertical sheeted dikes (see Figure 2-4 for a cross-section through the oceanic crust). These fluids can reach temperatures of over 400°C because of the heat transferred from the nearby magma body (**Figure 7-14**). The pressure on these fluids, because of the overlying column of seawater, is between approximately 300 and 500 bars. Under these physical conditions, the fluids can boil, producing a more dense chloride-rich fluid and less dense chloride-poor fluid. Up-flow of these fluids to the ocean floor appears to be along nearly vertical fractures between sheeted dike surfaces. These up-flow zones can tap different mixtures of the chloride-rich and chloride-poor fluids.

When these hydrothermal fluids mix with colder seawater at the ocean floor, metal sulfides are precipitated in a plume-like structure. The metal sulfides in the hydrothermal plume give it a dark color, producing what is termed a *black smoker*. MOR hydrothermal systems have a somewhat different character depending on the rate of seafloor spreading, producing differences in the thermal regimes present and in the permeability structure of the rocks. Judging from heat flow studies, the majority of heat at MOR is transferred conductively with the convective heat from MOR transferring only a small proportion of the heat (Hofmeister and Criss, 2005). Hydrothermal convection of seawater through the ocean crust is a relatively common feature along the approximately 55,000 km of MOR on earth.

The chemical interactions of seawater and basalt that occur in these hydrothermal systems are complex and change as a function of time and place because temperatures and the permeability of the crust change with time and place. Also, the degree of alteration and its character in rocks along the fluid

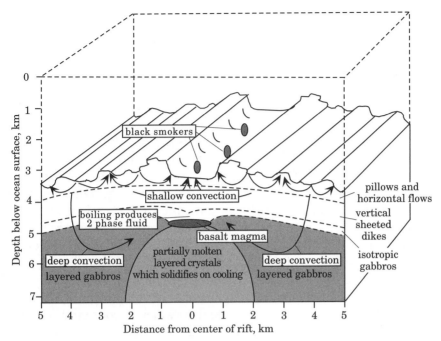

FIGURE 7-14 Hydrothermal vents at a MOR. Arrows give the path of seawater as it is heated and reacts with the basaltic rock before it emanates as hydrothermal solution, producing the indicated black smokers. Crystallization of cooling magma produces the layered gabbros that are present at depth. The cooling is rapid enough that there is likely only a transient presence of a magma chamber, but the hot rocks allow the convective system to remain active.

pathways change with the extent of reaction, adding to the complexity. The cumulative effect of these MOR hydrothermal systems is to transfer mass between seawater and oceanic crust, altering the composition of both significantly. As the hydrothermally altered crust is subducted into the mantle, it produces heterogeneity in mantle rocks.

Some general chemical trends appear to occur in most MOR hydrothermal systems. In the down-flow zones, seawater is heated. Because of its retrograde solubility, anhydrite precipitates from the heated seawater as it reaches temperatures of 150°C to 200°C. This modified seawater reacts with the basaltic rocks, lowering the modified seawater oxygen fugacity to near that of hematite plus magnetite by reacting Fe^{2+} in the basalt to produce Fe^{3+}. Oxidation of Fe^{2+} also reduces seawater SO_4^{2-} to H_2S. The *pH* of these hydrothermal fluids becomes somewhat acidic, probably reaching *pH* values of about 3.1 to 3.9. Mg^{2+} in the seawater reacts with olivine present in the basalt to produce chlorite and with the pyroxene to produce actinolite. These reactions deplete the hydrothermal solution of all of its Mg and allow the seawater component of sampled black smoker fluid to be determined from its Mg concentration. Plagioclase in the basalt reacts with Na^+ in seawater to produce albite. The Ca released contributes significantly to the Ca flux into the oceans. The albite produced can react to produce chlorite with fresh seawater or epidote with Mg-depleted seawater. Greenschist metamorphic facies assemblages of albite + actinolite + chlorite and albite + actinolite + epidote are therefore produced. Ferric micas and smectite are formed from a reaction of the basalt glass and the hydrothermal solution and can fix alkalis from solution in these phases.

It is thought that the MOR hydrothermal reactions remove approximately 50% of the annual river input of Mg into the oceans. The importance of magmatic volatile inputs to the hydrothermal systems is largely unknown. Black smoker CO_2 $\delta^{13}C$ values are identical to mantle CO_2 (see Chapter 11); however, whether the CO_2 came from a volatile magmatic phase or leaching of the MORBs is unknown. Hydrothermal vent fluids are enriched in methane and hydrogen, but the source of these volatiles could also be from leaching rather than a magmatic volatile phase. Alternatively, a biologic source is possible.

Chloride concentrations of the hydrothermal component in black smokers vary from about half to twice normal seawater in concentration. This is thought to indicate different mixtures of the more dense chloride-rich fluid and less dense chloride-poor fluid produced from boiling in these systems. The lower density, salt-depleted vapor escapes more readily to the seafloor, leaving a higher density brine phase at depths that later up-flow fluids can tap. Samples of the high-Cl fluid and Cl-poor vapor have been preserved in fluid inclusions in sampled rocks. Because the high-salinity fluid is denser than seawater, it may reside for extended periods of time in the deep levels of the convection systems before it is removed. Amphiboles produced in the hydration reactions that are taking place have high Cl contents consistent with being formed in the presence of a high Cl fluid.

The precipitation of sulfides from black smokers produces chimneys around the vent that can reach heights greater than 45 m. In other vents termed white smokers, calcium carbonate is precipitated that gives the fluid a whitish color, producing chimneys as high as 60 m. From the rate and amount of material deposited in observed hydrothermal systems, they must remain active for thousands of years at a particular location. Apparently, anhydrite ($CaSO_4$) is deposited first when Ca-rich hydrothermal solutions interact with SO_4^{2-} in seawater. Graham and coworkers (1988) argued the anhydrite reacts with vent fluid to precipitate marcasite, a dimorph of pyrite (FeS_2), and wurtzite, the high-temperature ZnS polymorph. The process is a balance between rates of anhydrite precipitation from cooling vent fluid and mineral dissolution during reaction with cooler seawater. This can be followed by pyrite deposition in the chimney and then the precipitation of the Cu sulfides bornite and chalcopyrite. With cooling of the fluid, zones of pyrite, marcasite, and then wurtzite are produced. **Figure 7-15** shows the character of a mound produced from the precipitating sulfide

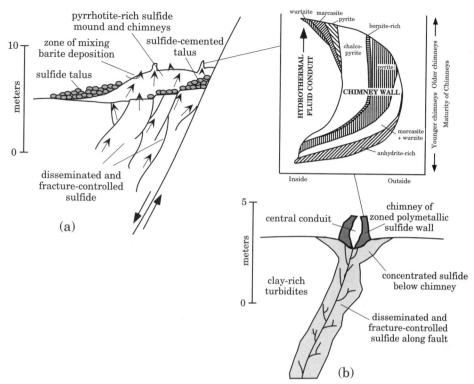

FIGURE 7-15 MOR hydrothermal mounds showing development of chimneys. (a) Diffuse flow system with slow discharge. (Modified from Koski, R.A., *et al.* The composition of massive sulfides from the sediment-covered floor of Escanaba Trough, Gorda Ridge: Implication for depositional processes. *Can. Mineral.* 26:655–673.) (b) Focused flow system with rapid discharge. (Modified from Graham, U.M., *et al.* Sulfides-sulfate chimneys on the East Pacific rise, 11° and 13°N latitude. Part I. Mineralogy and paragenesis. *Can. Mineral.* 26:487–504.)

phases. These mounds have been considered to be good analogues for the development of massive sulfide deposits found in rocks from oceanic environments.

■ Groundwater

The change in size of a particular groundwater reservoir is determined by the flux of water from precipitation in the recharge area and the flux out in discharge areas by using Equation 7.1. Groundwater has a highly variable composition depending on the following:

1. The composition of the precipitation recharging the groundwater reservoir.
2. Interaction with weathered soil and organic activity in the soil through which the water percolates before entering the groundwater reservoir.
3. The composition of subsurface rocks the water contacts in the groundwater reservoir.

4. The residence time, temperature, and the mineral surface area per unit volume of water in the reservoir.

From the numbers given in Figure 7-1, the average residence time for water in groundwater reservoirs is

$$t_R = \frac{8000 \text{ Eg}}{5 \text{ Eg yr}^{-1}} = 1600 \text{ years} \qquad [7.42]$$

This is an average time because some groundwater, like that in the Ogallala aquifer east of the Rocky Mountains, has remained there for over 10,000 years, whereas other water is present only since a recent rainstorm. Starting with the composition of meteoric water discussed above, reactions in the subsurface typically cause significant compositional changes.

Although the percentage of CO_2 in air from Table 7-1 is 350 ppm or 0.035%, the CO_2 in arable soils is typically between 0.15% and 0.65% and can reach values greater than 10%. This CO_2 is added by root respiration and microbial decay of the plant material that is present. According to Equation 7.15, this increased CO_2 causes $CaCO_3$ to dissolve. In carbonate terrains, if this elevated CO_2 is lost from groundwater to the atmosphere, supersaturation of solutions can result and lead to the formation of stalactites and stalagmites in caves.

The water's composition changes toward the solubilities of the minerals it contacts. The concentration of Ca^{2+} is controlled primarily by calcite and dolomite dissolution in limestone and plagioclase weathering in silicate rocks. The concentration of Na^+ in groundwater is determined by NaCl dissolution where it is present and the weathering of plagioclase where it is not. Mg^{2+} concentrations are controlled by dolomite weathering in limestones; amphibole, pyroxene, and biotite in felsic to intermediate silicate rocks; and olivine plus chlorite dissolution in ultramafic silicate rocks. K^+ is contributed by dissolution of biotite and K-feldspar.

Eight ions make up more than 90% of dissolved solids in most groundwater. These include the cations that typically have relative concentrations of $Ca^{2+} > Na^+ > Mg^{2+} > K^+$ and the anions with relative concentrations of $HCO_3^- > SO_4^{2-} > Cl^-$. Most groundwater is, therefore, prevalently a calcium bicarbonate solution. In limestones and granites, while HCO_3^- is typically the dominant anion, Cl^- is generally greater than SO_4^{2-}. At high pH, CO_3^{2-} can also become significant. In addition to these species, the neutral silica species, $SiO_{2(aq)}$, is generally important. **Table 7-10** lists the typical major solute compositions of water from sandstone, shale, limestone, and granite and indicates that waters from granites like those from other crystalline igneous rocks have significantly lower TDS than from sedimentary rock types. Waters from sandstones typically have higher Ca/Na ratios than waters from granites. Shales generally contain waters with higher TDS than other rock types, except for evaporites.

Table 7-10	Average major solute molalities multiplied by 10^3, pH, and ionic strength of groundwater from various rock types			
Species	Sandstone	Shale	Limestone	Granite
Cl^-	0.20	0.10	0.63	0.10
Na^+	0.50	2.51	0.10	0.40
Mg^{2+}	0.32	3.16	0.40	0.16
SO_4^{2-}	0.63	6.31	0.40	0.06
Ca^{2+}	1.00	3.16	2.00	0.32
K^+	0.10	0.06	0.20	0.10
HCO_3^-	2.51	7.94	5.01	1.26
$SiO_{2(aq)}$	0.20	0.20	0.10	0.20
pH	8.0	7.3	7.0	7.0
Ionic strength	4.00	20.0	6.3	1.6

From Stumm and Morgan (1981).

Limestones contain groundwaters with significant alkali elements and silica because, in addition to calcite, these rocks often have some clay, sulfate minerals, and dolomite.

Often, sedimentary rocks at great depths have saline pore waters. If they are encountered during drilling for petroleum, they are termed basinal or *oilfield brines*. These can be quite variable in composition, being anywhere from mildly saline to having salinities that are twice that of normal seawater. It is generally thought that these fluids obtain their salinity from dissolution of evaporites in the stratigraphic section. Typically, they have high Ca/Mg ratios relative to seawater. The Mg is lowered because of its reaction with $CaCO_3$ to produce dolomite in the subsurface. SO_4^{2-} is often low because of bacterial sulfate reduction to produce FeS_2 as well as from the precipitation of anhydrite. To understand how the composition of a groundwater develops, a model of water chemistry as it interacts toward equilibrium with the minerals it contacts can be constructed, as was done in Chapter 6. These models show that somewhat acidic soil waters entering an aquifer are neutralized at greater depth by reactions with silicate and carbonate and can increase their Cl^- and SO_4^{2-} concentrations if evaporite minerals are encountered.

■ Lakes

Lakes have similar compositions but tend to be less chemically variable than rivers or groundwater because of their large size relative to the surfaces of reacting mineral they contact. The chemistry of the nonconservative elements depends to a large extent on the lake's ability to mix surface and bottom waters.

Figure 7-16 shows the temperature profile of a typical temperate lake during summer and winter. In summer, the sun warms the top layer of the lake, the *epilimnion*, which makes it less dense than the bottom layer of the lake, the *hypolimnion*. The lower surface temperatures in the fall cause the epilimnion to cool to temperatures below the hypolimnion temperature. The epilimnion is then denser than the hypolimnion, which leads to overturning of the lake, bringing bottom waters to the surface. Because the densest water is at 4°C, it is also possible that on heating of the surface of the lake to 4°C in the spring, a second overturn of the lake can occur.

The *pH* of many lakes is controlled by biologic processes rather than with equilibrium with CO_2 in the atmosphere. *Photosynthesis* consumes dissolved $CO_{2(aq)}$. Bicarbonate in the lake water then reacts with H^+ by the reaction:

$$H^+ + HCO_3^- = CO_{2(aq)} + H_2O \qquad [7.43]$$

to restore the equilibrium $CO_{2(aq)}$, which increases *pH*. Respiration of organic matter, on the other hand, produces $CO_{2(aq)}$, which drives Reaction 7.43 to the left lowering the *pH*. The *pH* of lake surface water is typically higher during daylight hours and during the growing season, when photosynthesis is at a maximum. Because photosynthesis occurs in the top layer of a lake and respiration is often dominant at lake bottoms, the *pH* can change with depth in lakes when they have remained stratified for significant time periods. The *pH* of lakes can be lowered by the input of large amounts of low *pH* water (*acid rain*). This can be buffered by reactions with any carbonate sediments on the lake bottom. If buffering carbonates are not present then the lake can become quite acidic.

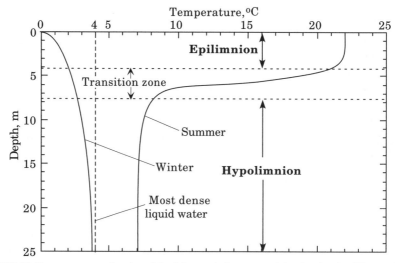

FIGURE 7-16 Temperature as a function of depth in a typical temperate lake showing the difference in the profile from winter to summer.

Lakes are divided into those that are "well fed" with plant nutrients, that is, *eutrophicated*, as opposed to those that are "poorly fed" or *oligotrophic*. In a eutrophicated lake, there is abundant production of organic matter by photosynthesis. The photosynthetic reaction for plankton growth at lake surfaces where sunlight can penetrate can be represented by (e.g., Stumm and Morgan, 1981)

$$106CO_2 + 16NO_3^- + HPO_4^{2-} + 122H_2O + 18\,H^+ + \text{trace elements} + h\nu \rightarrow$$
$$C_{106}H_{263}O_{110}N_{16}P_1 + 138O_2 \hspace{4cm} [7.44]$$

where $h\nu$ represents the sun's energy as given by Planck's constant, h, times the frequency of light, ν, and with $C_{106}H_{263}O_{110}N_{16}P_1$ representing the organic matter in plankton. Plankton, similar to most organisms, are considered to have the *Redfield ratio* (Redfield et al., 1963) of C:N:P to be 106:16:1. Writing the reaction in this manner shows the importance of N and P in combining with CO_2 and H_2O to produce life with energy from sunlight. It should be noted when considering N and P in organic matter more recent examination of marine planktonic organic matter (Hedges et al., 2002) suggests a composition that is less hydrogen and oxygen rich of $C_{106}H_{175-180}O_{35-44}N_{15-20}S_{0.3-0.5}$. Land plants have an estimated C:N:P of 822:9:1 according to Deevey (1973), whereas Likens et al. (1981) give a ratio of 2057:17:1; therefore land plants have a significantly higher concentration of carbon than the marine biomass. When N and P are not a concern, the photosynthetic reaction is often written in the simplified notation

$$CO_2 + H_2O + h\nu \rightarrow CH_2O + O_2 \hspace{4cm} [7.45]$$

which indicates the incorporation of the three dominant elements C, H, and O in their approximate ratio of 1:2:1 in organic matter.

When the organic matter produced in the photic zone at the surface sinks toward the bottom of the lake, the reverse of Reaction 7.44 or Reaction 7.45 occurs. This decreases the O_2 content of the lake so that fish and other organisms that need O_2 die, upsetting the ecologic balance in the lake. As shown by Reaction 7.44, it is the presence of nutrients N and P in the water that controls the process. Problems of eutrophication therefore occur with agricultural fertilizer runoff and disposal of N- and P-rich municipal wastes that are carried to lakes by streams.

Summary

Changes in composition and amount of natural waters can be modeled with a set of reservoirs with fluxes between the reservoirs. The residence time of a constituent in a reservoir at steady state is given by its concentration in the reservoir divided by the flux into or out of the reservoir.

Because of CO_2 in air, the *pH* of pure rainwater is calculated to be 5.66. This water can flux into a groundwater reservoir or can run off to rivers. Rivers transport dissolved, suspended, and bed loads of material. It is dominantly the dissolved load that fluxes into the oceans. Its composition is controlled by bedrock geology and ease of their erosion.

As a first approximation, seawater can be considered a 0.5 molal NaCl solution that contains an order of magnitude less $MgSO_4$ in solution. The elements in seawater are conservative, recycled, or scavenged. An oxygen minimum zone typically develops in water at a depth of about 1 km. Its position and the extent of O_2 depletion depends on the flux of organic matter from the photic zone above and the flux of cold oxygenated polar waters below. A CCD exists in the ocean below which carbonate material is not accumulating because of the increased solubility of carbonates with decreasing temperature and increased $CO_{2(gas)}$ concentrations with increased pressure. The *pH* of seawater is buffered between 7.8 and 8.4 primarily because of the presence of calcite and significant Ca^{2+} and HCO_3^- in solution. Evaporation of seawater precipitates calcite first followed by gypsum and halite. Further evaporation causes gypsum to be converted to anhydrite and the precipitation of a number of other sulfates and salts.

Seafloor hydrothermal systems at MORs heat seawater, precipitating anhydrite. Seawater oxidizes the Fe^{2+} in basalt to Fe^{3+} that lowers the *pH* of the modified seawater to about 3.1 to 3.9. Boiling occurs in the system that increases the Cl concentrations in the remaining brine. Black smokers can occur where the hydrothermal fluid vents into the ocean, producing chimneys from precipitated sulfides.

Groundwater composition depends on the nature of subsurface rocks and the precipitation recharging the reservoir. Reactions that occur in soils before fluxing to the groundwater reservoir, as well as the residence time are also important. Typically, the cation concentrations are given by $Ca^{2+} > Na^+ > Mg^{2+} > K^+$ with the anion concentrations of $HCO_3^- > SO_4^{2-} > Cl^-$, although significant variability exists. Lakes have similar compositions to rivers and groundwater but with less chemical variability. They can become eutrophic from the reaction of O_2 with the organic matter present if they do not experience convective overturn to incorporate additional O_2 into the lake from the atmosphere.

Key Terms Introduced

acid rain

aerobic

anaerobic

bed load

black smoker

carbonate compensation depth

cation exchange
conservative elements
coulometric titration
epilimnion
eutrophicated
flocculate
flux
hydrologic cycle
hypolimnion
lysocline
non–steady state
oil-field brine
oligotrophic

oxygen minimum zone
photic zone
photosynthesis
pycnocline
recycled elements
Redfield ratio
scavenged elements
steady state
sublimation
suspended load
thermocline
total dissolved solids

Questions

1. Explain the difference between steady-state and equilibrium.
2. How is the residence time of a substance in a reservoir calculated?
3. Why is the pH of pure uncontaminated rainwater 5.7 and not 7.0?
4. What controls the concentration of species in river water?
5. What is the difference between a conservative and a nonconservative component of seawater?
6. Why is the pH of the open surface layer of oceans between 8.0 and 8.35?
7. Describe the minerals that precipitate out of seawater as it evaporates.
8. What is a "black smoker"? What controls its chemical composition?
9. What is the anion species of greatest concentration in groundwater? Seawater?
10. What causes lakes to bring deep water to the surface?

Problems

1. If the flux of water into the ice reservoir were 4×10^{15} kg yr^{-1}, what would be the residence time of water at steady state? If because of global warming the flux of water out of the reservoir changed from the steady-state condition to 5×10^{15} kg yr^{-1}, how long would it take for all glaciers to melt?
2. From the molality of species in seawater, calculate its ionic strength.
3. From the values of Gibbs energy given in the back of the book, calculate the solubility of anhydrite in pure water. Will the solubility be greater or less in seawater? By using the values in Tables 7-7 and 7-8, if activity coefficients are unity and the relative ratio of species in seawater stays the same, to what percentage must seawater evaporate to start to

precipitate anhydrite? How does this compare when anhydrite actually starts to precipitate in evaporated seawater?

4. Surface water can show day to night as well as seasonal changes in *pH*. What causes this effect?

5. Explain how a lake becomes eutrophicated.

6. A measurement of seawater contains 41.3, 2.3, and 0.3 g m^{-3} of C, N, and P, respectively. With a consideration of the Redfield ratio, which nutrient would be biolimiting?

7. 14.4°C water issuing from a spring in serpentinized peridotite near John Day, Oregon has *pH* = 7.84 and a concentration of Ca^{2+} = 10 and HCO_3^- = 328.3 mg l^{-1} (Barnes and O'Neil, 1969).
   a. What is the saturation state of this water relative to calcite? At 14.4°C, the log activity product of calcite in water $(a_{Ca^{2+}}a_{HCO_3^-}a_{OH^-})$ = −12.354. Assume that activity coefficients are unity (log K_{H_2O} at 14.4°C = 14.36).
   b. With the high concentration of Mg^{2+} (58 mg l^{-1}) in the spring water, what would be the activity of calcite in a magnesium calcite in equilibrium with this water?

8. Consider the values in Table 7-10.
   a. By consulting Figure 6-7, is most groundwater saturated, supersaturated, or undersaturated with respect to quartz?
   b. As presented, how much is the average groundwater in shale out of charge equilibrium?
   c. Assume that standard conditions and activity coefficients of aqueous species are unity. For shale, calculate the partial pressure of CO_2 in equilibrium with the groundwater. Compare with the partial pressure of CO_2 in normal air, and explain the difference.

9. Given below is the composition of a flowing well water from a depth of 488 ft in the Lance Formation in Sheridan County, Wyoming (Hem, 1985).

Constituent	mg/L
Na$^+$ + K$^+$	661
Mg^{2+}	24
Ca^{2+}	37
HCO$_3^-$	429
Cl$^-$	82
SO$_4^{2-}$	1010
SiO$_2$	7.9
pH	7.3

a. Convert to moles per liter, and check the solutes for charge balance.
b. Calculate the ionic strength.
c. Determine f_{CO_2} in equilibrium with these samples, and compare with f_{CO_2} in air.

10. Given below is the mean composition of 283 water samples from a Columbia River Plateau basalt aquifer in the northwestern United States (Whiteman et al., 1994). Convert to moles per liter and calculate the charge imbalance. To bring composition into charge balance, change the concentration of the most concentrated solute (HCO_3^-). Now calculate the ionic strength, and determine f_{CO_2} in equilibrium with these samples and compare with f_{CO_2} in air.

Constituent	mg/L
Na^+	24.9
K^+	4.7
Mg^{2+}	10.7
Ca^{2+}	24.5
HCO_3^-	170
Cl^-	7.1
SO_4^{2-}	21.8
SiO_2	56.5
pH	7.6

Problems Making Use of SUPCRT92

11. Is seawater at 25°C and 1 bar undersaturated, saturated, or supersaturated with sepiolite ($Mg_4Si_6O_{15}(OH)_2(H_2O)_2 \bullet 4(H_2O)$)? Assume activity coefficients and the activity of H_2O and sepiolite are unity. Use values in Tables 7-7 and 7-8 and show your calculation.

12. Calculate the pH at 25°C and 1 bar as well as at 100°C and H_2O liquid–vapor saturation pressure for equal concentrations of Al^{3+} and $Al(OH)^{2+}$. Assume that the activity coefficient ratio and the activity of H_2O are unity. Does Al^{3+} or $Al(OH)^{2+}$ become more stable as temperature and pressure increase to 100°C?

References

Baes, C. F. and Mesmer, R. E., 1976, *The Hydrolysis of Cations*. Wiley Science, New York, 489 pp.

Berner, R. A., 1971, *Principles of Chemical Sedimentology*. McGraw-Hill, New York, 240 pp.

Berner, E. K. and Berner, R. A., 1996, *Global Environment: Water, Air, and Geochemical Cycles*. Prentice Hall, Upper Saddle River, NJ, 376 pp.

Bethke, C. M., 1996, *Geochemical Reaction Modeling*. Oxford University Press, New York, 397 pp.

Broecker, W. S. and Peng T.-H., 1982, *Tracers in the Sea*. Eldigio Press, Columbia University, New York, 690 pp.

Broecker, W. S., Takahashi, T., Simpson, H. J. and Peng T.-H., 1979, Fate of fossil fuel carbon dioxide and the global carbon budget. *Science*, v. 206, pp. 409–418.

Brown, J., Colling, A., Park, D., Phillips, J., Rothery, D. and Wright, J., 1989, *Ocean Chemistry and Deep-Sea Sediments*. Pergamon Press, Oxford, 134 pp.

Busey, R. H. and Mesmer, R. E., 1977, Ionization equilibria of silicic acid and polysilicate formation in aqueous sodium chloride solutions to 30°C. *Inorg. Chem.*, v. 16, pp. 2444–2450.

Chen, C-T. A. and Mong-Hsiu H., 1996, A mid-depth front separating the South China Sea water and the Philippine Sea water. *Jour. of Oceanogr.*, v. 52, pp. 17–25.

Deevey, E. S., Jr., 1973, Sulfur, nitrogen and carbon in the atmosphere. In G. M. Woodwell and E. V. Peacan (eds.) *Carbon and the Biosphere*. U. S. Atomic Energy Commission, CONF-720510, Washington, D.C., pp. 182–190.

Dickson, J. A. D., 2002, Fossil echinoderms as monitor of the Mg/Ca ratio of Phanerozoic oceans. *Science*, v. 298, pp. 1222–1224.

Drever, J. I., 1988, *The Geochemistry of Natural Waters*, 2nd edition. Prentice Hall, Englewood Cliffs, NJ, 437 pp.

Garrels, R. M., Mackenzie, F. T. and Hunt, C., 1975, *Chemical Cycles and the Global Environment Assessing Human Influences*. William Kaufmann, Inc., Los Altos, CA, 206 pp.

Gibbs, R. J., 1970, Mechanisms controlling world water chemistry. *Science*, v. 170, pp. 1088–1090.

Graham, U. M., Bluth, G. J. and Ohmoto, H., 1988, Sulfides-sulfate chimneys on the East Pacific rise, 11° and 13°N latitude. Part I. Mineralogy and paragenesis. *Can. Mineral.*, v. 26, pp. 487–504.

Hardie, L. A., 1967, The gypsum-anhydrite equilibrium at one atmosphere pressure. *American Mineralogist*, v. 52, pp. 171–185.

Havie, C. E., Weare, J. H., Hardie, L. A. and Eugster, H. P., 1980, Evaporation of seawater: Calculated mineral sequences. *Science*, v. 208, pp. 498–500.

Hedges, J. I., Baldock, J. A., Gelinas, Y., Lee, C., Peterson, M. L. and Wakeham, S. G., 2002, The biochemical and elemental composition of marine plankton: A NMR perspective. *Marine Chemistry*, v. 78, pp. 47–63.

Hem, J. D., 1985, Study and interpretation of the chemical characteristics of natural water, 3rd edition. *U.S. Geological Survey Supply Paper*, no. 2254, 264 pp + inset.

Hofmeister, A. M. and Criss, R. E., 2005, Earth's heat flux revised and linked to chemistry. *Tectonophysics*, v. 395, pp. 159–177.

Horita, J., Zimmermann, H. and Holland, H. D., 2002, Chemical evolution of seawater during the Phanerozoic: Implications from the record of marine evaporates. *Geochim. Cosmochim. Acta*, v. 66, pp. 3733–3756.

Koski, R. A., Shanks, III, W. C., Bohrson, W. A. and Oscarson, R. L., 1988, The composition of massive sulfides from the sediment-covered floor of Escanaba Trough, Gorda Ridge: Implication for depositional processes. *Can. Mineral.*, v. 26, pp. 655–673.

Langmuir, D., 1997, *Aqueous Environmental Geochemistry*. Prentice Hall, Upper Saddle River, NJ, 600 pp.

Likens, G. E., Bormann, H. F. and Johnson, N. M., 1981, Interactions between major biogeochemical cycles in terrestrial ecosystems. In Likens, G. E. (ed.) *Some Perspectives of the Major Biogeochemical Cycles*. SCOPE 17, Wiley, New York, pp. 93–112.

Mackenzie, F. T., 1998, *Our Changing Planet: An Introduction to Earth System Science and Global Environmental Change*, 2nd edition. Prentice Hall, Upper Saddle River, NJ, 486 pp.

Murray, J. W., 2001, Ocean Carbonate System: Control, Chapter 13. *Chemical Oceanography* OCN400 website, University of Washington, Seattle.

O'Neill, P., 1993, *Environmental Chemistry*, 2nd edition. Chapman & Hall, London, UK, 268 pp.

Pearson, P. N. and Palmer, M. R., 2000, Atmospheric carbon dioxide concentrations over the past 60 million years. *Nature*, v. 406, pp. 695–699.

Redfield, A. C., Ketchum, B. H. and Richards, F. A., 1963, The influence of organisms on the composition of seawater. In Hill, M. N. (ed.) *The Sea*. Wiley, New York, pp. 26–77.

Stallard, R. F. and Edmond, J. M., 1983, Geochemistry of the Amazon. 2. The influence of the geology and weathering environment on the dissolved load. *J. Geophys. Res.*, v. 88, pp. 9671–9688.

Stewart, F. H., 1963, Marine evaporties. *Data of Geochemistry*, 6th edition. U.S. Geol. Surv. Prof. Paper 440Y, 52 pp.

Stumm, W. and Morgan, J. J., 1981, *Aquatic Chemistry: An Introduction Emphasizing Chemical Equilibria in Natural Waters*, 2nd edition. Wiley-Interscience, New York, 780 pp.

Takahashi, T., Broecker, W. S., Bainbridge, E. and Weiss, R. F., 1980, Carbonate chemistry of the Atlantic, Pacific and Indian oceans: The results of the

GEOSECS expedition, 1972–1978, Technical Report No. 1, CU-1-80, Lamont-Doherty Geological Observatory, Palisades, NY, 15 pp.

Tyrrell, T. and Zeebe, R. E., 2004, History of carbonate ion concentration over the last 100 million years. *Geochim. Cosmoch. Acta*, v. 68, pp. 3521–3530.

Usiglio, M. J., 1849, Études sur la composition de l'eau de la Méditerranée et sur l'exploitation des sels qu'elle contient. *Ann. Chim. Phys.*, v. 27(3), pp. 172–191.

Van Andel, T. H., 1975, Mesozoic/Cenozoic calcite compensation depth and the global distribution of calcareous sediments. *Earth and Planetary Science Letters*, v. 26, pp. 187–194.

Whiteman, K. J., Vaccaro, J. J., Gontheir, J. B. and Bauer, H. H., 1994, The hydrologic framework and geochemistry of the Columbia River Plateau aquifer system, Washington, Oregon, and Idaho. *U.S. Geological Survey Professional Paper*, 1413-B, 73 pp.

Zimmerman, H., 2000, Tertiary seawater chemistry: Implications from primary fluid inclusions in marine halite. *American Jour. Science*, v. 300, pp. 723–767.

8 Chemistry of Igneous Rocks

Igneous rocks are those that have formed from a cooling magma produced from melting in the earth or other celestial bodies. On the earth, solidification of the magma can occur either at depth, producing an *intrusive* rock, or rise to the surface, producing an *extrusive* rock. The melting process and the composition of the rocks melted cannot be directly observed because the high temperatures needed to produce silicate melts occur at significant depths in the earth. Inferences about their origin must be made from the compositions of igneous rock samples from exposures at the surface and those available from drill cores. Direct samples of the mantle from which most magmas are thought to be produced are preserved in nodules in *basalt* and xenoliths from *kimberlite* bodies. Models are then constructed that explain the observed compositional trends found in a suite of igneous rocks. The process involves determining the composition of the parental material and magma, as well as the history of the magma from its place of origin to its site of final crystallization and cooling. This requires unraveling a number of interrelated chemical and physical processes. A number of chemical questions can be considered: What is the composition of the source rock for the magma? Did the magma's composition change from its source to its emplacement location, or is it a *primary magma*? What caused the trend in compositions observed in igneous rocks from field localities?

■ Composition of Igneous Rocks

All of the earth's solid core and mantle are composed of igneous rocks. Also, approximately 95% of the earth's crust is igneous. The main constituent of common igneous rocks is SiO_2 with concentrations generally between 40 and 78 wt%. Al_2O_3 is the next most abundant oxide component, typically consisting of 12 to 18 wt%. The nomenclature for igneous rocks is based on four major divisions according to their SiO_2 concentration. With increasing SiO_2, these are ultramafic, mafic, intermediate, and felsic, as outlined in **Figure 8-1**.

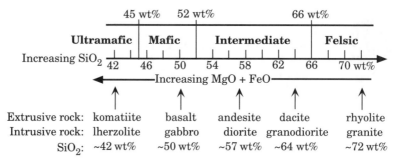

FIGURE 8-1 Major subdivisions of igneous rocks showing their range of SiO₂ in wt%. Also indicated is the most common extrusive and intrusive rock type in each subdivision.

The older terms ultrabasic, basic, and acidic can still be found in the literature but are being replaced because Si species in magmas are not the weak acids they are in aqueous solutions as was thought when these terms were first introduced. Figure 8-1 also gives the name and approximate silica content of the dominant extrusive and intrusive igneous rock type in each division. The major minerals that make up the indicated intrusive igneous rock type are shown in **Figure 8-2**. For extrusive rocks, the minerals of lower SiO₂ are typically present as *phenocrysts* in the rock, whereas the more SiO₂-rich phases appear in the *groundmass*. Phenocrysts are large crystals in a matrix of fine material termed the groundmass. Note the significant volume of feldspars in all of the

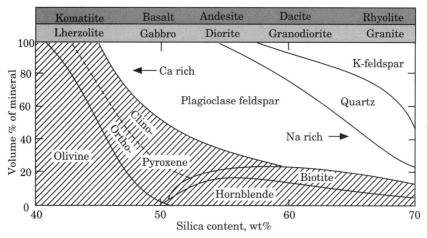

FIGURE 8-2 Volume percent of minerals in intrusive igneous rocks as a function of silica content. The lined region of the diagram indicates the mafic (dark-colored minerals), whereas the unlined area gives the felsic (light-colored) minerals in the rock. The light gray bar on the top of the diagram outlines the name of the intrusive igneous rock for the indicated range of SiO₂ content. The more dark gray bar specifies the equivalent extrusive rocks of similar composition.

rock types except ultramafic rocks. The plagioclase feldspars exhibit *solid solution* behavior where Ca-rich plagioclase (large *anorthite* component) exist in mafic rocks, whereas Na-rich plagioclase (large *albite* component) are found in rocks of greater SiO_2 content.

The most common rock type in the earth's crust is basalt, whose magma was produced from partial melting and *differentiation* in the mantle. The mantle has a *peridotite* composition (Chapter 2 has a more detailed discussion of the mantle's composition). Peridotite is a family of ultramafic rocks that is primarily made up of olivine with or without pyroxene. It contains an aluminum-rich phase as a common accessory. With increasing depth in the mantle, this is plagioclase, chrome spinel, and then garnet. Lherzolite is a type of peridotite that contains enstatite and diopside as its pyroxenes.

Representative compositions of some igneous rocks are given in **Table 8-1**. The variable amount of H_2O present in these rocks (up to 10 wt%) is not included. *Andesite* or its intrusive counterpart, diorite, has a higher Al content than other rock types. *Lherzolite*, on the other hand, has low Al_2O_3 but high MgO content as well as a significant amount of the oxides Cr_2O_3 and NiO. *Rhyolites* and *granites* typically possess high alkali (Na_2O and K_2O) as well as SiO_2 relative to the other rock types but lower ferromagnesium components (FeO and MgO). Finer divisions than shown in Figure 8-2 or Table 8-1 are often recognized. For instance, basalt can be termed *tholeiitic basalt* if it contains *normative hypersthene*, high-Al basalt if $Al_2O_3 > 17$ wt%, or *alkaline basalt* if it

Table 8-1   Average anhydrous compositions of common igneous rock types

	Lherzolite	Basalt	Gabbro	Andesite	Diorite	Rhyolite	Granite
SiO_2	44.2	50.8	51.1	54.2	58.6	73.7	72.0
TiO_2	0.1	2.0	1.2	1.3	1.0	0.2	0.3
Al_2O_3	2.1	14.1	15.9	17.2	17.0	13.5	14.4
Fe_2O_3	0.1	2.9	3.1	3.5	2.6	1.3	1.2
FeO	8.3	9.1	7.8	5.5	5.1	0.8	1.7
MnO	0.1	0.2	0.1	0.2	0.1	0.0	0.1
MgO	42.2	6.3	7.7	4.4	3.7	0.3	0.7
CaO	2.0	10.4	9.9	7.9	6.7	1.1	1.8
Na_2O	0.3	2.2	2.5	3.7	3.6	3.0	3.7
K_2O	0.1	0.8	1.0	1.1	1.8	5.4	4.1
P_2O_5	0.1	0.2	0.2	0.3	0.3	0.1	0.1
Cr_2O_3	0.4						
NiO	0.3						

From Nockolds (1954), LeMaitre (1976), and Maaloe and Aoki (1977).

contains both olivine and augite. Such compositional subdivisions are helpful in discriminating the environments and processes responsible for producing a particular rock.

What controls the composition of a rock crystallized from a magma or the composition of magma derived from melting of a rock? Equilibrium between the rock melted and magma occurs during most melting processes because of the elevated temperatures considered. From a thermodynamic standpoint, the controlling factors are then pressure, temperature, and the chemical composition of the system. Complexities typically arise after the initial melt has formed in equilibrium with its source because the magmas are not closed to mass or heat exchange and the processes that form them typically occur over a range of pressure and temperature. Of primary concern in changing the composition of a primary magma are magma recharge into a resident magma chamber, assimilation of surrounding rocks, fractional crystallization processes, and magmatic degassing. As a result, many different models have been developed to explain the occurrence and distribution of igneous rocks. To help understand these models, some of the basic processes involved can be considered in isolation.

■ Characteristics of Magma

The temperatures of magmas erupted at the earth's surface depend on their composition. This ranges from approximately 650°C for silicic rhyolite magmas to >1300°C for komatiite magmas. Water exhibits an important control on decreasing melting temperature relative to the anhydrous magma. Magmas typically contain dissolved H_2O in an amount that is both pressure and compositionally dependent. It ranges up to about 3 wt% in basalt, whereas intermediate and felsic magmas typically contain between 2 and 6 wt% H_2O. Magma densities are generally between 2.2 and 3.2 g cm^{-3}, with ultramafic melts having the greatest densities. Except for some mafic and ultramafic melts, magma is less dense than typical crustal rocks. This lower density relative to the enclosing solid rock makes melts buoyant in the earth's crust, causing them to rise toward the surface. Because the *compressibility* of magma is greater than that of the equivalent rock, the density contrast decreases with depth (Bottinga, 1985). This can cause ultramafic magmas to sink or not separate from the residual rock that produced it at deep levels in the earth's crust. Some have argued the sinking of ultramafic material into the mantle accounts for the fact that average continental crust is more SiO_2 rich than the basalt that is currently being derived from melting in the mantle.

Magmas have a broad range of viscosities. *Viscosity* is the response of a liquid to *shear stress*, determined from the tangential force needed to move a planar surface along its plane when separated from a similar surface by the liquid.

It is given as the ratio of shear stress-to-*shear strain rate* and has units of poise, which is equivalent to 0.1 Pa s. Viscosity can be thought of as a measure of a liquid's ability to restrain flow. The viscosity of a silicate magma increases with increased silica content. Low silica magmas of basalt composition have low viscosities and flow easily when erupted onto the earth's surface relative to high silica magmas of rhyolitic composition. This increased viscosity with increasing silica content means that silica-rich melts have more difficulty in rising toward the earth's surface after they are produced. This is one reason that there is a dominance of silica-rich igneous rocks that have crystallized in the crust (i.e., granites), whereas low silica rocks (i.e., basalts) more commonly reach the earth's surface as lava flows. Magma viscosity decreases with increasing temperature as well as alkali and water content. Viscosities can be as low as that of glycerol ($C_3H_8O_3$) at room temperature (approximately 10^{14} poise) for a high temperature ultramafic alkali-rich lava to that as high as glacial ice (approximately 10^{14} poise) for a dry rhyolitic magma near its crystallization temperature. The differences in viscosities reflect differences in the arrangement of atoms in the magma (Carmichael et al., 1974). The greater the interconnection of atoms (i.e., polymerization), the greater is the viscosity. Given their viscosity, silicate melts must have a significant interconnected structure that increases with increasing silica content but decreases with increasing H_2O.

Most silicate melts can be readily quenched to form glass, an amorphous metastable solid. *Raman spectroscopy* and x-ray analysis of glasses indicate they contain $(Si,Al)O_4$ tetrahedra that are polymerized into chains/networks, as shown schematically in **Figure 8-3**. The more Si in the glass, the more polymerized it is. Pure SiO_2 glass has a completely polymerized three-dimensional structure analogous to the regular crystal structures of quartz and feldspar minerals. Are

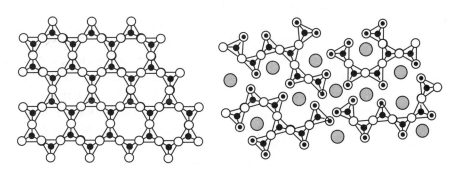

Pure SiO_2 glass SiO_2 glass with additional cations

FIGURE 8-3 Schematic diagrams of the structure of pure SiO_2 glass and glass that contains alkali and alkaline earth cations. Open circles are bridging oxygen atoms between silicon atoms, whereas open circles with an interior solid dot are nonbridging negatively charged oxygens. Solid circles in the tetrahedra are tetrahedrally coordinated Si and Al. The gray circles give the location of network modifying alkali and alkaline earth cations.

these chains/networks of (Si,Al)O$_4$ in the magma charged? Electrical conductance measurements indicate that magmas are somewhat ionic but much less charged than aqueous salt solutions. Magma can, therefore, be considered to be made up of negatively charged chains and networks of (Si,Al)O$_4$ tetrahedra together with alkalis, alkaline earth, and transition metal ions such as Na$^+$, K$^+$, Ca^{2+}, Mg^{2+}, and Fe^{2+} between the networks of (Si,Al)O$_4$ tetrahedra but in close association with them. This structure of magmas is confirmed by the low entropy of fusion of the glass, indicating little breakup of the structure of the glass on melting. Alkali and alkaline earth cations, therefore, are probably held in place by the networks of negatively charged (Si,Al)O$_4$. The degree to which they are fixed to particular sites in the melt is not known with any certainty.

■ Melting of Rocks

Some rocks made up of a single mineral phase and most rocks made up of more than one mineral phase melt *incongruently*. When incongruent melting occurs, the initial melt is different in composition than the average of the parent rock, the first *partial melt* typically being more silica rich. For instance, partial melting of a few percent of peridotite produces a basaltic magma. Because the earth's crust initially formed as a partial melt of the mantle, it is thought that the crust's original composition was basaltic. The basalt rock that was produced underwent later partial melting and produced the even more Si-rich rocks of granodiorite and granite found in the upper continental crust. The more mafic residual being denser than basalt is thought by many investigators to delaminate from the crust and sink into the mantle. Other investigators believe the presence of water during the initial melting of peridotite to form crust produced an original crust that was compositionally zoned with a more andesitic composition of upper crust and basaltic lower crust.

Anhydrous pressure–temperature melting curves for some common silicate minerals are shown in **Figure 8-4**. They all have positive slopes. Most of these pure-phase melting equilibria are congruent at low pressures, but K-feldspar melts incongruently to leucite plus melt. The lower slope of the quartz melting curve indicates quartz melts at a lower temperature than anorthite or forsterite at low pressures but at greater temperatures than these two minerals at high pressures. These slopes can be considered in the framework of the *Clapeyron equation* for the melting reaction, which can be written as

$$\frac{dP}{dT} = \frac{\Delta S_{fus}}{\Delta V_{fus}}$$

[8.1]

where ΔS_{fus} is the entropy of fusion and ΔV_{fus} is the volume of fusion to form the melt. For albite, the congruent melting reaction can be written as

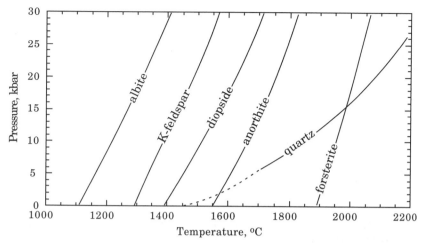

FIGURE 8-4 Pressure–temperature anhydrous melting curves for the indicated pure mineral. (Adapted from Boyd and England, 1963; Burnham, 1997.) The dashed line is the metastable extension of quartz melting into the cristobalite stability field.

$(NaAlSi_3O_8)_{solid} = (NaAlSi_3O_8)_{melt}$, where the subscripts solid and melt denote the solid mineral and melt phase, respectively. Because the melt phase is the high entropy phase, the positive slope shown for albite and the other minerals in Figure 8-4 implies

$$V_{solid} < V_{melt} \qquad\qquad [8.2]$$

This is typical for most types of solid–liquid equilibria (e.g., precipitates are observed to sink to the bottom of a beaker). The implication of the low slope of the quartz fusion *P-T* boundary is that the entropy of fusion for quartz is low. The three-dimensional structure in pure SiO_2 solid is preserved in the melt to a greater extent than is the structure in aluminosilicate melts that contain less silica. The equilibrium boundaries are not straight lines but have a small curvature because ΔV_{fus} decreases with increasing pressure. This occurs because the compressibilities of the melts are greater than the minerals.

Now consider melting in the presence of water. The melting relations for both anhydrous and H_2O-saturated conditions for quartz, diopside, and albite are shown in **Figure 8-5**. H_2O saturated means that the system is in equilibrium with an H_2O phase, and, therefore, the phases have their maximum H_2O content. This water enters the melt but not the mineral phase. Because the H_2O "dilutes" the activity of the mineral component in the melt, the introduction of H_2O lowers the temperature of the melting point of the mineral. Hydrous magmas are stable to lower temperatures at higher pressures because more H_2O can be dissolved in the melt with increased pressure. Unlike the anhydrous case, these equilibria have a negative *P-T* slope. This implies

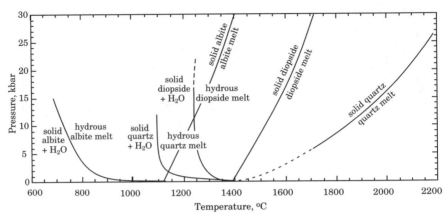

FIGURE 8-5 Melting curves as a function of pressure and temperature for the indicated mineral in anhydrous and H_2O-saturated systems. (Adapted from Burnham, 1997.)

$$V_{melt} < V_{solid} + V_{H_2O} \qquad [8.3]$$

Relative to a single mineral, the temperature of onset of melting is also lower for an assemblage of minerals. This occurs because there are *eutectics* between most mineral phases. That is, each successive mineral dissolves in the melt, diluting the activity of the other mineral phases in the melt and lowering its melting temperature. When three or more solid phases are involved, the lowest temperature that can be achieved is termed the *cotectic* temperature (see the discussion below about eutectics and cotectics). A familiar eutectic exists between ice and solid NaCl (halite). If salt is added to ice, the melting temperature of the ice is lowered. The lowest melting temperature that can be achieved with addition of NaCl is at the point where the liquid H_2O becomes saturated with hydrohalite, about $-21°C$ at 1 bar. This temperature is the eutectic temperature for ice–halite mixtures.

The melting of hydrous minerals is somewhat different from that for anhydrous minerals. At low pressure, the hydrous mineral dehydrates with increasing temperature before it melts. Consider the system quartz plus muscovite, as shown in **Figure 8-6**. At pressures greater than about 6 kbar, muscovite plus quartz reacts directly to an H_2O-rich melt with the precipitation of an aluminosilicate, sillimanite, at the muscovite + quartz eutectic according to the reaction

$$KAl_3Si_3O_{10}(OH)_2 + SiO_2 = Al_2SiO_5 + KAlO(OH)^+_{melt} \bullet Si_3O_6(OH)^-_{melt} \qquad [8.4]$$

muscovite   quartz   sillimanite

The equilibrium pressure and temperature is given by the short-dash lines in Figure 8-6. The reason for writing the melt species as $KAlO(OH)^+_{melt} \bullet Si_3O_6(OH)^-_{melt}$ is discussed below. At pressures less than about 6 kbar with increasing temperature, the dehydration reaction

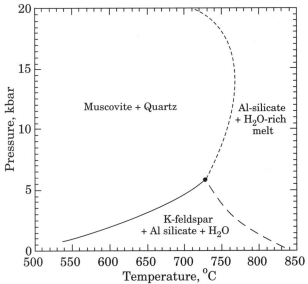

I FIGURE 8-6 Pressure–temperature phase relations for muscovite + quartz breakdown.

$$KAl_3Si_3O_{10}(OH)_2 + SiO_2 = KAlSi_3O_8 + Al_2SiO_5 + H_2O \tag{8.5}$$

occurs as given by the solid line in Figure 8-6. The aluminosilicate mineral produced can be determined by consulting Figure 5-9. The rock now consists of K-feldspar + Al-silicate together with an H_2O phase. If it continues to be heated and the H_2O remains with the rock, melting occurs by the reaction

$$KAlSi_3O_8 + H_2O = KAlO(OH)^+_{melt} \bullet Si_3O_6(OH)^-_{melt} \tag{8.6}$$

which has a negative slope given by the long-dash lines in Figure 8-6.

The P-T slopes of the muscovite + quartz melt equilibrium boundary increase with increasing pressure as the volume of the melt decreases (see Equation 8.1). When $\Delta V_{fus} = 0$, the slope of the boundary becomes infinite. At pressures greater than about 15 kbar, ΔV_{fus} becomes negative so that

$$V_{melt} < V_{solids} \tag{8.7}$$

and the melting boundary changes to a negative slope. This change from a positive to a negative slope occurs with melting of most hydrous solid phases with increased pressure.

The pressure–temperature stability relations for hornblende are given in **Figure 8-7**. Similar to muscovite, increasing the temperature of hornblende below about 1 kbar causes hornblende to react to form pyroxene plus an H_2O phase with a positive P-T slope. At higher pressures with increasing temperature,

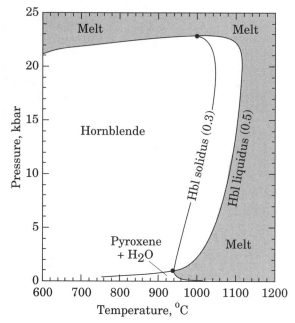

FIGURE 8-7 Pressure–temperature phase relations for hornblende stability. Because of solid solution in hornblende, the phase boundaries are not exact. The numbers in parentheses on the hornblende solidus and liquidus give the activity of H_2O along the boundary. (Adapted from Burnham, 1979.)

hornblende reacts directly to a melt. Because of the solid solution in hornblende and the fact that the melting is incongruent, the initial melt is not the same composition as the reacting hornblende. The hornblende, therefore, has separate *solidus* and *liquidus* equilibria. The solidus is the locus of points where melting begins, whereas the liquidus is the locus of points in which the solid is completely melted. The relationship between a liquidus and solidus is outlined below. Figure 8-7 indicates that at pressures of 21–22 kbar the melting of hornblende becomes nearly temperature independent as the solidus and liquidus nearly coincide. Hornblende, therefore, can melt by increasing pressure above 21–22 kbar at a constant temperature.

The addition of H_2O to the melt, like the addition of most new components, lowers the melt's Gibbs energy. The Gibbs energy of an H_2O-rich melt can be expressed as a mixture of anhydrous melt (An melt) and H_2O as

$$G_{H_2O\text{-rich melt}} = X_{An\ melt} G^o_{An\ melt} + (1 - X_{H_2O})G^o_{H_2O} +$$

mechanical mixture

$$X_{An\ melt}\ RT \ln X_{An\ melt} + (1 - X_{H_2O})\ RT \ln (1 - X_{H_2O}) + G_{nonideal}$$

ideal mixing   nonideal mixing

[8.8]

where the subscript, nonideal, indicates the contribution for the nonideal interactions during mixing in the melt. Because $V_{H_2O-rich\ melt}$ does not equal V_{solid} + V_{H_2O}, as must be the case for ideal mixing, nonideal effects need to be considered for hydrous melt equilibria. Experimental evidence indicates that $G_{nonideal}$ is a significant negative number in Equation 8.8, lowering the Gibbs energy of the hydrous melt and making it more stable relative to ideal mixing of the mineral + H_2O components. It is thought that H_2O reacts with bridging oxygen atoms between Si and Al atoms to produce OH functional groups that break apart the (Si,Al)–O chains in the melt.

Consider the reaction of water with albite component in the melt

$$n\text{NaAlSi}_3\text{O}_8(\text{An melt}) + \text{H}_2\text{O} = (n\text{NaAlSi}_3\text{O}_8\bullet\text{H}_2\text{O}) \qquad [8.9]$$

where n is the number of $NaAlSi_3O_8$ (= 262.2 g) molar units in a melt species relative to 1 mole of H_2O (= 18.01 g). Assume that n equals 1 and the mole fraction of H_2O in the melt is 0.5. On a wt% basis, the concentration of H_2O is then

$$\text{wt\% (H}_2\text{O)} = 18.01/(18.01 + 262.2) \times 100 = 6.4\ \text{wt\%} \qquad [8.10]$$

If n is 2 with a mole fraction of H_2O of 0.5,

$$\text{wt\% (H}_2\text{O)} = 18.01/(18.01 + 2 \times 262.2) \times 100 = 3.3\ \text{wt\%} \qquad [8.11]$$

Clearly, if component species are extended polymers, a few wt% of H_2O is equivalent to a large mole fraction of H_2O in the melt. A small wt% of H_2O, therefore, has a large effect on melting temperatures because it is present at a relatively large mole fraction (ideal mixing effects) and it depolymerizes the melt (nonideal effects), increasing the hydrous melt's entropy by creating more species and therefore lowering its Gibbs energy.

■ Melt Speciation Reactions

What are the mole fractions of melt species? A model of speciation in the melt must first be constructed. Relative to aqueous solutions, melt speciation models are in their infancy. One of the most widely cited is the Burnham model (1979, 1982, 1994), a quasicrystalline speciation model of aluminosilicate melts. Near their liquidus temperatures, the melt phase is considered to contain species that are similar in stoichiometry and structure to that found in the mineral phases on the liquidus based on an eight oxygen stoichiometry. These species are considered to mix ideally with each other in the melt. This is consistent with the measured low entropy of fusion of most aluminosilicate melts.

Melt species are considered to be neutrally charged with mineral-like structures but without the long range order found in mineral structures. With the Burnham model, the albite melting reaction is written

$$(NaAlSi_3O_8)_{solid} = (NaAlSi_3O_8)_{melt} \qquad [8.12]$$

where it is understood that $(NaAlSi_3O_8)_{solid}$ is part of the continuous albite mineral structure, whereas $(NaAlSi_3O_8)_{melt}$ is albite melt in units of $NaAlSi_3O_8$ that can polymerize and depolymerize over short time intervals and, therefore, does not have the three-dimensional structure found in albite crystals. Melts formed from multiphase silicate assemblages are similarly assumed to contain species that mimic those that crystallize from them. For incongruently melting minerals, such as with sanidine ($KAlSi_3O_8$), which produces the mineral leucite ($KAlSi_2O_6$) on melting, the Burnham melting reaction is

$$5KAlSi_3O_8 = KAlSi_2O_6 + 3(K_{1.33}Al_{1.33}Si_{2.67}O_8)_{melt} + 1.25(Si_4O_8)_{melt} \qquad [8.13]$$
$$\text{sanidine} \qquad \text{leucite}$$

The melt then contains a leucite-type species $(K_{1.33}Al_{1.33}Si_{2.67}O_8)_{melt}$. From mass balance a quartz-like species, $(Si_4O_8)_{melt}$, is also produced.

This model can be contrasted with the Ghiorso model (Ghiorso and Sack, 1995; Ghiorso et al., 1983) where component species in the melt are currently considered to be simple oxide phases for which Gibbs energy of mineral fusion are available. For instance, congruent albite melting is given by

$$NaAlSi_3O_8 = 0.5\ Na_2SiO_{3(melt)} + 0.5\ Al_2O_{3(melt)}) + 2.5\ SiO_{2(melt)} \qquad [8.14]$$
$$\text{albite}$$

where the energetics of any albite melt structure preserved in the melts is accounted for with regular solution mixing of these components species (see Chapter 4 for a discussion of regular solutions). Software packages, MELTS and pMELTS, are available to do the calculations (http://melts.ofm-research.org/) (Asimow and Ghiorso, 1998; Ghiorso and Sack, 1995; Ghiorso et al., 2002). The model is designed to calculate crystallization and melting in multicomponent natural magmas. It is less accurate in predicting crystallization and melting in simple experimentally determined binary and ternary systems across endmember compositions.

■ Water in Magmas

If melt species are some sort of silica polymer, then to understand the role of components such as H_2O and CO_2 in melts, their effects on the structure of the magma species must be determined. If CO_2 is added to H_2O, species such as H_2CO_3 and HCO_3^- are formed. How do CO_2 and H_2O added to the melt affect melt species? OH units are present in hydrous silica glasses, and the viscosity of melt decreases with the addition of water. Unlike other components in the melt, H_2O is therefore thought to break a bridging oxygen to silicon bond in the silica structure by a reaction such as

$$-Si-O-Al- + H_2O = -Si-OH + HO-Al-O- \qquad [8.15]$$

where the short horizontal lines (–) denote silicon/alumina–oxygen bonds. The silicate species denoted in Reaction 8.15 can be chains or perhaps three-dimensional extended polymer structures that decrease their polymerization by Reaction 8.15. Burnham and Davis (1971, 1974) determined the solubility of H_2O in albite melts. They noted that if aluminosilicate melt species were considered to be similar to mineral species from which they were formed written on an eight oxygen basis, the molar solubility of H_2O was similar in all aluminosilicate melts at the same pressure with very little temperature dependence. **Figure 8-8** shows the effect on melting for various activities of H_2O in an albite melt from anhydrous to H_2O saturated as calculated with the Burnham model.

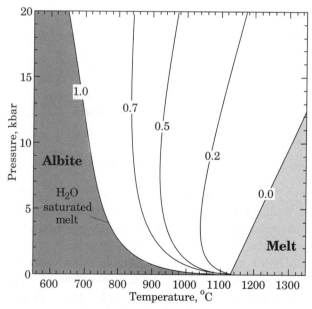

FIGURE 8-8 Pressure–temperature phase diagram for melting of albite plus H_2O contoured for H_2O activities in the magma. The standard state for H_2O stipulates that $a_{H_2O} = 1$ at H_2O saturation in the melt at the P and T of interest. (Adapted from Burnham, 1979.)

At elevated activities of H_2O in the melt, increasing pressure causes melting to occur, whereas at low a_{H_2O}, increasing pressure causes crystallization.

Standard states on which activities are defined can be chosen for the convenience of interpretation of the results. Burnham chose a standard state for H_2O in melts as pure H_2O at the mole fraction of water in the H_2O-saturated melt at the pressure and temperature of interest. At 800°C and 4 kbar, an H_2O-saturated melt contains $X_{H_2O}^{melt} = 0.55$ and $X_{Ab} = 0.45$; therefore, at 800°C and 4 kbar with $a_{H_2O} = \Gamma_{H_2O} X_{H_2O}^{melt} = 1$, $\Gamma_{H_2O} = 1/0.55 = 1.82$. If two melt species are produced, however, an expression like the following is more appropriate:

$$a_{H_2O} = k(X_{H_2O}^{melt})^2 \qquad \text{[8.16]}$$

Experiments indicate this is the case at $X_{H_2O} < 0.5$, where k is a constant independent of composition that depends mainly on pressure but is somewhat temperature dependent as well. At 800°C and 4 kbar, k = 3.31. Equation 8.16 is a Henry's law expression in the square of concentration, similar to the situation with NaCl activities in aqueous solutions that dissolve, producing Na^+ and Cl^- species (see Chapter 4 for a discussion of Henry's law for NaCl in water). Apparently, when H_2O is added to the melt, it disassociates to produce two species. Knowing the activity of H_2O in the melt allows melt–mineral equilibrium to be determined at H_2O-undersaturated conditions.

In the Burnham model for albite, Reaction 8.15 is given by

$$(NaAlSi_3O_8)_{melt} + H_2O = NaAlO(OH)^+_{melt} \bullet Si_3O_6(OH)^-_{melt} \qquad \text{[8.17]}$$

where $\bullet$ is used to indicate significant interaction between the positively and negatively charged species produced in the melt. At mole fractions of $H_2O > 0.5$, Burnham (1997) argued that associated H_2O begins to depolymerize the sheets in the melt by breaking the $(Al,Si)O_4$ networks into chains by a reaction such as

$$NaAlO(OH)^+_{melt} \bullet Si_3O_6(OH)^-_{melt} + H_2O = NaAlO(OH)^+_{melt} \bullet Si_3O_6(OH)^-_{melt} \bullet H_2O_{melt} \qquad \text{[8.18]}$$

This is similar to the reaction of H_2O proposed for Si_2O_8 melts. The model of aluminosilicate melts explains the lower molar solubility of H_2O in SiO_2 melts than in albite melts even when calculated on an Si_4O_8 basis. The pure SiO_2 melts lack the exchangeable sites found with the AlO_4 tetrahedra so that H_2O remains as a molecular species in the melt. CO_2 also appears to enter the melt as a molecular species and has little interaction with melt species. This is why CO_2 has a much lower solubility than H_2O in aluminosilicate melts.

Stolper (1982) and Silver and Stolper (1985, 1989) demonstrated that both hydroxyl (–OH) and molecular water (H_2O) are present in the melt, with molecular water dominating at high water concentrations and derived values of the equilibrium constant of Reaction 8.15. From this, for a particular melt composition, H_2O solubilities can be calculated. Their calculation procedure is quite accurate but does not allow the direct determination of H_2O solubilities of unmeasured compositions. It assumes a constant molar volume of H_2O in the melt as a function of pressure and temperature. The Ghiorso model uses the Nicholls (1980) regular solution model to calculate the activity of water in the melt from its mole fraction based on the activity of albite in the melt given by Burnham and Davis (1974) determinations from Reaction 8.16.

Figure 8-9 shows the possible albite–water melting relations at low, medium, and high pressure from Paillat et al. (1992). A mole of albite is given by $NaAlSi_3O_8$ for these determinations. At high enough temperature, there is complete solution between liquid albite and water. At low pressures (approximately 1 kbar), however, the coexisting magma–H_2O vapor region intersects anhydrous albite at its boiling temperature (Figure 8-9, point A). The maximum H_2O in the magma increases with decreasing temperature from its boiling temperature to the eutectic at point B. As an H_2O-saturated melt cools, it therefore becomes undersaturated with H_2O.

As pressure is increased (approximately 5 kbar), magma is produced at lower temperatures, and the coexisting magma–H_2O vapor region no longer extends to anhydrous albite compositions. This produces a critical point (Figure 8-9, point C) where magma and vapor are the same composition. Solubility of H_2O in the magma is a minimum at point E. A magma at point D with further cooling therefore produces a separate vapor phase that is reabsorbed at point F if it stays with the magma. At higher pressures (approximately 12 kbar) as an H_2O-saturated melt cools, it becomes undersaturated with H_2O similar to the situation

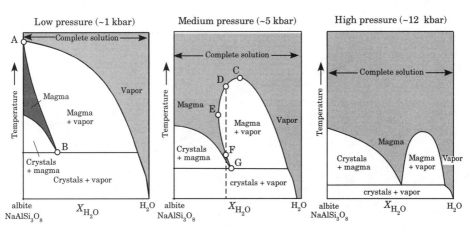

FIGURE 8-9 Possible high-temperature phase relations for albite–water at low, medium, and high pressure. (Adapted from Paillat et al., 1992.)

near 1 kbar. The relations given in Figure 8-9 are combined in **Figure 8-10** to show the pressure–temperature–mole fraction of H_2O stability of albite melts.

The effects of other components on the relations for hydrous albite melts shown in Figure 8-10 are to depress the melting temperatures. Volatile components (e.g., CO_2) preferentially partition into the vapor phase, expanding its field of stability. Nonvolatile components such as FeO and MgO expand the magma stability field. Using the Burnham model of ideal mixing of component magma species, the stability of olivine tholeiite basalt is given in **Figure 8-11**. Because all of the aluminosilicate melt species are assumed to mix ideally, the

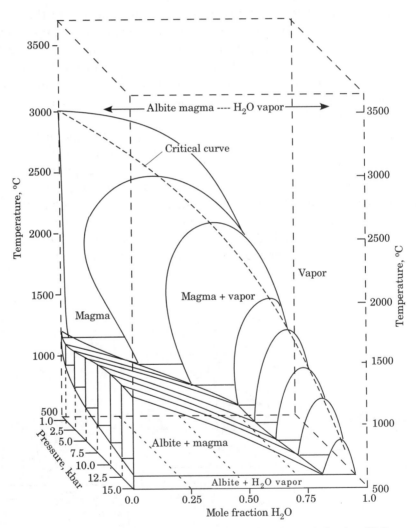

FIGURE 8-10 Schematic albite–water phase diagram as a function of the mole fraction of H_2O, pressure, and temperature. (Adapted from Paillat et al., 1992.)

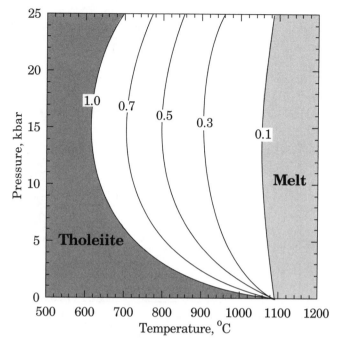

FIGURE 8-11 Pressure–temperature diagram for the beginning of melting (solidus) of olivine tholeiite at the indicated activity of H_2O in the melt. (Adapted from Burnham, 1979.)

melting of a complicated rock system that included water is similar to that shown in Figure 8-8. By adding additional mineral component species to the albite component to obtain a basalt melt, the temperature of initial melting is lowered by about 100°C. It is clear from Figure 8-11 that at pressures greater than about 15 kbar the isopleths of H_2O activity change from negative to positive. At these pressures, V_{H_2O} becomes very small because of its relatively large compressibility under pressure so that

$$V_{\substack{anhydrous \\ tholeiite}} + V_{H_2O} < V_{\substack{hydrous \\ tholeiite}} \qquad [8.19]$$

and the equilibrium phase boundary slopes become positive as calculated with Equation 8.1.

When considering other rock types in the earth, the lines in Figure 8-11 shift somewhat. The general relations outlined in Figure 8-11, however, still hold for any aluminosilicate rock system because of the ideal mixing of melt species in the Burnham model. The stability relations for hornblende from Figure 8-7 can be superimposed on Figure 8-11, as shown in **Figure 8-12**. Because metamorphosed hydrous olivine tholeiite becomes *amphibolite* with depth in the earth's crust, the hydrous olivine tholeiite field is given as amphibolite in the figure.

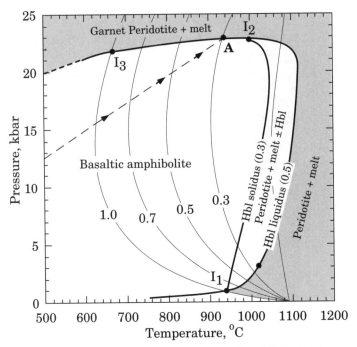

FIGURE 8-12 Pressure–temperature melting relations for amphibolite of olivine tholeiite composition from Figure 8-7 as heavy lines. Amphibolite with hornblende (Hbl) is stable on the lower temperature side of the pseudo-univariant boundary I_1–I_2–I_3. The lines in Figure 8-11 have been superimposed in light outline. (Adapted from Burnham, 1982.)

The diagram can then be used to understand reactions producing magmas in subduction zone environments from amphibolite melting. Oceanic lithosphere, whose top layer consists of hydrated olivine tholeiite basalt produced at mid-oceanic ridges, is subducted, together with some intercalated sediments, into the mantle at oceanic trenches. The angle of subduction varies from 20° to about 50° and occurs at rates between 2 and 10 cm/yr. Most of the water-rich sediments are "scraped off" the descending plate to build *accretionary prisms* (**Figure 8-13**).

Pressure and temperature increase in the hydrated tholeiite basalt as it is subducted. The rock first reacts to greenshist facies mineral assemblages. Hydration reactions absorb some of the pore fluid that is present, producing chlorite. The remaining water ascends into the accretionary prism above along fractures because of its lower density relative to the rock. At greater depths, the chlorite formed begins to break down, forming hornblende and releasing water. The water rises into the *mantle wedge* above, producing hornblende there as well. With further descent and heating of the subducted basaltic rock at the top of the descending plate, as temperatures increase to 550°C to 600°C,

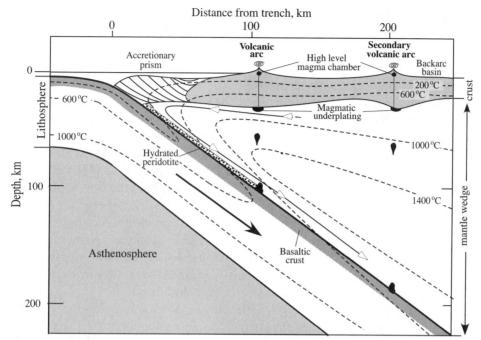

FIGURE 8-13 Cross-section through downgoing lithosphere indicating the location of basaltic crust and hydrated peridotite that can melt at the indicated locations, giving rise to volcanic arcs. The dashed lines give approximate isotherms in the system. (Adapted from Furukawa, 1993.) The open arrows give the likely movement of material as a function of time in the mantle wedge.

a hydrated amphibolite containing 2–3 wt% H_2O in hornblende is likely present. As the downgoing plate reaches melting conditions, a separate fluid phase is, therefore, not present, but rather it is an amphibolite with 2–3 wt% H_2O. Hornblende is also present in the ultramafic rocks in the wedge of mantle just above the downgoing plate, according to Burnham.

Shown as a dashed line with arrows in Figure 8-12 is a typical trajectory for the amphibolitic part of the downgoing plate. At point A, near 940°C and 23 kbar (approximately 78 km depth), the hornblende in the amphibolite melts, producing hydrous melt plus garnet peridotite. For melt to form at this pressure and temperature, a_{H_2O} in the melt must be greater than about 0.3, as given by the thin curved solid lines in Figure 8-12. This requires X_{H_2O} in the melt produced to be ≥0.5 (6.4 wt%). The maximum amount of melt that can be produced from an amphibolite is, therefore, about 2.5 wt% H_2O/6.4 wt% H_2O = 40% melt. For an amphibolite with less water, less melt can be produced. It is also less if the geothermal gradient is lower as higher activities of H_2O are required. For reasonable geothermal gradients for amphibolite in downgoing slabs, no matter what the H_2O content, melting occurs in the small pressure range between 22 and 23 kb (at depths of approximately 75–80 km).

The amount of melt produced is 5% to 40%, and the melt compositions range from *trondhjemite* (a quartz diorite or tonalite with minor mafic minerals) to diorite (Burnham, 1997). Production of the hydrous melt over a very small depth (= pressure) interval indicates why volcanic centers are at a uniform depth to the *Benioff zone* irrespective of the angle of subduction.

Rather than melting occurring in the downgoing plate, many investigators have argued that melting occurs primarily in the hydrated peridotite of the *mantle wedge* just above the subducting amphibolite. It is argued that dewatering phases in the downgoing plate enrich the mantle wedge with the isotopes ^{87}Sr, ^{210}Pb, and ^{10}B as well as with *large ion lithophile* (LIL) *elements*, Nd and B, found in erupted magmas. If melting occurs in the downgoing slab, these elements are thought to be added to the melt by incorporation of a few percent of subducted sediments together with amphibolite melting. In some arc magmas, the heavy rare earth elements (HREEs) and trace elements can be more depleted than in mid-oceanic ridge basalt. McCulloch and Gamble (1991) argued that this depletion is due to their continued extraction from the mantle wedge source as mantle flow is nearly in a closed cell.

After the magma is formed, whether in the downgoing plate or the mantle just above, it rises toward the surface because of its buoyancy. Its density is such, however, that it apparently reaches zero buoyancy at the base of the more silica-rich crust. Underplating of the crust by the magma occurs as shown in Figure 8-13. With the crystallization of more dense phases in this magma, a less dense residual magma is produced that now has a more hydrous and siliceous composition and can ascend into the crust, giving rise to the variable amounts of different rock types from various island arcs that generally have a large component of andesite (Gill, 1981). Because phlogopite is more stable at higher pressure than hornblende, a second volcanic arc can form with phlogopite dehydration at about 200 km above the Benioff zone if H_2O from phlogopite dehydration depresses the melting temperature sufficiently.

■ Eutectics and Melting

For any composition of silicate rock, if the temperature is high enough, it melts completely, producing a single melt phase. This implies that silicate magma at high enough temperature can have complete compositional variability between any endmembers, that is, total *miscibility* in the melt phase as implied by ideal mixing of melt species. Consider the temperature–composition phase diagram between the minerals diopside and anorthite at 1 bar shown in **Figure 8-14**. This is constructed with diopside on an eight oxygen basis to be consistent with the Burnham model.

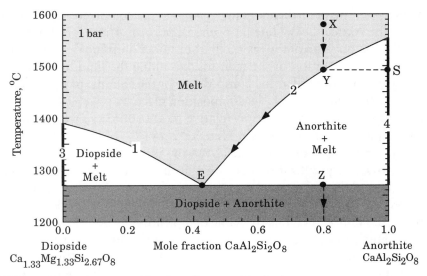

FIGURE 8-14 Temperature–composition phase diagram at 1 bar for the two-component system $Ca_{1.33}Mg_{1.33}Si_{2.67}O_8$ and $CaAl_2Si_2O_8$. (Adapted from Bowen, 1928.)

The compositions of the melt in equilibrium with pure diopside and pure anorthite are given by the lines labeled 1 and 2, respectively. These are termed liquidus lines, which give the composition of the liquid in equilibrium with the solid at the indicated temperature. Because of the significant difference between the crystal structure of diopside, a pyroxene, and anorthite, a feldspar, these minerals show no significant solid-solution behavior for each other, even at high temperatures. The compositions of the solids in equilibrium with the melt (solidus lines) are, therefore, given by the bold vertical line segments, labeled 3 and 4, along the edges of the diagram in Figure 8-14 at the composition of the pure diopside and anorthite phases, respectively.

Figure 8-14 indicates that the expected complete miscibility in the melt phase at temperatures of approximately 1560°C and above gives way to total crystallization of all compositions to pure diopside and anorthite crystals at temperatures below approximately 1270°C. Because of the lack of solution between the solid phases, this system displays a eutectic (given by point E). The eutectic point is the composition and lowest temperature where a melt can exist in this two-component system at a given pressure. The eutectic is also the composition and lowest temperature at which a melt is produced on heating a collection of diopside and anorthite crystals.

How is the crystallization and melting phenomena in this system described? Consider a melt at point X in Figure 8-14 where $X_{CaAl_2Si_2O_8} = 0.8$. As it cools, its composition remains the same until it encounters the liquidus at point Y.

From this point, with further cooling, crystals at the composition of the solidus given by the vertical heavy line (at point S), pure anorthite, are formed. This causes the composition of the melt to become more diopside rich. With further cooling, the composition of the melt follows along the liquidus as more and more crystals are formed at the composition of the solidus, of pure anorthite. At point E, the eutectic, with the composition given by $X_{CaAl_2Si_2O_8} = 0.42$, crystals of diopside in the proportion relative to anorthite as given by the *lever rule* are crystallized from the melt. That is, 42 mole % of the crystals are anorthite, and 58 mole % are diopside (given on an eight-oxygen basis as $Ca_{1.33}$ $Mg_{1.33}Si_{2.67}O_8$). Crystallization continues at the eutectic with this proportion of crystals until the last liquid disappears, which occurs when the total crystals produced become 80 mole % anorthite and 20 mole % of diopside, as given by point Z. Point Z then has the same composition as the starting magma at point X. Further cooling from Z does not affect the makeup of the system.

The phase diagram can also be used to analyze the composition of liquids produced on melting with increasing temperature. Consider a rock made up of 80% anorthite and 20% diopside ($Ca_{1.33}Mg_{1.33}Si_{2.67}O_8$) crystals. On heating, the first liquid to be formed is given by the eutectic composition that melts 42 mole % of anorthite and 58 mole % of diopside crystals at 1270°C. The heat supplied goes into the heat of fusion so that the temperature remains at the eutectic temperature. There is a greater ratio of diopside to anorthite crystals being melted (58 mole % diopside) than in the starting composition (20 mole % diopside), so, with continued melting at the eutectic, the diopside crystals are totally melted, whereas some anorthite crystals remain. When the last diopside crystals are melted, further melting causes the liquid to become more anorthite rich as it increases its temperature along the liquidus, following a path that is opposite to the direction of the arrows in Figure 8-14. At point Y, the last crystals of anorthite in the rock will be melted, and further heat causes the liquid to increase in temperature toward point X.

In many instances in the earth, a melt is separated from the crystals before all of the crystals are completely melted. Consider a rock of composition $X_{CaAl_2Si_2O_8} = 0.8$ in the diopside–anorthite system. As outlined above, this rock begins melting at point Z in Figure 8-14, producing a melt of composition E. If this melt is less dense than the surrounding rock, it can leave the melting region because of buoyancy forces. In this case, the remaining rock is more anorthite rich than the initial composition of $X_{CaAl_2Si_2O_8} = 0.8$. This rock would continue producing melt of the eutectic composition at the eutectic temperature until all of the diopside was melted. If this melt also left the melting region, then further melting of the remaining pure anorthite rock would occur only if the temperature increased to approximately 1560°C.

What determines the equilibrium melting relations for a phase that displays a separate liquidus and solidus such as that for anorthite in the diopside–anorthite system? The Gibbs energy of anorthite in the melt along the liquidus, $G_{An(melt)}$, must equal the Gibbs energy of anorthite in the solid, $G_{An(solid)}$ along the solidus, that is

$$G_{An(solid)} = G_{An(melt)} \qquad [8.20]$$

At constant pressure, the Gibbs energy of a pure solid is a function of temperature only, whereas the Gibbs energy of the melt is a function of both temperature and melt composition. For this calculation, the entropies of the pure solid and the pure anorthite melt, $S^{o}_{An(solid)}$ and $S^{o}_{An(melt)}$, respectively, are not considered temperature dependent. For the anorthite solid, this implies

$$G_{An(solid)} = G^{o}_{An(solid)} + (T - T_{fus})\, S^{o}_{An(solid)} \qquad [8.21]$$

where $G^{o}_{An(solid)}$ is the standard state Gibbs energy of pure anorthite at its fusion temperature for the given pressure and T_{fus} denotes the temperature of fusion. With mixing in the melt between the two components considered on an eight-oxygen basis taken as ideal, the Gibbs energy of the anorthite component in the melt is

$$G_{An(melt)} = G^{o}_{An(melt)} + (T - T_{fus})\, S^{o}_{An(melt)} + RT \ln X_{An} \qquad [8.22]$$

where $G^{o}_{An(melt)}$ denotes the standard state Gibbs energy of pure anorthite melt at its fusion temperature for the given pressure. Given these standard states,

$$G^{o}_{An(solid)} = G^{o}_{An(melt)} \qquad [8.23]$$

Equating $G_{An(solid)}$ in Equation 8.21 with $G_{An(melt)}$ in Equation 8.22 and taking account of Equation 8.23 result in

$$(T - T_{fus})\, S^{o}_{An(solid)} = (T - T_{fus})\, S^{o}_{An(melt)} + RT \ln X_{An} \qquad [8.24]$$

The standard entropy of fusion, ΔS^{o}_{fus}, is the entropy produced in going from the pure solid to the melt phase or $S^{o}_{An(melt)} - S^{o}_{An(solid)}$. Equation 8.24 can, therefore, be written as

$$RT \ln X_{An} = -\Delta S^{o}_{fus}(T - T_{fus}) \qquad [8.25]$$

Solving for X_{An} yields

$$X_{An} = \exp\left(-\frac{\Delta S^{o}_{fus}(T_{fus} - T)}{RT}\right) \qquad [8.26]$$

Equation 8.26 gives the mole fraction of anorthite in the melt along the liquidus at a particular temperature. To look at the change in temperature for the liquidus with a change in X_{An}, the series expansion of $\ln X_{An}$ needs to be considered. Because X_{An} varies between 0 and 1, $\ln X_{An}$ can be written as the series expansion

$$\ln X_{An} = (X_{An} - 1) - \frac{1}{2}(X_{An} - 1)^2 + \frac{1}{3}(X_{An} - 1)^3 \ldots \qquad [8.27]$$

If X_{An} is close to 1, it is clear that relative to the first term on the right side of Equation 8.27 all the other terms are small; therefore, near $X_{An} = 1$, Equation 8.25 can be written as

$$RT(X_{An} - 1) = -\Delta S^o_{fus}(T - T_{fus}) \qquad [8.28]$$

Taking the derivative of X_{An} with respect to T in Equation 8.28 with T approaching T_{fus} results in

$$\left(\frac{dT}{dX_{An}}\right) = \frac{RT_{fus}}{\Delta S^o_{fus}} \qquad [8.29]$$

Equation 8.29 then gives the slope of the liquidus near $X_{An} = 1$. The smaller the ΔS^o_{fus} of a component in the mixture relative to other components, the greater is the slope of the liquidus and the closer the eutectic is to this component's composition, as shown in **Figure 8-15**.

Unlike phases that contain alkali or alkaline earth elements, quartz retains much of its three-dimensional structure when melted as mentioned above. Its entropy of fusion is, therefore, relatively low. This means that the eutectic in a two-component system with Si_4O_8 as a component or the cotectic in a multicomponent system is toward the Si_4O_8-rich compositions. In general, when

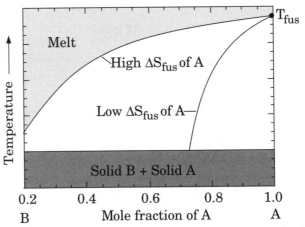

FIGURE 8-15 Temperature–composition phase diagram between two components showing the effects of changes in the entropy of fusion, ΔS_{fus}, on the position of the liquidus.

magmas cool and crystallize minerals, the crystals are less silica rich than the original melt, and the melt becomes more Si rich with increasing crystallization and decreasing T. This is why *Bowen's reaction series*, which gives the order of crystallization of silicate minerals from magmas, is olivine, pyroxene, hornblende, biotite, K-feldspar, and then quartz with decreasing temperature. That is, the concentration of silica in the minerals that crystallized relative to other oxides increases along the series because their entropy of fusion is decreasing as the melt becomes more polymerized.

■ Solid Solutions and Melting

The phase diagram for the melting of plagioclase at 1 bar is shown in **Figure 8-16**. It is a two-component system with albite and anorthite as endmembers. Unlike the system diopside and anorthite, the system albite and anorthite shows complete solution between the solids. The solidus for plagioclase, therefore, is a line from the albite to the anorthite endmember compositions. Because pure albite and pure anorthite melt congruently, the solidus and liquidus must converge at the pure endmember compositions, as shown in Figure 8-16.

Consider a melt at high temperature with 70 mole % of the anorthite component. As it cools, it begins to crystallize plagioclase at point W along the liquidus. The composition of the plagioclase crystallized is given by point X on the solidus. Because the crystals are more anorthite rich than the melt, the

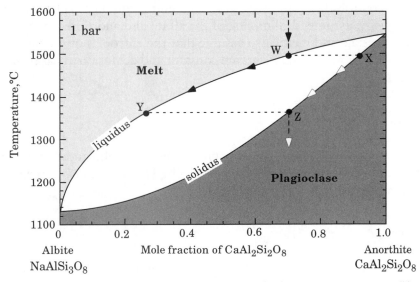

FIGURE 8-16 Temperature–composition phase diagram at 1 bar for the two-component system albite and anorthite. (Adapted from Bowen, 1915.)

residual melts become more albite rich. If equilibrium is maintained with the crystals, the composition of the melt follows along the liquidus, as shown by the solid arrows. The composition of the plagioclase crystals changes along the solidus, as given by the open arrows as temperature decreases. At point Y, the last of the melt will have crystallized with the composition and temperature of all of the plagioclase crystals given by point Z. Any further cooling occurs with crystals only and follows the vertical line down from point Z. Because no melt or crystals are lost from the system, the final plagioclase has the same composition as the original melt of 70 mole % of anorthite.

If crystals were separated from the melt after they crystallized, for instance, by settling to the bottom of a magma chamber, they would not change their composition continuously to retain equilibrium with the melt but would be isolated from the crystallizing melt. In this case, the first crystals formed would still be at the composition given by point X, but successive layers of plagioclase crystals would be produced at the bottom of the magma chamber that varied from composition X to pure albite. In this case, the average composition of the plagioclase crystals would still be 70 mole % of anorthite, but some of the crystals would be more anorthite rich and some less anorthite rich. This disequilibrium phenomena can also be used to explain Ca–Na zoning of plagioclase crystals. In this case, early formed cores are anorthite rich because they have formed from an early anorthite-rich melt. With continued precipitation of plagioclase, the early formed crystals often do not react with melt. Successive layers that precipitate on the crystals are more and more albite rich as the melt composition becomes more albite rich.

What controls the location of the liquidus and solidus in this phase diagram? For simplicity, assume ideal mixing of the albite and anorthite component in both the crystals and melt. Also assume that the entropies of the pure solids and endmember melt components are constant and do not vary near their melting points so that

$$\left(\frac{dS_{i,S}^{o}}{dT}\right)_P = \left(\frac{dS_{i,L}^{o}}{dT}\right)_P = 0 \qquad [8.30]$$

where $S_{i,S}^{o}$ and $S_{i,L}^{o}$ represent the entropies of the pure solid and liquid component i of either anorthite or albite, respectively. The change in Gibbs energy of a component i with T at constant P is given by

$$\left(\frac{dG_i}{dT}\right)_P = -S_i \qquad [8.31]$$

whereas the change in Gibbs energy per mole of component i with ideal mixing is

$$\left(\frac{dG_i}{dX_i}\right)_{P,T} = RT \ln X_i \qquad [8.32]$$

Therefore, the Gibbs energy of the albite, $G_{Ab,L}$, and anorthite, $G_{An,L}$, components in the melt are

$$G_{Ab,L} = G^o_{Ab,L} - (T - T_{Ab})\, S^o_{Ab,L} + RT \ln X_{Ab,L} \qquad [8.33]$$

and

$$G_{An,L} = G^o_{An,L} - (T - T_{An})\, S^o_{An,L} + RT \ln X_{An,L} \qquad [8.34]$$

respectively. $G^o_{i,L}$ is the standard state molar Gibbs energy and $S^o_{i,L}$ the entropy of the ith subscripted component in the melt. The standard state for albite and anorthite used is the pure endmember phases at the pressure of interest and their fusion temperatures, T_{Ab} and T_{An}, respectively. $X_{i,L}$ denotes the mole fraction of component i in the melt. Denoting the plagioclase solid phase with the subscript S, for the albite component

$$G_{Ab,S} = G^o_{Ab,S} - (T - T_{Ab})\, S^o_{Ab,S} + RT \ln X_{Ab,S} \qquad [8.35]$$

and for the anorthite component

$$G_{An,S} = G^o_{An,S} - (T - T_{An})\, S^o_{An,S} + RT \ln X_{An,S} \qquad [8.36]$$

With equilibrium between the albite component in the solid and melt

$$G_{Ab,S} = G_{Ab,L} \qquad [8.37]$$

For the standard state chosen of the pure solid and pure liquid at their melting temperature and the pressure of interest also requires

$$G^o_{Ab,S} = G^o_{Ab,L} \qquad [8.38]$$

Therefore, from Equations 8.33 and 8.35

$$- (T - T_{Ab})\, S^o_{Ab,L} + RT \ln X_{Ab,L} = - (T - T_{Ab})\, S^o_{Ab,S} + RT \ln X_{Ab,S} \qquad [8.39]$$

Because the standard entropy of fusion of albite, $\Delta S^o_{f,Ab}$, is equal to $S^o_{Ab,S} - S^o_{Ab,L}$, Equation 8.39 can be written as

$$RT \ln X_{Ab,S} = RT \ln X_{Ab,L} - (T - T_{Ab})\Delta S^o_{f,Ab} \qquad [8.40]$$

Dividing both sides of Equation 8.40 by RT and taking their exponential results in

$$X_{Ab,S} = X_{Ab,L}\exp(\Delta S^o_{f,Ab}(T_{Ab} - T)/RT) \qquad [8.41]$$

A similar analysis for the anorthite component gives

$$X_{An,S} = X_{An,L}\exp(\Delta S^o_{f,An}(T_{An} - T)/RT) \tag{8.42}$$

These two equations then give the solidus at a given temperature in terms of the melting temperature of the pure endmember and the composition of the liquid. Because

$$X_{Ab,S} + X_{An,S} = 1 \text{ and } X_{Ab,L} + X_{An,L} = 1 \tag{8.43}$$

the mole fraction of albite in the liquid is given by

$$X_{Ab,L} = \frac{1 - \exp(\Delta S^o_{f,An}(T_{An} - T)/RT)}{\exp(\Delta S^o_{f,An}(T_{An} - T)/RT) - \exp(\Delta S^o_{f,Ab}(T_{Ab} - T)/RT)} \tag{8.44}$$

Equation 8.44 gives the position of the liquidus at a given temperature. It depends on the entropy of fusion of the two pure components, as well as their fusion temperature. The greater the entropies of fusion, the larger is the temperature difference between the liquidus and solidus.

Eutectic Temperature as a Function of Pressure

Figure 8-17 shows the difference in the phase relations between diopside and anorthite as a function of pressure with no water present at 1 bar and 20 kbar and also under the conditions of H_2O saturation at 5 and 10 kbar. Increasing pressure moves the eutectic to more anorthite-rich compositions. Increasing pres-

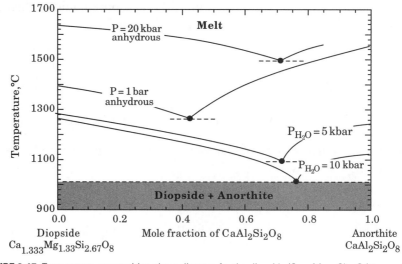

FIGURE 8-17 Temperature–composition phase diagram for the diopside ($Ca_{1.33}Mg_{1.33}Si_{2.67}O_8$) to anorthite ($CaAl_2Si_2O_8$) system without H_2O at 1 bar and 20 kbar and at H_2O saturation at 5 and 10 kbar. (Adapted from Yoder, 1965.)

sure in the dry system, however, increases the eutectic temperature, whereas in the H_2O-saturated system the temperature is decreased. This increased stability of the melt phase to lower temperatures occurs because the addition of H_2O creates more melt species, and therefore, the melt's entropy is increased, as discussed above.

More Complex Temperature–Composition Melting Phase Diagrams

In some binary and in more complicated systems, eutectics, solid solutions, and a *solvus* can be combined (see Chapter 4 for a discussion of a solvus). This is true for the albite–K-feldspar system, as shown in **Figure 8-18**. Consider a melt of 70 mole % $KAlSi_3O_8$ and 30 mole % $NaAlSi_3O_8$ that cools to its liquidus as given by point A in Figure 8-18. At this point, alkali feldspar of composition B along the solidus starts to crystallize. As the magma continues to cool, as shown by the solid arrows, its composition becomes more Na rich. To stay in equilibrium, the (K,Na)-feldspar must become more Na rich as its composition changes along the solidus, shown by the open arrows. Because 70 mole % $KAlSi_3O_8$ is at a greater concentration of $KAlSi_3O_8$ than point D, the magma will have completely crystallized as point C is reached, giving solid (K,Na)-feldspar of the composition of the starting magma at point G. Nothing happens on further cooling until the K-feldspar reaches point I. At this point, albite solid solution of the composition H starts to exsolve from the (K,Na)-feldspar. With further cooling, the (K,Na)-feldspar becomes more K rich as the albite becomes more Na rich during the exsolution process, as shown by the open arrows on both sides of the Ab + Ksp solvus.

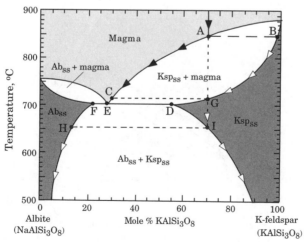

FIGURE 8-18 Phase diagram for the alkali feldspars $NaAlSi_3O_8$-$KAlSi_3O_8$ as a function of temperature at water saturation and 5 kbar. Ab_{ss} and Ksp_{ss} stand for Na + K solid solutions in albite and K-feldspar, respectively. (Adapted from Morse, 1970.)

Shown in **Figure 8-19** is the liquidus surface for the three-component system albite, anorthite, and diopside at 1 bar contoured for temperature. Cotectic crystallization is given by the line where the points A, B, C, D, and E are located. Temperature increases in the vertical direction out of the page. If a melt composition is at some temperature above the diopside field, it crystallizes diopside first on cooling. Plagioclase crystallizes first on cooling from above the plagioclase field.

Consider a melt of the composition given by point K. As it cools to 1300°C, it starts to crystallize diopside. With crystallization of diopside, the melt composition moves along the liquidus surface directly away from the diopside apex to point L on the cotectic line. At this point, with further cooling plagioclase starts to crystallize along with the diopside. The composition of the plagioclase is shown by the dashed tie line from the cotectic to the plagioclase solidus (point F). This composition is more anorthite rich than the Na/Ca ratio in the melt. As a result, with further cooling and crystallization, the melt composition moves along the liquidus away from both the diopside and anorthite apexes. If the crystals stay in equilibrium with the melt, the plagioclase changes its composition in the direction given by the open arrows on the albite–anorthite join. When the melt composition reaches point D, the melt will have com-

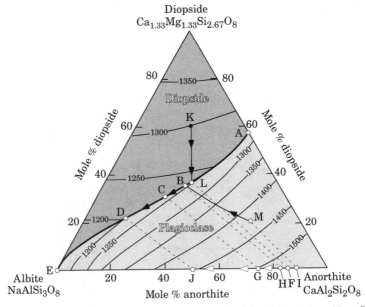

FIGURE 8-19 Liquidus surface in the system albite-anorthite-diopside at 1 bar. (Adapted from Bowen, 1915.) The field labeled as diopside is where diopside precipitates first, and the field labeled as plagioclase is where plagioclase precipitates first. The thick solid line along A-B-C-D-E is the cotectic line, where both diopside and plagioclase precipitate. The thin solid lines are isotherms labeled in °C. The dashed lines are tie lines to the solidus composition along the albite–anorthite join.

pletely solidified. At this point, there are 60 mole % diopside crystals and 40 mole % plagioclase crystals of 50 mole % anorthite composition, as given by point J. This is then the same average composition as given by the original melt composition at point K.

Now consider a melt of composition given by point M cooling to the liquidus at approximately 1420°C. This melt has a composition of 60 mole % anorthite, 20 mole % albite, and 20 mole % diopside component. At the liquidus, it starts to crystallize plagioclase of composition given by point I. The crystallization path of the liquidus follows along the curve to point B as plagioclase of higher anorthite composition than in the melt crystallizes. If the crystals of plagioclase stay in equilibrium with the melt, they change their composition to point H as the melt reaches point B. At this point, diopside begins to crystallize, and the melt changes its composition along the cotectic in the direction given by the solid arrow. At point C, the last of the melt is crystallized, and the final composition of the plagioclase is given by point G. When this occurs, a rock made up of 20 mole % of diopside crystals and 80 mole % of plagioclase crystals whose composition is 75 mole % anorthite and 25 mole % albite has been produced.

■ Distribution of Trace Components Between Rocks and Melts

As can be deduced from Table 2-2, the elements O, Fe, Si, Mg, Ca, Al, Na, Cr, Mn, Ti, Ni, and K each have average concentrations greater than 0.1 wt% in mantle + crust rocks. All of the other 80 elements combined generally contribute less than 0.5 wt%. These low-abundance elements are called *trace elements* because they typically are only incorporated in trace amounts in the common rock-forming silicates found in the crust and mantle. It should be noted, however, that trace elements can be major constituents of a variety of rare minerals containing sulfur, carbon, and phosphorous and in some less common simple oxides.

Changes in the concentration of trace elements in a suite of silicate rocks can be used to place constraints on igneous processes. The techniques used are discussed below. Many trace elements are incompatible in minerals and consequently increase in the melt fraction as the magma crystallizes. These elements typically have high charge or large size. Incompatible elements in common silicate minerals include the *high field strength* (HFS) *elements* and large ion lithophile (LIL) elements. The HFS elements form ions with higher charge than the +2 found for the Mg^{2+}, Fe^{2+}, and Ca^{2+} present in the octahedral sites common in ultramafic and mafic rock-forming minerals and are larger in size than the tetrahedral Si^{4+} and Al^{3+} sites in silicate minerals. HFS ions include Zr^{4+}, Hf^{4+}, Ta^{4+}, Nb^{5+}, Th^{4+}, U^{4+}, Mo^{6+}, W^{6+}, and the +3 valence *rare earth elements* (REEs).

LIL elements, which include Rb^+, Cs^+, Sr^{2+}, and Ba^{2+}, are typically too large to substitute in ultramafic and mafic rock-forming mineral structures.

REEs are particularly interesting in their behavior. REEs have a $5d^1\ 6s^2$ outer electron orbital arrangement. With increasing atomic number, the inner 4f shell is filled. The effects of these inner orbitals are blocked, however, by the outer $5d^1\ 6s^2$ orbitals. This makes REE's chemical behavior similar to one another and produces a +3 valence in these metals. There is systematic contraction of trivalent ionic radii with atomic number as positive charge is added to the nucleus. The +3 valence radii of the REEs are shown in **Figure 8-20**; however, Ce and Eu can have additional valence states. Under oxidized conditions, cerium becomes Ce^{4+}, and its ionic radii decreases. Under reducing conditions, europium becomes Eu^{2+}, and its ionic radii increases. Because oxidation states in the crust and mantle are reducing, europium occurs as Eu^{2+} and can exhibit anomalous chemical behavior relative to the other REE.

The REEs are divided into light and heavy groups based on their atomic weight. The light rare earth elements (LREEs) are from lanthanum, atomic number 57, to gadolinium, atomic number 64, whereas the HREEs are those from terbium, atomic number 65, to lutetium, atomic number 71. Increased nuclear stability for even-numbered atomic elements (*Oddo-Harkins rule*) results in a sawtooth REE abundance pattern for average CI chondritic meteorites, as shown in Figure 2-1 and **Figure 8-21**. Values for REEs measured in other rocks are typically normalized to chondritic values. This is done because other rocks

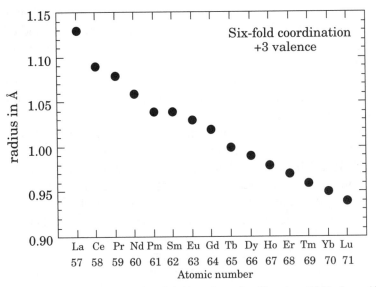

FIGURE 8-20 Radii of the REEs as a function of their atomic number. (Data from Whittacker and Muntus, 1970.)

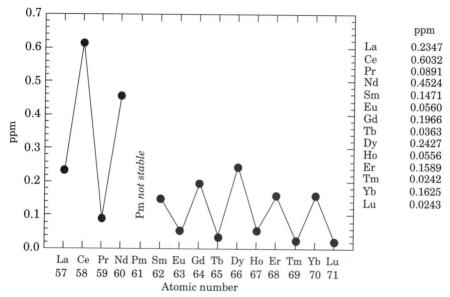

	ppm
La	0.2347
Ce	0.6032
Pr	0.0891
Nd	0.4524
Sm	0.1471
Eu	0.0560
Gd	0.1966
Tb	0.0363
Dy	0.2427
Ho	0.0556
Er	0.1589
Tm	0.0242
Yb	0.1625
Lu	0.0243

I FIGURE 8-21 REE concentrations in average CI meteorites as reported by Anders and Ebihara (1982).

also have abundance variation between odd and even atomic numbers, and thus, dividing by chondritic values smooths out this variation. Also, because chondritic values are considered to represent average solar nebular values, relative enrichment or depletion is given by normalized values that are greater or less than 1, respectively.

Figure 8-22 shows chondrite normalized REE concentrations for average upper continental crust, mid-oceanic ridge basalt, and depleted peridotite mantle. Consistent with the argument that the crust is a partial melt of the mantle, the mantle is depleted in the LREEs relative to chondritic ratios, and the upper crust is enriched. Mid-oceanic ridge basalt as a later partial melt from the depleted mantle has somewhat depleted LREEs. Both slopes and anomalies across a normalized series for a mineral or rock can provide information about geochemical process within the mantle or crust.

Why do minerals or melt incorporate or exclude certain trace components? Consider the chemical potential of trace component i in a melt given by

$$\mu_i^{melt} = (\mu_i^{melt})^\circ + RT \ln a_i^{melt} \qquad [8.45]$$

where $(\mu_i^{melt})^\circ$ is the standard state chemical potential of the ith trace element in the melt and a_i^{melt} is its activity. The chemical potential of i in a mineral phase can be expressed as

$$\mu_i^{min} = (\mu_i^{min})^\circ + RT \ln a_i^{min} \qquad [8.46]$$

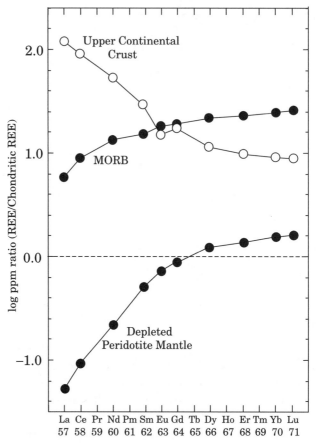

FIGURE 8-22 Typical REE abundance ratios for major reservoirs in the earth. MORB, mid-oceanic ridge basalt. (Data from Patchett, 1989.)

where $(\mu_i^{min})°$ is the standard state chemical potential of i in the mineral and a_i^{min} is its activity. At the low concentrations of trace components in minerals and melts, generally <0.1 wt%, numerous studies have indicated that Henry's law is obeyed. Consider a Henry's law standard state for i in each phase. That is, the standard state of i is at the pressure and temperature of interest with the fugacity of pure i in the phase being its Henry's law constant (see Chapter 4 for a discussion of the Henry's law standard state). For this standard state in the Henry's law region of concentration of trace element, i, in the mineral, its activity is given by

$$a_i^{min} = k_i^{min} X_i^{min} \qquad [8.47]$$

where k_i^{min} is the Henry's law constant for component i and X_i^{min} denotes its mole fraction in the mineral. If the mixing of the trace element in the mineral is ideal,

k_i^{min} is unity. Typically, the incorporation of trace components in minerals is highly nonideal with positive departure from ideality; k_i^{min} is, therefore, significantly larger than 1 as energy is required to incorporate the trace element in the mineral.

Now consider component i in the melt phase in the Henry's law concentration region. Its activity is given by

$$a_i^{melt} = k_i^{melt} X_i^{melt}$$

[8.48]

where k_i^{melt} denotes the Henry's law constant for i in the melt. At low concentrations of i, in the Henry's law region of trace component, Equations 8.45 and 8.46 can therefore be written as

$$\mu_i^{melt} = (\mu_i^{melt})^o + RT \ln (k_i^{melt} X_i^{melt})$$

[8.49]

and

$$\mu_i^{min} = (\mu_i^{min})^o + RT \ln (k_i^{min} X_i^{min})$$

[8.50]

respectively. At equilibrium, $\mu_i^{melt} = \mu_i^{min}$, and Equations 8.49 and 8.50 can be equated to give

$$(\mu_i^{min})^o - (\mu_i^{melt})^o = RT \ln \left(\frac{k_i^{melt} X_i^{melt}}{k_i^{min} X_i^{min}} \right)$$

[8.51]

Taking the exponential of both sides of Equation 8.51 results in

$$\exp \frac{((\mu_i^{min})^o - (\mu_i^{melt})^o)}{RT} = \left(\frac{k_i^{melt} X_i^{melt}}{k_i^{min} X_i^{min}} \right)$$

[8.52]

With the standard states chosen, the left side of Equation 8.52 is constant, independent of composition, but can be a function of pressure and temperature. If this constant is designated as k', Equation 8.52 becomes

$$\left(\frac{X_i^{min}}{X_i^{melt}} \right) = \left(\frac{k_i^{melt}}{k_i^{min}} \right) \frac{1}{k'}$$

[8.53]

Letting D_i, termed the *Nernst distribution coefficient* of i, be equal to the constants on the right side of Equation 8.53 implies

$$D_i = \left(\frac{k_i^{melt}}{k_i^{min}} \right) \frac{1}{k'} = \left(\frac{X_i^{min}}{X_i^{melt}} \right)$$

[8.54]

D_i depends on pressure and temperature and the identity of the trace element i but is independent of the amount of i in the Henry's law region of concentration. Generally, because of the low concentrations involved, values are determined in ppm by weight. At low concentrations,

$$\left(\frac{ppm_i^{min}}{ppm_i^{melt}}\right) = \left(\frac{X_i^{min}}{X_i^{melt}}\right) = D_i \qquad [8.55]$$

D_i is a constant in which the value is determined by values of both k_i^{min} and k_i^{melt}. k_i^{min} and k_i^{melt} depend on the nature of the mineral and melt phase, respectively. Not only do temperature and pressure affect the value of D_i, but also the melt composition and composition and structure of the mineral. Many different models have been constructed to give D_i over limited chemical and physical conditions. No model constructed to date appears to have general validity.

To quantify trace element behavior, all of the factors that control D_i must be determined. In most cases, it appears that lattice strain energy as the trace element substituted into the crystal structure is important in determining the contribution of k_i^{min} to D_i (Wood and Blundy, 2001). This strain energy increases with increasing charge difference between the "normal" and the substituted ion. With no strain energy around the trace element in the melt, its composition has little strain effect. D_i values for a given melt composition then appear to reflect the relative strain because of its substitution in the mineral and how this changes with composition, temperature, and pressure.

Although the melt composition is also important, in the present state of knowledge, the melt compositions are typically specified broadly based on silica content as peridotite, basalt, intermediate, or felsic for a given D_i. The temperatures considered are generally near the mineral-melt equilibrium temperature for a given melt composition, and their small range is generally not considered to be significant. Pressure effects can be significant but are generally ignored unless lower crust or mantle conditions are contrasted with upper crustal conditions.

The detailed knowledge of D_i is limited; however, in many cases, reported D_i values obtained from measurements of the trace component in glass and coexisting mineral from extrusive igneous rocks appear reasonable, and some general trends can be discerned. Averages of some values of D_i that have been reported in the literature for basalt to basaltic andesite and dacite to rhyolite rocks are given in **Table 8-2**. D_i for the REEs between orthopyroxene, clinopyroxene, garnet, and plagioclase and melt are greater in felsic than basaltic melts. Although Sr is incompatible in ultramafic minerals, it is clear from Table 8-2 that Sr substitutes readily for Ca in plagioclase, being of the correct size and charge (+2). HREEs are fractionated into garnet relative to the LREEs because their smaller size is more compatible. Problems can exist with the chang-

Table 8-2 D_i (mineral/melt) for the ith element and the indicated mineral in basaltic to basaltic andesite melts (labeled a) and in dacitic to rhyolitic melts (labeled b) near their melting temperature and at low pressure

	Oliv$_a$	Opx$_a$	Opx$_b$	Cpx$_a$	Cpx$_b$	Garnet$_a$	Garnet$_b$	Plag$_a$	Plag$_b$
Rb	0.0010	0.022	0.003	0.031	0.032	0.042	0.009	0.071	0.065
Sr	0.014	0.040	0.009	0.060	0.52	0.012	0.015	1.830	7.6
Ba	0.010	0.013	0.003	0.026	0.13	0.023	0.017	0.23	0.73
Ni	5.9-29	5		1.5-14					
Cr	0.70	10		34		1.0	3.7		
La	0.007		0.40	0.056	0.58	0.014	0.39	0.17	0.38
Ce	0.006	0.02	0.37	0.12	0.84	0.029	0.52	0.10	0.26
Nd	0.006	0.03	0.50	0.27	1.7	0.048	0.57	0.075	0.19
Sm	0.007	0.05	0.63	0.47	2.5	0.33	2.3	0.059	0.10
Eu	0.007	0.05	0.50	0.49	2.0	0.58	1.0	0.64	3.2
Gd	0.009	0.09	0.37	0.58	1.3	1.3	8.7	0.055	0.37
Dy	0.011	0.15	0.77	0.63	3.3	3.1	29	0.047	0.087
Er	0.018	0.23	0.36	0.62	1.2	5.6	43	0.047	0.070
Yb	0.032	0.34	1.06	0.58	2.9	18	42	0.049	0.072
Lu	0.031	0.42	1.10	0.52	2.7	20	35	0.044	0.067

Cpx, clinopyroxene; Oliv, olivine; Opx, orthopyroxene; Plag, plagioclase. Averaged values reported by Rollinson (1993).

ing composition of minerals. For instance, it has been demonstrated that for low-Ca pyroxene, intermediate melt D_i for some REEs increase significantly with increasing Ca concentration (Nielsen et al., 1992).

Figure 8-23 shows the measured distribution coefficients between clinopyroxene and melt for the REE, Ho, as a function of temperature for a number of different pressures labeled in kbar. The melts are of basalt and andesite composition with 50% and 60% SiO_2, respectively. D_{Ho} varies over nearly two orders of magnitude, depending on the physical and chemical conditions. Ho is incompatible, that is, partitioned into the melt ($D_{Ho} < 1$ or $\ln D_{Ho} < 0$) under most conditions in basaltic melts (50% SiO_2) but becomes compatible in clinopyroxene at high pressures and lower temperature in andesite type melts because k_{Ho}^{melt}, and, therefore, D_{Ho} increases (see Equation 8.54). The decrease in D_{Ho} as a function of increased temperature at constant pressure and melt composition as well as its decrease with decreasing pressure at constant temperature and melt composition is typical of all REEs. This allows other REEs to be extrapolated in pressure and temperature by considering the Ho data.

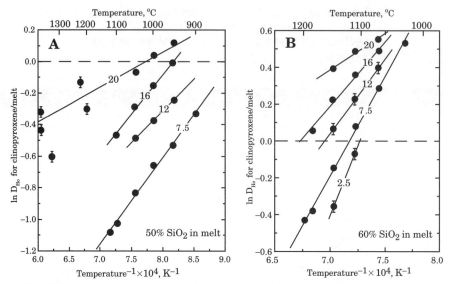

FIGURE 8-23 The partition coefficient for the REE, Holmium (Ho), between clinopyroxene and melt as a function of temperature at the labeled pressure in kbars. (A) 50% SiO_2 in the melt. (B) 60% SiO_2 in the melt. (Data from Green and Pearson, 1985.)

Dunn and Sen (1994) measured D_i for REEs between a basaltic melt and the indicated minerals as shown in **Figure 8-24**. Some important observations can be made. For each mineral, all of the REE are incompatible with $D_i < 1$ and increase in the melt with crystallization of the phases. Also, the D_i increases with increasing atomic number for olivine and orthopyroxene. This occurs because of the decreasing size of the REE with increasing atomic number. The sizes of REEs are larger than the Mg^{2+} and Fe^{2+} found in olivine and orthopyroxene in basalt. As REE size decreases, the incompatibility decreases. In the case of plagioclase with a larger octahedral site occupied by Ca^{2+}, the decrease in size makes REEs too small, and incompatibility increases with decreasing size. Europium is anomalous. As mentioned above, at reduced conditions found in the mantle, europium is present as Eu^{2+}. It has increased ionic size in the +2 oxidation state and has similar charge to the Ca^{2+} for which it is substituting. Eu, therefore, shows a positive spike relative to the other REEs.

Figure 8-25 shows measured values of D_i between plagioclase and its melt for the indicated fugacity of O_2 in the system. Only Eu is affected because it is the only REE that can exist in the +2 oxidation state as well as the typical +3 oxidation state. At the low oxidation states of rocks in the crust and mantle, Eu exists in the +2 state. This gives it a large D_i in plagioclase as it readily substitutes for Ca^{2+} in the structure. Rocks that crystallize plagioclase produce a negative Eu anomaly relative to other REEs in the remaining melt.

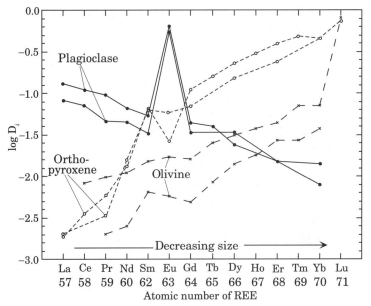

FIGURE 8-24 Base-10 logarithm of the measured distribution coefficient between the indicated mineral and alkali olivine basalt from two different samples for the indicated first row inner transition metals, REEs. (Data from Dunn and Sen, 1994.)

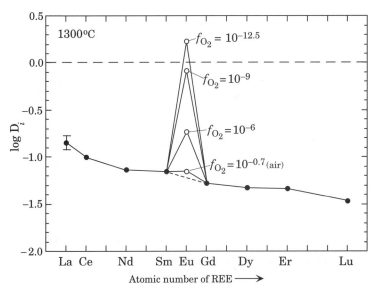

FIGURE 8-25 Measured D_i for the indicated REE between plagioclase and its melt at 1300°C and 1 bar at the indicated oxygen fugacity. (Data from Drake and Weill, 1975.)

Figure 8-26 gives distribution coefficients for the indicated REEs between hornblende and the indicated melt composition. The value of these data is that hornblende remains on the liquidus of the melt for a long time, and thus, its REE content can elucidate the crystallization sequence that has occurred. The D_i for hornblende-basalt are less than 1, as they are for other minerals crystallized from basalt melts. REE concentrations, therefore, increase in the melt. As the silica concentration of the melt phase increases, as occurs with increased crystallization, the D_i for the ith REE becomes greater. To model the change of concentration of REEs in melts, these changes with Si concentration must be included and lead to high concentrations of REEs in hornblende in silica-rich rocks.

For a given melt composition, the relative values of D_i behave the same. For instance, D_{Eu} is always anomalously low relative to the D_i of other REEs; therefore, Eu is excluded from hornblende relative to the trend of the other REEs. The behavior of D_i for REEs in hornblende appears to reflect behavior from the two different sizes of octahedral sites. The two large M4 sites behave similarly to the large octahedral sites in plagioclase, causing D_i to decrease with decreasing size of the REE. The five other smaller octahedral sites behave like

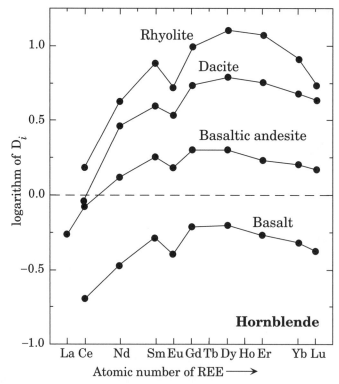

FIGURE 8-26 D_i for REEs between hornblende and the indicated composition of melt at low pressure near its melting temperature. (Data from Rollinson, 1993.)

the small sites in olivine or in clinopyroxene, causing D_i to increase with decreasing size of the REE. The combined effect is to produce lower D_i for LREEs and HREEs relative to intermediate values and lower Eu values.

■ Partial Melting and Crystallization Processes

The distribution coefficient, D_i, indicates how a trace component is distributed when a mineral is crystallized or precipitated from a melt. Trace elements can be used as tracers of certain processes if their D_i values are known. That is, they are present in such low concentrations that they do not affect the major crystallization or melting processes that may occur but can be used to put constraints on the extent of these processes.

Consider the concentration of trace element i in a melt, C_i^{melt}, relative to that in the original rock before it melted, $C_i^{o\ rock}$. If the concentration scale is on a mass basis such as ppm or wt%, then as a function of the mass fraction of a rock that has melted, F, the ratio of $C_i^{melt}/C_i^{o\ rock}$ is given by

$$\frac{C_i^{melt}}{C_i^{o\ rock}} = \frac{1}{D_i^{rock}(1 - F) + F} \qquad [8.56]$$

where D_i^{rock} is the distribution coefficient of i between the rock and coexisting melt. The value of D_i^{rock} can be computed from

$$D_i^{rock} = \sum_j D_i^{jth\ mineral} X_j \qquad [8.57]$$

where the summation is taken over all of the j minerals in the rock whose distribution coefficient i with the melt is $D_i^{jth\ mineral}$. The mole fraction of mineral j in the rock is given by X_j. **Figure 8-27a** indicates that if D_i^{rock} is small—that is, the element is incompatible, with a small degree of partial melting of the rock a large ratio of component i relative to its original concentration in the rock is incorporated in the magma. If D_i^{rock} is unity, then there is the same concentration of trace element i in both the melt and rock. Clearly, if D_i^{rock} is greater than 1, the melt has a lower concentration of i in it than in the rock from which it melted, as indicated in **Figure 8-27b**.

When changes in the concentration of a component with fractional crystallization are considered, a Rayleigh fractionation equation can be used:

$$C_i^{melt} = C_i^{o\ melt} F^{(D_i^{rock} - 1)} \qquad [8.58]$$

where C_i^{melt} is the concentration of i in the remaining melt fraction, F, and $C_i^{o\ melt}$ denotes the original concentration of i in the melt. The relations for a

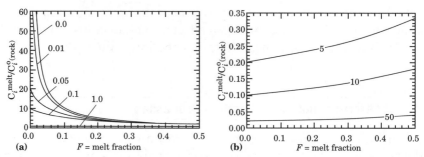

FIGURE 8-27 The concentration of trace element i in the melt relative to its concentration in the original rock as a function of the fraction of rock melted for the indicated distribution coefficient of i, D_i, between rock and melt.

given D_i^{rock} are shown in **Figure 8-28**. Starting with 100% melt, Figure 8-28 indicates that the effects of a small amount of crystallization are larger for a D_i^{rock} that is greater than 1 rather than less than 1; however, when small melt fractions are reached for $D_i^{rock} < 1$, large changes in $C_i^{melt}/C_i^{o \, melt}$ can occur.

Knowing the D_i for various trace elements and measuring the element concentrations in a suite of rocks allow inferences to be made about the percent of melting of the parental rock or the amount of crystallization of phases in a magma. For instance, partial melts derived from a source rock containing plagioclase that is not completely dissolved during melting contain less Eu than other REEs because D_{Eu} in plagioclase is larger than for the other REEs, leading to an "Eu anomaly" in the partial melt.

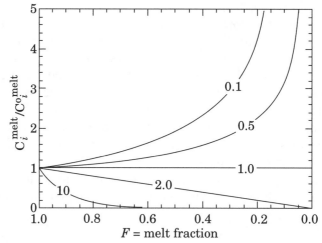

FIGURE 8-28 Relative ratio of the concentration of i in the melt to the original concentration of i in the melt as a function of the melt fraction during fractional crystallization. The numbers on the curve give the D_i^{rock} between the melt and the rock.

■ Mixing Relations

Mixing relations in igneous rocks are important because of the possibility of melting of two or more rock types to obtain the composition of magmas that produce the igneous rocks observed. Consider the production of a melt from rock 1 and rock 2 with a concentration of two different trace components **a** and **b** in each given on the same concentration scale (e.g., ppm or wt%) denoted as a_1 and b_1 and a_2 and b_2, respectively. If the magma mixture has a composition a_m and b_m, then

$$a_m = X_1 a_1 + (1 - X_1)a_2 \qquad [8.59]$$

and

$$b_m = X_1 b_1 + (1 - X_1)b_2 \qquad [8.60]$$

where X_1 is the mass fraction of rock 1 in the mixture. If the amount of **a** and **b** in each phase is known, then X_1 in the mixture can be computed from

$$X_1 = \frac{a_m - a_2}{a_1 - a_2} \qquad [8.61]$$

or

$$X_1 = \frac{b_m - b_2}{b_1 - b_2} \qquad [8.62]$$

The mole fraction of rock 1 contributed to the mixture can, therefore, be determined from either component, and these should agree. On a plot of the concentration of **a** versus **b**, all of the compositions of possible magmas, (a_m, b_m), from the melting of these two sources plot on the straight line segment from the (a_1, b_1) to (a_2, b_2) endmembers in the mixture. If a suite of related rocks does not plot on a straight line, then another process besides mixing must be important in determining their composition.

Instead of elements, consider ratios of components such as isotope or trace element ratios (e.g., $^{87}Sr/^{86}Sr$, $^{206}Pb/^{204}Pb$, Rb/Zr). Equation 8.59 can be divided by Equation 8.60 to obtain

$$\left(\frac{a}{b}\right)_m = \frac{a_1 X_1 + a_2(1 - X_1)}{b_1 X_1 + b_2(1 - X_1)} \qquad [8.63]$$

where $(a/b)_m$ is the ratio of **a** to **b** in the mixture. For the ratio of another set of components, **c** and **d**, a similar expression is

$$\left(\frac{c}{d}\right)_m = \frac{c_1X_1 + c_2(1 - X_1)}{d_1X_1 + d_2(1 - X_1)}$$ [8.64]

Combining Equations 8.63 and 8.64 to eliminate X_1 results in

$$A\left(\frac{a}{b}\right)_m + B\left(\frac{a}{b}\right)_m\left(\frac{c}{d}\right)_m + C\left(\frac{c}{d}\right)_m + D = 0$$ [8.65]

where

$$A = b_1c_2 - b_2c_1$$ [8.66]

$$B = d_1b_2 - d_2b_1$$ [8.67]

$$C = d_2a_1 - d_1a_2$$ [8.68]

$$D = a_2c_1 - a_1c_2$$ [8.69]

Equation 8.65 is a hyperbola rather than a straight line on a plot of (a/b) versus (c/d). Mixing should plot along this hyperbola if mixing of two endmembers is involved (**Figure 8-29**). If $B = 0$, that is, if $d_1b_2 = d_2b_1$, then all pairs of values

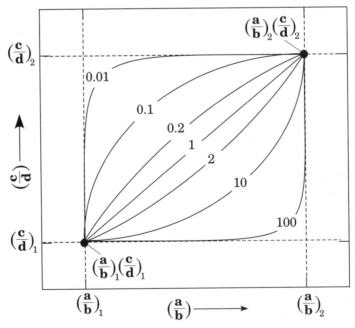

FIGURE 8-29 Ratio (a/b) – ratio (c/d) plot of endmember, 1, having values (a/b)₁ and (c/d)₁ mixing with another endmember, 2, with values of (a/b)₂ and (c/d)₂. The numbers on the lines give the values of r = d_1b_2/d_2b_1 (see text).

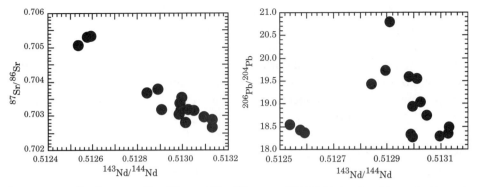

FIGURE 8-30 ^{87}Sr/^{86}Sr versus ^{143}Nd/^{144}Nd and ^{206}Pb/^{204}Pb versus ^{143}Nd/^{144}Nd plots for mid-oceanic ridge and ocean island basalts. (Data from Zindler et al., 1982.)

of $(a/b)_m$ and $(c/d)_m$ plot on the straight line segment between $(a/b)_1$, $(c/d)_1$ and $(a/b)_2$, $(c/d)_2$. If **b** and **d** are, therefore, the same component as in a plot of $(^{206}Pb/^{204}Pb)$ versus $(^{207}Pb/^{204}Pb)$, then mixing plots on a straight line. The deviation of Equation 8.65 from a straight line is given by departures of d_1b_2/d_2b_1 from unity. These are denoted as values of r in Figure 8-29.

Figure 8-30 displays ^{87}Sr/^{86}Sr, ^{143}Nd/^{144}Nd, and ^{206}Pb/^{204}Pb ratios for mid-oceanic ridge and ocean island basalts. Whereas a plot of ^{87}Sr/^{86}Sr versus ^{143}Nd/^{144}Nd could be interpreted as the mixing of components from two sources, the ^{206}Pb/^{204}Pb versus ^{143}Nd/^{144}Nd plot indicates at least three sources. This suggests that the upper mantle may be significantly heterogeneous.

Summary

Problems in understanding the genesis of igneous rocks occur because the physical conditions of melting cannot be directly observed and the source rocks are typically not exposed. Models, therefore, must be made based on the trends in compositions of rocks produced when the magmas cool. Processes can include partial melting and fractional crystallization where the concentrations in the melt and minerals are different and depend on the amount of melt or crystallization, respectively. Mixing of melts of different compositions is a process that also produces compositional trends in igneous rocks. It is generally accepted that low degrees of partial melting of the mantle yield low silica but high concentrations of incompatible elements like the HFS elements and LIL elements in the melt. Higher degrees of partial melting yield higher silica but lower concentrations of these incompatible elements in the melt.

Given the high temperatures involved, equilibrium can generally be assumed. Thermodynamics is, therefore, particularly helpful in understanding the changes in composition observed. Water has a particularly large effect on melting processes. Although H_2O is present as only a few wt% in most melts, this is a large mole fraction of melt species. It changes the energetics of the melt by reacting with bridging oxygen atoms in melt species and decreasing their size.

Trace elements are effective in monitoring igneous processes because they change their concentration in both the melt and mineral phases during melting or crystallization depending on the nature of the mineral and melt involved. They are present at such low concentrations, however, that they do not effect the chemical processes involved.

Key Terms Introduced

accretionary prism
albite
alkaline basalt
amphibolite
andesite
anorthite
basalt
Benioff zone
Bowen's reaction series
Clapeyron equation
compressibility
cotectic
differentiation
eutectic
extrusive
granite
groundmass
high field strength elements
incongruently
intrusive
kimberlite
large ion lithophile elements
lever rule

lherzolite
liquidus
mantle wedge
miscibility
Nernst distribution coefficient
normative hypersthene
Oddo-Harkins rule
partial melt
peridotite
phenocrysts
primary magma
Raman spectroscopy
rare earth elements
rhyolite
shear strain rate
shear stress
solid solution
solidus
solvus
tholeiitic basalt
trace elements
trondhjemite
viscosity

Questions

1. Describe the difference between a mafic and felsic magma. Give an example of each.
2. What is a liquidus? Solidus? Eutectic?
3. What is primary magma?
4. Explain what viscosity is. What controls it in a magma?
5. State the Clapeyron equation, and discuss the terms.
6. Explain why anhydrous minerals change the slope of their melting curves on a pressure–temperature phase diagram when water is added.

7. Describe the difference in melting of a hydrous mineral as opposed to an anhydrous mineral.
8. What is known of the structure and composition of species in a silica-rich magma?
9. Why is water so important in understanding melting or rocks in the earth?
10. Describe the sequence of crystallization of a binary melt system displaying a eutectic such as diopside-anorthite with cooling.
11. Describe the sequence of crystallization of a binary melt system such as albite-anorthite that displays complete solid solution with cooling.
12. What is the entropy of fusion of melting? What effect does it have on the liquidus in a system?
13. What is a high field strength element? LIL element? Rare earth element?
14. How is Henry's law used to describe trace element behavior?
15. What is a Nernst distribution coefficient?
16. Why does europium (Eu) behave differently during crystallization than other rare earth elements?
17. How does knowledge of Nernst distribution coefficients between minerals in a rock and the magma help determine the extent of melting?

Problems

1. With a mole of forsterite written on an eight-oxygen basis, what would be the mole fraction of H_2O in a 5 wt% H_2O plus 95 wt% forsterite melt?
2. Using the $NaAlSi_3O_8$-$KAlSi_3O_8$ temperature–composition phase diagram at water saturation and 5 kbar given in Figure 8-18, outline the crystallization sequence of a 40 mole % $KAlSi_3O_8$ and 60 mole % $NaAlSi_3O_8$ magma as it cools if the melt stays in equilibrium with the minerals.
3. Label the numbered fields and construct G-X_2 a diagram for temperature T_1 shown on the following constant pressure T-X diagram.

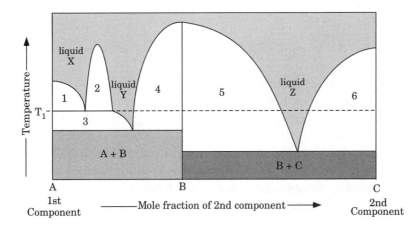

4. The liquid and solids fields α, β, and γ are shown in the phase diagram below.
   a. Label the numbered fields.
   b. Describe the fractional crystallization of composition X on cooling.

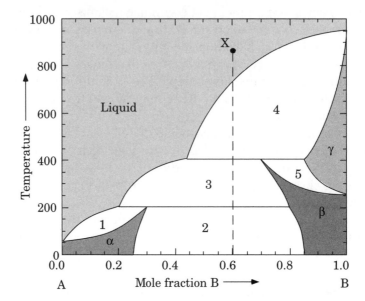

5. The element A is present in a silicate melt. The melt begins to crystal-lize in such a way that after solid crystals form, they do not react with the remaining liquid. If element A exhibits Henry's law behavior, where k_A^{solid} is the Henry's law constant in the solid and k_A^{melt} is the Henry's law constant in the melt, show that the ratio of the mole fraction of A in the residual melt, X_A^{melt}, to that in the starting melt, $X_A^{melt^o}$ is given by

$$\frac{X_A^{melt}}{X_A^{melt^o}} = F^{(k_A^{melt}/k_A^{solid} - 1)}$$

where F is the fraction of liquid remaining. (This is the Rayleigh distil-lation equation.)

6. How does the concentration of a trace element i in the melt change as a function of the melt fraction during the melting of a rock if $D_i = 2$ with an initial concentration of the trace element in the rock of 120 ppm? Plot your results.

7. How does the concentration of a trace element i in the melt change as a function of the melt fraction during the crystallization of a rock if $D_i = 5$ with an initial concentration of the trace element in the melt of 120 ppm? Plot your results.

8. Given below are concentrations from Sun and McDonough (1989) in ppm by weight for the indicated rare earths in N-type MORB (normal mid-oceanic ridge basalt), E-type MORB (from "enriched deeper tapped "plume" source mantle), and ocean island basalts (OIB). Construct a plot of the atomic numbers versus concentration for each of these reservoirs normalized to CI chondrite from Figure 8-21. Explain the likely cause of the difference between OIB behavior from those of N-type and E-type MORB.

Rare Earth	N-Type MORB	E-Type MORB	OIB
La	2.50	6.30	37.0
Ce	7.5	15.0	80
Pr	1.32	2.05	9.70
Nd	7.30	9.00	38.5
Sm	2.63	2.60	10.0
Eu	1.02	0.91	3.00
Gd	3.680	2.970	7.620
Tb	0.670	0.530	1.050
Dy	4.550	3.550	5.600
Ho	1.01	0.790	1.06
Er	2.97	2.31	2.62
Tm	0.456	0.356	0.350
Yb	3.05	2.37	2.16
Lu	0.455	0.354	0.300

9. Given below are REE determinations from Pride and Muecke (1982) for a granite sheet in the Scourian Complex, NW Scotland in ppm by weight. Plot a diagram of the atomic number versus log (ppm [granite sheet/chondrite]). Use chondrite values in Figure 8-21. What is the implication of the Eu value relative to other REE in the granite sheet?

Atomic Number	Granite Sheet
57	16.2
58	21.2
62	0.65
63	0.72
65	0.08
70	0.17
71	0.03

10. Determine the relative concentration (ratio) of trace element i in a partial melt to that in the rock at increments of $F = 0.1$ with $D_i = 2$ and $D_i = 0.2$. Plot your results. Why is the greatest departure from unity at the lowest degree of partial melting?

11. Determine the relative concentration (ratio) of trace element i in the melt as it crystallizes relative to its initial concentration in the melt, at increments of $F = 0.1$, with $D_i = 2$ and $D_i = 0.2$. Plot your results.

Problems Making Use of SUPCRT92

12. Plot the pressure–temperature equilibrium of Reaction 8.5

$$KAl_3Si_3O_{10}(OH)_2 + SiO_2 = KAlSi_3O_8 + Al_2SiO_5 + H_2O$$

between 500 and 5000 bars. First assume the phases are pure so that the equilibrium constant for the reaction is $K = 1$. Then plot the boundary with the activity of K-feldspar = 0.9. Compare with Figure 8-6.

References

Anders, E. and Grevesse N., 1989, Abundances of elements: Meteoritic and solar, *Geochem. Cosmochim. Acta*, v. 53, pp. 197–214.

Asimow, P. D. and Ghiorso, M. S., 1998, Algorithmic modifications extending MELTS to calculate subsolidus phase relations. *American Mineralogist.* v. 83, pp. 1127–1131.

Bottinga, Y., 1985, On the isothermal compressibility of silicate liquids at high pressure. *Earth and Planetary Sci. Let.*, v. 74, pp. 127–138.

Bowen, N. L., 1915, The crystallization of haplobasaltic, haplodiorite and related magmas. *Am. J. Sci.*, v. 190, pp. 161–185.

Bowen, N. L., 1928, *The Evolution of the Igneous Rocks*. Princeton Univ. Press, Princeton, N.J., 332 pp.

Boyd, F. R. and England, J. L., 1963, Effect of pressure on the melting of diopside CaMgSi$_2$O$_6$, and albite NaAlSi$_3$O$_8$ in the range up to 50 kilobars. *J. Geophys. Res.*, v. 68, pp. 311–323.

Broeker, W. S. and Oversby, V. M., 1971, Melting phenomena. In *Chemical Equilibria in the Earth*. McGraw-Hill, New York, pp. 219–235.

Burnham, C. W., 1979, The importance of volatile constituents. In *The Evolution of the Igneous Rocks: Fiftieth Anniversary Perspectives*. Princeton University Press, Princeton, N.J., pp. 439–482.

Burnham, C. W., 1982, The nature of multicomponent aluminosilicate melts. In Rickard, D. and Wickman, F. E. (eds.) *Chemistry and Geochemistry of*

Solutions at High Temperatures and Pressures. Pergamon Press, Oxford, pp. 197–227.

Burnham, C. W., 1994, Development of the Burnham model for prediction of H_2O solubility in magmas. *Rev. Minerol.*, v. 30, pp. 123–129.

Burnham, C. W., 1997, Magmas and hydrothermal fluids. In Barnes, H. L. (ed.) *Geochemistry of Hydrothermal Ore Deposits*, 3rd edition. John Wiley & Sons, New York, pp. 63–123.

Burnham, C. W. and Davis, N. F., 1971, The role of H_2O in silicate melts: I. P-V-T relations in the system $NaAlSiO_3$-H_2O to 10 kilobars and 1000 °C. *Amer. J. Sci.*, v. 270, pp. 54–79.

Burnham, C. W. and Davis, N. F., 1974, The role of H_2O in silicate melts: II. Thermodynamic and phase relations in the system $NaAlSiO_3$-H_2O to 10 kilobars, 700° to 1100 °C. *Amer. J. Sci.*, v. 274, pp. 902–940.

Carmichael, I. S. E., Turner, F. J. and Verhoogen, J., 1974, *Igneous Petrology.* McGraw-Hill, New York, 739 pp.

Cox, K. G., Bell, J. D. and Pankhust, R. J., 1979, *The Interpretation of Igneous Rocks.* Allen and Unwin, London, 450 pp.

Drake, M. J. and Weill, D. F., 1975, Partition of Sr, Ba, Ca, Y, Eu^{2+}, Eu^{3+}, and other REE between plagioclase feldspar and magmatic liquid: an experimental study. *Geochim. Cosmochim. Acta*, v. 39, pp. 689–712.

Dunn, T. and Sen, C., 1994, Mineral/matrix partition coefficients for orthopyroxene, plagioclase, and olivine in basaltic to andesitic systems: a combined analytical and experimental study. *Geochim. Cosmochim. Acta*, v. 58, pp. 717–733.

Furukawa, Y., 1993, Magmatic processes under arcs and formation of the volcanic front. *J. Geophys. Res.*, v. 98, pp. 8309–8319.

Ghiorso, M. S., Carmichael, I. S. E., Rivers, M. L. and Sack, R. O., 1983, The Gibbs free energy of mixing of natural silicate liquids: An expanded regular solution approximation for the calculation of magmatic intensive variables. *Contrib. Mineral. Petrol.*, v. 84, pp. 107–145.

Ghiorso, M. S., Hirschmann, M. M., Reiners, P. W. and Kress, V. C. III, 2002, The pMELTS: A revision of MELTS aimed at improving calculation of phase relations and major element partitioning involved in partial melting of the mantle at pressures up to 3 GPa. *Geochemistry, Geophysics, Geosystems,* v. 3(5), p. 10.

Ghiorso, M. S. and Sack, R. O., 1995, Chemical mass transfer in magmatic processes. IV. A revised and internally consistent thermodynamic model for the interpolation and extrapolation of liquid-solid equilibria in magmatic systems at elevated temperatures and pressures. *Contrib. Mineral. Petrol.*, v. 119, pp. 197–212.

Gill, J. B., 1981, *Orogenic Andesites and Plate Tectonics.* Springer-Verlag, Berlin, 390 pp.

Green, T. H. and Pearson, N. J., 1985, Experimental determination of REE partition coefficients between amphibole and basaltic to andesitic liquids at high pressure. *Geochim. Cosmochim. Acta,* v. 49, pp. 1465–1468.

LeMaitre, R. W., 1976, The chemical variability of some common igneous rocks. *J. Petrol.,* v. 17, pp. 589–637.

Maaloe, S. and Aoki, K., 1977, The major element chemistry of the upper mantle estimated from the composition of lherzolites. *Contrib. Mineral. Petrol.,* v. 63, pp. 161–173.

McCulloch, M. T. and Gamble, J. A., 1991, Geochemical and geodynamical constraints on subduction zone magmatism. *Earth Planet. Sci. Lett.,* v. 102, pp. 358–374.

Morse, S. A., 1970, Alkali feldspar with water at 5 kb pressure. *J. Petrol.,* v. 11, pp. 221–251.

Nielsen, R. L., Gallahan, W. E. and Newberger, F., 1992, Experimentally determined mineral-melt partition coefficients for Sc, Y and REE for olivine, orthopyroxene, pigeonite, magnetite and ilmenite. *Contrib. Mineral. Petrol.,* v. 110, pp. 488–499.

Nicholls, J., 1980, A simple model for estimating the solubility of H_2O in magmas. *Contrib. Mineral. Petrol.,* v. 74, pp. 211–220.

Nockolds, S., 1954, Average chemical compositions of some igneous rocks. *Geo. Soc. Am. Bull.,* v. 65, pp. 1007–1032.

Paillat, O., Elphick, S. C. and Brown, W. L., 1992, The solubility of water in $NaAlSi_3O_8$ melts: A re-examination of Ab-H_2O phase relations and critical behavior at high pressures. *Contrib. Mineral. Petrol.,* v. 112, pp. 490–500.

Patchett, P. J., 1989, Radiogenic isotope geochemistry of rare earth elements. *Rev. Mineral.,* v. 21, pp. 25–44.

Pride, C. and Muecke, G. K., 1982, Geochemistry and origin of granite rocks, Scourian Complex, NW Scotland. *Contrib. Mineral. Petrology,* v. 80, pp. 379–385.

Rollinson, H. R., 1993, *Using Geochemical Data: Evaluation, Presentation, Interpretation.* Harlow, Essex, England, 352 pp.

Silver, L. A. and Stolper, E. M., 1985, A thermodynamic model for hydrous silicate melts. *J. Geol.,* v. 93, pp. 161–178.

Silver, L. A. and Stolper, E. M., 1989, Water in albitic glasses. *J. Petrol.,* v. 30, pp. 667–709.

Stolper, E. M., 1982, The speciation of water in silicate melts. *Geochim. Cosmochim. Acta,* v. 46, pp. 2609–2620.

Sun, S.-S. and McDonough, W. F., 1989, Chemical and isotopic systematics of ocean basalts: Implications for mantle composition and processes. In Saunders, A. D. and Norry, M. J. (eds.) *Magmatism in the Ocean Basins.* Geol. Soc. Special Pub. No. 42, pp. 313–345.

Whittacker, E. J. W. and Muntus, R., 1970, Ionic radii for use in geochemistry. *Geochim. Cosmochim. Acta*, v. 34, pp. 945–956.

Winter, J. D., 2001, *An Introduction to Igneous and Metamorphic Petrology.* Prentice Hall, Upper Saddle River, N.J., 697 pp.

Wood, B. J. and Blundy, J. D., 2001, The effect of cation charge on crystal-melt partitioning of trace elements. *Earth Planet. Sci. Lett.*, v. 188, pp. 59–71.

Yoder, H. S., Jr., 1965, Diopside-anorthite-water at five and ten kilobars and its bearing on explosive volcanism. *Carnegie Inst. Washington Yearbook*, v. 64, pp. 82–89.

Zindler, A., Jagoutz, E. and Goldstein, S., 1982, Nd, Sr and Pb isotopic systematics in a three-component mantle: A new perspective. *Nature*, v. 298, pp. 519–523.

Chemical Controls on Soil Formation, Diagenesis, Metamorphism, and Hydrothermal Ore Deposition

Below the soil, sedimentary rocks cover about two-thirds of the continental crust, and sediments cover most of the oceanic crust. Because igneous and metamorphic rocks are unstable at the earth's surface, they break down and produce soils and sediments. Reactions with rainwater on land surfaces produce soils on surface rocks. Most sediments, however, are eroded from land surfaces and transported to the ocean, where they accumulate near the edge of the continent in *geoclines*. Over time, these sediments and any additional volcanics form layers that sink because the crust of the earth, to a large extent, floats on the mantle, being in *isostatic equilibrium* with it, similar to an iceberg in the ocean. If more ice is added to the top of the iceberg, the layers below sink. Thus, as more sediment is deposited, the previously deposited layers subside. As sediments are buried deeper and deeper in the crust, they heat up and undergo reactions that turn them into sedimentary rocks. With deeper burial, reactions continue to occur that produce minerals stable at the new pressures and temperatures encountered. The processes responsible for these changes are termed diagenesis and metamorphism.

Diagenetic and metamorphic processes differ from igneous ones in that mineralogy and texture change while the rock remains in a solid state. The presence of an aqueous fluid between the mineral grains facilitates these changes. At low temperatures, roughly below 200°C to 300°C, the processes are typically referred to as "diagenetic," whereas above these temperatures, they are termed "metamorphic." A sedimentary rock that has experienced regional metamorphic changes has previously undergone diagenetic processes when it was initially buried at the top of the earth's crust. Consider first the formation of soil.

■ Soil Formation

Soils are complex mixtures with a wide range of compositions derived both from the underlying bedrock and from material that is transported to the site of soil formation by floods, wind, ice, and landslides. A mixture of solid, liquid, and gas occurs wherein the spatial arrangement of these phases gives the soil its texture, or *soil fabric*. The soil gas phase is present in the 25% to 50% void space of a well-developed soil. Plants growing in the soil absorb O_2 to oxidize sugars producing CO_2 in a process called *respiration*. Aerobic bacteria and fungi break down organic material by reaction with O_2 producing additional CO_2. *Soil air* typically has a lower O_2 and greater CO_2 concentration than normal atmospheric air. The CO_2 concentration of soil air can reach 12% in the summer under moist soil conditions. Waterlogged soils block air infiltration and often have abundant CH_4 and significant H_2S trapped in their pores produced from organic matter decay, resulting in an *anoxic* environment.

The liquid phase in soils is termed the *soil solution* and occupies 35% to less than 1% of the soil. It is typically derived from atmospheric precipitation and modified by reaction with the minerals and organic matter it encounters. As infiltrating water moves downward, it dissolves material from higher in the soil and deposits it at a lower level in the soil layers. Some of this material is carried in *chelates*. Chelates are organic molecules derived from soil organic matter that behave as weak acids in aqueous solutions and can bond metals at two or more molecular sites. The order of highest to lowest stability of metal chelates is typically Fe^{3+}, Fe^{2+}, Ca^{2+}, and Mg^{2+}. Chelating organic molecules are referred to as bidentate or polydentate depending on whether the metal is chemically bonded at two or at more than two molecular sites in the chelate. A common bidentate chelator is the oxalate ion, which has two negatively charged oxygen sites that can coordinate with a positively charged metal ion:

$$\begin{array}{c} \\ \overset{\displaystyle O}{\underset{\displaystyle O^-}{}}\!\!C\!-\!C\!\!\overset{\displaystyle O}{\underset{\displaystyle O^-}{}} \end{array}$$

Peptides and sugars (see Chapter 15) in soils have function groups that act as polydentate chelates.

Soils form compositional layers because of the downward *chromatographic* action of reactive water. These layers are designated O, A, E, B, C, and R, as shown in **Figure 9-1**. Soils can be immature or mature and often contain a variable amount of organic material, depending on the nature of the parental material that is weathered, the climate, and the extent of vegetation present. Humic or organic-rich soils are common in bog and swamp areas, but soils in most landscapes are dominated by silicate minerals. The procedure for naming the common types of soils is outlined in **Figure 9-2**. Most soils are destroyed over long time frames by glaciation and other types of erosion. When they are preserved in the geologic record, however, they give important information about past climatic conditions.

	Soil layer	General Detailed		Properties
O	Organic layer	**O**		Typically a thin black or dark brown layer of organic plant residue, but can be quite thick if a peat or muck layer is present
	Leaf layer		Oi	Plant fibers recognizable
	Humus		Oa	*Saprophytes* (organisms) have decomposed the organic fibers
A	Topsoil	**A**		A dark gray mixture of mineral particles and organic matter
			Ap	Darkened, plowed, humus rich layer
Subsoil: **E**	Leached layer	**E**		Light gray colored due to low organics and fine particles being eluviated (washed downward), contains primary minerals resistant to chemical weathering
B	Accumulation zone	**B**		Zone where material leached and transported downward from above collects, rich in clays ± calcite ± gypsum. It can be from ~2 cm to 2 m thick.
			Bt	Clay layer (German = Ton)
			Bk	Carbonate layer
			By	Gypsum layer
C	Parent material	**C**		Little changed by soil formation, but generally partially weathered. Can be 30 m thick in tropical soils
R	Bedrock	**R**		The unweathered parent material

I FIGURE 9-1 General properties of soil horizons.

Unlike soils, sediments accumulate in bodies of water or river floodplains. Most sedimentary material is transported to the ocean where it is transformed into sedimentary rocks; however, significant lake, floodplain, and windblown deposits are also common. After deposition, the sediments are preserved when they are buried by a later input of sediments, preventing their erosion.

■ Diagenesis

Diagenetic reactions are those that transform unconsolidated sediments deposited at the earth's surface—sand, mud, carbonate, and organic matter—into coherent lithified rock of sandstone, shale, limestone, and coal, respectively, as they are buried. The sediments then undergo both cementation and compaction. The cementation can start early, as in *beachrock*, rock formed at beaches from cementation of sand grains by precipitation of calcium carbonate. In the early stage of *diagenesis*, the cements are typically aragonite or Mg-calcite, whereas silica cements are common in late (deeper) diagenesis. During diagenesis, there is an overall reduction of porosity and breakdown of organic material.

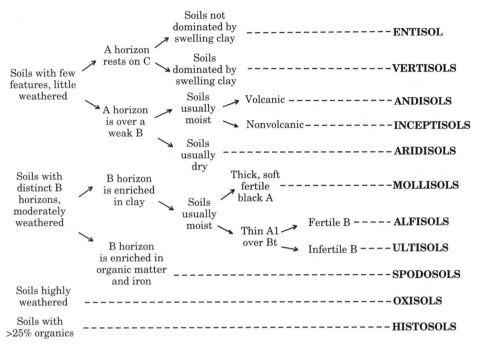

I FIGURE 9-2 General classification scheme for soils.

Generally, sandstone retains a greater porosity with depth than shale or limestone. Because of the greater solubility of carbonates than aluminosilicate phases in water, limestones can recrystallize to massive units with low porosity but can also develop a secondary porosity due to later fracturing and dissolution or by dolomitization of the original limestone. Typically, during diagenesis pores are filled with water, and it is through this water that the diagenetic reactions occur. This water is present because it is

1. Trapped at the time of deposition of the sediments
2. Transported through the sediments from elsewhere due to a fluid pressure gradient
3. Produced by mineral dehydration reactions, mainly in clays, during diagenesis

Formation waters in marine sediments typically increase in salinity to values exceeding that of seawater at depth in sedimentary basins. This increased salinity occurs because of dissolution of evaporite minerals in shales and salt layers. Also, charged clay mineral surfaces act to retain salts while allowing water molecules to escape upward. This then increases the salinity of the residual water. In general, relative to seawater, K in formation waters is lower and Ca higher because of reactions with minerals during burial.

Consider a marine mud. When it is deposited, it can have a porosity of up to 80%. Two possibilities exist depending on the amount and makeup of its organic matter: (1) Organic matter reacts with oxygenated seawater so the seawater in the pores and above the sediment–water interface becomes anoxic, or (2) organic matter in the mud is completely oxidized by diffusion of oxygen from the overlying water mass, and formation water remains oxidized. In this latter case, organic matter is oxidized to CO_2, which is incorporated into the pore water. Reactions of clays with K^+ reduce the potassium concentration in the pore water and increase the illite content at the expense of other clay minerals in the clay assemblage. Just below the first few centimeters of clay, the SO_4^{2-} from seawater in the pore fluid is reduced with the aid of bacteria, such as *Desulphovibrio*, to H_2S by the reaction

$$2H^+ + 2CH_2O + SO_4^{2-} \rightarrow 2CO_2 + H_2S + 2H_2O. \qquad [9.1]$$
$$\text{organic matter}$$

This reaction can continue to a number of meters deep, but the limited organic matter is usually exhausted in the upper one-half meter in normal marine environments, resulting in an organic free mud. In anoxic waters, the reaction can start in the water column above the top clay layer. In this case, the reaction is typically limited by the amount of seawater-derived SO_4^{2-} available, and the mud can become organic rich.

The H_2S produced in normal marine environments by Reaction 9.1 escapes into the overlying seawater or becomes oxidized to elemental S by a reaction like

$$2H_2S + O_2 \rightarrow 2S + 2H_2O. \qquad [9.2]$$

Alternatively, in anoxic waters, it reacts with the Fe^{2+} that is present in solution from the clays and precipitates amorphous $FeS \bullet nH_2O$, as given by

$$Fe^{2+} + H_2S + nH_2O \rightarrow FeS \bullet nH_2O + 2H^+ \qquad [9.3]$$

The S produced by Reaction 9.2 and the $FeS \bullet nH_2O$ produced by Reaction 9.3 typically react during early diagenesis to produce pyrite, FeS_2. *Fermentation* of the organic matter (see Chapter 15) also proceeds during early diagenesis down to a burial depth of about 1 km.

Much of the interstitial pore water in muddy sediments is expelled during compaction, but sediments at 1 km depth can contain approximately 30% H_2O (**Figure 9-3a**). Much of this is in interlayers of clays and adsorbed on clay surfaces. **Figure 9-3b** shows the composition of typical shale in terms of Al, Si, and alkali + alkaline earth elements, giving its concentration of illite + chlorite + Si phases. The compositions of some phases that are produced during diagen-

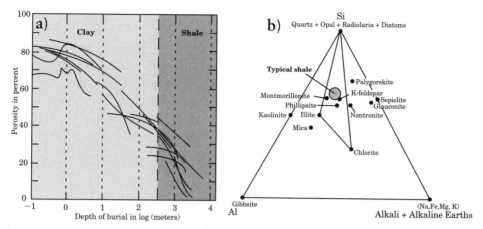

FIGURE 9-3 (a) Porosity of clay/shale with depth from a variety of sources. (Adapted from Baldwin, 1971.) (b) Compositions of original minerals and phases produced during diagenesis in shale in terms of Si, Al, and alkali + alkaline earth elements. The gray circle gives the bulk composition of typical shale. (Adapted from Garrels, 1986.)

esis are also shown. Clays and other silicates (e.g., quartz) dominate shales, but carbonates (calcite and dolomite) are often also present. Total organic matter in most shales is less than 1%. Shales that are more organic rich than this are called *black shales* and can have greater than 15% total organic carbon.

Clay Diagenesis

Hower et al. (1976) analyzed the change in composition with depth of Oligocene-Miocene clay-rich sediments from the U.S. Gulf Coast. Changes with increasing depth in these sediments can be considered to be those that are produced in clay with increasing diagenesis. **Figure 9-4** shows that the calcite content in progressively larger size fractions decreases to zero at greater depths; however, perhaps the most extensive changes occur in mixed layer illite–smectite clay, the dominant mineral in the shales. Mixed layer clays have interwoven sheets, typically illite–montmorillonite and chlorite–montmorillonite. During diagenesis, smectite layers are slowly transformed to illite layers, termed the *I/S transformation*. Both smectite and illite are three-layered clays with one layer of alumina octahedra sandwiched between two layers of silica tetrahedra (see Chapter 5 for a further discussion of compositional changes in clays). One of the most common smectites is montmorillonite. Its endmember composition is

$$Al_2Si_4O_{10}(OH)_2 \cdot nH_2O \qquad [9.4]$$

but Fe^{2+} and Mg^{2+} are typically substituted for some Al^{3+} in the structure. The charge difference is generally balanced by substitution of Ca^{2+} and Na^+ in an

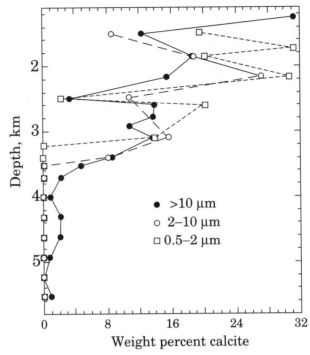

FIGURE 9-4 Calcite content of the indicated size fraction of shale with depth in Oligocene-Miocene sediments from the U.S. Gulf Coast. (Adapted from Hower et al., 1976.)

interlayer structural position. Illites have Al^{3+} substituting for Si^{4+} with K^+ added to the interlayer position for charge balance, which can produce an end-member illite composition of

$$KAl_2(AlSi_3)O_{10}(OH)_2 \qquad [9.5]$$

The I/S transformation is then the substitution of Al^{3+} plus K^+ for Si^{4+} in the clay mineral's structure.

As shown in **Figure 9-5a,** the increased diagenesis with depth causes mixed layer montmorillonite with about 20% illite (at approximately 2 km and 55°C to 80°C) to be converted to mixed layer clay with about 80% illite (at approximately 4 km and 120°C to 140°C). **Figure 9-5b** indicates that K-feldspar is lost from the rock as illite layers are produced. The illite-forming reaction is, therefore, likely to be

$$Al_2Si_4O_{10}(OH)_2 \cdot nH_2O + KAlSi_3O_8 \rightarrow KAl_2(AlSi_3)O_{10}(OH)_2 + 4SiO_{2(aq)} + nH_2O \quad [9.6]$$
$$\text{montmorillonite} \qquad \text{K-feldspar} \qquad \text{illite}$$

This reaction conserves Al between the solid phases. Also, quartz (or at lower temperatures amorphous silica) precipitates from the $SiO_{2(aq)}$ that is produced

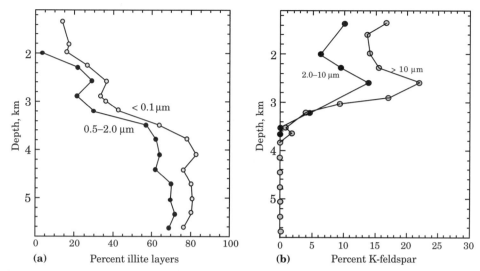

FIGURE 9-5 (a) Percent illite layers with depth in illite-montmorillonite clay. (b) Percent K-feldspar in the indicated clay fractions contained in Oligocene-Miocene sediments from the U.S. Gulf Coast. (Adapted from Hower et al., 1976.)

if the solution becomes supersaturated with these phases. The smectite to illite reaction is influenced by temperature, time, and the availability of K-containing phases.

At a 2.5-km depth, chlorite appears as outlined in **Figure 9-6a**. Many investigators believe that chlorite forms as a byproduct of the conversion of smectite to illite with Mg derived from solution. Others believe chlorite forms directly from montmorillonite and Mg-rich solutions. Because the kaolinite content decreases below 3.4 km, as shown in **Figure 9-6b**, chlorite may also be formed by reaction of kaolinite with the quartz and available Mg^{2+} and Fe^{2+} in solution. For Mg endmember chlorite, the reaction would be

$$7H_2O + Al_2Si_2O_5(OH)_4 + 5Mg^{2+} + SiO_2 \rightarrow Mg_5Al(AlSi_3)O_{10}(OH)_8 + 10H^+ \quad [9.7]$$
$$\text{kaolinite} \qquad\qquad\qquad \text{quartz} \qquad \text{clinochlore}$$

At temperatures of approximately 300°C at depths of >5.5 km, illite is recrystallized to sericite, a fine-grained mica, and then to the muscovite structure with the reaction considered to be metamorphic.

Sandstone Diagenesis

Source rocks and the extent of chemical weathering determine the nature of material in a particular sand sediment. After deposition, reactions in the sandstone together with pore-solution migration can modify the composition of sediments significantly. This occurs because a well-sorted sandstone can have

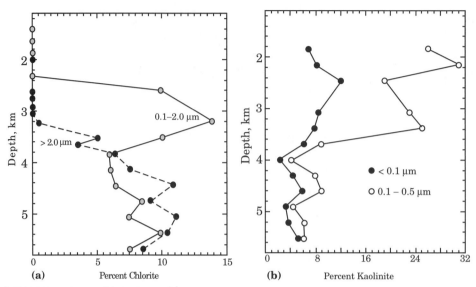

FIGURE 9-6 Percent (a) chlorite and (b) kaolinite as a function of depth in the indicated size fractions contained in Oligocene-Miocene sediments from the U.S. Gulf Coast. (Adapted from Hower et al., 1976.)

a high *permeability*, allowing the passage of a large volume of solution. These solutions are commonly saline and typically somewhat alkaline (significant CO_3^{2-}) with lower Na, Mg, sulfate, and K relative to Cl but higher Ca, Sr, and Si than seawater. These fluids can then precipitate significant quantities of $CaCO_3$ and SiO_2 cements that decrease the sandstone's porosity. Because calcite has *retrograde solubility*, its ability to form cement at high temperatures is diminished, and SiO_2 cements are the norm. **Figure 9-7** shows the relationship between increased clay content of a sandstone and depth of burial with its decrease in porosity. The higher the clay content of the sandstone, the earlier during diagenesis it loses its porosity. Because of the rigid sand grains present, *pressure solution* can occur in burial. That is, the greater pressure at grain–grain contacts as opposed to grain–fluid contacts causes the mineral to be nonhydrostatically stressed. Because solubility of silicates increases with pressure, the more stressed regions of the mineral have greater solubility in the fluid. Dissolution can then occur, supersaturating the fluid relative to normally stressed mineral surfaces. Precipitation in a normally stressed environment then occurs, helping to cement the sand grains.

Carbonate Diagenesis

Shallow-water sedimentary carbonate that is forming now is about two-thirds aragonite and one-third Mg-calcite together with minor Ca-rich dolomite. These are precipitated by organisms and accumulate in carbonate muds. In

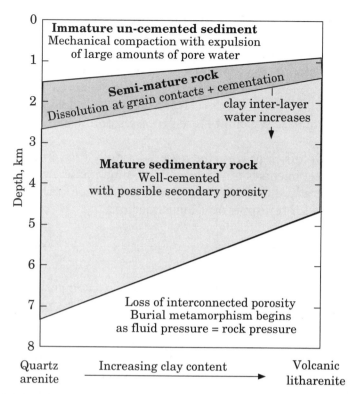

FIGURE 9-7 Schematic diagram indicating the relation between clay content and porosity loss for sediment buried at a rate of about 30 m per million years.

pelagic sediments, calcite dominates the carbonate component. The aragonite and Mg-calcite in carbonates are thermodynamically unstable in the initial seawater pore fluid. The finer size fractions are especially reactive because their large surface area gives them additional surface energy making them even less stable. Dissolution of these less stable carbonates supersaturates the pore solutions relative to average-sized carbonates. Carbonate cements precipitate that with further compaction produce low permeability carbonate units. A *secondary porosity* can be created by dissolution reactions from the passage of a carbonate-undersaturated groundwater.

Under some conditions in the geologic past, dolomite has been precipitated directly out of seawater. Most dolomite strata, however, form by dolomitization of previously precipitated calcite layers. This requires reaction with a fluid with a higher Mg/Ca than presently found in average seawater or temperatures greater than 40°C. Dolomites are presently forming in *sabkha*-type surface environment. Aragonite/anhydrite precipitates from seawater when it becomes supersaturated during evaporation of H_2O from the supratidal pools that are present. This then increases the Mg/Ca ratio of the fluid, enabling dolomitization. It has also been argued that because mixtures of 70% groundwater and 30% normal seawater are undersaturated with calcite but supersaturated with dolomite that dolomitization occurs at the seawater–meteoric water interface. Others have argued that compacting basinal mudstones produces a Mg-rich fluid derived from clays that can cause dolomitization of the carbonates it contacts. Which of these processes dominate is still an area of active research.

There are two different ways the material required to form new minerals and cement the sediments is transported in the fluid phase, by *diffusion* or by *advection*. Generally, diffusion operates over short distances, whereas advection is important for transport over longer distances.

■ Diffusion

Diffusion is the process of material transport along a compositional or *chemical potential gradient* in a medium that is stationary. This medium can be a gas, liquid, or solid. In diagenesis, it is diffusion in the H_2O found in the pores and along grain boundaries that is most important. The change in the concentration of component or species i with time, dm_i/dt, due to diffusion, which is often given in mol cm^{-2} s^{-1}, is termed the diffusional flux of i, J_i^{dif}. Diffusion in one direction is given by

$$J_i^{dif} = -L_i \left(\frac{\partial \mu_i}{\partial x} \right)_t \qquad [9.8]$$

L_i is the "phenomenologic coefficient" of i in units of mol^2 $joule^{-1}$ cm^{-1} s^{-1}. It depends on the nature of the medium through which diffusion occurs and gives the ability of i to diffuse. μ_i is the chemical potential of i in units of joules mol^{-1}, whereas x denotes the distance along the diffusion direction in cm. The derivative $(\partial \mu_i/\partial x)$ gives the Gibbs energy "drive" for diffusion. Because the chemical potential gradient can vary with time, the derivative of μ_i is taken at a particular time, t. If $\partial \mu_i/\partial x = 0$, the diffusional flux is zero and no diffusion occurs.

The quantification of diffusion was first outlined by Adolf Fick in 1855 nearly 30 years before J. Willard Gibbs began to publish his "thermodynamic" model of chemical systems. Fick proposed that diffusion was the "simple spreading of a soluble substance in its solvent." It occurred because of the attraction and repulsion of molecules, and thus an analogy was made with Fourier's and Ohm's laws; therefore,

$$J_i^{dif} = -\frac{D_i}{1000} \left(\frac{\partial c_i}{\partial x} \right)_t \qquad [9.9]$$

where D_i represents the diffusion coefficient in units of cm² s⁻¹ and c_i is concentration of i in units of molarity, that is, moles per 1000 cm³. If $dc_i/dx = 0$,
this equation implies that no diffusion occurs. Equations 9.8 and 9.9 are related. As outlined in Chapter 4, where the thermodynamics of mixtures was
discussed, the chemical potential of i can be written as

$$\mu_i = \mu_i^\circ + RT \ln \gamma_i + RT \ln c_i \qquad [9.10]$$

where μ_i°, γ_i, and c_i represent the standard state chemical potential, activity
coefficient, and concentration of i, respectively.

Consider a standard state for determining μ_i° of unit concentration referenced to infinite dilution at the pressure and temperature of interest. The derivative of Equation 9.10 at constant pressure and temperature with respect
to c_i is then

$$\left(\frac{\partial \mu_i}{\partial c_i}\right)_{T,P} = RT\left(\frac{\partial \ln \gamma_i}{\partial c_i}\right)_{T,P} + \frac{RT}{c_i} \qquad [9.11]$$

If γ_i is considered constant with respect to concentration, the first term on the
right side of Equation 9.11 is zero, and together with Equations 9.8 and 9.9,
the relationship between L_i and D_i is given by

$$L_i = \frac{D_i c_i}{1000 \, RT} \qquad [9.12]$$

Thus, the diffusion and phenomenologic coefficients are directly related.

Problems arise, however, when γ_i cannot be taken as being independent of
concentration. Recall that for charged species, their activity coefficients can
be strong functions of concentration as given by the Debye-Hückel theory.
Also, as is outlined in Chapter 12, ions can increase in concentration near a
charged mineral surface because of electrostatic attraction. One could ask:
"Why doesn't diffusion cause the dissipation of this increased concentration
of ions?" The reason this does not occur is because the energetics caused by
surface charge decrease with distance from the surface. Both c_i and γ_i change
as a function of distance, but their product, a_i, the activity of i, and therefore,
μ_i is constant as a function of distance. With constant μ_i, there is no diffusion.
This is why Equation 9.8 rather than Equation 9.9 is strictly true. It is generally better to consider changes in μ_i rather than c_i when considering diffusion
phenomena. Despite these concerns, Fick's law in terms of concentration gradients is most often used in diffusion calculations. Because Equations 9.8 and
9.9 refer to fluxes at constant time, they are very helpful in steady-state situations, that is, for time-independent diffusion profiles.

If the concentration at a location changes with time, Equations 9.8 and 9.9 still hold at a particular time; however, the description becomes more complex. Consider a diffusional flux as a function of time in a medium as shown in **Figure 9-8**, where J_1^{dif} at x_1 is different from J_2^{dif} at x_2. If the system is not at steady-state clearly, the concentration at points between x_1 and x_2 changes with time. If Δx, the difference between x_2 and x_1 is small, J_2^{dif} can be calculated by adding to J_1^{dif} the change in the flux with distance times the change in distance

$$J_2^{\text{dif}} = J_1^{\text{dif}} + \Delta x \left(\frac{dJ^{\text{dif}}}{dx} \right) \qquad [9.13]$$

or rearranging

$$J_2^{\text{dif}} - J_1^{\text{dif}} = \Delta x \left(\frac{dJ^{\text{dif}}}{dx} \right) \qquad [9.14]$$

In Figure 9-8, for a unit cross-sectional area perpendicular to the page, the volume between x_1 and x_2 equals Δx. The amount of material that accumulates in this volume is equal to the difference of the flux into the volume minus the flux out as given by the left side of Equation 9.14. The change in the amount of material that diffuses is also equal to its volume, Δx, times its change in chemical potential with time so that the right side of Equation 9.14 is

$$\Delta x \left(\frac{dJ^{\text{dif}}}{dx} \right) = \Delta x \left(\frac{d\mu_i}{dt} \right) \quad \text{or} \quad \left(\frac{dJ^{\text{dif}}}{dx} \right) = \left(\frac{d\mu_i}{dt} \right) \qquad [9.15]$$

Substituting Equation 9.15 into Equation 9.8 gives

$$\left(\frac{d\mu_i}{dt} \right) = \frac{d}{dx} \left(L \frac{d\mu_i}{dx} \right) \qquad [9.16]$$

This is Fick's second law of diffusion in one dimension. Equation 9.16 indicates that μ_i can be a function of both distance and time. Solutions to this second-

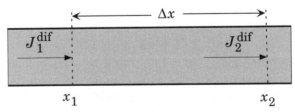

FIGURE 9-8 Control volume for diffusional flux of species through a unit cross-sectional area at x_1 of J_1^{dif} and at x_2 of J_2^{dif}.

order differential equation with various initial conditions are given in Crank (1975). If L is not a function of distance, Equation 9.16 becomes

$$\left(\frac{d\mu_i}{dt}\right) = L\left(\frac{d^2\mu_i}{dx^2}\right)$$

[9.17]

This form of Fick's second law is most often used when diffusional fluxes are time dependent because L is generally considered to be constant.

■ Advection of Fluid

Material can also be transported by movement of the medium in which it resides. In the case of diagenetic processes, this again is H_2O. Before considering fluid flow in the earth's crust, it is necessary to understand how fluid pressure in pores and fractures changes with depth in the crust. Differences in fluid pressure and elevation drive fluid flow (e.g., water flows downhill). Fluid pressure in interconnected pores near the earth's surface is given by a *hydrostatic pressure* gradient. This is the pressure caused by an overlying column of fluid from the location of interest in the crust to the top of the *water table*. If the fluid is not moving and is therefore static, the fluid pressure, P_f, as discussed in Chapter 2, is given by

$$P_f = \rho_f\, g\, z$$

[9.18]

where ρ_f in this case is the average density of the fluid above the location of interest, g is the acceleration of gravity, and z is the height of the fluid in the gravitational field above the location of interest. Although g depends on the position in the earth, for the small distances considered relative to the radius of the earth, g can be considered a constant.

Figure 9-9 shows the H_2O fluid pressure in pores as a function of depth determined from down-hole fluid pressure measured in a well from Brazoria County, Texas in the U.S. Gulf Coast. At depths of 3 km or more below the surface, fluid pressure is no longer given by the hydrostatic pressure gradient. It increases downward along a greater gradient because the interconnection to surface fluid is destroyed on compaction and the much greater density of the overlying rocks is loaded on the fluid. That is, as the pores close and become disconnected with depth, the pressure on the fluid changes from hydrostatic to lithostatic. Lithostatic pressure is the pressure exerted by the overlying column of rocks from the location of interest to the earth's surface. In the case of Brazoria County, Texas, fluids are at lithostatic pressure at depths of approximately 6 km or greater.

This transition from hydrostatic to lithostatic pressure on the fluid occurs because with increasing depth there is a greater difference between hydrostatic

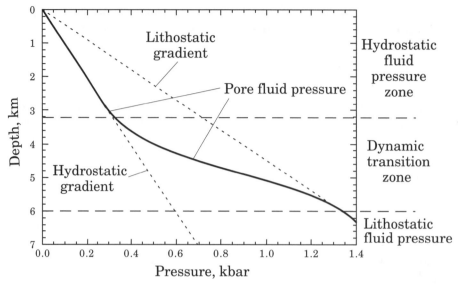

FIGURE 9-9 Fluid pressure as a function of depth, given as the solid line, derived from downwell fluid pressure measurements in Brazoria County, Texas. The dashed lines give the calculated pressure for a column of fluid (hydrostatic) and rock (lithostatic). (From Walther, 1990.)

pressure on the fluid in the pores and the lithostatic pressure on the minerals. To maintain the pore space near the earth's surface, the mineral grains must be strong enough to support the stress difference on different surfaces of the grains (**Figure 9-10**). At some depth, with increasing temperature and greater difference between fluid and lithostatic pressure, the effective crushing strength of the rock grains is exceeded, and the interconnected porosity collapses. At what depth this occurs depends on a number of factors. One of the most important is the composition of the rocks. Rocks with a large clay component as opposed to a sand component undergo the transition closer to the earth's surface (Figure 9-6). The transition has been observed to start at depths of between 1 and 10 km and to occur over a depth interval of approximately 100 meters to a few km.

■ Fluid Flow

Fluid movement in the hydrostatic fluid pressure regime near the earth's surface is by the flow of fluid through a rigid framework of pores and fractures in the rock, termed fluid flow through a porous medium. If the properties of this porous medium can be averaged by considering a representative volume, the flux of fluid, q, defined as the volume of fluid that moves through a unit cross-section in the volume in a given amount of time (cm^3 cm^{-2} s^{-1}), is described by *Darcy's law*. In one dimension Darcy's law is

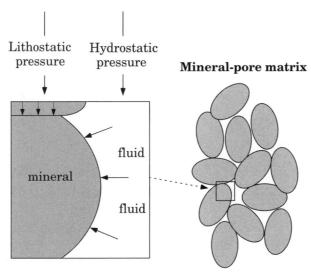

FIGURE 9-10 Diagram indicating the differential stress between lithostatic pressure developed at mineral–mineral contacts and the hydrostatic pressure at fluid–mineral contacts. This situation occurs near the earth's surface where the differential stress between the fluid and interior of mineral grains can be maintained by the mineral's crushing strength.

$$q = -K \left(\frac{dh}{dx} \right) \qquad [9.19]$$

where

> q = fluid flow volume flux (cm^3 cm^{-2} s^{-1}). This is the volume of fluid transported through a unit area in unit time.
>
> K = *hydraulic conductivity* in the representative volume (cm s^{-1}). It is a constant that depends on the properties of both the fluid and the rock that relates fluid flux to the hydraulic gradient.
>
> h = *hydraulic head* in flow direction (cm) giving the relative energy for flow as a length measurement.
>
> x = distance in the flow direction (cm).

The hydraulic head is a measure of fluid potential energy given as a length. It consists of both a *pressure head*, h_p, and an *elevation head*, h_e. The pressure head measures the potential at a location for pressure–volume work and is given by

$$h_p = \frac{\rho_f g}{P_f} \qquad [9.20]$$

where P_f and ρ_f are the fluid pressure and density, respectively. The elevation head, h_e, gives the gravitational potential at a distance above a reference height (h_e = height).

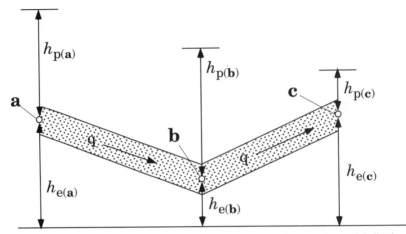

FIGURE 9-11 Locations **a**, **b**, and **c** in fluid-saturated rock at different depths in the earth indicating the direction of fluid flux, q, is controlled by the sum of the pressure head, h_p, and the elevation head, h_e.

Figure 9-11 shows the values of h_p and h_e at three different locations, **a**, **b**, and **c**. The total head is given by the sum of the pressure head, h_p, plus the elevation head, h_e, at a particular location. The total head at **a** is greater than the total head at **b**. The fluid flux, q, therefore, is from **a** to **b** as shown by the arrow. Fluid also flows from **b** to **c** even though it flows "uphill" because the greater pressure head at **b** than at **c** causes the total head at **b** to be greater than the total head at **c**.

At depths where fluid pressure is above hydrostatic pressure, the pore size is controlled by the amount of fluid present, and Darcy's law must be modified (see below). At these depths, the pressure head is not independently variable from the elevation head. This situation starts to occur at a depth of about 3 km in Brazoria County, Texas, as shown in Figure 9-9. Below the 3-km depth, downward flow of fluids does not occur. Regional metamorphic processes occur under these conditions. If fluids do not flow downward, where do the fluids necessary for metamorphic reactions and to produce the veins that are observed come from? To understand the nature of fluid flow during metamorphism at depth in the earth's crust, where fluid pressure equals lithostatic pressure, an understanding of the processes that produce fluid during metamorphism is needed.

■ Metamorphism

Metamorphic reactions are those that occur at greater temperatures than diagenesis, roughly above 200°C to 300°C. On the upper end of the metamorphic temperature scale is the melting of rocks, which ranges from about 650°C for water-saturated silica-rich rocks to about 1100°C for dry basalt. Pressures during metamorphism in the crust are those developed near the earth's surface for some reactions during *contact metamorphism* by an igneous intrusion to those greater than 20 kbar for *regional metamorphism* at the base of thickened continental crust.

Consider the regional metamorphism of a sequence of three layers of sediment and pyroclastics/lava in a geocline that were deposited at successively later times in the order 1, 2, and 3 from early to late. Each layer experiences a different pressure and temperature history as it is buried, heats up, and is eventually exhumed. The first layer experiences the greatest increase in pressure and temperature as the other layers are loaded on top of it. The highest pressure experienced by a layer occurs before the highest temperature because loading rock on top of a layer immediately increases the pressure, whereas the increase in temperature occurs over a longer time period as heat transport in the earth is a slow process. Each layer has a distinct pressure–temperature path, exhibited as a loop, through its burial and exhumation history. Schematic pressure and temperature paths for rock units 1, 2, and 3 are shown in **Figure 9-12**. This analysis is a simplification of the metamorphic process because many loadings of sediments and rocks and their subsequent loss by erosion with a variety of heat fluxes can be involved in a single regional metamorphic event. This is further complicated by deformation, which can produce continental thickening or thinning.

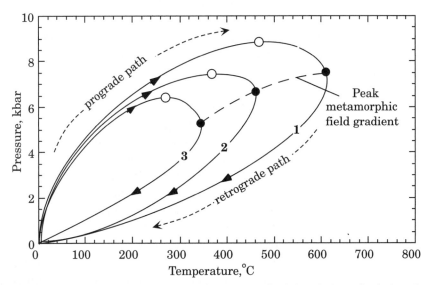

FIGURE 9-12 Pressure–temperature trajectories of three layers of rock deposited over time in the order **1**, **2**, and **3**. The arrows give the direction of increased time, with the open circles indicating the highest pressure and the filled circles indicating the highest temperature experienced by each of the three layers of rock.

It is also important to note that pressure–temperature paths for contact metamorphism are different from those given in Figure 9-12, and some regional-appearing metamorphism is actually contact in nature.

It is metamorphism at the highest temperature, shown by the solid circles in Figure 9-12, that is generally preserved in rocks recovered in the field at the earth's surface (see below). *Prograde reactions* are those that occur during the change in temperature and pressure up to the peak temperature experienced by a particular rock. Any reactions that occur during cooling from the peak temperature are termed *retrograde reactions*. Generally, retrograde reactions only occur if a fluid phase is present during cooling along a retrograde P-T path. The points of highest temperature, shown by filled circles in Figure 9-12, define an arc that gives the *metamorphic field gradient*. This is concave toward the temperature axis, whereas the geothermal gradient occurring at an instant in time is convex. This is because the rocks in each layer did not reach their maximum metamorphic grade, that is, greatest temperature, at the same time. The metamorphic field gradient, therefore, is not a geothermal gradient at a particular instant in time.

The highest pressure occurs at the same time in each layer, when the maximum load of sediments in the geocline occurs. At later times, overlying rocks are being eroded from the earth's surface above the layer. The highest temperature in a layer occurs later than peak pressure and depends on the influx of heat during thermal relaxation toward a new steady-state temperature gradient. Rocks, therefore, continue to heat up after unroofing starts and pressure begins to decrease. For rocks now observed at the earth's surface, typically those that have been most deeply buried reach their temperature maximum at a later time than rocks that have been less deeply buried.

Devolatilization reactions that produce H_2O and CO_2 during regional metamorphism are fairly pressure insensitive at mid- to lower-crustal levels. **Figure 9-13** shows a series of devolatilization reactions. It is clear that the last devolatilization reaction occurs at the greatest temperature experienced by a rock. Because of solid solutions in minerals, these reactions that produce fluid are continuously taking place in many rocks in the sequence with increasing temperature. This fluid allows reactions between minerals to occur. When the rocks begin to cool, retrograde reactions can occur only if a fluid phase is present to produce the lower temperature volatile-containing mineral phase, that is, the reactions need a H_2O and/or CO_2 phase to proceed in the opposite direction.

What is the composition of fluid in metamorphic rocks? In most metamorphic terrains, the dominant rock type is *pelite* to semipelite. A pelite is a mudstone, an argillaceous rock that after diagenesis becomes a shale. Semipelite is a pelite that contains significant quartz or carbonate. **Table 9-1** lists the compositions of average unmetamorphosed pelitic rock and average high-grade metamorphosed pelitic rock. Given the variability in sampling, within the uncertainty of the analysis, no significant difference in composition occurs during meta-

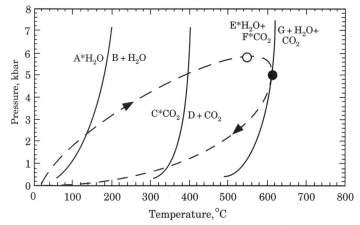

FIGURE 9-13 Pressure–temperature diagram showing some model univariant devolatilization reactions as solid lines with the mineral phases denoted by letters. The dashed line is the *P–T* trajectory of a packet of rock as a function of time with the highest pressure given by an open circle and the highest temperature by a filled circle. The asterisk after a letter indicates that H_2O or CO_2 is bound in the mineral such as in the minerals biotite and calcite, respectively. Continuous release of H_2O and CO_2 occurs during the prograde path to the greatest temperature experienced by the packet of rock.

morphism except for loss of H_2O and CO_2. That is, metamorphism on the large scale is isocompositional except for the loss of H_2O and CO_2. Table 9-1 indicates that on average, 2.60 wt% of H_2O and 2.42 wt% of CO_2 are released during metamorphism of pelitic rocks. This is equivalent to 1.44 moles of H_2O and 0.55 moles of CO_2 (= 2 total moles) per 1000 g of rock. If retained in the rock, this fluid would occupy a volume of 12% at 500 °C and 4 kbar. Because

Table 9-1   Average composition of low-grade shale and high-grade pelitic rocks

Oxide	Low Grade (wt%)	High Grade (wt%)
SiO_2	58.38	63.51
TiO_2	0.65	0.79
Al_2O_3	15.47	17.35
Fe_2O_3	4.03	2.00
FeO	2.46	4.71
MgO	2.45	2.31
CaO	3.12	1.24
Na_2O	1.31	1.96
K_2O	3.25	3.35
H_2O	5.02	2.42
CO_2	2.64	0.22
Total	98.78	99.86

From Walther and Orville (1982).

the porosity of metamorphic rocks is 0.2% or less, this implies that the fluid escapes during metamorphism.

Transport of Fluid at Mid-Crustal and Deeper Levels

Consider a volume of fluid produced by the destruction of volatile-bearing minerals and existing along grain boundaries at mid levels of the earth's crust. A small amount of fluid can be retained at mineral triple junctions. An interconnected fluid space, however, occurs with only a small amount of the fluid released during pelite metamorphism. This interconnected fluid volume could have a complicated three-dimensional geometry depending on the local conditions, but it does have some vertical extent. The vertical extent of interconnected fluid produced is shown schematically by the vertical shaded fluid-filled space depicted in **Figure 9-14**.

At some vertical location, z_o, in the fluid, the pressure in the static fluid, P_{sf}, is exactly equal to the pressure in the rock, P_R. A change of Δz in the vertical direction with z increasing downward results in a change in pressure in the rock, ΔP_R, of

$$\frac{\Delta P_R}{\Delta z} = g\,\rho_R \qquad\qquad [9.21]$$

where ρ_R is the density of the rock. The corresponding change in the static fluid pressure, ΔP_{sf}, with ρ_f equal to the fluid density, is given by

$$\frac{\Delta P_{sf}}{\Delta z} = g\,\rho_f \qquad\qquad [9.22]$$

Because $\rho_R > \rho_f$, these equations indicate $P_{sf} > P_R$ at all points above z_o and $P_{sf} < P_R$ for all points below z_o. The rock in contact with the fluid, therefore, is under directed nonhydrostatic stress.

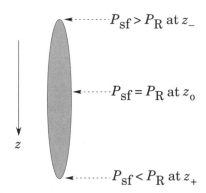

FIGURE 9-14 Static fluid-filled space at depth in the earth's crust. If pressure in the static fluid, P_{sf}, equals pressure in the rock, P_R, at some elevation, z_o, then at all points above this elevation the fluid pressure is greater than rock pressure and at all points below fluid pressure it is less than rock pressure.

Given the elevated temperatures and long time frames of interest at mid-crustal levels, these wall rocks have vanishingly small strength to support this nonhydrostatic stress. The strength of the rocks, therefore, is not enough to prevent mechanical failure. This instability causes the fluid-filled fracture to propagate upward by wedging open the rock at the top where $P_{sf} > P_R$ and closing the fracture at the bottom where $P_{sf} < P_R$. This is how magmas, with their lower density than the surrounding rocks, ascend in the crust. In the case of H_2O + CO_2 fluids in rocks, the density and viscosity contrasts are significantly greater so the rocks fracture. This fracturing is referred to as *hydrofracturing*. The mechanically unstable situation can be made mechanically stable by flowing a fluid through the fracture. The pressure change in a steady-state flowing fluid, ΔP_{ff}, is given by

$$\frac{\Delta P_{ff}}{\Delta z} = \frac{\Delta P_{sf}}{\Delta z} + \frac{\Delta P_{vis}}{\Delta z}$$ [9.23]

where ΔP_{vis} is the change in the *viscous pressure* in the fluid. Because of its viscosity, a fluid will flow with greater velocity at its center, with the velocity decreasing to near zero at the walls of the fracture. Because of this, there will be a tangential force on the walls of the fracture that decreases in the direction of flow. The situation is outlined in **Figure 9-15**.

To keep the fracture open requires

$$\frac{\Delta P_R}{\Delta z} = \frac{\Delta P_{ff}}{\Delta z}$$ [9.24]

If the fracture is not oriented vertically but is at some angle θ with the vertical along the x direction, then $\Delta z = \Delta(x \cos \theta)$, where θ is the angle of the fracture with the vertical (**Figure 9-16**); therefore, to keep the fracture open

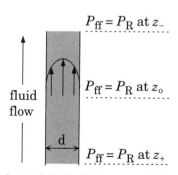

$P_{ff} = P_R$ at z_-

$P_{ff} = P_R$ at z_0

fluid flow

d

$P_{ff} = P_R$ at z_+

FIGURE 9-15 Fluid pressure in a flowing fluid, P_{ff}, which generates fluid pressure equal to rock pressure along the walls of a fracture of width d. The vertical arrows in the fracture indicate that due to fluid viscosity, the fluid flow is more rapid in the center and decreases as the walls of the fracture are approached, creating a tangential pressure on the walls of the fracture.

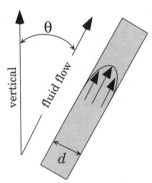

I FIGURE 9-16 Fluid in a layer flowing at some angle θ to the vertical.

$$\frac{\Delta P_{vis}}{\Delta z} = \frac{\Delta P_{vis}}{\Delta(x \cos \theta)} = (\rho_R - \rho_f)\, g \qquad [9.25]$$

or with θ constant

$$\frac{\Delta P_{vis}}{\Delta x} = (\rho_R - \rho_f)\, g \cos \theta \qquad [9.26]$$

This is the viscous pressure gradient that maintains fluid pressure just equal to rock pressure along the walls of the fracture.

The flow of fluid in a crack or grain boundary can be modeled as laminar flow between two parallel plates with some *tortuosity*, τ. The average fluid velocity, v_{av}, is given by

$$v_{av} = \frac{2}{3} v_{max} = \frac{\tau d^2}{12\eta}\left(\frac{\Delta P}{\Delta x}\right) \qquad [9.27]$$

where v_{max} is the maximum velocity at the center of the fracture, d is the width of the opening of the fracture, η denotes the viscosity of the fluid, and $(\Delta P/\Delta x)$ gives the fluid pressure gradient in the x direction. The tortuosity, τ, is a nondimensional parameter accounting for the orientation of the fractures/grain boundaries relative to the flow path. For single fracture flow τ = 1, whereas for grain boundary flow τ is between 0.5 and 0.8 (Bear, 1975).

To determine the extent of flow needed to produce the required $\Delta P_{vis}/\Delta x$, the viscosity of the fluid is required. The viscosity, η, of supercritical aqueous fluids is not very sensitive to changes in temperature and pressure. For H_2O + CO_2 fluids along geotherms in the earth's crust of 15 to 45°C km^{-1}, η are typically between 0.1 and 0.2 centipoise. No significant error is introduced in the analysis here if a value of η = 0.15 centipoise is used (Walther and Orville, 1982). The fluid flux volume (cm^3 per cm^2 per s) is equal to the cross-sectional area of the fracture perpendicular to the flow direction times the average fluid velocity

$$q = v_{av} d \, l = \frac{\tau d^3 l}{12\eta}\left(\frac{\Delta P}{\Delta x}\right) \qquad [9.28]$$

where l (cm^{-1}) is the length of fracture opening per unit area perpendicular to the flow direction. With the fluid pressure gradient equal to the viscous pressure gradient required to keep the fracture open, Equation 9.28 becomes

$$q = v_{av} d \, l = \frac{d^3 l \tau \, g \, (\rho_R - \rho_f)\cos\theta}{12\eta} \qquad [9.29]$$

Linking Fluid Flow in the Crust to Darcy's Law

Fluid flow in the middle to lower crust can be either along grain boundaries or in more widely spaced fractures. To determine the extent of flow with a continuum model like Darcy's law, a representative volume is considered. It is not clear, however, that a representative volume exists at deep levels in the crust. That is, the scale of heterogeneity in the crust likely expands along with the size of the representative volume. No matter what the size of the volume considered, there can exist larger fractures that are not considered in the model volume. To get around this problem, a continuum model can be considered between more widely spaced fractures. In this case, with laminar flow, the flux can be determined with the Darcy flow equation as given in Equation 9.19:

$$q = -K\left(\frac{db}{dx}\right) \qquad [9.30]$$

Under the conditions that fluid pressure is equal to rock pressure, the change in energy of the fluid must be by buoyancy forces so that the energy of the fluid depends only on its position in the gravitational field and the density difference between the fluid and rock. That is, fluid is transported toward the earth's surface because it is less dense than the surrounding rocks. Tectonic stresses in the solid rock are not important as an energy source because they operate over much longer time frames and the fluid itself is in *hydrostatic equilibrium*. Thus, the fluid maintains no internal stresses. The loss, therefore, in gravitational energy of the fluid because of its transport toward the earth's surface is what drives fluid flow. The tectonic stresses become important in how they affect the hydraulic conductivity in the representative volume.

To understand the hydraulic head as a measure of fluid energy where $P_f = P_R$, consider the dimensions of energy (units $= ML^2/t^2$, where M = mass, L = length, and t = time) of the fluid per unit volume (L^3) that become $(ML^2/t^2)/(L^3)$ or $(L)(L/t^2)(M/L^3)$. These energy units are dimensionally satisfied using elevation, z (L), the acceleration of gravity, g (L/T^2), and density difference between the rock and fluid of $(\rho_r - \rho_f)$ with dimensions of (M/L^3); therefore, the energy of transport per unit volume, E_v (viscous pressure) is equal to

$$E_v = \Delta P_{vis} = z\, g\, (\rho_R - \rho_f). \tag{9.31}$$

The expression on the right side of Equation 9.31 is the pressure difference between the rock and fluid, and this is what causes the fluid to flow. If Equation 9.31 is divided by $(\rho_R - \rho_f)$, the result is energy of transport per unit mass

$$E_m = \frac{E_v}{(\rho_R - \rho_f)} = z\, g \tag{9.32}$$

Dividing Equation 9.32 by g, energy per unit weight E_w, is produced or energy calibrated in terms of length, that is, elevation, z,

$$E_w = \frac{E_v}{g\,(\rho_R - \rho_f)} = z \tag{9.33}$$

The change in energy per unit weight along the fluid flow path is then

$$\left(\frac{dE_w}{dx}\right) = \left(\frac{dz}{dx}\right) \tag{9.34}$$

Darcy's law then states that fluid flow along the flow direction is proportional to this energy change or

$$q = K\left(\frac{dE_w}{dx}\right) = K\left(\frac{dz}{dx}\right) \tag{9.35}$$

where

$$K = \frac{d^3 l\, \tau\, g\, (\rho_R - \rho_f)}{12\,\eta} \tag{9.36}$$

Because

$$\left(\frac{dz}{dx}\right) = \cos\theta \tag{9.37}$$

substituting Equations 9.36 and 9.37 into Equation 9.35 gives Equation 9.29.

Rate of Fluid Production

How much fluid is produced during metamorphism as a function of time? As stated above, about 2 moles of fluid is released per 1000 g of pelitic rock during the metamorphic process. This process of metamorphosing a rock from low grade to high grade takes place over about 200°C, roughly from 350°C to 550°C. The length of time it takes to release the 2 moles of fluid is the time needed to increase the temperature of the rock 200°C. Heat must be added. **Figure 9-17**

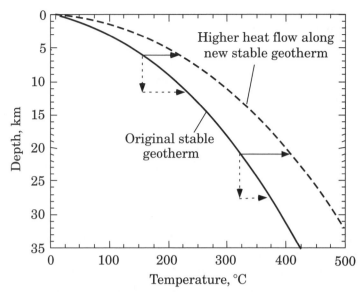

FIGURE 9-17 A typical geotherm as a function of depth in the crust given by the solid line. The solid horizontal arrows give perturbation of the temperature in a rock caused by increased heat flux from the base of the crust. The dashed arrows give the perturbation with loading and subsequent heating of a rock along the geotherm.

shows a typical steady-state geotherm for the crust (solid line). For metamorphism to occur, the pressure in the rock volume or the heat flux in or out of a rock must be changed. Metamorphism occurs during the time interval that pressure and temperature are changing before a new steady-state geotherm is established. That is, at steady state, the amount of heat that enters the bottom of the crust plus that produced in the crust is equal to the amount that leaves the top of the crust, and no metamorphism occurs.

How can this stable geothermal gradient be perturbed? Possibilities include rapid thickening by depositing sediment on a sedimentary pile. This causes colder rocks to sink into hotter areas of the earth, and heating of these rocks as the system relaxes to new pressure and temperature conditions causes metamorphism along a new geotherm, as indicated by the dashed arrows in Figure 9-17. Overthrust thickening of the sedimentary pile has the same effect. It is also possible that rise of magma to the upper mantle or lower crust could increase the heat flow into the crust. Rocks are progressively metamorphosed as temperature increases, as shown by the solid horizontal arrows in Figure 9-17, until a new steady-state geothermal gradient is established at a higher steady-state heat flow.

Present-day heat flow varies at the earth's surface from about 0.8 μcal cm^{-2} s^{-1} (0.8 *heat flow units* [HFU]) at the bottom of ocean trenches and on old continental shield areas to 2.5 μcal cm^{-2} s^{-1} (2.5 HFU) in tectonically active areas such as the East Pacific rise. The value 1 µcal cm^{-2} s^{-1} is referred to as a HFU (heat flow is discussed in more detail in Chapter 2).

To consider metamorphism, a simple thermal model can be constructed where the steady-state heat flow in the crust is 0.8 HFU, and it is instantaneously changed at the base of the crust from 0.8 to 2.5 HFU. The additional heat causes the temperature of rocks above to increase until a new steady-state geotherm is established, where 2.5 HFU are lost at the earth's surface. Instead of determining the actual change in the geotherm with time, assume that all of the heat added goes into metamorphosing the rocks in the crust. That is, by increasing the heat flow by 1.7 HFU instantaneously and assuming that all of this extra heat is used to cause metamorphism as the rocks heat up, the calculation puts an upper bound on the rate of temperature increase. How much heat is needed to metamorphose 1000 g of rock from 350°C to 550°C?

The average heat capacity of minerals like quartz, mica, and feldspar between 350°C and 550°C is approximately 0.25 cal/g °C. In the 200°C temperature rise that occurs during metamorphism, 50 kcal per kg of average rock are, therefore, required. A significant amount of heat (enthalpy of reaction) is also needed to release H_2O and CO_2 from the minerals during the metamorphism. **Table 9-2** lists the heats required to liberate a mole of H_2O or CO_2 by the indicated reactions. As a first approximation, the heat of reaction needed to liberate the 2 moles of H_2O plus CO_2 in a kilogram of average rock is then about 40 kcal. To metamorphose a kilogram of rock therefore requires 90 kcal, of which 50 kcal is required to increase the temperature and 40 kcal to release the H_2O and CO_2 from the minerals.

What is the average rate of advancement of isotherms through the crust during metamorphism? With a density of rock of 2.6 g cm^{-3}, the volume of a kg of rock is 385 cm^3. With heat supplied for metamorphism of 1.7 HFU and with 90×10^3 cal needed to metamorphose a kilogram of rock, the average rate of advancement of the isotherms is

| Table 9-2 | Enthalpy of reaction per mole of H_2O and CO_2 released, ΔH_R, for the indicated metamorphic reactions at 5 kbar and 500 °C as calculated from SUPCRT92 |

Reaction	ΔH_R per mole of fluid (kcal/mole)
Pyrophylite = andalusite + 3 quartz + H_2O	14.3
Muscovite + quartz = K-feldspar + kyanite + H_2O	14.4
3 Tremolite + 5 calcite = 11 diopside + 2 forsterite + 5 CO_2 + 3 H_2O	17.6
Calcite + quartz = wollastonite + CO_2	20.6
4 Zoisite + quartz = grossular + 5 anorthite + 2 H_2O	25.6

$$\frac{1.7 \times 10^{-6} \text{ cal cm}^{-2} \text{ s}^{-1} \ 385 \text{ cm}^3 \text{ kg}^{-1}}{90 \times 10^3 \text{cal kg}^{-1}} = 7.18 \times 10^{-9} \text{ cm s}^{-1} \qquad [9.38]$$

This rate indicates isotherms advance 2.3 mm in 1 year or can metamorphose a 23-km thick crust in 10 million years. Because regional prograde metamorphic events typically last a few tens of millions of years, this would be near the upper limit for the rate of prograde metamorphism. To consider a slow rate of metamorphism, somewhat arbitrarily, a rate that is an order of magnitude slower can be considered. This would take 100 million years to metamorphose 23 km of crust.

The average rate of volatile release from the volatile bearing minerals throughout the metamorphic pile for the rapid (maximum) rate of metamorphism, q_{max}, would be

$$q_{max} = \frac{1.7 \times 10^{-6} \text{ cal cm}^{-2} \text{ s}^{-1}}{90 \times 10^3 \text{ cal kg}^{-1}} \times 2 \text{ mol kg}^{-1} = 3.8 \times 10^{-11} \text{ mol cm}^{-2} \text{ s}^{-1} \qquad [9.39]$$

Because there are 50 g in the 2.0 moles of volatiles, q_{max} can also be given as 1.0×10^{-9} g cm^{-2} s^{-1}. With an order of magnitude slower (minimum) metamorphic event the fluid flux, q_{min}, would be 1.0×10^{-10} g cm^{-2} s^{-1}. That is, q, the mass flux of fluid out of the metamorphic pile if all of the fluid produced is transported upward is on average between 10^{-10} and 10^{-9} g cm^{-2} s^{-1}, as shown by the vertical extent of the hatched area in **Figure 9-18**. With a knowledge of the likely range of q, everything except d and l are known in Equation 9.29 above for approximately vertical flow. Log q can be plotted versus log d and contoured for l as given by the diagonal dashed lines in Figure 9-18 labeled with the value of l in cm of fracture per cm^2.

Width of Fractures and Quartz Veins

What is the width of the fractures that transport fluid upward? In metamorphic quartz, planes of fluid inclusions are observed as shown in the photomicrograph given in **Figure 9-19**. These fluid inclusion planes formed when a fluid-filled microfracture in the mineral sealed and the fluid in the fracture migrated to lower its surface area of contact by forming a plane of fluid inclusions. The minimum thickness of fractures, which occurred just before the fracture closed down, can be calculated by redistributing the fluid in fluid inclusion planes over the total area of the healed fracture. When this is done, the width of these fractures is about 0.02 µm. This minimum width of the fracture is shown by a vertical line in Figure 9-18. This width is over an order of magnitude larger than the width of a monomolecular absorbed layer on opposing surfaces of the fracture. This indicates a true fluid phase is transported during metamorphism rather than the flux due to an adsorbed phase on mineral surfaces.

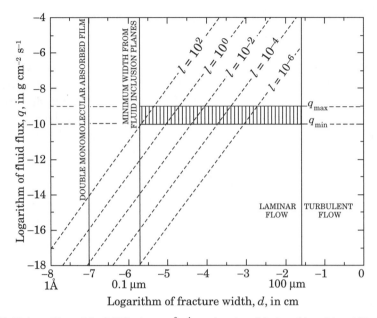

FIGURE 9-18 Logarithms of the fluid flux in g cm^{-2} s^{-1} as a function of the logarithm of the width of the fracture contoured as a function of length of fractures, l, in cm per cm^2 given as diagonal dashed lines. The hatched area gives the range of fluid flux and fracture width appropriate for the earth's crust. The constraints on fracture width given by the vertical solid lines are discussed in the text. (From Walther and Orville, 1982.)

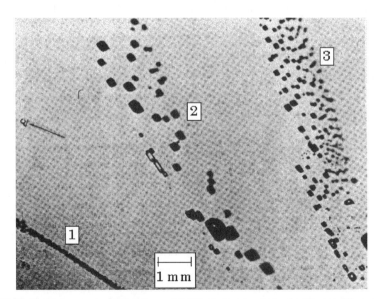

FIGURE 9-19 Photomicrograph of three planes of fluid inclusions in quartz orientated at different angles. These have formed from fluid trapped along a sealing fracture. The trapped fluid minimizes its contact area with the mineral by fluid migrating to form isolated inclusions. The larger the inclusions, the greater is the distance they are apart. (From E. Roedder, Evidence from fluid inclusions as to the nature of the ore-forming fluids. In Symposium-Problems of Postmagmatic Ore Deposition, Prague, 1963. *Prague Geol. Survey*, 1965.)

The fluid velocity will increase as the width of the fracture increases. At some point, the fluid will no longer be laminar in nature but become turbulent. Turbulent flow occurs in a tabular fracture if the *Reynolds number* exceeds about 2300. For a Reynolds number of 2300, the computed width of the fracture is 0.025 cm. This is shown by the vertical line on the right side of Figure 9-18. Using this width with q_{max} from Equation 9.39, a fluid density of 0.9 g cm^{-3} and $\eta = 0.15$ centipoise for a vertical fracture gives l from Equation 9.29 of

$$l = \frac{1.0 \times 10^{-9}\text{g cm}^{-2}\text{ s}^{-1} \times 12 \times 0.0015 \text{ poise}}{0.9\,\text{g cm}^{-3} \times (0.025 \text{ cm})^3 \times 1.9 \times 10^3 \text{ dynes cm}^{-3}} = 6.7 \times 10^{-10} \text{ cm per cm}^2 \qquad [9.40]$$

This would be equivalent to a length of fracture of only 6.7 cm in each km^2 of rock. Because the number of fractures appears to be much greater than this, the assumption of laminar flow used to obtain Equation 9.29 appears reasonable.

Fluid is produced throughout the metamorphic pile at each volatile-containing mineral undergoing reaction. This fluid produces grain boundary channels and microfractures in minerals. At increased distances along the flow path, the fluid pathways likely coalesce to form larger through-going fractures. Quartz solubility decreases both with decreasing P and decreasing T; therefore, quartz precipitates as the fluid is transported upward along the fractures, producing quartz veins. Fractures appear to have opened and sealed numerous times during metamorphic devolatilization. The fractures keep forming at the site of quartz veins because as a monomineralic material, quartz propagates fractures much more readily than the heterogeneous surrounding rock. Also, any fracture that propagates outside the quartz vein carries H_2O, which hydrates the surrounding minerals, causing an increased volume in the solids that can seal the fracture. Consistent with this analysis is the observation that quartz veins seen at mid and lower crustal depths have hydrated mineral selvages around them. These quartz veins then give the path of the main flux of the fluid that drained the progressive metamorphic terrain. It is not unlike a dendritic stream pattern of fluid transport on the earth's surface, as shown in **Figure 9-20.**

In summary, one concludes that the rate of reactions decreases dramatically if no fluid phase is present. Other than next to major channel ways, little fluid is present unless it is actively being produced. This is why high-grade minerals (e.g., garnets) can be observed on the earth's surface. After the fluid leaves the immediate area where it is produced, reactions to form new minerals do not occur at measurable rates. One then talks about "metamorphic facies" in a "petrogenetic grid." These then give the assemblages of minerals stable at the highest temperature a particular rock was subjected to that which typically occurs at a different time than the highest pressure experienced by the rock and at a different time than nearby rocks that are experiencing different metamorphic facies conditions.

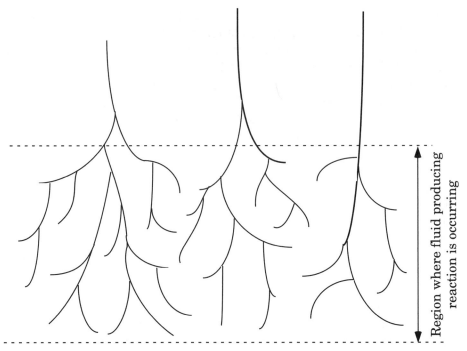

Region where fluid producing reaction is occurring

FIGURE 9-20 Pattern of quartz veins in the earth's crust that mark the site of major channel ways for the volatile release that occurs during progressive regional metamorphism.

■ Ore Deposits

Mineral deposits form where normal geological processes occur long enough and in the correct sequence to concentrate an element of economic value. What processes lead to the deposition of the ore minerals? The primary ones are as follows:

1. Magmatic segregation (e.g., layer and podiform chromite from basaltic magma)
2. Magmatic-hydrothermal differentiation (e.g., statiform igneous Pt + Pd ± Au reefs)
3. Weathering or sediment–seawater interaction (e.g., banded iron)
4. Hydrothermal fluid/mineral reactions (e.g., porphyry Cu + Mo)

Secondary enrichment of ore can occur if deposits have crumbled and are eroded by physical sorting during the flow of water, giving rise to what are termed *placer deposits*. Placers can also form in soil by physical and chemical weathering of rock with high starting concentrations of an element of economic value. For instance, the weathering of ultramafic rocks can produce placers of Pt and/or Pd.

The chemical mechanism for concentration of the element of interest results from dissimilar solubilities of the element in minerals of a residual rock as opposed to a fluid phase it reacts with, whether this is a magma or aqueous solution. In some ores such as *bauxite deposits* produced by the weathering of Al-rich rocks, the residual solid phases concentrate the element of interest, as the weathering fluids remove much of the other non-ore material. Examples are the deposits of New Caledonia and Arkansas.

In most deposits, however, the magma or aqueous solution is where the concentrating occurs. In this case, a process develops that supersaturates the metals in the fluid phase and allows the precipitation of an ore mineral at a locality. The precipitation process is often a change in the physical state of the system that rapidly changes the pressure and temperature of the fluid. For instance, a rapid decrease in pressure could be triggered by fracturing of the rocks, changing aqueous fluid pressure from lithostatic to hydrostatic and leading to fluid boiling. The boiling or pressure change supersaturates the fluid with respect to the minerals of interest. Alternatively, the temperature could cool by proximity to the surface or mixing with cooler groundwater or seawater. A fluid can also supersaturate with the elements of interest by changing its chemistry (often increasing its pH) due to reaction with minerals it contacts or mixing with seawater.

In magmatic segregation, the ore metals are less concentrated in crystallizing non-ore minerals than melt. They, therefore, concentrate in the residual melt upon cooling. The residual melt typically concentrates sulfur along with these ore-forming metals. Metal sulfides are crystallized, producing a sulfide-rich deposit of the metal. In magmatic-hydrothermal differentiation deposits, such as the Bushveld of South Africa or the Skaergaard intrusion of east Greenland, platinum group elements and Au are concentrated in a late stage magmatic aqueous-rich melt produced during the final stages of crystallization. A metal-rich aqueous fluid can also separate from the magma. This residual fluid phase then precipitates ore minerals by one of the processes outlined above.

In some settings, weathering leaches and transports an ore metal to its site of deposition. This process has upgraded copper ore in many porphyry Cu-Mo deposits (see below) where it is termed *supergene* enrichment. Similarly, hydrothermal fluids consisting of heated groundwater leach minerals from source rocks and transport them to a site of ore deposition. Because of the high temperatures and the large flux of water in hydrothermal systems, hydrothermal reactions are the most effective ways to concentrate many ore metals (Reed, 1997). Hydrothermal ore deposits are the most important type of metallic ore deposits and occur throughout the cordillera of the western hemisphere and other mountain ranges at convergent plate boundaries. To understand the formation of hydrothermal ore deposits, the source of metals, its transport in hydrothermal fluids, and the mechanism of metal deposition must be considered.

Porphyry Ore Deposits

Consider *porphyry* copper-molybdenum ore deposits, one of the largest sources of Cu and Mo. They occur in and around felsic to intermediate composition porphyritic *stocks* and dikes. These are igneous intrusions with abundant plagiocase, K-feldspar, and quartz and with lesser amounts of hornblende and biotite and minor amounts of the oxide phases magnetite (Fe_3O_4) and ilmenite ($FeTiO_3$). The porphyritic texture refers to conspicuous feldspar and quartz phenocrysts set in a fine-grained groundmass. Porphyry ore deposits contain large tonnages of Cu and Mo sulfide at low grade. Coeval precious metal deposits (*epithermal deposits*) occur above some porphyry ore deposits in veins, stockworks, breccia pipes, and hot-spring deposits. The porphyry ores and associated epithermal deposits include many of the world's largest accumulations of Cu, Mo, Au, Ag, Sn, and W. Large porphyry Cu-Mo deposits occur at El Salvador; Chile; Butte, Montana; and Bingham, Utah. Porphyry Mo deposits include those at Climax, Colorado and Buckingham, Nevada.

The melt from which the porphyry rocks were crystallized was probably produced in the mantle wedge above subducted oceanic lithosphere. Rather than ascending to the earth's surface to produce a volcanic eruption, the magma ascends to a depth of 1 to 2.5 km where it crystallizes in place as a stock. The decreasing pressure on its ascent saturates or nearly saturates the magma with water at lithostatic pressure because H_2O is less soluble in magmas at lower pressures (see Chapter 8). Having been intruded into cooler rocks, the margins of the magma body crystallize rapidly. With continued crystallization of anhydrous minerals, the concentration of H_2O and ore metals in the remaining magma increases until a separate water-rich phase is produced. The pressure on this fluid can increase to greater than lithostatic. When the tensile strength of the rock that confines this fluid is exceeded, the rock is hydrofractured. Aqueous fluids containing ore metals then move upward into overlying rocks where hydrostatic pressures conditions prevail. Metaliferous veins form along the fluid flow path. The decompression release of water from the magma after fracturing causes rapid crystallization, producing the porphyritic texture that gives the ore deposits their name. The magmatic water released, which contains the ore metals and sulfur moves into *crackle*, *stockwork*, and *brecciated zones* where it cools and reacts with the rock and precipitates sulfide minerals.

Because of density differences in and therefore pressure differences at the bottom of equal columns of hot and cold water, heated groundwater convects around the cooling porphyry stock. To understand this circulation, a simple fluid convective cooling model of an igneous intrusion can be constructed. **Figure 9-21** shows the calculated temperatures and stream lines for H_2O flow around a model stock after 20,000 years whose initial temperature was 750°C. The stream lines, which show the fluid pathways, are spaced further apart where the fluid descends than where fluid ascends along the sides to the top of

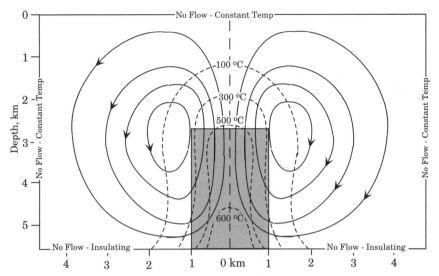

FIGURE 9-21 A model of H_2O circulation and temperature distribution around an igneous pluton given by the shaded area intruded with an initial temperature of 750°C after 20,000 years of cooling. (Adapted from Norton and Cathles, 1979.)

the cooling stock. The closely spaced lines imply the fluid flux is largest along the pluton margin, which is where cooling of the fluid starts to occur, thereby removing heat from the system which causes ore metals to precipitate from the magmatic fluids.

The convecting water generally contains chloride from the magmatic vapor phase (**Table 9-3**) and can also react with NaCl in country rocks to become Cl rich. These metals dissolve as Cl complexes in the fluid. *Skarn* deposits in porphyry Cu + Mo systems form by contact metasomatic mineralization where silica-rich metal-bearing fluids from cooling igneous plutons react with limestone of dolostone. Ca-rich garnet and clinopyroxene are produced. During this process, metal sulfides are precipitated from the increase in *pH* of the fluid by reaction with carbonate. Cu skarns associated with porphyry Cu deposits typically contain 1 to 2 wt% Cu. The Cu-bearing minerals are chalcopyrite ($CuFeS_2$) and bornite (Cu_4FeS_5). A schematic cross-section through a porphyry Cu deposit that includes a skarn is shown in **Figure 9-22**. It gives the alteration zones produced and the location of ore.

Metal Sulfide Solubilities

How are metals transported and precipitated in ore deposits? Consider Zn in a hydrothermal porphyry Cu + Mo ore deposit. Zn occurs predominately in the mineral sphalerite, ZnS. Because Zn^{2+} is the dominant Zn species in dilute aqueous solutions, a sphalerite solubility reaction can be written as

Table 9-3	Temperature, oxygen fugacity, and mole percent of constituents in volcanic gas from Mt. Momotombo, Nicaragua

Temp, °C = 820	log f_{O_2} = −13.55
Constituent	**Mole %**
H_2O	97.11
H_2	0.70
CO_2	1.44
CO	0.0096
SO_2	0.50
H_2S	0.23
S_2	0.0003
HCl	2.89
HF	0.259

Reported in Symonds et al. (1994).

$$2H^+ + 0.5O_2 + ZnS = Zn^{2+} + 0.5S_2 + H_2O \qquad [9.41]$$

The equilibrium constant for Reaction 9.41 is

$$\log K_{(9.41)} = \log \frac{a_{Zn^{2+}} f_{S_2}^{0.5} a_{H_2O}}{a_{H^+}^2 f_{O_2}^{0.5} a_{ZnS}} \qquad [9.42]$$

A reasonable average pressure in a hydrothermal porphyry Cu ore deposit is 1 kbar. This corresponds to an average depth somewhat under 4 km under a lithostatic pressure gradient. Although the hydrothermal system is cooling with time from its intrusion temperature of 650°C to a background geothermal gradient that, at 4 km, is about 100°C, a reasonable average temperature is 350°C. Using SUPCRT92, log $K_{(9.41)}$ at 1 kbar and 350°C is calculated to be 10.11. Although the activity of H_2O and ZnS can be taken as unity, what are the fugacities of O_2 and S_2 and the activity of H^+ needed to determine the concentration of Zn^{2+} at sphalerite saturation from Equation 9.42?

Given in **Figure 9-23** is a fugacity of O_2–S_2 diagram calculated for the S-O-Fe-Cu system at 350°C and 1 kbar. It is constructed similarly to the activity–activity diagrams introduced in Chapter 6 where aqueous solutions were discussed. Chalcopyrite, Cp, is the dominant ore mineral in hydrothermal ore deposits, and the range of f_{O_2} and f_{S_2} where it is stable is given by the cross-hatched area in Figure 9-23. Pyrite + magnetite ± pyrrhotite are also present in many deposits along with the chalcopyrite. Note the stable coexistence of pyrite + magnetite + pyrrhotite in the chalcopyrite stability field. With these phases present according to Figure 9-23, f_{O_2} and f_{S_2} are buffered to −30.22 and −9.38, respectively.

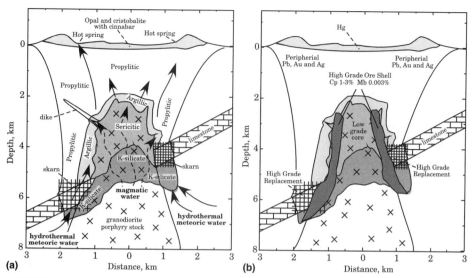

FIGURE 9-22 Schematic diagrams showing the development of a porphyry ore deposit. (a) Alteration zones observed in the field with possible fluid pathways shown by arrows. (b) Areas where ore is deposited in the porphyry ore deposit.

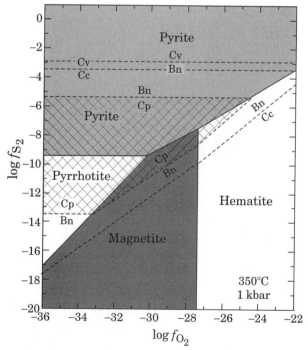

FIGURE 9-23 Mineral stability is the system O-S-Fe-Cu at 350°C and 1 kbar. Bn, bornite (Cu_5FeS_4); Cc, chalcocite (Cu_2S); Cv, covellite (CuS); Cp, chalcopyrite ($CuFeS_2$). The field where Cp is stable is given by the cross-hatched area.

Using these values for f_{O_2} and f_{S_2}, Equation 9.42 at 1 kbar and 350°C becomes

$$-0.31 = \log a_{Zn^{2+}} + 2pH \qquad [9.43]$$

The amount of Zn that can therefore be transported in solution before it precipitates sphalerite increases as pH decreases. Even if the pH is low, however, the amount of Zn that can be carried in solution as Zn^{2+} is very small. For instance, at $pH = 4$, $a_{Zn^{2+}} = 10^{-8.3}$, a value too small to carry enough Zn in solution to produce an ore deposit. What is the minimum concentration of Zn or any other element of interest that needs to be carried in a hydrothermal solution to form an ore deposit? This is difficult to determine exactly, but with $m_{Zn} = 10^{-8}$, to produce an average-sized ore deposit of 1 million tons of Zn (approximately 1.4×10^{10} moles) with all of the Zn precipitated from the fluid requires about 1.4×10^{18} liters of H_2O. The ocean contains 1.4×10^{21} liters of H_2O. This deposit would require an integrated flux of 1/1000 of the H_2O found in the ocean. This appears significantly too large, and most investigators put the minimum Zn molality of an ore-forming solution at 10^{-6} m.

How can the concentration of Zn in solution be increased? **Figure 9-24** presents a photomicrograph of a fluid inclusion trapped in beryl from a hydrothermal ore deposit. Because it was formed at higher temperature, on cooling to earth surface conditions, it produces a H_2O vapor bubble and precipitates a number of minerals, including halite (NaCl) and sylvite (KCl). Like most fluid inclusions observed in hydrothermal ore deposits, there is a large concentration of Cl in the trapped aqueous fluid. Zn can also be carried as Zn-Cl species

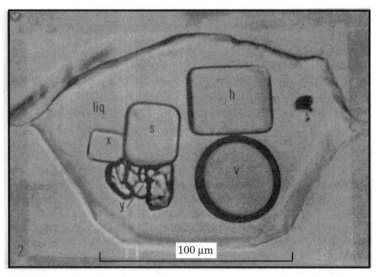

FIGURE 9-24 Fluid inclusion in beryl from the Muzo mine, Colombia viewed at earth surface conditions containing a vapor bubble (v), halite crystal (h), sylvite crystal (s), and two unidentified solid phases (x and y). (Courtesy of Roedder, 1972.)

in aqueous solutions with up to four Cl^- in the species. The reactions and their equilibrium constant expressions calculated at 1 kbar and 350°C from SUPCRT92 are

$$2H^+ + Cl^- + 0.5\,O_2 + ZnS = ZnCl^+ + 0.5\,S_2 + H_2O \tag{9.44}$$

$$\text{Log } K_{(9.44)} = 16.620 = \log\left(\frac{a_{ZnCl^+}\,f_{S_2}^{0.5}\,a_{H_2O}}{a_{H^+}^2\,a_{Cl^-}\,f_{O_2}^{0.5}\,a_{SPH}}\right) \tag{9.45}$$

$$2H^+ + 2Cl^- + 0.5\,O_2 + ZnS = ZnCl_2 + 0.5\,S_2 + H_2O \tag{9.46}$$

$$\text{Log } K_{(9.46)} = 17.420 = \log\left(\frac{a_{ZnCl_2}\,f_{S_2}^{0.5}\,a_{H_2O}}{a_{H^+}^2\,a_{Cl^-}^2\,f_{O_2}^{0.5}\,a_{SPH}}\right) \tag{9.47}$$

$$2H^+ + 3Cl^- + 0.5\,O_2 + ZnS = ZnCl_3^- + 0.5S_2 + H_2O \tag{9.48}$$

$$\text{Log } K_{(9.48)} = 16.385 = \log\left(\frac{a_{ZnCl_3^-}\,f_{S_2}^{0.5}\,a_{H_2O}}{a_{H^+}^2\,a_{Cl^-}^3\,f_{O_2}^{0.5}\,a_{SPH}}\right) \tag{9.49}$$

and

$$2H^+ + 4Cl^- + 0.5\,O_2 + ZnS = ZnCl_4^{2-} + 0.5\,S_2 + H_2O \tag{9.50}$$

$$\text{Log } K_{(9.50)} = 17.442 = \log\left(\frac{a_{ZnCl_4^{2-}}\,f_{S_2}^{0.5}\,a_{H_2O}}{a_{H^+}^2\,a_{Cl^-}^4\,f_{O_2}^{0.5}\,a_{SPH}}\right) \tag{9.51}$$

With the log f_{O_2} and log f_{S_2} buffered to −30.22 and −9.38, respectively, the equilibrium constant expressions for the Zn-Cl aqueous species become

$$\text{Log } K_{(9.44)} = 6.20 = \log\left(\frac{a_{ZnCl^+}}{a_{H^+}^2\,a_{Cl^-}}\right) \tag{9.52}$$

$$\text{Log } K_{(9.46)} = 7.00 = \log\left(\frac{a_{ZnCl_2}}{a_{H^+}^2\,a_{Cl^-}^2}\right) \tag{9.53}$$

$$\text{Log } K_{(9.48)} = 5.965 = \log\left(\frac{a_{ZnCl_3^-}}{a_{H^+}^2\,a_{Cl^-}^3}\right) \tag{9.54}$$

and

$$\text{Log } K_{(9.50)} = 7.022 = \log\left(\frac{a_{ZnCl_4^{2-}}}{a_{H^+}^2\,a_{Cl^-}^4}\right) \tag{9.55}$$

Figure 9-25 shows a plot of sphalerite solubility from each of the Zn-Cl aqueous species with log $f_{O_2} = -30.22$ and log $f_{S_2} = -9.38$ at 1.0 kbar and 350°C as a function of Cl⁻ in solution at neutral pH (= 5.33 at 1 kbar and 350°C). The solubility increases significantly with increasing Cl⁻ in solution. The species ZnCl⁺ dominates at low Cl⁻ concentrations, but species with a higher ratio of Cl to Zn become important as Cl⁻ concentrations are increased. Because Equations 9.52 to 9.55 have a_{H^+} in the dominator raised to the second power, a decrease in pH of 1 unit increases Zn in solution by 2 orders of magnitude.

The pH of solution during the formation of porphyry ore deposits likely becomes somewhat acid rather than the neutral pH used in the calculation of Figure 9-25. This occurs because SO₂ and HCl are present in the fluid expelled from the intrusions at high temperature. Table 9-3 gives the measurements of the constituents of volcanic gas from a stratovolcano from a convergent plate boundary at Momotombo, Nicaragua. Because 1000 g = 55.51 moles of H₂O, the gas on cooling produces a 0.8 molal SO₂ and 5 molal HCl fluid. The cooling also disproportionates SO₂ in the presence of H₂O to the acids, H₂SO₄ and H₂S. Disassociation of these acids occurs decreasing the solution pH as cooling continues.

The HCl also disassociates on cooling by the reaction

$$HCl = H^+ + Cl^- \tag{9.56}$$

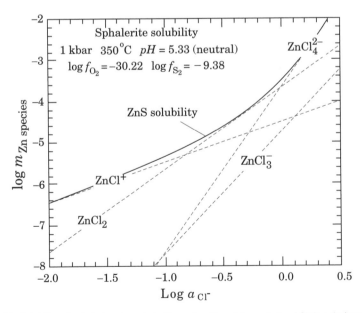

FIGURE 9-25 Solubility of sphalerite at neutral pH as a function of the activity of Cl⁻ in solution at 350°C and 1 kbar where log $f_{O_2} = -30.22$ and log $f_{S_2} = -9.38$.

Shown in **Figure 9-26** is the log of the equilibrium constant for Reaction 9.56 plotted at 0.5 and 1.0 kbar. Note the increased disassociation of HCl, that is, the production of acid as temperature decreases. Rocks react with the acid produced, neutralizing it, and cause rock alteration. The ore minerals are precipitated because the solution supersaturates from increased *pH* as acids interact with the host rock. For this reason, ore-forming solutions are thought to be from somewhat acidic to near neutral under most conditions.

Rock Alteration

The acidic solutions produced in porphyry Cu-Mo deposits as hydrothermal fluids cool cause a series of alterations in the rock that are termed potassic, advanced argillic, sericitic, argillic, and propylitic alterations with decreasing alteration intensity (Figure 9-22). The alteration observed is the combined result of fluid–rock reactions taking place in an environment where temperature and, therefore, the *pH* of solutions changes as a function of both time and distance from the intrusion.

Potassic alteration occurs early and at the highest temperatures in the central portions of the deposit with K introduced from the fluid to produce new and recrystallized K-feldspar and biotite. Anhydrite or apatite is commonly found as an alteration phase. Ca and Na are leached from the rock. A characteristic reaction is

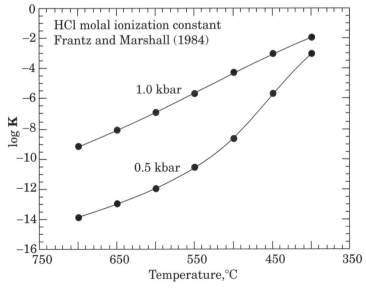

I FIGURE 9-26 Molal equilibrium constant of HCl disassociation to H$^+$ and Cl$^-$.

$$K^+ + NaAlSi_3O_8 \rightarrow KAlSi_3O_8 + Na^+ \qquad \text{[9.57]}$$
$$\text{albite} \qquad \text{K-feldspar}$$

which is shown on the activity of Na^+/H^+ versus activity of K^+/H^+ diagram given in **Figure 9-27a**. The potassic alteration zone also contains magnetite and pyrite together with the primary Cu and Mo ore minerals (e.g., chalcopyrite [$CuFeS_2$], bornite [Cu_5FeS_4], and molybdenite [MoS_2]).

At lower temperature and increased distance from the intrusion core, sericitic alteration can be identified. Sericite is a fine-grained white mica (muscovite to paragonite or phengite). In the sericite zone, all of the minerals in the rock are altered to quartz, sericite, and pyrite. Cu ore minerals, typically chalcopyrite, are found here in veins with pyrite-rich halos. A common sericite producing reaction is

$$2H^+ + 3KAlSi_3O_8 \rightarrow KAl_3Si_3O_{10}(OH)_2 + 6SiO_2 + 2K^+ \qquad \text{[9.58]}$$
$$\text{K-feldspar} \qquad \text{muscovite} \qquad \text{quartz}$$

On **Figure 9-27a** and **Figure 9-27b** the location in the general alteration path with time on cooling is shown by the shaded arrows. In some deposits, advanced argillic alteration occurs.

Advanced argillic alteration is a high-temperature, intense alteration caused by low K^+/H^+ and Na^+/H^+ activity ratios in the hydrothermal fluid. This fluid leaches the alkalis from the rock. The only remaining primary igneous mineral is quartz. The original Al-containing minerals (feldspars, mica) are altered to

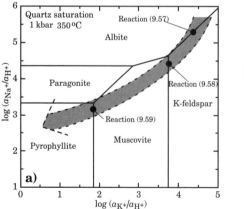
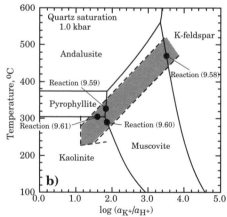

FIGURE 9-27 (a) Log activity of K^+ over activity of H^+ versus log activity of Na^+ over H^+ diagram for the $Na_2O–K_2O–Al_2O_3–SiO_2–H_2O–HCl$ system at 350°C and 1 kbar with quartz present. (b) Temperature–log activity of K^+ over activity of H^+ phase diagram at 1 kbar and with quartz present for the $K_2O–Al_2O_3–SiO_2–H_2O–HCl$ system (stability boundaries calculated with SUPCRT92). The shaded arrows give the general alteration path with time for an acidic cooling fluid.

aluminosilicates (kaolinite, pyrophyllite, andalusite, dickite, topaz, and alunite). The reactions include

$$2H^+ + 2KAl_3Si_3O_{10}(OH)_2 + 6SiO_2 \rightarrow 3Al_2Si_4O_{10}(OH)_2 + 2K^+ \qquad [9.59]$$

$$\text{muscovite} \qquad \text{quartz} \qquad \text{pyrophyllite}$$

$$2H^+ + 3H_2O + 2KAl_3Si_3O_{10}(OH)_2 \rightarrow 3Al_2Si_2O_5(OH)_4 + 2K^+ \qquad [9.60]$$

$$\text{muscovite} \qquad \text{kaolinite}$$

and

$$Al_2Si_2O_{10}(OH)_2 + H_2O \rightarrow Al_2Si_2O_5(OH)_4 + 2SiO_2 \qquad [9.61]$$

$$\text{pyrophyllite} \qquad \text{kaolinite} \qquad \text{quartz}$$

as given in Figures 9-27a and 9-27b.

Argillic alteration is less intense than advanced argillic alteration and is produced by a higher K^+/H^+ in the fluid. The reactions occur at relatively low temperature with the remaining plagioclase altered to kaolinite ± chlorite. The amphibole present is altered to a smectite clay phase. The alteration of albite to kaolinite is given by

$$2H^+ + H_2O + 2NaAlSi_3O_8 \rightarrow Al_2Si_2O_5(OH)_4 + 4SiO_2 + 2Na^+ \qquad [9.62]$$

$$\text{albite} \qquad \text{kaolinite} \qquad \text{quartz}$$

Propylitic alteration is the lowest temperature alteration. Plagioclase, hornblende, and biotite are altered to chlorite, epidote, carbonates, and clays. It is similar to greenshist facies metamorphism. H_2O, H^+, CO_2, and S are typically added to the rock from the fluid. Generally, either small amounts of pyrite are present or no sulfide minerals are present at all. A typical endmember reaction is the alteration of calcium–iron hornblende + anorthite to iron chlorite + epidote + quartz

$$CaFe_5Al_2Si_7O_{22}(OH)_2 + 3CaAl_2Si_2O_8 + 4H_2O \rightarrow$$

$$\text{Ca-Fe amphibole} \qquad \text{anorthite}$$

$$Fe_5Al_2Si_3O_{10}(OH)_8 + 2Ca_2Al_3Si_3O_{12}OH + 4SiO_2 \qquad [9.63]$$

$$\text{chlorite} \qquad \text{epidote} \qquad \text{quartz}$$

Summary

This chapter outlines some of the chemical changes observed in rocks because of reaction with aqueous fluids. Soils form from the action of groundwater as it percolates to the water table. This fluid can leach material high in the soil profile and deposit it with changed chemical conditions at depth, producing a series of layers in the soil. Most sediments, however, are transported to bodies of water; lakes, and the ocean, where they are deposited. The reactions that go on in these sediments with burial are termed diagenetic.

In normal marine environments, organic matter is oxidized by bacteria to CO_2 in the water column or at the sediment–water interface. Seawater sulfate trapped in the sediments is reduced to H_2S, which typically escapes upward. In anoxic marine environments, however, organic matter can accumulate with seawater sulfate being reduced to amorphous $FeS \bullet nH_2O$. This is transformed during diagenesis to pyrite.

During diagenesis of clay material, any calcite present is generally lost during an early stage by dissolution in undersaturated pore solutions and removal by fluid advection during sediment compaction. Smectite–group minerals in the clay fraction are transformed to illite. With later diagenesis when the temperature reaches 300°C, illite is transformed to sericite. Further alteration at high temperatures to muscovite is generally considered a metamorphic reaction.

Currently, the sedimentary marine carbonate is about two-thirds aragonite and one-third Mg-calcite, with calcite dominant during clay deposition. Dolomite appears to form during diagenesis by solutions that have a higher Mg/Ca ratio than present-day seawater.

Because a pore fluid is present during diagenesis and metamorphism, material is transported by diffusion and/or advection of this fluid. Pressure in fluid filled pores and fractures in the earth is hydrostatic near the surface but becomes lithostatic at depth when the mass of overlying rocks is loaded on the fluid filled pores. Fluid flow, at depths where fluid is under a lithostatic gradient, is not described by Darcy's law. It always moves with a transport component toward the earth's surface because the fluid is much less dense (more buoyant) than the surrounding rocks. This fluid is produced by destruction of CO_2- and H_2O-containing minerals to the maximum temperature experienced by the rock.

When fluid is present, new minerals can form from the destruction of unstable phases. If no fluid is present, the unstable high-temperature minerals are preserved within rocks as they are brought to the earth's surface. Fluid is transported away from the sites where it is produced by microfractures developing in the rocks. Coalescence of fluids from these fractures during transport produces quartz-rich veins in the rocks.

Ore deposits are unusual events that require a method to concentrate the element of interest. In hydrothermal ore deposits, hot groundwater can scavenge metals from accessory minerals in granitic host rocks and contribute their concentrations to those from a water-rich magmatic phase. The metals are typically carried as chloride species at somewhat acid to neutral pH and are deposited as metal sulfides. Deposition occurs on supersaturation as the fluid reacts with rocks, increasing its pH, and on cooling. This occurs most significantly on the sides and tops of porphyry stocks.

The arguments presented here were mainly based on equilibrium constraints. In the chapters that follow, isotopic techniques to understand these processes are introduced, and time is brought into the analysis by considering the kinetics of reactions.

Key Terms Introduced

advection
anoxic
bauxite deposits
beachrock
black shales
brecciated zones
chelates
chemical potential gradient
chromatographic
contact metamorphism
crackle
Darcy's law
Desulphovibrio
devolatilization reaction
diagenesis
diffusion
elevation head
epithermal deposits
fermentation
geocline
geothermal gradient
heat flow unit
hydraulic conductivity
hydraulic head
hydrofracturing
hydrostatic equilibrium
hydrostatic pressure
isostatic equilibrium

I/S transformation
lithostatic pressure
metamorphic field gradient
pelite
permeability
placer deposits
porphyry
pressure head
pressure solution
prograde reactions
regional metamorphism
respiration
retrograde reactions
retrograde solubility
Reynolds number
sabkha
secondary porosity
skarn
soil air
soil fabric
soil solution
stock
stockwork
supergene
tortuosity
viscous pressure
water table

Questions

1. Why does soil typically form in layers?
2. How does soil air differ from air in the atmosphere?
3. Describe a mollisol.
4. What is diagenesis as opposed to metamorphism?
5. What happens to sulfate in seawater that is retained in pores during sedimentation?

6. How does a black shale differ from a normal shale?
7. Describe the I/S transformation.
8. Describe the ways dolomite can form.
9. Give the equation for diffusional transport. Define the terms.
10. Why is diffusional transport better described by chemical potential gradients than concentration gradients?
11. What is Fick's second law? What does it describe?
12. How is fluid pressure calculated at depth in the earth?
13. Why do lithostatic and hydrostatic pressure differ?
14. Give Darcy's law. Define the terms.
15. If a rock has experienced peak metamorphic conditions, what can be said about the pressure and temperature?
16. At mid levels of the earth's crust, in regionally metamorphosed areas, veins containing quartz are observed. Where does the water and silica come from?
17. Describe the movement of fluids at mid-crustal conditions. How does this vary from fluid transport near the earth's surface?
18. What is hydrofracturing?
19. How does heat flow vary on the surface of the earth? Why?
20. Besides raising its temperature, how else is heat used in a rock?
21. What is the approximate width of fluid transporting fractures during regional metamorphism of rocks? What controls this width?
22. What processes need to happen to form an ore deposit?
23. Describe how a porphyry rock forms.
24. Where do the metals deposited in porphyry Cu deposits come from? What causes them to be deposited?
25. What species carry Zn in aqueous solutions? What controls their stability?
26. List the types of alteration patterns that can occur in porphyry Cu deposits.
27. Why are ore-forming solutions typically somewhat acidic?

Problems

1. Using the thermodynamic data given in the back of the book, calculate the oxygen fugacity in equilibrium with 0.1 m H_2S and elemental sulfur at 25°C and 1 bar (Reaction 9.2). Assume that activity coefficients are unity. Because O_2 and S are the elements in their most stable state at 25°C and 1 bar, their standard state Gibbs energy is zero. Will Fe^{2+} or Fe^{3+} be stable in the solution at the calculated oxygen fugacity (compare Figure 14-8 with 14-3)?
2. At 25°C and 1 bar, the diffusion coefficient of NaCl in a 1 molar aqueous solution is 1.47×10^{-5} cm^2 s^{-1}. If a steady-state gradient exists with

NaCl at 1.1 molar at x and 0.9 molar at $x + 2$ cm, how much NaCl will diffuse through 1 cm^2 perpendicular to the diffusion direction in 1 hour?

3. Consider a tanker filled with cesium waste that is illegally dumped onto a floodplain by a river. The waste spreads out in a thin layer so that there are 100 mg of cesium per cm^2. The river then rapidly deposits sediment on the floodplain, covering the thin layer so that diffusion is both up and down. What is the concentration of cesium because of diffusion after 10 months (need to convert to s) as a function of distance up and down from the layer? The diffusion coefficient of Cs, $D = 2.4 \times 10^{-5}$ cm^2 s^{-1}. The solution to Equation 9.17 for these conditions is

$$c(x,t) = \frac{b}{2\sqrt{\pi Dt}} \exp(-x^2/4Dt)$$

with the concentration, c, in mg per cm^3 and where b is the total Cs deposited per cm^2 before diffusion starts. Plot enough c as a function of x to get an idea of the profile at both positive and negative x. What will happen to the profile as a function of increased time?

4. At a 6-km depth with average fluid density of 1.0 g cm^{-3} and rock density of 2.7 g cm^{-3}, what would be the difference between hydrostatic and lithostatic pressure?

5. If fluid pressure in a silty sand 1 km below the surface was 100 bars at one location and 120 at another location 1 km away, what is the fluid flux if the hydraulic conductivity is 1×10^{-5} cm s^{-1} and the fluid density is 1.0 g cm^{-3}?

6. At a point, Z°, in a fracture filled with stationary fluid, fluid pressure equals the pressure on the fracture. At a point 20 meters above Z°, how much greater would fluid pressure be than rock pressure? Give the answer in bars (g cm^{-1} s^{-2} = dyne cm^{-2} = 10^{-6} bar).

7. a. Do a dimensional analysis of Equation 9.29 in CGS units (see Appendix B) to indicate that it is dimensionally correct and indicate the dimensions of q.
   b. If q is a volume flux, how can it have these units?
   c. When a flux in g is required, how should Equation 9.29 be modified?

8. At mid levels of the earth's crust, one encounters a granite of density = 2.7 g cm^{-3} with a fracture density of 1 meter per square meter orientated at 30° from the vertical. If fluid flux of 5×10^{-10} g cm^{-2} s^{-1} is flowing through the fractures with a viscosity of 0.15 centipoise and density of 0.9 g cm^{-3}, what would be the width of the fracture?

9. The Gibbs energy of formation of zincite $(ZnO) = -76,600$ cal mol^{-1} at 25°C and 1 bar. Use the thermodynamic data for Zn^{2+} in the back of the book to plot the log concentration of Zn^{2+} as a function of pH between 1 and 10 at zincite saturation. Assume that activity coefficients of aqueous species are unity. The plot indicates that the solubility of zincite is very low and decreases with pH at high pH, but this is not true. Give a reason why.

10. Pure chalcocite (Cu_2S), chalcopyrite $(CuFeS_2)$, and pyrite (FeS_2) are found together at the earth's surface.
    a. Use the values in the back of the book to determine the fugacity of sulfur gas that would be in equilibrium with this assemblage.
    b. Determine ΔS_R° of a reaction involving the phases and using this value argue if the fugacity of sulfur gas increases or decreases with increasing temperature.

11. Penniston-Dorland and Ferry (2008) analyzed quartz veins in quartz bearing wall rocks. They argue that if the vein contained a fluid of lower solubility of SiO_2 than the wall rock because of a lower X_{H_2O} in the vein ". . . considering Fick's first law, the gradient in dissolved silica between vein and wall rock would drive silica diffusion from wall rock to vein. . . ." Explain why this would not occur.

Problems Making Use of SUPCRT92

12. At 350°C and 500 bars, calculate the sulfur and oxygen fugacity and the activity of Fe^{2+} over the activity of H^+ squared in equilibrium with pyrite + magnetite + pyrrhotite. Assume pure minerals and H_2O. If $pH = 5$, determine the activity of Fe^{2+} in solution.

13. Calculate $\log (a_{Mg^{2+}}/(a_{H^+})^2)$ for equilibrium between pure kaolinite, quartz, and clinochlore, 14Å (see Equation 9.7), at 500 bars for 100°C, 150°C, 200°C, 250°C, and 300°C. If pH is constant, how does the equilibrium activity of Mg^{2+} change in the solution with increasing temperature? Does this favor the production of kaolinite or 14Å clinochlore?

14. Calculate the standard enthalpy (H) of both quartz and K-feldspar at 400°C and 600°C at 1 and 4 kbar.
    a. What is the standard enthalpy per mole required to raise the temperature of these two minerals by 200°C at 1 kbar? By 4 kbar?
    b. Why are the enthalpies different but the changes in enthalpy the same at the two pressures?
    c. What are these enthalpy changes over the 200°C interval per gram-atom for the two minerals?

References

Badwin, B., 1971, Ways of deciphering compacted sediments. *J. Sediment. Petrol.*, v. 41, pp. 293–301.

Beane, R. E. and Titley, S. R., 1981, Porphyry copper deposits. Part II. In Skinner, B. J. (ed.) *Economic Geology Seventy-Fifth Anniversary, Volume 1905–1980*. Economic Geology Pub., New Haven, pp. 235–269.

Bear, J., 1975, *Dynamics of Fluid in Porous Media*. American Elsevier, New York, 764 pp.

Berner, R. A., 1980, *Early Diagenesis: A Theoretical Approach*. Princeton Univ. Press, Princeton, 241 pp.

Cox, D. P. and Singer, D. A. (eds.), 1986, Mineral deposit models. *U.S. Geological Survey Bulletin,* No. 1693, 379 pp.

Crank, J., 1975, *The Mathematics of Diffusion*, 2nd ed., Oxford Press, London, 347 pp.

Garrels, R. M., 1986, Sediment cycling and diagenesis. In *Studies in Diagenesis. U.S. Geological Survey Bulletin*, No. 1578, pp. 1–11.

Guilbert, J. M. and Park, C. F., Jr., 1986, *The Geology of Ore Deposits*. W. H. Freeman and Company, New York, 985 pp.

Hower, J., Elinger, E. V., Hower, M. E. and Perry, E. A., 1976, Mechanism of burial metamorphism of argillaceous sediment. 1. Mineral and chemical evidence. *Geol. Soc. Am. Bull.*, v. 87, pp. 725–737.

Norton, D. and Cathles, L. M., 1979, Thermal aspects of ore deposition. In Barnes, H. (ed.) *Geochemistry of Hydrothermal Ore Deposits*. John Wiley & Sons, New York, pp. 611–631.

Norton, D. and Knapp, R., 1977, Transport phenomena in hydrothermal systems: The nature of porosity. *Am. J. Sci.*, v. 277, pp. 913–936.

Norton, D. and Knight, J., 1977, Transport phenomena in hydrothermal systems: Cooling plutons. *Am. J. Sci.*, v. 277, pp. 937–981.

Reed, M.N., 1997, Hydrothermal alteration and its relationship to ore fluid composition. In Barnes, H. L. (ed.) *Geochemistry of Hydrothermal Ore Deposits*, 3rd edition. Wiley, pp. 303–366.

Roedder, E., 1972, Composition of fluid inclusions: Data of geochemistry. *U.S.G.S. Prof. Paper,* 440-JJ, 164 pp.

Roedder, E., 1984, Fluid inclusions. *Rev. Mineral.*, v. 12, Mineralogy Soc. Amer., Washington, D.C., 644 pp.

Spear, F. S., 1993, *Metamorphic Phase Equilibria and Pressure–Temperature– Time Paths*. Mineralogical Society of America, Washington, D.C., 799 pp.

Symonds, R. B., Rose, W. I., Bluth, G. J. S. and Gerlach, T. M., 1994, Volcanic-gas studies: Methods, results, and applications. *Rev. Mineral.*, v. 30, pp. 1–66.

Walther, J. V., 1990, Fluid dynamics during progressive regional metamorphism. In Bredehoeft, J. D. and Norton, D. (eds.) *The Role of Fluids in Crustal Processes*. National Academy of Sciences Press, Washington, DC, pp. 64–71.

Walther, J. V. and Orville, P. M., 1982, Rates of metamorphism and volatile production and transport in regional metamorphism. *Contrib. Mineral. Petrol.*, v. 79, pp. 252–257.

10 Radioactive Isotope Geochemistry

In 1896, the Frenchman Antoine Becquerel discovered radioactivity. The recognition that certain elements undergo spontaneous decay at a constant rate led to the development of techniques to obtain the age of a variety of rocks. Investigators first tried to measure the helium produced by uranium decay to obtain an age; however, it was not until 1907 that the American Bertram Boltwood developed a technique based on the U/Pb ratio in rocks that accurate radioactive dating could be done. This allowed important constraints to be put on the age of the earth and our solar system. The widely held view at the time, based on the calculations of Lord Kelvin, was that the earth was no more than 40 million years old. That this time was much too short was demonstrated by radioactive age dating. *Radioactive isotope* investigations also provided absolute ages for successions of strata and, therefore, the periods of the geologic time scale. A comparison of radioactive ages of the ocean crust as opposed to those found on continents helped to refine and extend the plate tectonic model of the movement of the earth's crust. Radioactive isotope geochemistry's ability to quantify rates and fluxes has led to increased understanding of many geologic processes. For example, isotope geochemical studies have allowed sources of magmas, metals in ore deposits, and masses of ocean water to be determined. In addition, our knowledge of the sources and residence times of species in the atmosphere and groundwaters is due primarily to isotopic studies.

Each element has a specified number of protons in its nucleus given by its atomic number. The chemical behavior of an element is, however, due primarily to the electrons orbiting around the nucleus of the atom. For instance, salt, $NaCl$, is put on food, which results in the consumption of the ions Na^+ and Cl^-; in fact, our blood contains substantial Cl^-. If one breathes Cl_2 gas, however, it is deadly. The different behavior of the two species, Cl^- and Cl_2, in our body is due to the difference in their electron structure.

Consider the mass of an atom, which is often given in atomic mass units. Atomic mass units, also termed *daltons*, given by the symbol Da, are units of g(N°)$^{-1}$. The electron has a mass of 0.000549 Da, the proton 1.00728 Da, and the neutron 1.00866 Da. The mass of an atom, therefore, rests primarily in its nucleus. The Bohr-type representation of an atom, giving the relative locations of negatively charged electrons, positively charged protons, and uncharged neutrons, is shown in **Figure 10-1**. An isotope of an element is distinguished by differences in the number of neutrons in the nucleus. This retains the charge characteristics but gives it a different mass and, therefore, a difference in atomic weight.

Most elements in nature have more than one isotope. For instance, oxygen has three naturally occurring isotopes denoted by ^{16}O, ^{17}O, and ^{18}O with abundances of 99.759%, 0.037%, and 0.204%, respectively. The superscripted numbers on the chemical symbol give the *atomic mass number*, the number of protons plus neutrons in the nucleus. Because oxygen has 8 protons in its nucleus, ^{16}O, ^{17}O, and ^{18}O indicate oxygen with 8, 9, and 10 neutrons, respectively.

The technique to measure the ratio of isotopes in a sample is outlined in **Figure 10-2**. Differences in the mass of different isotopes result in slightly different chemical properties; however, as mentioned above, the major chemical differences among atoms are due to the differences in bonding of electrons orbiting around the nuclei. Different isotopes of the same element have nearly identical bonding electrons and, therefore, bond similarly.

Isotopes are either *stable* or radioactive. Stable isotopes are those that do not undergo radioactive decay. In contrast, radioactive isotopes undergo spontaneous time-dependent decay to form new elements or isotopes. Radioactivity occurs when the *coulombic* repulsion of positive charges between protons becomes greater than the local binding forces in the nucleus. **Figure 10-3** shows the

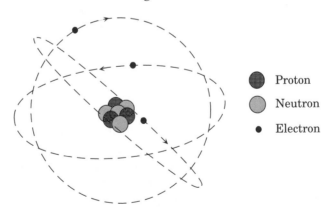

FIGURE 10-1 Simple Bohr-type model of a lithium atom. Atomic nuclei are between 10^{-12} and 10^{-13} cm in size and are made up of protons of positive charge and neutrons of neutral charge. At distances on the order of 10^{-8} cm, electrons of negative charge move in circular orbits around the nucleus. The size of the electrons relative to protons and neutrons and the distance of the electron orbits from the nucleus are exaggerated in the diagram. If the neutrons and protons were the size of air rifle pellets, the electron orbitals would be at football field distances.

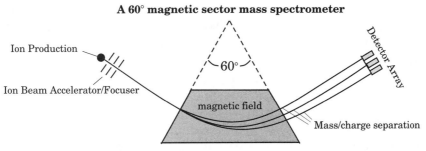

A 60° magnetic sector mass spectrometer

FIGURE 10-2 Isotopic abundances are measured with a mass spectrometer. First atoms or molecules are ionized. In solid source instruments the sample is heated to very high temperatures by a filament and a vapor of species is produced. The temperature is high enough that the species become ionized. In gas-source instruments a stream of electrons is used to bombard a stream of species in a gas and knock out electrons. The ions produced are accelerated, enter a magnetic field and their path is altered depending on their mass to charge ratio.

$$\frac{\text{mass}}{\text{charge}} = \frac{(\text{magnetic field strength})^2\ (\text{path radius})^2}{2 \times (\text{accelerating potential})}$$

An array of detectors then measures the ions from each mass to charge stream and from this the relative isotopic abundances are determined.

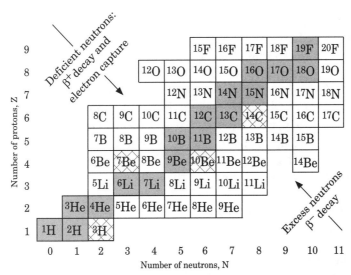

FIGURE 10-3 The low atomic weight part of the chart of the nuclides giving the number of protons and neutrons in each isotope. The stable nuclides are shown in gray, whereas the unstable nuclides have a white background. The hatch marks through an unstable nuclide indicate that it has sufficient stability to be useful in geologic dating. Isotopes with excess neutrons decay by β^- emission, leading to an increase in atomic number, whereas those with deficient neutrons relative to the stable isotopes decay by β^+ emission or electron capture, leading to a decrease in atomic number.

low proton number part of the chart of the *nuclides*, which displays the arrangement of both stable and radioactive isotopes. Each particular nuclide, that is isotope, contains a specific number of both protons and neutrons in its nucleus. There are about 1700 known nuclides, with about 430 naturally occurring. Of these, about 200 are stable.

The stable nuclides of the low atomic weight elements fall in a band at or just above a 1:1 ratio of neutrons to protons. Isotopes with a greater number of neutrons per proton than the stable isotopes become unstable and decay by β^- decay, and those with fewer neutrons are also unstable and undergo β^+ decay or electron capture (see below). With an increasing number of protons, the stable isotopes have an increasing ratio of neutrons to protons in their nucleus. For instance, the stable lead isotopes ^{206}Pb, ^{207}Pb, and ^{208}Pb have 82 protons but 124, 125, and 126 neutrons, respectively. This occurs because the binding forces in the nucleus are short range, acting between nearest neighbors, whereas the electrostatic repulsion between protons acts across the nucleus. As the number of protons increases, the increased repulsion in the nucleus is balanced by holding them further apart by increased numbers of neutrons.

In Figure 10-3, no stable nuclides exist with atomic numbers (protons + neutrons) of 5 or 8. This occurs because of the particular stability of the helium nucleus of 2 protons and 2 neutrons. There is not enough binding energy to overcome the charge repulsion needed to add a proton to this nucleus. Likewise, no stabilization energy exists to retain an additional neutron in a helium nucleus. The logical nucleus of atomic number 8 would exist as 2 helium nuclei.

Because the differences in chemical properties between isotopes of the same element are very small, the ratio of isotopes in a mass of the element is nearly fixed in nature. The atomic weight of an element reflects this common ratio of the isotopes. For instance, carbon, wherever it is found on the earth, is made up of 98.93% ^{12}C and 1.07% ^{13}C, giving it an atomic weight of 12.0107 g mol^{-1}; however, the sum of the parts of an isotope (electrons + protons + neutrons) does not add up to the measured weight of the isotope.

Consider a mole of ^{12}C, with a mass of 12.0000 g. The mass of six electrons, six protons, and six neutrons, the constituents of ^{12}C, is

$$6\,(0.000549\text{ g }(N^\circ)^{-1} + 1.00728\text{ g }(N^\circ)^{-1} + 1.00866\text{ g }(N^\circ)^{-1}) = 12.0989\text{ g }(N^\circ)^{-1} \quad [10.1]$$

$$\text{electron} \qquad\qquad \text{proton} \qquad\qquad \text{neutron}$$

This is greater than the mass of a mole of ^{12}C because in ^{12}C, as in any stable atom, there are strong molecular binding forces lowering the energy of the nucleus and making it a thermodynamically stable configuration. That is, when protons and neutrons are bonded to form the nucleus, the decrease of the internal energy (referred to as the binding energy) is so great that this energy difference is measured in terms of a difference of mass.

Thermodynamics, therefore, indicates that a collection of six electrons, six protons, and six neutrons is unstable relative to a carbon-12 atom. A difference in binding energies of the nuclei of different isotopes means that the mass of electrons, protons, and neutrons cannot be added to obtain the molecular weight of an isotope; however, consider normal hydrogen, ^{1}H. An atom of hydrogen consists of one electron and one proton. The sum of a mass of a mole of electrons and protons is

$$\underset{\text{electron}}{0.000549 \text{ g } (N°)^{-1}} + \underset{\text{proton}}{1.00728 \text{ g } (N°)^{-1}} = 1.00783 \text{ g } (N°)^{-1} \qquad [10.2]$$

which is equal to the measured molecular weight of ^{1}H. This indicates that the long-range electrostatic energy between electrons and protons is much less than the binding energy between protons and neutrons.

The binding energy per mole of nucleon, $E_{binding}$, for ^{12}C from Equation 10.1 is $(12.0989 – 12.0000)/12 = 0.00824$ g mol^{-1}. Using Einstein's special law of relativity, this mass change, Δm, is equal to energy in Joules:

$$E_{binding} = \Delta m \, c^2 = 8.24 \times 10^{-6} \text{ kg mol}^{-1} (3.00 \times 10^8 \text{ m s}^{-1})^2 = 7.42 \times 10^{11} \text{ J mol}^{-1} \qquad [10.3]$$

where c denotes the speed of light (3.00×10^8 m s^{-1}). This is a very large amount of energy compared with normal chemical reactions involving electron transfers during bond formation between atoms where the energy of reaction is generally a few thousand J mole^{-1} or less. Reactions that produce free electrons are therefore common and result in little energy exchange, whereas reactions that separate neutrons from protons are less common and are confined to particular high atomic weight isotopes. These latter reactions include nuclear fission reactions and release a large amount of energy during their decay reactions, as indicated by Equation 10.3.

■ Radioactive Decay

Investigations of cosmic rays and scattering processes in particle accelerators demonstrated that dozens of subatomic particles existed besides protons, electrons, and neutrons. *Antiparticles* were also observed. An antiparticle has the same mass as a particle but opposite charge. Interaction of antiparticles and particles leads to their destruction giving rise to high-energy photons (*gamma rays*) or other particle–antiparticle pairs equal to the energy difference of their rest masses. The subatomic particles (antiparticles) that are important in radioactive decay, along with their charge and rest mass, are given in **Table 10-1**.

Radioactive isotopes decay by one of three processes: *beta decay*, *alpha decay*, or *nuclear fission*. They produce new nuclides that also decay if

| Table 10-1 | Subatomic particles involved in radioactive decay |

Particle	Symbol	Charge	Mass (g)	Mass (amu)
alpha	α^{2+}	+2	6.645×10^{-24}	4.0015
proton	p^+	+1	1.673×10^{-24}	1.00727
neutron	n	0	1.675×10^{-24}	1.00866
electron	β^-	−1	9.109×10^{-28}	5.485×10^{-4}
positron	β^+	+1	9.109×10^{-28}	5.485×10^{-4}
neutrino		0	$<10^{-32}$	$<5 \times 10^{-9}$
photon		0	0	0

amu = atomic mass unit

radioactive. The decay series stops when a stable isotope, that is, one that does not undergo decay, is produced. The most common decay process is beta decay, which is either negative or positive.

Beta Decay

In negative beta decay, an electron (β^- particle) is spontaneously emitted from the nucleus. This causes the nucleus of the isotope to decrease its number of neutrons by one, producing a proton according to the reaction:

$$\text{"neutron"} \rightarrow \text{"proton"} + \beta^- + \textit{antineutrino} + \text{gamma ray} \qquad [10.4]$$

This reaction produces a new element whose atomic number is greater than its parent. Because of the strong energy of cohesion (which is measured in mass changes), the change of mass of a "neutron" to a "proton" in the nucleus cannot be determined from mass changes of unbonded neutrons and protons. The neutron and proton as they exist in a nucleus are, therefore, indicated with quotation marks, and the energy of the gamma ray produced depends on the particular reaction. The gamma ray (energy) is dissipated by collisions with other atoms and converted into heat energy. The antineutrinos released are antiparticles of *neutrinos* that have kinetic energy but near zero mass, zero electric charge, interact sparingly with matter, and rotate clockwise along their direction of transport into space with spin 1/2. These particles then conserve angular momentum for the reaction. Antiparticles have the same mass as the normal particle but have opposite values for other properties, including charge, *baryon number*, and *strangeness*. Energy and charge are conserved in the reaction, as they must be in any chemical reaction.

In positive beta decay, a positron, designated as a β^+ particle, that has the same mass as an electron but is positively charged is spontaneously released from decay of a proton to form a neutron in the nucleus. This decay also produces a new element but with an atomic number that is one less than its parent according to the reaction

$$\text{``proton''} \rightarrow \text{``neutron''} + \beta^+ + \text{neutrino} + \text{gamma ray} \qquad [10.5]$$

Besides a β^+ particle, a neutrino and gamma ray are also released from the nucleus. A neutrino is a stable subatomic particle of spin 1/2 with kinetic energy that conserves the angular momentum of the reaction but has near zero mass and zero electric charge.

In a third type of beta decay, a new element is produced when a proton is converted to a neutron by electron capture. That is, an electron orbiting close to the nucleus is spontaneously captured by the positive charge in the nucleus. This decay reaction is represented as

$$\text{``proton''} + \beta^- \rightarrow \text{``neutron''} + \text{neutrino} + \text{gamma ray} \qquad [10.6]$$

Alpha Decay

As opposed to beta decay, in alpha decay, an alpha particle, α^{+2}, consisting of two "protons" and two "neutrons," which has a charge of +2, is ejected from the nucleus. The atomic number of the isotope is, therefore, decreased by two and the mass by about 4 Da. As given in **Table 10-2**, a number of natural radioactive *isotopes* of atomic number equal to or greater than 58 (cerium) can decay by emitting an alpha particle. They do this by *quantum mechanical tunneling* to overcome the coulomb energy barrier faced by alpha particle emission from the nucleus. For instance, ^{238}U (uranium) undergoes spontaneous decay to ^{234}Th (thorium) by alpha decay according to the reaction

$$^{238}U^{+6} \rightarrow {}^{234}Th^{+4} + \alpha^{+2} + \text{gamma ray} + \Delta E \qquad [10.7]$$

where ΔE is the kinetic energy of the ejected alpha particle and nucleus recoil. To conserve charge for this reaction where ^{238}U is bound in a mineral, its valence is taken as +6. This implies the valence of the product ^{234}Th becomes +4, and no electrons are transferred in the decay reaction. The charged alpha particle typically escapes the isotope where it is produced and bonds with electrons from the electron cloud of other atoms it traverses. It becomes an uncharged He atom, and its kinetic energy is converted into heat energy. This electron transfer to form He causes oxidation of the electron donating element (e.g., Fe^{2+} to Fe^{3+}) and local charge imbalance on the atomic scale as is the case with the Th^{+4} produced.

When isotopes undergo radioactive decay, their products can also be unstable and undergo decay. As a result, some isotopes undergo an entire series

Table 10-2 Decay processes and half-lives of naturally occurring radioactive isotopes

Radioactive isotope	Decay process	Half-life (yr)
^{3}H	β^-	12.26
^{7}Be	EC	0.146
^{10}Be	β^-	1.5×10^6
^{14}C	β^-	5730
^{26}Al	β^+, EC	7.16×10^5
^{32}Si	β^-	276
^{36}Cl	β^+, EC, β^-	3.08×10^5
^{37}Ar	EC	0.0953
^{39}Ar	β^-	269
^{40}K	β^-, EC, β^+	1.25×10^9
^{50}V	β^-, EC	6×10^{15}
^{53}Mn	EC	3.7×10^6
^{59}Ni	EC, β^+	8×10^4
^{81}Kr	EC	2.13×10^5
^{85}Kr	β^-	10.6
^{87}Rb	β^-	4.88×10^{10}
^{113}Cd	β^-	9×10^{15}
^{115}In	β^-	5.1×10^{14}
^{123}Te	EC	1.2×10^{13}
^{138}La	EC, β^-	2.69×10^{11}
^{142}Ce	α	$\sim 5 \times 10^{16}$
^{144}Nd	α	2.1×10^{15}
^{147}Sm	α	1.06×10^{11}
^{148}Sm	α	7×10^{15}
^{149}Sm	α	$>1 \times 10^6$
^{152}Gd	α	1.1×10^{14}
^{174}Hf	α	2.0×10^{15}
^{176}Lu	β^-	3.57×10^{10}
^{187}Re	β^-	4.23×10^{10}
^{190}Pt	α	6×10^{11}
^{232}Th	Decay chain	1.40×10^{10}
^{234}U	Decay chain	2.44×10^5
^{235}U	Decay chain	7.04×10^8
^{238}U	Decay chain	4.47×10^9

The half-life is the time it takes for half the atoms of the isotope to decay. EC, electron capture.

From Ottonello (1997).

of decay reactions. These include the isotopes ^{238}U, ^{235}U, ^{234}U, and ^{232}Th that decay through a chain of decay products before finally being transformed to stable Pb isotopes that do not undergo decay. For instance, the ^{238}U decay product, ^{234}Th, is radioactive and undergoes β^- decay to ^{234}Pa (protactinium), which also decays. The decay chain of ^{238}U involves 11 different elements and terminates when stable ^{206}Pb is produced. Decay chains for the isotopes ^{238}U, ^{235}U, and ^{232}Th are given in Appendix H.

Spontaneous Fission

Spontaneous fission is the third type of natural decay. During this decay, the nucleus breaks into two or more unequal fragments. Artificial isotopes with atomic numbers above about 100 are most susceptible to spontaneous fission. The natural isotopes ^{232}Th, ^{235}U, and ^{238}U are known to undergo natural spontaneous fission. The rates are extremely slow. The fastest natural fission decay is found in ^{238}U. With a *half-life* of 8.2×10^{15} years, its decay rate is insignificant relative to the alpha and beta particle decay chain and can be safely ignored for most purposes, as it is in Table 10-2. The half-life is the amount of time it takes for half of the number of isotopes present to decay.

■ Energy of Isotope Decay

Radioactive decay occurs at a constant rate unaffected by temperature, pressure, or nonradioactive isotope chemical changes involving the isotope. The internal energy changes for a system caused by temperature, pressure, or nonradioactive chemical state changes are trivially small compared with changes caused by a radioactive decay reaction, which are measured as changes in mass. All radioactive decay results in mass loss that is converted into nonmass energy (e.g., kinetic and heat) and is, therefore, exothermic. For instance, consider the negative beta decay reaction of ^{40}K to ^{40}Ca:

$$^{40}K^+ \quad \rightarrow \quad ^{40}Ca^{2+} \quad + \quad \beta^- \quad + \quad \Delta E \qquad [10.8]$$

daltons: 39.963999   39.962591   0.000549

where ΔE is the nonmass energy released that includes an antineutrino and gamma ray. The mass in daltons of each particle is given below its symbol. The mass loss for the reaction that is converted into energy, ΔE, is

$$39.963999 \text{ Da} - (39.962591 \text{ Da} + 0.000549 \text{ Da}) = 0.000859 \text{ Da} \qquad [10.9]$$

This change in mass per mole (6.02×10^{23} atom mol^{-1}) of ^{40}K atoms is related to a change in internal energy using Einstein's relationship between mass and energy. Because a dalton is equivalent to 1.66×10^{-27} kg, the energy is

$$\Delta E = \Delta mc^2$$
$$= 6.02 \times 10^{23} \text{ mol}^{-1} \times 0.000859 \text{ Da} \times 1.66 \times 10^{-27} \text{ kg Da}^{-1}$$
$$\times (3.00 \times 10^8 \text{ m s}^{-1})^2 \quad\quad\quad\quad\quad\quad [10.10]$$
$$= 7.73 \times 10^{10} \text{ J mol}^{-1}$$

where Δm is the mass change per mole and c is the speed of light. This energy change per mole is extremely large compared with possible internal energy changes from temperature, pressure, or nonradioactive chemical reactions in the earth. These typically produce changes of a few thousand J mol^{-1} (see Chapters 3 and 4 where the thermodynamic model of energy transfer is outlined). As a result, radioactive decay serves as a reliable clock for measuring geologic time because it is unaffected by other chemical processes or changes in pressure and temperature. After a radioactive isotope of an element is fixed in the structure of a growing crystal, the isotope decays to its daughter isotope at a fixed rate.

■ Determining Isotope Decay Times

The Decay Time Equation

The total number of radioactive isotopes that decay is proportional to the number present. That is, radioactive decay as a function of time, t, is described by

$$-\left(\frac{dP}{dt}\right) = \lambda P \quad\quad\quad\quad\quad\quad [10.11]$$

where P is the number of "parent" radioactive isotopes remaining at any time, t, and λ stands for the decay constant, which is different for each radioactive isotope. Because each decay of a parent isotope produces a daughter isotope,

$$-\frac{dP}{dt} = \frac{dD}{dt} \quad\quad\quad\quad\quad\quad [10.12]$$

where D is the number of daughter isotopes produced. Rearranging decay Equation 10.11 gives

$$-\frac{1}{P}dP = \lambda dt \quad\quad\quad\quad\quad\quad [10.13]$$

Taking the indefinite integral of both sides of this rate equation results in

$$-\ln P = \lambda t + C \quad\quad\quad\quad\quad\quad [10.14]$$

where C is a constant of integration. Evaluating Equation 10.14 at time = 0 indicates that $C = -\ln P^o$, where P^o is the number of parent isotopes present at time = 0. This substitution gives

$$\ln P - \ln P^o = -\lambda t \qquad\qquad [10.15]$$

Taking the exponential of both sides of Equation 10.15 results in

$$P = P^o \, e^{-\lambda t} \qquad\qquad [10.16]$$

Equation 10.16 rearranged to give the number of original parent isotopes is

$$P^o = P e^{\lambda t} \qquad\qquad [10.17]$$

The number of daughter isotopes produced in time interval t, D*, is the difference between P^o and P. From Equation 10.17, this is

$$D^* = P^o - P = P e^{\lambda t} - P = P(e^{\lambda t} - 1) \qquad\qquad [10.18]$$

Clearly, these daughter isotopes are added to any daughter isotopes that were there originally. The total daughter isotopes present, D, is the sum of the number of original daughters, D^o, and those produced by radioactive decay, D*,

$$D = D^o + D^* \qquad\qquad [10.19]$$

Substituting Equation 10.18 into Equation 10.19 yields

$$D = D^o + P(e^{\lambda t} - 1) \qquad\qquad [10.20]$$

Solving Equation 10.20 for t, the time since decay started, gives

$$t = \frac{1}{\lambda} \ln\left(\frac{D - D^o}{P} + 1\right) \qquad\qquad [10.21]$$

This is the *decay time equation*. The decay constant, λ, is found by laboratory determination. Equation 10.21 is used to obtain a time since a rock or mineral became a closed system to the decaying isotope. A measurement of the concentration of parent isotopes remaining, P, and daughter isotopes, D, present in the sample is required. The original daughter isotopes present, D^o, at the start of the process must also be determined by some other independent means (see below).

λ is related to the half-life, the time it takes for half of the parent isotopes to decay. From Equation 10.16, this half-life time, $t_{1/2}$, when $P = 0.5\ P^o$ is given by

$$0.5P^o + P^o e^{-\lambda t_{1/2}} \qquad\qquad [10.22]$$

or by

$$t_{1/2} = \ln\left(\frac{P^o}{0.5P^o}\right) \times \frac{1}{\lambda} = \frac{\ln 2}{\lambda} \cong \frac{0.693}{\lambda} \qquad\qquad [10.23]$$

For example, λ for $^{238}U = 1.537 \times 10^{-10}$ yr^{-1}. Its half-life is, therefore, 4.47×10^9 years, as given in Table 10-2.

Closure Temperature

There are various ways to estimate D^o needed to evaluate Equation 10.21. For noble gas daughters (e.g., $^{40}K \rightarrow {}^{40}Ar$ dating), it is often assumed that any daughter isotopes produced in the melt are not incorporated into the growing crystal during crystallization. The noble gas escapes the local melt environment before the melt is incorporated into the mineral; therefore, D^o is taken as zero at the time the crystal starts to grow. After the growth of the mineral, the noble gas daughter produced as a function of time, D^*, is typically caged in the crystal and is measured to obtain a date of crystallization. Because the gas is caged rather than bonded in the crystal, a later reheating event that exceeds the *closure temperature* for Ar gas in the mineral allows the diffusion of Ar out of the crystal. The closure temperature is then the temperature below which the daughter isotope is retained in the mineral. If a later reheating event above the closure temperature has occurred, the time measured from the concentration of D would be the time since the sample cooled below the closure temperature after reheating.

Because Ar loss is a diffusion process, there is not a unique closure temperature, as it is dependent on the rate of cooling through the window of temperature where diffusion can occur. The closure temperature is different for each mineral depending on its structure and cooling rate. Closure temperatures also depend on the diffusion distance and, therefore, on a mineral's grain size. For Ar, with cooling at 5°C per million years (m.y.), the closure temperature apparently decreases from about 700°C to 200°C in the relative order hornblende > biotite > K-feldspar > plagioclase (Berger and York, 1981). These temperature-dependent closure temperatures have been used to place constraints on rates of uplift in tectonically active regions where it is called *thermochronology* (Reiners and Ehlers, 2005).

Isocrons

Another way to estimate D^o is to note that Equation 10.20 is a straight line when D is plotted against P. The slope of this line is equal to $(e^{\lambda t} - 1)$ and has an intercept of D^o. Consider a suite of minerals in a rock or rocks crystallized from an initially homogeneous magma. Because different minerals crystallize with different affinities for a particular element, the amounts of radioactive parent isotope, P, are different in each mineral. The same should be true of any daughter isotope present originally as well; however, the ratio of the daughter isotope, D^o, to a different nonradioactive isotope of the same daughter element, D^{n-r}, should be the same throughout the magma during crystallization. This occurs because isotopes of the same element have the same electron configuration and, therefore, very similar chemical properties and are not fractionated. Their ratios are then the same in all phases that crystallize from the magma.

A plot of D/D^{n-r} versus P/D^{n-r} for a suite of samples crystallized at the same time should fall on the same straight line as given by Equation 10.20, where the ratios are used rather than the isotopes themselves. A straight line plot of D/D^{n-r} versus P/D^{n-r} for samples formed at the same time is referred to as an *isochron* plot. The time since the mineral formed is obtained from the slope of the line, $(e^{\lambda t} - 1)$. If the samples do not fall on a straight line, then the system was not completely closed (isolated) since the initial resetting, the daughters were not completely homogenized, or there were different closure temperatures for the samples. With these ideas in mind, particular age dating schemes are considered.

■ Rubidium-Strontium Systematics

Age Dating

Consider crystallization during cooling of an intrusive magma body, say a granite stock. Because of their difference in size and valence (see Chapter 5), Rb and Sr in the magma body show different preferences for incorporation into minerals depending on the mineral structure. K-feldspar crystallizes first with a high Rb/Sr ratio because the substitution of Rb^+ for K^+ in the K-feldspar structure increases its concentration while a smaller concentration of Sr^{2+} is incorporated. When plagioclase begins to crystallize, it has a lower Rb/Sr ratio because Sr^{2+} is incorporated into the Ca^{2+} site, but little Rb^+ is accommodated in the structure. Finally, biotite begins to crystallize with a higher Rb/Sr than K-feldspar.

Although the concentration of Rb and Sr is likely to be similar throughout the original magma body because of mixing by thermal convection, the chemical and mechanical *fractionation* during crystallization causes different regions in the solidified granite to have different relative abundance of minerals. Different Rb/Sr ratios both in whole rock samples taken from different parts of the intrusion and differences in Rb/Sr for various mineral phases are therefore produced. As a function of time after the magma crystallized, the ^{87}Rb decays to ^{87}Sr. Minerals/rocks with different Rb concentrations produce different amounts of ^{87}Sr with time. If this produced ^{87}Sr is determined, an age can be calculated.

^{87}Rb decays by spontaneous emission of a β^- particle from its nucleus to form stable ^{87}Sr. The rate Equation 10.20 is written as

$$^{87}\text{Sr} = {}^{87}\text{Sr}_0 + {}^{87}\text{Rb}(e^{\lambda t} - 1) \qquad [10.24]$$

where $\lambda = 1.42 \times 10^{-11}$ yr^{-1}. ^{87}Sr$_0$ denotes the original amount of ^{87}Sr in the sample. Because ratios of isotopes rather than absolute concentrations are more likely to be homogenized to a common starting value, Equation 10.24 is divided through by the concentration of the stable nonradiogenic ^{86}Sr isotope to give

$$\underset{\text{measured}}{\frac{^{87}\text{Sr}}{^{86}\text{Sr}}} = \underset{\text{initial}}{\frac{^{87}\text{Sr}_0}{^{86}\text{Sr}}} + \underset{\text{produced}}{\frac{^{87}\text{Rb}}{^{86}\text{Sr}}(e^{\lambda t} - 1)} \qquad [10.25]$$

Figure 10-4 shows a plot of ^{87}Rb/^{86}Sr versus ^{87}Sr/^{86}Sr for three rocks that are plagioclase rich, K-feldspar rich, and biotite rich. The initial homogenized values of ^{87}Rb/^{86}Sr and ^{87}Sr/^{86}Sr in each rock are indicated with the points A*, B*, and C*. Arrows between initial and measured values in the figure give the evolution of the isotopes as a function of time for each rock. ^{87}Sr$_0$/^{86}Sr is the value of ^{87}Sr/^{86}Sr along the isochron at ^{87}Rb/^{86}Sr = 0.0. The slope of the isochron line on which the measured samples plot equals $e^{\lambda t} - 1$. The age of the samples, therefore, is calculated knowing λ. One assumes that the samples all started with the same initial ^{87}Sr to ^{86}Sr ratio, ^{87}Sr$_0$/^{86}Sr, and did not lose or

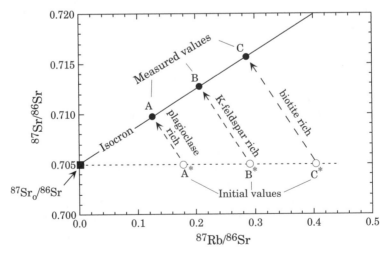

FIGURE 10-4 Rb-Sr isochron plot of ^{87}Rb/^{86}Sr versus ^{87}Sr/^{86}Sr for three rocks, A, B, and C. A*, B*, and C* give their initial values at time 0, and the long dashed arrows indicate the change in the values with time.

gain Rb or Sr except by radioactive decay. That is, the system remained closed. For the isochron shown in Figure 10-4, its slope is $(0.720 - 0.705)/0.4 = 3.75 \times 10^{-2}$ for an age of 2.6 billion years (b.y.).

Often, assuming that the system is closed is not valid because of a later re-heating event during a subsequent metamorphism that tends to homogenize the isotopes as temperatures above the Rb/Sr closure temperature are reached and diffusion of Rb and/or Sr isotopes occurs. If the isotopes are completely homogenized by the latter event, then the measured time would be of the latter event. If homogenization was only partial, then the values of $^{87}Sr/^{86}Sr$ would not plot on a straight isochron line.

Determining Magma Sources

Rb-Sr isotope systematics are also used to help determine whether magmas originated in the mantle, crust, or had a mixed source. For this analysis, the initial ratio of $^{87}Sr/^{86}Sr$ at the time the earth formed, $^{87}Sr_o/^{86}Sr$, is required. The earth and meteorites are thought to have formed at the same time from the solar nebula with a constant $^{87}Sr/^{86}Sr$ ratio. The $^{87}Sr/^{86}Sr$ ratio of the original earth is, therefore, obtained from meteorites. Meteorites with very low Rb have not had their $^{87}Sr/^{86}Sr$ changed significantly with time. The small change that has occurred can be determined by extrapolating back to the time of the earth's formation using the decay equation. If $^{87}Sr/^{86}Sr$ from these meteorites at the time the earth formed are similar to those of terrestrial silica-rich rocks at this time, then $^{87}Sr_o/^{86}Sr = 0.6989$ for terrestrial rocks at their formation.

Sometime after its formation, the earth's crust differentiated from the mantle. Rb, as an alkali, fractionated preferentially with the alkali K into the silica-rich minerals that dominate the crust. The alkaline earth, Sr, on the other hand, behaves similarly to the alkaline earth, Ca, and, therefore, more was preferentially retained in the mantle relative to K. The differentiation of the core does not play a part because of its lack of significant Rb or Sr. The $^{87}Sr/^{86}Sr$ ratio in the crust, because of greater ^{87}Rb and lower ^{86}Sr content, therefore, should have increased more rapidly than in the mantle with increasing time since crustal differentiation. This difference then gives a means of distinguishing rocks formed from partial melting of rocks in the mantle as opposed to those from the crust.

The present ratio of $^{87}Sr/^{86}Sr$ in the mantle is determined by measuring recently erupted basalts from oceanic environments where continental crustal rock contamination ($^{87}Sr/^{86}Sr \sim 0.720$) and seawater contamination ($^{87}Sr/^{86}Sr = 0.709$) are not a problem. Such measurements give $^{87}Sr/^{86}Sr$ ratios in the range of 0.7025–0.7065 (see Figure 10-8). The range in values indicates that there is some heterogeneity in the mantle. Taking an average present day value of 0.704 for the mantle, the $^{87}Sr/^{86}Sr$ in the mantle at any time in the past is determined by a linear extrapolation between 0.699 and 0.704 and the time interval between them. $^{87}Sr/^{86}Sr$ ratios are measured for a suite of rocks and $^{87}Sr_o/^{86}Sr$ determined. If the value is lower than 0.706, one surmises that the rocks are mostly from a mantle source. If it is significantly higher, it must have a crustal component.

Strontium Isotopes and Global Tectonic Events

Strontium isotopes have also been used to understand global tectonic processes. For instance, consider the change in the ratio of $^{87}Sr/^{86}Sr$ determined for seawater as a function of time, as shown in **Figure 10-5**. Sr behaves chemically the same whether it is ^{87}Sr or ^{86}Sr. The change of the ratio $^{87}Sr/^{86}Sr$ in seawater is controlled by the amount of Sr that fluxes in and its $^{87}Sr/^{86}Sr$ ratio minus the amount of Sr that fluxes out and its $^{87}Sr/^{86}Sr$ ratio. As argued by Veizer (1989), the dominate Sr fluxes into seawater are from river suspended and dissolved loads, groundwater discharge, ocean crust–seawater interactions, and diagenetic reflux from buried pore water. The ocean crust–seawater interactions are divided into those at mid-oceanic ridges, those along ridge flanks, and interactions with old oceanic crust. Removal of Sr is by sedimentation and exchange of high $^{87}Sr/^{86}Sr$ in seawater with lower $^{87}Sr/^{86}Sr$ in newly injected basalts on the ocean floor during hydrothermal alteration.

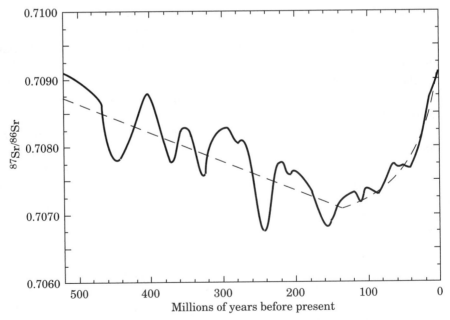

FIGURE 10-5 Variations in $^{87}Sr/^{86}Sr$ of seawater during the Phanerozoic. (Adapted from Veizer, 1989.) The solid line is the measured change with the dashed line giving the average trend of the data with time.

Note the general decrease in $^{87}Sr/^{86}Sr$ of seawater during the *Paleozoic* with an increase occurring in the Cretaceous and Tertiary. On shorter time intervals on the order of 10 m.y., there are clearly significant oscillations in the $^{87}Sr/^{86}Sr$ ratio. The reasons for the longer term trends and the shorter term variability are not understood with any certainty. It has been argued that the dominate control is the changing flux of Sr and $^{87}Sr/^{86}Sr$ from river input. These reflect tectonic and climatic changes. For instance, if tectonic forces expose old granitic rocks with high $^{87}Sr/^{86}Sr$ to weathering, the $^{87}Sr/^{86}Sr$ of seawater increases. Understanding of these linkages should lead to better models of tectonic and erosional processes through time. It has been argued that the rapid rise in $^{87}Sr/^{86}Sr$ over the last 20 m.y. is due to the contribution from weathered granitic basement rocks exposed during the Himalayan uplift.

■ Samarium-Neodymium Systematics

Age Dating

The systematics of Sm-Nd dating are similar to those of Rb-Sr dating. Sm-Nd isotopic measurements are generally used to date mafic and ultramafic rocks where the lower rubidium concentrations make Rb-Sr dating difficult. Sm and Nd exist at ppm levels in most common silicate minerals. ^{147}Sm decays to ^{143}Nd by the reaction

$$^{147}Sm^{+3} \rightarrow \ ^{143}Nd^{+1} + \alpha^{+2} + \Delta E \tag{10.26}$$

where Sm and Nd are in the +3 and +1 valence state, respectively, and ΔE is the nonmass energy released. The rate equation is

$$^{143}Nd = \ ^{143}Nd_o + \ ^{147}Sm \ (e^{\lambda t} - 1) \tag{10.27}$$

with $\lambda = 6.54 \times 10^{-12}$ yr^{-1} or a half-life of 106 b.y. With this slow rate of decay, Sm-Nd dating has an uncertainty on the order of 20 m.y. and, therefore, is primarily used to date Precambrian rocks.

Ratios are determined relative to the nonradiogenic isotope ^{144}Nd so the rate equation becomes

$$\underbrace{\frac{^{143}Nd}{^{144}Nd}}_{\text{measured}} = \underbrace{\frac{^{143}Nd_o}{^{144}Nd}}_{\text{initial}} + \underbrace{\frac{^{147}Sm}{^{144}Nd}}_{\text{produced}}(e^{\lambda t} - 1) \tag{10.28}$$

Sm and Nd are light rare earth elements with ionic radii of 1.04 and 1.08 Å, respectively. They have an electron structure that varies only in the number of 4f orbital electrons (see Chapter 5). They, therefore, have very similar chemical properties and are not reset as easily as in Rb-Sr dating. **Figure 10-6** shows an Sm-Nd isochron for gabbro from the Stillwater complex.

Chondritic Uniform Reservoir and Epsilon Values

As with $^{87}Sr/^{86}Sr$, $^{143}Nd/^{144}Nd$ is used to draw inferences about whether igneous rocks are from mantle or crustal sources. Judging from values of $^{143}Nd/^{144}Nd$ in *chondritic meteorites*, the $^{143}Nd/^{144}Nd$ of the bulk earth, if it had not differentiated into core, mantle, and crust, would be 0.5126 at the present time. Knowing the age of the earth and taking the average ^{147}Sm concentration of these meteorites, the decay equation is used to calculate $^{143}Nd/^{144}Nd$ at any time in the past. The value of $^{143}Nd/^{144}Nd$ at the time of the earth's formation is calculated to be about 0.5058. The model value, for the evolution of $^{143}Nd/^{144}Nd$ as a function of time for an undifferentiated earth, is referred to as the chondritic uniform reservoir (CHUR) value and is plotted as a dashed line as a function of time in **Figure 10-7.**

Sm and Nd are both trace elements. (The behavior of trace elements in crystallizing igneous melts is discussed in Chapter 8.) Nd is more incompatible in most minerals crystallizing from a mafic melt than Sm because of its larger ionic radius. Nd is, therefore, more concentrated in the melt. If the crust of the earth formed from a partial melt extracted from the CHUR reservoir at 3.5 b.y., it would evolve along a shallower trajectory than CHUR, as shown in Figure 10-7, because the partial melt would be lower in ^{147}Sm. The residual mantle would have a greater concentration of ^{147}Sm than CHUR and would evolve along the steeper trajectory as shown. If at a later time a partial melt was extracted from the residual mantle, its $^{143}Nd_0/^{144}Nd$ would be greater than CHUR because the residual mantle has a higher Sm/Nd ratio.

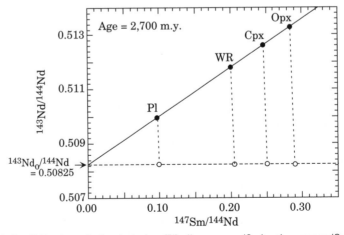

FIGURE 10-6 Sm–Nd isochron plot for plagioclase (Pl), clinopyroxene (Cpx), orthopyroxene (Opx), and the whole rock (WR) for the Stillwater complex gabbro giving a value of $^{143}Nd_0/^{144}Nd = 0.50825$ at 2700 m.y. (Adapted from DePaolo and Wasserburg, 1979.)

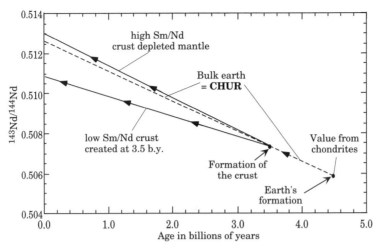

FIGURE 10-7 Model of the evolution of ^{143}Nd to ^{144}Nd in rocks formed from a CHUR that partially melted at 3.5 b.y., producing reservoirs of ^{147}Sm depleted crust and ^{147}Sm enriched mantle.

The difference in a rock's ^{143}Nd$_o$/^{144}Nd from CHUR is typically given as a ratio, termed its *epsilon value*, defined as

$$\varepsilon_{Nd} = \left[\frac{(^{143}Nd_o/^{144}Nd)_{Rock}}{(^{143}Nd_o/^{144}Nd)_{CHUR}} - 1 \right] \times 10^4 \qquad [10.29]$$

ε_{Nd} of a rock then gives the extent of difference from the CHUR reservoir at the time a particular sample formed. A positive value of ε_{Nd} indicates the rock came from a depleted residual rock (i.e., mantle). A negative value indicates that the rock was formed from an enriched source with lower Sm/Nd (i.e., continental crust). A ^{143}Nd$_o$/^{144}Nd value for a rock that plots on the CHUR line would have an $\varepsilon_{Nd} = 0$. Values of ε_{Nd}, therefore, help one determine the evolution of various rock reservoirs in the earth. For instance, 3.8 b.y. rocks have ε_{Nd} between +2 and +4. This implies that a ^{147}Sm depleted partial melt (continental crust) was extracted from this source at least 2×10^5 years earlier to produce the positive ε_{Nd} values.

There is still a large debate about whether the continental crust reached its present size early in earth history or has grown in size as a function of time. The decay Equation 10.28 is used to back calculate the ^{143}Nd/^{144}Nd ratio with time of a crustal rock from its measured ^{143}Nd/^{144}Nd and ^{147}Sm/^{144}Nd ratios. Extrapolation backward of this trajectory to its intersection with the CHUR model trajectory gives the crustal rock's Sm-Nd model age. This has been done for crustal rocks in western North America. These Sm-Nd model ages are greater than ages measured by isochrons in these rocks. These rocks at the time they

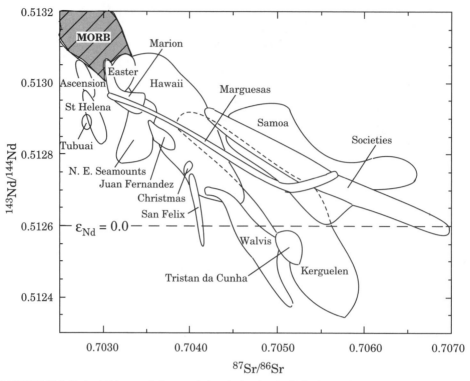

FIGURE 10-8 Ratio of Nd versus Sr isotopes in basalt showing the limited range for MORBs and ocean island basalts. (Adapted from Hart et al., 1986.)

formed then plot above the CHUR line in Figure 10-7. This indicates that some
differentiation of crustal material must have occurred earlier to produce these
values, and therefore, the crust has grown at least somewhat with time.

Combining Neodymium and Strontium Isotope Analysis

If the mantle is not well mixed, it would be reasonable to assume that the deep
mantle, which geologists believe was not involved in crust formation, is undifferentiated relative to CHUR and is considered a CHUR reservoir. The upper
mantle would then show depletion of the elements extracted when the crust
formed. Consider also the behavior of Rb/Sr isotope ratios. With the formation of the crust, the Rb/Sr ratio of the residual mantle would decrease so the
amount of radiogenic Sr produced there would be less. Thus, depleted mantle
would have a lower $^{87}Sr/^{86}Sr$ than undepleted mantle. On the other hand, formation of depleted mantle increases its Sm/Nd ratio so that depleted mantle
would have a higher $^{143}Nd/^{144}Nd$ than undepleted mantle. With greater depletion and more time, $^{87}Sr/^{86}Sr$ increases more slowly, and $^{143}Nd/^{144}Nd$ increases
more rapidly compared with those ratios in an undepleted mantle reservoir.

Oceanic basalts are used as samples of the mantle. Again, although elements fractionate between melt and residual rock when basalt is produced from the mantle, the ratio of isotopes of the element does not. Uncontaminated basalts directly from the mantle should sample its isotopic ratios. **Figure 10-8** shows measurements of basalts from *mid-oceanic ridge basalts (MORBs)* and those from the indicated ocean islands. These are considered to be direct mantle samples. A general inverse correlation exists between $^{87}Sr/^{86}Sr$ and $^{143}Nd/^{144}Nd$, as predicted for crustal extraction from the mantle. This general correlation is termed the *mantle array*. MORB values are well confined and sample the most depleted mantle. The volcanic hot spot islands appear, in general, to have trends that span some distance between undepleted mantle, given by the dashed line labeled $\varepsilon_{Nd} = 0.0$, and MORB values. This is consistent with MORB sampling of depleted upper mantle and the oceanic islands sampling a deeper less depleted source. The trends for a particular ocean island suggest that contamination with upper depleted mantle has occurred in many cases. Alternatively, multiple mantle endmembers produced by recycling of crustal material back in the mantle during subduction could be playing a part.

Figure 10-9 shows Nd versus Sr isotopic ratios measured in Paleozoic S-type and I-type granites from Australia. S-type granites are dominantly from sedimentary

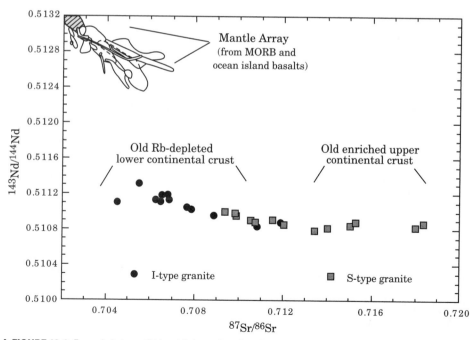

FIGURE 10-9 Expanded view of Nd and Sr isotopic ratios given in Figure 10-8 so that ratios from Paleozoic granitic batholiths from southeastern Australia are plotted. (Adapted from McCulloch and Chappell, 1982.)

sources, and I-type granites are from igneous sources with little or no sedimentary component. As anticipated because the granites have had low Sm for an extended period, their $^{143}Nd/^{144}Nd$ values are low compared with the current mantle values given by the mantle array. The S-type granites from Australia likely formed high in the crust with a component of melted Rb-rich illites producing their high $^{87}Sr/^{86}Sr$ ratio. The I-type granites were probably produced from melting of the lower crust or upper mantle where Rb concentrations are low. The low Rb led to a low production of ^{87}Sr and, therefore, low $^{87}Sr/^{86}Sr$.

■ Uranium-Thorium-Lead Systematics

Age Dating

Natural uranium consists dominantly of ^{238}U with a small amount of ^{235}U (0.72%) and an even smaller amount of ^{234}U (0.0055%). ^{238}U and ^{235}U decay to ^{234}Th and ^{231}Th, respectively, by alpha decay, and the thorium isotopes produced decay by a series of shorter lived intermediate radioactive isotopes to form the stable isotopes of ^{206}Pb and ^{207}Pb, respectively. The decay series are outlined in Appendix H. The overall reactions are

$$^{238}U^{+6} \rightarrow {}^{206}Pb^{+2} + 8\alpha^{+2} + 6\beta^- + 6\,e^- + \Delta E \qquad \text{half-life} = 4.51 \times 10^9 \text{ years} \qquad [10.30]$$

and

$$^{235}U^{+6} \rightarrow {}^{207}Pb^{+2} + 7\alpha^{+2} + 7\beta^- + 3\,e^- + \Delta E \qquad \text{half-life} = 0.71 \times 10^9 \text{ years} \qquad [10.31]$$

where, as indicated above, α^{+2} denotes an alpha particle, e^- an electron lost from the element's electron cloud, β^- an electron lost from its nucleus, and ΔE gives the nonmass energy produced. Electrons are lost from the element's electron cloud during its transformation to Pb to maintain charge balance. Uranium containing minerals may also contain ^{232}Th, which undergoes spontaneous decay to ^{208}Pb by the overall reaction

$$^{232}Th^{+4} \rightarrow {}^{208}Pb^{+2} + 6\alpha^{+2} + 4\beta^- + 6\,e^- + \Delta E \qquad \text{half-life} = 14.01 \times 10^9 \text{ years} \qquad [10.32]$$

The overall Reactions 10.30, 10.31, and 10.32 also produce neutrinos and gamma rays but no other particles with mass, and this nonmass energy is included in ΔE.

Three dating schemes are then possible: ^{238}U to ^{206}Pb, ^{235}U to ^{207}Pb, and ^{232}Th to ^{208}Pb. A good mineral to date using U-Th-Pb techniques is zircon ($ZrSiO_4$), a common accessory mineral in igneous rocks, particularly granites. It is resistant to both physical and chemical weathering, and thus, it is likely to remain a closed system. Also, it tends to concentrate U and to a lesser extent Th while excluding Pb during growth; therefore, it has the high original parent-to-daughter ratio necessary for accurate dating. Typically, the measured radioactive isotopes of U and Th as well as the daughter product Pb isotopes are expressed as ratios with the nonradiogenic isotope of Pb, ^{204}Pb. Thus, for $^{238}U \rightarrow {}^{206}Pb$ dating

$$\underbrace{\frac{^{206}Pb}{^{204}Pb}}_{\text{measured}} = \underbrace{\frac{^{206}Pb_0}{^{204}Pb}}_{\text{initial}} + \underbrace{\frac{^{238}U}{^{204}Pb}(e^{\lambda_{238}t} - 1)}_{\text{produced}} \qquad [10.33]$$

The radioactive age is then given by

$$t = \frac{1}{\lambda_{238}} \ln \frac{\left(\frac{^{206}Pb}{^{204}Pb}\right) - \left(\frac{^{206}Pb_0}{^{204}Pb}\right)}{\left(\frac{^{238}U}{^{204}Pb}\right)} + 1 \qquad [10.34]$$

where $\lambda_{238} = 1.55 \times 10^{-10}$ yr^{-1}. The equation characterizing $^{235}U \rightarrow {}^{207}Pb$ decay is given by

$$\underbrace{\frac{^{207}Pb}{^{204}Pb}}_{\text{measured}} = \underbrace{\frac{^{207}Pb_0}{^{204}Pb}}_{\text{initial}} + \underbrace{\frac{^{235}U}{^{204}Pb}(e^{\lambda_{235}t} - 1)}_{\text{produced}} \qquad [10.35]$$

where $\lambda_{235} = 9.85 \times 10^{-10}$ yr^{-1}. An age can be determined by an expression similar to Equation 10.34. The age can also be determined for decay of ^{232}Th in a similar manner; however, the three age determination schemes used on the same sample often do not agree because the minerals in a rock are not entirely closed to these isotopes. Loss of any of the intermediate isotopes as well as Pb produces erroneous ages; however, Pb reacts chemically the same whether it is ^{206}Pb or ^{207}Pb. An age, therefore, can be calculated from Equations 10.33 and 10.35 even if lead is lost. That is, because the rates of decay are different, the ratio of daughter isotopes is a measure of time, and no parent isotopic values are needed.

Concordia Diagrams

From Equation 10.33, the radiogenic lead produced by ^{238}U decay, $^{206}Pb^*$, is

$$^{206}Pb^* = {}^{206}Pb - {}^{206}Pb_0 = {}^{238}U(e^{\lambda_{238}t} - 1) \qquad [10.36]$$

From Equation 10.35, the radiogenic lead produced by ^{235}U decay, $^{207}Pb^*$, is

$$^{207}Pb^* = {}^{207}Pb - {}^{207}Pb_o = {}^{235}U(e^{\lambda_{235}t} - 1)$$ [10.37]

Equations 10.36 and 10.37 cast in terms of ratios become

$$\frac{^{206}Pb^*}{^{238}U} = e^{\lambda_{238}t} - 1$$ [10.38]

and

$$\frac{^{207}Pb^*}{^{235}U} = e^{\lambda_{235}t} - 1$$ [10.39]

Plotting one of these ratios against the other as a function of time produces a *concordia* diagram, as shown in **Figure 10-10**. The concordia curve is the locus of all points for samples that satisfy both isochron Equations 10.33 and 10.35 and, therefore, have remained closed since their formation. Ages are indicated by the numbers along the curve in billions of years. If the system loses lead at some later time, the $^{206}Pb/^{207}Pb$ ratio of the rocks departs from the concordia at the age of the lead loss and subsequently plots below the concordia. In fact, if a set of samples lose different amounts of Pb at the same time, then these sam-

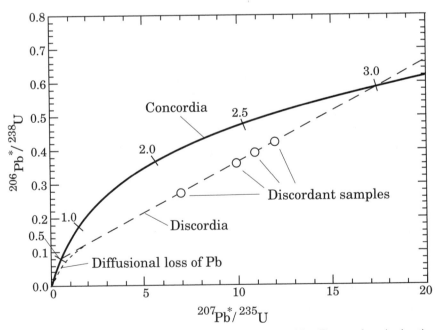

FIGURE 10-10 Concordia for U-Pb systematics given as the solid curved line. The ages from the time the system is closed are labeled in billions of years along the line. Discordant samples are shown that give an age of 3.0 b.y. and a latter event where each of the samples have lost a different amount of lead at 0.5 b.y. These data can also be interpreted as a diffusional loss of Pb.

ples plot as a linear array below the concordia along what is termed a *discordia*. Extension of the discordia up to where it intersects the concordia gives the age of initial formation of the sample suite—about 3.0 b.y. in the case given in Figure 10-10. Downward extension would represent the age of the Pb loss event if only one loss event has occurred. This is about 500 m.y. for the samples in Figure 10-10. Arguments have been made that lead loss is better modeled as a continuous diffusion process. In this case, a linear discordia array is also produced, but the downward extrapolation is meaningless. Diffusional Pb loss would increase with decreasing pressure during uplift of crystalline rocks and also with age due to increased damage of the mineral from the particles released during the radioactive decay.

Holmes-Houterman Common Pb Model

A model has been produced to help understand the development of Pb isotopic ratios through time. The Holmes-Houterman model assumes reservoirs with different concentrations of Pb and U develop at different times during the earth's history. Consider $^{206}Pb/^{204}Pb$ and $^{207}Pb/^{204}Pb$, and remember that the decay of ^{238}U adds uranogenic ^{206}Pb and the decay of ^{235}U uranogenic ^{207}Pb as a function of time. Because ^{204}Pb is not produced from radioactive decay, these ratios increase with time. Mixing processes appear to have caused these ratios to be uniform throughout the solar nebula at the time the earth formed. Isotopic measurements of troilite meteorites, which contain insignificant U and Th but significant Pb, yield primordial ratios of $^{206}Pb/^{204}Pb$ and $^{207}Pb/^{204}Pb$ of 9.307 and 10.294, respectively. Based on the age of the earth and present ^{235}U and ^{238}U abundances, primordial $^{235}U/^{238}U$ was 0.316. Of the primordial ^{238}U with a half-life of 4.51 b.y., about half remains today. Of the primordial ^{235}U with a half-life of 0.71 b.y., a little more than 1% remains. Because of the shorter half-life of ^{235}U, $^{207}Pb/^{204}Pb$ increases more rapidly than $^{206}Pb/^{204}Pb$ at first; however, after several half-lives, there is little ^{235}U left, slowing the rate of increase of $^{207}Pb/^{204}Pb$. The rate of $^{206}Pb/^{204}Pb$ increase is, therefore, greater closer to present time, as shown by the decreasing slopes for the younger isochrons in **Figure 10-11**.

Because the primordial $^{235}U/^{238}U$ is known, the relative rate of $^{207}Pb/^{204}Pb$ to $^{206}Pb/^{204}Pb$ produced depends only on the amount of U to Pb. The variable μ gives $^{238}U/^{204}Pb$ in the reservoir as it exists today. If $\mu = 9.1$, the estimated average bulk value for the present earth (Allègre et al., 1988), then the Pb isotope ratios for the earth would develop along a $\mu = 9.1$ growth curve just above the $\mu = 9$ growth curve shown in Figure 10-11. If different reservoirs were produced from the solar nebula, they would have different μ and would now all plot on the 0 isochron, the geochron. This is true of meteorites that have different μ. They all plot in a linear array along the geochron. This is a one-stage model in which growth is along a single growth curve. No terrestrial Pb appears

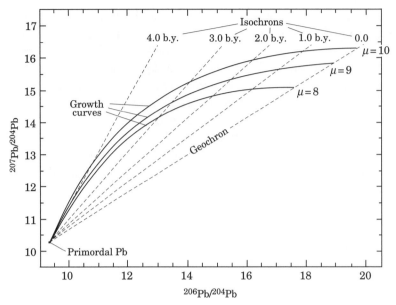

FIGURE 10-11 Holmes-Houterman diagram indicating the growth of common Pb isotope ratios from primordial values to predicted present day values for three different present day values of $\mu = {}^{238}U/{}^{204}Pb$.

to be exactly single stage but has existed in different reservoirs, with different values of μ for some time.

If Pb isotope ratios plot to the right of the geochron line, they cannot have a one-stage history but must have a more complicated history. Consider a two-stage model. The mantle reservoir with $\mu = 9.1$ develops from primordial Pb at the age of the earth as shown in **Figure 10-12**. At 2.7 b.y., crustal granitic material is derived from the mantle, fractionating U and other incompatible elements into the granite. This granite now has $\mu = 25$ and evolves its Pb isotopes along the line marked $\mu = 25$ in Figure 10-12. Because the amount of granitic crust fractionated is small relative to the size of the mantle, the mantle's Pb isotopic ratios remain unchanged. The mantle continues to evolve along the $\mu = 9.1$ line. If at 0.5 b.y. basalt from the mantle intrudes the crust, causing local melting of the granite, the rocks formed will be a mixture of mantle and crustal Pb. Any K-feldspar crystallized incorporates this mixed Pb but no U. If Pb in these K-feldspars in crustal granite are measured, they would plot as the open squares in Figure 10-12. Even more complex models have also been devised to interpret Pb isotope ratios found in major reservoirs in the earth.

U and Th Decay Series Dating

Radium-226 Dating

Dating is also undertaken with daughter radioactive isotopes produced during the U and Th isotope decay series. Although most of these isotopes have

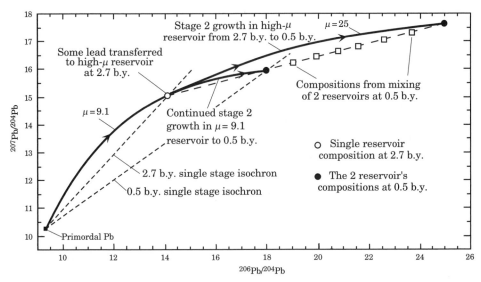

FIGURE 10-12 A two-stage Holmes-Houterman diagram for the development of common lead with stage 1 to 2.7 b.y. and stage 2 to 0.5 b.y. The measured values occur to the right of a single stage 0.5 b.y. isochron. (Long, L. E., 1999, Lead: Stable isotopes. In Marshall, C. and Fairbridge, R. (eds.) *Encyclopedia of Geochemistry.* Kluwer, Dordrecht, pp. 363–368. Reprinted with kind permission from Springer Science and Business Media.)

half-lives of minutes or less (see Appendix H), the ones given in **Table 10-3** have half-lives longer than a year and, therefore, have potential use in dating geologic processes. For instance, radium-226 with a half-life of 1622 years is produced by alpha decay of thorium-230 with a half-life of 80,500 years. The change in the number of ^{226}Ra isotopes is given by the number produced minus the number that decay or

$$\frac{dN_{226_{Ra}}}{dt} = \lambda_{230_{Th}}N_{230_{Th}} - \lambda_{226_{Ra}}N_{226_{Ra}} \qquad [10.40]$$

Table 10-3   Long-lived U and Th radioactive daughter isotopes

Nuclide	Half-life (yr)	Ultimate parent
^{234}U	245,000	^{238}U
^{230}Th	80,500	^{238}U
^{231}Pa	34,300	^{235}U
^{226}Ra	1622	^{238}U
^{210}Pb	22.3	^{238}U
^{227}Ac	22.0	^{235}U
^{228}Ra	6.7	^{232}Th
^{228}Th	1.91	^{232}Th

where N is the number of isotopes and λ the decay constant of the subscripted isotope. At steady state, the number of ^{226}Ra atoms is constant so that

$$0 = \lambda_{230\text{Th}} N_{230\text{Th}} - \lambda_{226\text{Ra}} N_{226\text{Ra}} = \frac{dN_{230\text{Th}}}{dt} - \frac{dN_{226\text{Ra}}}{dt} \qquad [10.41]$$

The concentration of ^{226}Ra at steady-state is termed the *supported concentration* because it is supported by production from ^{230}Th. Because Ra and Th have different chemical properties, they are fractionated in a chemical reaction. If the contribution from ^{230}Th decay or loss from ^{226}Ra decay changes from this steady-state condition, then

$$\frac{dN_{226\text{Ra}}}{dt} = \left(\frac{dN_{226\text{Ra}}}{dt}\right)_s + \left(\frac{dN_{226\text{Ra}}}{dt}\right)_u \qquad [10.42]$$

where the subscripts s and u stand for the supported and unsupported production of ^{226}Ra, respectively. If the system remains closed, the unsupported ^{226}Ra is given by

$$\left(\frac{dN_{226\text{Ra}}}{dt}\right)_u = \left(\frac{dN^o_{226\text{Ra}}}{dt}\right)_u e^{-\lambda t} \qquad [10.43]$$

where the superscript o indicates the value at the time the steady state was disturbed. Note the exponential decay as a function of time of the unsupported ^{226}Ra. Consider a system in which an initial value of ^{226}Ra is equal to the steady-state value produced by ^{230}Th decay. If the system then excludes ^{230}Th, measuring the present value of ^{226}Ra concentration produced by unsupported decay allows the time since loss of ^{230}Th event to be determined from Equation 10.43. With ^{230}Th in seawater excluded from deposited sediments, the deposition rate of slowly accumulating sediments has been determined by measuring ^{226}Ra as a function of depth in a sediment core.

Lead-210 Dating

^{210}Pb with its 22.3-year half-life has been used to determine sedimentation rates of recent rapidly accumulating sediment in a similar way as with radium-226 dating. **Figure 10-13** gives ^{210}Pb and ^{226}Ra measurements in a mud sediment core. The ^{226}Ra gives the ^{210}Pb supported by the ^{238}U decay chain. The unsupported product of ^{210}Pb from ^{226}Ra in the atmosphere is absorbed on clay particles. The unsupported ^{210}Pb is given by

$$\left(\frac{dN_{210\text{Pb}}}{dt}\right)_u = \left(\frac{dN^o_{210\text{Pb}}}{dt}\right)_u e^{-\lambda t} \qquad [10.44]$$

This curve is fit to the profile to obtain an age with depth as shown on the right side of Figure 10-13, assuming a constant sedimentation rate.

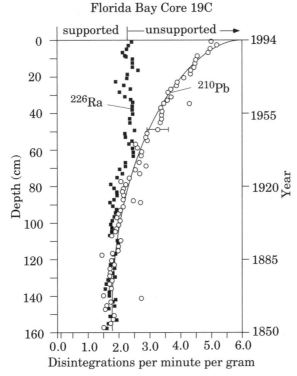

Florida Bay Core 19C

FIGURE 10-13 ^{210}Pb and ^{226}Ra in a core with depth from a mudbank in Florida Bay. The line gives the fit of Equation 10.44 to the data (U.S.G.S. FS-073-98).

Another uranium decay series dating technique involves the measurement of ^{234}U concentrations. It has been used to obtain the age of *Quaternary* corals to clarify the chronology of the period. ^{234}U determinations have also been used to date changes in the natural rate of ^{14}C production in the atmosphere. This has allowed better calibration of ^{14}C ages.

■ Potassium-Argon Systematics

Age Dating

^{40}K makes up $0.01167 \pm 0.00004\%$ of all K isotopes. It undergoes two principal types of decay. One is β^- decay (88.8%) with the emission of an electron, antineutrino, and gamma ray from its nucleus by the reaction

$$^{40}K^+ \rightarrow {}^{40}Ca^{2+} + \beta^- + \text{antineutrino} + \text{gamma ray} + \Delta E \qquad [10.45]$$

This is generally not used for dating because ^{40}Ca is the common Ca isotope, and thus, the amount produced relative to the background level is difficult to quantify. ^{40}K also undergoes decay by electron capture to produce a neutron from a proton (11.2%) according to the reaction

$$^{40}K^+ + \beta^- \rightarrow {}^{40}Ar + \text{neutrino} + \text{gamma ray} + \Delta E \qquad [10.46]$$

This decay reaction is used for K-Ar dating (Dalrymple and Lanphere, 1969). The amount of ^{40}Ar present in a sample as a function of time is given by

$$^{40}Ar = {}^{40}Ar_o + \frac{\lambda_a}{\lambda} {}^{40}K(e^{\lambda t} - 1) \qquad [10.47]$$

where $^{40}Ar_o$ is the initial ^{40}Ar in the sample at time 0, whereas λ denotes the total decay constant of ^{40}K ($\lambda = 5.543 \times 10^{-10}$ yr^{-1}). λ_a denotes the constant for decay of ^{40}K to ^{40}Ar ($\lambda_a = 0.581 \times 10^{-10}$ yr^{-1}).

The method allows dating between hundreds of thousands of years and the age of the earth, depending on the K concentration in the sample. With the diffusional escape of the noble gas Ar on heating of a sample, no original ^{40}Ar is present for dating an event that heated the rock above the Ar closure temperature. This makes $^{40}K/^{40}Ar$ for a sample high so that even though the decay has a long half-life (1.25 b.y.) young ages are precisely determined. With $\lambda_a/\lambda = 0.105$ and no initial $^{40}Ar_o$, Equation 10.47 becomes

$$^{40}Ar = 0.105 \, {}^{40}K \, (e^{\lambda t} - 1) \qquad [10.48]$$

Only measurements of ^{40}Ar and ^{40}K in a sample are, therefore, needed to determine the time since it became a closed system. All of the Ar is released by fusing the sample in a vacuum, and the concentration of ^{40}Ar is determined with a mass spectrometer. The amount of ^{40}K in the sample is determined by measuring the concentration of K in the sample by a standard laboratory method such as *atomic absorption* or *x-ray fluorescence* analysis. The ^{40}K is then $0.01167 \pm 0.00004\%$ of total K. The amount of any ^{40}Ar present because of air trapped in the sample is determined by measuring nonradiogenic ^{36}Ar in the sample and subtracting ^{40}Ar according to the ratio of $^{40}Ar/^{36}Ar = 295.5$ found in air. Care must be taken in the analysis because of possible partial loss of ^{40}Ar. The Ar produced from decay increases the Ar concentration in the mineral. A gradient then exists in the concentration of Ar in the mineral relative to its surface. Because the Ar is not bound in the mineral structure, it can diffuse down the concentration gradient to the surface, resulting in loss of Ar. This makes resetting and partial resetting of K-Ar systematics relatively common.

^{40}Ar/^{39}Ar Dating

When partial loss of Ar is suspected, the $^{40}Ar/^{39}Ar$ method is used (McDougall and Harrison, 1988). ^{39}K in a sample is converted to ^{39}Ar by irradiating it in a nuclear reactor. The efficiency of the irradiation to produce ^{39}Ar is determined by irradiating a sample standard of known K concentration and age and, therefore, known ^{39}K concentration. Corrections are needed for reactor produced Ar isotopes from Ca, K, and Cl in the sample. From the measured concentration of ^{39}Ar and the efficiency of irradiation, the concentration of ^{39}K and finally ^{40}K in a sample is determined. By knowing ^{40}K and ^{40}Ar*, the radiogenically produced Ar, an age is obtained. The relationship is

$$t = \frac{1}{\lambda} \ln \left(\frac{^{40}\text{Ar}^*}{^{39}\text{Ar}} J + 1 \right)$$
[10.49]

where J is given by

$$J = \frac{e^{\lambda t_m} - 1}{(^{40}\text{Ar}^* / {}^{39}\text{Ar})_m}$$
[10.50]

In Equation 10.50, t_m and $(^{40}\text{Ar}^*/{}^{39}\text{Ar})_m$ are the known age of the flux monitor and its radiogenic Ar 40:39 ratio, respectively. Because only ratios of radiogenic ^{40}Ar*/^{39}Ar determined in a mass spectrometer in an irradiated sample and monitor are required to obtain an age, this technique is helpful when only a small sample size is available, such as with extraterrestrial rocks. Also, a series of age measurements from a single sample that is step heated to release the Ar from different crystallographic locations can be made. If the sample has remained a closed system since its formation, the dates determined at each step should be the same. If dates are different as a function of temperature, inferences are made on the nature of a later heating event during the geologic history of the sample. Typically, Ar obtained at lower step heated temperatures gives younger ages, indicating a loss of radiogenic Ar. A laser can be used to release Ar from a spot on the irradiated sample. By comparing ages as a function of distance in a mineral from laser released Ar, diffusional effects are evaluated.

■ Artificial Isotopes

Because of atmospheric testing of thermonuclear explosions in the 1950s and 1960s, a number of artificial radioactive isotopes were added to the atmosphere. One of these was ^{137}Cs with a half-life of 30.3 years. ^{137}Cs was rapidly incorporated into sediments by rainfall. For sediments deposited in the 1950s and 1960s, the fallout from particular tests is identified from the known production by profile matching. ^{137}Cs profiles have been used to date mine waste in lacustrine sediments. **Figure 10-14** gives the ^{137}Cs profile from Biscayne Bay mud near Miami, Florida.

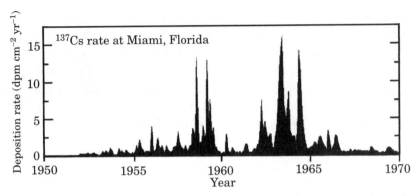

FIGURE 10-14 Deposition rate of ^{137}Cs at Miami, Florida at the indicated time as measured from sediment cores. dpm denotes ^{137}Cs concentration in disintegrations per minute. (Adapted from Holmes, 1998.)

■ Cosmogenic Isotopes

Cosmic rays from the sun contain high energy protons and other particles that collide with the nuclei of elements in the earth's atmosphere and with minerals at the earth's surface. These reactions can produce radioactive isotopes or release particles from the nuclei that collide with other nuclei to produce radioactive isotopes. Of importance for geologic analysis are ^{3}H, ^{7}Be, ^{10}Be, ^{14}C, ^{26}Al, and ^{36}Cl.

Carbon-14 Dating

A flux of neutrons, ^{1}n, is produced from cosmic ray proton bombardment of the nuclei of atmospheric gases. ^{14}N can capture a neutron, releasing a proton, ^{1}H, and heat energy according to the reaction

$$^1n + {}^{14}N \rightarrow {}^{14}C + {}^1H + \Delta E \tag{10.51}$$

where ΔE is nonmass energy released of 1.01×10^{-19} J per atom. ^{14}C rapidly reacts with the O_2 in the atmosphere to produce CO_2. The CO_2 is dispersed evenly throughout the atmosphere because of the rapid mixing of air masses. The content of ^{14}C in CO_2 is, therefore, similar throughout the atmosphere. Its production rate has been fairly constant over the last 70,000 years except during the 1950s and 1960s when above-ground nuclear tests produced a spike in concentration by doubling its production rate.

Once produced, ^{14}C undergoes spontaneous decay by the conversion of a neutron to a proton with the loss of an electron according to the reaction

$$^{14}C \rightarrow {}^{14}N + \beta^- + \text{antineutrino} + \Delta E \qquad \text{half-life} = 5730 \text{ years} \tag{10.52}$$

The concentration of ^{14}C is, therefore, nearly in steady state in the atmosphere with the amount produced by cosmic ray–induced neutrons balanced by its spontaneous decay.

Plants incorporate a $^{14}C/^{12}C$ ratio while alive, which depends on the atmospheric $^{14}C/^{12}C$, by continued photosynthetic fixation of C from atmospheric CO_2. The ratio incorporated in a plant is slightly dependent on changes in amount of sunlight received and water stress it experiences; however, $^{14}C/^{12}C$ in plants, in general, reflects the $^{14}C/^{12}C$ in the atmosphere at the time the C was fixed because these environmental factors generally are second order. At the plant's death, the source of ^{14}C is no longer available, and the concentration of ^{14}C then decreases with time. This implies the concentration of ^{14}C is a measure of time since death of the plant. The method differs from other isotope methods in that the radioactivity is generally measured rather than ratio of the concentration of parent to daughter. In this way, the ^{14}N produced is not needed. As a gas, ^{14}N is easily lost and is contaminated with the large concentration of ^{14}N in air.

The rate equation for carbon-14 dating is

$$^{14}C_{mes} = {}^{14}C_{eq}\, e^{-\lambda_{14C}t} \tag{10.53}$$

where $\lambda_{14C} = 1.209 \times 10^{-4}$ yr^{-1}, $^{14}C_{eq}$ is the concentration of ^{14}C in equilibrium with the atmosphere, and $^{14}C_{mes}$ is the amount of ^{14}C measured after isolation time t. Solving for t and converting to a base-10 logarithm gives

$$t = 19{,}035 \log_{10} \left({}^{14}C_{eq}/{}^{14}C_{mes} \right) \tag{10.54}$$

Because of the rapid decay of ^{14}C, even using accurate measurements, the technique is limited to time periods no greater than 12 half-lives or about 70,000 years. Recently, for very precise measurements or for very small samples, *accelerator mass spectrometry* has been used to measure $^{14}C/^{12}C$ ratios and extend the measurement's precision limits. Accelerator mass spectrometry has also been used to obtain precise Be and Cl isotope ratios needed for dating.

Carbon-14 dating is used extensively in archeology to date skeletal remains and campfires as well as volcanic eruptions, volcanic ash layers, and glacial events. Carbon-14 has also been used to obtain the age of waters to determine rates of deep ocean circulation and groundwater transport. The presence or absence of a spike in ^{14}C produced by the above-ground atomic bomb testing is often a helpful marker. When considering groundwaters, account must be taken of carbon from the dissolution of carbonates because carbon from carbonates contains no ^{14}C, and therefore, groundwaters may appear much older than they really are.

Other Cosmogenic Isotope Dating Techniques

^{10}Be is a radioactive isotope of beryllium created in the atmosphere by cosmic ray *spallation* reactions of N and O nuclei. Spallation occurs when an atomic nucleus splits into three or more fragments due to collision with cosmic rays. The production of ^{10}Be depends on latitude because of the latitudinal dependence of the cosmic ray flux. Unlike ^{14}C, which is incorporated into a gas that remains in the atmosphere and therefore becomes well mixed, ^{10}Be concentration in the atmosphere is not well mixed because it is absorbed on dust particles and is rapidly removed, mainly by raindrops; therefore, its flux to the earth's surface depends on location. With a half-life of 1.5 m.y., it is useful in dating slowly accumulating sediments. Typically, a profile of ^{10}Be in sediments is obtained, and under the assumption of a constant sedimentation rate, the profile is fit to obtain a time and average ^{10}Be flux. ^{10}Be has been observed in recently erupted lavas from volcanic arcs and is believed to reflect ^{10}Be from subducted marine sediments (Tera et al., 1986).

Another isotope of beryllium, ^{7}Be, is also radioactive and formed by cosmic ray spallation of N and O nuclei. It has similar chemical properties to ^{10}Be but has a half-life of only 53 days. It has been used in environmental studies in which dating of sediments in time frames of less than a year are required.

The cosmogenic isotope, ^{36}Cl, is also of interest for dating. ^{36}Cl is produced by cosmic ray spallation of ^{39}K and ^{40}Ca in rocks at the earth's surface to a depth of about 1 m. It is radioactive and decays with a half-life of $3.01 \pm 0.04 \times 10^5$ years. This decay is used to date surface exposure ages of rocks and meteorites. ^{36}Cl is also produced in the atmosphere by cosmic ray spallation of ^{40}Ar and ^{36}Ar and, like ^{10}Be, is incorporated into falling precipitation. Its flux also increased from above-ground nuclear testing in the 1950s and 1960s. If this peak is identified, it is used as a marker. Its flux at a particular location is dependent on latitude, distance from the coast, winds, and *orographic* effects. When averaged over sufficient time periods, the atmospheric flux is nearly constant. Weathering of exposed rocks releases ^{36}Cl to surface water.

Surface waters, therefore, have a flux of ^{36}Cl both from meteoric and weathering sources. Once in groundwater, a small flux of ^{36}Cl is produced from ^{35}Cl by the capture of neutrons from the decay of U and Th. This production is difficult to quantify and constitutes a background level of ^{36}Cl. If the ^{36}Cl concentration or more commonly the ^{36}Cl/Cl of recharge waters to an aquifer is determined at a particular location and ^{36}Cl/Cl in discharge measured, the residence time of the water in the aquifer can be determined. Waters as old as 1 m.y. have been dated by the ^{36}Cl technique (Torgersen et al., 1991).

Tritium, ^{3}H, dating is based on the formation of tritium in the stratosphere where fast neutrons cause fission of ^{14}N to produce ^{12}C + ^{3}H. The ^{3}H is oxidized and incorporated into H$_2$O. Because of its upper atmosphere production, its concentration in H_2O varies with time of year, being greater in springtime when stratosphere–troposphere mixing is greatest. It also varies with location, being lower in ocean precipitation. ^3H decays by β^- emission to ^3He. A measure of the concentration of ^3H in water gives the age relative to the time the water precipitated out of the atmosphere. With a half-life of 12.43 years, groundwaters of ages of up to about 70 years are determined. The lack of measurable tritium, therefore, indicates an age of greater than 70 years for groundwater. Because of hydrogen bomb testing, anthropologic input of ^3H to the atmosphere occurred between 1952 and 1963, with a peak in 1963. This makes accurate dating in this time frame more difficult; however, like ^{36}Cl, this "bomb spike," if found in groundwater, provides a time tracer during fluid flow.

■ Fission Track Dating

A useful dating technique has been developed based on the tracks left by particles in minerals and glasses produced when isotopes undergo spontaneous fission (Wagner and Van Den Haute, 1992). ^{238}U, ^{235}U, and ^{232}Th decay primarily by emitting an alpha particle with rate constants of 1.551×10^{-10}, 9.849×10^{-10}, and 4.948×10^{-11} yr^{-1}, respectively; however, they also undergo much slower spontaneous fission. Only the natural fission of ^{238}U is significant with a rate constant, λ_f, of $8.46 \pm 0.06 \times 10^{-17}$ yr^{-1}. During ^{238}U fission, two nuclei are released in opposite directions with nearly equal and high enough momentum to produce damage channels and areas of atomic misalignment around the channels in SiO_2-rich crystals or glasses before they come to rest. These fission tracks are made optically visible for a distance of up to about 8 μm in each direction by using an etchant acid on a polished mineral or glass surface. The density and length of tracks are then determined with an optical microscope. The number of tracks produced depends on time and the amount of uranium in the sample.

Because of the slow rate of fission, only minerals that have a significant amount of U, such as apatite, zircon, titanite, mica, and allanite, are typically used for dating. The concentration of ^{238}U is determined by subjecting the sample surface to a neutron flux from an atomic reactor to induce fission. The induced track density (tracks/area) can then be determined. The fission track age, t_{ft}, is given by

$$t_{ft} = \frac{1}{\lambda_\alpha} \ln \left(\lambda_\alpha \frac{\rho_s}{\rho_i} \rho_d \zeta g + 1 \right)$$

[10.55]

where λ_α is the alpha decay constant for ^{238}U, ρ_s denotes the spontaneous track density, ρ_i gives the induced track density, and ρ_d represents the *dosimeter* track

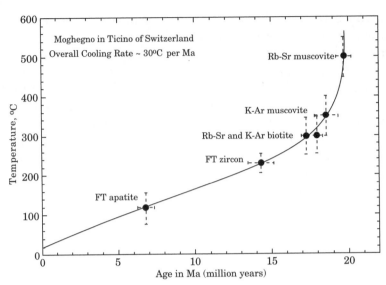

FIGURE 10-15 Temperature versus age of cooling from Bosco Gurin in the Lepontine Alps determined from Rb-Sr, K-Ar, and fission track (FT) measurements. (Adapted from Hurford, 1986.)

density. The dosimeter tracks are those produced in a glass with known U concentration that is irradiated along with the sample. The *g* specifies the geometry factor (which accounts for the fact that the sample's fission tracks come from twice the volume of the induced surface tracks) and ζ, the zeta factor, is a constant that includes the fission decay constant and neutron capture cross-section. ζ is determined by calculation from Equation 10.55 with a sample of known age. Typically, 30–50 calibrations with a sample of known age are made for a particular reactor flux and dosimeter to determine ζ (Gallagher et al., 1998).

After they are produced, fission tracks undergo significant annealing even at low temperatures if long enough time frames are considered. They do so by shortening their lengths. The rate of shortening is time and temperature dependent and is different for different minerals. It also appears to be somewhat compositionally dependent. The ability to retain tracks with rising temperature increases in the order biotite, apatite, muscovite, allanite, and titanite. Tracks in apatite anneal rapidly above 120°C but slow to nearly an imperceptible rate as temperatures decrease to 60°C. Zircon tracks anneal rapidly above about 350°C and slows to imperceptible rates below about 230°C. Because a surface density of randomly oriented tracks is made, the shortening of tracks during annealing decreases the observed surface density and results in a younger fission track age.

Two types of track densities are typically determined: surface and confined. Both track types are made visible by etching. Surface tracks are those that intersect the surface. Confined tracks are those entirely below the surface. These tracks are made visible by some imperfection in the sample, generally created during polishing, allowing etchant to reach the track. The horizontal tracks (up to approximately 15° of horizontal) tend to be simultaneously in focus along their entire length. The distribution of lengths of these horizontal tracks is measured to ±0.2 μm. Using the distribution of horizontal confined track lengths among samples from different localities, thermal uplift histories in sedimentary basins are determined from changes in track lengths. **Figure 10-15** shows ages versus blocking temperatures measured by various techniques for a sample from the Lepontine Alps. These then give the cooling history of the area from peak metamorphic conditions as it was uplifted during denudation.

Summary

The recognition that a number of naturally occurring isotopes are radioactive and undergo spontaneous time-dependent decay has had a profound effect on our understanding of the earth, allowing rates and fluxes of geologic processes to be determined. These geochronometers include the decay of isotopes ^{40}K, ^{87}Rb, ^{235}U, ^{238}U, ^{232}Th, and ^{147}Sm. Additionally, shorter lived radioactive isotopes produced in the chain of decay products of ^{238}U and ^{235}U to their final stable isotopes, ^{206}Pb and ^{207}Pb, respectively, are used to determine fluxes in sediments and water masses. These include ^{226}Ra and ^{210}Pb with half-lives of 1622 and 22.3 years, respectively. Profiles of these isotopes in sediments allow rates of deposition to be determined.

Cosmogenically produced radionuclides, which include ^{3}H, ^{7}Be, ^{10}Be, ^{14}C, ^{26}Al, and ^{36}Cl, have been measured to obtain dates and fluxes of earth surface processes occurring in the last million years. Tracks in minerals and glasses produced from the spontaneous fission of ^{238}U are used to determine uplift rates in mountain building processes. Other dating techniques of earth material not mentioned above have been and are being developed based on the half-lives and chemical behavior of the radioactive isotopes given in Table 10-1 as well as their daughter isotopes.

Radioactive isotopes and their daughters are used to discern where material originated. In particular, whether magmas originated in the crust or mantle is determined from Rb/Sr and Sm/Nd isotopic systematics. This analysis is predicated on the fact that although elements are fractionated and therefore change their concentration because of chemical reactions isotope ratios of the same element do not. As discussed in Chapter 11, when the behavior of stable isotopes is considered, this is not strictly true, and under some conditions, isotopes are fractionated during a reaction to a very limited extent. Ratios of some isotopes, therefore, can change slightly in geologic processes.

Key Terms Introduced

accelerator mass spectrometry
alpha decay
antineutrino
antiparticle
atomic absorption
atomic (mass) number
atomic weight
baryon number
beta decay
chondritic meteorites
closure temperature
concordia
coulombic
daltons (Da)
decay time equation
discordia
dosimeter
epsilon value (ε_{Nd})
fractionation
gamma rays

half-life
isochron
isotope
mantle array
mid-oceanic ridge basalts (MORB)
neutrino
nuclear fission
nuclide
orographic
Paleozoic
quantum mechanical tunneling
Quaternary
radioactive isotope
spallation
stable isotope
strangeness
supported concentration
thermochronology
x-ray fluorescence

Questions

1. What is an isotope?
2. Define half-life.
3. What are the units of a radioactive decay constant? How is its value determined?
4. How does an alpha particle differ from a helium atom?
5. Does a proton or a neutron have a greater mass?
6. What are the different processes by which an isotope can undergo spontaneous decay?
7. What is a neutrino?
8. How do particles and antiparticles differ?
9. What does a closure temperature give?
10. Explain how measurements of $^{87}Sr/^{86}Sr$ are used to determine whether a rock was formed in the mantle as opposed to the crust.
11. What kind of rocks are generally dated with samarium decay to neodymium?
12. What does a CHUR value of +3 indicate?
13. What is the correlation shown in the mantle array for $^{143}Nd/^{144}Nd$ versus $^{87}Sr/^{86}Sr$?
14. What is a Pb concordia diagram?
15. In decay series dating, what is the "supported concentration"?

Problems

1. How many protons, electrons, and neutrons are present in $^{87}Sr^{2+}$, ^{14}C, and $^{37}Cl^-$?

2. Is the following reaction endothermic or exothermic? Why?

   proton + electron $\rightarrow$ neutron

3. Name the modes of spontaneous natural radioactive isotope decay found in minerals. Describe how the atomic number and atomic mass number change from parent to daughter in each.

4. Calculate the mass change for the following nuclear reaction involving hydrogen colliding with lithium (Li) to produce two helium (He) atoms

   $$^{1}H + {}^{7}Li \rightarrow 2\,{}^{4}He$$

   where the rest mass of the atoms is as follows: $^{1}H = 1.007825$, $^{7}Li = 7.016004$, $^{4}He = 4.002603$ Da (1 Da = 931.6 MeV [million electron volts] and 1 MeV = 1.602×10^{-13} J). Compare the molar energy released by the reaction to the energy available by drinking one 250-ml serving of fruit juice (180 kJ per 100 ml).

5. The $^{14}C/^{12}C$ of the Dead Sea Scrolls measured in 2008 was 0.784 of what are found in living plants today. What is its age?

6. Calculate the binding energies (in MeV) of the isotopes from their masses in **Table 10-4**, where

   $$\Delta E_{binding}(in\ MeV) = \frac{931.6\ \Delta m\ (in\ Da)}{A}$$

| Table 10-4

Element	Z	A	Atomic weight (Da)
He	2	4	4.0026
B	5	8	8.0246
N	7	14	14.0032
O	8	16	15.9949
Ne	10	20	19.9924
Al	13	27	26.9815
Fe	26	56	55.9349
Sr	38	86	85.9092
Ag	47	107	106.9050
Au	79	197	196.9665
Pb	82	206	205.9744
U	92	235	235.0439

where A is the atomic mass number (neutrons + protons) and Δm represents the difference in mass between the atomic weight of the element and the mass of a hydrogen (1.007825 Da) times Z plus the mass of a neutron (1.008665) times N. Z denotes the atomic number (number of protons) and N the neutron number of the isotope. Plot $\Delta E_{binding}$ against A. What isotope has the lowest binding energy? Use the graph to explain why energy is released if low atomic mass isotopes are combined (fusion) and high atomic number isotopes are split (fission).

7. **Table 10-5** shows the isotope measurements from minerals in the Godthaab area of Greenland. Plot an Rb/Sr isocron diagram with these values. Determine the age of the Godthaab complex and the initial ratio of $^{87}Sr/^{86}Sr$.

Table 10-5

Mineral	$^{87}Sr/^{86}Sr$	$^{87}Rb/^{86}Sr$
1	0.715	0.20
2	0.718	0.25
3	0.746	0.85
4	0.760	1.10
5	0.805	1.95
6	0.826	2.60
7	0.880	3.32
8	0.908	3.85

8. The measurements in **Table 10-6** were obtained from minerals in the Lizard complex in Cornwall, Great Britain (Cook et al., 2000). The

Table 10-6

Mineral	$^{143}Nd/^{144}Nd$	$^{147}Sm/^{144}Nd$
LZ01	0.51300	0.175
2555(a)	0.51310	0.198
2555(b)	0.51310	0.205
2559	0.51305	0.250
2560	0.51330	0.260
2553	0.51385	0.431
2553 cpx	0.51400	0.490
2554 sp	0.51425	0.729

complex is thought to be a piece of old ocean floor. Plot an Sm/Nd isocron diagram with these values. Determine the age of the Lizard complex and the initial ratio of $^{143}Nd/^{144}Nd$. Indicate which mineral is not in equilibrium with the others. All oceanic rocks are younger than a few hundred million years. What do the data imply?

9. Ion microprobe U-Pb measurements on a fossil tooth gave the values in **Table 10-7** (Sano and Terada, 1999). What is the age of this tooth?

Table 10-7

Sample	$^{206}Pb/^{204}Pb$	$^{238}U/^{206}Pb$
1.1	36.7	8.82
2.1	26.4	5.95
7.1	57.8	13.99
12.1	78.8	17.30

10. The following values (**Table 10-8**) were measured in a glass from tephra at Pianico-Sellere in the southern Alps of Italy (Pinti et al., 2001). What are the number of moles of ^{40}K and $^{40}Ar^*$ and the age of this tephra? (Atomic weight of $^{40}K = 39.964$ and ^{40}K is 0.01167% of the total K.)

Table 10-8

Sample	Weight (g)	K (wt %)	Total ^{40}Ar (10^{13} atoms/g)	$^{40}Ar^*$ (%)
PNCO-V1	0.60373	4.806	2.47	15.8

$^{40}Ar^*$, radiogenic Ar.

11. An age was obtained from a basalt from the Devil's Causeway, Colorado (Kunk et al., 2001) by the $^{40}Ar/^{39}Ar$ technique from the measurements in **Table 10-9**. $^{40}Ar^*$ stands for radiogenic Ar. $^{39}Ar_K$ is the Ar produced from radiation of ^{40}K, and J is defined in Equation 10.50. What is the age of the Devil's Causeway?

Table 10-9

Sample	$^{40}Ar^*/^{39}Ar_K$	J
3BBS13-9-99	2.443	0.005021

12. **Table 10-10** shows the U-Pb isotopic measurements of a zircon in the Littlewood rhyolite from Coats Land, Antarctica (Gose et al., 1997). What are the ^{238}U to ^{206}Pb, ^{235}U to ^{207}Pb, and $^{207}Pb/^{206}Pb$ ages of this sample?

Table 10-10

Sample	Weight (mg)	U (ppn)	Pb rad (ppm)	Total common Pb (pg)					
Z1	0.062	154	29.5	6	18712	0.1147	0.18727	1.9800	0.07668

Ratios are corrected atomic ratios.

13. ^{14}C in a piece of charcoal from an archaeological site has 60% of that found in a living tree. What is the age of the site?

14. Well-preserved collagen was extracted from bones of wooly mammoths (Mammuthus primigenius Blum.) from the permafrost area of Siberia. The ^{14}C activity of the youngest specimens gave 4.5 disintegrations per minute per g C. If this youngest date is the age when wooly mammoth became extinct, when did this happen (^{14}C activity of living tissue gives 15.3 disintegrations per minute per g C)?

15. The isotope ^{90}Sr with a half-life of 28.78 years is produced in nuclear explosions. Its chemical behavior is similar to Ca and is incorporated into growing bones. Suppose a 3-year-old child drank milk from cows that had incorporated ^{90}Sr during feeding. If 1.5 µg was absorbed through the milk by the child, how much would remain when the child was 21?

References

Allègre, C. J., Lewin, E. and Dupré, B., 1988, A coherent crust-mantle model for the uranium-thorium-lead isotopic system. *Chem. Geol.*, v. 70, pp. 211–234.

Berger, G. W. and York, D., 1981, Geothermometry from $^{40}Ar/^{39}Ar$ dating experiments. *Geochim. Cosmochim. Acta*, v. 45, pp. 795–811.

Cook, C. A., Holdsworth, R. E., Styles, M. T. and Pearce, J. A., 2000, Preemplacement structural history recorded by mantle peridotites: An example from the Lizard Complex, SW England. *J. Geol. Soc. Lond.*, v. 157, pp. 1049–1064.

Dalrymple, G. B. and Lanphere, M. A., 1969, *Potassium Argon Dating*. W. H. Freeman, San Francisco, 257 pp.

DePaolo, D. J. and Wasserburg, G. J., 1979, Sm-Nd age of the Stillwater complex and the mantle evolution curve for neodymium. *Geochim. Cosmochim. Acta*, v. 43, pp. 999–1008.

Gallagher, K., Brown, R. and Johnson, C., 1998, Fission track analysis and its application to geological problems. *Annu. Rev. Earth Planet. Sci.*, v. 26, pp. 519–572.

Gose, W. A., Helper, M. A., Connelly, J. N., Hutson, F. E. and Dalziel, I. W. D., 1997, Paleomagnetic data and U-Pb isotopic age determinations from Coats Land, Antarctica: Implications for late Proterozoic plate reconstructions. *J. Geophys. Res.*, v. 102, pp. 7887–7902.

Hart, S. R., Gerlach, D. C. and White, W. M., 1986, A possible new Sr-Nd-Pb mantle array and consequences for mantle mixing. *Geochim. Cosmochim. Acta*, v. 50, pp. 1551–1557.

Holmes, C. W., 1998, Short-lived isotopic chronometers—a means of measuring decadal sedimentary dynamics: FS–073-98. *U.S. Geological Survey Fact Sheet*, 2 pp.

Hurford, A. J., 1986, Cooling and uplift patterns in the Lepontine Alps South Central Switzerland and an age of vertical movement on the Insubric fault line. *Contrib. Mineral. Petrol.*, v. 92, pp. 413–427.

Kunk, M. J., Winick, J. A. and Stanley, J. O., 2001, ^{40}Ar/^{39}Ar age-spectrum and laser fusion data for volcanic rocks in west central Colorado. *Open-File Report 01-472*, U.S. Geological Survey, Denver, CO, 94 pp.

Long, L. E., 1999, Lead: Stable isotopes. In Marshall, C. and Fairbridge, R. (eds.) *Encyclopedia of Geochemistry*. Kluwer, Dordrecht, pp. 363–368.

McCulloch, M. T. and Chappell, B. W., 1982, Nd isotopic characteristics of S- and I-type granites. *Earth Planet. Sci. Lett.*, v. 96, pp. 256–268.

McDougall, I. and Harrison, T. M., 1988, *Geochronology and Thermochronology by the ^{40}Ar/^{39}Ar Method*. Oxford University Press, New York, 212 pp.

Ottonello, G., 1997, *Principles of Geochemistry*. Columbia University Press, New York, 894 pp.

Pinti, D. L., Quidelleur, X., Chiesa, S., Ravazzi, C. and Gillot, P.-Y., 2001, K-Ar dating of an early Middle Pleistocene distal tephra in the interglacial varved succession of Piànico-Sèllere (Southern Alps, Italy). *Earth Planet. Sci. Lett.*, v. 188, pp. 1–7.

Reiners, P. W. and Ehlers, T. A. (eds.), 2005, *Low-Temperature Thermochronology: Techniques, Interpretation, and Applications*. Reviews in Mineralogy & Geochemistry, v. 58, Mineralogical Society America, Chantilly, Virginia, 622 pp.

Sano, Y. and Terada, K., 1999, Direct ion microprobe U-Pb dating of fossil tooth of a Permian shark. *Earth Planet. Sci. Lett.*, v., 174, pp. 75–80.

Tera, F., Brown, L., Morris, J., Sacks, I. S., Klein, J. and Middleton, R., 1986, Sediment incorporation in island-arc magmas: Inferences from ^{10}Be. *Geochim. Cosmochim. Acta*, v. 50, pp. 535–550.

Torgersen, T., Habermehl, M. A., Phillips, F. M., Elmore, D., Kubik, P., Jones, B. G., Hemmick, T. and H. E. Gove, 1991, Chlorine 36 dating of very old groundwater, 3: Further studies in the Great Artesian Basin, Australia. *Water Resour. Res.*, v. 27, pp. 3201–3213.

Veizer, J., 1989, Strontium isotopes in seawater through time. *Annu. Rev. Earth Planet. Sci.*, v. 17, pp. 141–168.

Wagner, G. and Van Den Haute, P., 1992, *Fission-Track Dating*. Kluwer, Dordrecht, 285 pp.

11

Stable Isotope Geochemistry

Stable isotope ratios of low atomic number elements vary in the earth's atmosphere, ocean, surface fresh water, and mantle lithosphere reservoirs both at the present time and as documented in the geologic record (Criss, 1999). Judging from oxygen isotope ratios of zircons, these distinct reservoirs have existed for at least the last 4.3 billion years (Mojzsis et al., 2001). Variations are produced by important fluxes between reservoirs. Active tectonic processes brought on by plate tectonics together with the existence of a large reservoir of H_2O at the earth's surface induced stable isotope heterogeneity of low atomic number elements between high- and low-temperature reservoirs in and on the earth. Basalt injection at mid-oceanic ridges (*MORBs*) fluxes material with mantle lithospheric signatures to the surface reservoirs, whereas surface material formed at low temperatures is fluxed back to the mantle during the subduction of lithospheric plates. Typical time scales of these fluxes are short compared with the age of the earth so that virtually all of the major chemical cycles are near steady state. The boundary conditions on the steady states, however, are changing slowly over geologic time as the earth continues to evolve.

Because isotopes of a given element vary only in their number of neutrons, they have nearly identical electronic structure. The types of bonds they form in any compound, therefore, are identical. This is why, except for changes due to decay of radioactive isotopes, isotopic ratios remain nearly constant for elements during a chemical reaction or phase change. The statement "nearly identical electronic structure," however, implies some small differences exist. Because different isotopes of the same element have a different number of neutrons in their nuclei, they change their ratios to each other slightly during certain reactions. The energy well for bonding with distance between atoms is slightly deeper for the heavier isotope, as shown in **Figure 11-1**. Heavier isotopes, therefore, form slightly stronger bonds.

With vibrational and rotational oscillation about the mean bond length, less potential energy is stored in the lighter isotope's bond. With less energy in

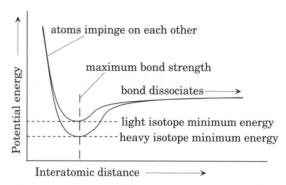

the bond, the lighter isotope bonds are more easily broken. In general, the effect is largest for the isotopes of an element with the greatest difference in bond strengths and, therefore, with the largest relative mass differences. Because there are only a few mass unit differences between the isotopes of a given element, light elements show greater isotopic *fractionation* effects. This is because of the larger relative mass differences between isotopes of light elements compared with isotopes of heavier elements for the same absolute mass difference. Isotopes of elements of low atomic mass, therefore, have significant effects, whereas isotopes of elements of high atomic mass have much smaller, less easily measurable differences.

The fractionation effect is important in covalently bonded species because of the mass-dependent vibrational and rotational modes developed in these bonds when electrons are shared between elements. Ionic and metallic bonds do not share electrons, so they do not develop mass-dependent effects in their bonds and show little fractionation of isotopes. For instance, the isotopes ^{48}Ca and ^{40}Ca, with a mass difference of about 20%, have relatively small fractionation effects because of the ionic nature of Ca bonding in species. ^{13}C and ^{12}C, on the other hand, with a mass difference of less than 10%, have much greater effects because they develop covalent bonds with other elements. The fractionation effects are temperature dependent and get smaller at higher temperatures. This occurs because the relative difference in the depth of the bonding energy wells of isotopes of the same element gets smaller with higher temperatures and the excited modes of vibration and rotation of the bonds become similar. That is, bonding energies become less mass dependent with increased temperature.

Advances in measuring techniques and equipment development for determining isotope ratios of elements continue to be made to increase relative abundance and absolute sensitivity. Of particular importance is counting statistics of the rarer isotope. Much of the advancement has centered on improved ionization techniques. In traditional thermal ionization mass spectrometry, a sample is ionized by placing it on a metal strip heated by an electric current. Instrument discrimination can be improved by using a double-spike technique (Johnson and Beard, 1999). Secondary ionization mass spectrometers use a primary ion beam that is accelerated and focused onto the surface of a sample and sputters material into the gas phase with about 1% coming off as ions. The low yield of the ions limits the method when isotopic ratios are considered. With resonance ionization mass spectrometry, photons are used to ionize the sample. Laser ionization mass spectrometry uses a laser pulse to ablate material from the surface of a solid sample and ionize it.

| Table 11-1 | **Elements Whose Stable Isotope Ratios Are Typically Used To Investigate Earth Processes** |

Element	Isotopic Abundances	Ratio Used	Standard (*Ratio*) Standard
Hydrogen	^1H (99.985%) and ^2H (0.015%) also represented as D and called deuterium. A small amount of ^3H, tritium, also exists but it is radioactive.	$D/^1H$	SMOW (Standard Mean Ocean Water) $(^2D/^1H)_{SMOW} = 1.557 \times 10^{-4}$
Boron	^{10}B (19.8%) and ^{11}B (80.2%)	$^{11}B/^{10}B$	SRM (Searless Lake CA boric acid) $(^{11}B/^{10}B)_{SRM} = 4.04558$
Carbon	^{12}C (98.90%) and ^{13}C (1.10%). A small amount of ^{14}C is also present but it is radioactive.	$^{13}C/^{12}C$	PDB (Peedee formation belemnite, South Carolina) $(^{13}C/^{12}C)_{PDB} = 1.122 \times 10^{-2}$
Nitrogen	^{14}N (99.63%) ^{15}N (0.37%)	$^{15}N/^{14}N$	N_2 in atmosphere $(^{15}N/^{14}N)_{atm} = 3.613 \times 10^{-3}$
Oxygen	^{16}O (99.762%) ^{17}O (0.038%) ^{18}O (0.200%)	$^{18}O/^{16}O$	SMOW (PDB is often used for carbonates) $(^{18}O/^{16}O)_{SMOW} = 2.0052 \times 10^{-3}$ $(^{18}O/^{16}O)_{PDB} = 3.76 \times 10^{-4}$ $\delta^{18}O_{SMOW} = 1.03086\, \delta^{18}O_{PDB} + 30.86\,‰$
Sulfur	^{32}S (95.02%), ^{33}S (0.75%), ^{34}S (4.21%), and ^{36}S (0.02%)	$^{34}S/^{32}S$	Canyon Diablo meteorite (CDM), Meteor Crater, Arizona $(^{34}S/^{32}S)_{CDM} = 4.43 \times 10^{-2}$

By convention, ratios of isotopes are reported as heavy-to-light.

Multicollector inductively coupled plasma mass spectrometers are perhaps the greatest recent advance in technology. An inductively coupled plasma source produces a plasma to gain high ionization efficiency in the sample. A multicollector allows simultaneous collection of isotope production. These advances have opened to analysis more massive elements with much smaller relative mass isotope fractionations (Anbar, 2004; Larson et al., 2003; Welch et al., 2003; Wombacher et al., 2003). Differences in the ratios of isotopes in natural samples can now be measured for many elements, including H, He, Li, B, C, N, O, Si, S, Cl, Ca, Fe, Mg, Pd, Ba, Cr, Se, Ni, Mo, Ti, and Cd. Other elements will undoubtedly be added in the future. The isotopes that have been most extensively analyzed to help understand earth processes are given in **Table 11-1**. Significant differences in ion ratios of these light mass elements can be measured in natural samples using a traditional gas-source mass spectrometer.

■ Types of Isotope Fractionation

Two types of effects, equilibrium and kinetic, lead to stable isotope ratio fractionations. *Equilibrium isotope effects* involve changes that occur when one phase is transformed to another, such as with the evaporation of seawater to H_2O-saturated air or precipitation of H_2O from an H_2O-saturated air mass. In these equilibrium exchanges, the light isotope is fractionated into the phase in which the element is less strongly bound on average. Evaporation/precipitation equilibrium, therefore, leads to a greater concentration of the light isotope in the vapor relative to the liquid. Equilibrium effects also occur when two species mix in the same phase. During this process, exchange occurs between species to bring the isotopic ratio to equilibrium. For instance, if CO_2 containing only ^{16}O is mixed with water containing only ^{18}O, an exchange occurs between the two species to bring them to isotopic equilibrium. Written for the exchange of a single oxygen, this is characterized by the exchange reaction

$$\frac{1}{2}C^{16}O_2 + H_2^{18}O = \frac{1}{2}C^{18}O_2 + H_2^{16}O \qquad [11.1]$$

The equilibrium constant for this reaction can be determined with a standard state taken as the pure single isotopic composition species at the temperature and pressure of interest. Activity coefficient ratios of ^{18}O in CO_2 to ^{18}O in H_2O and ^{16}O in CO_2 to ^{16}O in H_2O are taken as unity. The equilibrium constant of Reaction 11.1, therefore, becomes

$$K_{(11.1)} = \frac{X_{C^{18}O_2}^{1/2} \, X_{H_2^{16}O}}{X_{C^{16}O_2}^{1/2} \, X_{H_2^{18}O}} \qquad [11.2]$$

where X_i indicates the mole fraction of the ith subscripted species. Given the abundance of ^{16}O (99.76%) and its small relative difference in species, the ratio of $X_{H_2}^{16O}$ to $X_{C^{16}O_2}^{1/2}$ is approximated as unity and

$$K_{(11.1)} = \frac{X_{C^{18}O_2}^{1/2}}{X_{H_2^{18}O}^{1/2}} = \frac{\dfrac{n_{^{18}O}^{CO_2}}{n_{^{18}O}^{CO_2} + n_{^{16}O}^{CO_2}}}{\dfrac{n_{^{18}O}^{H_2O}}{n_{^{18}O}^{H_2O} + n_{^{16}O}^{H_2O}}} = \frac{\dfrac{n_{^{18}O}^{CO_2}}{n_{^{16}O}^{CO_2}}}{\dfrac{n_{^{18}O}^{H_2O}}{n_{^{16}O}^{H_2O}}} = \frac{R_{CO_2}}{R_{H_2O}} = \alpha_{H_2O}^{CO_2} \qquad [11.3]$$

where n_j^i designates the number of moles of isotope j in species i and R_{CO_2} denotes the molar ratio of ^{18}O to ^{16}O in CO_2, whereas R_{H_2O} is this ratio in H_2O. Because R_{CO_2} and R_{H_2O} are nearly the same, their ratio, the *fractionation factor* of oxygen isotopes between CO_2 and H_2O, $\alpha_{H_2O}^{CO_2}$, is close to one. The value of $K_{(11.1)} = \alpha_{H_2O}^{CO_2}$ at 25°C and 1 bar is 1.0412 so that $\ln K_{(11.1)}$ equals 0.0404. The standard-state Gibbs energy of this reaction, ΔG_R^o, is

$$\Delta G_R^o = -R\,T \ln K_{(11.1)} = -8.314\,\text{J mol}^{-1}\,\text{K}^{-1} \times 298\,\text{K} \times 0.0404 = -100\,\text{J mol}^{-1} \qquad [11.4]$$

This is the Gibbs energy difference for reacting the standard-state pure endmember isotope phases. The actual difference in isotopic ratios for species in a typical reaction is generally more than 1000 times less because of the similarity of isotopic ratios, that is, concentrations, between species. The Gibbs energy change of a typical natural isotope exchange reaction is, therefore, less than 0.1 J mol^{-1}. This is generally a very small energy difference compared with the ΔG_R for most reactions observed on the earth (see Chapters 3 and 4). Isotopic exchange, therefore, does not significantly affect the elemental equilibrium between species in a reaction. This is why isotope composition is not considered when normal chemical equilibrium between species is considered.

Kinetic isotopic effects due to differences in the rate of diffusion and vibration among isotopes also come into play for isotopic exchange during a chemical reaction. With its greater mass, the heavier isotope has a slower rate of diffusion and vibration. The lighter isotope, therefore, comes into contact with other reactants more often than the heavier isotope and becomes more concentrated in the product species. The effect is important for reactions with significant reaction times. One of these is bacterial reduction of sulfate to sulfide, which is measurably faster for the lighter ^{32}S isotope than ^{34}S.

■ Determining δ Values

Because of the similarity of isotopic ratios in various phases and species, values are typically given in "*delta*" notation. That is, rather than reporting values of the isotopic ratios in a phase or species A as R_A, isotopic ratios are expressed as differences relative to a standard times 1000:

$$\delta^{(\text{heavy})}A = \left[\frac{R_A - R_{\text{standard}}}{R_{\text{standard}}}\right] \times 1000 = \left[\frac{R_A}{R_{\text{standard}}} - 1\right] \times 1000 \qquad [11.5]$$

where $\delta^{(\text{heavy})}A$ denotes the "delta" value of A for the indicated heavy isotope in the sample, whereas R_A specifies the ratio of the heavy isotope to the light isotope in the Ath sample. R_{standard} gives the ratio in a standard. Because δ values are multiplied by 1000, they are expressed in units of "per mil" designated with the symbol ‰. Ratios between two phases, A and B, that are in isotopic equilibrium are related to their fractionation factor, α_B^A, by

$$\alpha_B^A = \frac{R_A}{R_B} = \frac{\delta^{(\text{heavy})}A + 1000}{\delta^{(\text{heavy})}B + 1000} \qquad [11.6]$$

For O and H, the standard for isotopic analysis is typically a sample of ocean water, referred to as *standard mean ocean water* (*SMOW*). Because the original volume of SMOW used as a standard is nearly gone, a new sample of seawater of similar isotopic constituents V-SMOW (Vienna SMOW) is often used. The isotopic ratio of oxygen in a sample is then given by

$$\delta^{18}O(in\ ‰) = \left[\frac{(^{18}O/^{16}O)_{\text{sample}} - (^{18}O/^{16}O)_{\text{SMOW}}}{(^{18}O/^{16}O)_{\text{SMOW}}}\right] \times 1000 \qquad [11.7]$$

where the subscript outside the parentheses indicates the phase considered. For hydrogen isotopes, values are reported as

$$\delta D(in\ ‰) = \left[\frac{(D/H)_{\text{sample}} - (D/H)_{\text{SMOW}}}{(D/H)_{\text{SMOW}}}\right] \times 1000 \qquad [11.8]$$

where D and H denote deuterium and normal hydrogen, respectively.

Positive $\delta^{18}O$ or δD values indicate the samples are enriched in the heavy isotope relative to seawater (SMOW). Most meteoric waters have negative values of $\delta^{18}O$ and δD. The evaporation process fractionates the light isotope into water vapor, creating negative values for H_2O in the atmosphere. Because it is generally well mixed, the deep ocean reservoir has $\delta^{18}O \cong \delta D \cong 0.0‰$ nearly everywhere; however, surface waters may deviate significantly because of local evaporation and dilution processes by rain and river water (Delaygue et al., 2000).

Because O and H isotopes in water participate in the same bonds, the values of δD vary more than $\delta^{18}O$ because the relative mass difference is greater between ^2D and ^1H than between ^{18}O and ^{16}O. Being derived from the oceanic reservoir of a fixed isotopic composition by evaporation and precipitation processes, the values of $\delta^{18}O$ and δD in meteoric water are temperature dependent but related. When δD is plotted against $\delta^{18}O$ for meteoric water derived from the ocean, the values are clustered around the relation

$$\delta D = 8\,\delta^{18}O + 10 \qquad\qquad [11.9]$$

In plots of δD against $\delta^{18}O$, the line produced by Equation 11.9 is referred to as the *global meteoric water line* (GMWL), as shown in **Figure 11-2**.

Most water vapor condenses to produce precipitation under near equilibrium conditions from vapor-saturated air masses. It evaporates under nonequilibrium conditions, however, where the air mass is vapor undersaturated. The nonequilibrium evaporation process produces the intercept value of +10‰ rather than the zero value that would be produced in an equilibrium process and is the reason that seawater does not plot on the line. Variability in evaporation also produces the small deviation of precipitation at a particular location from the GMWL. In some areas, more refined lines have been constructed. For instance, a local meteoric water line for North America has been developed (Yurtsever and Gat, 1981) that is

$$\delta D = 7.95\,\delta^{18}O + 6.03 \qquad\qquad [11.10]$$

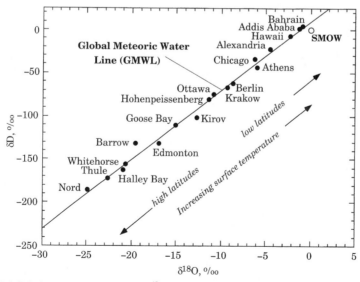

FIGURE 11-2 Relationship between δD and $\delta^{18}O$ for meteoric water that is ultimately derived from ocean water (SMOW). (From data compiled by Rozanski et al., 1993.)

Again, atmospheric water derived from evaporation of the ocean has negative δD and $\delta^{18}O$ values given by the meteoric water line. Because of the Rayleigh fractionation effect (see below), atmospheric water, as it continues to lose precipitation, becomes even more negative along the meteoric water line. At high latitudes (i.e., near the poles), where the amount of water in the atmosphere is lower, greater fractionation effects lowers δD and $\delta^{18}O$ to a larger extent than in the larger amount of water derived near the equator. The isotopic fractionation effect also has some effect as the lower the temperature of evaporation, the greater the lowering of δD and $\delta^{18}O$.

After a mass of water vapor is in the atmosphere, its isotopic composition is modified by precipitation processes. The initial precipitation of atmospheric H_2O fractionates the heavier isotope into the precipitated H_2O, leaving the remaining atmospheric H_2O vapor isotopically lighter. Because this process is the opposite of the evaporation process, both the remaining atmospheric water and the precipitated water still have isotopic ratio values given by the meteoric water line but are on opposite sides of the initial atmospheric water isotopic ratio. The remaining atmospheric H_2O becomes isotopically lighter and lighter as more and more precipitation occurs the farther inland an air mass is from its oceanic source. In **Figure 11-3**, isopleths of δD and $\delta^{18}O$ become more

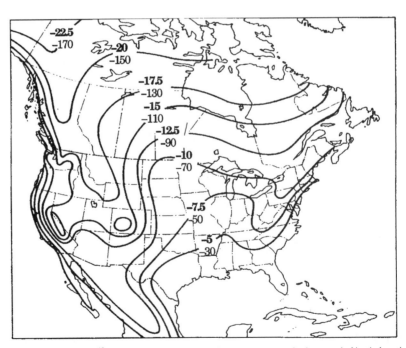

FIGURE 11-3 Isopleths of $\delta^{18}O$ and δD for the average of measurements of rainwater in North America. (Modified after Sheppard et al., 1969.)

negative as atmospheric water from the Pacific Ocean loses mass because of precipitation as it moves eastward. This is also true to some extent for evaporated ocean water moving westward from the East Coast and northward from the Gulf of Mexico. Again, the relative ratio of δD and $\delta^{18}O$ continues to be given by the meteoric water line. The isotopic composition of precipitation, however, becomes lighter because the mass of water in the air mass decreases with distance from the ocean, as shown in **Figure 11-4**.

The change in isotopic ratios during precipitation events is described by *Rayleigh distillation*. The isotopic ratio, R_a, of a mass of H_2O, a, remaining in the atmosphere after precipitation is related to the original isotopic ratio in the atmospheric H_2O mass, R_a^o, by

$$R_a = R_a^o (f_a)^{\alpha-1} \quad \text{or} \quad \delta_a = (\delta_a^o + 1000)\, f_a^{\alpha-1} - 1000 \qquad [11.11]$$

where α specifies the fractionation factor between liquid and vapor at the temperature considered (α_{vapor}^{liquid} for $^{18}O/^{16}O = 1.0092$ at 25°C) and f_a denotes the fraction of atmospheric H_2O remaining in the atmosphere. With the definition of α given in Equation 11.5 combined with Equation 11.10, an expression for the isotopic composition of the precipitated phase, R_p, is

$$R_p = \alpha R_a^o (f_a)^{\alpha-1} \quad \text{or} \quad \delta_p = \alpha(\delta_a^o + 1000)\, f_a^{\alpha-1} - 1000 \qquad [11.12]$$

Shown in **Figure 11-5** are the changes due to Rayleigh distillation for a closed system between reactant phase A and product phase B with $\alpha_A^B = 0.990$ and initial $\delta^{heavy}A = 0.0$. As the reactant disappears, the isotopic ratio in the average product B becomes equal to that in the initial reactant. If the system were open to a continuous supply of new reactant with $\delta^{heavy}A = 0.0$, then $\delta^{heavy}B$ would be fixed at -10, independent of the extent of reaction, and Rayleigh distillation would not occur.

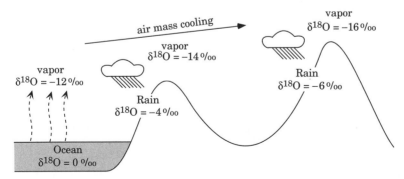

FIGURE 11-4 Schematic diagram showing possible changes in $\delta^{18}O$ for atmospheric water vapor evaporated from the ocean and falling as rain with the cooling of the air mass with increased elevation as it moves further inland.

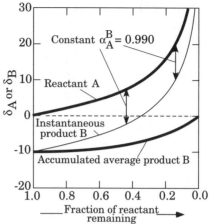

FIGURE 11-5 Change in δ value of the reactant A with initial $\delta_A = 0.0$ and product B as a function of the fraction remaining for Rayleigh distillation. With a decrease in α, the difference between δ of residual reactant A and product B increases.

Table 11-2   **Some Relations Between Variables Used in Isotopic Analysis**

Expression	Isotope ratios	Delta notation
Fractionation factor between 1 and 2	$\alpha_2^1 = \dfrac{R_1}{R_2}$	$\alpha_2^1 = \dfrac{\delta_1 + 1000}{\delta_2 + 1000}$
Change of standard from i to j	$\alpha_{ik} = \alpha_{ij} \cdot \alpha_{jk}$	$\delta_{ik} = \delta_{ik} + \delta_{jk} + \dfrac{\delta_{ik} \cdot \delta_{jk}}{1000}$
Isotope mass balance of system	$R_{sys} = \displaystyle\sum_{i=1}^{n} X_i R_i$	$\delta_{sys} = \displaystyle\sum_{i=1}^{n} X_i \delta_i$
Fractionation between species	$1000 \ln\alpha_2^1 \approx \delta_1 - \delta_2$ $1000(\alpha_2^1 - 1) \approx \delta_1 - \delta_2$	$\Delta_2^1 = \delta_1 - \delta_2$
Equation of line on isothermal δ_2–δ_1 plot	$\delta_1 = \dfrac{R_1}{R_2} \cdot \delta_2 + 1000 \cdot \left(\dfrac{R_1}{R_2} - 1\right);$ $\delta_1 = \alpha_2^1 \cdot \delta_2 + 1000 \cdot \alpha_2^1 - 1;$	$\delta_1 = \delta_2 - \Delta_{12}$
Rayleigh fractionation law	$R_1 = R_1^o \cdot f^{\alpha - 1}$	$\delta_1 = (\delta_1^o + 1000) f^{\alpha - 1} - 1000$
Slope of the GMWL	$\dfrac{\alpha_D - 1}{\alpha_O - 1}$	$\dfrac{\Delta_D}{\Delta_O}$
GMWL intercept: (From Rayleigh law = column 2, Closed system = column 3)	$\delta D^o - \dfrac{\alpha_D - 1}{\alpha_O - 1} \cdot \delta^{18}O^o$	$\delta D^o - \dfrac{\Delta_D}{\Delta_O} \cdot \delta^{18}O^o$

As an example, consider a mass of atmospheric water with $\delta^{18}O = -10‰$ at 25°C. With $R_{SMOW} = 2.0052 \times 10^{-3}$ from Equation 11.6, $R_a^o = 1.98515 \times 10^{-3}$. If 40% of this water precipitates at 25°C using Equation 11.11,

$$R_a = 1.98515 \times 10^{-3}(0.6)^{1.0092} = 1.97584 \times 10^{-3} \qquad [11.13]$$

or the remaining atmospheric water has $\delta^{18}O = -14.6‰$. The isotopic composition of the precipitation $R_p = 1.99402 \times 10^{-3}$ or $\delta^{18}O = -5.6‰$.

The Rayleigh distillation model helps explain much about the changes in delta values that are observed in isotope hydrology. This includes the slope of the GMWL, the latitudinal depletions in heavy H and O isotopes of precipitation, the elevation effect and the changes in delta values of precipitation as a storm proceeds. Some helpful relations used in isotopic analysis are given in **Table 11-2.**

■ Isotope Exchange Between Minerals and Water

Significant deviations for surface water and groundwater from the meteoric water line are generally interpreted as the interaction with the oxygen and hydrogen reservoir in rocks; however, in some instances, surface waters also deviate from the meteoric water line as a result of local evaporation/precipitation processes.

Calcite and aragonite precipitated from the ocean are enriched in $^{18}O/^{16}O$ relative to ocean water. This enrichment is temperature dependent. Cast in terms of temperature, the relationship is fit to an equation (Craig, 1965):

$$T\,(°C) = 16.9 - 4.2(\delta_c - \delta_w) + 0.13(\delta_c - \delta_w)^2 \qquad [11.14]$$

where δ_c specifies the $\delta^{18}O$ of the CO_2 obtained from acid digestion of the carbonate and δ_w denotes $\delta^{18}O$ of seawater. When using Equation 11.14, a change in δ_c of 0.05‰, the accuracy of a typical measurement, results in a change in temperature of about 0.2°C. The $\delta^{18}O$ of seawater is included because the isotopic composition of surface seawater varies somewhat locally when there is a significant addition of river water or where local evaporation and precipitation processes are important.

When using Equation 11.14 to reconstruct temperatures in the geologic past, care must be taken as most preserved oceanic sediments are deposited on continental platforms where river water can alter δ_w of the limited ocean water present in the shallow seas during deposition of the carbonate. Also, the change in δ_w of the ocean because of the ice produced during ice ages must be considered. Continental ice is derived from the precipitation of meteoric water. This incorporates depleted ^{18}O relative to seawater into the ice, making δ_w of the ocean significantly more positive during ice ages. The $\delta^{18}O$ of seawater has varied from about +1.0‰ relative to SMOW at the maximum extent of continental glaciation during the last ice age, to 0‰ at present and would be about −0.7‰ if all the ice on earth were melted back into the ocean.

It is difficult to calculate a precise paleotemperature of the ocean. Aragonite and calcite, because of their somewhat different structures, give slightly different values. Most carbonates are precipitated from supersaturated seawater, and Equation 11.14 assumes an equilibrium distribution of isotopes between carbonate and seawater. Also, certain organisms do not precipitate carbonate in equilibrium with seawater because of vital effects. Vital effects refer to biological characteristics unique to each organism that can influence isotopic fractionation in their shells. This is because organisms are not passive recorders of their biogeochemical environment but complex living things. In addition, there are $\delta^{18}O$ and temperature differences between surface and bottom waters, growth layers produced by different annual temperatures, and seasonal variations in chemistry and temperature.

Figure 11-6 gives measurements of δD in an ice core from Vostok, Antarctica (78.4° S, 106.9° E) as a function of its age. Also shown are measured $\delta^{18}O$ of plankton and estimated subpolar sea surface temperatures determined by Martinson et al. (1987) from tuning within the age uncertainties of measured $\delta^{18}O$ of the radiolarian *Cycladophora davisiana* from the North Pacific with $\delta^{18}O$ of total planktonic and limited benthic $\delta^{18}O$, total $\delta^{13}C$, and percent of $CaCO_3$ present. The temperature record derived from the marine sediments appears to correlate with the ice core record. For instance, the peak of the last glaciation at about 18 thousand years shows up as a strong decrease in δD in the ice record and $\delta^{18}O$ in the planktonic record.

$\delta^{18}O$ of Marine Carbonates

It has been argued that average $\delta^{18}O$ of Phanerozoic marine carbonate rocks decreases with age based on the values reported in **Figure 11-7** (Veizer and Hoefs, 1976). There is a large scatter, however, in the calculated mean that depends on age. For instance, the standard error for the middle Permian is shown as a vertical dashed line in Figure 11-7. If the trend is real, it may be due to exchange of meteoric water after deposition, higher ocean temperatures in the past, or lower $\delta^{18}O$ values of seawater with increasing age. The curve is constructed from different carbonate secreting organisms over time: brachiopods in the Paleozoic, belemnites in the Mesozoic, and benthic forams in the Cenozoic. Recently, some investigators argued that the Paleozoic data are biased by samples from the large abundance of epicontinental seas present during this era. These samples would be prone to mixing of isotopically lighter meteoric water from river input. There is, at present, no general consensus on which process accounts for the observed trend of the data.

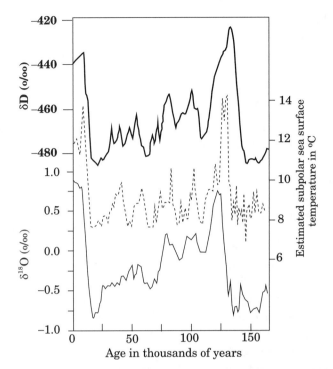

FIGURE 11-6 Variation of δD (bold line) as a function of time in the Vostok ice core compared with $\delta^{18}O$ of plankton (normal line) and estimated summer sea surface temperature (dashed line). (Adapted from Jouzel et al., 1987; and from *Quat. Res.* 27, D. G. Martinson, N. G. Pisias, J. D. Hayes, J. Imbrie, T. C. Moore, Jr. and N. J. Shackleton, Age dating and the orbital theory of the ice ages: development of a high-resolution 0 to 300,000-year chronostratigraphy, pp. 1–29, 1987, with permission from Elsevier.)

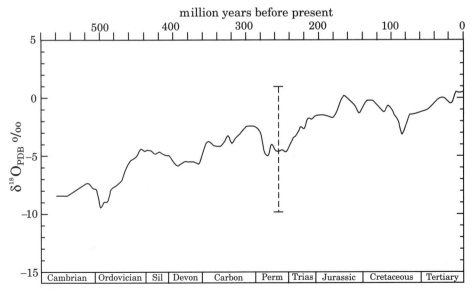

FIGURE 11-7 The running mean of $\delta^{18}O$ relative to the PDB standard of shells of marine organisms as a function of their age. The error bar for the middle Permian gives the 95% confidence limit for this time period. (Adapted from Veizer et al., 1999.)

Just as there is an ice volume effect for modern pelagic carbonates, there should be an epicontinental sea volume effect as well. This effect is small in the open ocean but potentially large in epicontinental seaways where $\delta^{18}O$ of "sea-water" might be as low as -4 to -6. This has large consequences for Equation 11.14.

$\delta^{18}O$ and δD of Groundwaters

Shallow groundwaters tend to be a reasonably good average of local $\delta^{18}O$ and δD of meteoric waters. In some cases, the groundwater $\delta^{18}O$ is enriched somewhat in ^{18}O by evaporation from standing water and wet surfaces produced during a rain event. In other cases, it is somewhat ^{18}O depleted by selective recharge of groundwater during winter storm events; however, large changes in $\delta^{18}O$ of surface waters typically occur only with exchange of oxygen from the rock reservoir. Although shallow groundwaters are out of equilibrium with the rocks they contact, it is not until reaction rates increase with increasing temperature that significant exchange occurs. The increased temperatures and larger mineral surface area to water volume that develops at depths of approximately 100 m and below increase the reaction rate sufficiently such that most meteoric waters exchange their oxygen to a measurable degree with the $\delta^{18}O$-enriched minerals that are present.

In **Figure 11-8**, therefore, water from rivers, lakes, and snow from a particular area plot close to the meteoric water line. Deep groundwaters and those from geothermal systems and hot springs at elevated temperatures from these same

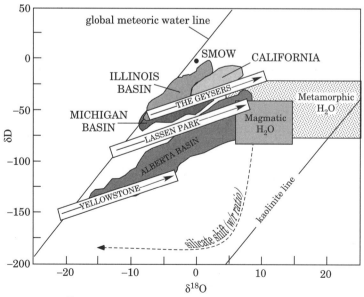

FIGURE 11-8 δD versus $\delta^{18}O$ plot showing departures from the meteoric water line of water from various environments. (Adapted from Gregory, 2002.)

areas plot to the right of the meteoric water line along the trends given by the arrows because of isotopic exchange with rocks. In general, the higher the temperature and, therefore, the deeper the formation water, the greater are its $\delta^{18}O$ and δD along the trends shown. Interaction with rocks in these areas increases the $\delta^{18}O$ of the water dramatically but has less of an effect on the δD values. This occurs because rocks have a large reservoir of oxygen to react with, but there is less hydrogen in minerals to alter the δD values of the water. Figure 11-8 also indicates that these deep groundwaters and geothermal waters do not appear to derive their $\delta^{18}O$ and δD values from magmatic or metamorphic waters. Magmatic or metamorphic waters appear to be in isotopic equilibrium with the local host rock. In fact, δD of unaltered igneous rocks, metamorphic rocks, and marine sediments are reasonably similar, suggesting water sampled by all these rocks has a surface water source.

$\delta^{18}O$ of Rocks

What is the reservoir of oxygen in rocks composed of? Carbonates typically have a $\delta^{18}O$ between 20‰ and 30‰. Carbonates precipitate in seawater at earth surface conditions where isotopic fractionations are large. For instance, with $\alpha_{water}^{calcite} = 1.0288$ at 25°C, the isotopic composition of calcite that would precipitate at equilibrium with ocean water (SMOW) where $\delta^{18}O = 0.0‰$ is calculated from Equation 11.6:

$$1.0288 = \frac{\delta^{18}O_{calcite} + 1000}{0.0 + 1000} \qquad [11.15]$$

This calcite would then have a $\delta^{18}O = 28.8‰$. Given the differences in temperature, aragonite, or dolomite precipitation rather than calcite and the possible changes in the isotopic composition of seawater with time, the given range of $\delta^{18}O$ is not surprising.

Unaltered igneous rocks have a $\delta^{18}O$ between 5.5 and 11‰. The vast bulk of the mantle has values near 5.5‰, and isotopic fractionation during melting is small because of the high temperatures involved. MORB is, therefore, also near 5.5‰. Values higher than 7‰ observed for igneous rocks would suggest a surface component is present. Hydrothermally altered basalt is enriched in ^{18}O with $\delta^{18}O$ equal to approximately 9‰, which is the average value for greenstones. Shales and pelites have a range of $\delta^{18}O$ values commonly between 8‰ and 25‰. Many detrital minerals in these rocks gained their $\delta^{18}O$ at high temperatures and preserve these low values in their new environment. In contrast, minerals precipitated during diagenetic reactions at lower temperature produce the higher $\delta^{18}O$ values.

To develop a better understanding of isotopic exchange between minerals and water, consider the quartz–water exchange reaction

$$\frac{1}{2}Si^{16}O_{2(qtz)} + H_2^{18}O = \frac{1}{2}Si^{18}O_{2(qtz)} + H_2^{16}O \qquad [11.16]$$

The equilibrium constant with a standard state of the pure single isotope phase at the pressure and temperature of interest and assuming activity coefficient ratios are unity is

$$K_{(11,16)} = \frac{X^{1/2}_{Si^{18}O_2}X_{H_2^{16}O}}{X^{1/2}_{Si^{16}O_2}X_{H_2^{18}O}} = \frac{\left[\dfrac{n^{quartz}_{18O}}{n^{quartz}_{16O}}\right]^{1/2}}{\left[\dfrac{n^{H_2O}_{18O}}{n^{H_2O}_{16O}}\right]} \qquad [11.17]$$

The equilibrium constant of an isotope exchange reaction written for the exchange of one atom between species is equal to α^A_B; therefore,

$$K_{(11.16)} = \alpha^{quartz}_{H_2O} \qquad [11.18]$$

As an equilibrium constant, $K_{(11.16)}$ (i.e., $\alpha^{quartz}_{H_2O}$) can possibly be pressure and temperature dependent. From the thermodynamic model, the pressure dependence of ln K at constant temperature is

$$\left(\frac{d\ln K}{dP}\right)_T = \frac{\Delta V^o_R}{RT} \qquad [11.19]$$

where ΔV^o_R is the standard-state volume of reaction. The volumes of pure $Si^{18}O_2$ and $Si^{16}O_2$ and pure $H_2^{18}O$ and $H_2^{16}O$ are virtually identical so that ΔV^o_R is nearly zero. There is, therefore, no significant pressure dependence of $\alpha^{quartz}_{H_2O}$ for isotopic exchange. This is generally true for α^A_B of any two species or phases.

There is, however, a temperature dependence of α^A_B as indicated above that causes α^A_B to decrease as temperature increases, as shown in **Figure 11-9** for mineral–water exchange. Because of the small energy drive for isotopic exchange reactions, investigators use elevated temperatures to increase the rate of reaction and obtain equilibrium. The isotopic exchange equilibria are linear with $1/T^2$ at high temperature but start to curve at near surface temperatures where the dependence becomes proportional to $1/T$. This behavior is due to the fact that the energy differences between isotopes are due to their differences in mass-dependent vibrational frequencies that are themselves temperature dependent. This dependency is calculated from statistical mechanics by considering the reduced *partition function* ratio for each phase in an exchange pair.

Figure 11-10 gives the ln $\alpha^{quartz}_{mineral}$ calculated by Keiffer (1982) from reduced partition function ratios. Values are positive except for calcite at high temperature. This is because quartz has a highly polymerized structure, and for silicates, the more polymerized the silica tetrahedra in the mineral the higher is its isotopic ratio,

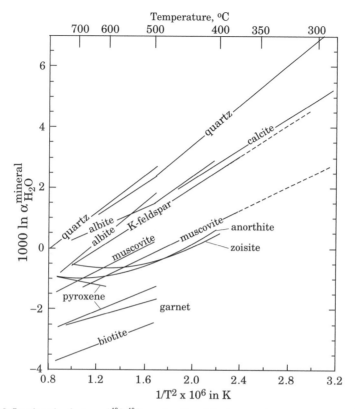

FIGURE 11-9 Fractionation factors of $^{18}O/^{16}O$ as a function of the inverse of temperature squared measured between the indicated mineral and H$_2$O. (Adapted from Kyser, 1987.)

$R_{mineral}$, and therefore the lower will be ln $\alpha_{mineral}^{quartz}$. The fractionation factor between quartz and calcite is small and between quartz and rutile is large because of the greater differences in bond strengths between quartz and the lower bond strengths in rutile than the difference in quartz and the high bond strengths found in calcite. In all cases, however, as infinite temperature is approached, all of the fractionation factors tend toward zero. In theory, the equilibria shown can be used to determine the temperature of last equilibrium between quartz and the indicated mineral as ln $\alpha_{mineral}^{quartz}$ is a function of temperature but not pressure.

Problems exist in using these isotopic fractionations to obtain a temperature of last equilibrium for minerals in a rock. For instance, the isotopic equilibrium between quartz and forsterite can be calculated. They are not, however, in chemical equilibrium for the temperatures shown and therefore do not coexist. If reaction to equilibrium occurs, it will always produce a magnesium silicate with an Si/Mg ratio greater than forsterite. Also, there are problems of achieving isotopic equilibrium solely by exchange reactions because of the small Gibbs energy drive of these reactions.

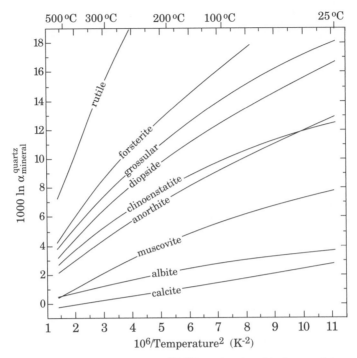

FIGURE 11-10 Calculated fractionation factors for $^{18}O/^{16}O$ as a function of the inverse of the square of temperature between the indicated mineral and quartz. (Adapted from Kieffer, 1982.)

Consider a system containing a fluid that has reached chemical and isotopic equilibrium with a mineral phase. If the fluid is replaced by another fluid of identical elemental composition but with a different isotopic composition, there is little drive for isotopic exchange toward equilibrium of the fluid with the minerals. The Gibbs energy drive was calculated above in Equation 11.4. An isotopic disequilibrium can persist for an extended time until the elemental components of the minerals are dissolved in the fluid due to a chemical reaction between minerals and fluid with a significant Gibbs energy drive.

■ Changes of $\delta^{18}O$ During Open System Mass Transfer

A model was constructed to look at changes in $\delta^{18}O$ between two minerals through a fluid in a rock (Criss et al., 1986; Gregory and Criss, 1986; Gregory et al., 1989). Consider a rock made of equal oxygen moles of feldspar and quartz equilibrated at a high (roughly magmatic) temperature. At these temperatures, $\delta^{18}O_{quartz} - \delta^{18}O_{feldspar} = \Delta^{18}O^{quartz}_{feldspar} \sim 1.0‰$. Reasonable model values of $\delta^{18}O_{quartz}$ and $\delta^{18}O_{feldspar}$ consistent with this value are 10‰ and 9‰, respectively (**Figure 11-11**, point [10, 9]). If the rock cools to a temperature where equilibrium between quartz and feldspar is given by $\Delta^{18}O^{quartz}_{feldspar} = 2‰$, the minerals would now be out of isotopic equilibrium. $\delta^{18}O_{quartz}$ and $\delta^{18}O_{feldspar}$ should now have values that reside on the line labeled $\Delta^{18}O^{quartz}_{feldspar} = 2‰$ on Figure 11-11. If the rock made up of feldspar and quartz reacts to isotopic equilibrium at the lower temperature, the path to this line for this closed system is given by the short arrow from (10, 9) to the line labeled $\Delta^{18}O^{quartz}_{feldspar} = 2‰$.

Instead of a closed system, consider the possibility that the new equilibrium is attained through a meteoric water with $\delta^{18}O_{water} = -6‰$. The rock is considered to have a water filled porosity of 0.1%. Introducing this water into the system starts isotopic exchange of the quartz and feldspar through the fluid to eventual isotopic equilibrium at the lower temperature. With $\delta^{18}O_{water} = -6‰$, equilibrium occurs when $\delta^{18}O_{feldspar} = -2‰$ and $\delta^{18}O_{quartz} = 0‰$ (Figure 11-11, point [0, -2]). The trajectories to equilibrium depend on the relative exchange rates between minerals and fluid, k, and the fluid flow rate, u. These trajectories are given by the solid curved lines with arrows in the figure assuming the feldspar reacts 100 times faster than the quartz. As the ratio of u, the reciprocal of the time needed to exchange the water in the pores, to

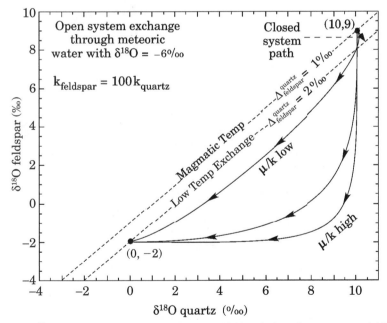

FIGURE 11-11 $\delta^{18}O$ changes for two solid phases (quartz and feldspar) of equal oxygen content initially out of equilibrium which react through a fluid to equilibrium at a lower temperature as a function of time. The fluid fluxes into the system in a porosity of 0.001 with $\delta^{18}O = -6‰$. Quartz and feldspar are present with initial $\delta^{18}O = 10‰$ and 9‰, respectively. The feldspar reacts 100 times faster than the quartz. The solid curves are trajectories to equilibrium with the $\delta^{18}O = -6‰$ fluid of $\delta^{18}O_{quartz}$ and $\delta^{18}O_{feldspar}$ for different fluid flux, μ, to rate constant, k, ratios. (Adapted from Gregory and Criss, 1986.)

the rate constant, $k_{feldspar}$, increases—the lines showing isotopic exchange trajectories depart to a greater extent from the closed system situation.

Measurements of $\delta^{18}O$ values of coexisting quartz and feldspar where they are the dominant minerals in a rock are used with the aid of the model to distinguish closed system from open system conditions. **Figure 11-12** plots $\delta^{18}O_{quartz}$ and $\delta^{18}O_{feldspar}$ pairs measured in granite plutons. Note the disequilibrium fractionations between the feldspar and quartz. Values are incompatible with closed system exchange and indicate the positively sloped disequilibrium arrays that are the result of hydrothermal exchange. As a function of cooling time, the slopes of the arrays become shallower, allowing a length of exchange time to be estimated. Arrays as given in Figure 11-12 occur for nearly all plutonic igneous rocks in the upper 10 km of the crust, indicating reaction with circulating heated meteoric water. At MORBs, seawater, circulating through the hot ocean crust that is present there, has shifted $\delta^{18}O$ of the basalts to values below those of MORBs.

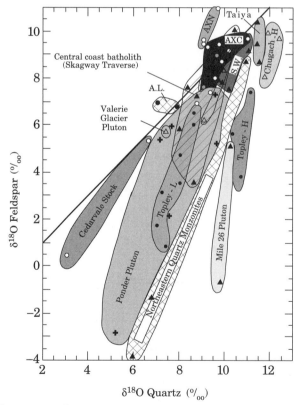

FIGURE 11-12 $\delta^{18}O_{quartz}$ versus $\delta^{18}O_{feldspar}$ plots for the indicated granitic plutons in Cordilleran batholiths. Note the positive steep slope of the arrays characteristic of subsurface hydrothermal exchange. (Adapted from Gregory et al., 1989.)

■ Carbon Isotopes

Carbon fluxes readily between the earth's atmosphere, ocean, and crust. As the basic building block of life, knowledge of carbon's behavior in the environment is a key to understanding interactions in the biosphere. Two stable isotopes of carbon, ^{12}C and ^{13}C, are used to help understand the carbon cycle (see Chapter 15). Values of δ^{13}C are generally reported relative to PDB (Peedee formation belemnite from South Carolina).

The main constituents of plants formed during photosynthesis are carbohydrates. Production of carbohydrates fractionates ^{13}C/^{12}C relative to atmospheric CO_2. Although the details of the reactions that occur are complex, the overall reaction for oxygenic photosynthesis to form carbohydrates is described by the reaction

$$m\text{CO}_2 + n\text{H}_2\text{O} + \overset{\text{solar}}{\underset{\text{energy}}{}} \rightarrow \text{C}_m(\text{H}_2\text{O})_n + m\text{O}_2 \qquad [11.20]$$

Different reaction pathways exist for Reaction 11.20 depending on the type of plant.

The three-carbon, C_3, *Calvin cycle* is used by marine plants and most terrestrial plants. Maize, sugarcane, sorghum, and hot-weather grasses, however, use the four-carbon, C_4, *Hatch-Slack cycle* to produce carbohydrates during photosynthesis. One difference between C_3 and C_4 plants is that C_4 plants do not change their rates of growth with changes in atmospheric CO_2 as readily as C_3 plants. This is because C_4 plants preconcentrate CO_2 in their leaf cells, which allows metabolism to remain high while stomata remain closed to reduce water loss.

A third reaction pathway is *crassulacean acid metabolism* (CAM). It is more similar to C_4 but tends to enrich the plant in deuterium relative to the water used in the reaction. This occurs because the CAM process is used by succulents from water-limited environments. One of the distinguishing characteristics between pathways is that the C_4 and CAM plants produce a smaller carbon isotopic fractionation (δ^{13}C from −9‰ to −16‰ with most values near −13‰ at 25°C) for the incorporation of atmospheric C from CO_2 into carbohydrates than the C_3 pathway (δ^{13}C from −23‰ to −33‰ with most values near −26‰ at 25°C). The fractionation difference has implications for using carbon isotopic systematics for determining paleoenvironmental conditions. Mammalian herbivores produce about a +11‰ fractionation in the δ^{13}C value of their tooth apatite relative to what they eat. Analysis of δ^{13}C in teeth, therefore, is used to help put constraints on their relative diet of C_3 and C_4 plants (Koch et al., 1994).

Figure 11-13 shows the range of values of δ^{13}C for various carbon reservoirs on earth. In Figure 11-13, for organic matter and its products, δ^{13}C is more

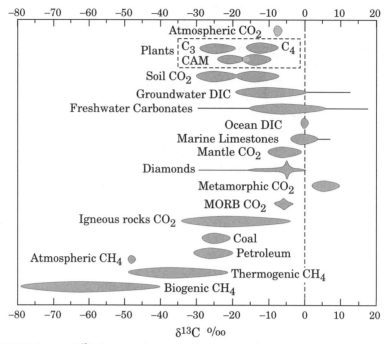

FIGURE 11-13 Range of $\delta^{13}C$ of some carbon reservoirs on the earth relative to the Peedee formation belemnite (PDB) standard. (Adapted from Hoefs, 1996.)

negative than atmospheric CO_2 ($\delta^{13}C \sim -7‰$) and marine carbonates ($\delta^{13}C \sim 0‰$), irrespective of age. Fossil fuels have lower $\delta^{13}C$ than CO_2 in air because plants, which produced the precursors of the fossilized organic carbon compounds, prefer ^{12}C containing species over ^{13}C in photosynthesis.

Temporal Changes in $\delta^{13}C$

$\delta^{13}C$ in the atmosphere and ocean surface waters has declined over the past decades, whereas the concentration of CO_2 has increased (**Figure 11-14**). This fits a fossil fuel CO_2 source and argues against a dominant oceanic CO_2 source for the increased CO_2 in the atmosphere. CO_2 degassed from the ocean has a $\delta^{13}C$ value close to that of atmospheric CO_2 and would not cause such a shift.

Figure 11-15 displays the Tertiary $\delta^{13}C$ record of marine carbonates from Shackleton (1987). Note the general decrease in $\delta^{13}C$ during the last 20 million years. Most researchers believe it reflects an increase in the abundance of calcite-secreting planktonic foraminifera, which adds limestone to the carbonate reservoir. This causes the atmosphere–ocean system to become lighter so that $\delta^{13}C$ of newly deposited ocean sediments decreases. The trend could also be due to a general decrease in sea level toward the present. This would lead

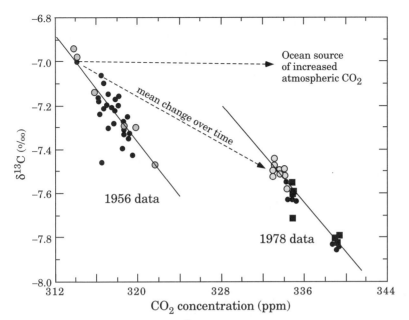

FIGURE 11-14 Change in the relation between $\delta^{13}C$ and CO_2 concentration between 1956 and 1978. The symbols give three different data sets from the western United States. The dashed line is the mean translation of the mixing trend over this time period. (Adapted from Keeling et al., 1979.)

to a less efficient burial of light organic matter. The decay of this organic matter makes the atmosphere–ocean system lighter.

Figure 11-15 also indicates a sharp decrease in $\delta^{13}C$ at the Cretaceous–Tertiary boundary and near the Paleocene–Eocene boundary both followed by a slower increase. The rapid decreases are consistent with a rapid decrease in the organic carbon production and burial rate. If the normal drawdown of

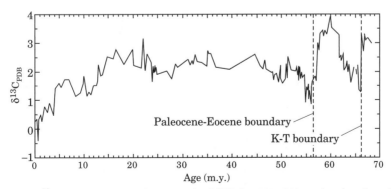

FIGURE 11-15 $\delta^{13}C_{PDB}$ in bulk sediments (~carbonate) at DSDP sites 525 + 528 as a function of age. (Adapted from Shackleton, 1987.)

light carbon in the atmosphere by organic matter suddenly stops, the atmosphere–ocean system will rapidly get lighter, and this is recorded in the carbonate record. The slower subsequent increases in isotopic values reflect the slow return to more normal organic carbon production and burial rates.

To examine the cause of the rapid decrease in $\delta^{13}C$ in more detail, consider the sediments from a *DSDP core* deposited in a time interval that extends across the Cretaceous–Tertiary (K-T) boundary at approximately 65 million years ago (**Figure 11-16**). Note the decrease in the amount of $CaCO_3$ and $\delta^{13}C$, whereas Ir and $\delta^{18}O$ increased. It is generally accepted that the increased Ir content of K-T boundary sediments represents material from the impact of a large meteorite or comet. The $\delta^{18}O$ increase was not likely due to "flash heating" of the atmosphere but rather to greenhouse heating brought on by the death of planktonic foraminifera and calcareous nanoplankton in the surface oceans. The boundary layer clay is similar to silica-rich material above and below the boundary but lacks the carbonate material. Hsü et al. (1982) argued that the observed 3‰ decrease in $\delta^{13}C$ of carbonate occurred after the K-T boundary event, followed by an increase reaching a maximum 40,000 to 50,000 years later. $\delta^{13}C$ returned to more Cretaceous-like values 300,000 to 400,000 years after the maximum. The evidence suggests that at the end of the Cretaceous, the carbonate compensation depth rapidly shallowed so that little carbonate was deposited in the boundary layer clay. The light carbon that would have gone into the organic carbon instead remained in the shallow ocean–atmosphere system, decreasing its $\delta^{13}C$ and leading to a doubling to quadrupling of atmospheric CO_2 because of the decreased photosynthesis.

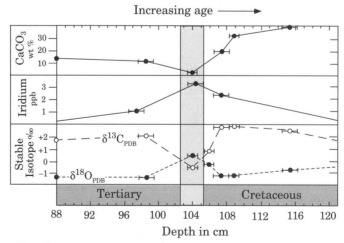

FIGURE 11-16 $\delta^{18}O$, $\delta^{13}C$, Ir, and $CaCO_3$ concentrations across the K-T boundary layer clay (light gray) in the Deep Sea Drilling Project core at site 524 (29°29' S, 3°31' E). (Adapted from Hsü et al., 1982.)

Other causes of the changes in the isotopic record have been proposed. If the Chicxulub cratering of the carbonate shelf on the Yucatan Peninsula was responsible for the K-T event, then the rapid decrease in $\delta^{13}C$ could have been caused by the CO_2 released from any organic material present. As they appear to be of the correct age, another possibility is release of light mantle CO_2 from the Deccan Traps flood basalts of west-central India. Other scenarios are possible, and this is still a matter of active debate.

Also present in the sedimentary record are globally widespread events on the order of 100,000 to a million years in length where anonymously organic-rich sediments have been deposited and buried in oceanic sediments. These have been referred to as *oceanic anoxic events* (OAEs). The ocean must have been anoxic during these events to allow the organic carbon produced by organisms in the surface ocean layer to be deposited and buried. One of the most prominent OAEs occurred in the middle Cretaceous at the Cenomanian–Turonian stage boundary (**Figure 11-17**). An understanding of the process involved in OAEs in the Cretaceous is important because Cretaceous rocks are a major source of hydrocarbons used by humankind.

The organic-rich sediments were deposited in shallow continental seaways and at continental margins (water depth < 1–2 km). A positive excursion of $\delta^{13}C$ of pelagic carbonates occurs that indicates an increase in the burial of light organic carbon, increasing the $\delta^{13}C$ of the atmosphere/ocean reservoir. The increase of $\delta^{13}C$ of +2 to +3‰ is calculated to require an organic carbon burial of 1.5 times the rate in sediments outside the event. Apparently, a maximum in sea level induced a maximum in oceanic shelf area. The shallow continental seas warmed, increasing the production rate of organic carbon that increased the extent of the O_2 minimum zone (for a further discussion, see Arthur et al., 1987). The warmer seas produce the small decrease in $\delta^{18}O$ recorded at the

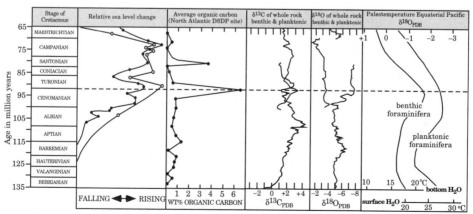

FIGURE 11-17 Sea level change, organic carbon, $\delta^{13}C$, $\delta^{18}O$, and paleotemperature of the equatorial Pacific derived from $\delta^{18}O$ of benthic and planktonic foraminifera. (Adapted from Scholle and Arthur, 1980.)

boundary. As shown in Figure 11-17, a less dynamic OAE appears to have occurred at the Aptian–Albian boundary.

Figure 11-18 displays a well-recorded spike in $\delta^{13}C$ in sediments at the Cenomanian–Turonian boundary from Othfresen, Germany. These spikes in $\delta^{13}C$ are a common feature of sediments at this boundary that are interpreted to represent a global rise and fall in sea level. It is not completely clear why sea level rose in the Cenomanian and then fell in the Turonian, but it likely is due to tectonic events. These could then change climate, oceanic circulation, location of the O_2 minimum zone, and nutrient supply from rivers and the deep ocean. Because the carbon isotopic fluctuations are widespread and independent of environment and latitude, they are used as a stratigraphic tool to correlate beds over large distances. That is, both negative and positive spikes in $\delta^{13}C$ are used as time markers in sediments if correctly assigned. This is particularly important in the late Precambrian where index fossils are rare and many sedimentary sections have undergone tectonic or metamorphic alteration. In addition, $\delta^{13}C$ is used to correlate from marine to terrestrial sections where index fossils are often rare or lacking.

Carbon isotopes have also been used to determine the timing of the evolution of C_4 type plants, its relation to the uplift of the Himalayas, and its effect on climate. $\delta^{13}C$ determinations are used as a discriminator on paleodiet reconstructions. Measurements of $\delta^{13}C$ in goethite have been used to place constraints on paleoatmospheric CO_2 concentrations throughout the Phanerozoic. Oil companies use $\delta^{13}C$ to help distinguish which rocks produced a particular oil or gas reservoir. Studies in which stable isotopic information such as $\delta^{13}C$ is useful have been rapidly increasing.

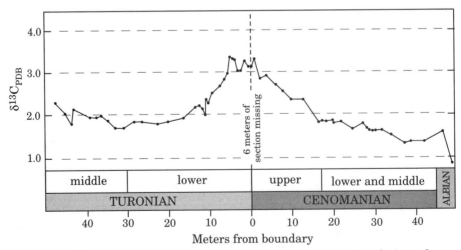

FIGURE 11-18 $\delta^{13}C_{PDB}$ of carbonate from the Cenomanian–Turonian boundary section at Othfresen, Germany showing the spike in $\delta^{13}C$ that is also present in many other sections in Europe. (Adapted from Schlanger et al., 1987.)

■ Sulfur Isotopes

As with carbon, systematics of sulfur isotopes are also used extensively to understand earth processes. They are particularly helpful when considering the cycling of sulfur between the ocean, crust, and atmosphere. **Figure 11-19** shows the present day global sulfur cycle. Most of the sulfur exists in sedimentary rocks, with the amount of oxidized sulfur somewhat greater than reduced sulfur; 9% of the total mass of sulfur in the sulfur cycle is in the ocean reservoir. Its concentration is 0.028 molal, existing mainly as the ions SO_4^{2-} and $NaSO_4^-$. The removal of sulfur in normal marine environments is generally by precipitation of gypsum in sediments; however, in *euxinic* basins bacterial reduction of sulfate occurs in the water column so that reduced sulfur is precipitated instead.

Sulfur isotopic ratios are generally expressed as

$$\delta^{34}S(in\ ‰) = \frac{(^{34}S/^{32}S)_{sample} - (^{34}S/^{32}S)_{CDM}}{(^{34}S/^{32}S)_{CDM}} \times 1000 \qquad [11.21]$$

where the subscript CDM indicates the isotopic ratio for the Canyon Diablo iron meteorite standard that has a $^{32}S/^{34}S$ ratio of 22.52. Observed values of $\delta^{34}S$ range from –55‰ for biogenic pyrite in some coal samples to values greater than +35‰ for some sulfate minerals, as indicated in **Figure 11-20**.

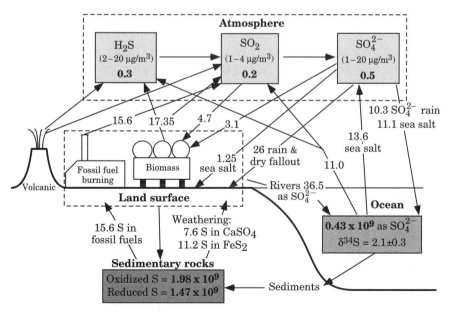

FIGURE 11-19 Present-day global sulfur cycle. Reservoirs are shown in boxes in units of 10^{11} moles. The arrows give fluxes in 10^{11} moles per year. The dashed boxes give the atmospheric and land surface reservoirs. For the land surface reservoir, the total flux in equals the total flux out. (Values mainly from Garrels et al., 1975.)

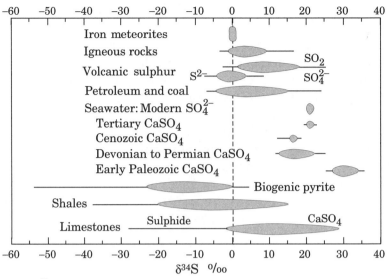

I FIGURE 11-20 $\delta^{34}S$ for various reservoirs of sulfur in the earth. (Adapted from Hoefs, 1996.)

Values of $\delta^{34}S$ in sulfide and sulfate minerals are used to help put constraints on the fluids from which they precipitated and, in many hydrothermal ore deposits, their temperature of formation; however, like oxygen isotopes in silicates, sulfur isotopic ratios in sulfides and sulfate are often not precipitated with equilibrium isotopic ratios at low temperatures. This appears to be a particular problem with sulfur because of the differences in oxidation state of sulfur in different phases.

Consider the oxidation-reduction reaction (see Chapter 13):

$$SO_4^{2-} + CH_4 + 2H^+ = H_2S + CO_2 + 2H_2O \tag{11.22}$$

For SO_4^{2-} and H^+ a standard state of a 1 molal solution having the properties of the species at infinite dilution at the pressure and temperature of interest is considered. For CH_4, H_2S, and CO_2, a standard state of the pure ideal gas at 1 bar and the temperature of interest is used, and for H_2O, a pure phase at the pressure and temperature of interest standard state is employed. The standard-state Gibbs energy of the reaction at 25°C and 1 bar in cal mol^{-1} from SUPCRT92 data is

$$\Delta G_R^o = (-8,021 + -94,254 + [2 \times -56,688]) - (-177,930 + -12,122.4 + 2 \times 0.0)$$
$$= -25,599 \text{ cal mol}^{-1} \tag{11.23}$$

The value of the logarithm of the equilibrium constant of Reaction 11.22 is

$$\log K_{(11.22)} = \frac{-25,599 \text{ cal mol}^{-1}}{2.303 \times 1.987 \text{ cal mol}^{-1}\text{K}^{-1} \times 298.15 \text{ K}} = 18.76 \qquad [11.24]$$

The equilibrium constant expression for Reaction 11.22 is, therefore, written as

$$K_{(11.22)} = \frac{f_{H_2S} \, f_{CO_2} \, a_{H_2O}^2}{a_{SO_4^{2-}} \, f_{CH_4} \, a_{H^+}^2} = 10^{18.76} \qquad [11.25]$$

Assume $a_{H_2O} = 1$ and stipulate a maximum fugacity of H_2S and CO_2 of 1 bar. With a pH of 3 or greater, the expression given in Equation 11.25 becomes

$$a_{SO_4^{2-}} \, f_{CH_4} < 10^{-24.76} \qquad [11.26]$$

This implies $a_{SO_4^{2-}}$ or f_{CH_4} must be at least as low as 10^{-12} molal or bars, respectively. In other words, the reaction uses virtually all of at least one of these reactants as it reacts to equilibrium.

Because of the large change in oxidation state of sulfur during SO_4^{2-} reduction to H_2S by this reaction, it is extremely sluggish at earth surface conditions. Its rate is controlled by mediation of anoxic sulfur-reducing bacteria, *Desulfovibrio* in marine and brackish waters and *Desulfatomaculum* in freshwater. These organisms use the energy change of the reaction to sustain their metabolism, increasing the rate of the reaction. The H_2S produced in normal marine environments escapes to the overlying seawater, becomes oxidized in the local oxidizing environment to elemental S by a reaction like

$$2H_2S + O_2 \rightarrow 2S + 2H_2O \qquad [11.27]$$

or reacts with Fe^{2+} in anoxic waters and precipitates amorphous $FeS \bullet nH_2O$ by a reaction like

$$Fe^{2+} + H_2S + nH_2O \rightarrow FeS \bullet nH_2O + 2H^+ \qquad [11.28]$$

The S produced by Reaction 11.27 and the $FeS \bullet nH_2O$ produced by Reaction 11.28 typically react during diagenesis in sediments to produce pyrite, FeS_2.

Three factors control values of $\delta^{34}S$ in the pyrite produced. First is the value of $\delta^{34}S$ of the seawater. Although its present value is $+20 \pm 1‰$, it appears seawater $\delta^{34}S$ has varied from $+10.5$ to $+31‰$ over the last 1.8 billion years. This can be determined because there is little fractionation of ^{34}S when gypsum precipitates from seawater (see Equation 11.33). The $\delta^{34}S$ of unaltered gypsum of a particular age, therefore, records $\delta^{34}S$ of seawater of that age, as outlined in **Figure 11-21**. Gypsum deposits from the Permian Period (~250 million years ago) have $\delta^{34}S$ of approximately $+10.5‰$. Devonian Period (~350 million years ago) gypsum has $\delta^{34}S$ of approximately $+25‰$.

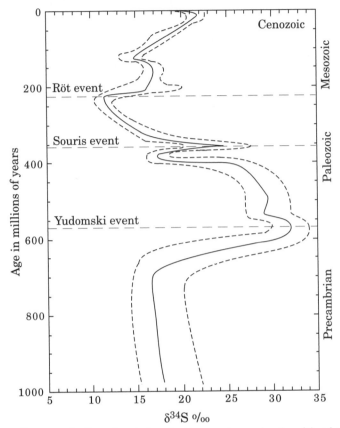

FIGURE 11-21 δ^{34}S values of sulfate minerals deposited in seawater as a function of time from the present. The solid line gives the estimated value. The dashed lines give the range of uncertainty. Because the Δ^{34}S fractionation between sulfate minerals and seawater is only +1.65, the difference is nearly lost in the noise of the measurements. In this case, the measured δ^{34}S sulfate values are taken as those in seawater. The events denote times of significant changes in δ^{34}S in the geological record. (Adapted from Claypool et al., 1980.)

From Figure 11-19, the average residence time of sulfur in seawater calculated by the flux into the ocean is

$$\frac{0.43 \times 10^9 \times 10^{11} \text{mol}}{(36.5 + 10.3) \times 10^{11} \text{mol yr}^{-1}} = 9.2 \text{ million years} \qquad [11.29]$$

These changes, therefore, likely occurred on time scales at least as great as this. The factors affecting these changes in δ^{34}S are reduction of seawater sulfate by bacteria and fractionation of light S into pyrite, causing pyrite and any precipitated sulfate minerals to increase their δ^{34}S values; oxidation during weathering of sedimentary pyrite, causing the seawater and precipitated seawater sulfate $\delta^{34}S$ values to decrease; and weathering of old evaporites with the contribution of its S with a given $\delta^{34}S$ to the ocean reservoir.

Kinetic isotopic effects during the reactions to reduce sulfur are also important. These depend on the temperature, pH, the amount of sulfur present, and time. Bacterial-mediated reduction of sulfate causes the rate of reaction to increase, which causes the difference between $\delta^{34}S_{H_2S}$ and $\delta^{34}S_{SO_4^{2-}}$, that is, $\Delta^{34}S_{SO_4^{2-}}^{H_2S}$ to decrease from its equilibrium value at 25°C of –75‰. With bacterial mediation, the lowering of $\Delta^{34}S_{SO_4^{2-}}^{H_2S}$ for this reaction is anywhere from –75‰ at slow rates of reaction to 0‰ at rapid reaction rates. This produces a positive correlation between $\delta^{34}S_{pyrite}$ and organic content of marine sediments. The third variable is the extent the reactions are open to H_2S escape into the water column. A completely open system can be modeled with Rayleigh fractionation, as was done for $\delta^{18}O$ in the atmosphere, as outlined in the discussion of Equations 11.11 to 11.13. The result of all of these effects is that $\delta^{34}S$ of 90% of pyrite in recent marine sediments is –25 ± 20‰.

What is the implication of gypsum and, therefore, seawater $\delta^{34}S$ of +10‰ in the Permian? The Permian was a time when large amounts of gypsum were formed. Can precipitation of gypsum lower $\delta^{34}S$ of seawater? Today, the average $\delta^{34}S$ of river water is approximately +7‰. If SO_4^{2-} ($\delta^{34}S$ = +20‰) and oxidized FeS_2 ($\delta^{34}S$ = –25‰) are the only two inputs to sulfur in river water, then the amount of sulfur transported by rivers to the oceans from FeS_2, x, is determined from 7‰ = $(1 - x)$ 20‰ + x (–25‰) or $x = 29\%$. This is consistent with the global sulfur cycle shown in Figure 11-19 as the flux of S due to FeS_2 weathering is 7.6×10^{11} mol/yr in a river flux of 36.5×10^{11} mol/yr or 30%. If the percentage of S that fluxes out of seawater as FeS_2 is also 30%, with 70% from gypsum deposition, then seawater $\delta^{34}S$ would remain at +20‰. If gypsum deposition increased relative to pyrite but the sulfate-to-sulfide ratio of the flux into seawater remained the same, then the gypsum deposited would get lighter with time. More of the isotopically light S from pyrite would be incorporated into the sulfate in gypsum.

This implies that SO_4^{2-} from oxidized pyrite is not converted back to FeS_2. Four moles of O_2 are removed from the atmosphere for each mole of FeS_2 oxidized. Modeling of the system is complicated, but the excess gypsum formed during the Permian stored anywhere from one to three times as much oxygen as is currently in the atmosphere. Clearly, O_2 in the atmosphere did not approach zero during the Permian because aerobic organisms did not die out. A feedback that prevented the decrease in oxygen to significantly lower levels must have operated. One such mechanism is that precipitation of gypsum requires calcium, and the calcium needed far exceeds that in the ocean so that it must be supplied by a calcium containing phase like calcite. The dissolution of calcite releases CO_2 to the atmosphere. If the CO_2 is photosynthesized into organic carbon and buried, then O_2 would be released to the atmosphere

$(CO_2 + H_2O \rightarrow CH_2O + O_2)$. Abundant gypsum deposition does appear to correlate with increased organic carbon deposition in the sedimentary record (Garrels and Lerman, 1984). Only further study will determine whether other feedbacks are important.

Geothermometry Using Sulfur Isotopic Ratios

Differences in equilibrium fractionation of isotopes between different sulfide minerals have been measured. If sulfides crystallized at the same time in equilibrium with the same reservoir of sulfur, then the $\delta^{34}S$ difference between them can be used to obtain a temperature of crystallization. For instance, Kajiwara and Krouse (1971) proposed the following relation between the isotopic ratios in pyrite and galena at equilibrium:

$$10^3 \ln \alpha_{Gn}^{Py} = \frac{1.1 \times 10^6}{T^2}$$

[11.30]

where T is temperature in K. If measurements of $\delta^{34}S$ for pyrite and galena from an ore deposit are made and it is assumed they formed in equilibrium, a value of α_{Gn}^{Py} is determined from the sulfur isotope equivalent of Equation 11.6. The temperature of formation is then computed from Equation 11.30. For instance, if the measured $\alpha_{Gn}^{Py} = 1.00537$, the temperature of formation would be 453 K or 180°C. Like the problems discussed above for oxygen isotope equilibrium between silicates or carbonates, problems exist with the assumption of equilibrium being reached between pyrite and galena. The temperatures determined, therefore, need to be corroborated by other means.

■ Link of Sulfur-Carbon-Isotopes Through Iron in Sedimentary Rocks

The iron and sulfur present in sedimentary rocks in the crust has its ultimate source from the earth's mantle where they both existed in a reduced oxidation state. When transported to the earth's crust and crystallized in igneous rocks, the ferrous mantle iron of valance +2 is incorporated in iron silicates. The reduced mantle sulfur of −1 valence is often crystallized with reduced iron in pyrite (FeS_2). In sedimentary rocks, pyrite is present in abundance in organic-rich shales; however, abundant oxidized Fe^{+3} occurs in iron oxides and hydroxides in organic-free shales and sandstones (e.g., hematite). Also, large quantities of oxidized +6 sulfur are present in sedimentary sulfates, particularly in evaporite deposits (e.g., gypsum, $CaSO_4 \cdot 2H_2O$). Why are these oxidized minerals present given the low oxidation state of their initial mantle source?

To oxidize ferrous iron in iron silicates and pyrite to form the ferric iron oxides/hydroxides present as well as the oxidation of reduced sulfide in pyrite to sulfates of sedimentary rocks requires an oxidant to remove the electrons in the reactants (see Chapter 14). The large quantities of ferric oxides/hydroxides and sulfates present in sedimentary rocks indicate that significant amounts of oxidant must have been reduced to receive these electrons. O_2 is an oxidant because it can take electrons in a reaction by producing an O^{2-} species such as is present in H_2O and oxide minerals. The amount of oxidant required for all of the iron oxides/hydroxides and sulfates present in crustal rocks is, however, much larger than all the O_2 in the atmosphere. The dominant oxidant responsible for the oxidized iron and sulfur, therefore, must be carbon because it is the only other element of abundance on earth that undergoes oxidation/reduction.

Mantle-derived carbon is introduced as CO_2 into the surface crustal environment through degassing during volcanic eruptions. This carbon and the large amount of carbon in carbonates as CO_3^{2-} are in an oxidized state (valence $= +4$). In order for carbon to absorb the electrons from iron and sulfur, a more reduced carbon species must be produced. This reduced carbon species is present in organic compounds where carbon generally has a valence of 0; however, organic carbon is not spontaneously formed from atmospheric CO_2 or carbonate minerals. It requires added energy in order for the reaction to occur. Not until the advent of a photosynthetic pathway, which uses the additional energy from the sun to reduce carbon in CO_2 to organic carbon, could significant oxidized iron and sulfur species develop on the earth's surface. In fact, the photosynthetic reaction is what produced the O_2 present in our atmosphere.

Because of this transfer of electrons, pyrite, carbonate, gypsum, and organic carbon are linked in the sedimentary rock record through oxidation/reduction (redox) reactions. The redox reaction involving calcite with pyrite can be written as

$$4FeS_2 + 15CaCO_3 + 7SiO_2 + 31H_2O \rightarrow$$

pyrite calcite quartz

$$8CaSO_4 \bullet 2H_2O + 2Fe_2O_3 + 15CH_2O + 7CaSiO_3 \qquad [11.31]$$

    gypsum hematite organic C Ca-silicate

The redox reaction involving the reduction of carbon in dolomite with pyrite is

$$8FeS_2 + 15CaMg(CO_3)_2 + 14SiO_2 + CaSiO_3 + 62H_2O \rightarrow$$

pyrite dolomite   quartz Ca-silicate

$$16CaSO_4 \bullet 2H_2O + 4Fe_2O_3 + 30CH_2O + 15MgSiO_3 \qquad [11.32]$$

    gypsum hematite organic C Mg-silicate

The extent of these redox reactions is influenced by the need of life to obtain new carbon from the oxidized carbon reservoir in carbonates/CO_2 and the degree to which the produced compounds containing oxidized or reduced elements are buried and preserved in sedimentary rocks. Sedimentary rocks contain significant quantities of oxidized and reduced Fe, S, and C; therefore, Reactions 11.31 and 11.32 should help in understanding the changes in the relative abundance of pyrite and carbonates as opposed to sulfates and organic matter in the sedimentary rock record as a function of time.

Reactions Occurring Through the Ocean

As observed today, the precipitation of pyrite is typically in anoxic, organic carbon-rich aqueous environments, which prevents pyrite from undergoing oxidation; however, Reactions 11.31 and 11.32 indicate that pyrite should disappear as organic carbon is produced. Consistent with Reactions 11.31 and 11.32, evaporites containing gypsum with its oxidized sulfur are also deposited in some anoxic seawater environments where organic matter is present. The reason for this seemingly conflicting behavior is that Reactions 11.31 and 11.32, as applied to deposition of sediments, are really shorthand for a set of reactions linked through the ocean that occur in different surface and crustal environments on the earth and are generally not in equilibrium. These include reduction of oxidized carbon, organic matter oxidation, sulfate reduction, pyrite oxidation, iron reduction, and iron oxidation. Reactions 11.31 and 11.32 should really be viewed as electron mass balance equations when considering the sedimentary record of deposition of sediments.

Isotopic Fractionation of Sulfur and Carbon

Significant fractionation of sulfur isotopes occurs as sulfate from gypsum is biologically reduced to sulfide in pyrite. The amount of fractionation depends on environmental factors and varies with $\delta^{34}S_{pyrite} - \delta^{34}S_{gypsum} = -46‰$ to $-75‰$. The $\delta^{34}S$ values of the pyrite formed decreases (becomes lighter) as bacteria preferentially use ^{32}S over ^{34}S in the reaction. The $\delta^{34}S$ values in the remaining gypsum increase because of isotope mass balance in the process. Likewise, the biological reduction of carbon in carbonate (carb) produces organic matter (org) with a lower $\delta^{13}C$. The fractionation varies with $\delta^{13}C_{org} - \delta^{13}C_{carb} = -24‰$ to $-35‰$ for the more common C3 plants and averages $-12‰$ for C4 plants. Mass balance requires the $\delta^{13}C$ values of the remaining carbonate to become heavier; therefore, if Reactions 11.31 and 11.32 proceed in the forward direction because of biological activity, $\delta^{13}C$ values of carbonate will increase, and $\delta^{34}S$ values of sulfate will decrease.

Given in **Figure 11-22** are the averages of $\delta^{34}S$ of sulfates and $\delta^{13}C$ of carbonates precipitated in evaporites as a function of their age for the Phanerozoic.

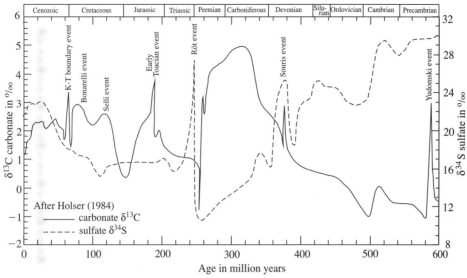

FIGURE 11-22 Averages of $\delta^{34}S$ values of sulfates precipitated from seawater relative to the Canyon Diablo troilite (CDT) and $\delta^{13}C$ values of carbonates precipitated from seawater relative to Peedee belemnite (PDB) as a function of their age. Also shown are carbon and sulfate isotope events recorded in the rocks.

In most cases, these values can be considered to be in equilibrium with seawater at the time they were precipitated. The inorganic fractionation of sulfur isotopes between the seawater sulfate ion (SO_4^{2-}) and sulfate minerals ($CaSO_4$) can be described by

$$\delta^{34}S_{SO_4^{2-}} - \delta^{34}S_{CaSO_4} = 1-1.6‰ \qquad [11.33]$$

The inorganic fractionation of carbon between seawater bicarbonate (HCO_3^-) and carbonate minerals (carb) is given by

$$\delta^{13}C_{HCO_3^-} - \delta^{13}C_{carb} = \sim 1‰ \qquad [11.34]$$

Therefore, these sulfates and carbonates record the $\delta^{34}S$ and $\delta^{13}C$ values of seawater at the time they were precipitated except for a $\sim 1‰$ shift in their delta values. Resetting of the values during diagenesis can be checked because there should be little change in $\delta^{34}S_{SO_4^{2-}} - \delta^{34}S_{CaSO_4}$ and $\delta^{13}C_{HCO_3^-} - \delta^{13}C_{carb}$ even if $\delta^{34}S$ values of sulfates and $\delta^{13}C$ values of carbonates change significantly.

The values shown in Figure 11-22 are averages and carry significant uncertainty as the measurements are typically done on evaporite deposits where sulfates and carbonates can be found together. These environments can have poor circulation relative to open seawater, and mixing of the limited seawater present with river water of different isotopic composition can alter values somewhat. The ages assigned to the rocks are probably accurate to $\sim 5\%$ in most cases, although a lack of fossils in evaporites means that biostratigraphic correlations cannot be made.

Note in Figure 11-22, the general inverse correlation between $\delta^{34}S$ and $\delta^{13}C$ values as a function of age from ~580 to 250 m.y. A general inverse correlation also exists during the past 120 m.y., although in the opposite sense. This behavior is predicted by Equations 11.31 and 11.32 because the oxidized species of sulfur and carbon are on opposite sides of the reaction. From the beginning of the Cambrian to the end of the Permian, the $\delta^{13}C$ values of carbonate, on average, increased and the $\delta^{34}S$ values of sulfate decreased indicating Reactions 11.31 and 11.32 were proceeding in the forward direction toward present time. What has driven these reactions in the forward direction?

The general trend in the curves is likely due to the demand for organic carbon from the evolution of life with photosynthesis exceeding respiration by some small amount. The radiation of species became significant in the Cambrian with photosynthetic organisms transferring carbon from the oxidized reservoir in carbonates + atmospheric CO_2 to organic carbon as new environmental niches were exploited through time. Some of this carbon was then buried in sedimentary rocks. Reactions 11.31 and 11.32 would then be driven to the right, stripping electrons from the surface environment. The conversion of carbonate to organic carbon increases the $\delta^{13}C$ values of carbonate. The loss of electrons causes −1 sulfur in pyrite to be converted to +6 sulfur in sulfates. Sulfates then incorporate lighter sulfur so that their $\delta^{34}S$ values decrease.

This behavior is particularly true for the late Paleozoic when land and swamp plants were evolving weather resistant *lignin*. Because of resistance to its biological and chemical decomposition, lignin is more likely to be buried. The electrons lost when oxidized carbon is transferred to lignin also stabilize zero valance O_2 over the −2 valance of oxygen found in H_2O and in minerals; therefore, the concentration of oxygen in the atmosphere increased at this time as well (see Figure 16-2).

Reduced Sulfur Isotope Record

To understand the isotope changes of sulfur in more detail, consider the isotopic composition of reduced sulfur in oceanic sediments from the Archean. Given in **Figure 11-23** are determinations of the average $\delta^{34}S$ values of pyrite in sediments and their metamorphic equivalents as a function of time. Sulfur enters the surface reservoir from the mantle through volcanic input of reduced sulfur in SO_2 gas with subordinate H_2S gas. Both have $\delta^{34}S$ values of 0‰, reflecting their mantle source. Sulfur reacts with the Fe^{2+} present in rocks and generally produces FeS_2 with no detectable isotopic fractionation as the mantle derived sulfur is completely used to produce pyrite in most cases. Over time, sediments containing FeS_2 with $\delta^{34}S$ values equal to 0‰ are subducted back to the man-

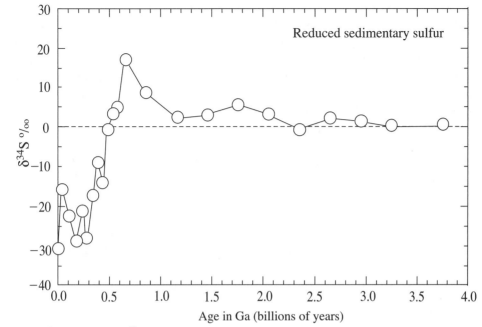

FIGURE 11-23 Average $\delta^{34}S$ values of sedimentary sulfides relative to Canyon Diablo troilite within the indicated time periods. (Adapted from Canfield, 2004.)

tle at convergent plate margins, causing significant overturn. It is estimated between 4 and 10 times the current surface sulfur inventory has been returned to the mantle during the history of the earth (Canfield, 2004).

Early on the earth's atmosphere was in a reduced state and the earth's surface became oxidized with the evolution of photosynthetic life forms. With this oxidation at the surface, oxidized sulfur precipitated as gypsum in shallow marginal basins where Ca^{2+} and SO_4^{2-} were concentrated by H_2O evaporation while deep waters remained anoxic and pyrite continued to precipitate. Both the gypsum and pyrite formed from mantle derived sulfur have $\delta^{34}S$ values of 0‰. Organisms then evolved to use the oxidized sulfur as an energy source through its reduction. Biological reduction of sulfate to sulfide partitions heavier sulfur into the gypsum and lighter sulfur into pyrite. Because the sulfate is not completely converted to sulfide, significant isotopic fractionation occurs. The lighter reduced sulfur is stable in anoxic bottom waters and is returned to the mantle by subduction, making the surface reservoir of sulfur increasingly heavier than the mantle reservoir ($\delta^{34}S = 0$‰) as a function of time. The average $\delta^{34}S$ of pyrite increased to more positive values with decreasing age as shown by the averages given in Figure 11-23.

As oxygen continued to increase in the atmosphere the entire water column in the ocean becomes oxidized similar to its present state. It is argued that this occurred at about 0.7 Ga (Canfield, 2004), although OAEs have occasionally occurred in the Phanerozoic (see discussion of Figure 11-17). Seawater sulfate could then collect in pores on the ocean floor, be fixed in sediments, and subducted back into the mantle. Because of this, isotopically heavy sulfur was lost from the surface sulfur reservoir; therefore, both sulfates and sulfides have become isotopically lighter into the Phanerozoic, as indicated in Figures 11-22 and 11-23, respectively. At 600 m.y., sulfates had an average $\delta^{34}S$ value of about +30‰, and sulfides had an average $\delta^{34}S$ value of about +3‰. Given these positive values, the surface sulfur reservoir was isotopically heavy relative to the mantle ($\delta^{34}S \sim 0$‰) at 600 m.y. According to Figure 11-22, the increase in $\delta^{13}C$ values and decrease in $\delta^{34}S$ values was brought to an abrupt end near the Permian-Triassic boundary (~251.4 m.y.). Apparently, the mass extinction that occurred at this time halted the production of a large amount of organic matter, and oxidation of recently deposited organic carbon ensued, driving Reactions 11.31 and 11.32 in the reverse direction.

Permo-Triassic Extinction Event

The most extensive extinction of species documented in the rock record occurred at the Permian-Triassic boundary. Although it is still a matter of debate, it has been argued that sea level fell in the Permian before the extinction event (Algeo and Seslavinsky, 1995). Alternatively, evidence exists for widespread uplift of the southern margins of Gondwana (Faure, deWit, and Willis, 1995). Late Permian global coal hiatus linked to either scenario is consistent with $\delta^{13}C$ values of carbonates decreasing at this time as Carboniferous coal swamps became exposed, decayed, and released lighter carbon to the atmosphere. This light atmospheric carbon was absorbed by the ocean decreasing $\delta^{13}C$ of the carbonates precipitated; however, the $\delta^{13}C$ value of carbonates rose to +3‰ to +4‰ at the very end of the Permian (see **Figure 11-24**). This likely occurred by increased storage of organic carbon in sediments and gas clathrates because of the evolution of vascular land plants with their weather-resistant lignin.

To understand the isotopic changes involved, consider the total carbon in the Atmosphere-Terrestrial biosphere-Ocean reservoir, the *ATO*, at the end of the Permian. This carbon was partitioned between organic and inorganic reservoirs; therefore, a mass balance for carbon isotopes in the ATO can be written as

$$\delta^{13}C_{org} [Y] + \delta^{13}C_{carb} [X] = \delta^{13}C_{ave}[C_{ATO}] \tag{11.35}$$

where [Y] gives the amount of organic carbon, [X] represents the amount of inorganic carbon, and [C_{ATO}] stands for the total carbon in the atmosphere-

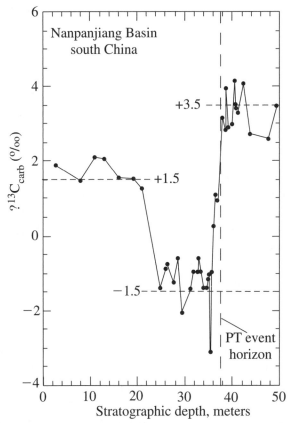

FIGURE 11-24 $\delta^{13}C$ of carbonates across the PT event horizon. The long dashed line gives the contact of Upper Permian skeletal packstone and the base of the calcmicrobial framework of the Triassic Griesbachian. (Adapted from Krull et al., 2004.)

terrestrial biosphere-ocean reservoir. If carbon in the ATO at this time, including that in clathrates, is similar in size to that determined for the present, then $[C_{ATO}]$ was about 50,000 Pg C (see Figure 16-1). The average $\delta^{13}C$ values of the ATO in Permo-Triassic times, $\delta^{13}C_{ave}$, likely reflected a primordial mantle out gassing value near $-5.5‰$ on the PDB scale as little carbon is returned to the mantle. Assume before the end of Permian organic carbon storage event, the oxidized carbon in carbonates had an average $\delta^{13}C_{carb}$ value of $1.5‰$. This is the average of the $\delta^{13}C$ values of carbonate after the isotopic change event associated with the P-T boundary event as given in Figure 11-24 and values determined for the early Permian. If the carbon is partitioned between the organic and inorganic carbon reservoir with an average $\delta^{13}C_{carb} - \delta^{13}C_{org} = -28‰$ (Hayes et al., 1989), then before the organic carbon storage event, $\delta^{13}C_{org} = -26.5‰$. Substituting the values outlined into Equation 11.35 gives

$$-26.5\text{‰} \times (50,000 \text{ Pg C} - [X]) + 1.5\text{‰} \times [X] = -5.5\text{‰} \times (50,000 \text{ Pg C}) \qquad [11.36]$$

or $[X] = 37,500$ Pg C with $\delta^{13}C$ values averaging $+1.5$‰; therefore, the size of the organic carbon reservoir is 12,500 Pg C with $\delta^{13}C$ values averaging -26.5‰.

If carbon is transferred from the inorganic to the organic reservoir during the carbon storage event to make $\delta^{13}C$ of the inorganic carbon reservoir $+3.5$‰, then with $\delta^{13}C_{carb} - \delta^{13}C_{org} = -28$‰ isotopic mass balance implies

$$-24.5\text{‰} \times (50,000 \text{ Pg C} - [X]) + 3.5\text{‰} \times [X] = -5.5\text{‰} \times (50,000 \text{ Pg C}) \qquad [11.37]$$

In this case, $[X] = 33,930$ Pg C with average $\delta^{13}C = +3.5$‰, and the size of the organic carbon reservoir becomes 16,070 Pg C with an average $\delta^{13}C = -24.5$‰.

The organic carbon storage event, therefore, increased the organic carbon reservoir by approximately 30%. The $\delta^{13}C$ value of carbonate spiked downward to a minimum of between -1‰ to -2‰ (see Figure 11-24) sometime after the Permo-Triassic boundary event defined by the appearance of a new fauna. The minimum likely occurs where the decrease in biomass because of the extinction of the old fauna is balanced by the increase in biomass due to the radiation of the new fauna. The increase in $\delta^{13}C$ values after the minimum occurred appears to show two peaks separated by a minimum at about 30 meters in the section given in Figure 11-24 before it restabilized to a $\delta^{13}C$ value of near $+1.5$ at about 20 meters height; therefore, the event, manifested in changes in $\delta^{13}C$ values, is likely more complicated in detail.

Assume during the Permo-Triassic event a change of $\delta^{13}C$ from $+3.5$‰ to -1.5‰ with $\delta^{13}C_{carb} - \delta^{13}C_{org} = -28$‰ occurred in the carbonate in the ATO. This requires a shift of the reservoirs of

$$-29.5\text{‰} \times (50,000 \text{ Pg C} - [X]) + -1.5\text{‰} \times [X] = -5.5\text{‰} \times (50,000 \text{ Pg C}). \qquad [11.38]$$

or $[X] = 42,860$ Pg C with $\delta^{13}C$ averaging -1.5‰ for the carbonate and the size of the organic carbon reservoir is 7140 Pg C with $\delta^{13}C$ averaging -29.5‰. This change is then

$$+3.5\text{‰} \times (33,930 \text{ Pg}) - 1.5\text{‰} \times (42,860 \text{ Pg C}) = 1.8 \times 10^5 \text{ Pg C‰ units} \qquad [11.39]$$

Organic carbon of $\delta^{13}C = -24.5$‰ oxidized from the ATO to the inorganic reservoir could possibly cause the shift. The amount needed is

$$\frac{1.8 \times 10^5 \text{ Pg C‰}}{(-1.5\text{‰}) - (-24.5\text{‰})} = 7826 \text{ Pg C of organic carbon} \qquad [11.40]$$

This is 7826 Pg C/12,500 Pg C = 63% the total organic reservoir in the ATO calculated above. Because by most investigators' estimates most of the organic carbon at present in the ATO is in clathrates, a clathrate destabilization event could possibly cause the shift. In fact, because the $\delta^{13}C$ of methane clathrate is typically about $-65‰$ this requires a mass of

$$\frac{1.8 \times 10^5 \text{ Pg C‰}}{(-1.5‰) - (-65‰)} = 2,835 \text{ Pg C} \qquad [11.41]$$

The present reservoir of methane clathrate is ~10,000 Pg C. The isotopic shift would require release of about 28% of the present gas clathrate reservoir. Because of the ease of releasing gas clathrate formed on continental shelves with sea level lowering or continental uplift and the reasonableness of this value, many scientists believe the release of methane clathrates was responsible for the Permo-Triassic extinction event. The methane added to the atmosphere and its oxidized product, CO_2, causing global warming. This then becomes a positive feedback situation where increased warming causes the release of more methane from clathrates (see discussion of Figure 16-4 on methane clathrate stability). It is argued this would lead to a loss of vertical circulation and stagnation in the ocean, causing it to become eutrophic. The anoxia produced together with changing weather patterns from global warming lead to the mass extinction event observed.

The isotopic shift, however, could also have been produced by adding additional CO_2 from mantle outgassing with $\delta^{13}C = -5‰$ to the ATO. The Siberian flood basalts appear to be of the correct age (Reichow et al., 2002). There appears to be a good correlation of flood basalts and extinction events (Erwin, 2006, Figure 8-1), suggesting a link. The amount of carbon needed is

$$\frac{1.8 \times 10^5 \text{ Pg C‰}}{(-1.5‰) - (-5‰)} = 51,400 \text{ Pg C from } CO_2 \text{ gas} \qquad [11.42]$$

The present mantle flux of carbon from CO_2 to the atmosphere is ~0.05 Pg C yr^{-1}. If the CO_2 flux were twice as much as this at the Permo-Triassic boundary, perhaps because of the eruption of the Siberian flood basalts, the required amount of CO_2 could be supplied in 51,400 Pg C/0.10 Pg C yr^{-1} = 0.51 m.y. Because this time frame is in the window given for the length of the P-T event, this is also a possible mechanism. As the Siberian flood basalts erupted onto Permian coal deposits, it is also possible that part of the decrease in $\delta^{13}C$ values occurring at the Permo-Triassic boundary was due to vaporization of the light carbon in the coals by the erupted lavas.

Changes of $\delta^{13}C$ and $\delta^{34}S$ from the Mid-Cretaceous to the Present

Note in the past ~120 m.y. Figure 11-22 indicates, in a general sense, that Reactions 11.31 and 11.32 proceed in the reverse direction as present time is approached. Why has this occurred? The isotopic record appears to be the result of decrease in CO_2 contributed to the atmosphere by decreased volcanic eruptions toward the present. This is consistent with the documented decrease in seafloor spreading (Conrad and Lithgow-Bertelloni, 2006) determined from analysis of the age of the ocean floor. If CO_2 is used rather than carbonate as an oxidant of pyrite to be reduced to organic carbon, Reaction 11.31 becomes

$$4FeS_2 + 15CO_2 + 8CaSiO_3 + 15H_2O \rightarrow 8CaSO_4 + 2Fe_2O_3 + 15CH_2O + 8SiO_2 \quad \text{[11.43]}$$

Alternatively, the introduction of CO_2 into the equation can be used to eliminate the connection to silicate rocks by removing the phases $CaSiO_3$ and SiO_2 while retaining carbonate as an oxidant:

$$4FeS_2 + 7CO_2 + 8CaCO_3 + 15H_2O \rightarrow 8CaSO_4 + 2Fe_2O_3 + 15CH_2O \quad \text{[11.44]}$$

If seawater is brought into the picture as another phase then the calcium phases can be eliminated to give

$$4FeS_2 + 15CO_2 + 23H_2O \rightarrow 2Fe_2O_3 + 15CH_2O + 8SO_4^{2-} + 16H^+ \quad \text{[11.45]}$$

In Reactions 11.43, 11.44, and 11.45, CO_2 is on the opposite side of the reaction relative to organic matter, CH_2O; therefore, a decrease in CO_2 causes these reactions to proceed in the reverse direction. This decrease in CO_2 toward the present, during the past ~120 m.y., is consistent with a general decrease in temperature brought on by decreased greenhouse warming of CO_2 from the mid-Cretaceous to the present (the hothouse to ice house transition), as has been shown by a number of proxies.

It has been argued (Holser et al., 1988, see also Mackenzie and Agegian, 1989, Figure 2) that sea level plays an important role in the directions of Reactions 11.31 and 11.32. Falling sea level causes Reactions 11.31 and 11.32 to proceed in the forward direction with $\delta^{34}S$ of sulfates decreasing and $\delta^{13}C$ of carbonates increasing as exposed pyrite sulfide is oxidized and exposed carbonates are weathered. This is, however, opposite to what has happened during the fall in sea level from the mid-Cretaceous to the present. One of the problems is that Reaction 4.19 of Holser et al. (1988)

$$CO_{2(g)} + 2Fe_2O_3 + 8CaSO_4 + 8MgSiO_3 + 15CH_2O \rightarrow$$
$$4FeS_2 + 8CaMg(CO_3)_2 + SiO_2 + 15H_2O \quad \text{[11.46]}$$

has the oxidized species CO_2 on the opposite side of the reaction as the oxidized carbon in carbonates. They did this because they surmised periods of elevated CO_2 in the atmosphere were periods of carbon transfer from the reduced organic carbon to the oxidized carbonate reservoir; however, increased atmospheric CO_2 produces a greater amount of organic matter as a function of time not less, as has been shown at least for plants on laboratory time scales.

Local Maxima and Minima in $\delta^{13}C$ and $\delta^{34}S$

The local maxima and minima, that is, events in $\delta^{13}C$ and $\delta^{34}S$ existing on the general trend lasting tens of million years, could be due to any number of processes. In analyzing the measurements from a particular location, it needs to be kept in mind that measurements may represent local rather than global oceanic values; however, most investigators believe the decrease in the $\delta^{13}C$ values of carbonate at the K-T boundary found in many studied sections was caused by a meteorite impact. The positive spikes of $\delta^{13}C$ values in the Mesozoic are generally attributed to OAEs as lighter organic carbon is deposited on the seafloor, causing the $\delta^{13}C$ values of carbonates to increase. Negative excursions in $\delta^{13}C$ are generally attributed to release of isotopically lighter methane from clathrates with low $\delta^{13}C$ values to the atmosphere. Carbon with low $\delta^{13}C$ values is then incorporated into carbonates. Events like the Röt, Souris, and Yudomski shown in Figures 11-21 and 11-22 are still a matter of active debate.

■ Nitrogen Isotopes

There are two stable isotopes of nitrogen ^{14}N and ^{15}N that change their relative abundance in many earth processes. The $^{15}N/^{14}N$ ratio for substance A is reported as

$$\delta^{15}N_A(in\ \text{‰}) \equiv \left[\frac{(^{15}N/^{14}N)_A - (^{15}N/^{14}N)_{standard}}{(^{15}N/^{14}N)_{standard}}\right] \times 1000 \qquad [11.47]$$

where $(^{15}N/^{14}N)_{standard}$ is the isotopic ratio in atmospheric N_2 of 0.0036765; therefore, $\delta^{15}N$ of air, $\delta^{15}N_{air} = 0.0$. The range of $\delta^{15}N_A$ values measured in terrestrial material is generally between –20‰ and +30‰; however, $\delta^{15}N_A$ in carbonaceous chondrites and other stony meteorites have a much broader range, suggesting the earth's mantle has heterogeneous $^{15}N/^{14}N$ values. MORBs have a $\delta^{15}N$ between –4.5 and +7.5, and igneous rocks in general span the range from –12 to +22. The nitrogen in igneous rocks generally resides in ammonium ions substituting for alkalis in feldspars. $\delta^{15}N$ of various nitrogen reservoirs on earth are shown in **Figure 11-25**. They often trace quantities in samples, the problems of contamination of N_2 (mass = 14 +14 = 28) with CO (mass = 12 + 16 = 28), and the difficultly in transporting N_2 in evacuated mass spectrometer lines did not allow the routine measurement of nitrogen isotopes until continuous flow in helium gas instruments became available about 25 years ago.

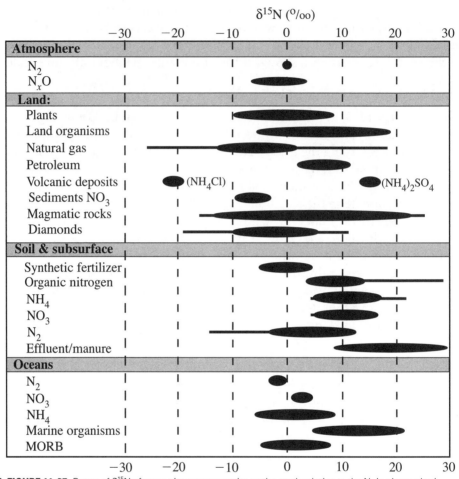

FIGURE 11-25 Range of $\delta^{15}N$ of some nitrogen reservoirs on the earth relative to the N_2 in air standard. (Adapted from SAHRAs [Sustainability of semi-Arid Hydrology and Riparian Areas] website.)

A change in oxidation state of nitrogen during a reaction generally changes the ratio of $^{15}N/^{14}N$ between the reactants and products—the more oxidized species incorporating a greater concentration of ^{15}N. Because of the large number of different oxidation states of nitrogen compounds, earth materials can change their ratios of $^{15}N/^{14}N$ significantly. This is compounded by the fact that nitrogen is involved in one-way life processes, and most of these show kinetic isotopic effects, which are concentration and time dependent, with the slowest step typically having the greatest effect. α_B^A for these processes are less than unity because life processes are typically reductive. The fluxes involved occur between the reservoirs outlined in the global nitrogen cycle, as shown in **Figure 11-26**. Nitrogen isotope analysis can help in understanding the fluxes be-

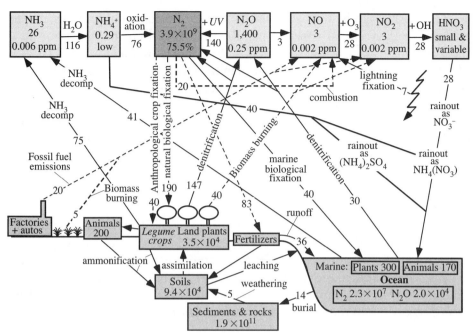

FIGURE 11-26 Global nitrogen cycle with reservoir abundances in Tg (Teragrams = 10^{12} g) and fluxes (given by arrows) in Tg per year. The solid lines are natural fluxes and the dashed lines are anthropogenic. (Adapted from Garrels et al., 1975, considering Galloway, 1998; Söderlund and Svensson, 1976; Stedman and Shetter, 1983; and others.)

tween the reservoirs; however, nitrogen fractionation reactions can often proceed to near completion because of the small amounts involved so that the overall α_B^A approaches 1.0 (see Figure 11-5). This often makes interpretations of the isotopic ratios difficult.

Global Nitrogen Cycle

The majority of the earth's nitrogen resides in sediments and rocks; however, there are orders of magnitude more in the atmosphere than in soils, the ocean, or continental and marine biota. Atmospheric nitrogen exists as seven different species: N_2, NH_3, NH_4^+, N_2O, NO, NO_2, and HNO_3, which are derived from a number of different sources. Because of concerns about the production of nitric acid (acid rain) and ammonium nitrate (smog particles) from the oxidized nitrogen species, $NO + NO_2$, these are often designated collectively as NO_x. As shown in Figure 11-26, substantial fluxes exist between the atmosphere and the biosphere, land surface, and ocean. Oxidized nitrogen species can also be produced in the atmosphere by the reaction of N_2 and O_2 in the intensive heat produced by lightning.

Organisms require nitrogen as an essential component of proteins, nucleic acids, and other cellular constituents. Relative to carbon and phosphorous in most organisms the nitrogen needed is close to the *Redfield ratio* of (C:N:P) of (106:16:1) that is present in phytoplankton. Given the large amount of carbon in organic matter, this implies a substantial need for nitrogen in organisms. Because of the need for nitrogen and the large reservoir of N_2 in the atmosphere, a number of organisms have evolved to reduce N_2 to a form that can be used in organic compounds. There is a small equilibrium fractionation in the reaction

$$N_{2(gas)} = N_{2(aq)} \qquad [11.48]$$

of $\alpha^{N2(aq)}_{N2(gas)} = 1.00085$ at 0°C. A group of photosynthetic cyanobacteria of both free-living aerobic (e.g., *Azotobacter*) and free-living anaerobic (e.g., *Desulfovibrio*) as well as symbiotic forms (e.g., *Nostoc*, a common photosynthetic partner of lichens; *Rhizobium*, in root nodules of legumes, termites, protozoa) has evolved that can convert $N_{2(aq)}$ into nitrogen in organic tissue, $NH_{3(org)}$, with the release of O_2, according to the *nitrogen-fixation* reaction

$$2N_{2(aq)} + 6H_2O + \substack{solar \\ energy} \rightarrow 4NH_{3(org)} + 3O_2 \qquad [11.49]$$

$NH_{3(org)}$ stands for nitrogen of the correct oxidation state to attach to a carbon in an organic molecule as $C-N{<}^H_H$. The triple bond between N atoms in $N_{2(aq)}$ that needs to be broken by Reaction 11.49 is very strong. The large solar energy needed relative to the difference in energy states between ^{15}N and ^{14}N isotopes results in little detectable fractionation with $\alpha^{NH3(org)}_{N2(aq)} = 0.997$ to 1.001 reported. With decay of the organism ammonia, NH_3 gas, is released in an *ammonification* reaction

$$NH_{3(org)} - decay \rightarrow NH_{3(gas)} \qquad [11.50]$$

The $NH_{3(gas)}$ is quite soluble and becomes hydrated in water producing the ammonium ion

$$NH_{3(gas)} + H_2O \rightarrow NH_4^+ + OH^- \qquad [11.51]$$

Because Reactions 11.50 and 11.51 typically go to completion, neither fractionates nitrogen isotopes significantly.

The ammonium ion is consumed by nitrifying bacteria (*Nitrosomonas*) to produce nitrites by the oxidation reaction

$$2NH_4^+ + 3O_2 \rightarrow 2NO_2^- + 4H^+ + 2H_2O \qquad [11.52]$$

with $\alpha_{NH_4^+}^{NO_2^-} = 1.02$ as the product is more oxidized. Bacteria of the genus *Nitrobacter* then oxidize the nitrites to nitrates

$$2NO_2^- + O_2 \rightarrow 2NO_3^- \qquad\qquad [11.53]$$

These last two reactions are termed bacterial *nitrification* reactions. More complex organisms will then undergo an *assimilation* reaction of nitrate to obtain their needed nitrogen. The overall assimilation reaction can be written as

$$NO_3^- + 2H_2O \rightarrow NH_{3(org)} + 2O_2 + OH^- \qquad\qquad [11.54]$$

The nitrogen has been reduced from a +5 to −3 oxidation state in the reaction. A kinetic fractionation favoring ^{14}N occurs. Some plants assimilate $NH_{3(gas)}$ or organic N rather than NO_3^-. $\alpha_{NO_3^-}^{NH_{3(org)}}$ of 0.974 to 1.000 have been reported from field and laboratory measurements (Fogel and Cifuentes, 1993).

Nitrogen can also be returned to the atmosphere by the process of *denitrification* with the aid of the bacterium *Thiobacillus denitrificans* using the energy of organic matter oxidation, as given by

$$4NO_3^- + 5CH_2O \rightarrow 2N_2 + 5HCO_3^- + H^+ + 2H_2O \qquad\qquad [11.55]$$

or by anaerobic bacteria (e.g., *Pseudomonas*, a genus of *gram-negative* bacteria capable of degrading a variety of compounds) in the absence of organic carbon

$$4NO_3^- + 2H_2O \rightarrow 2N_2 + 5O_2 + 4OH^- \qquad\qquad [11.56]$$

The fractionation factor for denitrification has been reported to be approximately 0.980 as lighter nitrogen fractionates into N_2.

The ratio of nitrogen to phosphorus in open ocean water is given by the Redfield ratio 16:1. This suggests that although a great deal of energy is required organisms have evolved to just keep up with the need for fixed nitrogen by organisms and its loss by denitrification and burial in sediments but produce no extra. In this case, the amount of organic matter in the ocean is controlled to a first approximation by the availability of phosphorous with nitrogen-fixing bacteria producing just enough fixed nitrogen needed by organisms.

Nitrogen Isotopes in the Food Chain

As outlined above, the kinetic assimilation of nitrogen fractionates ^{14}N preferentially into the organism; however, $\delta^{15}N$ of animals increase up the food chain (*trophic effect*), as shown in **Figure 11-27**. This occurs because animal waste material (e.g., urea) has $\delta^{15}N$, which is 2‰ to 4‰ less than that obtained from the food they consume.

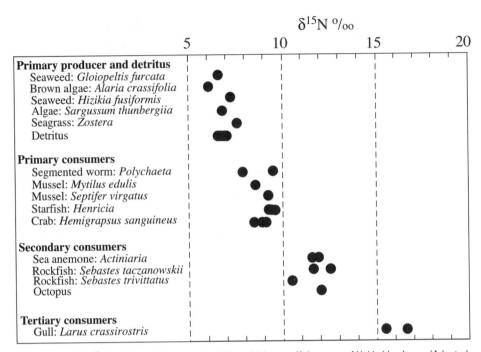

FIGURE 11-27 $\delta^{15}N$ (‰) of biota from the Usujiri intertidal zone off the coast of Hokkaido, Japan. (Adapted from Minagawa and Wada, 1984.)

Because the amount of nitrogen fixed in fertilizer production plus anthropological crop fixation is of the same order as natural biological fixation, the bacteria in soils responsible for denitrification in the global nitrogen cycle are presently not keeping up; therefore, the concentration-dependent nitrogen assimilation in land and marine organisms is increasing. This is in despite of the fact that humankind biomass burning is transferring a significant amount of nitrogen to the atmosphere as nitric oxide, NO, and nitrogen dioxide, NO_2 (see Figure 11-26).

In addition to anthropogenic changes, the $\delta^{15}N$ of soils is affected by many factors, including the soil depth and therefore its interaction with atmospheric nitrogen, type, and extent of vegetation present, climate, and particle size. For natural soils the two factors, soil drainage and extent of organic litter are particularly important (Shearer and Kohl, 1988). Measurements of $\delta^{15}N$ in soil and water can be used to pinpoint the source of nitrogen (**Figure 11-28**). Animal manure, which includes nitrate, has $\delta^{15}N$ values in the range of +10‰ to +25‰ (*tropic effect*). Septic tank waste has $\delta^{15}N$ of approximately +10. Artificial fertilizers produced from atmospheric nitrogen have $\delta^{15}N$ of 0‰ ± 3‰. Also, ^{15}N-labeled fertilizer can be applied to a soil and the portion of fertilizer uptake in plants, as opposed to the portions remaining in the soil, lost by denitrification into the atmosphere and leached into runoff waters can be estimated.

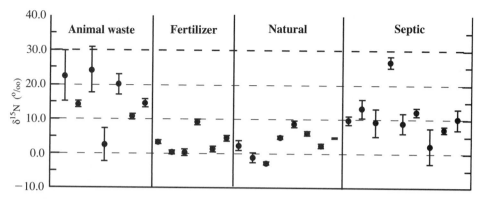

FIGURE 11-28 Literature values of $\delta^{15}N$ in vadose zone soil water beneath the indicated N sources. Dots give means and error bars are 1 standard deviation of the mean. (Adapted from Fogg et al., 1998.)

■ Boron Isotopes and *pH*

The isotopic composition of boron in marine carbonates apparently is a tracer of seawater *pH*. The concentration of boron in the ocean is 4.52 ppm with a residence time of about 16 million years. There are two stable isotopes of boron ^{10}B and ^{11}B (see Figure 10-3) making up 19.8% and 80.2% of oceanic boron, respectively. Relative to the National Bureau of Standards boric acid, SRM 957, where $^{11}B/^{10}B$ is 4.04558, $^{11}\delta B_{seawater} = +39.5‰$. In seawater, boron exists dominantly as the two species boric acid, $B(OH)_3^o$, and borate ion, $B(OH)_4^-$, with less than 10% as Na, Mg, and Ca borate species. Ignoring the smaller effects of the Na, Mg, and Ca borate species ^{11}B abundance of $B(OH)_3^o$ and $B(OH)_4^-$ in seawater are related by

$$\delta^{11}B_{seawater} = X_{B(OH)_3^o}\,\delta^{11}B_{B(OH)_3^o} + X_{B(OH)_4^-}\,\delta^{11}B_{B(OH)_4^-} \qquad [11.57]$$

where X stands for the mole fraction of the subscripted species. $B(OH)_3^o$ and $B(OH)_4^-$ abundance are related through *pH* by

$$B(OH)_3^o + H_2O = B(OH)_4^- + H^+ \qquad K_{(11.58)} = \frac{\Gamma_{B(OH)_4^-}\,X_{B(OH)_4^-}\,a_{H^+}}{\Gamma_{B(OH)_3^o}\,X_{B(OH)_3^o}\,a_{H_2O}} \qquad [11.58]$$

where $\log K_{(11.58)} = -8.830$ at 25°C. With the ratio of the Raoult's law activity coefficients $\Gamma_{B(OH)_4^-}/\Gamma_{B(OH)_3^o}$ and the activity of H_2O equal to unity the *pH* of the solution is given by

$$pH = \log\left(\frac{X_{(BOH)_4^-}}{1 - X_{B(OH)_4^-}}\right) - \log K_{(11.58)} \qquad [11.59]$$

Given in **Figure 11-29** is the speciation of boron as a function of pH at 25°C from Equation 11.59. With a pH of 8.3 at 25°C, Equation 11.59 indicates $X_{B(OH)_4^-} = 0.228$, and therefore, $X_{B(OH)_3^o} = 0.772$. For these conditions, Equation 11.57 can therefore be written as

$$+39.5\text{‰} = 0.772\ \delta^{11}B_{B(OH)_3^o} + 0.228\ \delta^{11}B_{B(OH)_4^-} \tag{11.60}$$

An isotopic exchange reaction can be written between the two boron species:

$$^{10}B(OH)_3^o + {}^{11}B(OH)_4^- = {}^{11}B(OH)_3^o + {}^{10}B(OH)_4^- \tag{11.61}$$

where

$$K_{(11.61)} = \alpha_{B(OH)_4^-}^{B(OH)_3^o} = \frac{X_{{}^{11}B(OH)_3^o} X_{{}^{10}B(OH)_4^-}}{X_{{}^{10}B(OH)_3^o} X_{{}^{11}B(OH)_4^-}} \tag{11.62}$$

$K_{(11.61)}$ has a value of 1.0194 at 25°C, 1.0206 at 0°C, and 1.0177 at 60°C, as calculated by Kakihana et al. (1977).

Because $\alpha_{B(OH)_4^-}^{B(OH)_3^o}$ has a value near 1, it can be represent by $1.0 + \varepsilon$ where $\varepsilon = 0.0N$. N is, therefore, the number of hundredths of the value above 1. A good approximation is then

$$100\ \ln\left(\alpha_{B(OH)_4^-}^{B(OH)_3^o}\right) = N \tag{11.63}$$

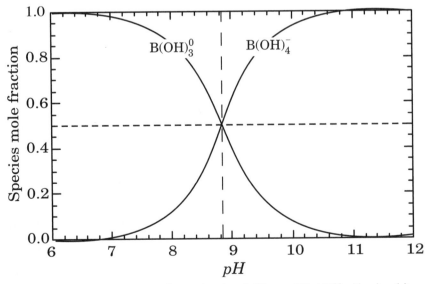

FIGURE 11-29 Distribution of boron species as a function of pH in pure H_2O at 25°C with unit activity coefficients. This can be used to approximate their relative abundance in seawater using an activity coefficient model.

Also knowing that $1.0N = \alpha^{B(OH)_3^o}_{B(OH)_4^-}$, the expression $100\,(1.0N - 1) = N$ can be written as

$$100\left(\alpha^{B(OH)_3^o}_{B(OH)_4^-} - 1\right) = N \qquad [11.64]$$

Equating N in Equations 11.63 and 11.64 gives

$$\ln\left(\alpha^{B(OH)_3^o}_{B(OH)_4^-}\right) = \left(\alpha^{B(OH)_3^o}_{B(OH)_4^-} - 1\right) \qquad [11.65]$$

The definition of α for boron isotopes, $\alpha^{B(OH)_3^o}_{B(OH)_4^-}$ given in Equation 11.6 implies

$$\alpha^{B(OH)_3^o}_{B(OH)_4^-} - 1 = \frac{R_{B(OH)_3^o} - R_{B(OH)_4^-}}{R_{B(OH)_4^-}} \qquad [11.66]$$

Also, Equation 11.5 for boron isotopes can be rearranged to give

$$R_A = \frac{R_{standard}(\delta^{11}B_A + 1000)}{1000} \qquad [11.67]$$

Taking note of Equation 11.67 for both boron species, Equation 11.66 can be written as

$$\alpha^{B(OH)_3^o}_{B(OH)_4^-} - 1 = \frac{(\delta^{11}B_{B(OH)_3^o} = 1000) - (\delta^{11}B_{B(OH)_4^-} + 1000)}{\delta^{11}B_{B(OH)_4^-} + 1000} = \frac{\delta^{11}B_{B(OH)_3^o} - \delta^{11}B_{B(OH)_4^-}}{\delta^{11}B_{B(OH)_4^-} + 1000}$$

$$[11.68]$$

Making the reasonable approximation that the denominator on the right side of the last term in Equation 11.68 can be approximated as 1000 yields

$$1000\left(\alpha^{B(OH)_3^o}_{B(OH)_4^-}\right) - 1 = \delta^{11}B_{B(OH)_3^o} - \delta^{11}B_{B(OH)_4^-} \qquad [11.69]$$

Equation 11.60 constrained at $pH = 8.3$, and Equation 11.69 can be solved to give $\delta^{11}B_{B(OH)_3^o} = 43.9\,‰$ and $\delta^{11}B_{B(OH)_4^-} = 24.5‰$. Given in **Figure 11-30** is this calculation for seawater from Hemming and Hanson (1992).

Solving Equation 11.68 for $\delta^{11}B_{B(OH)_3^o}$ and substituting it in Equation 11.57 give

$$\delta^{11}B_{seawater} = X_{B(OH)_3^o}\left[1000\left(\alpha^{B(OH)_3^o}_{B(OH)_4^-} - 1\right) + \delta^{11}B_{B(OH)_4^-}\right] + X_{B(OH)_4^-}\,\delta^{11}B_{B(OH)_4^-} \qquad [11.70]$$

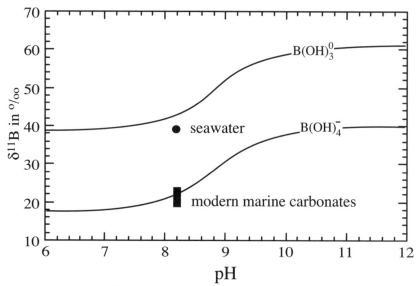

FIGURE 11-30 $\delta^{11}B_{B(OH)_3^0}$ and $\delta^{11}B_{B(OH)_4^-}$ as a function of *pH* in seawater calculated from equations in the text with log $K_{(11.58)} = -8.830$ and $K_{(11.61)} = 1.0194$. Also shown is $\delta^{11}B_{seawater}$ and the range of measured $\delta^{11}B_{carb}$ plotted at *pH* = 8.2. (Adapted from Hemming and Hanson, 1992.)

or because $X_{B(OH)_3^0} = (1 - X_{B(OH)_4^-})$ Equation 11.70 can also be written as

$$\delta^{11}B_{seawater} = (1 - X_{B(OH)_4^-})1000(\alpha^{B(OH)_3^0}_{B(OH)_4^-} - 1) +$$

$$(1 - X_{B(OH)_4^-})\delta^{11}B_{B(OH)_4^-} + X_{B(OH)_4^-} \delta^{11}B_{B(OH)_4^-} \qquad [11.71]$$

Simplifying and solving for $X_{B(OH)_4^-}$ yields

$$X_{B(OH)_4^-} = \frac{\delta^{11}B_{B(OH)_4^-} - \delta^{11}B_{seawater}}{1000\left(\alpha^{B(OH)_3^0}_{B(OH)_4^-} - 1\right)} + 1 \qquad [11.72]$$

The boron isotopic ratio in carbonate minerals, $\delta^{11}B_{carb}$, can be measured. The 10- to 70-ppm boron in carbonate (carb) is incorporated solely from the charged species $B(OH)_4^-$ in seawater substituting for CO_3^{2-}, by the reaction,

$$CaCO_{3(carb)} + B(OH)_4^- = Ca(HBO_3)_{carb} + HCO_3^- + H_2O \qquad [11.73]$$

The isotopic fractionation of carbonate/seawater $= -16‰$ to $-25‰$ depending on the carbonate secreting fauna analyzed. $\delta^{11}B_{B(OH)_4^-}$ in seawater can be determined from $\delta^{11}B_{carb}$ as this is equal to $\delta^{11}B_{B(OH)_4^-}$ in the carbonate. $\delta^{11}B_{seawater}$ can either be taken as equal to the present value +39.5 or a model using the differ-

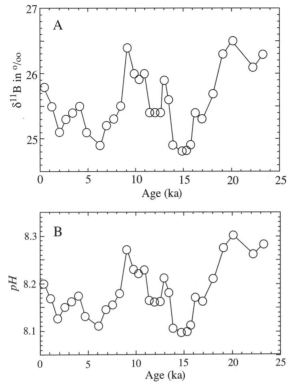

FIGURE 11-31 (A) $\delta^{11}B$ as a function of age measured on planktonic foraminifer shells of *G. sacculifer*. (B) Calculated *pH* of surface water as a function of age calculated from the measurement in (A) according to the procedure outlined in the text. (Adapted from Palmer and Pearson, 2003.)

ence between surface and deep planktonic $\delta^{11}B$ is used to model $\delta^{11}B_{seawater}$ at a particular site at a particular time. Because $\alpha^{B(OH)_3^0}_{B(OH)_4^-}$ is a function of temperature, a sea surface temperature is assumed (typically 27°C for tropical conditions independent of age). Knowing $\alpha^{B(OH)_3^0}_{B(OH)_4^-}$ at temperature, $\delta^{11}B_{B(OH)_4^-}$ in seawater and $\delta^{11}B_{seawater}$, $X_{B(OH)_4^-}$ can be calculated from Equation 11.72.

Because *pH* is a function of $X_{B(OH)_4^-}$ (Equation 11.59) the *pH* of seawater can be determined. Given in **Figure 11-31** are determinations of $\delta^{11}B_{seawater}$ from Palmer and Pearson (2003) and their calculated *pH* for surface waters. The *pH* of surface water is one of two variables required to define the state of inorganic carbon species in seawater and the CO_2 concentration in the atmosphere in equilibrium with it (see Chapter 7).

Summary

The variability in atomic weights of isotopes of the same element leads to very small differences in their chemical behavior when they are involved in a reaction.

As a result, stable isotopes fractionate to a small degree between species and phases when they undergo a change from one phase to another. For most isotopes, the difference is so small as to be undetectable in natural samples, although recent experimental advances allow these differences to be determined in an increasing number of elements. Because the energetics of isotopic exchange are so small, they do not, in general, significantly affect the rate or equilibrium of normal chemical reactions; however, when the mass difference of the isotopes is large and they participate in covalent bonding, the small differences are readily determined. Differences in stable isotope ratios of H, C, O, S, and N for a large variety of natural samples are routinely measured and place limits on various mass-dependent physical processes that occur on the earth. Because different processes fractionate the isotopes differently, measurements of isotopic ratios, in particular chemical species, phases, or reservoirs, are used to put some constraints on the magnitude of the processes responsible for the chemical fluxes between these entities.

Stable isotopic variations are characterized as a ratio of the heavy-to-light isotope reported as a difference between the ratio of the entity of concern, R_A, relative to the ratio in a standard, $R_{standard}$, times 1000, giving the "per mil" value as

$$\delta^{(heavy)} A = \left[\frac{R_A - R_{standard}}{R_{standard}} \right] \times 1000$$

Oxygen and hydrogen isotopes of water fractionate to a measurable extent during evaporation of the ocean, during precipitation from the atmosphere, and exchange between water and minerals. Because the degree of fractionation is temperature dependent, knowledge of $\delta^{18}O$ of seawater and the carbonates precipitated from the seawater is used to back calculate paleotemperatures of the ocean. $\delta^{18}O$ and δD of minerals and rocks in sedimentary basins and geothermal areas have been used to determine the extent of fluid–rock reactions in these environments. Extensive measurements of $\delta^{18}O$ and δD for igneous rocks have shown that nearly all of these rocks in the upper 10 km of the earth's crust have reacted with circulating meteoric water.

Carbon isotopes undergo significant fractionation during photosynthesis. Because different plants fractionate carbon in atmospheric CO_2 to different extents, measurements of $\delta^{13}C$ are used to place constraints on paleoenvironmental conditions. $\delta^{13}C$ of organic matter ($-30‰ < \delta^{13}C < -7.5‰$) is more negative than atmospheric CO_2 ($\delta^{13}C \sim -8‰$) and marine carbonates ($\delta^{13}C \sim 0‰$). This produces distinct $\delta^{13}C$ of different fossil fuel accumulations that are used as an atmospheric tracer when they are burned. $\delta^{13}C$ of deposited carbonate decreased sharply at the Cretaceous–Tertiary and near the Paleocene–Eocene boundary followed by slow increases. It is argued that this is due to a rapid decrease of production and burial of organic carbon and a slow return to normal productivity.

$\delta^{34}S$ of sulfide and sulfate minerals have been measured to put constraints on the fluid from which they precipitated and their temperature of formation. Differences in $\delta^{34}S$ during oxidation/reduction reactions have also been considered to help understand the reduction of sulfate that occurs in the subsurface when pyrite is formed.

Sulfur and carbon isotopes in minerals precipitated from the ocean appear to be linked as a function of geological time. This is due to the transfer of electrons between pyrite, carbonate, gypsum, and organic matter. Changes in carbon isotopes in rocks near the Permo–Triassic boundary can be used to place constraints on possible processes that may have occurred.

Nitrogen isotopes of earth material can be viewed in reference to the nitrogen cycle to place constraints on biological processes, including changes in the food chain. Because of the fractionations involved, boron isotopes in carbonate minerals have been used to decipher the *pH* of surface seawater back in geological time.

Key Terms Introduced

ammonification
assimilation
ATO
carbonate compensation depth
Calvin cycle
crassulacean acid metabolism
delta notation
denitrification
DSDP core
equilibrium isotope effects
euxinic
fractionation
fractionation factor
global meteoric water line

gram negative
Hatch-Slack cycle
kinetic isotope effects
lignin
MORB
nitrification
nitrogen fixation
oceanic anoxic events
partition function
Rayleigh distillation
Redfield ratio
standard mean ocean water (SMOW)
stable isotope
trophic effect

Questions

1. What is a stable isotope?
2. Why do some processes fractionate stable isotopes?
3. When a liquid is evaporated, why is there a slight preference for the lighter isotope to go into the vapor?
4. Which factors lead to the largest fractionation of stable isotopes?
5. How are isotopic ratios in a phase typically reported?

6. What is "delta" notation? Give the definition of α_B^A and $\Delta^{18}O_B^A$ between phase A and B?

7. What is the significance of the meteoric water line on a $\delta^{18}O$ versus δD plot?

8. What causes departures of a $\delta^{18}O$ and δD in groundwater from the meteoric water line?

9. What is Rayleigh distillation, and how is it used to explain why δ^{18} values of rainwater decrease with increasing distance inland from the ocean?

10. Explain how measurements of the isotopic ratios of oxygen in carbonates precipitated from the ocean are used to determine past oceanic temperature.

11. Why is isotopic exchange between minerals dependent on temperature but not pressure?

12. How does the C_3 Calvin cycle differ from the C_4 Hatch-Slack cycle as far as isotopic fractionation is concerned?

13. What is the record of $\delta^{13}C$ across the Cretaceous–Tertiary (K-T) boundary? Why?

14. What controls the values of $\delta^{34}S$ of sedimentary pyrite?

15. Explain how nitrogen isotopic ratios change in the food chain.

16. How can boron isotope analysis of carbonates be used to determine the pH of seawater?

Problems

1. Write the equilibrium isotopic exchange reaction for one mole of ^{34}S and ^{32}S between pyrite (FeS_2) and galena (PbS).

2. List the possible distinguishable stable isotopic species that exist for H_2O. Which have the same atomic mass?

3. A sign at a local farm stand reads this: "All produce is grown with organic fertilizers." How could you verify this by an isotopic measurement?

4. Calculate the value of $\delta^{18}O$ for meteoric water of $\delta D = -66‰$ according to the GMWL and the local meteoric water line for North America.

5. Calculate the $\delta^{18}O$ for atmospheric water at 25°C that originally had $\delta^{18}O = -5.0‰$ after 50% has precipitated. Also, determine $\delta^{18}O$ of the precipitated phase.

6. Clay minerals formed during weathering in humid environments have $\delta^{18}O$ and δD values that are largely determined by the isotopic composition of local meteoric water. Lines showing the distribution of $\delta^{18}O$ and δD in different clay minerals, when plotted on a $\delta^{18}O$ versus δD diagram, are parallel to the meteoric water line but have different

intercepts. For kaolinite, this line is $\delta D = 8\delta^{18}O - 220$, and the fractionation of oxygen between kaolinite and water is $\alpha = 1.027$. With this information, calculate $\delta^{18}O_{kaolinite}$ and $\delta D_{kaolinite}$ formed by weathering of feldspar at Manama, Bahrain and Thule, Greenland (see Figure 11-2). Construct a diagram showing the meteoric water line, the line for kaolinite passing through the two sites, and tie lines connecting the isotopic compositions of rainwater and kaolinite at Bahrain and Thule. Using the results of your computations, what is the fractionation factor for the distribution of hydrogen isotopes between kaolinite and water (from Krauskopf and Bird, 1995)?

7. Starting with the expressions for δ_A and α_B^A given in Equations 11.5 and 11.6, respectively, derive the equation $10^3(\alpha_B^A - 1) \approx \delta_A - \delta_B$ making use of the approximation $\delta_B \ll 10^3$.

8. Starting with the approximation $10^3(\alpha_B^A - 1) \approx \delta_A - \delta_B$ derived in Problem 11-7 and using the series expansion of $\ln x$ for $2 \geq x > 0$ given by $\ln x = (x-1) - \frac{1}{2}(x-1)^2 + \frac{1}{3}(x-1)^3 - \ldots$, explain why $10^3 \ln \alpha_B^A \approx \delta_A - \delta_B$ is also a good approximation.

9. $\delta^{18}O$ values of quartz and rutile of 9.63‰ and 3.17‰, respectively, have been measured in a quartz vein. With the fractionation of quartz–water of

$$1000 \ln \alpha_{qtz-water} = \frac{2.51 \times 10^6}{T^2} - 1.96 \text{ (T in degrees K)}$$

and for rutile–water of

$$1000 \ln \alpha_{rutile-water} = \frac{4.1 \times 10^6}{T^2} + 0.96 \text{ (T in degrees K)}$$

What is the temperature of oxygen isotope equilibration for quartz–rutile in the vein?

10. Three SiO_2 polymorphs are coesite, α-quartz, and stishovite. What is the likely order of fractionation of ^{18}O to ^{16}O between these minerals? Why?

11. Similar to Equation 11.30, the temperature dependence of the fraction factor of oxygen between calcite and pure water is given by O'Neal et al. (1969)

$$10^3 \ln \alpha_{water}^{calcite} = 2.78 \times 10^6/T^2 - 2.89 \text{ where T is in kelvins.}$$

(a) If calcite precipitated at 20°C in equilibrium with pure meteoric water having $\delta D = 74$‰ what is its $\delta^{18}O$ value? You can make use of the approximation introduced in Problem 11-7.

(b) Use Equation 11.14 to calculate the precipitation temperature of calcite. Why is this number lower than 20°C?

12. (a) From Figure 7-1, determine the mass percent of surface water that is ice. Assume it has an average $\delta^{18}O = -30$‰ (a reasonable approximation, see Figure 11-3). If the amount of ice on earth was doubled, what would $\delta^{18}O$ of the ocean become? (Because oxygen is approximately 99.8% ^{16}O, $\delta^{18}O$ is considered to change directly with mass changes.)

(b) If a marine carbonate produced under these conditions had a $\delta^{18}O = 32.1$‰, at what temperature would it have been formed?

(c) If today's value of $\delta^{18}O$ of the ocean was mistakenly used, what would be the calculated temperature?

13. Measured values of $\delta^{18}O$ for a reaction involving quartz and rutile from a hydrothermal ore deposit are +8.0 and –7.0‰, respectively. If the minerals formed in equilibrium and were not altered, at what temperature did the reaction occur? What was the $\delta^{18}O$ of the water present during the reaction?

14. Consider a rock composed of 35 mole % quartz with a value of $\delta^{18}O = +6.0$‰, 55 mole % alkali feldspar with a value of $\delta^{18}O = -4.0$‰ and 10 mole % mica with a value of $\delta^{18}O = -6.0$‰. Using the following expressions from Bottinga and Javoy (1975)

$$\Delta^{18}O_{feldspar}^{quartz} = \delta^{18}O_{quartz} - \delta^{18}O_{feldspar} = \frac{0.97 \times 10^6}{T^2} \quad \text{and}$$

$$\Delta^{18}O_{mica}^{quartz} = \delta^{18}O_{quartz} - \delta^{18}O_{mica} = \frac{3.69 \times 10^6}{T^2}$$

(a) Calculate an expression for $\Delta^{18}O_{feldspar}^{mica}$.

(b) If the rock is heated and isotopically equilibrates at 500°C in a closed system, what are the $\delta^{18}O$ values of the quartz, feldspar, and mica?

References

Algeo, T. J. and Seslavinsky, K. B., 1995, The Paleozoic world: continental flooding, hypsometry, and sealevel. *Amer. Jour. Sci.*, v. 295, pp. 787–822.

Anbar, A. D., 2004, Iron stable isotopes: beyond biosignatures. *Earth Planet Sci. Lett.*, v. 217, pp. 223–236.

Arthur, M. A., Schlanger, S. O., and Jenkyns, H. C., 1987, The Cenomanian-Turonian ocean anoxic event. II. Palaeoceanographic controls on organic-matter production and preservation. In Brooks, J. and Fleet, A. J. (eds.) *Marine Petroleum Source Rocks*. G.S.A. Special Pub. No. 26, Blackwell Scientific, Oxford, pp. 371–400, 401–420.

Bottinga, Y. and Javoy, M., 1975, Oxygen isotope partitioning among the minerals in igneous and metamorphic rocks. *Rev. Geophysics Space Physics*, v. 13, pp. 401–418.

Canfield, D. E., 2004, The evolution of the earth surface sulfur reservoir. *Amer. Jour. Sci.*, v. 304, pp. 839–861.

Claypool, G. E., Holser, W. T., Kaplan, I. R., Sakai, H., and Zak, I., 1980, The age curves for sulfur and oxygen isotopes in marine sulfate and their mutual interpretation. *Chem. Geol.* v. 28, pp. 199–260.

Conrad, C. P. and Lithgow-Bertelloni, C., 2006, Faster seafloor spreading and lithosphere production during the mid-Cenozoic, Paper No. 142-14, *Geological Society of America, Philadelphia Annual Meeting* (22–25 October, 2006).

Craig, H., 1965, The measurement of oxygen isotope paleotemperatures. In *Stable Isotopes in Oceanographic Studies and Paleotemperatures*. Spoleto Conference in Nuclear Geology, July 26–27, 1965. Consiglio Nazionale delle Richerche, Laboratorio di Geologia Nucleare, Pisa, pp. 1–24.

Criss, R. E., 1999, *Principles of Stable Isotope Distribution*. Oxford University Press, New York, 254 pp.

Criss, R. E., Gregory, R. T., and Taylor, H. P., 1986, Kinetic theory of oxygen isotope exchange between minerals and water. *Geochim. Cosmochim. Acta*, v. 51, pp. 1099–1108.

Delaygue, G., Jouzel, J., and Dutay, J.-C., 2000, Oxygen 18-salinity relationship simulated by an oceanic general circulation model. *Earth Planet. Sci. Lett.*, v. 178, pp. 113–123.

Erwin, D. H., 2006, *Extinction: How Life on Earth Nearly Ended 250 Million Years Ago*. Princeton University Press, Princeton, 296 pp.

Faure, G., 1986, *Principles of Isotope Geology*. John Wiley & Sons, New York, 589 pp.

Faure, K., deWit, M. J., and Willis, J. P., 1995, Late Permian global coal hiatus linked to ^{12}C-depleted CO_2 flux into the atmosphere during the final consolidation of Pangaea. *Geology*, v. 23, pp. 507–510.

Fogel, M. L. and Cifuentes, L. A., 1993, Isotope fractionation during primary production. In Engel, M. H. and Macko, S. A. (eds.) *Organic Geochemistry*. Plenum, New York, pp. 73–98.

Fogg, G. E., Rolston, D. E., Decker, D. L., Louie, D. T., and Grismer, M. E., 1998, Spatial variation in nitrogen isotope values beneath nitrate contaminated sources. *Ground Water*, v. 36, pp. 418–426.

Galloway, J. N., 1998, The global nitrogen-cycle: Changes and consequences. Environmental. *Pollution*, v. 102(S1), pp. 15–24.

Garrels, R. M. and Lerman, A., 1984, Coupling of the sedimentary sulfur and carbon cycles: An improved model. *Amer. Jour. Sci.*, v. 284, pp. 989–1007.

Garrels, R. M., Mackenzie, F. T., and Hunt, C. A., 1975, *Chemical Cycles and the Global Environment: Assessing Human Influences*. William Kaufmann, Inc., Los Altos, CA, 206 pp.

Gregory, R. T., 2002, Stable isotopes as tracers of global cycles. In *Encyclopedia of Physical Science and Technology*, 3rd edition, v. 15. Academic Press, San Diego, pp. 695–713.

Gregory, R. T. and Criss, R. E., 1986, Isotopic exchange in open and closed systems. *Rev. Mineral.*, v. 16, pp. 91–125.

Gregory, R. T., Criss, R. E., and Taylor, H. P. Jr., 1989, Oxygen isotope exchange kinetics of mineral pairs in closed and open systems: Application to problems of hydrothermal alteration of igneous rocks and Precambrian iron formations. *Chem. Geol.*, v. 75, pp. 1–42.

Hayes, J. M., Popp, B. N., Takigiku, R., and Johnson, M. W., 1989, An isotopic study of biogeochemical relations between carbonates and organic carbon in the Greenhorn Formation. *Geochim. Cosmochim. Acta*, v. 53, pp. 2961–2972.

Hemming, N. G. and Hanson, G. N., 1992, Boron isotope composition and concentration in modern marine carbonates. *Geochem. Cosmochim. Acta*, v. 56, pp. 537–543.

Hoefs, J., 1996, *Stable Isotope Geochemistry*, 4th edition. Springer-Verlag, Berlin, 212 pp.

Holser, W. T., 1984, Gradual and abrupt shifts in ocean chemistry during phanerozoic time. In Holland, H. D. and Trendall, A. F. (eds.) *Patterns of Change in Earth Evolution*. Report of the Dahlem Workshop on Patterns of Change in Earth Evolution, Berlin, 1983, May 1–6 (Physical, Chemical, and Earth Sciences Research Reports). Springer-Verlag, New York, pp. 123–143.

Holser, W. T., Schidlowski, M., Mackenzie, F. T., and Maynard, J. B., 1988, Geochemical cycles of carbon and sulfur. In Gregor, C. B., Garrels, R. M., Mackenzie, F. T., and Maynard, J. B. (eds.) *Geochemical Cycles in the Evolution of the Earth*. John Wiley and Sons, New York, pp. 105–173.

Hsü, K. J., He, Q., McKenzie, J. A., Weissert, H., Perch-Nielsen, K., Oberhänsli, H., Kelts, K., LaBrecque, J., Tauxe, L., Krähenbühl, U., Percival, S. F., Jr., Wright, R., Karpoff, A. M., Petersen, N., Tucker, P., Poore, R. Z., Gombos, A. M., Pisciotto, K., Carman, M. F., Jr. and Schreiber, E., 1982, Mass mortality and its environmental and evolutionary consequences. *Science*, v. 216, pp. 249–256.

Johnson, C. M. and Beard, B. L., 1999, Correction of instrumentally produced mass fraction during isotopic analysis of Fe by thermal ionization mass spectrometry. *Int. J. Mass Spectrom.*, v. 193, pp. 87–99.

Jouzel, J., Lorius, C., Petit, J. R., Genthon, C., Barkov, N. I., Kotlyakov, V. M., and Petrov, V. M., 1987, Vostok ice core: A continuous isotope temperature record over the last climatic cycle (160,000 years). *Nature*, v. 329, pp. 403–407.

Kajiwara, Y. and Krouse, H. R., 1971, Sulfur isotope partitioning in metallic sulfide systems. *Can. J. Earth Sci.*, v. 8, pp. 1397–1408.

Kakihana, H., Kotaka, M., Satoh, S., Nomura, M., and Okamoto, M., 1977, Fundamental studies on the ion exchange separation of boron isotopes. *Bulletin Chemical Society Japan*, v. 50, pp. 158–163.

Keeling, C. D., Mook, W. C., and Tans, P. P., 1979, Recent trends in the $^{13}C/^{12}C$ ratio of atmospheric carbon dioxide. *Nature*, v. 277, pp. 121–123.

Kieffer, S. W., 1982, Thermodynamics and lattice vibrations of minerals. 5. Application to phase equilibria, isotopic fractionation, and high-pressure thermodynamic properties. *Rev. Geophys. Space Phys.*, v. 20, pp. 827–849.

Krauskopf, K. B. and Bird, D. K., 1995, *Introduction to Geochemistry*, 3rd edition. McGraw-Hill, New York, 647 pp.

Krull, E. S., Lehrmann, D. J., Druke, D., Kessel, B., Yo, Y., and Li, R., 2004, Stable carbon isotope stratigraphy across the Permian-Triassic boundary in shallow marine carbonate platforms, Nanpanjiang Basin, south China. *Palaeogeography, Palaeoclimatology, Palaeoecology*, v. 204, pp. 297–315.

Koch, P. L., Fogel, M. L., and Tuross, N., 1994, Tracing diets of fossil animals using stable isotopes. In Lajtha, K. and Michner, R. H. (eds.) *Stable Isotopes in Ecology and Environmental Science*, Blackwell, Oxford, pp. 63–92.

Kyser, T. K., 1987, Equilibrium fractionation factors for stable isotopes. In Kyser, K. K. (ed.) *Stable Isotope Geochemistry of Low Temperature Fluids*, Short-Course v. 13 Mineralogical Association of Canada, Ontario, pp. 1–84.

Larson, P. B., Maher, K., Ramos, F. C., Chang, Z., Gaspar, M., and Meinert, L. D., 2003, Copper isotope ratios in magmatic and hydrothermal ore-forming environments. *Chem. Geol.*, v. 201, pp. 337–350.

Mackenzie, F. T. and Agegian, C. R., 1989, Biomineralization and tentative links to plate tectonics. In Crick, R. E. (ed.) *Origin, Evolution, and Modern Apects of Biomineralization in Plants and Animals*. Plenum Press, New York, pp. 11–27.

Martinson, D. G., Pisias, N. G., Hayes, J. D., Imbrie, J., Moore, T. C., Jr., and Shackleton, N. J., 1987, Age dating and the orbital theory of the ice ages: development of a high-resolution 0 to 300,000-year chronostratigraphy. *Quat. Res.*, v. 27, pp. 1–29.

Minagawa, M., and Wada, E., 1984, Stepwise enrichment of 15N along food chains: Further evidence and the relation between δ15N and animal age. *Geochim. Cosmochim. Acta*, v. 48, pp. 1135–1140.

Mojzsis, S. J., Harrison, T. M., and Pidgeon, R. T., 2001, Oxygen isotope evidence from ancient zircons for liquid water at the Earth's surface 4,300 Myr ago. *Nature*, v. 409, pp. 178–181.

O'Neil, J. R., Clayton, R. N., and Mayeda, T. K., 1969, Oxygen isotope fractionation in divalent metal carbonates. *J. Chem. Phys.*, v. 51, pp. 5547–5558.

Palmer, M. R. and Pearson, P. N., 2003, A 23,000-year record of surface water pH and P_{CO_2} in the Western Equatorial Pacific Ocean. *Nature*, v. 300, pp. 480–482.

Reichow, M. K., Saunders, A. D., White, R. V., Pringle, M. S., Al'Mukhamedov A. I., Medvedev, A. I. and Kirda, N. P., 2002, $^{40}Ar/^{39}Ar$ Dates from the West Siberian Basin: Siberian flood basalt province doubled. *Science* 7 June 2002, v. 296. no. 5574, p. 1846 – 1849.

Rozanski, K., Araguás-Araguás, L., and Confiantini, R., 1993, Isotopic patterns in modern global precipitation. In Swart, P. K., Lohmann, K. C., McKenzie, J., and Savin, S. (eds.) *Climate Change in Continental Isotopic Records*. Geophysical Monograph 78, *Amer. Geophy. Union*, Washington, DC, pp. 1–15.

Schlanger, S. O., Arthur, M. A., Jenkyns, H. C., and Scholle, P. A., 1987, The Cenomanian-Turonian ocean anoxic event. I. Stratigraphy and distribution of organic carbon-rich beds and the marine $\delta^{13}C$ excursion. In Brooks, J. and Fleet, A. J. (eds.) *Marine Petroleum Source Rocks*. G.S.A. Special Pub. No. 26. Blackwell Scientific, Oxford, pp. 371–400.

Scholle, P. A. and Arthur, M. A., 1980, Carbon isotope fluctuations in Cretaceous pelagic limestones: Potential stratigraphic and petroleum exploration tool. *Amer. Assoc. Petrol. Geol. Bull.*, v. 64, pp. 67–87.

Shearer, G. and Kohl, D. H., 1988, Nitrogen isotopic fractionation and ^{18}O exchange in relation to the mechanism of denitrification of nitrite by *Pseudomonas stutzeri. J. Biol. Chem.* v. 263, pp. 13231–13245.

Stedman, D. H. and Shetter, R. E., 1983, The global budget of atmospheric nitrogen species. In Schwartz, S. E. (ed.) *Trace Atmospheric Constituents*. Adv. Environ. Sci. Technol., v. 12, Wiley, New York, pp. 411–454.

Shackleton, N. J., 1987, The carbon isotope record of the Cenozoic: History of organic carbon burial and of oxygen in the ocean and atmosphere. In Brooks, J. and Fleet, A. J. (eds.) *Marine Petroleum Source Rocks*. Geol. Soc. Spec. Publ. no. 26. Blackwell Scientific, Oxford, pp. 423–434.

Sheppard, S. M. F., Nielsen, R. L., and Taylor, H. P., Jr., 1969, Oxygen and hydrogen isotopic ratios of clay minerals from porphyry copper deposits. *Econ. Geol.*, v. 64, pp. 755–777.

Söderlund, R. and Svensson, B. H., 1976, The global nitrogen cycle: Nitrogen, phosphorus and sulphur—*Global Cycles*. SCOPE 7 Report. *Ecol. Bull. (Stockholm)*, v. 22, pp. 23–73.

Veizer, J. and Hoefs, J., 1976, The nature of $^{18}O/^{16}O$ and $^{13}C/^{12}C$ secular trends in sedimentary carbonate rocks. *Geochim. Cosmochim. Acta*, v. 40, pp. 1387–1395.

Veizer, J., Ala, D., Azmy, K., Bruckschen, P., Buhl, D., Bruhn, F., Carden, G. A. F., Diener, A., Ebneth, S., Godderis, Y., Jasper, T., Korte, C., Pawellek, F., Podlaha, O. G., and Strauss, H., 1999, $^{87}Sr/^{86}Sr$, $\delta^{13}C$ and $\delta^{18}O$ evolution of Phanerozoic seawater. *Chem. Geol.*, v. 161, pp. 59–88.

Welch, S. A., Beard, B. L., Johnson, C. M., and Braterman, P. S., 2003, Kinetic and equilibrium Fe isotope fractionation between aqueous Fe(II) and Fe(III). *Geochim. Cosmochim. Acta*, v. 67, pp. 4231–4250.

Wombacher, F., Rehkämper, M., Mezger, K., and Münster, C., 2003, Stable isotope compositions of cadmium in geological materials and meteorites determined by multiple-collector ICPMS. *Geochim. Cosmochim. Acta*, v. 67, pp. 4639–4654.

Yurtsever, Y. and Gat, J. R., 1981, Atmospheric waters. In Gat, J. R. and Gonfiantini, R. (eds.) *Stable Isotope Hydrology, Deuterium and Oxygen-18 in the Water Cycle*. Technical Reports Series No. 219, International Atomic Energy Agency, Vienna.

12

Surface Sorption Geochemistry

At the earth's surface and in the crust, whenever chemical reactions are occurring, that is, when one mineral or group of minerals react to form other minerals, the reactions occur at the mineral surfaces with the aid of a fluid phase. This can either be a magma or aqueous phase. When aqueous fluids are involved, these reactions describe dissolution of the unstable mineral and then precipitation of new stable minerals out of the aqueous phase. Sorptive species at the mineral–water interface control the rate of these aqueous surface reactions. *Surface sorption* on minerals also influences the rate at which material is transported by fluid flow and diffusion in groundwater. For instance, if some contaminant species in a fluid are bound to a mineral surface, their ability to be transported in the fluid is prevented. It is for this reason that the distribution of pesticides and fertilizers in soils depends on mineral surface reactions; therefore, to understand changes in the quality of groundwater and the extent of migration of contaminants and what controls the rate of dissolution and growth of minerals, the extent of sorptive species reactions on organic matter and mineral surfaces must be characterized.

Mineral surface sorption involves the removal of aqueous species from solution and their attachment to the mineral's surface. Three types of surface sorption are identified: *absorption*, which implies the species is incorporated inside the solid; *adsorption*, where the species is held on the surface either electrostatically or through sharing of electrons with atoms in the mineral; and a precipitate from solution forming on the mineral's surface.

H^+ and other positively charged solution species adsorb to negative surfaces and vice versa (see below). At constant pressure, temperature, and *pH*, sorption typically increases with increasing concentration of the *sorbate* in solution, the sorbate being the species that is removed from the solution. The term *sorbent* refers to the phase to which the sorbate enters or attaches to. The reactions between sorbate and sorbent are characterized by *sorption isotherms*, as shown in **Figure 12-1**. These give the amount of sorbate attached to the sorbent per unit concentration of sorbate in solution. Figure 12-1 indicates that

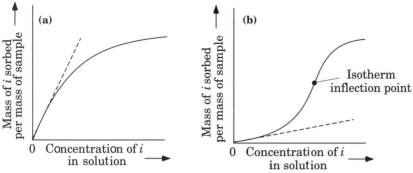

FIGURE 12-1 Amount of substance *i* attached to or incorporated in a solid as a function of its concentration in solution at constant pressure, temperature, and *pH* for substances that display two types of behavior: (a) Freundlich sorption isotherm behavior or (b) S-type sorption isotherm behavior.

at low concentrations there is nearly a linear relation between the amount added to solution and the amount incorporated in and/or on the mineral surface, as given by the dashed line. This occurs because the chemical environment on the surface and in solution is largely unaffected when trace sorption occurs in dilute solutions at constant *pH*. At higher concentrations of *i*, the mass of *i* adsorbed per unit increase in solution concentration either decreases or increases. Most often the slope on a sorption isotherm decreases with increasing concentration in solution, as shown in Figure 12-1a. This is attributed to a decrease in the number of unoccupied sorption sites on the solid surface. Also, repulsive interactions occur between attached charged species and those in solution with charge of the same sign.

In S-type sorption, as given in Figure 12-1b, the sorbate is likely interacting with other sorbate species on the surface or *ligands* in solution, causing more sorption per unit increase in solution, at least initially. Generally, no matter what behavior occurs initially, the sorption isotherm at high enough solution concentrations flattens to horizontal, and no more sorption occurs as the solution concentration increases. The exception is when more than one layer of sorption is possible. In this case, the mass of *i* sorbed increases to large values before the solution concentration of *i* becomes independent of the amount of *i* sorbed. If a surface precipitate is formed, the sorbent concentration is controlled by the precipitate's solubility.

■ Freundlich Sorption Isotherm

A simple way to describe the common sorption isotherm behavior given in Figure 12-1a and 12-1b up to its inflection point is to use the *Freundlich sorption* isotherm equation,

$$C_i^{\text{sample}} = K_f \, (C_i)^m \qquad\qquad\qquad [12.1]$$

where C_i^{sample} specifies the mass of i sequestered in or on the sample per unit mass of solid sample. C_i denotes the concentration of i in solution, and m represents an exponent on C_i. K_f is the Freundlich sorption coefficient. K_f and m are generally considered to be constants at constant temperature, pressure, and pH. Taking the logarithm of both sides of Equation 12.1 yields

$$\log C_i^{\text{sample}} = m \log C_i + \log K_f \qquad\qquad\qquad [12.2]$$

Plots of $\log C_i^{\text{sample}}$ as a function of $\log C_i$ are, therefore, straight lines with a slope m and intercept of $\log K_f$, as shown in **Figure 12-2**. Linear plots of measurements of $\log C_i^{\text{sample}}$ versus $\log C_i$ are used to determine K_f and m. Note that Freundlich sorption displays no a priori limit to the amount of possible sorption.

To help understand the mechanisms by which sorption between solids and aqueous species in solution occur, consider the interface region. At the interface, discontinuities in density, composition, and structure occur in both the mineral and aqueous solution. In the mineral, the repeating arrangement of atoms that defines the structure of the mineral is terminated. At the interface, therefore, part of the coordination environment of the atoms is either missing or different from the interior of the mineral. The mineral surface, therefore, has different properties, including thermodynamic properties, from the bulk of the mineral and generally manifests a surface charge. **Figure 12-3** displays the surface charge of a number of solids in solution as a function of pH at earth surface conditions. The reason the surface charge is a function of pH is that H^+ and OH^- are adsorbed to the surface and change the surface charge (see below). Silicate min-

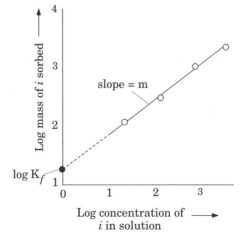

FIGURE 12-2 Logarithm of mass of sorbent versus the log concentration in solution for a substance that displays Freundlich sorption behavior.

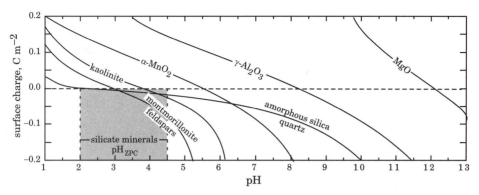

FIGURE 12-3 Surface charge at 25°C and 1 bar in coulombs per meter squared of some solids as a function of *pH*. Because the curves but not the *pH* where the surface has zero charge are dependent on ionic strength, they are meant to exemplify trends in a semiquantitative way. (Adapted from Stumm and Morgan, 1981.)

erals tend to have a zero point of charge at *pH* 4.5 or less. At *pH* above 4.5, therefore, silicate minerals in aqueous solutions manifest a negative charge.

Because of surface charge, oppositely charged aqueous species are adsorbed from the solution to the mineral surface. For this reason, water near the surface has a different concentration of solutes from the bulk solution. The surface, therefore, has a volume, that is, a mass in both the mineral and aqueous solution. The thickness of this surface layer is given by the distance over which the repeating crystalline structure of the mineral stops, through the layer of adsorbed species, to the bulk solution unaffected by the mineral surface. Because the surface involves a layer on the mineral and a layer in the solution, it is referred to as a *double layer*, as shown in **Figure 12-4**. In fresh water at earth surface conditions, the fluid part of the double layer is from about 5 to 10 nm = 50–100 Å thick.

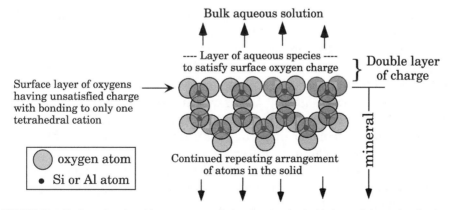

FIGURE 12-4 Surface of a mineral in an aqueous solution showing the double layer of charge that develops when the surface oxygens of the mineral manifest a charge that is neutralized by a layer of charged aqueous species.

■ Thermodynamics of Surfaces

Despite the force of gravity, a droplet of water placed on a table does not flatten out to a single layer of H_2O molecules but rather retains a flattened spherical shape. This is an equilibrium shape because if an additional force is put on the droplet, it deforms. When the force is removed, the droplet returns to its original shape. The droplet of water, therefore, has a *surface energy* that balances the gravitational energy to flatten it. The Gibbs energy per unit surface area, γ_s, is defined as

$$\gamma_s = \left(\frac{\partial G}{\partial A_s} \right)_{P, T, n} \qquad [12.3]$$

where A_s is the area of the surface, whereas P, T, and n indicate that pressure, temperature, and the chemical potential of all of the n components in the phase are kept constant. Because the droplet of H_2O tends to a state of minimization of surface area rather than increasing its surface area by spreading out under the force of gravity, γ_s must be positive. If the Gibbs energy of a system depends on the surface area, the total differential of G needs to reflect this dependence. When there are j such different surfaces of molar surface area, A_j, the total differential of G depends on T, P, chemical potential of the systems components, and the j surface areas present so that

$$dG = -SdT + VdP + \sum_i \mu_i dn_i + \sum_j \gamma_{s_j} dA_j \qquad [12.4]$$

Equation 12.4 indicates that for a phase of a given mass as particle size decreases, that is, its surface area per mole increases, the molar Gibbs energy increases because the γ_s are positive.

Consider a mole of 1-cm cubes of quartz at earth surface conditions. With a density of 2.65 g cm^{-3} and a molecular weight of 60.08 g mol^{-1}, the surface area of these cubes is 136 cm^2 mol^{-1}. If these cubes are broken into 1-µm cubes, the surface area is four orders of magnitude greater, that is, 1.36×10^6 cm^2 mol^{-1}. This increased surface area with a positive surface Gibbs energy causes small particles to have a greater Gibbs energy than large particles and therefore to have a greater solubility in solution. The effect generally becomes significant when the diameter of particles is about 100 µm or less. Many clay minerals, therefore, exhibit the effect.

The changes in the Gibbs energy of a surface layer, dG_s, depend on changes in both chemical composition and surface area. At constant pressure and temperature, the relationship is

$$dG_s = \sum_i \mu_i dn_i + \gamma_s dA_s \qquad [12.5]$$

By definition, the G_s of the surface phase is given by the sum of the Gibbs energy of each component in the surface phase plus the total surface Gibbs energy. The Gibbs energy for each component is equal to its chemical potential, μ_i, times the number of moles of that component, n_i, in the surface. The surface Gibbs energy is given by the Gibbs energy per unit surface, γ_s, times the amount of surface, A_s, so that

$$G_s = \sum_i \mu_i n_i + \gamma_s A_s \qquad [12.6]$$

Taking the total differential of both sides of Equation 12.6 yields

$$dG_s = \sum_i \mu_i dn_i + \sum_i n_i d\mu_i + \gamma_s dA_s + A_s d\gamma_s \qquad [12.7]$$

Subtracting Equation 12.5 from 12.7 and solving for $d\gamma_s$ gives

$$d\gamma_s = -\sum_i \frac{n_i}{A_s} d\mu_i \qquad [12.8]$$

Because of the minus sign, Equation 12.8 indicates that the adsorption of component i at the surface lowers the surface phase's chemical potential, that is, its Gibbs energy. This is the thermodynamic argument as to why adsorption occurs.

Figure 12-5 shows the types of species that are sorbed to the solid surface. These are called surface complexes. If water molecules are adsorbed between the metal center sorbed and the surface, the complex is termed an *outer sphere complex*, as shown in Figure 12-5a. If the metal center is bound directly to the surface, the complex is denoted as *inner sphere*, as shown in Figure 12-5b. As with functional groups on organic molecules, these inner sphere complexes are understood in terms of bonding to specific sites on the mineral surface.

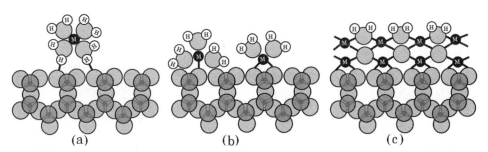

(a)   (b)   (c)

FIGURE 12-5 Three kinds of adsorption of a species at a surface of a mineral oxide that occur in an aqueous solution. The gray circles represent oxygen and the small back circles Si and Al; the circles labeled with an M denote metal centers, whereas the circles with an H stand for hydrogen atoms bound to the oxygens. (a) An outer sphere complex with H_2O between the metal center and the mineral surface. (b) Inner sphere complexes with the metal atom bonded directly to the surface. (c) A surface precipitate. (Adapted from Brown, 1990.)

Surface precipitates also form if sorption complexes cover the surface and start to bond together, as shown in Figure 12-5c.

When a mineral phase with significant surface area is added to pure water, the pH of the water invariably increases in less than a minute to a new nearly steady-state value. At longer times, on the order of tens of minutes, the pH begins to slowly drift lower from the value obtained after the rapid initial pH shift. This initial shift is due to rapid adsorption of H^+ to the mineral surface. The drift of pH at greater times is due to mineral dissolution and diffusion of hydrogen atoms into the mineral surface phase. Reactions can be written to characterize this initial H^+ adsorption.

Silicate Surfaces

To assess these reactions, consider the surface of a quartz grain as shown schematically in **Figure 12-6**. Si is tetrahedrally bonded to oxygen atoms in quartz, with each oxygen bound between two Si. If an oxygen resides at the surface, its coordination is reduced because it is typically bonded to only one Si within the interior. In water, the surface coordinates H^+ from solution to balance this negative charge if this lowers the Gibbs energy of the surface double layer. That is, protons are adsorbed to balance the negative valence charge on the surface oxygen. Depending on Gibbs energy constraints, the surface undergoes reactions with H^+ in solution to become neutral or positively charged. More H^+ is adsorbed as the H^+ concentration in the solution increases and the pH of the solution decreases, forming a positively charged surface at low pH.

Reactions can be written between the surface species that develop. If H^+ is the only species adsorbed, the reactions that occur and their equilibrium constants are expressed as

$$\equiv SiO^- + H^+ = \equiv SiOH \qquad K_{(12.9)} = \frac{\gamma_{\equiv SiOH}}{\gamma_{\equiv SiO^-} a_{H^+}} \frac{[\equiv SiOH]}{[\equiv SiO^-]} \qquad [12.9]$$

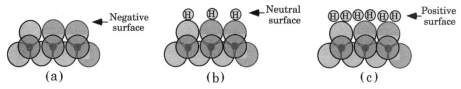

FIGURE 12-6 Schematic diagram of the surface of quartz in contact with an aqueous solution. Dark gray circles represent oxygen, black circles denote silicon, and the H-labeled circles indicate adsorbed H^+ ions. (a) High pH solution where H^+ in solution is low and little adsorption occurs and the surface is negative. (b) As solution pH decreases, some H^+ is adsorbed and the charge of the surface becomes neutral. (c) At low pH, the surface becomes positive as more H^+ from the solution is adsorbed.

and

$$\equiv SiOH + H^+ = \equiv SiOH_2^+ \qquad K_{(12.10)} = \frac{\gamma_{\equiv SiOH_2^+}}{\gamma_{\equiv SiOH}\, a_{H^+}} \frac{[\equiv SiOH_2^+]}{[\equiv SiOH]} \qquad [12.10]$$

In Equations 12.9 and 12.10, the species $\equiv SiOH$, $\equiv SiOH_2^+$, and $\equiv SiO^-$ represent the neutrally, positively, and negatively charged surface complexes, respectively, as shown in Figure 12-6. The $\equiv$ is used to represent the valence bonds of a surface Si to three oxygen atoms in the mineral. a_{H^+} denotes the activity of H^+ in the aqueous solution. [ ] indicates concentration in sites per unit surface area. γ denotes the activity coefficient of the subscripted surface species based on a hypothetical standard state of 1 mole of surface species per unit surface area behaving as though they are neutrally charged with no interactions with other species on the surface or in solution.

As an ion, i, in solution approaches a charged surface, its chemical environment changes. The Gibbs energy of i includes an electrical field energy term because of the charged surface. The added energy is the work necessary to bring the charged species from the bulk solution to its position in the electric field. At constant pressure and temperature, this energy can be characterized by

$$\mu_i = RT\ln a_i = RT\ln \gamma_i + RT\ln m_i \qquad [12.11]$$

where γ_i must account for interactions with other species and the extra energy because of its position in the electric field. This extra energy per mole of species with a valence, z, is given by the *electrostatic potential* of the field, Ψ, times a mole of charge, that is, a faraday, F, times z, so that

$$RT\ln\gamma_i = RT\ln\gamma_i^\infty + \Psi F z \qquad [12.12]$$

where γ_i^∞ is the activity coefficient of i in the bulk solution (at infinite distance from the surface).

A potential is a value at a location in a field that gives the amount of a variable at the location relative to other locations in the field. For instance, the change in gravitational potential gives the change in gravitational energy with location for a unit mass. Electrostatic potential is the potential due to being in an electrical field for a unit charge.

The energy difference of the activity coefficient of a charged and uncharged surface site is given by the charge difference times Faraday's constant, F, times the electrostatic potential at the surface, Ψ^o, as the γ_i^∞ cancels. In this case, the ratio of activity coefficients of the charged and uncharged surfaces is given by

$$\frac{\gamma_{\equiv SiO^-}}{\gamma_{\equiv SiOH}} = \frac{\gamma_{\equiv SiOH}}{\gamma_{\equiv SiOH_2^+}} = e^{(-F\Psi^o/RT)} \qquad [12.13]$$

where Ψ° is the electrostatic potential at the surface relative to the bulk solution. The electrostatic potential at the surface is a scalar voltage, which indicates the strength of the electric field. The energy needed to bring a mole of charges from the bulk solution to the surface is then $F\Psi^\circ$. This charging is non-*Nernstian*, and Ψ° is not directly measurable and depends on the solution's *ionic strength*. A model must therefore be used to determine the activity coefficient ratios in Equations 12.9, 12.10, and 12.13. The values of $\log K_{(12.9)}$ and $\log K_{(12.10)}$ are experimentally determined. At 25°C and 1 bar for quartz surfaces, they are 6.8 and –1.8, respectively.

Aluminum Surfaces

A similar analysis of corundum and gibbsite gives the charging reactions for an aluminum surface species as

$$\equiv\!AlO^- + H^+ = \equiv\!AlOH \qquad K_{(12.14)} = \frac{\gamma_{\equiv AlOH}}{\gamma_{\equiv AlO^-}\, a_{H^+}} \frac{[\equiv\!AlOH]}{[\equiv\!AlO^-]} \qquad [12.14]$$

and

$$\equiv\!AlOH + H^+ = \equiv\!AlOH_2^+ \qquad K_{(12.15)} = \frac{\gamma_{\equiv AlOH_2^+}}{\gamma_{\equiv AlOH}\, a_{H^+}} \frac{[\equiv\!AlOH_2^+]}{[\equiv\!AlOH]} \qquad [12.15]$$

Log $K_{(12.14)}$ and log $K_{(12.15)}$ at 25°C and 1 bar on corundum have been experimentally determined to be 8.0 and 9.0, respectively. These reactions give the *pH*-dependent charge on the surface.

■ *pH* of Zero Point of Charge

The *pH* of zero point of charge, pH_{zpc}, of a simple single cation oxide or hydroxide is obtained by combining the equilibrium constant expressions. Taking the logarithm of these equilibrium constants for a Si—O surface and combining them gives

$$\log K_{(12.9)} + \log K_{(12.10)} = \log\left[\frac{\gamma_{\equiv SiOH_2^+}}{\gamma_{\equiv SiO^-}\, a_{H^+}^2} \frac{[\equiv\!SiOH_2^+]}{[\equiv\!SiO^-]}\right] \qquad [12.16]$$

Because the activity coefficients of the charged surface species are considered to be of similar magnitude, their ratio is taken as unity. Remembering that $pH = -\log a_{H^+}$, the *pH* where $[\equiv\!SiO^-] = [\equiv\!SiOH]$, that is, the *pH* where the surface has no net charge, pH_{zpc}, is obtained by solving Equation 12.16 for a_{H^+} to give

$$pH_{zpc} = 0.5(\log K_{(12.9)} + \log K_{(12.10)}) \qquad [12.17]$$

At solution pH less than the pH_{zpc}, the surface is positively charged, and above the pH_{zpc}, it has a negative charge.

■ Surface Charge Measurements

The charge on a surface as a function of pH is commonly obtained by two different types of techniques: by applying an electric field to the surface and by titrating the surface with acid and base. With the *electrophoresis* technique, the pH of the isoelectric point, pH_{iep}, where the surface has no charge is determined by preparing solutions of a given pH that include fine particles of high surface area of the solid. When an electric potential field is applied, if the mineral particles have a surface charge, they move toward the oppositely charged electrode in the field. At pH_{iep}, the mineral particles do not move in the field. Their density then remains constant between the electrodes. Rather than having particles move in a stationary fluid with the streaming potential technique, a solid of interest is kept stationary, and an ionic fluid is put in motion over the surface. From changes in the applied electric field potential, the pH_{iep} is determined.

With the surface titration technique, an aqueous solution of fine particles of high surface area of the solid is prepared and the pH measured. Acid or base is titrated into the solution, and the pH is measured again. From the amount of acid or base added and the difference in pH of the solution before and after the addition, the amount of H^+ adsorbed to the surface or released to solution can be calculated. At the pH_{zpc}, the addition of acid or base changes the solution pH with no H^+ adsorbed or released. In theory, pH_{iep} and pH_{zpc} should be equal. In practice, at times they are somewhat different. **Table 12-1** lists a compilation of some pH_{zpc} determined by various investigators. Difficulties arise because of the sensitivity of the measurements to small amounts of contamination and interactions with ions contributed from dissolution during the course of the measurements.

Figure 12-7 gives the surface charge per unit surface area as a function of pH determined by surface titration measurements of quartz and corundum. In this figure, surface charge is given in moles of electron charges per cm^2. There are 6.24×10^{18} electron charges (e.c.) per coulomb (C) and 6.02×10^{23} charges per mole so that

$$10^{-10} \text{ mol cm}^{-2} \times \frac{6.02 \times 10^{23} \text{ e.c. mol}^{-1}}{6.24 \times 10^{18} \text{ e.c. C}^{-1} \times 10^{-4} \text{ m}^2 \text{ cm}^{-2}} = 0.0965 \text{ C m}^{-2} \qquad [12.18]$$

Using this relation, Figure 12-3 can be directly compared with Figure 12-7. The Si surface sites on quartz remain negative throughout the pH region shown in Figure 12-7, as its pH_{zpc} is 2.5 ± 0.6.

Table 12-1 | *pH* of zero points of charge determined for some minerals in aqueous solution at 25°C and 1 bar

Mineral	pH_{zpc}
Periclase	12.4
Chrysotile	>10
Gibbsite	10
Calcite	9.5
Corundum	8.5
Hematite	8.5
Goethite	7.3–7.8
Magnetite	6.5–6.6
Anatase	5.8
Zircon	5
Kaolinite	2–4.6
Quartz	2–3
Feldspars	2–2.4
Montmorillonite	2–3
Albite	2.0

From Parks (1967) and Stumm and Morgan (1981).

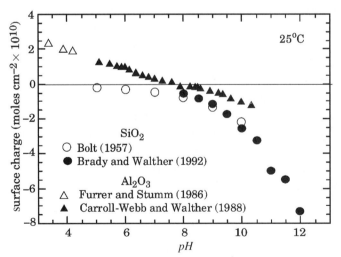

FIGURE 12-7 Surface charge on SiO_2 (circles) in 0.001 M NaCl and Al_2O_3 (triangles) in 0.1 M $NaNO_3$ determined from surface titration as a function of *pH* at 25°C. (Walther, 1996.)

■ Aluminosilicate Surface Charge

With knowledge of the behavior of Si and Al sites on simple oxides, the charge on an aluminosilicate mineral surface is considered. First, surface charge occurs from imperfections in the mineral. Cations of one valence replace cations of another valence because of mineral lattice imperfections. Those that occur at the surface produce a permanent surface charge, σ_p. Perhaps the most common is Al replacement of Si in surface Si—O tetrahedra in clay mineral sheets, producing a negative surface charge. Surface charge also develops because surface cations are lost when a mineral is immersed into an aqueous solution. That is, if alkalis or alkaline earths are present in the mineral structure, pH-dependent alkali and alkaline earth-leaching reactions occur on the mineral surface. The loss of these cations at the surface produces a negative surface layer, whose charge is designated as σ_c. While the leaching is occurring, the charged mineral surface also undergoes *coordinative reactions* with the aqueous solution as discussed above for H^+ on quartz and corundum. These reactions are pH dependent and occur between aqueous species and surface oxygens that are in coordination with cations within the mineral. This charge is denoted as σ_{cr}. The total net charge on a mineral surface, σ_o (charge per cm^2), is then

$$\sigma_o = \sigma_p + \sigma_c + \sigma_{cr} \qquad [12.19]$$

This surface charge is balanced by a charge, σ_d, as a result of a diffuse layer of oppositely charged counter ions accumulating in solution near the solution–mineral surface so that overall charge neutrality is maintained. These are any cation or anion in solution including H^+ and OH^-. As pointed out above, this creates an electric double layer where the surface charge plus the layer of oppositely charged ions in solution sums to zero. In other words, there exists a localized disturbance of electroneutrality, but the system as a whole is in electrical balance.

Different models have been developed to explain the behavior of charged mineral surface layers in solution. In a Gouy-Chapman type surface charging model

$$\sigma_o + \sigma_d = 0 \qquad [12.20]$$

Often, a more complex surface charging model is used containing an additional surface layer, termed a *Stern layer*, to account for differences between charges of inner and outer sphere surface complexes. In this case

$$\sigma_o + \sigma_\beta + \sigma_d = 0 \qquad [12.21]$$

where σ_β is the charge in the Stern or β layer. This charging model is often referred to as a *triple-layer model* because of the additional layer of charge. In these one-dimensional models, the charge, q, is related to the electrostatic potential, Ψ, by *Poisson's equation* of electrostatics,

$$\frac{d^2\Psi}{dx^2} = -\frac{q}{\varepsilon\varepsilon_o} \tag{12.22}$$

where q is the volumetric charge density, ε denotes the dielectric constant of the solution (taken as pure water = 78.3 at 25°C), and ε_o is the *permittivity* in a vacuum (8.854×10^{-14} C V^{-1} cm^{-1}). The dielectric constant and permittivity are discussed in Chapter 6 where the properties of water molecules are considered.

In both charging models, the charge density is integrated from the start of the diffuse layer to infinity to give the total charge in the diffuse layer of

$$\sigma_d = (8RT\varepsilon\varepsilon_o I \times 10^3)^{1/2}\sinh\left(\frac{zF\Psi_d}{2RT}\right) \tag{12.23}$$

where I signifies the solution's ionic strength (M), z represents the ionic charge at the surface, and Ψ_d denotes the electrostatic potential (V) at the start of the diffuse layer. At 25°C, Equation 12.23 becomes

$$\sigma_d = 0.1174I^{1/2} \sinh(19.46z\Psi_d) \tag{12.24}$$

In the Gouy-Chapman model, $\Psi^o = \Psi_d$, and in a triple-layer model, $\Psi^o = \Psi_\beta + \Psi_d$. **Figure 12-8** shows charge distributions as a function of distance from the surface for the two models. The width of the diffuse layer, w_d, is equal to

$$w_d = \left(\frac{\varepsilon\varepsilon_o RT}{2F^2 10^3 I}\right)^{1/2} \tag{12.25}$$

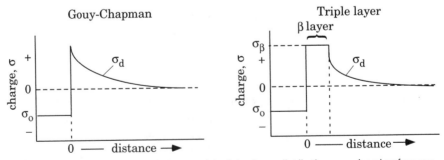

FIGURE 12-8 Gouy-Chapman and triple layer models of the charge distribution on a mineral surface as a function of distance from the surface. The negative internal surface charge is balanced by a positive charge from adsorbed and diffuse aqueous species that decrease at increasing distance from the surface.

In water at 25°C and 1 bar, this width is computed to be $2.8 \times 10^{-8} \, I^{-1/2}$ cm. The higher the ionic strength, the more compact is the diffuse layer because a greater concentration of counter ions is near the surface. In 0.1 M electrolyte solution, the diffuse layer is computed to be 8.9×10^{-8} cm = 8.9 Å, or only about three molecules of H_2O in thickness on average.

■ Metal Ion Adsorption

Consider the adsorption of a metal, M^{2+}, from a liter of solution initially at a concentration of 10^{-5} M onto 1 g of silica. Further assume the silica has 10^{-4} mol of surface sites per gram. (With a surface area = 100 m² g, there are approximately 6 sites nm^{-2}.) This adsorption reaction and its equilibrium constant are

$$\equiv SiOH + M^{2+} = \equiv SiOM^+ + H^+ \quad K_{(12.26)} = \left[\frac{\gamma_{\equiv SiOM^+}}{\gamma_{\equiv SiOH} \, \gamma_{M^{2+}}} \frac{[\equiv SiOM^+] \, a_{H^+}}{[\equiv SiOH][M^{2+}]} \right] \quad [12.26]$$

The amount of M^{2+} adsorbed, therefore, depends on pH. With the surface constants for H^+ adsorption given in Equations 12.9 and 12.10 and assuming activity coefficients are unity, the distribution of solution and surface species in the system is determined by knowing $K_{(12.26)}$.

From Equations 12.9, 12.10, and 12.26, the total number of silica sites, [$\equiv Si$], that is, the sum of the number of each of the types of surface sites, $\equiv SiO^-$, $\equiv SiOH$, $\equiv SiOH_2^+$, and $\equiv SiOM^+$, is calculated from

$$10^{-4} = [\equiv Si] = [\equiv SiOH] \left(\frac{1}{[H^+] \, K_{(12.9)}} + 1 + [H^+] \, K_{(12.10)} + \frac{K_{(12.26)} \, [M^{2+}]}{[H^+]} \right) \quad [12.27]$$

The total moles of metal, [total M^{2+}], are the sum of the moles in solution and those on the surface so that

$$10^{-5} = [\text{total } M^{2+}] = [M^{2+}] \left(1 + \frac{K_{(12.26)} \, [\equiv SiOH]}{[H^+]} \right) \quad [12.28]$$

For a particular pH, Equations 12.27 and 12.28 are used to determine the concentration of each species. With log $K_{(12.26)}$ = 2, the results are plotted in **Figure 12-9**. M^{2+} adsorption changes from nearly zero to total adsorption over a limited pH range.

Figure 12-10 shows the adsorption of M^{2+} on a silica surface as a function of M^{2+} concentration in solution at constant pH. With decreasing pH, the surface adsorption is less for the same concentration of M^{2+} in solution. This occurs because as pH is lowered M^{2+} must compete with increasing concentration of H^+ for the available sites. The tendency of a cation to adsorb on a silicate

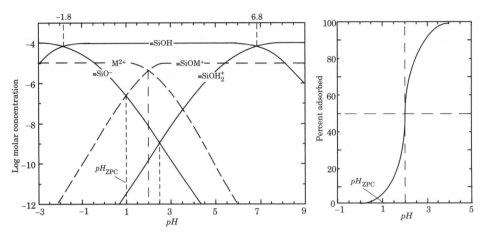

FIGURE 12-9 The diagram on the left gives the distribution of species as a function of pH for M^{2+} binding to a silica surface. The diagram on the right plots the M^{2+} adsorption percentage as a function of pH.

at acid pH appears to increase with an ion's radius so that for doubly charged ions relative adsorption order is

$$Ba^{2+} > Sr^{2+} > Ca^{2+} > Mg^{2+}$$

For a singly charged ion the order is

$$Cs^+ > Rb^+ > K^+ > Na^+ > Li^+$$

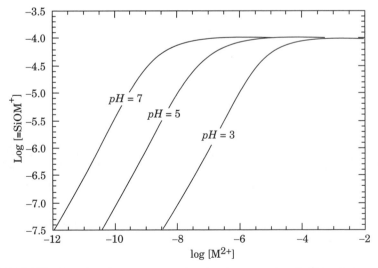

FIGURE 12-10 M^{2+} cation adsorption on quartz as a function of concentration of M^{2+} in solution at the indicated pH with log $K_{(12.26)} = 2$ and the total surface sites available of 10^{-4}.

Relative silicate adsorption for the first-row transition elements increases with the inverse of the ion radius with atomic number as given by the high-spin octahedral electronic configuration (*Irving-Williams series*) so that

$$Cu^{2+} > Ni^{2+} > Co^{2+} > Fe^{2+} > Mn^{2+}$$

Figure 12-11 shows the relative adsorption for a number of metals onto silica as a function of *pH*.

With unit activity coefficients, Equation 12.26 is recast as

$$k = \frac{K_{(12.26)}}{a_{H^+}} = \frac{[\equiv SiOM^+]}{[\equiv SiOH][M^{2+}]} \qquad [12.29]$$

where k is a constant at constant *pH*, temperature, and pressure. The total number of surface sites per unit surface area, [≡Si], with Equation 12.27 in mind is written as

$$[\equiv Si] = [\equiv SiOH]\left(\frac{1}{[H^+]\,K_{(12.9)}} + 1 + [H^+]\,K_{(12.10)}\right) + [\equiv SiOM^+] \qquad [12.30]$$

With *pH* constant, the terms in parentheses on the right side of Equation 12.30 are constant. Denoting this constant as *k*, Equation 12.30 becomes

$$[\equiv Si] = [\equiv SiOH]k + [\equiv SiOM^+] \qquad [12.31]$$

Solving for [≡SiOH] and substituting the result into Equation 12.29 gives

$$k = \frac{[\equiv SiOM^+]}{[M^{2+}]\frac{1}{k}([\equiv Si] - [\equiv SiOM^+])} \qquad [12.32]$$

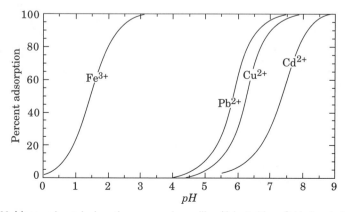

I FIGURE 12-11 Measured metal adsorption on amorphous silica. (Adapted from Schindler et al., 1976.)

With $K_L = k/k$ and by algebra manipulation, Equation 12.32 becomes

$$\frac{[M^{2+}]}{[\equiv SiOM^+]} = \frac{[M^{2+}]}{[\equiv Si]} + \frac{1}{K_L[\equiv Si]} \qquad [12.33]$$

Langmuir Adsorption Isotherm

Equation 12.33 has the form of a *Langmuir adsorption* isotherm equation for element M. Originally developed for gas adsorption onto solid surfaces, the model reduces to *Henry's law* at low concentrations. The Langmuir isotherm is the most commonly used *adsorption isotherm* equation. Its success reflects its ability to fit a wide variety of adsorption data and the ease with which its adjustable parameters are estimated. Langmuir adsorption for aqueous species to a mineral is given by

$$\frac{[M^{2+}]}{[M_{solid}]} = \frac{[M^{2+}]}{\beta_2} + \frac{1}{\beta_1\beta_2} \qquad [12.34]$$

where β_1 denotes a constant related to the binding energy for adsorption, β_2 is the adsorption maximum for the sample, and $[M_{solid}]$ gives the mass of M adsorbed per mass of solid. For the silica surface case outlined above, $\beta_1 = K_L$ and $\beta_2 = [\equiv Si]$. In addition, with Reaction 12.26 as the only M adsorption reaction $[\equiv SiOM^+] = [M_{solid}]$. Making these substitutions, Equation 12.34 is equivalent to Equation 12.33.

Inspection of Equation 12.34 indicates a plot of $[M^{2+}]/[M_{solid}]$ as a function of $[M^{2+}]$ is a straight line with slope equal to $1/\beta_2$ and intercept of $1/(\beta_1\beta_2)$. Thus, the Langmuir equation is a linearization of surface adsorption that is used to determine K_L and $[\equiv Si]$. If M^{2+} adsorption is not near its adsorption maximum so that $\beta_2 >> [M^{2+}]$, the first term on the right side of Equation 12.34 is ignored. The equation then reduces to a Henry's law type equation of

$$[M_{solid}] = \beta_1\beta_2 [M^{2+}] \qquad [12.35]$$

This is similar to a Freundlich adsorption isotherm with $m = 1$, where $K_f = \beta_1\beta_2$ (Equation 12.1). If, on the other hand, $[M^{2+}]$ is large, then the last term in Equation 12.34 is taken as zero. In this case

$$[M_{solid}] = \beta_2 \qquad [12.36]$$

This corresponds to the maximum amount adsorbed that becomes independent of $[M^{2+}]$, as shown at high $[M^{2+}]$ in Figure 12-10.

Rearranging Equation 12.34 yields

$$\frac{\beta_2}{[M_{solid}]} = 1 + \frac{1}{\beta_1[M^{2+}]} \qquad [12.37]$$

Because $\beta_1[M^{2+}]/\beta_1[M^{2+}] = 1$, the right side of Equation 12.37 is equal to

$$\frac{\beta_1[M^{2+}]}{\beta_1[M^{2+}]} + \frac{1}{\beta_1[M^{2+}]} = \frac{\beta_1[M^{2+}] + 1}{\beta_1[M^{2+}]} \qquad [12.38]$$

Denoting the fraction of total capacity that is adsorbed, that is, $[M_{solid}]/\beta_2$, as χ, the left-hand side of Equation 12.37 becomes $1/\chi$. With this in mind, Equation 12.37 is rewritten as

$$\frac{1}{\chi} = \frac{\beta_1[M^{2+}] + 1}{\beta_1[M^{2+}]} \qquad [12.39]$$

or inverting both sides gives

$$\chi = \frac{\beta_1[M^{2+}]}{1 + \beta_1[M^{2+}]} \qquad [12.40]$$

Equation 12.40 is another form of the Langmuir adsorption isotherm equation, and χ is plotted as a function of log $[M^{2+}]$ in **Figure 12-12** for β_1 of 1 and 100. (Note the similarity of Figures 12-11 and 12-12.) When $[M^{2+}]$ approaches 0 so that $\beta_1[M^{2+}]$ is small compared with 1, χ approaches $\beta_1[M^{2+}]$. This case usually occurs for trace components in solution adsorbed to a surface.

Another way to derive the Langmuir equations is to consider the rate at which adsorption occurs. This rate, R_{ad}, is equal to the fraction of unadsorbed sites, $(1 - \chi)$, times a rate constant, k_{ad}, times the amount in solution able to adsorb, $[M^{2+}]$ or

$$R_{ad} = (1 - \chi)\, k_{ad}\, [M^{2+}] \qquad [12.41]$$

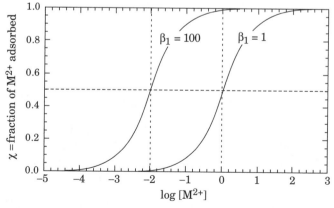

FIGURE 12-12 Langmuir adsorption behavior for species M^{2+} at constant *pH* as a function of concentration with $\beta_1 = 1$ and 100.

The rate at which desorption, R_{ds}, occurs is equal to the fraction adsorbed, χ, times a rate constant, k_{ds}, or

$$R_{ds} = \chi k_{ds} \qquad [12.42]$$

At equilibrium, the two rates are equated, giving

$$(1 - \chi) \, k_{ad} \, [M^{2+}] = \chi k_{ds} \qquad [12.43]$$

or solving for χ

$$\chi = \frac{k_{ad}[M^{2+}]}{k_{ds} + k_{ad}[M^{2+}]} \qquad [12.44]$$

This has the same form as Equation 12.40.

Bidentate Adsorption

At times, rather than a single sorption complex, a metal sorbs by both a *monodentate* and *bidentate* complex, where monodentate and bidentate refer to a species with one and two bonds to the sorbate, respectively. As an example, consider lead adsorption on amorphous silica. Pb^{2+} appears to adsorb by bonding both to one and two surface oxygens, as shown in **Figure 12-13**. Pb^{2+} adsorption as a monodentate complex is described by

$$Pb^{2+} + {\equiv}SiOH = {\equiv}SiOPb^+ + H^+ \qquad K_{(12.45)} = \frac{[{\equiv}SiOPb^+][H^+]}{[Pb^{2+}][{\equiv}SiOH]} \qquad [12.45]$$

For the bidentate complex, its adsorption reaction and equilibrium constant are written as

$$Pb^{2+} + 2 \ {\equiv}SiOH = ({\equiv}SiO)_2Pb^\circ + 2H^+ \qquad K_{(12.46)} = \frac{[({\equiv}SiO)_2Pb^\circ][H^+]^2}{[Pb^{2+}][{\equiv}SiOH]^2} \qquad [12.46]$$

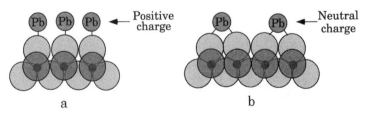

FIGURE 12-13 Pb^{2+} adsorption as (a) monodentate Pb surface species with a positive charge and (b) bidentate surface species with zero surface charge.

The percent of each species adsorbed is given by

$$\%(\equiv SiOPb^+) = \frac{100K_{(12.45)}}{K_{(12.45)} + K_{(12.46)}[\equiv SiOH][H^+]^{-1}}$$ [12.47]

and

$$\%(\equiv SiO_2Pb^\circ) = \frac{100K_{(12.46)}[\equiv SiOH][H^+]^{-1}}{K_{(12.45)} + K_{(12.46)}[\equiv SiOH][H^+]^{-1}}$$ [12.48]

The calculation of the amount of Pb^{2+} adsorbed to amorphous silica at a given *pH* becomes more complicated than that for the single adsorption site. Pb^{2+} adsorption is shown along with some other cations in Figure 12-11. Below *pH* 4, little adsorption of Pb^{2+} occurs, but by *pH* 7, virtually all of the Pb adsorption sites are occupied.

■ Colloids

Colloids are very small particles of one phase dispersed in another phase. Generally, they have diameters between 10 and 10^5 Å. These diameters are greater than the width of a unit cell of a mineral (approximately 10 Å) or a true solution species (typically a couple of Å). They are smaller than those that quickly settle out of solution when diameters become greater than about 10 μm (**Figure 12-14**). If the colloids are a liquid in a liquid, they are called an *emulsion*, as occurs with petroleum in water. Most chert appears to be precipitated from a colloidal suspension of SiO_2 in water. The colloids have very large surface area per mole and, therefore, adsorb significant concentrations of ions onto their surface.

The positive Gibbs surface energy present would decrease if colloids amalgamated during random collision due to *Brownian motion*; however, the colloid's

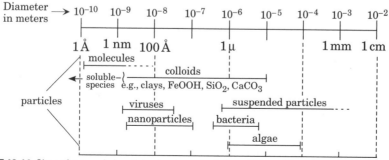

I FIGURE 12-14 Sizes of suspended particles in aqueous solutions. (Adapted from Stumm, 1977.)

surface charge keeps them dispersed. Particles with a strong charge of the same sign stay suspended for a long time. *Flocculation* occurs with decrease of charge with changing *pH* toward the colloid's pH_{zpc} and also with increasing ionic strength because this compresses the electrical double layer and allows closer approach of two colloids. Flocculation is thermodynamically favored because the surface area, and therefore, the Gibbs energy per mole decreases. Small particles in river water, therefore, flocculate and settle out when mixed with ocean water because its higher ionic strength allows the particles to approach close enough to combine and decrease their total surface Gibbs energy.

■ Hydrophobic Sorption

Many compounds, particularly organics, are hydrophobic—that is, they do not dissolve readily in water but are lipophilic, dissolving readily in organic solvents and have high affinity for sorption on organic matter surfaces. Organic chemicals are, therefore, retained on organic matter surfaces in soils. Generally, the greater the concentration of organic matter in the soil is, the longer the organics are retained. At concentrations of organic carbon less than about 0.01 g kg^{-1}, sorption of hydrophobic organic chemicals appears to be controlled by surface properties of the soil clay minerals present. These charged surfaces become sorption sites by attaching to polar functional group on the organic molecules.

A measure of the hydrophobic tendency is through the octanol-water *distribution coefficient*, K_{ow}. K_{ow} is the ratio of a compound's solubility in *n*-octanol

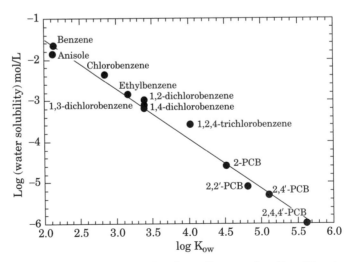

FIGURE 12-15 Logarithm of aqueous solubility in moles per liter versus logarithm of the octanol-water distribution coefficient as reported by Chiou et al. (1983).

to that in water. As the name indicates, *n*-octanol is an eight-carbon alcohol with the chemical formula $C_8H_{17}OH$ (see Chapter 15). This long-chain alcohol mimics lipid (fat) tissue behavior and provides a useful way to measure a compound's ability to bioaccumulate in an animal's fat from uptake in water. Distribution coefficients vary over more than 12 orders of magnitude so logarithmic values are typically considered. As shown in **Figure 12-15**, log K_{ow} is inversely proportional to a compound's solubility in water. Values of K_{ow} are also useful for predicting adsorption of fertilizers, pesticides, and industrial chemicals in soils.

Summary

Surface properties of solid phases in aqueous solutions become important when the solid particles are small. Most surface effects stem from the fact that solid surfaces in aqueous solutions are charged. Surface charge also plays an important role in the kinetics of mineral dissolution and precipitation reactions in aqueous solutions. This is addressed in Chapter 13.

The surface charge is either positive or negative and depends on solution *pH*. Because either H^+ or OH^- is adsorbed to the surface, changing its charge, there exists a *pH* where the solid surface has no charge, the *pH* of zero point of charge, pH_{zpc}. At *pH* above pH_{zpc}, the surface is negatively charged, and below the pH_{zpc}, the surface is positively charged. Models have been developed to characterize the charge as a function of *pH*. Because the charge on the surface layer produces a layer of opposite charge in solution, these models are often called double layer models. More sophisticated models that have two charged layers in the solution have been developed and are referred to as triple layer models.

For most silicate minerals in the earth's crust, their pH_{zpc} are between 2 and 4.5. Because the surface charge causes ions of opposite charge to be attached to or incorporated in the solid surface, for most silicate minerals in solutions of *pH* 4.5 or greater, cations are attached. One way to describe this sorption is with a Freundlich sorption isotherm, where the log of sorption concentration is related to the log of its concentration in solution. A more quantitative model, however, is to use a Langmuir adsorption isotherm that considers a fixed number of possible adsorption sites on the surface.

Surface charge is important in understanding the stability of colloids. Octanol-water distribution coefficients are used to help understand adsorption of organics in soils.

Key Terms Introduced

absorption	bidentate
adsorption	Brownian motion
adsorption isotherm	colloids

coordinative reactions
dielectric constant
distribution coefficient
double layer
electrophoresis
electrostatic potential
emulsion
flocculation
Freundlich sorption
Henry's law
inner sphere complex
ionic strength
Irving-Williams series
Langmuir adsorption

ligands
monodentate
Nernstian
outer sphere complex
permittivity
Poisson's equation
sorbate
sorbent
sorption isotherms
Stern layer
surface energy
surface sorption
triple-layer model

Questions

1. How does absorption differ from adsorption?
2. What is a sorption isotherm? Type S isotherm?
3. What are the properties of Freundlich sorption?
4. How does Langmuir adsorption differ from Freundlich sorption?
5. Why do mineral surfaces manifest a charge?
6. Why does mineral surface charge change with pH?
7. What are inner sphere complexes as opposed to outer sphere complexes?
8. How does the Gibbs energy of an adsorbed species on a mineral surface change because of its electrical field energy from a charged surface?
9. Explain the nature of the chemical environment at the pH of zero point of charge.
10. Describe the electrical double layer on a mineral surface.
11. What is a Gouy-Chapman type surface charging model? Triple-layer model?
12. What is a colloid?
13. Explain why increasing ionic strength decreases the thickness of the layer of diffuse ions next to a mineral surface.
14. By what observation or experiment could you establish that particles of an Au sol carry a positive charge?

Problems

1. On a charged mineral surface in water, the diffuse layer of charge imbalance is greater in fresh water than in seawater. Why?

2. Figure 12-3 plots the surface charge in C m^{-2} from -0.2 to 0.2 ($\times 10^{-10}$). Recast these numbers to give surface charge in terms of moles cm^{-3} to compare the values in Figure 12-7.

3. At 25°C and 1 bar, 10^{-3} moles of the mineral goethite, FeOOH, is added to a liter of solution containing an inert background electrolyte ($I = 10^{-2}$ M) and 10^{-4} M of Zn^{2+} at $pH = 5.0$. The goethite adsorbs 25% of the Zn^{2+} from solution and has 0.2 moles of active sites per mole of the mineral and a surface area of 500 m^2 g^{-1}. A titration of the goethite with acid and base gives the total charge in the suspension at $pH = 5$ of $[\equiv Fe-OH_2^+] - [\equiv Fe-O^-] = 2 \times 10^{-5}$ M. Determine the following:

   (a) Charge density in μC cm^{-2}
   (b) Electrical potential on the goethite surface in volts and
   (c) Thickness of the double layer, κ^{-1}, in nm (**Table 12-2** may be helpful).

Table 12-2

x	sinh(x)
2.80	8.19192
2.81	8.27486
2.82	8.35862
2.83	8.44322
2.84	8.52867
2.85	8.61497
2.86	8.70213
2.87	8.79016
2.88	8.87907
2.89	8.96887
2.90	9.05956

4. Why are alkaline solutions of concentrations of metal ions not stored in glass bottles?

5. To separate aliquots of hematite suspension (10 mg/L) at 25°C, the listed quantities of phthalic acid, H_2P_T, in **Table 12-3** are added at $pH = 6$ with a constant ionic strength of 10^{-3} M and the remaining amount in solution determined (from Stumm, 1992).

   (a) Fit the data in terms of a Freundlich and a Langmuir isotherm.
   (b) Is one of the measurements suspect? Which one and why?
   (c) Estimate the maximum adsorption capacity of H_2P_T by hematite after excluding this measurement.

Table 12-3

H$_2$P$_T$ added [M]	H$_2$P$_T$ in solution [M] $\times$ 10^7
1 $\times$ 10^{-7}	0.43
3 $\times$ 10^{-7}	1.4
1 $\times$ 10^{-6}	4.8
3 $\times$ 10^{-6}	18.5
1 $\times$ 10^{-5}	78
3 $\times$ 10^{-5}	280
1 $\times$ 10^{-4}	890

6. 20.42 g of kaolinite were added to a series of 200 mL arsenate (AsO$_4^{3-}$) solutions of the indicated initial concentrations that resulted in the final equilibrium concentrations in **Table 12-4** (from Evangelou, 1998).
   (a) Fit a Freundlich and Langmuir isotherm to the data.
   (b) If you wanted to characterize adsorption at low concentration of arsenate in solution, which isotherm would you use?
   (c) What is the maximum arsenate that is adsorbed by the 20.42 g of kaolinite as determined from the Langmuir isotherm?

Table 12-4

Initial concentration (mg l^{-1})	Equilibrium concentration (mg l^{-1})
4.88	1.20
10.09	3.56
15.36	6.78
20.11	10.1
30.87	17.6
40.62	25.0
51.78	33.4
80.96	58.4
117.24	90.7
138.28	109
157.91	128
180.53	150
205.73	175
222.83	192

7. Muhammad et al. (1998) measured metal adsorption from solution on 1 g of sand at 20°C in 250 ml of solution at pH 7.4 and obtained the equilibrium values in solution in mg/l and amount adsorbed on the sand in mg/g of sand given in **Table 12-5**. Fit a Langmuir isotherm to each set of data and determine the maximum adsorption per gram of sand.

|Table 12-5

Cu solution	Cu adsorb	Pb solution	Pb adsorb	Cd solution	Cd adsorb
940	145.0	68.0	15.0	115.8	14.3
1860	250.0	96.0	20.5	256.3	29.3
2750	312.5	260.0	54.0	412.9	40.4
3200	332.5	355.0	71.5	543.6	49.2

8. For standard conditions plot, the fraction of Pb adsorbed to 1 kg of gibbsite, $Al(OH)_3$, with a surface area of 4 m^2 per g as a function of pH from 4 to 7. Determine the moles of adsorption sites present. Total concentration of Pb added to solution $= 10^{-9}$ moles. Assume 8.5 adsorption sites per nm^2 and Pb^{2+} adsorbs by a monodentate complex with log $K_{Pb\ ad} = 5.45$ (White and Driscoll, 1987) and site activity coefficients are unity. The amount of Pb adsorbed is small compared with the amount of H$^+$ and OH.

9. Cu^{2+} has leached from underground copper pipes where goethite is present in the soil. H$^+$ adsorption to the goethite surface is given by

$$\equiv FeOH_2^+ = H^+ + \equiv FeOH \qquad \log K_1 = -6.0$$
$$\equiv FeOH = H^+ + \equiv FeO^- \qquad \log K_2 = -8.8$$

Copper adsorption by

$$\equiv FeOH + Cu^{2+} = H^+ + \equiv FeOCu^+ \qquad \log K_3 = 8.0$$

(a) What is the pH of the zero point of charge of the goethite?

(b) Calculate an adsorption isotherm for Cu^{2+} showing the mole fraction of goethite sites occupied per concentration of Cu in solution at $pH = 7$ if the total number of goethite adsorption sites is 10^{-4} moles per liter of soil fluid and assuming electrostatic effects are not important.

References

Bolt, G., 1957, Determination of the charge density of silica soils. *J. Physical Chem.*, v. 61, pp. 1166–1170.

Brady, P. and Walther, J. V., 1992, Surface chemistry and silicate dissolution at elevated temperature. *Am. J. Sci.,* v. 292, pp. 639–659.

Brown, G. E., Jr., 1990, Spectroscopic studies of chemisorption reaction mechanisms at oxide-water interfaces. In Hochella, M. F. and White, A. F. (eds.) *Mineral Water Interface Geochemistry, Reviews in Mineralogy,* v. 23. Mineralogical Society of America, Washington, D.C., pp. 309–363.

Carroll-Webb, S. A. and Walther, J. V., 1988, A surface complex reaction model for the *pH* dependence of corundum and kaolinite dissolution. *Geochim. Cosmochim. Acta,* v. 52, pp. 2609–2623.

Chiou, C. T., Porter, P. E. and Schmedding, D. W., 1983, Partition equilibria of nonionic organic compounds between soil organic matter and water. *Environ. Sci. Technol.,* v. 17, pp. 227–231.

Evangelou, V. P., 1998, *Environmental Soil and Water Chemistry: Principles and Applications.* John Wiley & Sons, New York, 564 pp.

Furrer, F. and Stumm, W., 1986, The coordination chemistry of weathering. I. Dissolution kinetics of γ-Al_2O_3 and BeO. *Geochim. Cosmochim. Acta,* v. 50, pp. 1847–1860.

Hochella, M. F., Jr. and White, A. F., 1990, Mineral–water interface geochemistry. *Rev. Mineral.,* v. 23, 603 pp.

Huang, C. P. and Stumm, W., 1973, Specific adsorption of cations on hydrous α-Al_2O_3. *J. Colloid. Interface Sci.,* v. 22, pp. 231–259.

Manahan, S. E., 2000, *Environmental Chemistry.* Lewis Publishers, Boca Raton, FL, 898 pp.

Muhammad, N., Parr, P., Smith, M. D. and Wheatly, A. D., 1998, *Adsorption of Heavy Metals in Slow Sand Filters.* Abstracts for the 24th WEDC Conf., Sanitation and Water for All, Islamabad, Pakistan, pp. 346–349.

Parks, G. A., 1967, Aqueous surface chemistry of oxides and complex oxide minerals: isoelectric point and zero point of charge. *Adv. Chem.,* v. 67, pp. 121–160.

Schindler, P. W. and Kamber, H. R., 1968, Die aciditat von silanolgruppen. *Helv. Chim. Acta,* v. 51, pp. 1781–1786.

Schindler, P. W., Fürst, B., Dick, R. and Wolf, P. U., 1976, Ligand properties of surface silanol group d. I. Surface complex formation with Fe^{3+}, Cu^{2+}, Cd^{2+}, and Pb^{2+}. *J. Colloid Interface Sci.,* v. 55, pp. 469–475.

Stumm, W., 1977, Chemical interaction in particle separation. *Environ. Sci. Technol.,* v. 11, pp. 1066–1070.

Stumm, W., 1992, *Chemistry of the Solid-Water Interface.* Wiley, New York, 428 pp.

Stumm, W. and Morgan, J. J., 1981, *Aquatic Chemistry: An Introduction Emphasizing Chemical Equilibria in Natural Waters*. John Wiley and Sons, New York, 780 pp.

Walther, J. V., 1996, Relation between rates of aluminosilicate mineral dissolution, *pH*, temperature and surface charge. *Am. J. Sci.*, v. 296, pp. 693–728.

White J. R. and Driscoll, C. T., 1987, Manganese cycling in an acidic Adirondack Lake. *Biogeochemistry*, Vol. 3, No. 1/3, Acidification of the Moose River System in the Adirondack Mountains of New York State, pp. 87–103.

13 | Chemical Kinetics

Chemical kinetics considers the rate of chemical processes. These are represented as balanced reactions with equilibrium occurring when the forward and reverse reaction rates are equal. That is, reactions proceed from unstable states toward equilibrium, and when equilibrium is approached, the back reaction occurs, slowing the rate of the overall reaction to zero at equilibrium. At equilibrium then, the rate of reaction is not zero, but rather, the forward rate equals the reverse rate of reaction. In general, a reactant in a reaction consists of a set of molecules or species with a range of energies distributed according to a *Boltzmann distribution*. For reactions to occur, some of these species need enough energy to make it across a reaction threshold and proceed toward the more stable state of lower energy. The threshold may be the energy needed to break or rearrange bonds. This energy is denoted as an *activation energy*, E_a.

Consider the balls shown in **Figure 13-1**. Ball C has the lowest energy and is, therefore, in the most stable state. Ball B is clearly unstable because with time it decreases its energy by moving to position A or C. Ball A is also unstable but must overcome an energy barrier, E_a, to reach the more stable state of C. Imagine that the position of A represents the average energy of a mole of species in a reaction. This mole of species has a distribution of energies, some of which have greater than the average energy of A. By reacting, that is, interacting, with other species, a transfer of energy occurs so that a species develops enough energy to exceed the energy barrier, E_a, at the B position. It can then progress to the more stable C position. If E_a is large, it would be unlikely that many species could develop the necessary energy, and a long time would be needed for a noticeable number of species of A to overcome the energy barrier. The size of E_a therefore determines the reaction rate. If E_a is sufficiently large, no species has enough energy to cross the energy barrier, and no reaction occurs. One refers to this state in which no reactions occur, but a more stable state exists as a *metastable* state. Metastable states then have significant activation energy barriers separating them from more stable states that keep reactions from happening in observable time. Because this designation requires time, it is outside

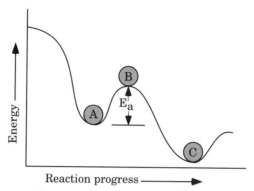

FIGURE 13-1 Relative energy of balls A, B, and C in a gravitational field. Ball C is most stable, being at the lowest energy. For ball A to reach this position, it must overcome the activation energy, E_a, by increasing its energy to the position of ball B.

the model of thermodynamics. In a thermodynamic context, however, the metastable state is considered in a similar way as the stable equilibrium state if the time frame is short relative to rates along possible reaction paths to more stable states.

Consider aragonite, a form of $CaCO_3$ (e.g., pearl). It is unstable at the earth's surface compared with calcite, but its rate of reaction to calcite is extremely slow. Such a transformation requires a change in orientation of the atoms from pseudohexagonal in aragonite to perfect rhombohedrons in calcite. Aragonitic shells from the Mesozoic era are known, but none from the Paleozoic survived. Aragonite is considered metastable. On laboratory time scales, aragonite solubility is measured and its equilibrium constant, K_{arag}, determined

$$CaCO_3 = Ca^{2+} + CO_3^{2-}$$
aragonite

[13.1]

Calcite solubility is also measured and its equilibrium constant, K_{cal}, determined. Because aragonite at earth surface conditions is unstable with respect to calcite, its Gibbs energy is greater; therefore, $K_{arag} > K_{cal}$. Where both aragonite and calcite are present, waters have generally reached the greater solute concentrations consistent with aragonite equilibrium, being supersaturated with respect to calcite. The rate of aragonite precipitation is, therefore, more rapid than the rate of calcite precipitation. Other important metastable phases include amorphous silica (opal) and chalcedony. Again, it is a matter of time.

Most waters at the earth's surface are supersaturated with respect to quartz, the stable SiO_2 phase, and Si in solution appears to be controlled by the dissolution of other silicates and the growth of siliceous diatoms. This occurs because of the slow rate of the quartz precipitation reaction. Equilibrium between these metastable phases and aqueous solutions is determined in the laboratory and observed in nature on significantly longer time scales; however, when *catalytic reaction* paths are available for these metastable phases, they are observed to revert to the more stable phase much more rapidly. These catalytic paths then have lower activation energy barriers for reaction. Again, although the equilibrium state is well described by thermodynamics and thermodynamics also indicates in which direction a reaction should proceed, it does not indicate the reaction path or rate of the reaction as it proceeds from an unstable state to equilibrium. This is the realm of kinetic studies.

There are *homogeneous reactions* that occur in one phase, for instance, a reaction between gases in a gas phase or between dissolved species in an aqueous solution. There are also *heterogeneous reactions* that occur between phases, for instance, the dissolution of a mineral into an aqueous solution or the conversion of aragonite to calcite. As expected, heterogeneous reaction rates are generally controlled by processes occurring at the interface between the phases.

■ First-Order Decay and Growth Equations

The kinetics of chemical reactions is characterized in terms of its *reaction order*. The order of a reaction is the power to which concentrations of a component in a rate law are raised. The overall reaction order equals the sum of the powers of all of the concentrations. Consider a rate law for reactants A and B going to products C and D given by

$$\frac{d[A]}{dt} = k_r^+ [A]^2 [B] \qquad [13.2]$$

where k_r^+ is the forward reaction rate constant, t denotes time, and $[x]$ gives the concentration of reactant x. This rate equation is determined by changing the concentration of A and B in the reaction. The rate with these changed concentrations must also be given by Equation 13.2. For instance, if one doubles the concentration of B, the rate of change of A with time should double. This rate is second order in [A] and third order overall. Given in **Table 13-1** are the characteristics of first and second order rate laws.

A type of reaction important to many geologic processes is a reaction that is written as A → C + D with a rate of reaction given by

$$\frac{d[A]}{dt} = -k_r^+ [A] \qquad [13.3]$$

This is a first-order reaction in A (see Table 13-1). The minus sign is used to indicate that the amount of A typically decreases with time. The rate does not

Table 13-1   Rate laws

	Differential rate law	Integrated rate law	Linear plot	Half-life
First order	$\dfrac{d[A]}{dt} = -k[A]$	$[A] = [A^\circ]\exp(-kt)$	$\ln[A]$ vs. t	$\dfrac{\ln 2}{k}$
Second order	$\dfrac{d[A]}{dt} = -k[A]^2$	$\dfrac{1}{[A]} - \dfrac{1}{[A^\circ]} = kt$	$\dfrac{1}{[A]}$ vs. t	$\dfrac{1}{k[A^\circ]}$

depend on the concentration of C or D and is not zero at equilibrium but decreases to become equal to the reverse reaction at equilibrium. Many simple reactions are first order. These include anaerobic decay of organic material as well as radioactive decay. Rearranging Equation 13.3 gives

$$\frac{d[A]}{[A]} = -k_r^+ \, dt \qquad\qquad [13.4]$$

The integral of Equation 13.4 is written as

$$\int_{[A^\circ]}^{[A]} \frac{d[A]}{[A]} = -k_r^+ \int_0^t dt \qquad\qquad [13.5]$$

Performing the integration from time 0 with initial concentration $[A^\circ]$ to time t with concentration $[A]$ results in

$$\ln[A] - \ln[A^\circ] = -k_r^+ (t - 0) \ \text{ or } \ \ln\frac{[A]}{[A^\circ]} = -k_r^+ t \qquad\qquad [13.6]$$

Exponentiation of both sides of Equation 13.6 yields

$$[A] = [A^\circ] e^{-k_r^+ t} \qquad\qquad [13.7]$$

This is a decay-rate equation with rate constant k_r^+. In Chapter 10, this equation is discussed in terms of radioactive decay of isotopes. If the exponent is positive, that is, if k_r^+ is a negative number, the concentration of A increases with time, and it is called a growth-rate equation. Note the exponential change in the concentration of A as a function of time given by this equation. This is the overall reaction given by a macroscopically observable rate expression. What are the elementary molecular steps through which this reaction proceeds?

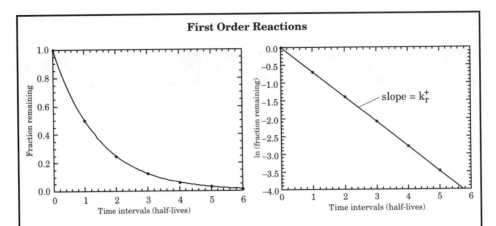

First Order Reactions

First order reactions are given by a rate of change of reactant of concentration [A] of

$$\text{rate} = -\frac{d[A]}{dt} = k_r^+[A]$$

This can be integrated in time to give

$$\ln\frac{[A]}{[A^o]} = k_r^+ t$$

where $[A^o]$ is the initial concentration. Therefore, the slope on a natural logarithm of the fraction of the initial amount versus time gives k_r^+. A half-life is the amount of time it takes for half the material to be reacted.

■ Elementary Reactions

Elementary reactions characterize the reaction on the molecular scale and are typically unimolecular and first order with respect to a single reactant (spontaneous decay) or bimolecular depending on the first order of two reactant concentrations (interaction of two molecules). There appear to be some elementary reactions in the atmosphere, however, that are first order with respect to three reactants (see Chapter 16). The observable overall reaction is made up of a series of elementary reactions that occur one after another in a chain. For instance, consider the chain of elementary reactions for the macroscopic reaction A → C given by

$$A \xrightarrow{\ R_A\ } B \xrightarrow{\ R_B\ } C \qquad\qquad\qquad [13.8]$$

where R_i stands for the rate of reaction of reactant i. B is termed a *reaction intermediary*. This is a configuration of molecules through which the reacting molecules, A, must pass to become products, C. If the molecular reactions of A to B to C are first order, the rate of reaction of A is characterized by

$$R_A = \frac{d[A]}{dt} = -k_A^+[A] \qquad\qquad [13.9]$$

with the minus sign indicating the concentration of A is decreasing as a function of time.

As given by Equation 13.7, the concentration of A as a function of time is

$$[A] = [A^o]e^{-k_A^+ t} \qquad\qquad [13.10]$$

The concentration of B is equal to the amount produced from A minus the amount that reacts to C or

$$\frac{d[B]}{dt} = k_A^+[A] - k_B^+[B] \qquad\qquad [13.11]$$

If there is no B or C to start with but only A at $[A^o]$, from expressions of Equation 13.10, the concentration of B as a function of time is given by

$$[B] = [A^o]\left(\frac{k_A^+}{k_B^+ - k_A^+}\right)(e^{-k_A^+ t} - e^{-k_B^+ t}) \qquad\qquad [13.12]$$

Because $[A] + [B] + [C] = [A^o]$, the concentration of C as a function of time by substitution of Equations 13.10 and 13.12 is

$$[C] = [A^o]\left(1 + \frac{1}{k_A^+ - k_B^+}\right)(k_B^+ e^{-k_A^+ t} - k_A^+ e^{-k_B^+ t}) \qquad\qquad [13.13]$$

When the rate constant k_B^+ is much larger than k_A^+, then $e^{-k_B^+ t}$ is small compared with $e^{-k_A^+ t}$ and can be taken as zero. Equation 13.13 then becomes

$$[C] = [A^o]\left(1 + \frac{k_B^+}{k_A^+ - k_B^+}\right)e^{-k_A^+ t} = [A^o](1 - e^{-k_A^+ t}) \qquad\qquad [13.14]$$

In contrast, if k_A^+ is much larger than k_B^+, then Equation 13.13 becomes

$$[C] \cong [A^o](1 - e^{-k_B^+ t}) \qquad\qquad [13.15]$$

The concentration of C is, therefore, dominated by the step with the smaller rate constant. This step is termed the rate-limiting elementary reaction step. That is, if k_B^+ is much larger than k_A^+, when B is formed, it reacts quickly to C. The rate of formation of C is then determined almost solely by the rate the intermediary B is formed. If instead the rate constant k_A^+ is much larger than k_B^+, then a large concentration of B is produced, and its breakdown to C is the rate limiting step. Again, it is the smallest rate constant in the reaction series that dominates the rate. The key to molecular kinetics is then identifying the rate-determining elementary reaction and the rate at which it reacts.

Generally, rates are considered under the conditions when the reaction intermediaries are at a steady-state concentration as a function of time. This is termed *secular equilibrium*. In a reaction, this steady state is established and propagated through the entire reaction chain such that $k_A^+[A] = k_B^+[B] = k_C^+[C]$. A steady-state ratio of reactants to products is achieved for each elementary reaction in the chain because the amount produced at any instant in time becomes equal to the amount reacted to the next species in the chain.

Hydrogen Iodine

Consider the simple homogeneous reaction between hydrogen gas (H_2) and iodine gas (I_2) to produce HI gas at temperatures below about 350°C (Gimblett, 1970) of

$$H_2 + I_2 \rightarrow 2HI \qquad [13.16]$$

Because of conservation of mass

$$\frac{d[HI]}{dt} = -2\frac{d[H_2]}{dt} = -2\frac{d[I_2]}{dt} \qquad [13.17]$$

where [ ] indicates the moles of the gas species in a unit volume of the gas and t is time. The rate of reaction has been measured and is given by

$$R_+ = \frac{d[HI]}{dt} = k_r^+[H_2][I_2] \qquad [13.18]$$

where R_+ denotes this forward rate of Reaction 13.16. k_r^+ again is the forward rate constant that depends on temperature and pressure but is independent of the concentration of H_2 or I_2. The rate depends only on the product of the concentration of the reactants and not on the concentration of the products. The rate of this reaction is then second order overall and first order with respect to H_2 or I_2.

Because the HI formation reaction is first order in two reactant species, the reaction as written probably also describes the simple rate–limiting elementary molecular reaction of the collision of two molecules. If the macroscopic rate equation depended on concentration raised to a fractional power, it cannot be characterized by an elementary molecular reaction. Also, if the overall rate is greater than third order, it is probably not an elementary reaction because rates that depend on simultaneous reaction of four or more reactant entities are not common.

Consider an elementary molecular–scale reaction process in which two molecules collide as characterized by a rate equation such as Equation 13.18. Whether the collision produces the product species depends on both the geometry of approach of the molecules and their internal energy. Given the large number of molecules in random orientation in a homogeneous gas or liquid phase, even with unsymmetric molecules, the geometric effects typically average out. That is, no factors that affect the orientation of the molecules should appear in the rate equation. As mentioned above, the reactant molecules in a gas have a range of internal energies given by Boltzmann's law. The molecules with higher energy tend to have higher velocity because some of this internal energy is kinetic. When the reactant molecules collide, the kinetic energy released on collision has the potential to reconfigure the bonding of the molecules to a *transitory state*. A transitory state is a state of some bonding between the reactant molecules that has a higher internal energy than the sum of the internal energies of the average H_2 plus average I_2 in the gas phase. This would be position B in Figure 13-1.

This bonding is denoted by writing the species in this transition state as $H_2 \bullet I_2$ or $HI \bullet HI$. This species is termed the *activated complex* for the reaction. From the rate equation, however, there must be an equal number of H and I in the molecular transitory state (activated complex) for the rate-determining elementary reaction. With the molecule of the activated complex written as $HI \bullet HI$, its formation is given by

$$H_2 + I_2 \rightarrow HI \bullet HI \qquad \text{[13.19]}$$

The average molar Gibbs energy change of this forward reaction to form the activated complex, ΔG_{act}^+, would be positive at constant temperature and pressure. Some of the unstable transitory state molecules decay to produce the final HI molecules according to the reaction

$$HI \bullet HI \rightarrow 2HI \qquad \text{[13.20]}$$

This has a negative molar Gibbs energy of decay of the activated complex, $-\Delta G_{act}^-$. Other activated complexes revert back to the original molecules by the reaction

$$HI \bullet HI \rightarrow H_2 + I_2 \qquad \text{[13.21]}$$

which also has a negative molar Gibbs energy of reaction given by $-\Delta G_{act}^+$.

Along a reaction coordinate, that is, sequence, these reactions are represented as

$$H_2 + I_2 \leftrightarrow HI \bullet HI \xrightarrow{\text{rate determining}} 2HI \qquad \text{[13.22]}$$

Statistical thermodynamics shows that by assuming equal a priori probabilities of energy levels in the reactants and activated complexes, the rates of the forward and reverse reactions, ↔, can be considered an equilibrium. The reaction intermediaries H_2 and I_2 are then modeled as being in equilibrium with the activated complex of the rate-limiting reaction HI•HI → 2 HI, as shown in **Figure 13-2**.

The description of the molecular dynamics in the reaction are then a series of steps in an elementary reaction sequence along a reaction coordinate. This coordinate is not time because some of the activated complexes revert back to the reactant molecules. The difference in energy between the beginning and final states is ΔG_R, but the reaction must first go through a transitory state of higher energy. If the breakdown of the activated complex, HI•HI, is assumed to be the slowest and therefore rate limiting, then it is in equilibrium with the reactants to form it. An equilibrium, therefore, can be written as

$$H_2 + I_2 = HI\bullet HI \qquad \text{with} \qquad K_{act}^+ = \frac{\gamma_{HI\bullet HI}}{\gamma_{H_2}\gamma_{I_2}}\frac{[HI\bullet HI]}{[H_2][I_2]} \qquad [13.23]$$

where K_{act}^+ is the equilibrium constant for the forward reaction to form the rate-limiting activated complex and γ denotes the activity coefficient of the subscripted species whose concentration is given in brackets. K_{act}^+ is related to the *standard-state* Gibbs energy of the forward reaction to form the activated complex, ΔG_{act}^{o+}, by

$$-RT\ln K_{act}^+ = \Delta G_{act}^{o+} \qquad [13.24]$$

Solving for the concentration of activated complexes from the equilibrium constant expression given in Equation 13.23 yields

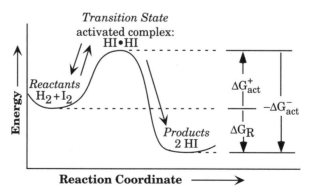

FIGURE 13-2 The rate-limiting reaction $H_2 + I_2 \rightarrow 2$ HI along the reaction coordinate, indicating the energy changes that occur.

$$[\text{HI•HI}] = K_{act}^+ \frac{\gamma_{H_2}\gamma_{I_2}}{\gamma_{HI•HI}}[H_2][I_2] \qquad [13.25]$$

The rate or frequency of decomposition of the activated complexes is calculated from the *partition functions* of the distribution of energies of the molecules in the system over the allowable energy states. From this statistical mechanical analysis, the rate of activated complex decomposition is given by kT/h, where k is *Boltzmann's constant*, h is *Planck's constant*, and T is the temperature in K.

The probability of breakdown of the rate-limiting activated complex to products rather than reactants, the transmission coefficient, is denoted by κ. In most reactions, κ has a value near 1. That is, when the activated complex breaks down, it nearly always does so to the product species. The forward rate is then equal to the number of activated complexes produced times the rate of decomposition of the activated complex times the transmission coefficient:

$$R_+ = [\text{HI•HI}]\frac{kT}{h}\kappa \qquad [13.26]$$

This rate is equated with the rate given by Equation 13.18 to yield

$$[\text{HI•HI}]\frac{kT}{h}\kappa = k_r^+[H_2][I_2] \qquad [13.27]$$

With consideration of Equations 13.24 and 13.25, the rate constant, k_r^+, is calculated to be

$$k_r^+ = \kappa\,\frac{kT}{h}\,K_{act}^+\frac{\gamma_{H_2}\gamma_{I_2}}{\gamma_{HI•HI}} = \kappa\,\frac{kT}{h}\frac{\gamma_{H_2}\gamma_{I_2}}{\gamma_{HI•HI}}e^{-(\Delta G_{act}^{o+}/RT)} \qquad [13.28]$$

knowing that

$$\Delta G_{act}^{o+} = G_{act}^o - G_{H_2}^o - G_{I_2}^o \qquad [13.29]$$

where G_i^o is the standard-state Gibbs energy of the *i*th species.

Now consider the reverse of Reaction 13.20 written to form an activated complex from the product species in the overall reaction:

$$2\text{HI} \leftrightarrow \text{HI•HI} \quad \text{with} \quad K_{act}^- = \frac{\gamma_{HI•HI}}{\gamma_{HI}^2}\frac{[\text{HI•HI}]}{[\text{HI}]^2} \qquad [13.30]$$

where K_{act}^- is the equilibrium constant for this reverse activated complex-forming reaction from the products of the overall reaction. By a similar analysis with that above

$$R_- = k_r^-[\text{HI}]^2 \qquad [13.31]$$

with a rate constant, k_r^-, given by

$$k_r^- = -\kappa' \frac{kT}{h} K_{act}^- \frac{\gamma_{HI}^2}{\gamma_{HI \cdot HI}} = \kappa' \frac{kT}{h} \frac{\gamma_{HI}^2}{\gamma_{HI \cdot HI}} e^{-(\Delta G_{act}^{o-}/RT)} \qquad [13.32]$$

ΔG_{act}^{o-} designates the Gibbs energy change for the reverse activated complex-forming reaction from the overall products so that

$$\Delta G_{act}^{o-} = G_{act-}^o - 2G_{HI}^o \qquad [13.33]$$

With the transmission coefficient similar in both directions, the ratio of k_r^+ to k_r^- from Equations 13.28 and 13.32, respectively, is written as

$$\frac{k_r^+}{k_r^-} = \frac{\gamma_{H_2} \gamma_{I_2}}{\gamma_{HI}^2} e^{-(\Delta G_{act}^{o+} - \Delta G_{act}^{o-})/RT} \qquad [13.34]$$

The standard-state Gibbs energy change of the overall reaction, ΔG_R^o, is the standard-state Gibbs energy of the forward activated complex-forming reaction minus the reverse-activated complex-forming reaction so that

$$\Delta G_{act}^{o+} - \Delta G_{act}^{o-} = (G_{act+}^o - G_{H_2}^o - G_{I_2}^o) - (G_{act-}^o - 2G_{HI}^o)$$
$$= (G_{act+}^o - G_{act-}^o) + (2G_{HI}^o - G_{H_2}^o - G_{I_2}^o) \qquad [13.35]$$

If the activated complex for the forward and reverse reactions is the same, then

$$(G_{act+}^o - G_{act-}^o) = 0 \text{ and } \Delta G_{act}^{o+} - \Delta G_{act}^{o-} = \Delta G_R^o \qquad [13.36]$$

where ΔG_R^o is the standard-state Gibbs energy difference between the products and reactants in the overall reaction.

Up to this point, standard-state properties are used rather than those of the actual reaction. With similar forward and reverse activated complexes, Equation 13.34 yields

$$\frac{k_r^+}{k_r^-} = \frac{\gamma_{H_2} \gamma_{I_2}}{\gamma_{HI}^2} e^{-\Delta G_R^o/RT} \qquad [13.37]$$

From Equations 13.26 and 13.31, the ratio of the forward and reverse reaction is

$$\frac{R_+}{R_-} = \frac{k_r^+ [H_2][I_2]}{k_r^- [HI]^2} = \frac{\gamma_{H_2} \gamma_{I_2}}{\gamma_{HI}^2} \frac{[H_2][I_2]}{[HI]^2} e^{-\Delta G_R^o/RT} \qquad [13.38]$$

$$= \frac{a_{H_2} a_{I_2}}{a_{HI}^2} e^{-\Delta G_R^o/RT} = e^{-\Delta G_R/RT} \qquad [13.39]$$

where ΔG_R is the total Gibbs energy change of the overall reaction as written. At equilibrium, $\Delta G_R = 0$ and, therefore, $R_+ = R_-$. Because from Equation 13.39 $R_- = R_+ e^{\Delta G_R/RT}$, the net forward reaction $R_{net} = R_+ - R_-$ is given by

$$R_{net} = R_+ (1 - e^{\Delta G_R /RT}) \qquad [13.40]$$

From infinite series analysis, the series expansion of e^x is written as

$$e^x = 1 + x + \frac{x^2}{2!} + \frac{x^3}{3!} + \cdots \qquad [13.41]$$

If $x \ll 1$, the higher order terms are ignored, and e^x is given by

$$e^x = 1 + x \qquad [13.42]$$

If, therefore, the reaction is close to equilibrium so that $\Delta G_R < 0.4RT$, the series expansion of $e^{\Delta G_R/RT}$ gives

$$e^{\Delta G_R/RT} = 1 + \frac{\Delta G_R}{RT} \qquad [13.43]$$

In this case

$$R_{net} = -R_+ \frac{\Delta G_R}{RT} \qquad [13.44]$$

That is, close to equilibrium at constant temperature, the net rate of reaction is a linear function of ΔG_R with a slope of $-R_+/RT$ on a rate versus ΔG_R plot. Although Equation 13.44 was derived for elementary molecular reactions assuming that the activated complex is identical in the forward and reverse direction, it appears to be a reasonable approximation for many heterogeneous reactions of geochemical interest given the present state of knowledge. Many reactions at depth in the earth's crust are considered to be near equilibrium so that the relation in Equation 13.44 is a reasonable way to analyze their rates.

How far from equilibrium can Equation 13.44 be used? At constant pressure with $\Delta C_P = 0$, $\Delta G_R = \Delta S_R(T - T_{eq})$, where T_{eq} represents the reaction's equilibrium temperature and ΔS_R is the reaction entropy. If Equation 13.44 is used up to $\Delta G_R = 0.4RT$, the temperature is

$$T = \frac{T_{eq} \Delta S_R}{\Delta S_R - 0.4R} \qquad [13.45]$$

For a typical dehydration reaction to produce 1 mole of H_2O, ΔS_R is about 20 cal mol^{-1} K^{-1}. With an equilibrium temperature of 773 K (500°C) and remembering that $R = 1.987$ cal mol^{-1} K^{-1}, the maximum overstep temperature would be 805 K (532°C), or 32°C greater than the equilibrium temperature. In some cases, the assumption of $\Delta C_P = 0$ may not be a good approximation, but for overstepping of dehydration reactions for up to 10°C, little error is generally incorporated if Equation 13.44 is used.

Hydrogen Bromide

Consider another reaction between gas phase molecules:

$$H_2 + Br_2 = 2HBr \tag{13.46}$$

The rate of this reaction has been measured and is

$$-\frac{d[HBr]}{dt} = \frac{k'[H_2][Br_2]^{1/2}}{1 + k''([HBr]/[Br_2])} \tag{13.47}$$

where k' and k'' are constants independent of composition at a fixed temperature and pressure (Gimblett, 1970). Equation 13.47 indicates that Reaction 13.46 is not an elementary reaction as written. Among other things, the rate depends on the concentration of products of the reaction, which is not the case for an elementary reaction.

This observed rate expression has been determined to be a combination of elementary reaction steps that proceed in two reaction chains to produce HBr. These are

$$H_2 + Br \rightarrow HBr + H \tag{13.48}$$

and

$$H + Br_2 \rightarrow HBr + Br \tag{13.49}$$

Each reaction produces reactants for the other reaction. Thus, the reaction depends on an endless chain of the two reactions. Where do the atoms of Br and H come from? Br comes from the chain-initiating reaction

$$Br_2 \rightarrow Br + Br \tag{13.50}$$

The reaction of H_2 to form H, however, is energetically unfavorable, and Reaction 13.48 takes its place. Also, because [HBr] is in the denominator, it must have an inhibiting effect. This occurs by reducing [H] according to the reaction

$$H + HBr \rightarrow H_2 + Br \tag{13.51}$$

Finally, the chain is terminated by the reaction

$$Br + Br \rightarrow Br_2 \tag{13.52}$$

The combination of these five different elementary reactions gives the overall rate expression in Equation 13.47. It is clear that even in the case of a seemingly simple homogeneous reaction between gas molecules the elementary reaction steps can be complicated. One would surmise that most heterogeneous reactions of geochemical interest are even more complicated. For instance, the reaction that leads to the dissolution of multication silicates in H_2O is viewed as chains of elementary reactions involving each cation where the rate-controlling elementary reaction step in each chain is dependent on the products of other chains in the overall dissolution reaction.

■ Temperature Dependence of Rates

From the expression for the rate constant of the forward elementary Reaction 13.28 discussed above

$$k_r^+ = \kappa \, \frac{kT}{h} \frac{\gamma_{H_2}\gamma_{I_2}}{\gamma_{HI \cdot HI}} e^{-\Delta G_R^{o+}/RT} \qquad [13.53]$$

Because

$$\Delta G_{R+}^o = \Delta H_{R+}^o - T\Delta S_{R+}^o \qquad [13.54]$$

the logarithm of Equation 13.53 becomes

$$\ln k_r^+ = \ln \left(\kappa \, \frac{kT}{h} \frac{\gamma_{H_2}\gamma_{I_2}}{\gamma_{HI \cdot HI}}\right) - \frac{\Delta H_{R+}^o}{RT} + \frac{\Delta S_{R+}^o}{R} \qquad [13.55]$$

As the standard heat capacity of reaction, ΔCp_{R+}^o is often close to zero. ΔH_{R+}^o and ΔS_{R+}^o generally are then taken as independent of temperature. With activity coefficients and κ constant, the derivative of both sides of Reaction 13.55 with respect to temperature is

$$\frac{d \ln k_r^+}{dT} = \frac{1}{T} + \frac{\Delta H_{R+}^o}{RT^2} \qquad [13.56]$$

This gives the temperature dependence of k_r^+.

Before the advent of the transition-state theory of kinetics and activated complexes at the molecular level, the change in the rate constant with temperature was equated with a macroscopically measured value, k_{mes}^+. An activation energy for the reaction, E_a, was determined by using the Arrhenius relation for the temperature dependence of the overall measured rate from

Table 13-2   Activation energies of some typical reactions of geochemical interest

Reaction	E_a (kcal/mol)	Conditions	Reference
Aqueous species diffusion	~3–4	Tracer ionic diffusion in aqueous solutions to 600°C and 5 kbar	Calculated from Oelkers and Helgeson (1988)
Solid-state diffusion in garnet	60	Mg-Fe solid-state interdiffusion 500–700°C	Lasaga et al. (1977)
Solid-state diffusion in fayalite	49.8	Mg tracer diffusion down c-axis 900–1100°C	Misener (1974)
Kaolinite surface dissolution		Aqueous solution 25–80°C	
	16.0 ± 0.6	pH = 1	
	13.3	pH = 2	
	10.3	pH = 3	Carroll and Walther
	7.7	pH = 4	(1990)
	4.9	pH = 5	
	2.3	pH = 6	
	1.7	pH = 7	
	3.4	pH = 8	
	5.3	pH = 9	
	7.0	pH = 10	
	8.5	pH = 11	
	9.9	pH = 12	
Quartz surface dissolution		Aqueous solution 25–60°C	
	10.5	pH = 4	Brady and Walther (1990)
	11.0	pH = 6	
	13.0	pH = 8	
	22.0	pH = 10	
	23.3	pH = 11	
Silica glass surface dissolution	22.7	Aqueous solution 40–85°C pH = 4.1	Mazer and Walther (1994)
Calcite surface dissolution	8.4	Aqueous solution 5–50°C	Sjoberg (1976)

$$\frac{d \ln k_{mes}^+}{dT} = \frac{E_a}{RT^2} \quad \text{or} \quad \frac{d(\ln k_{mes}^+)}{d\left(\dfrac{1}{T}\right)} = \frac{E_a}{R}$$ [13.57]

Equating Equations 13.56 and 13.57 at constant volume gives

$$E_a = RT + \Delta H_{R+}^o$$ [13.58]

Because RT is generally small compared with ΔH_{R+}^o, E_a is often equated with the standard heat of the overall forward rate limiting reaction, ΔH_{R+}^o.

Consider a reaction with $E_a = 50$ kJ mol^{-1} (approximately 12 kcal mol^{-1}) at 25°C. With this activation energy, Equation 13.57 gives $d(\ln k_{mes}^+)/dT = 0.068$, or the reaction rate constant, k_{mes}^+, doubles with a 10°C change in temperature. Some values of E_a reported for various reactions are given in **Table 13-2**. A process in which the rate-controlling step is aqueous diffusion has an E_a on the order of 3 to 4 kcal mol^{-1}. Solid-state diffusion processes of cations in minerals have E_a that are on the order of 50 to 60 kcal mol^{-1}. E_a for surface-controlled dissolution reactions are generally at intermediate values. Because the E_a of surface-controlled mineral dissolution in aqueous solutions is pH dependent, the nature of the rate-limiting reaction must also be pH dependent.

■ Rates from Irreversible Thermodynamics

Although thermodynamics considers energy changes between states at equilibrium, the second law of thermodynamics with its consideration of entropy allows thermodynamics to be linked to irreversible processes. This linkage is due to the fact that certain restrictive relations must operate near equilibrium for thermodynamics to be valid at equilibrium. Irreversible thermodynamics then gives a fundamental understanding of the kinetics of transport processes by setting up near to equilibrium steady states of energy and mass transport. This section outlines these states. A more detailed presentation is given in de Groot and Mazur (1962) and Katchalsky and Curran (1967).

Thermodynamics stipulates that at constant pressure and temperature reactions proceed in a direction that lowers the Gibbs energy of the system. If this is considered in terms of entropy production for a *closed system*, the change in entropy in going from one equilibrium state to another, dS_{heat}, is defined as

$$dS_{heat} \equiv \frac{dq_{rev}}{T}$$ [13.59]

where dq_{rev} is the reversible heat adsorbed from the surroundings through the boundaries of the system during the process. Now consider a spontaneous, that is, irreversible, change in an *isolated system*. An isolated system is one that does not allow heat or mass transfer across its boundaries, so $dq_{rev} = 0$. In this case, the entropy change, dS_{irr}, must be such that

$$dS_{irr} \geq 0 \tag{13.60}$$

with the equal sign referring to equilibrium in the system. dS_{irr} is then a measure of irreversibility.

The change in entropy for this isolated irreversible reaction is not due to a heat exchange with the surroundings as for an equilibrium change of state. One, therefore, has two sources of entropy production, reversible and irreversible. The total entropy change for a change in state in an *open system* with an irreversible reaction taking place is written as the sum of these two contributions:

$$dS = dS_{heat} + dS_{irr} \tag{13.61}$$

where dS_{heat} is caused by heat and mass flux with the surroundings just as in the equilibrium change in state, and dS_{irr} is the irreversible change of entropy due to internal processes in the system when going from one state to another. The entropy change for an irreversible process, dS_{irr}, therefore, depends on *how* one goes from one state to another.

The heat and mass transfer in irreversible processes is cast in terms of fluxes, that is, the amount transferred through a cross-sectional area of the boundary of the system in a given time. With the cross-sectional area of A and a heat flux, J_q,

$$J_q = -\frac{1}{A}\frac{dq}{dt} \tag{13.62}$$

The mass flux of the ith component, J_i, is given by

$$J_i = -\frac{1}{A}\frac{dn_i}{dt} \tag{13.63}$$

Entropy that is irreversible is produced by the chemical reactions going on in the system, of which the amount of Gibbs energy drive available is given by the *chemical affinity* of the reaction, **A**, as introduced in Chapter 3. The entropy produced at constant volume in the system by chemical reactions, dS_{irr}, is then the rate of reaction, r, times **A**/T or

$$dS_{irr} = r\frac{\mathbf{A}}{T} \tag{13.64}$$

The entropy production in the system can be made into an intensive variable, σ, that is, one that does not depend on mass by considering entropy production per unit of infinitesimally small volume of the system. σ is defined as

$$\sigma \equiv \frac{1}{A}\frac{dS}{dx}\frac{}{dt}$$

[13.65]

where dx is the infinitesimal length of system volume normal to surface **A**.

To consider this further, remember that the change in internal energy of a volume is characterized by

$$dU = TdS - PdV + \sum_i \mu_i dn_i$$

[13.66]

or solving for dS

$$dS = \frac{dU + PdV}{T} - \sum_i \frac{\mu_i}{T} dn_i$$

[13.67]

From the first law of thermodynamics, $dq_{rev} = dU + PdV$, and Equation 13.67 then becomes

$$dS = \frac{dq_{rev}}{T} - \sum_i \frac{\mu_i}{T} dn_i$$

[13.68]

There are three ways to produce entropy in the system volume:

1. Heat transfer into the volume.
2. Mass transfer into the volume.
3. Reaction in the volume.

The first term on the right of Equation 13.68 is entropy from heat and the second term from mass transfer. Using the equations for fluxes given in Equations 13.62 and 13.63 and adding to this, the entropy produced by chemical reactions from Equation 13.64 and Equation 13.65 becomes (Groot and Mazur, 1962)

$$\sigma = -J_q \frac{d\left(\frac{1}{T}\right)}{dx} + \sum_i J_i \frac{d\left(\frac{\mu_i}{T}\right)}{dx} + r\left(\frac{A}{T}\right)$$

[13.69]

$$\underset{\text{transfer}}{\underset{\text{heat}}{}} \qquad \underset{\text{transfer}}{\underset{\text{mass}}{}} \qquad \underset{\text{reactions}}{\underset{\text{chemical}}{}}$$

Each term in the entropy production expression is the product of a flux and a force, as given in **Table 13-3**. It is clear that the temperature is the force that causes heat to flow. If the temperature gradient is zero, heat does not flow. Similarly, the gradient in the chemical potential, μ_i, is the force that causes a flow

Table 13-3	Sources of entropy production as a function of time		
Entropy production by		**Flux**	**Force**
Heat transfer		J_q	$-d(1/T)/dx$
Mass transfer		J_i	$d(\mu_i/T)/dx$
Chemical reactions		r	A/T

or flux of matter. For a chemical reaction the affinity change, **A**, must be positive for the reaction to occur; therefore, **A** is the driving force for the reaction.

The terms in Equation 13.69 for σ, therefore, are cast in terms of the flux of i, J_i, times its conjugate force, X_i:

$$\sigma = \sum_i J_i X_i \qquad [13.70]$$

It must be true that each J_i is a function of each force X_i. For instance, a gradient in chemical potential causes heat to be transferred. Because the force terms are zero or positive if entropy is produced, a *Taylor's series* expansion about zero for J_i in these forces can be set up:

$$J_i = 0 + \sum_j \frac{dJ_i}{dX_j}(X_j - 0) + \sum_i \sum_j \frac{d^2 J_i}{dX_j dX_i} \frac{(X_j - 0)(X_i - 0)}{2} + \ldots \qquad [13.71]$$

where j counts on the forces. Truncating the series after the first two terms gives

$$J_i = \sum_j L_{i,j} X_j \qquad [13.72]$$

where the constants $L_{i,j} = dJ_i/dX_j$ are termed the *phenomenological coefficients*. The truncation implies that the fluxes are considered to be linearly related to the forces with slope $L_{i,j}$. This has been shown to be reasonable for both heat and mass diffusive flux, Darcy flow, and for chemical reactions near equilibrium.

Diffusion

Consider a system at constant temperature and pressure open to diffusional mass transfer of two components, **a** and **b**, in the x direction with no heat or chemical reactions occurring. From Equation 13.72, the flux of **a** is written as

$$J_a = L_{aa} X_a + L_{ab} X_b \qquad [13.73]$$

and for **b** as

$$J_b = L_{ba} X_a + L_{bb} X_b \qquad [13.74]$$

where

$$L_{aa} = \frac{dJ_a}{dX_a}, \; L_{ab} = \frac{dJ_a}{dX_b}, \; L_{bb} = \frac{dJ_b}{dX_b} \quad \text{and} \quad L_{ba} = \frac{dJ_b}{dX_a} \qquad [13.75]$$

For a diffusional flux the forces are

$$X_a = \frac{1}{T}\frac{d\mu_a}{dx} \quad \text{and} \quad X_b = \frac{1}{T}\frac{d\mu_b}{dx} \qquad [13.76]$$

If there is no change in μ_i with distance, the forces are zero and there is no diffusion of i.

As discussed in Chapter 12, in a solution next to a charged mineral surface, a concentration gradient of ions of opposite charge develops. Why does diffusion in the solution not occur to destroy the gradient? The fact that diffusion does not occur implies that although there is a concentration gradient, no chemical potential (i.e., no activity gradient, which is the force necessary to drive diffusion) exists. No chemical potential gradient with distance x implies

$$\frac{d\mu_i}{dx} = \frac{d(\mu_i^o + RT \ln a_i)}{dx} = RT\frac{d \ln a_i}{dx} = 0 \qquad [13.77]$$

At constant temperature as a function of distance from the surface

$$\frac{d \ln(\gamma_i m_i)}{dx} = 0 \qquad [13.78]$$

Because the surface is charged, there are strong interactions between the surface and charged species in solution. Clearly, these increase the stability of the ion near the surface, thereby decreasing γ_i. With increasing distance from the surface, γ_i increases as the interactions with the surface decrease. Equation 13.78 states that the activity of the ion, that is $\gamma_i m_i$, must be constant as a function of distance. Therefore, the activity of i in the double layer in the solution next to the mineral is equal to that in the bulk solution. With γ_i increasing away from the mineral surface, therefore, m_i must decrease. This gradient in the concentration of i is, therefore, a thermodynamically stable feature.

A relation between the phenomenological coefficients and diffusion coefficients is generated by first considering the Onsager relation. It was shown by Onsager (1931) that if J_i and X_i are independent, then the relation

$$L_{ab} = L_{ba} \qquad [13.79]$$

must hold; therefore, the number of independent coefficients is greatly reduced. With

$$\mu_i = \mu_i^o + RT \ln\gamma_i m_i \qquad [13.80]$$

and assuming γ_i is a constant, independent of m_i

$$\frac{d\mu_i}{dm_i} = \frac{RT}{\gamma_i m_i} \qquad [13.81]$$

This is the case if species i mixes ideally in solution or in the case of neutral species that are dilute enough to be in the Henry's law region. Under these conditions, the force of diffusion is

$$\frac{1}{T}\frac{d\mu_i}{dx} = \frac{1}{T}\frac{d\mu_i}{dm_i}\frac{dm_i}{dx} = \frac{R}{\gamma_i m_i}\frac{dm_i}{dx} \qquad [13.82]$$

so that J_a is written as

$$J_a = L_{aa}\frac{R}{\gamma_a m_a}\frac{dm_a}{dx} + L_{ab}\frac{R}{\gamma_b m_b}\frac{dm_b}{dx} \qquad [13.83]$$

Because the molal diffusion equation for J_a is

$$J_a = D_{aa}\frac{dm_a}{dx} + D_{ab}\frac{dm_b}{dx} \qquad [13.84]$$

The relation between D_{ij} and L_{ij} is given by

$$D_{aa} = L_{aa}\frac{R}{\gamma_a m_a} \quad \text{and} \quad D_{ab} = L_{ab}\frac{R}{\gamma_b m_b} \qquad [13.85]$$

Care must be taken when using equations such as Equation 13.84 because the driving force for diffusion is $1/T\,(d\mu_i/dx)$ and not dm_i/dx. With these ideas relating Gibbs energy and entropy in linear irreversible reactions to the rate of elementary reaction, heterogeneous reactions between minerals and fluids of importance in studying the earth can be analyzed; however, at greater distances from equilibrium, nonlinear processes can occur. These nonlinear, far from equilibrium processes can lead to ordered structures and pattern formation even though they increase entropy of the system with time (see the section on deterministic chaos in Chapter 1).

Autocatalytic Processes

Consider heat transfer through a pot of water in a temperature gradient. At small temperature gradients and therefore close to equilibrium, *Fourier's law* holds, and the entropy production is given by the linear heat flux times the force of heat transfer, $d(1/T/dx)$; however, at higher temperature gradients, the system becomes nonlinear as fluid transport occurs. The positive nonlinear feedback between fluid transport and diffusion causes large-scale pattern formation by self-organization resulting in the formation of stable convection cells. Pattern formation also occurs in other natural far from equilibrium processes.

The change in entropy of a system, dS, can be defined as the sum of that produced internally, d_iS, and that transported through the boundaries of the system, d_eS. Close to equilibrium d_iS dominates, and the system reaches a state of maximum entropy at equilibrium; however, when d_eS is large, it can overwhelm d_iS. In this case, the system may not proceed to equilibrium, but spontaneous self-organization can occur. One can think of this as the system being caught in a state of local entropy maximum in its attempt to converge on the state of overall maximum entropy. What is required is a nonequilibrium *autocatalytic process*. An autocatalytic process is one in which the rate increases by increasing the results of the process. In these systems, small perturbations are amplified (*positive feedback*) rather than converging on a stable value.

Consider the following feedback between dissolution and flow processes. Initially, a volume of unaltered rock in a porous medium contains a fluid at equilibrium. A nonequilibrium undersaturated fluid, which can cause dissolution of the rock, is introduced into the rock volume. A reaction front will develop as the flowing nonequilibrium fluid reacts with the unaltered rock before it reaches equilibrium. This front is also a porosity/permeability front where the porosity/permeability is greater behind the front because of mineral dissolution. Because of small initial perturbations of porosity/permeability along the front, at some locations, greater dissolution will occur while at other locations along the front less dissolution occurs.

Dissolution-flow is then an autocatalytic process. More fluid will flow where greater dissolution has produced greater porosity/permeability, which in turn increases the extent of flow. As flow is diverted by capture from the areas initially of lower porosity/permeability, these areas will show less advancement of the reaction front. As a function of time, the flow diversion and, therefore, the extent of reaction will produce a steady state in which a regular banded patterned of altered and unaltered rock is produced.

As another example of pattern formation, consider the common oscillatory zoning of anorthite-rich and albite-rich layers that can occur in plagioclase crystallizing from a magma (Haase et al., 1980). If the rate of crystallization of a unit of anorthite is promoted by the presence of already precipitated anorthite and the rate of crystallization of albite increases if already precipitated albite is present, this is an autocatalytic process in each component. They are linked if the growth of anorthite or albite on the plagioclase depletes the nearby melt of its anorthite and albite components, respectively.

Starting at point **a** in **Figure 13-3**, the incorporation of anorthite component increases the rate of anorthite component incorporation and this causes the composition to move to point **b**. At point **b**, all of the anorthite in the melt is swept from the vicinity of the growing crystal and anorthite incorporation

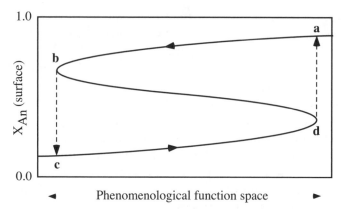

FIGURE 13-3 Functional relation between the mole fraction of anorthite incorporated relative to albite on a plagioclase crystal given a positive feedback autocatalytic process that is limited by the diffusion of the component in the melt to the growing plagioclase surface. (Adapted from Hasse et al., 1980.)

stops. At this point, the albite component is high enough in the nearby melt that it starts to crystallize on the plagioclase and the reaction jumps from point **b** to **c**. With the incorporation of albite, its rate of crystallization increases. Albite then dominates on the growing plagioclase until its incorporation stops as point **d** is reached as the albite component is depleted in the nearby melt. The melt then starts to crystallize anorthite at point **a**. An oscillatory growth pattern in the feldspar is produced.

■ Solid-Fluid Reactions

Consider the heterogeneous reaction of quartz dissolution at 25°C and 1 bar in pure H_2O. The dominant Si species in solution at a pH below 8 can be written as $Si(OH)_4^o$. The overall dissolution reaction is

$$2H_2O + SiO_{2(qtz)} \rightarrow Si(OH)_4^o \qquad [13.86]$$

The rate of this reaction is measured by monitoring the increase in $Si(OH)_4^o$ molality undersaturated far from equilibrium solutions as a function of time, as this must be equal to the amount of quartz that dissolves. In strongly undersaturated solutions, no significant back reaction occurs. The rate is observed to be constant as a function of time so that

$$\frac{dm_{Si(OH)_4^o}}{dt} = k^+ \qquad [13.87]$$

where k^+ is the rate constant for the dissolution reaction that is dependent on pressure and temperature but independent of the concentration of $Si(OH)_4^o$ in solution. It is also observed that if the quartz is ground, producing twice the surface area and this quartz is reacted under the same conditions, the rate is twice as high, although the amount of quartz is the same. The rate of reaction between silicate minerals and aqueous solutions is, therefore, controlled by reactions on the mineral surface. The activated complex controlling the reaction rate must be in equilibrium with a surface species being a molecular configuration at the interface between the quartz and H_2O. What is the nature of this surface species?

If quartz reacts in solutions of decreasing pH, the dissolution rate constant, k^+ of quartz, decreases. k^+ of quartz reaches a minimum at a pH of approximately 2.5 at 25°C, as shown in **Figure 13-4**. Below as well as above this pH, the rate increases. This indicates that the concentration of the rate-controlling activated complex is affected by pH and decreases to a minimum at a pH of approximately 2.5. Because the aqueous Si species in solution remains $Si(OH)_4^0$, it is not affected by changes in pH. The quartz surface, however, has a pH-dependent charge. The quartz surface exhibits a large negative charge at high pH and decreases its charge to zero at a pH of approximately 2.5, its zero point of charge, pH_{zpc} (see Chapter 12). The equilibrium between this charged surface species and H^+ in the solution is written as

$$\equiv Si-O^- + H^+ = \equiv Si-OH \qquad [13.88]$$

where $\equiv Si-O^-$ and $\equiv Si-OH$ represent the negatively and neutrally charged surface species, respectively, and $\equiv$ denotes the bonding of Si with three oxygens in the mineral.

There is some surface Si that is bound to less than three oxygens in the mineral at corners and along kinks at surface ledges and defects, but bonding to three

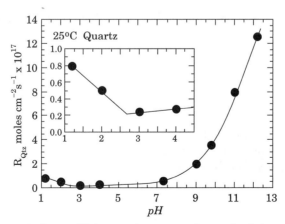

FIGURE 13-4 Long-term, far from equilibrium, steady-state rate of quartz dissolution at 25°C as a function of *pH*. The inset is a blowup of the low *pH* region. (Adapted from Wollast and Chou, 1986.)

interior oxygens dominates surface Si. A one-to-one correspondence has been shown between the decrease in the rate and decreasing concentration of Si–O⁻ on the surface. One can, therefore, conclude that the surface species Si–O⁻ controls the dissolution rate from the *pH* of the zero point of charge (approximately 2.5) to high *pH*. The overall dissolution reaction is, therefore, described by

$$\begin{array}{cc} \equiv\!Si\!-\!O \\ \equiv\!Si\!-\!O\!-\!\!\!-\!Si\!-\!O^- + 3H_2O \rightarrow & \begin{array}{c} \equiv\!Si\!-\!OH \\ \equiv\!Si\!-\!O^- + Si(OH)_4^o \\ \equiv\!Si\!-\!OH \end{array} \\ \equiv\!Si\!-\!O \end{array} \qquad [13.89]$$

where the lines between Si and O represent Si–O bonds. The left side of the surface species denoted with $\equiv$ is bonded to the mineral, and the right side is in contact with the solution. Reaction 13.89 indicates three bonds into the mineral must be broken to release an Si from the surface into solution. The sequence of doing this on a molecular scale is not known. From Reaction 13.89, the rate of quartz dissolution above its *pH* of zero point of charge is given by

$$R_+ = k^+ [\equiv\!Si\!-\!O^-][H_2O]^3 \qquad [13.90]$$

While holding the *pH* of solution constant, if the reaction is observed long enough, the overall rate slows as quartz approaches equilibrium with $Si(OH)_4^o$ in solution. That is, the reverse reaction with a rate constant of

$$R_- = k^-[\equiv\!Si\!-\!O^-][Si(OH)_4^o] \qquad [13.91]$$

becomes significant. At equilibrium, $R_+ = R_-$ so that

$$k^+ [\equiv\!Si\!-\!O^-][H_2O]^3 = k^-[\equiv\!Si\!-\!O^-][Si(OH)_4^o] \qquad [13.92]$$

With a solution of nearly pure H_2O, $[H_2O]$ is considered a constant in the reaction and its value taken as unity. Therefore,

$$\frac{k^+}{k^-} = m^{eq}_{Si(OH)_4^o} \qquad [13.93]$$

where $m^{eq}_{Si(OH)_4^o}$ denotes the equilibrium molality of $Si(OH)_4^o$ in solution. With a standard state for H_2O and quartz of the pure phase and for $Si(OH)_4^o$ a 1 molal solution whose properties are at infinite dilution, the equilibrium constant for this quartz dissolution Reaction 13.89 is

$$K_{qtz} = m^{eq}_{Si(OH)_4^o} \qquad [13.94]$$

as the activity coefficient of $Si(OH)_4^o$ and the activity of H_2O and quartz are taken as unity.

Combining Equations 13.93 and 13.94, K_{qtz} is also equal to the ratio of the rate constants of the forward and reverse reactions

$$K_{qtz} = \frac{k^+}{k^-}$$

[13.95]

K_{qtz} is independent of pH, but both the forward and reverse rates are pH dependent because the number of Si—O$^-$ on the quartz surface depends on pH. Like quartz, most reactions between aluminosilicate minerals and aqueous fluids are also influenced by the nature of charges on the mineral's surface. Rate expressions are very complicated in a complex aluminosilicate because the charging at all of the different types of cation sites present on the surface must be considered.

Feldspar Dissolution in H$_2$O

Consider the reactions that occur when albite, $NaAlSi_3O_8$, is put in an aqueous solution at 25°C. Albite has a structure of corner sharing Al—O and Si—O tetrahedra linked in a three-dimensional array with charge balancing Na$^+$ in large irregular cavities in the tetrahedral framework. When immersed in water, in less than a minute, a loss of Na$^+$ from the albite surface occurs that is represented by the reaction

[13.96]

Again, the short straight lines stand for oxygen–cation bonds. The symbols $\equiv$ and > indicate that three and two oxygen bonds, respectively, exist in the feldspar structure. The left side of the surface species, therefore, is bonded into the structure of the mineral; the right side is considered to be in contact with an aqueous solution. The (–1) above the surface species on the right indicates that the surface has a unit negative charge spread over the four oxygens.

In more compact notation, Reaction 13.96 is represented as

$$\left(\frac{\equiv Al}{\equiv Si}{>}O\right)_4{-}Na \longrightarrow 4\left(\frac{\equiv Al}{\equiv Si}{>}O\right)^{-0.25} + Na^+$$

[13.97]

where each of the four surface O has a –0.25 charge for charge counting purposes. Because each oxygen is bonded between two cations in different tetrahedra, a single negative valence charge is considered to be contributed by each oxygen to a tetrahedral cation bond. An Al tetrahedra has a total -1 charge with a $-1/4$ charge on each oxygen as indicated in Reaction 13.97. The negative surface charge produces bonds with H^+ (i.e., H_3O^+) from solution so that the reaction

$$
\begin{array}{c}
\overset{(-1)}{\equiv Si} \\
\quad\equiv Si\diagdown \\
\qquad O \\
\quad >Al\diagdown \\
\qquad O \\
\quad >Si\diagdown + mH_3O^+ \\
\qquad O \\
\quad >Al\diagdown \\
\qquad O \\
\quad\equiv Si
\end{array}
\longrightarrow
\begin{array}{c}
\overset{(m-1)}{\equiv Si} \\
\quad\equiv Si\diagdown \\
\qquad O \\
\quad >Al\diagdown \\
\qquad O \\
\quad >Si\diagdown mH_3O \\
\qquad O \\
\quad >Al\diagdown \\
\qquad O \\
\quad\equiv Si
\end{array}
\qquad\text{[13.98]}
$$

also occurs rapidly, coming to equilibrium in a couple of minutes. This may also be written in more compact notation as

$$4(\overset{\equiv Al}{\underset{\equiv Si}{}} > O)^{-0.25} + mH_3O^+ \rightarrow (\overset{\equiv Al}{\underset{\equiv Si}{}} > O)_4 - H_3O_m^{(m-1)} \qquad \text{[13.99]}$$

If m is less than 1, the surface remains negative, whereas if m is greater than 1, the surface is positively charged. As discussed in Chapter 12, the mineral surface is usually charged in aqueous solutions in which the charge depends on the solution pH. When albite is added to an aqueous solution, the pH increases as a result of this H_3O^+ adsorption. The combination of Reactions 13.97 and 13.99 can be viewed as an exchange of Na^+ for H_3O^+. This exchange, however, is not complete because more Na^+ is released than H_3O^+ absorbed. The resulting albite surface in solution is then negatively charged. The albite mineral surface remains negatively charged until the concentration of H^+ (i.e., H_3O^+) in solution increases to about $10^{-2}m$—that is, a pH of 2 or lower is reached, at which point the albite surface becomes positively charged.

Also occurring relatively rapidly on addition of albite to solution is hydration of the surface by breaking of a surface Si—O bond, which is represented by the reaction

$$\overset{\equiv Si}{\underset{\equiv Si}{}} > O + H_2O \rightarrow 2 \equiv Si-OH \qquad \text{[13.100]}$$

for an oxygen bound between two Si. For the oxygen bound between an Si and Al, the species produced by Reaction 13.99 with $m = 1$ is considered to be hydrated by the reaction

$$(\overset{\equiv Al}{\underset{\equiv Si}{}} > O)_4 - H_3O + 3H_2O \rightarrow 4(\equiv Al-OH)^{-1/4} + 4 \equiv Si-OH + H^+ \qquad \text{[13.101]}$$

After this relatively rapid surface exchange and hydration, the mineral begins to dissolve on a significantly longer time frame. This dissolution is pH dependent. At acidic conditions above a pH of approximately 2, the negatively charged Si surface species is modeled as dissolving by the reaction

$$\begin{matrix} \equiv Si-O \searrow \\ \equiv Al-O - Si-O^- \\ \equiv Si-O \nearrow \end{matrix} + H^+ + 3H_2O \longrightarrow Si(OH)_4^0 + \begin{matrix} \equiv Si-OH \\ \equiv Al-OH \\ \equiv Si-OH \end{matrix} \qquad [13.102]$$

whereas the Al species dissolution is represented by the reaction

$$\begin{matrix} \equiv Si-O \searrow \\ \equiv Si-O - Al-OH^- \\ \equiv Si-O \nearrow \end{matrix} + 4H^+ \longrightarrow Al^{3+} + \begin{matrix} \equiv Si-OH \\ \equiv Si-OH \\ \equiv Si-OH \end{matrix} + H_2O \qquad [13.103]$$

As pH decreases, the forward reactions are favored in both cases. Also, the reactions occur in parallel on the albite surface with the production of one reaction being a reactant in the other after a surface charge is produced. The dissolution is nonstoichiometric with the Al detachment Reaction 13.103 being faster than the Si detachment Reaction 13.102, producing an Si-rich surface at low pH.

At alkaline pH, the two parallel dissolution reactions are represented by

$$\begin{matrix} \equiv Si-O \searrow \\ \equiv Al-O - Si-O^- \\ \equiv Si-O \nearrow \end{matrix} + 2OH^- + H_2O \longrightarrow Si(OH)_4^0 + \begin{matrix} \equiv Si-O^- \\ \equiv Al-O^- \\ \equiv Si-O^- \end{matrix} \qquad [13.104]$$

and

$$\begin{matrix} \equiv Si-O \searrow \\ \equiv Si-O - Al-OH^- \\ \equiv Si-O \nearrow \end{matrix} + 3OH^- \longrightarrow Al(OH)_4^{-1} + \begin{matrix} \equiv Si-O^- \\ \equiv Si-O^- \\ \equiv Si-O^- \end{matrix} \qquad [13.105]$$

In this case, as pH increases, the reactions are driven to the right, and the dissolution rate increases. This dissolution is also nonstoichiometric as a result of different rates of parallel chain reactions, but with the Si detachment, Reaction 13.104 is greater than the Al detachment Reaction 13.105 so that the surface becomes Al rich. Again, with time, the cation that is released at a greater than stoichiometric rate slows as its surface exposure decreases, and the rates become stoichiometric. The time frame for this to occur at 25°C is about a week at acid or alkaline pH but is months at near-neutral pH where reaction rates are slower.

Rates of dissolution increase in acidic and alkaline solutions relative to near-neutral solutions because albite forms increasing concentrations of charged surface complexes as pH is lowered or raised relative to near-neutral conditions. These charged surface complexes focus the bonding electron cloud, thereby polarizing the surface cation–oxygen bonds and weakening them (Zinder et al., 1986). The weakened bond has a lower activation energy for reaction and, therefore, a greater probability of surface detachment, or rate of dissolution, than unpolarized cation–oxygen bonds (Brady and Walther, 1989). **Figure 13-5** shows the measurements of the long-term, far from equilibrium, steady-state rate of albite dissolution after stoichiometric dissolution has been established.

Most aluminosilicates dissolve in a similar manner to albite with initial alkali for H_3O^+ exchange and the early rate of release of Si greater than Al relative to stoichiometric proportions at alkaline *pH* and Al greater than Si at acid *pH*. When long-term stoichiometric dissolution is eventually established, the mineral surface is Si rich at low *pH* and Al rich at high *pH*. This stoichiometric dissolution rate increases in acid solution with decreasing *pH* and increases in alkaline solutions with increasing *pH* for most aluminosilicate minerals, as shown for albite in Figure 13-5.

Relative Rates of Silicate Dissolution

To consider the relative dissolution rates between silicate minerals, the relative strength of the cation–oxygen bonds in silicates needs to be considered because this is proportional to the activation energy of the rate-limiting elementary reaction. A measure of bond strength between cations and oxygen is the *mean electrostatic site potential* of the oxygen bound to the cation in the structure. Electrostatic potential is computed by assuming each atom is a point charge and summing valence charge to bond distance ratios. Oxygen electrostatic site potential increases with the increased sharing of an oxygen site by additional silicon and is minimal for oxygen sites not bound to other silicon atoms (Smyth, 1989).

If the breaking of the Si—O bond is rate limiting, then there should be a relation between bond strength and rate of reaction. Because rates are *pH* depen-

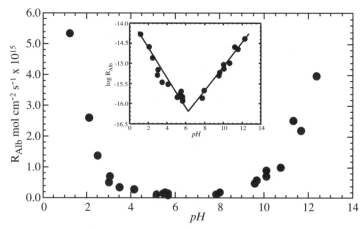

FIGURE 13-5 Long-term, far from equilibrium, steady-state 25°C and 1 bar albite dissolution rate as a function of *pH*. The insert is the data plotted as a function of log rate. (Adapted from Chou and Wollast, 1985.)

dent, rates need to be compared at the same *pH*. This was done by Brady and Walther (1989) at *pH* 8, as shown in **Figure 13-6**. Silicate minerals that form at higher temperatures, like forsterite and enstatite, have a smaller site potential and therefore dissolve more rapidly than silicate minerals stable at lower temperatures, like quartz and kaolinite. The reason nepheline dissolution rates are so high is not clear.

Reaction Rates Between Minerals at Elevated Temperatures and Pressures

Consider the mechanisms that produce a new mineral during progressive metamorphism in the earth's crust, as outlined in Chapter 9. As the temperature increases toward peak metamorphic conditions, lower grade mineral assemblages become unstable relative to higher grade assemblages. The Gibbs energy of the higher grade assemblage minus the Gibbs energy of the lower grade assemblage becomes negative. With this chemical affinity drive, the reaction proceeds by a number of sequential steps, with the slowest being rate limiting.

The product minerals in the reaction do not typically grow at the site of the reacting phases but occur several grain lengths away. This happens because of the difficulty of transporting Al, as discussed below. It is assumed that a fluid phase is present because, as argued in Chapter 9, insignificant reactions to form new minerals occur during regional metamorphism if a fluid phase is not present. The higher Gibbs energy of the reactant grains relative to fluid-phase species causes the reactant grains to dissolve, reacting toward an equilibrium between the grain boundary fluid and the reactant mineral surface. This fluid becomes supersaturated with respect to the potentially stable higher grade phases that have yet to nucleate. The reactant material in the fluid migrates to an area that minimizes the activation energy for *nucleation*. The higher grade phases then nucleate from the fluid when the activation energy

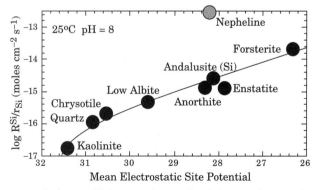

FIGURE 13-6 Long-term, far from equilibrium, steady-state dissolution rates for the indicated minerals at 25°C, 1 bar, and *pH* 8 as a function of site potentials in volts of oxygen bonds connecting detaching silicon atoms to the mineral surface. r_{Si} is the ratio of SiO_2 to total non-alkali oxides in the mineral formula. (Reprinted from *Geochim. Cosmochim. Acta* 53, P. V. Brady and J. V. Walther, Controls of silicate dissolution rates in neutral and basic pH solutions at 25°C, pp. 2823–2830, 1989, with permission from Elsevier.)

barrier for nucleation in the fluid is exceeded. After the higher grade phase has nucleated, it grows from elements obtained from the fluid transported from the dissolving reactant grains.

Any of these sequential processes can be rate limiting. For a given overstep of the Gibbs energy of reaction, these processes are

1. Dissolution of the reactant mineral
2. Transport of material to the site of product-phase nucleation
3. Nucleation
4. Growth of the product phase

as illustrated in **Figure 13-7.**

Nucleation

Nucleation is either homogeneous, nucleating in the fluid phase, or heterogeneous, nucleating from the fluid on a surface of a preexisting mineral grain. The Gibbs energy of an embryo of the solid phase relative to the fluid phase, ΔG_e, is equal to the difference in Gibbs energy per molecular unit volume of the nucleating phase to that in the fluid phase, ΔG_V (a negative number if supersaturated), times the total volume V of the embryo. To this must be added the *surface energy* per unit surface area of the embryo in the fluid, γ_s, times the embryo's surface area, A_s (see Chapter 12).

$$\Delta G_e = \Delta G_V V + \gamma_s A_s \tag{13.106}$$

$d(\Delta G_e)/dr$ where r is the radius of the embryo must be negative for the new embryo to be stable.

As the overstep temperature increases, ΔG_V becomes a larger negative number as more of the reactant is dissolving in the fluid, but the surface energy, γ_s, is positive. If the embryo is approximated as a sphere, which would minimize the surface area per unit volume, then

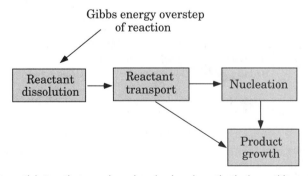

FIGURE 13-7 Sequential steps that occur in a mineral–mineral reaction in the earth's crust for a Gibbs energy of overstep of the reaction.

$$A_s = 4\,\pi r^2 \tag{13.107}$$

Because the volume of a sphere is $4/3\pi r^3$, the total energy change to form the embryo from reactant species is

$$\Delta G_e = \frac{4}{3}\pi r^3 \Delta G_V + 4\pi r^2 \gamma_s \tag{13.108}$$

The surface energy term dominates at small r, whereas the volume term dominates at large r; therefore, $d(\Delta G_e)/dr$ changes sign from positive to negative with increasing r, as shown in **Figure 13-8**.

The critical radius where $d(\Delta G_e)/dr = 0$ for the embryo to form a stable crystal is denoted by r_c. Taking the derivatives of both sides of Equation 13.108 with respect to r at r_c where $d(\Delta G_e)/dr = 0$, yields

$$r_c = \frac{-2\gamma_s}{\Delta G_V} \tag{13.109}$$

Equation 13.109 indicates that when ΔG_V is zero, the critical embryo radius would need to be infinite. As the solution becomes supersaturated relative to embryo formation, ΔG_V becomes negative, and for the same surface energy, the required radius of the embryo decreases. With decreasing critical radius, the probability of forming the critically sized embryo increases, and thus, the nucleation rate increases.

Consider a solution with a molality of $SiO_2 = 7.2929 \times 10^{-2}$ at 500°C and 2 kbar (this is a quartz saturated solution at 501°C and 2 kbar). Quartz solubility at 500°C and 2 kbar computed from SUPCRT92 = 7.2594×10^{-2} molal. What would be the critical radius for homogeneous nucleation for this slightly supersaturated solution if quartz was not present? The excess Gibbs energy drive for nucleation would be

$$\Delta G_{ex} = -2.303\,RT \log\left(\frac{Q}{K}\right) = -2.303 \times 8.314 \times 773.15 \times \log\left(\frac{7.2929 \times 10^{-2}m}{7.2594 \times 10^{-2}m}\right)$$
$$= -30.0\ \text{J mol}^{-1} \tag{13.110}$$

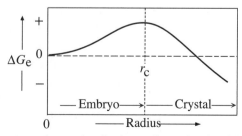

FIGURE 13-8 The Gibbs energy to form a mineral embryo, ΔG_e, as a function of the embryo radius. r_c denotes the critical radius above which the embryo is stable.

or per cm^3

$$\Delta G_V = \frac{-30.0 \text{ J mol}^{-1}}{22.688 \text{ cm}^3 \text{ mol}^{-1}} = -1.32 \text{ J cm}^{-3} \qquad [13.111]$$

The surface Gibbs energy of SiO_2 (amorphous) in water at earth surface conditions $= 46$ erg cm$^{-2} = 4.6 \times 10^{-6}$ J cm^{-2} (Berner, 1980, Table 5-1). Taking this value for quartz in water at 500°C and 2 kbar, from Equation 13.109, the critical radius is

$$r_c = \frac{-2 \times 4.6 \times 10^{-6} \text{ J cm}^{-2}}{-1.32 \text{ J cm}^{-3}} = 6.97 \times 10^{-6} \text{ cm} = 697\text{Å} \qquad [13.112]$$

With the molar volume of quartz $= 22.688$ cm^3 mol^{-1}, the volume of a unit of SiO_2 of quartz can be obtained by dividing by $N°$

$$\frac{22.688 \text{ cm}^3 \text{ mol}^{-1}}{6.022 \times 10^{23} \text{units mol}^{-1}} = 3.77 \times 10^{-23} \text{ cm}^3 \text{ per SiO}_2 \text{ unit} \qquad [13.113]$$

If one assumes a spherical SiO_2 unit cell, its radius is

$$r_{\text{unit}} = \left(\frac{3.77 \times 10^{-23}}{4/3\pi}\right)^{1/3} = 2.08 \times 10^{-8} \text{ cm} = 2.08 \qquad [13.114]$$

which is somewhat less than the measured half lengths of the unit cell parameters of quartz (a = 4.913 Å, c = 5.405 Å).

Given the two orders of magnitude difference in the computed critical radius and that of a unit cell, the probability that units of SiO_2 will congregate to homogeneously nucleate an embryo of critical radius is small; however, given the large number of SiO_2 units in an interstitial metamorphic fluid and the 100,000-year time frame available during a 1°C increase in temperature for regional metamorphism, its probability increases significantly.

For reactions producing new minerals in the earth's crust, nucleation most likely occurs by heterogeneous reactions. Heterogeneous nucleation is dominant because r_c of heterogeneous nucleation is lower than for the homogeneous case of fluid-phase nucleation. r_c of heterogeneous nucleation is decreased relative to fluid-phase nucleation by a lowering of the interfacial energy, γ_s, between atomic clusters of the newly formed embryo and the mineral substrate on which it nucleates. This occurs by matching lattice and bond types rather than from similarities in chemical composition on the mineral surface substrate. For most prograde metamorphic reactions, r_c for heterogeneous nucleation is small, and the nucleation rate does not appear to control the reaction rate. This is consistent with the fact that newly nucleated index minerals at isograd reactions observed in the field appear in a regular manner, whereas the disappearance of the reactant phase is often irregular. The product mineral phases, therefore, are observed to be readily nucleated.

If nucleation is not generally a rate-limiting factor, then either breakdown and growth of the mineral surface or material transport from the reactant to the product mineral must be. Because these are sequential processes, the slowest of the two is rate limiting. The question is whether reactions at a surface, which require bonds to be broken and rearranged in a more stable configuration, or transport of material by diffusion or fluid advection is slower.

Surface Reaction

Consider first the rate of surface reaction at medium grades of metamorphism (approximately 500°C). This rate is influenced by the extent of Gibbs energy overstepping of the reaction. The rate is calculated from experiments designed to determine the equilibrium position in pressure and temperature of reactions like

$$Al_2Si_4O_{10}(OH)_2 = Al_2SiO_5 + 3SiO_2 + H_2O \qquad [13.115]$$

pyrophyllite   andalusite   quartz

A common way to determine the equilibrium temperature for this reaction at a specified pressure is to prepare reaction capsules of all the phases in the reaction, including H_2O with all of the minerals ground up except one, which is left as a single crystal. The single crystal is carefully weighed before it is loaded into the capsule. After the capsule is sealed, it is heated. A set of capsules is brought to different temperatures for a fixed amount of time (e.g., 2 weeks) at the same pressure. After 2 weeks, the capsules are opened, and the single crystals are again weighed. The weight change is plotted as a function of temperature as shown for single crystals of quartz for Reaction 13.115 from Kerrick (1968) in **Figure 13-9**.

As expected, the quartz crystal gains weight at temperatures above the equilibrium temperature for Reaction 13.115 as the reaction goes to the right. The quartz crystal loses weight below the equilibrium temperature as Reaction 13.115 goes to the left and is zero at the equilibrium temperature. As the reactions were all run for the same amount of time, the greater weight change as one gets further from the equilibrium temperature is an effect of the chemical affinity, that is, Gibbs energy change of the reaction. Because the reaction is near equilibrium, the relation in Equation 13.44 is used. With m as the mass of material lost or gained per unit surface area,

$$\frac{dm}{dt} = k_r^+ \frac{\Delta G_R}{RT} \qquad [13.116]$$

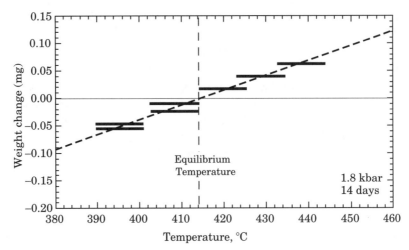

FIGURE 13-9 Thick horizontal line segments bracket temperature and weight change errors for measurements of a single crystal of quartz in the reaction pyrophyllite = andalusite + 3 quartz + H_2O after 14 days. The diagonal dashed line gives the best fit to the weight change as a function of temperature. The vertical line drawn through zero weight change gives the equilibrium temperature of the reaction. (Adapted from Wood and Walther, 1983.)

The change in ΔG_R with temperature is given by

$$\left(\frac{d\Delta G_R}{dT}\right)_P = -\Delta S_R \tag{13.117}$$

If $\Delta C_p = 0$ or is small, ΔS_R is considered to be constant, and ΔG_R is explicitly expressed as a function of temperature at constant pressure as

$$\Delta G_R = -(T - T_{eq}) \, \Delta S_R \tag{13.118}$$

Being near the equilibrium temperature, the values of ΔG_R from Equation 13.118 can be substituted into Equation 13.116 to give

$$\frac{dm}{dt} = \frac{k_r^+ (\Delta S_R (T - T_{eq}))}{R T_{eq}} \tag{13.119}$$

Equation 13.119 for a constant time interval gives the linear relationship shown by the heavy dashed line in Figure 13-9.

In the single-crystal experiments, reactions at the surface of the single crystals are likely rate limiting. With knowledge of ΔS_R for the reaction and dm/dt from the single-crystal weight loss, k_r^+ is calculated for the reaction on the single crystal normalized per unit of its surface area. To compare k_r^+ for simple minerals, such as quartz, with those of more complicated minerals, such as grossular, the rates are normalized to *gram atoms of oxygen*, g.a.o. (i.e., per mole of oxygen). For instance, there are 2 g.a.o. in quartz (SiO_2) but 12 g.a.o. in grossular ($Ca_3Al_2Si_3O_{12}$) per mole. When these normalized k_r^+ are plotted as a function of inverse temperature, **Figure 13-10** is obtained.

Also plotted on this diagram are rate studies from lower temperature rate investigations at near-neutral pH. It appears that these pH dependent rates are greater and therefore overwhelm any pH independent rate mechanism that may occur at these lower temperatures. The near linearity when plotted as a function of $1/T$ (an *Arrhenius plot*) stems from the fact that

$$\left(\frac{d \ln k_r^+}{d\frac{1}{T}} \right)_P \cong \frac{\Delta H_{R+}^o}{R} \qquad [13.120]$$

where ΔH_{R+}^o is the standard-state heat of the overall forward reaction (see Equation 13.56). If the heat capacity of the reaction is close to zero, then ΔH_{R+}^o does not change significantly with temperature, and Equation 13.120 produces a straight line when $\ln k_r^+$ is plotted against $1/T$. The line shown in Figure 13-10 is given by

$$\log k_r^+ = -\frac{2900}{T} - 6.85 \qquad [13.121]$$

with T in Kelvin and k_r^+ in g.a.o. cm^{-2} s^{-1}.

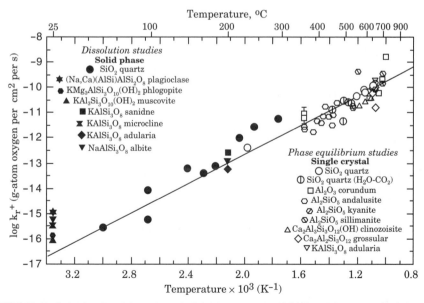

FIGURE 13-10 Arrhenius plot of the reciprocal of absolute temperature (1/T) in Kelvin versus the base-10 logarithm of the rate constant, in gram atoms of oxygen per cm^2 per second, determined from both phase equilibrium investigations and dissolution measurements at near-neutral pH. (Adapted from Wood and Walther, 1983.)

As an example of calculating the time for a surface-controlled reaction to occur, consider a dehydration reaction at 500°C (773 K). Hydration reactions written for 1 mole of H_2O released typically have a ΔS_R of about 20 cal °C^{-1} mol^{-1}. k_r^+ is calculated from Equation 13.121 and for 773 K is 2.5×10^{-11} g.a.o. cm^{-2} s^{-1}. With a 1°C overstep of the equilibrium temperature, the reaction rate from Equation 13.119 is

$$\frac{dm}{dt} = \frac{2.5 \times 10^{-11} \text{ g.a.o. cm}^{-2}\text{s}^{-1} \times 20 \text{ cal}\,°\text{C}^{-1}\text{ mol}^{-1} \times 1.0°\text{C}}{1.987 \text{ cal K}^{-1}\text{mol}^{-1} \times 773 \text{ K}} \quad [13.122]$$

$$= 3.26 \times 10^{-13} \text{ g.a.o. cm}^{-2} \text{ s}^{-1}$$

A crystal of surface area, s, containing n g.a.o. then gains or loses material at a rate given by

$$\frac{1}{s}\frac{dn}{dt} = 3.26 \times 10^{-13} \text{ g.a.o. cm}^{-2} \text{ s}^{-1} \quad [13.123]$$

Assuming a spherical crystal of radius r containing N g.a.o. per cm^3, the total g.a.o. in the crystal is given by

$$n = \frac{4}{3}\pi r^3 N \quad [13.124]$$

Taking the derivative of both sides of Equation 13.124 with respect to r results in

$$\frac{dn}{dr} = 4\pi r^2 N \quad [13.125]$$

Combining the two derivative expressions in Equations 13.123 and 13.125 gives

$$\frac{1}{s}\frac{dn}{dt}\frac{dr}{dn} = \frac{3.26 \times 10^{13}\text{g.a.o. cm}^{-2} \text{ s}^{-1}}{4\pi r^2 N} \quad [13.126]$$

Because the surface area is related to the radius of a sphere by

$$s = 4\pi r^2 \quad [13.127]$$

the rate of change of radius of the crystal is given by

$$\frac{dr}{dt} = \frac{3.26 \times 10^{-13}}{N} \quad [13.128]$$

Integrating the rate of change of radius in time from $r°$ at $t = 0$ to r at time t gives

$$r - r° = \frac{3.26 \times 10^{13}\text{g.a.o. cm}^{-2} \text{ s}^{-1} \ t}{N \text{ g.a.o. cm}^{-3}} \quad [13.129]$$

Consider the dissolution of a quartz crystal (where $N = 0.09$ g.a.o. cm^{-3}) of radius $r^\circ = 0.1$ cm. The time needed to dissolve it completely for a 1°C overstep is $t = 2.8 \times 10^{10}$ s, or 875 years. Regional metamorphism typically occurs with temperature increases of 10°C per million years and contact events with 1°C increases in 1000 years. Surface reactions are relatively rapid with respect to these temperature changes, and reactions should generally not be surface reaction limited.

■ Rates of Material Transport

The reaction rate could be limited by the rate of material transport from the reactant mineral to the site of new mineral growth. This transport is either by diffusion or flow. These processes are parallel, and thus, the most rapid transport process is rate determining. For a diffusional flux, J^{dif} (mol cm^{-2} s^{-1}), Fick's first law of diffusion (see Equations 13.84 and 13.86) for component a is

$$J_a^{dif} = -D_a \left(\frac{dc_a}{dx}\right)_t = -L_{aa}\left(\frac{d\mu_a}{dx}\right)_t \qquad [13.130]$$

where J_a^{dif} gives the flux of a along a composition/chemical potential gradient at a particular time, whereas D_a and L_{aa} represent the diffusion coefficient and phenomenological coefficient, respectively, of a because of a change in a. If D_a has units of cm^2 s^{-1}, then c_a is in mol cm^{-3}. Three types of diffusion are possible at medium grades of metamorphism (approximately 500°C).

Diffusion through the mineral grains themselves could conceivably occur. The diffusion coefficient for this solid-state diffusion, D^{ss}, however, is on the order of 10^{-21} cm^2 s^{-1} (Walther and Wood, 1984) for most elements, although hydrogen diffusion is significantly more rapid. This value is too small to allow significant diffusion over the distance needed in the time frames of millions of years or less that are available. For instance, using Equation 13.130 with a large concentration gradient of a, say a difference of 1 mole per cm^3, over a distance of 1 cm gives

$$J_a^{dif} = -10^{-21}\ \text{cm}^2\ \text{s}^{-1}\ \frac{1\ \text{mol/cm}^3}{1\ \text{cm}} = -10^{-21}\text{mol cm}^{-2}\ \text{s}^{-1} \qquad [13.131]$$

To, therefore, move 0.001 mole of material through a 1 cm^{-2} area would take 10^{18} s, or 32 billion years. This slow rate of solid-state diffusion is confirmed by studies of compositional gradients in minerals that indicate large gradients as a function of distance of opposite slope persist in mineral grains. For instance, patterns of oscillatory zoning are observed to survive peak metamorphic conditions. This implies that solid-state diffusion is not important for transporting material the distances of a number of mineral grains in length required to make new minerals.

Another possibility is grain boundary diffusion. That is, diffusion through the disordered structure of a number of unit cell depths that exists on the surface of a typical mineral. The grain boundary diffusion coefficient, D^{gb}, is about four orders of magnitude more rapid than solid-state diffusion or about 10^{-17} $cm^2\,s^{-1}$ at 500°C (Joesten, 1983). To determine the surface of grain boundaries available for grain boundary diffusion, consider a rock made up of 1 mm mineral grains so that there are about 20 cm of grain boundary length per cm^2 of rock. With a width of the grain boundary of 20 Å (2×10^{-7} cm), which is equal to about a 2 unit cell thickness of the mineral, and with two minerals at each boundary, the area through which the flux occurs is 4×10^{-7} cm^2 per cm^2 of rock. The four orders of magnitude greater diffusion coefficient of grain boundary diffusion relative to solid-state diffusion is balanced by the six orders of magnitude less area available to diffuse through relative to solid-state diffusion. This is also a very inefficient process.

The final diffusion process to consider is diffusion in a discrete fluid phase along gain boundaries and in microfractures. In this case, the diffusion coefficient, D^{aq}, at 500°C is about 10^{-4} $cm^2\,s^{-1}$. What is the cross-sectional area of this grain boundary/microfracture fluid through which true fluid diffusion operates? This likely has a minimum width of approximately 200 Å because this is the width of fluid-filled fractures and grain boundaries at the time they begin to seal, as determined in Chapter 9 from a variety of fluid inclusion planes in regionally metamorphosed rocks (see the discussion for Figure 9-19). Because of the 13 orders of magnitude greater values of the diffusion coefficient for fluid-phase diffusion, as opposed to grain boundary diffusion, even if fluid is present only intermittently, fluid-phase diffusion is the dominant diffusion mechanism.

From this discussion, it is clear that if no fluid is present, no significant diffusion occurs, but if a fluid phase is present, there is a potential for fluid-phase diffusional flux. For the case of fluid diffusion, D_a^{aq} in Fick's law is used to determine L_{aa}^{aq}, the phenomenological coefficient of diffusion of **a** in the fluid due to a change in **a** from

$$L_{aa}^{aq} = \frac{D_a^{aq} C_a\ \phi\ \tau}{1000\ RT} \tag{13.132}$$

where

D_a^{aq} = aqueous diffusion coefficient of **a** ($cm^2\,s^{-1}$).

C_a = concentration of **a** (moles per liter). The 1000 in the denominator is to convert this to moles per cm^3.

ϕ = area of fluid-filled fracture normal to the diffusion direction per cm^2 (dimensionless).

τ = tortuosity of the flow (dimensionless).

The *tortuosity*, τ, is a parameter accounting for the orientation of the fractures relative to the flow path. From Fick's law, therefore, the flux of **a** due to aqueous phase diffusion, J_a^{dif}, is

$$J_a^{dif} = -\frac{D_a^{aq} C_a\ \phi\ \tau}{1000\ RT}\frac{d\mu}{dx}\ \text{mol cm}^{-2}\ \text{s}^{-1}$$

[13.133]

Because the flux of **a** depends on fluid-filled channels, it is possible that the fluid is moving as discussed in Chapter 9 and the transport of component **a** by fluid flow may also be important.

Transport by Fluid Flow

In Chapter 9, when metamorphism of the crust was discussed, a model of fluid flow during a regional metamorphic event was constructed. The fluid flux, q, was given in Equation 9.29 as

$$q = \frac{d^3 l\,\tau\,g(\rho_R - \rho_f)\ \cos\theta}{12\,\eta}$$

[13.134]

The symbols for and range of values of variables for Equation 13.134 are given in **Table 13-4**. For grain boundary flow at 45° to the vertical with a tortuosity,

Table 13-4	Parameters and their values for considering fluid flow during regional metamorphism in the earth's crust where fluid pressure equals rock pressure, as discussed in Chapter 9

Parameter	Symbol	Value
Fluid flux	q	10^{-10} to 10^{-9} cm^3 per cm^{-2} s^{-1} for regional metamorphism
Grain boundary or fracture density	l	20 cm per cm^2 (1-mm grains) to 10^{-5} cm per cm^2 (1 km of fracture per km^2)
Width of grain boundary or fracture	d	(calculated) can range from 0.02 μm to more than 10 μm
Density of fluid	ρ_f	~0.9 g cm^{-3} (supercritical H_2O–CO_2 fluid)
Density of rock	ρ_R	~2.8 g cm^{-3} (average for crust)
Viscosity of fluid	η	1 to 2×10^{-3} poise (dyn s cm^{-2}) (supercritical H_2O–CO_2 fluid)
Tortuosity	τ	(dimensionless) 1.0 for a straight fracture from 0.5 to 0.8 for grain boundary flow
Acceleration of gravity	g	981 cm s^{-2}
Angle to the vertical of flow	θ	0° for vertical flow to 90° for horizontal flow

$\tau = 0.7$ grain boundary/fracture width using Equation 13.134 for values consistent with those in Table 13-4 of $q = 5 \times 10^{-10}$ cm^3 per cm^{-2} s^{-1} and $\eta = 1.5 \times 10^{-3}$ poise d can be calculated. In this case, the width of the fluid in the grain boundary/fracture is

$$d = \left(\frac{q \, 12 \, \eta}{l \tau \, g(\rho_R - \rho_f) \cos \theta} \right)^{1/3}$$

$$= \left(\frac{5 \times 10^{10} \text{cm s}^{-1} \times 12 \times 1.5 \times 10^{-3} \text{ dyn s cm}^{-2}}{20 \text{ cm} \times 0.7 \times 981 \text{ cm s}^{-1} \times (2.8 - 0.9) \text{ g cm}^{-3} \times 0.707} \right)^{1/3} \qquad [13.135]$$

$$= 7.9 \times 10^{-6} \text{ cm} = 790 \text{ Å}$$

The width (aperture) of the fracture, d, is above the 200 Å fracture width at which fractures seal as determined from fluid-inclusion planes.

The velocity, v, of the fluid along the flow path is equal to

$$v = \frac{q \cos \theta}{dl} \qquad [13.136]$$

or

$$v = \frac{5 \times 10^{-10} \text{cm s}^{-1} \times 0.707}{7.9 \times 10^{-6} \text{cm} \times 20 \text{ cm}^{-1}} = 2.2 \times 10^{-6} \text{cm s}^{-1} = 0.69 \text{m yr}^{-1} \qquad [13.137]$$

With a grain boundary flow at 45° to the vertical, therefore, its velocity is on the order of 0.7 meters a year. If the flow is nearly horizontal, for example, at 86° to the vertical, the flow would be about 10 times slower (cos 86° = 0.070). Because the flow is buoyancy driven, if the potential flow direction was completely horizontal with cos $\theta = 0$, the fluid would not flow.

Instead of being channeled along grain boundaries, consider a fluid flux that is accommodated by a representative volume with say 100 m of fracture per 100 m^2 area. This is then equivalent to a set of fractures 1 m apart across a 10×10 m area so that

$$l = \frac{100 \text{ m}}{100 \text{ m}^2} = 10^{-2} \text{ cm}^{-1} \qquad [13.138]$$

If these are oriented at 45° to the vertical with $\tau = 1$ and with the flow rate given above, the width of the fracture is calculated to be

$$d = \left(\frac{5 \times 10^{10} \text{ cm s}^{-1} \times 12 \times 1.5 \times 10^{-3} \text{ dyn s cm}^{-2}}{10^{-2} \text{ cm} \times 981 \text{ cm s}^{-1} \times (2.8 - 0.9) \text{ g cm}^{-3} \times 0.707} \right)^{1/3} = 8.8 \times 10^{-5} \text{cm} \qquad [13.139]$$

This fracture aperture is greater than an order of magnitude wider than calculated above for grain boundary flow, but it is still quite a narrow fracture (0.88 μm). The fluid would have an average velocity of

$$v = \frac{5 \times 10^{-10} \ 0.707}{8.8 \times 10^{-5} \ 10^{-2}} = 4.0 \times 10^{-4} \text{ cm s}^{-1} = 126 \text{ m yr}^{-1} \qquad [13.140]$$

This is over two orders of magnitude greater than the grain boundary flow calculated in Equation 13.137.

Solute Transport in Flowing Fluid

The flux of solute, **a,** in the flowing fluid is equal to $C_a q$, where C_a is the concentration of solute in the fluid in moles per cm^{-3}. From Equation 13.134, therefore

$$C_a q = \frac{C_a d^3 l \tau g (\rho_R - \rho_f) \cos \theta}{12 \eta} \text{ mole cm}^{-2} \text{ s}^{-1} \qquad [13.141]$$

The ratio of transport of **a** by diffusion (Equation 13.133) relative to fluid flow (Equation 13.141) is given by

$$\frac{J_a^{\text{dif}}}{C_a q} = \frac{12 \eta \ D_a \phi (d\mu/dx)}{d^3 l RT \ g \ (\rho_R - \rho_f) \cos \theta} \qquad [13.142]$$

Because d times l is equal to ϕ

$$\frac{J_a^{\text{dif}}}{C_a q} = \frac{12 \eta \ D_a (d\mu/dx)}{d^2 RT \ g \ (\rho_R - \rho_f) \cos \theta} \qquad [13.143]$$

With values at 500°C for H_2O of $\eta = 1.5 \times 10^{-3}$ poise (Walther and Orville, 1982), $D_a = 10^{-4}$ cm^2 s^{-1} (Nigrini, 1970), $\rho_R = 2.8$ g cm^{-3}, and $\rho_f = 0.9$ g cm^{-3}, Equation 13.143 becomes

$$\frac{J_a^{\text{dif}}}{C_a q} = \frac{12 \times 1.5 \times 10^{-3} \text{poise } 10^{-4} \text{cm}^2\text{s}^{-1} \ d\mu/dx}{d^2 \ 8.31 \text{J mol K}^{-1} \ 773 \text{K } 981 \text{ cm s}^{-2} (2.8 - 0.9) \text{ g cm}^{-3} \cos \theta}$$

$$\qquad [13.144]$$

$$= 1.5 \times 10^{-13} \frac{d\mu/dx}{d^2 \cos \theta} (\text{J mol}^{-1} \text{ cm}^{-3})$$

The relative effectiveness of flow over diffusion is dependent on the width of the fractures/grain boundaries squared. With increasing width of fractures/grain boundaries, therefore, flow becomes the dominant mechanism of transport. Diffusion is favored by increasing the chemical potential gradient between reactants and products and by increasing the angle of flow relative to the vertical. For grain boundary transport whose width is given in Equation 13.135,

$$\frac{J_a^{dif}}{C_a q} = \frac{1.5 \times 10^{-13} \, d\mu/dx}{(7.9 \times 10^{-6})^2 \, 0.707} \, (\text{J mol}^{-1} \, \text{cm}^{-3})$$

$$= 3.4 \times 10^{-3} \, d\mu/dx \, (\text{J mol}^{-1} \, \text{cm}^{-1})$$

[13.145]

Unless the chemical potential gradient is very large, that is, a number of kJ per mole per cm, as is the case at compositional boundaries between layers, fluid flow appears to be the dominant mechanism of transport.

Summary

Although thermodynamics indicates the net direction of a reaction, the rates of reaction are modeled by considering that the reactants must proceed over an energy barrier to form product species. This energy is termed the activation energy for the reaction. The greater the energy barrier, the slower the reaction is. For the reaction

$$A + B \rightarrow C + D$$

with rate expression given by

$$\frac{d[A]}{dt} = k_r^+ [A]^2 [B]$$

where k_r^+ is the forward rate constant, the rate is termed third order overall because this is given by the sum of the exponents on the concentration terms given by the brackets.

Of particular importance are first-order reactions given by

$$[A] = [A_o] e^{-k_r^+ t}$$

where $[A_o]$ is the concentration of A at time 0. This is because at the molecular level the reactions are often first order with respect to a single reactant or bimolecular, depending on the first order of two reacting concentrations.

The macroscopically observed reaction is made up of a chain of elementary molecular reactions, with the slowest being the rate-limiting elementary reaction step. The reactants for this limiting step are considered to be in equilibrium with an activated complex of higher energy, and it is the breakdown of this complex that determines the rate.

Because at equilibrium the forward and reverse rate of reaction are equal, the relative rate of reaction is given by

$$R_+/R_- = e^{-\Delta G_R/RT}$$

where ΔG_R is the Gibbs energy of the rate-determining reaction. The net rate of reaction becomes

$$R_{net} = R_+(1 - e^{-\Delta G_R/RT})$$

and near equilibrium

$$R_{net} = \frac{-R_+\Delta G_R}{RT}$$

The temperature dependence of the rate constant is given by

$$\left(\frac{d \ln k_r^+}{dT}\right) = \left(\frac{1}{T} + \frac{\Delta H_R^o}{RT^2}\right)$$

where ΔH_R^o denotes the standard enthalpy of the rate-determining elementary reaction and is often equated with the activation energy of the overall reaction.

Irreversible thermodynamics through the production of nonreversible entropy during heat transfer, mass transfer, and chemical reactions are used to place constraints on reaction rates. Near-equilibrium entropy production from each is considered as the linear product of a flux and a force with the constant that relates the two, termed the phenomenological coefficient. For chemical diffusion rates, the driving force is the change in chemical potential rather than concentration of the species. Far from equilibrium, the increased irreversible entropy production can lead to pattern formation.

Heterogeneous mineral dissolution reactions in aqueous solutions are controlled by charged activated mineral surface complexes, and therefore, the rates are *pH* dependent. Rates appear to increase from the *pH* of zero surface charge both with increasing and decreasing *pH*. The rate-limiting complexes appear to be different for breaking oxygen–silica and oxygen–alumina bonds. Initial overall rates are, therefore, nonstoichiometric with a greater release of Al at acid *pH* and greater Si release to solution at alkaline *pH*. The overall rates become stoichiometric when the surface exposure of the more slowly reacting cation increases with removal of the initially more rapidly reacting cation, which then decreases.

At elevated temperatures in the crust, it appears that aqueous diffusion and flow processes control the rate of new mineral formation. Where a fluid phase is available reaction rates typically do not allow overstepping of equilibrium by more than 1°C or so.

Hopefully an appreciation of elementary molecular processes as rate-limiting processes in the complex heterogeneous reactions common on and in the earth has been developed. It should be apparent that most reactions that occur between minerals on and in the earth occur through a fluid-phase intermediary. During weathering, diagenesis, and metamorphism, this is an H_2O-rich phase, whereas with the formation of igneous rocks it involves a silica-rich magma phase. All of these reactions occur as the systems evolve toward a state of chemical equilibrium with increased overall entropy.

Key Terms Introduced

activated complex
activation energy
Arrhenius plot
autocatalytic process
Boltzmann distribution
Boltzmann's constant (k)
catalytic reaction
chemical affinity
closed system
elementary reaction
gram atoms of oxygen
heterogeneous reaction
homogeneous reaction
isolated system
mean electrostatic site potential

metastable
nucleation
open system
partition function
phenomenological coefficients
Planck's constant (h)
positive feedback
reaction intermediary
reaction order
secular equilibrium
standard state
surface energy
Taylor's series
tortuosity
transitory state

Questions

1. When the rate of the forward reaction is equal to the rate of the reverse reaction, what is the Gibbs energy of the overall reaction?
2. What is the difference between a homogeneous and a heterogeneous reaction?
3. Why is a rate expression of rate $= k_r^+[A]^{1/2}$, where $[A]$ is the concentration of reactant A, not the characterization of an elementary reaction?
4. Describe the transition state theory of chemical kinetics.
5. How does the Gibbs energy of an overall reaction affect its rate? Near equilibrium?
6. What factors determine the rate of dissolution of a mineral in solution?
7. What does nonstoichiometric early dissolution of an aluminosilicate mineral imply about the long-term stoichiometric rate of dissolution?
8. What is reversible as opposed to irreversible heat of a reaction?
9. Define chemical affinity.
10. What is a phenomenological coefficient?

11. What factors control the precipitation of a mineral from solution?

12. Describe how a feldspar dissolves to equilibrium in an aqueous solution.

13. Why do reactions of one mineral assemblage to another generally occur in a fluid phase?

14. What is an Arrhenius plot?

Problems

1. Specify the reaction order of the following rate expressions where k_r^+ is the forward rate constant. Indicate which ones could be elementary reactions.
   (a) Rate = $k_r^+[A]$
   (b) Rate = $k_r^+[A][B]^{1/2}$
   (c) Rate = $k_r^+ [A]^2$
   (c) Rate = $k_r^+[A][B][C]$

2. If the reaction is zeroth order with respect to [A], which of the following quantities when plotted versus time should be a straight line: ln [A], 1/[A], [A], or $[A]^2$?

3. Given the rate expression in Equation 13.10, determine the time when [A] = 0.5[A°].

4. The rate constant of a first-order decay reaction is $3.5 \times 10^{-4}\,s^{-1}$. What is its half-life? What percentage would be left after 10 minutes?

5. For the second-order reaction, 2 A = B + C, [A] decreases from a concentration of 30 to 15 in 45 minutes. What time is needed for [A] to reduce from a composition 30 to 3.75?

6. Assume the elementary reaction for precipitation of calcite in aqueous solutions is $Ca^{2+} + CO_3^{2-} \rightarrow CaCO_3$ and the reverse reaction is also elementary. Write an equation for the net reaction rate, R_{net}, near equilibrium.

7. When nitric oxide (NO) is reacted with oxygen, it appears to display third-order kinetics with the rate given by $k_r^+[NO]^2[O_2]$, and the rate can decrease with increasing temperature. (Elementary reactions always increase with temperature.) If NO_3 is formed as an intermediary,
   (1) What could the elementary reactions be?
   (2) What are their orders?
   (3) What would be the apparent rate order of the sum of the elementary reactions?

8. A is involved in a first-order reaction, and [A] is measured as a function of time, as given in **Table 13-5**.

Table 13-5

Hours	0	24	48	96
Molality	1.0	0.905	0.819	0.670

(a) Determine the value of the rate constant.
(b) Determine the half-life of A.
(c) Calculate [A] when time = 120 hours.
(d) Calculate the time when [A] is reduced to 10%.

9. For the reaction $2 NO + O_2 \rightarrow 2 NO_2$, the results of three experiments outlined in **Table 13-6** were obtained. What is the order of the reaction with respect to
(a) [NO] and
(b) [O_2]?

Table 13-6

Run	[NO]	[O_2]	Rate
1	1.2	0.5	1.2
2	1.2	1.0	2.4
3	2.4	0.5	4.8

10. Pearson et al. (1995) estimated that a grain of diamond at 1 bar and 1000°C would convert to graphite in a billion years and at 1200°C would require a million years. What is the activation energy of the transition?

11. The data in **Table 13-7** give the increase in molality of Al with time at 25°C in a pH 11.2 solution containing ground corundum, Al_2O_3 (from Carroll-Webb and Walther, 1988).

Table 13-7

Time (h)	Al molality (in 10^{-5})
0	0.0
24	1.1
48	2.0
96	2.4
144	3.5
240	4.2
408	5.7
528	6.8

(a) Why is the rate of dissolution greater initially than when state-state conditions are achieved?

(b) Determine the steady-state corundum dissolution rate in moles Al released per hour.

(c) If 1.32 g of corundum was used in the experiment with a surface area of 3.7 m^2/g, what is the rate of dissolution in moles Al cm^{-2} s^{-1}?

(d) If the rate is surface controlled and 35 times lower in a *pH* 9.5 solution, what does this say about the charge of the rate-determining surface complex?

12. At 300°C, a solution is supersaturated to twice the concentration of equilibrium with a potential mineral that has a Gibbs surface energy of 100 erg cm^{-2} in the solution. Plot the radius of an embryo of the mineral versus its Gibbs energy, ΔG_e, from 0 to 50 Å. Indicate the critical radius above which the embryo becomes stable. (Molar volume of mineral is 50 cm^3.)

13. At midcrustal conditions, what would be the vertical flux of fluid per cm^2 along grain boundaries driven by buoyancy if they are modeled as squares of 1 mm on a side and the width of the grain boundaries was 10^{-5} cm? Give the answer in liters per cm^{-2} per thousand years.

14. What is the average velocity of the fluid flux in Problem 13?

Problems Making Use of SUPCRT92

15. Using the expression in Equation 13.116, calculate the net rate of reaction at 2 kbar and 300°C, 400°C, and 500°C in moles per unit surface area for

(a) Andalusite → kyanite, limited by the surface area of the andalusite

(b) Pyrophyllite + 6 diaspore → 4 andalusite + 4H$_2$O, limited by the surface area of pyrophyllite

(c) Would the amount of andalusite increase or decrease with reaction at 500°C and 2 kbar? Pyrophyllite?

(d) Why are the rates of reaction of pyrophyllite so much greater than andalusite?

(e) With a surface area of pyrophyllite of 1 cm^2, how many years will it take for 1 mole of it to react at 400°C and 2 kbar?

References

Bear, J., 1975, *Dynamics of Fluids in Porous Media*. Elsevier, New York, 764 pp.

Berner, R. A., 1980, *Early Diagenesis: A Theoretical Approach*. Princeton University Press, Princeton, 241 pp.

Brady, P. V. and Walther, J. V., 1989, Controls of silicate dissolution rates in neutral and basic pH solutions at 25°C. *Geochim. Cosmochim. Acta*, v. 53, pp. 2823–2830.

Brady, P. V. and Walther, J. V., 1990, Kinetics of quartz dissolution at low temperatures. *Chem. Geol.*, v. 82, pp. 253–264.

Carroll, S. A. and Walther, J. V., 1990, Kaolinite dissolution at 25°, 60°, and 80°C. *Am. J. Sci.*, v. 290, pp. 797–810.

Carroll-Webb, S. A. and Walther, J. V., 1988, A surface complex reaction model for the pH-dependence of corundum and kaolinite dissolution rates. *Geochim. Cosmochim. Acta*, v. 52, pp. 2609–2623.

Chou, L. and Wollast, R., 1985, Steady-state kinetics and dissolution mechanisms of albite. *Am. J. Sci.*, v. 285, pp. 963–993.

de Groot, S. R. and Mazur, P., 1962, *Non-Equilibrium Thermodynamics*. North-Holland, Amsterdam, 510 pp.

Gimblett, F. G. R., 1970, *Introduction to the Kinetics of Chemical Chain Reactions*. McGraw-Hill, New York, 199 pp.

Joesten, R., 1983, Grain growth and grain boundary diffusion in quartz from the Christmas Mountains (Texas) contact aureole. *Am. J. Sci.*, v. 283-A (Orville vol.), pp. 233–254.

Katchalsky, A. and Curran, P. F., 1967, *Nonequilibrium Thermodynamics in Biophysics*. Harvard University Press, Cambridge, 248 pp.

Kerrick, D. M., 1968, Experiments on the upper stability limit of pyrophyllite at 1.8 kilobars and 3.9 kilobars water pressure. *Am. J. Sci.*, v. 266, pp. 204 –214.

Lasaga, A. C., Richardson, S. M. and Holland, H. D., 1977, The mathematics of cation diffusion and exchange between silicate minerals during retrograde metamorphism. In Saxena, S. K. and Bhattacharji, S. (eds.) *Energetics of Geological Processes*. Springer-Verlag, New York, pp. 353–388.

Mazer, J. J. and Walther, J. V., 1994, Dissolution kinetics of silica glass as a function of pH between 40 and 85°C. *J. Non-Cryst. Sol.*, v. 170, pp. 32–45.

Misener, D. J., 1974, Cationic diffusion in olivine to 1400°C and 35 kbar. In Yoder, H. S. (ed.) *Geochemical Kinetics*. Carnegie Institution, Washington, DC, pp. 117–129.

Nigrini, A., 1970, Diffusion in rock alteration systems. I. Predictions of limiting equivalent ionic conductances at elevated temperatures. *Am. J. Sci.*, v. 269, pp. 65–91.

Oelkers, E. H. and Helgeson, H. C., 1988, Calculation of the thermodynamic and transport properties of aqueous species at high pressures and temperatures: Aqueous tracer diffusion coefficients of ions to 1000°C and 5 kb. *Geochim. Cosmochim. Acta*, v. 52, pp. 63–85.

Onsager, L., 1931, Reciprocal relations in irreversible processes. II. *Phys. Rev.*, v. 38, pp. 2265–2279.

Pearson, D. G., Davies, G. R. and Nixon, P. H., 1995, Orogenic ultramafic rocks of UHP (diamond facies) origin. In: Coleman, R. G. and Wang, X. (eds.) *Ultrahigh Pressure Metamorphism*. Cambridge University Press, Cambridge, England, pp. 456–510.

Penniston-Dorland, S. C. and Ferry, J. M., 2008, Element mobility and scale of mass transport in the formation of quartz veins during regional metamorphism of the Waits River Formation, east-central Vermont. *Amer. Mineral.*, v. 93, pp. 7–21.

Sjoberg E. L., 1976, A fundamental equation for calcite dissolution kinetics. *Geochim. Cosmochim. Acta*, v. 40, pp. 441–447.

Smyth, J., 1989, Electrostatic characterization of oxygen sites in minerals. *Geochim. Cosmochim. Acta*, v. 53, pp. 1101–1110.

Walther, J. V. and Orville, P. M., 1982, Volatile production and transport in regional metamorphism. *Contrib. Mineral. Petrol.*, v. 79, pp. 252–257.

Walther, J. V. and Wood, B. J., 1984, Rate and mechanism in prograde metamorphism. *Contrib. Mineral. Petrol.*, v. 88, pp. 246–259.

Wollast, R. and Chou, L., 1986, Process, rate, and proton consumption by silicate weathering. Presented at the 13th International Society of Soil Science Congress, Hamburg, Germany, pp. 127–136.

Wood, B. J. and Walther, J. V., 1983, Rates of hydrothermal reactions. *Science*, v. 222, pp. 413–415.

Zinder, B., Furrer, G. and Stumm, W., 1986, The coordination chemistry of weathering. II. Dissolution of Fe(III) oxides. *Geochim. Cosmochim. Acta*, v. 50, pp. 1861–1869.

14 Oxidation and Reduction

Neutrons are not stable outside the nucleus and undergo spontaneous decay, as discussed in Chapter 10. Protons and electrons are, however, exchanged between species. Chemical reactions that cause a change in pH relate to movement of protons between species in acid–base reactions. The other important group of reactions relates to changes in electrons between species.

A reaction involving the transfer of electrons from one species to another is called an *oxidation–reduction* (redox) reaction. For instance, the reaction

$$4Fe^{2+} + 4H^+ + O_2 \rightarrow 4Fe^{3+} + 2H_2O \qquad [14.1]$$

is a redox reaction. In this reaction, Fe is considered to be oxidized because it reacts with O_2. This oxidation of Fe causes electrons to be lost from the iron species as Fe^{2+} is transformed into Fe^{3+}. Fe^{2+} is referred to as the reduced species of Fe in this reaction and Fe^{3+} the oxidized species. Because the reaction must be balanced, where do the electrons lost by Fe^{2+} when Fe^{3+} is produced go? Oxygen has gained electrons because it has zero valence charge in O_2 but has a valence of –2 in H_2O. Because oxygen has gained electrons, Reaction 14.1 has reduced O_2. Whenever a species is oxidized in a reaction, a species must also be reduced. These reactions then change the *valence* of species.

As discussed in Chapter 5 and given in Appendix C, many elements can have multiple valence states. A change in an element's valence between species is then a change in its oxidation state. Consider uncharged elemental sulfur that has six electrons in orbitals above the inert Ne gas electron configuration. It can donate up to six electrons (oxidize) when bonding with another element, producing S^{+6} such as in the species SO_4^{2-}. Sulfur can also gain two electrons (reduce) to reach the inert Ar gas electron configuration of S^{2-} in H_2S. Clearly, oxidation increases the valence of a species, and reduction decreases it. Besides Fe, S, and O, elements that commonly change their oxidation state at earth surface conditions include Mn, C, N, and H. Together with oxygen and sulfur, the last three of these are major building blocks of life. The energy needed for most life processes involves oxidation reactions. That is, most oxidation reactions are exothermic and drive metabolic reactions.

Redox reactions do not have to involve oxygen. The reaction

$$2Fe_{(solid)} + 3Cl_{2(gas)} \rightarrow 2Fe^{3+} + 6Cl^- \qquad [14.2]$$

is a redox reaction where Fe is oxidized, changing its valence state from 0 in solid Fe metal to +3 in the aqueous Fe^{3+} species; however, the oxidizing agent is $Cl_{2(gas)}$ rather than O_2. Again, when electrons are lost the species is oxidized, and when a species gains electrons, it is reduced. Although a reaction that oxidizes a species must also reduce a species, the effects of oxidation or reduction can be analyzed separately by writing *half-reactions* that show the explicit role of electrons.

Consider the reaction that causes electricity to flow in a Zn-Cu battery, as outlined in **Figure 14-1**, that can be represented by the reaction

$$Zn + Cu^{2+} \rightarrow Cu + Zn^{2+} \qquad [14.3]$$

The oxidation half-reaction for this redox reaction occurring at the Zn electrode is

$$Zn \rightarrow Zn^{2+} + 2e^- \qquad [14.4]$$

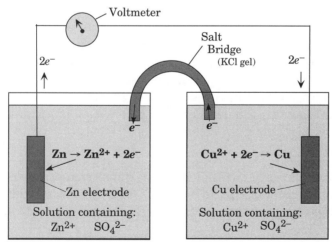

FIGURE 14-1 The Zn-Cu battery split into its two half-reactions, with the reaction $Zn = Zn^{2+} + 2e^-$ occurring at the Zn electrode and the reaction $Cu^{2+} + 2e^- = Cu$ occurring at the Cu electrode. The measured voltage times the 2 moles of electrons that are passed through the voltmeter per mole of reaction gives the energy change per mole of Reaction 14.3. The salt bridge allows SO_4^{2-} to diffuse from the Cu half-reaction to the Zn half-reaction to maintain charge balance in the two reaction vessels.

where e^- denotes an electron. The reduction half-reaction is

$$Cu^{2+} + 2e^- \rightarrow Cu \qquad [14.5]$$

which occurs at the Cu electrode.

Electrons are then required to produce solid Cu. Because "free" electrons do not exist, half-reactions of oxidation and reduction must always be combined in a real reaction. The electrons that are transferred between species, however, can be transferred by a wire between electrodes to produce the overall reaction. The potential to transfer electrons (voltage) is measured along the wire with a voltmeter as shown. If the voltage for Reaction 14.3 is zero so that there is no potential for electron transfer, the redox reaction would be at equilibrium.

■ Gibbs Energy of Redox Reactions

Remember from the first law of thermodynamics that for a closed system the energy change for a reaction is characterized by work plus heat changes. At constant temperature and pressure, the energy change is the Gibbs energy of reaction, ΔG_R, given by

$$\Delta G_R = dq_{rev} + dw \qquad [14.6]$$

for a reversible reaction in which dq_{rev} denotes the reversible heat adsorbed and dw gives the amount of work done on the system. In the case of electron transfer, the work is electrical.

The electrical work per mole done during a redox reaction is that needed to move a mole of charges (i.e., electrons), that is, a faraday of charge given by *Faraday's constant*, F (96,485 coulombs per mole), through a potential difference, ΔE, in volts. This energy is expressed in joules per mole of charge because coulombs × volts = joules. For a mole of reversible reaction, this is

$$dw = nF\Delta E \qquad [14.7]$$

where n is the number of moles of electrons transferred per mole of reaction. For a reversible reaction with no heat exchange with the surroundings, $dq_{rev} = 0$. The Gibbs energy change per mole of this reaction is then

$$\Delta G_R = nF\Delta E \qquad [14.8]$$

As with other reactions, a *standard state* is specified to help calculate values of ΔG_R. The standard-state change of Gibbs energy is then given by

$$\Delta G_R^{\circ} = nF\Delta E^{\circ} \qquad [14.9]$$

where $\Delta E°$ is the voltage of the reaction under standard-state conditions. Because volts, ΔE, times coulombs per mole, F, equals joules per mole if calories are considered, the conversion 1 cal = 4.1840 J is needed.

It would be helpful to have a standard voltage assigned to each half-reaction because any two half-reactions could then be combined to obtain the standard voltage of a redox reaction. To do this, all of the half-reactions are written as oxidation half-reactions to produce electrons in the reaction as it proceeds from left to right. $\Delta E°$ of a redox reaction is then obtained by subtracting $n\Delta E°$ of one half-reaction (to make it a reduction half-reaction) from $n\Delta E°$ of another (the oxidation reaction). A reference, however, from which to tabulate the voltage changes in half-reactions is required. This is similar to the problem of determining the thermodynamic properties of a charged species in a solution that must be electrically neutral. In the case of charged species in solution, the thermodynamic values are referenced to H^+. With oxidation-reduction half-reactions, they are referenced to the standard hydrogen gas half-reaction written as a reduction reaction (**Figure 14-2**).

Consider the redox reaction

$$Zn + 2H^+ \rightarrow Zn^{2+} + H_{2(g)} \qquad [14.10]$$

This is separated into an oxidation half-reaction:

$$Zn \rightarrow Zn^{2+} + 2e^- \qquad [14.11]$$

with a voltage $n\Delta E_{Zn}$ and a reduction half-reaction

$$2H + 2e^- \rightarrow H_{2(g)} \qquad [14.12]$$

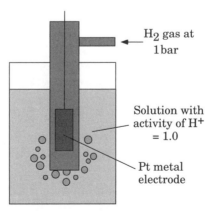

FIGURE 14-2 Standard hydrogen gas half-cell used to help measure the $Eh°$ of a solution at 25°C and 1 bar. H_2 gas at 1 bar is bubbled past a platinum electrode in contact with a solution of $pH = 0$. The reaction $2H^+ + 2e^- = H_{2(g)}$ occurs at this electrode when electrons are transferred along the wire attached to the Pt electrode.

with voltage designated as $n\Delta E_H$. n is the number of electrons transferred in the reaction; in this case, $n = 2$. The only measurable voltage, ΔE_{cell}, is that of the reaction cell as a whole. ΔE_{cell} must be related to the oxidation and reduction half-reaction voltages by

$$\Delta E_{cell} = n\Delta E_{Zn} + n\Delta E_H \qquad [14.13]$$

The standard convention is to take the Half-reaction 14.12 for transfer of 1 mole of electrons at standard conditions of $a_{H^+} = 1.0$ with a fugacity of H_2 of 1.0 bar and arbitrarily assign its voltage change, ΔE_H, at the pressure and temperature of interest or as represented in the chemical literature, Eh, a value of zero (**Table 14-1**). Generally, values of 25°C and 1 bar are considered. If the voltage of Reaction 14.10 is measured, therefore, under standard conditions written as an oxidation of Zn, the cell voltage, ΔE_{cell}, is assigned to $n\Delta E_{Zn}$ for the Zn half-reaction.

Because the determination is relative to the standard hydrogen half-reaction, the voltage assigned to the Zn oxidation half-reaction is referred to as its Eh. If the Zn oxidation half-reaction is at standard-state conditions, the voltage is designated as $Eh°$. Some $Eh°$ values at 25°C and 1 bar are given in Table 14-1. $Eh°$ of half-reactions not listed are calculated by rearranging Equation 14.9 to give

| Table 14-1 | At 25°C and 1 bar standard oxidation potentials, $Eh°$, in volts for the indicated half-reaction |

Half-reaction	$Eh°$ (V)	Half-reaction	$Eh°$ (V)
$1/4CH_2O_{(s)} + 1/4H_2O = 1/4CO_{2(g)} + H^+ + e^-$	−8.20	$H_{2(g)} = H^+ + e^-$	0.000
$K_{(s)} = K^+ + e^-$	−2.93	$H_2S_{(aq)} = S_{(s)} + 2H^+ + 2e^-$	+0.145
$Ca_{(s)} = Ca^{2+} + 2e^-$	−2.86	$Cu^+ = Cu^{2+} + e^-$	+0.162
$Mg_{(s)} = Mg^{2+} + 2e^-$	−2.35	$Cu_{(s)} = Cu^{2+} + 2e^-$	+0.340
$Be_{(s)} = Be^{2+} + 2e^-$	−1.81	$Ag_{(s)} = Cl^- + AgCl + e^-$	+0.604
$Zn_{(s)} = Zn^{2+} + 2e^-$	−0.763	$FeCl^+ = Fe^{3+} + Cl^- + e^-$	+0.76
$Fe_{(s)} = Fe^{2+} + 2e^-$	−0.474	$Fe^{2+} = Fe^{3+} + e^-$	+0.77
$Pb_{(s)} + SO_4^{2-} = PbSO_{4(s)} + 2e^-$	−0.356	$2H_2O = O_{2(g)} + 4H^+ + 4e^-$	+1.229
$Pb_{(s)} = Pb^{2+} + 2e^-$	−0.124	$PbSO_{4(s)} + 2H_2O = PbO_{2(s)} + SO_4^{2-} + 4H^+ + 2e^-$	+1.685
$Fe_{(s)} = Fe^{3+} + 3e^-$	−0.060	$O_{2(g)} + 2H_2O = O_{3(g)} + 2H^+ + 2e^-$	+2.07

Increasing reduction (left column, upward); *Increasing oxidation* (right column, upward)

$Eh°$ values are relative to the standard H_2-H^+ half-reaction (Equation (14.12) for the transfer of 1 mole of electrons). Subscripts (s), (g), and (aq) indicate the species are in a solid, gaseous, and aqueous state, respectively. H_2O is always in a liquid state. Voltages are calculated with $\Delta G°$ of species from SUPCRT92 and also taken from literature sources. All substances and species are at unit activity.

$$Eh^\circ = \frac{\Delta G_R^\circ}{Fn}$$ [14.14]

ΔG_R° denotes the standard-state Gibbs energy of the reaction in joules per mole at the temperature and pressure of interest, which includes the oxidation half-reaction of interest written to the hydrogen reduction half-reaction.

Consider the oxidation half-reaction between Fe metal and the Fe^{2+} aqueous species

$$Fe \rightarrow Fe^{2+} + 2e^-$$ [14.15]

Combining this with the hydrogen reduction half-reaction produces the complete reaction

$$Fe + 2H^+ \rightarrow Fe^{2+} + H_2$$ [14.16]

From SUPCRT92 data, where the standard-state Gibbs energy of the species at 25°C and 1 bar is given in calories, the standard-state Gibbs energy of Reaction 14.16 in joules per mole with 4.184 J cal^{-1} is calculated to be

$$\Delta G_R^\circ = (-21{,}870 + 0.0) - (0.0 + 2 \times 0.0) = -21{,}870 \text{ cal mol}^{-1}$$
$$ Fe^{2+} \quad\;\; H_2 \qquad\; Fe \qquad\; H^+$$ [14.17]
$$= -91{,}504 \text{ J mol}^{-1}$$

Therefore,

$$Eh^\circ = \frac{-91{,}505 \text{ J mol}^{-1}}{96{,}485 \text{ C} \times 2} = -0.474 \text{ V}$$ [14.18]

which is the value given in Table 14-1.

As an example of combining voltages of half-reactions to determine the voltage of another half-reaction, consider from Table 14-1 the half-reaction Fe $\rightarrow$ Fe^{2+} + 2e$^-$, with $nEh^\circ = -0.948$ V, and the half-reaction Fe $\rightarrow$ Fe^{3+} + 3e$^-$, with $nEh^\circ = -0.180$ V. The Eh° of the half-reaction Fe^{2+} $\rightarrow$ Fe^{3+} + e$^-$ is then $(-0.18$ V$) - (-0.95$ V$) = 0.77$ V. This Eh° is also given in Table 14-1.

■ Calculating the *Eh* of a Non–Standard-State Reaction

Consider the generalized redox reaction

$$aA + bB \rightarrow cC + dD$$ [14.19]

Redox reactions are often difficult to balance. A procedure for balancing these reactions is given in Appendix I. The Gibbs energy of Reaction 14.19 is

$$\Delta G_R = nFEh \qquad \qquad [14.20]$$

Eh represents the voltage of the reaction that, at least in principal, is measurable, and knowing the number of electrons per mole of reaction transferred ΔG_R is determined. From the definition of standard states and *activity product*, Q, ΔG_R is given by

$$\Delta G_R = \Delta G_R^o + RT \ln Q \qquad \qquad [14.21]$$

where R denotes the universal gas constant ($8.3145 \text{ J mol}^{-1} \text{ K}^{-1}$) and T is temperature in Kelvin.

The activity product is the product of all of the species activities raised to the power of the species' coefficient in the reaction. For Reaction 14.19, this is

$$Q = \frac{a_C^c \, a_D^d}{a_A^a \, a_B^b} \qquad \qquad [14.22]$$

With the relations in Equations 14.9 and 14.20, Equation 14.21 for an oxidation-reduction reaction is written as

$$nFEh = nFEh^o + RT \ln Q \qquad \qquad [14.23]$$

where F denotes Faraday's constant ($96,485 \text{ C mol}^{-1}$ or because a joule is a volt-coulomb, $96,485 \text{ J V}^{-1} \text{ mol}^{-1}$) and n (mol^{-1}) is the number of electrons transferred per mole of reaction. Solving for Eh gives

$$Eh = Eh^o + \frac{RT}{nF} \ln Q \qquad \qquad [14.24]$$

Equation 14.24 is the *Nernst equation*, which relates voltages of oxidation-reduction reactions to the reaction's activity product. If the reaction is at equilibrium so that $Eh = 0$, then the activity product becomes the equilibrium constant, **K**. Written in terms of a base-10 logarithm, Equation 14.24 becomes

$$Eh = Eh^o + 2.303 \frac{RT}{nF} \log Q \qquad \qquad [14.25]$$

At 25°C and 1 bar, the Nernst equation reduces to

$$Eh = Eh^o + \frac{0.05927}{n} \log Q \qquad \qquad [14.26]$$

As an example of the use of the Nernst equation, consider water at 25°C and 1 bar containing both Fe^{2+} and Fe^{3+}. This is a redox couple for which the oxidation half-reaction

$$Fe^{2+} \rightarrow Fe^{3+} + e^- \qquad [14.27]$$

is written. If the molality of $Fe^{2+} = 10^{-4}$ and that of $Fe^{3+} = 10^{-2}$, the Q of the half-reaction is $10^{-2}/10^{-4} = 100$. Eh^o of the reaction is obtained from values in Table 14-1. The Eh^o of the reaction $Fe = Fe^{2+} + 2e^-$ is -0.474 V. This is subtracted from the Eh^o of the reaction $Fe = Fe^{3+} + 3e^-$ of -0.060 V to obtain Eh^o $= +0.414$ V for Reaction 14.27. In the water from Equation 14.26, therefore, the $Eh = 0.414 + 0.05927 \times \log(100) = 0.532$ V. Alternately, if the voltage of a reaction is measured, the value of the activity product is determined.

■ Stability of H_2O and *Eh–pH* Diagrams

Consider the stability of liquid H_2O relative to H_2 and O_2 gas. This breakdown reaction of H_2O is written as

$$2H_2O = 2H_2 + O_2 \qquad K_{(14.28)} = \frac{f_{H_2}^2 f_{O_2}}{a_{H_2O}^2} \qquad [14.28]$$

where f_i represents the fugacity of gas i. This is a redox reaction because oxygen has been oxidized and hydrogen reduced. The equilibrium constant for Reaction 14.28 is determined from

$$\log K_{(14.28)} = -\frac{\Delta G_R^o}{2.303RT} \qquad [14.29]$$

at 25°C, and 1 bar calculated from SUPCRT92 data is

$$\Delta G_R^o = (2 \times 0 + 0) - (2 \times (-56,688)) = 113,376 \text{ cal mol}^{-1} \qquad [14.30]$$
$$\quad\;\; H_2 \; O_2 \qquad\qquad H_2O$$

For the nearly pure H_2O in most aqueous solutions, the activity of H_2O is taken as 1.0. With $R = 1.987$ cal^{-1} mol^{-1} K^{-1} at 25°C (298.15 K), Equations 14.28, 14.29, and 14.30 give

$$-83.1 = \log (f_{H_2}^2 f_{O_2}) \qquad [14.31]$$

This indicates the degree of H_2O disassociation to H_2 and O_2 is very small at earth surface conditions.

An upper limit on the fugacity of O_2 in the earth's atmosphere in equilibrium with H_2O is when the atmosphere is nearly pure O_2 or the fugacity of O_2 is about 1 bar. Any higher value of f_{O_2} could not exist in the approximately 1 bar atmosphere at the earth's surface. The fugacity of H_2 in equilibrium with a pure O_2 atmosphere in equilibrium with liquid H_2O from Equation 14.31 is

$$f_{H_2} = (10^{-83.1})^{1/2} = 10^{-41.55} \text{ bar} \qquad [14.32]$$

A lower limit on O_2 produced under reducing conditions in equilibrium with H_2O is in a pure H_2 atmosphere. In this case, the fugacity, f_{H_2}, is taken as 1.0. The *oxygen fugacity* in such an atmosphere from Equation 14.31 is

$$f_{O_2} = 10^{-83.1} \text{ bar} \qquad [14.33]$$

Therefore, f_{O_2} limited by the breakdown of H_2O in a nearly pure O_2 and nearly pure H_2 atmosphere is between 1 and $10^{-83.1}$ at the earth's surface. Remember that the number $10^{-83.1}$ is a "model" value of the concentration of O_2 in the atmosphere. It is so small that it implies that a small fraction of a single molecule of O_2 would exist in such an H_2 atmosphere if it had the same mass as that of the earth's atmosphere. This is clearly a physically unreal situation; however, model oxidation-reduction values, although not real, help in the understanding of relations between the variables used to describe the state of oxidation of the oxidation-reduction pairs in a system.

Now consider the stability of water in terms of redox half-reactions. The breakdown of H_2O to O_2 is characterized by the oxidation half-reaction

$$2 H_2O \rightarrow O_2 + 4 H^+ + 4e^- \qquad [14.34]$$

Reaction 14.34 is often referred to as the *water electrode* reaction. For this reaction, at 25°C and 1 bar, the Nernst equation is

$$Eh_{(14.34)} = Eh^o_{(14.34)} + \frac{0.0591}{4} \log \frac{a^4_{H^+} f_{O_2}}{a^2_{H_2O}} \qquad [14.35]$$

Assuming $a_{H_2O} = 1.0$ and with the highest possible fugacity of O_2 in our atmosphere of unity (i.e., pure O_2), Equation 14.35 becomes

$$Eh_{(14.34)} = Eh^o_{(14.34)} + \frac{0.0591}{4} \log a^4_{H^+} \qquad [14.36]$$

Because $Eh^o_{(14.34)}$ at 25°C and 1 bar from Table 14-1 is 1.23 V,

$$Eh_{(14.34)} = 1.23 + 0.059 \log a_{H^+} = 1.23 - 0.059 \, pH \qquad [14.37]$$

Equation 14.37 gives the stability of H_2O in terms of Eh and pH limited by 1 bar of oxygen fugacity. That would be the maximum oxidation potential in equilibrium with liquid water at 1 bar and 25°C. Equation 14.37 depends on both Eh and pH. Both Eh and pH are, therefore, important for considering the chemical character of various redox environments. It is informative to look at diagrams involving Eh and pH as independent variables. The line produced by Equation 14.37 is shown in **Figure 14-3**. Along this line, $f_{O_2} = 1$ and $f_{H_2} = 10^{-41.55}$, as indicated in the figure.

The stability of H_2O at low oxidation potentials where H_2 breaks down is given by the oxidation half-reaction

$$H_2 \rightarrow 2\,H^+ + 2\,e^- \qquad\qquad\qquad [14.38]$$

This is the standard hydrogen electrode reaction where $Eh°$ at 25°C and 1 bar = 0.0. The Nernst equation for Reaction 14.38 is

$$Eh_{(14.38)} = 0.0 + \frac{0.0591}{2} \log \frac{a_{H^+}^2}{f_{H_2}} \qquad\qquad [14.39]$$

The lowest oxidation potential would be when the fugacity of H_2 in the atmosphere is 1. Equation 14.39 then becomes

$$Eh = -0.059\,pH \qquad\qquad\qquad [14.40]$$

Equation 14.40 is also shown as a solid line in Figure 14-3 giving the stability of H_2O at 25°C in equilibrium with $f_{H_2} = 1$. Along this line, $f_{O_2} = 10^{-83.1}$

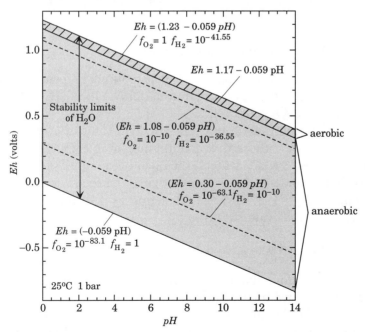

FIGURE 14-3 *Eh*–*pH* diagram at 25°C and 1 bar showing the stability of H₂O as limited by the equations given in the text (shaded area). A given value of oxygen fugacity in equilibrium with liquid H₂O becomes a line on an *EH*–*ph* diagram.

bar. Equations 14.37 and 14.40 then limit the *Eh-pH* equilibrium of H_2O at the earth's surface as given by the shaded area in Figure 14-3. Between these limits, the oxidation potential in equilibrium with H_2O is contoured.

For instance, at an $f_{O_2} = 10^{-10}$ from Equation 14.31, $f_{H_2} = 10^{-36.55}$, and with $a_{H_2O} = 1$, Equation 14.39 becomes

$$Eh = 0.0 + 0.05917 \log a_{H^+} + \frac{0.05917}{2} 36.55 = 1.08 - 0.059\, pH \qquad [14.41]$$

With $f_{H_2} = 10^{-10}$ so that $f_{O_2} = 10^{-63.1}$, the relationship between *Eh* and *pH* is given by

$$Eh = 0.30 - 0.059\, pH \qquad [14.42]$$

Equation 14.42 is plotted as a dashed line in Figure 14-3 and labeled with the fugacities of O_2 and H_2.

■ *Eh* of Natural Waters

In natural waters, atmospheric O_2 is the major oxidant and organic matter the major reductant. Life through photosynthesis uses energy from the sun to drive the reaction

$$nH_2O + nCO_2 + solar\ energy \rightarrow (CH_2O)_n + nO_2 \qquad [14.43]$$

In this reaction oxygen is oxidized, carbon is reduced, and energy is added to the organic matter; however, the organic matter produced, $(CH_2O)_n$, is out of equilibrium with the O_2 produced.

With this dichotomy, natural waters are divided into *aerobic* and *anaerobic*. Aerobic waters are those that contain measurable concentrations of O_2 ($P_{O_2} > 10^{-4}$ bar). All surface waters, the open ocean at all depths, and lakes at all levels if they undergo yearly convective overturn are aerobic. With $f_{O_2} = 10^{-4}$ and taking $a_{H_2O} = 1$, according to Reaction 14.35, the boundary between aerobic and anaerobic waters is given by

$$Eh_{(14.34)} = 1.23 + 0.059 \log a_{H^+} - 0.059 = 1.17 - 0.059\, pH \qquad [14.44]$$

Aerobic waters, therefore, should plot between Equations 14.37 and 14.44. This limited region is shown by the ruled area at the top of the region where H_2O stability is plotted in Figure 14-3.

Anaerobic waters occur because organic matter and other reduced species react with O_2, lowering the oxygen fugacity. Although organic matter was represented as $(CH_2O)_n$ in Reaction 14.43, in reality, it is more complex. The number of redox reactions, therefore, that can be written with organic matter is large. The oxidation half-reaction for formaldehyde, CH_2O, is

$$CH_2O + H_2O \rightarrow CO_2 + 4H^+ + 4e^- \qquad [14.45]$$

This reaction is used as a proxy for all organic matter reactions.

With $Eh° = -8.20$ V and assuming $a_{H_2O} = a_{CH_2O} = 1$, the presence of formaldehyde (organic matter) should have an Eh given by

$$Eh_{(14.45)} = -8.20 + 0.059 \log (a_{H^+}) + 0.059 \log f_{CO_2}^{0.25} \qquad [14.46]$$

If $f_{CO_2} = 1$, the maximum f_{CO_2} possible in a 1-bar atmosphere, Equation 14.46 plots along the line

$$Eh_{(14.45)} = -8.20 - 0.59 \, pH \qquad [14.47]$$

in Figure 14-3. This Eh is significantly below the most reducing values shown in the figure. This implies CH_2O, and for that matter, any organic matter is never in equilibrium with H_2O at earth surface conditions. Aqueous environments that therefore contain organic matter are not at equilibrium.

As a function of time, the organic matter reacts with oxidized species catalyzed by bacteria. Different bacteria have evolved to obtain energy from oxidation of specific reduced species. For instance, *Thiobacillus ferroxidans*, an acidophilic bacteria, oxidizes Fe^{2+}, elemental sulfur, and reduced inorganic sulfur compounds to obtain energy. *Thiobacillus thiooxidans*, however, only oxidizes elemental sulfur and reduced inorganic sulfur compounds and not Fe^{2+}. These reactions cause the Eh of the water to decrease. Only if the water absorbs O_2 from the atmosphere faster than reactions of organic matter in the water consumes it will it maintain an aerobic state.

In some instances, oxygenated surface waters become nutrient rich (*eutrophic*). The nutrients cause the rapid growth of algae. Bacterial oxidation of the organic matter produced can cause more rapid depletion of oxygen than is supplied to the water body from the atmosphere. Thus, eutrophicated waters become anaerobic.

The range of measured Eh and pH for natural environments is given in **Figure 14-4** by the darkest shaded area (Baas-Becking et al., 1960). The most common measurements are inside the hatched-line loop. None of the measured Eh-pH pairs plot in the area of the diagram defined as aerobic water. This appears to occur because there is great difficulty in measuring Eh in equilibrium with O_2. These waters have a small concentration of H_2O_2 in them. Bockris and Oldfield (1954) suggested that in the presence of H_2O_2 at concentrations greater than 10^{-6} M, platinum electrodes in Eh meters obey a relationship between Eh and pH of

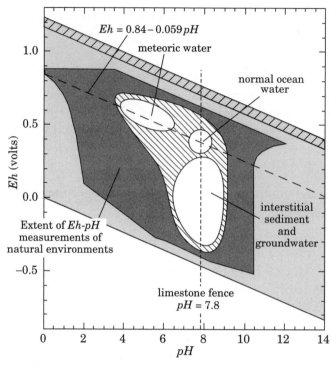

FIGURE 14-4 *Eh–pH* diagram indicating the limit for natural water modified after Baas-Becking et al. (1960) shown in the heavier shaded area relative to the stability of H₂O as given in Figure 14-3. The most common values are within the ruled area. Also shown are the *Eh–pH* characteristics of meteoric, open seawater, and interstitial waters. The carbonate fence is after Krumbein and Garrels (1952).

$$Eh = 0.84 - 0.0592\ pH \tag{14.48}$$

due to adsorption of OH° or H_2O_2 covering the electrode surface and poisoning it. Thus, the electrode behaves as a *pH* electrode according to the reaction

$$H_2O \rightarrow OH^\circ + H^+ + e^- \tag{14.49}$$

In Figure 14-4, Equation 14.48 plots as a line that runs through measured aerobic waters.

In addition to problems of electrode poisoning, in most instances, *Eh* values determined from two different oxidation-reduction couples in the same water give different values. **Figure 14-5** shows field-measured values of *Eh* plotted against the *Eh* computed from the indicated redox couples in the water. Note the generally poor agreement between the *Eh* computed for a redox couple and the measured *Eh* for the water. This discrepancy is due to *Eh* measurement errors and mixed potentials existing in the waters.

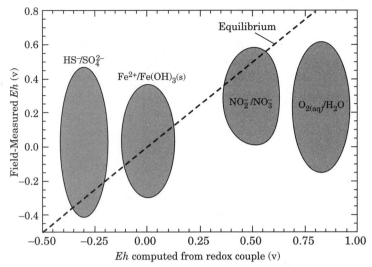

FIGURE 14-5 Shaded areas indicate where field measurements of *Eh* plot as a function of *Eh* calculated from concentrations of species in the indicated redox couples. For the $O_{2(aq)}/H_2O$ couple, there is never agreement between the two. (Modified after Lindberg and Runnells, 1984.)

Berner (1971) argued that the species O_2, CH_4, N_2, HCO_3^-, and SO_4^{2-} are electroinactive, in that oxidation-reduction couples made up of one species are different from that for the system as a whole. Waters that contain significant concentrations of Fe, Mn, and reduced sulfur species appear to behave better with respect to *Eh*.

▪ *Eh* and *pH* in Sedimentary Fluids

The *pH* of many fluids in sedimentary rocks is near 7.8. This is because acid in these waters is consumed by reaction with silicate minerals, releasing cations to the pore waters. As *pH* increases to 7.8, precipitation of calcium carbonate occurs. Environments of *pH* lower than 7.8 are unfavorable to calcite deposition. Above *pH* 7.8, carbonates are abundant in sediments. This boundary is shown as a vertical dashed line in Figure 14-4 and is referred to as the "limestone fence" for *pH*. Calcite precipitation does not depend on *Eh* but is highly *pH* dependent. An increase in *pH* tends to cause calcite to precipitate and limits how alkaline the waters become.

An exception to this scenario is peat deposits in which silicate material is limited. The acids produced by organic matter decay are not buffered by silicates and remain acidic, and oxidation of the carbon compounds drives the *Eh* to negative values.

How does *Eh* change as aerobic surface water is trapped in sediments? Aerobic bacteria live at the sediment–water interface and feed on organic matter falling through the water column. That is, they use the energy released by oxidizing the carbon in the organic matter. Bacteria, however, need a sink for electrons in these oxidation reactions. With O_2 present, the electron sink is the reduction of oxygen. Using the standard convention of writing, this as an oxidation half-reaction (the backward reaction being the reduction reaction required); the electron sink is

$$2H_2O \rightarrow O_2 + 4H^+ + 4e^- \tag{14.50}$$

The *Eh* of this reaction in equilibrium with the atmosphere where $f_{O_2} = 0.21$ at 25°C and 1 bar and at *pH* 7.8 with $a_{H_2O} = 1.0$ is calculated from the Nernst Equation 14.26 as

$$Eh_{O_2} = Eh^o_{(14.50)} + \frac{0.05917}{n} \log \frac{a^4_{H^+} f_{O_2}}{a^2_{H_2O}} \tag{14.51}$$

Thus, the *Eh* of O_2 reduction is

$$Eh_{O_2} = 1.23 \ V + \frac{0.05917}{4} \log[(10^{-7.8})^4 (0.21)] = 1.23 \ V - 0.47 \ V = 0.76 \ V \tag{14.52}$$

Because of the lack of significant transport of water in sediments and with the bacteria consuming O_2 at the sediment–water interface, interstitial waters in sediments typically remain closed to exchange with O_2 in the water column above.

After the O_2 is consumed in the interstitial waters, anaerobic bacteria use other oxidized species to take up electrons during the oxidization of any organic material present. The predominant nitrogen, sulfur, and carbon species in near-neutral *pH* aerobic surface waters are NO_3^-, SO_4^{2-}, and HCO_3^-, respectively. Oxidation of organic matter with these three oxidized species consuming the electrons occurs in succession. After O_2 reduction, electrons are consumed first by NO_3^- in the bacteria-mediated denitrification half-reaction. Written in terms of oxidation of N, this is

$$\frac{1}{10}N_2 + \frac{3}{5}H_2O \rightarrow \frac{1}{5}NO_3^- + \frac{6}{5}H^+ + e^- \tag{14.53}$$

The ΔG^o_R of this reaction at 25°C and 1 bar from SUPCRT92 data is

$$\Delta G^o_R = \frac{1}{5}(-7700) + \frac{6}{5}(0.0) - \frac{1}{10}(0.0) - \frac{3}{5}(-56{,}688) = 32{,}473 \ cal \ mol^{-1} \tag{14.54}$$

which is 135,932 J and results in Eh° of $+1.41$ V. Bacteria produce N_2 as a waste product from the NO_3^- electron sink half-reaction. To determine the Eh conditions produced, the concentration of N_2 in water, $N_{2(aq)}$, is needed. As a first approximation, it is assumed that the $N_{2(aq)}$ present is in equilibrium with N_2 in the atmosphere. Henry's law constant, K_H, for the partitioning of gases in H_2O is given by

$$K_{H(gas)} = \frac{m_{gas}}{f_{gas}} \qquad [14.55]$$

for the subscripted gas. m_{gas} and f_{gas} denote the molality and fugacity of the gas, respectively (see Chapter 4 for a further discussion of the implications of Henry's law behavior).

Values of $K_{H(gas)}$ at 25°C and 1 bar for some gases are tabulated in **Table 14-2**. Because 80% of dry air is N_2, $P_{N_2} = f_{N_2}$ in H_2O-saturated air at 1 bar is about 0.77 bar. The molality of N_2 in water, m_{N_2}, from Equation 14.55 is then 4.9×10^{-4} m. The Eh where $m_{N_{2(aq)}} = m_{NO_3^-}$ in equilibrium with normal air is given by

$$Eh = Eh^\circ + 0.05917 \log\left(\frac{(10^{7.8})^{6/5}(4.9 \times 10^4)^{1/5}}{(0.8)^{1/10}}\right) \qquad [14.56]$$

The Eh of nitrate reduction is then

$$Eh_{NO_3^-} = 1.41 - 0.59 = 0.82 \text{ V} \qquad [14.57]$$

After the denitrification, half-reaction is used as an electron sink for oxidizing carbon; the water contains the species $N_{2(aq)}$, SO_4^{2-}, and HCO_3^-. With

Table 14-2   **Henry's law constants for the indicated gas in H_2O at 25°C and 1 bar**

Gas	$K_{H(gas)}$
CH_4	1.32×10^{-3}
CO_2	3.34×10^{-2}
H_2	7.8×10^{-4}
N_2	6.40×10^{-4}
NO	2.0×10^{-3}
NH_3	57
O_2	1.26×10^{-3}
SO_2	1.2

Values in molality bar^{-1} mainly from Manahan (2000).

this water, the electrons produced by bacteria feeding on organic matter are absorbed by the reduction direction of the HS^- oxidation half-reaction:

$$\frac{1}{2}H_2O + \frac{1}{8}HS^- = \frac{1}{8}SO_4^{2-} + \frac{9}{8}H^+ + e^- \quad\quad [14.58]$$

The ΔG_R^o of this reaction at 25°C and 1 bar is

$$\Delta G_R^o = \frac{1}{8}(-177{,}930) + \frac{9}{8}(0.0) - \frac{1}{2}(-56{,}688) - \frac{1}{8}(2{,}860) = 5{,}745 \text{ cal mol}^{-1} \quad [14.59]$$

which is 24,050 J and results in $Eh_{SO_4^{2-}}^o$ of 0.249 V. The Eh at non–standard-state conditions is calculated from

$$Eh_{SO_4^{2-}} = Eh_{SO_4^{2-}}^o + 0.05917 \log\left(\frac{(a_{SO_4^{2-}})^{1/8}(a_{H^+})^{9/8}}{(a_{HS^-})^{1/8}(a_{H_2O})^{1/2}}\right) \quad [14.60]$$

Assuming $a_{H_2O} = 1$ at pH 7.8, the Eh of equal activities of SO_4^{2-} and HS^- for sulfate reduction is

$$Eh_{SO_4^{2-}} = Eh_{SO_4^{2-}}^o + 0.05917 \log(10^{-7.8})^{9/8} = 0.249 - 0.519 = -0.270 \text{ V} \quad [14.61]$$

After SO_4^{2-} is reduced, bacteria use the reduction of nitrogen to ammonia given by the reduction direction of the oxidation half-reaction

$$\frac{1}{3}NH_4^+ = \frac{1}{6}N_{2(aq)} + \frac{4}{3}H^+ + e^- \quad\quad [14.62]$$

The ΔG_R^o of this reaction at 25°C and 1 bar is

$$\Delta G_R^o = \frac{1}{6}(0) + \frac{4}{3}(0) - \frac{1}{3}(-18{,}990) = 6300 \text{ cal mol}^{-1} \quad [14.63]$$

which is 26,371 J and results in $Eh_{N_2}^o = 0.273$ V. The Eh at nonstandard conditions is given by

$$Eh_{N_2} = Eh_{N_2}^o + 0.05917 \log\left(\frac{(f_{N_2})^{1/6}(a_{H^+})^{4/3}}{(a_{NH_4^+})^{1/3}}\right) \quad [14.64]$$

As calculated above, the concentration of $N_{2(aq)}$ in equilibrium with the $f_{N_2} = 0.77$ bar in the atmosphere is 4.9×10^{-4} m. In equilibrium with the atmosphere, the Eh of equal activities of $N_{2(aq)}$ and ammonia is

$$Eh_{N_2} = Eh_{N_2}^o + 0.05917 \log\left(\frac{(0.8)^{1/6}(10^{-7.8})^{4/3}}{(4.9 \times 10^{-4})^{1/3}}\right) \quad [14.65]$$

The Eh of N_2 reduction under these conditions is

$$Eh_{N_2} = 0.273 - 0.551 = -0.278 \ V \qquad \text{[14.66]}$$

With the conversion of N_2, the water contains the species NH_4^+, HS^-, and HCO_3^-. HCO_3^- is then reduced to methane by the reduction direction of the oxidation half-reaction

$$\frac{1}{8}CH_4 + \frac{3}{8}H_2O = \frac{1}{8}HCO_3^- + \frac{9}{8}H^+ + e^- \qquad \text{[14.67]}$$

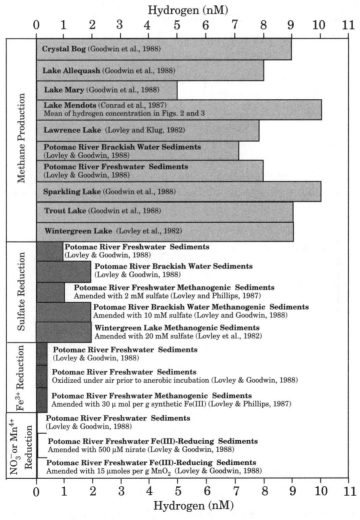

FIGURE 14-6 Concentrations of H_2 in the indicated sediments grouped according to their predominant electron accepting processes. (Modified after Lovley and Goodwin, 1988.)

The ΔG_R^o of this reaction at 25°C and 1 bar is

$$\Delta G_R^o = \frac{1}{8}(-140{,}282) + \frac{9}{8}(0) - \frac{1}{8}(-12{,}122) - \frac{3}{8}(-56{,}688) = 5238 \text{ cal mol}^{-1} \qquad [14.68]$$

which is 21,926 J and results in $Eh_{HCO_3^-}^o = 0.227$ V. The Eh at nonstandard conditions is

$$Eh_{HCO_3^-} = Eh_{HCO_3^-}^o + 0.05917 \log\left(\frac{(a_{HCO_3^-})^{1/8}(a_{H^+})^{9/8}}{(f_{CH_4})^{1/8}(a_{H_2O})^{3/8}}\right) \qquad [14.69]$$

Assuming $a_{H_2O} = 1$ at $pH = 7.8$ with $a_{HCO_3^-} = 10^{-4}$ m, the fugacity of CH_4 at equal molalities of HCO_3^- and $CH_{4(aq)}$ is 0.076. Under these conditions, the Eh of bicarbonate reduction is

$$Eh_{HCO_3^-} = Eh_{HCO_3^-}^o + 0.05917 \log\left(\frac{(10^4)^{1/8}(10^{-7.8})^{9/8}}{(0.076)^{1/8}}\right) \qquad [14.70]$$

or

$$Eh_{HCO_3^-} = 0.227 - 0.541 = -0.314 \qquad [14.71]$$

After bicarbonate is reduced, oxidized organic species in the water are used to consume electrons. The sequence of reactions indicates that after oxygen is

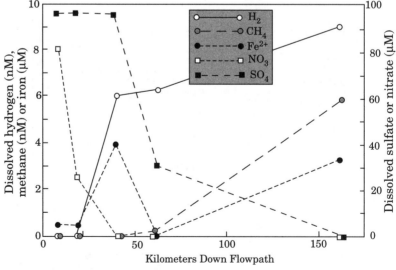

FIGURE 14-7 Distribution of redox species along the flow path of the Middendorf aquifer, South Carolina. (Modified after Lovley and Goodwin, 1988.)

consumed the most oxidized redox couple in turn is used to supply electrons needed as given by lower and lower values of its Eh as calculated above.

As outlined above, there are problems in measuring and interpreting Eh because most environments are not at redox equilibrium. H_2 is an important intermediate in most bacterially mediated redox reactions. Lovley and Goodwin (1988) argued that a better way to characterize redox environments is by H_2 contents. **Figure 14-6** gives H_2 gas concentrations measured in waters from sediments for the indicated dominant electron accepting process. The greater the Eh of the electron acceptor reaction from organic matter oxidation, the lower are the H_2 concentrations.

Figure 14-7 shows the distribution of redox species along the flow path of the Middendorf aquifer, South Carolina. The aerobic water in the recharge area becomes increasingly more reduced as it reacts with the reducing components encountered in the rocks along its flow path. Such changes in the redox state of the water have important consequences for the stability of Fe and other species that change oxidation state in solution, as discussed below.

■ Fe Species

Figure 14-8 is a plot indicating the range of stability of the dominant Fe species in solution in terms of Eh-pH at 25°C and 1 bar. The lines indicate equal activities of the ions in the solution. These lines were calculated from the

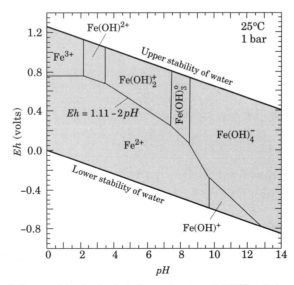

I FIGURE 14-8 Eh–pH diagram giving the dominate Fe species present at 25°C and 1 bar.

thermodynamic data for Fe^{2+} and Fe^{3+} given by SUPCRT92 together with values for equilibrium constants for the *hydrolysis* of these species from Baes and Mesmer (1976). (See Nordstrom and Munoz [1985] for a more detailed discussion of iron species in solution.)

Consider the oxidation half-reaction:

$$Fe^{2+} = Fe^{3+} + e^- \qquad Eh^{\circ}{}_{(14.72)} = 0.77 \text{ V} \qquad [14.72]$$

This reaction does not depend on *pH* but plots as a horizontal line at 0.77 V between the Fe^{2+} and Fe^{3+} fields in Figure 14-8. To consider the boundary between the species Fe^{2+} and $Fe(OH)_2^+$, consider first the hydrolysis of Fe^{3+} to $Fe(OH)_2^+$:

$$Fe^{3+} + 2H_2O = Fe(OH)_2^+ + 2H^+ \qquad [14.73]$$

With log $K_{(14.73)} = -5.67$ at 25°C and 1 bar (Baes and Mesmer, 1976, Table 10-18), the ΔG_R^o of Reaction 14.73 is

$$\Delta G_R^o = -2.303RT \log K_{(14.73)} = 3.24 \times 10^4 \text{ J mol}^{-1} \qquad [14.74]$$

Eh° for Reaction 14.73 is then

$$Eh^o_{(14.73)} = \frac{3.24 \times 10^4}{96,482} = 0.34 \text{ V} \qquad [14.75]$$

Combining Reaction 14.72 with Reaction 14.73 gives

$$Fe^{3+} + 2H_2O + (Fe^{2+}) = Fe(OH)_2^+ + 2H^+ + (Fe^{3+} + e^-) \qquad [14.76]$$

or

$$Fe^{2+} + 2H_2O = Fe(OH)_2^+ + 2H^+ + e^- \qquad [14.77]$$

Adding $Eh^o_{(14.72)}$ and $Eh^o_{(14.73)}$ results in

$$Eh^o_{(14.77)} = 0.34 + 0.77 = 1.11 \text{ V} \qquad [14.78]$$

The Nernst equation for Reaction 14.77 with equal activities of Fe^{2+} and $Fe(OH)_2^+$ is, therefore, written as

$$Eh = 1.11 - 2\,pH \qquad [14.79]$$

This is the boundary line plotted between Fe^{2+} and $Fe(OH)_2^+$, as indicated in Figure 14-8. The other boundaries are constructed in a similar manner using the hydrolysis constants reported by Baes and Mesmer (1976).

Examination of Figure 14-8 indicates that in deep ground water the Eh is low enough that for $pH < 8$ Fe is present as Fe^{2+}; however, Fe^{3+} is the dominant species at low pH and high Eh. In addition, the oxidation state of all of the Fe species in contact with the atmosphere is 3+. Because these species have limited stability relative to Fe mineral phases, the concentration of Fe in equilibrium with waters in contact with the earth's atmosphere is very small in contrast to the greater concentrations of Fe that exist in solution as Fe^{2+} at depth.

Eh-pH diagrams are also used to consider the stability of iron oxide minerals. In the presence of H_2O at 25°C and 1 bar, the stable oxides of Fe are hematite, Fe_2O_3, and magnetite, Fe_3O_4. Iron exists in Fe_2O_3 as Fe^{3+}, and in Fe_3O_4, a mixed valence with two Fe^{3+} and one Fe^{2+} is present. Magnetite is present, therefore, in lower oxidation state environments than hematite.

Consider the oxidation-reduction reaction between these two iron oxide minerals:

$$2Fe_3O_4 + H_2O = 3Fe_2O_3 + 2H^+ + 2e^-$$

[14.80]

Using the SUPCRT92 data

$$\Delta G_R^o = 3\,(-178,155) - (2(-242,574) + (-56,688))\text{ cal mol}^{-1} = 30,840\text{ J mol}^{-1}$$
$$\text{hematite} \qquad \text{magnetite} \qquad H_2O$$

[14.81]

so that

$$Eh^o_{14.80} = 0.160\text{ V}$$

[14.82]

The Nernst equation for Reaction 14.80 at 25°C and 1 bar is

$$Eh_{(14.80)} = 0.160 + \frac{0.05917}{2}\,\log\frac{a^3_{Fe_2O_3}a^2_{H^+}}{a^2_{Fe_3O_4}a_{H_2O}}$$

[14.83]

With nearly pure magnetite, hematite, and water, their activities are taken as unity. Equation 14.83 then reduces to

$$Eh_{(14.80)} = 0.16 + \frac{0.05917}{2}\,\log a^2_{H^+} = 0.16 - 0.05917\,pH$$

[14.84]

This is shown as a solid line in **Figure 14-9**. The stability boundary between magnetite and hematite is parallel to the water stability boundaries, and all but the most reducing environments are in equilibrium with hematite.

The stability of aqueous species relative to hematite and magnetite is also considered. The equilibrium between hematite and Fe^{3+} is given by

$$Fe_2O_3 + 6H^+ = 2Fe^{3+} + 3H_2O$$

[14.85]

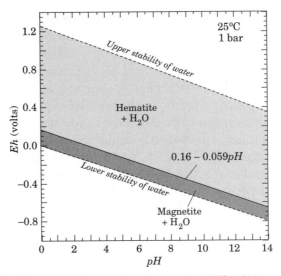

I FIGURE 14-9 *Eh–pH* stability diagram for hematite and magnetite at 25°C and 1 bar.

From SUPCRT92 data log $K_{(14.85)}$ at 25°C and 1 bar is calculated to be 0.109. Assuming the activities of Fe_2O_3 and H_2O are unity, the equilibrium constant expression becomes

$$\log K_{(14.85)} = \log \left(\frac{a_{Fe^{3+}}^2}{a_{H^+}^6} \right) = 0.109 \qquad [14.86]$$

or

$$\log a_{Fe^{3+}} = 0.055 - 3\,pH \qquad [14.87]$$

Because Fe did not change its oxidation state, the *Eh* of Reaction 14.85 does not change, but $a_{Fe^{3+}}$ is a function of *pH*. To display this equilibrium, an activity of Fe^{+3} is stipulated and the equilibrium *pH* for this plotted. This is shown in **Figure 14-10** for $a_{Fe^{3+}} = 10^{-6}$ and $10^{-4}\ m$, as indicated by the vertical dashed lines.

The relative stability of hematite and Fe^{2+} in solution is considered with the aid of the reaction

$$2Fe^{2+} + 3H_2O = Fe_2O_3 + 6H^+ + 2e^- \qquad Eh^\circ = +0.773\ V \qquad [14.88]$$

From the Nernst equation

$$Eh_{(14.88)} = 0.773 + \frac{0.05917}{2} \log \left(\frac{a_{H^+}^6}{a_{Fe^{2+}}^2} \right) \qquad [14.89]$$

$$= 0.773 - 0.059 \log a_{Fe^{2+}} - 0.178\,pH \qquad [14.90]$$

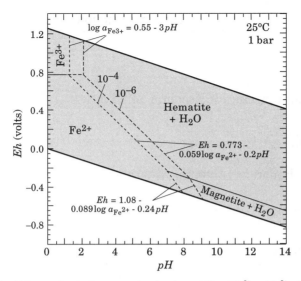

FIGURE 14-10 *Eh–pH* diagram for the Fe system showing the stability of Fe^{2+} and Fe^{3+} with hematite and magnetite in water at 25°C and 1 bar.

The relative stability of magnetite and Fe^{2+} is given by

$$3Fe^{2+} + 4H_2O = Fe_3O_4 + 8H^+ + 2e^- \qquad Eh^o = 1.08 \text{ V} \tag{14.91}$$

so that Eh is given by

$$Eh_{(14.91)} = 1.08 + \frac{0.05917}{2} \log\left(\frac{a_{H^+}^8}{a_{Fe^{2+}}^3}\right) \tag{14.92}$$

or

$$Eh_{(14.91)} = 1.08 - 0.089 \log a_{Fe^{2+}} - 0.233 \, pH \tag{14.93}$$

Equation 14.87 for Fe^{3+} and Equations 14.90 and 14.93 for Fe^{2+} activity in solution of 10^{-4} and 10^{-6} *m* are shown in Figure 14-10. To have appreciable Fe in solution at high Eh, the solution must be very acidic ($pH < 1.35$ for $> 10^{-4}$ *m*) or must have a low Eh where Fe^{2+} is stable.

Activity of Electrons in Solution

Besides using Eh to monitor the oxidation state of a system, the activity of the electron is also used. That is, similar to pH, pe is defined as

$$pe \equiv -\log a_{e^-} \tag{14.94}$$

where a_{e^-} is the activity of an electron in a reaction.

Consider the reaction

$$Fe^{3+} + e^- = Fe^{2+}$$ [14.95]

The equilibrium constant for Reaction 14.95 is

$$K_{(14.95)} = \frac{a_{Fe^{2+}}}{a_{Fe^{3+}} a_{e^-}}$$ [14.96]

or considering Equation 14.94, this is written as

$$\log K_{(14.95)} = \log \frac{a_{Fe^{2+}}}{a_{Fe^{3+}}} + pe$$ [14.97]

Solving for pe gives

$$pe = \log K_{(14.95)} - \log \frac{a_{Fe^{2+}}}{a_{Fe^{3+}}}$$ [14.98]

A standard state is assigned to pe, denoted as pe°. This is done by stipulating that all of the other species are in their standard state, which implies $a_i = 1$, where i in this case is Fe^{2+} or Fe^{3+}. pe° is then given by

$$pe^\circ \equiv \frac{1}{n} \log K_{(14.95)}$$ [14.99]

where n is the number of moles of electrons transferred per mole of reaction. In the case of Reaction 14.95, $n = 1$. Combining Equations 14.98 and 14.99 results in

$$pe = pe^\circ - \log \frac{a_{Fe^{2+}}}{a_{Fe^{3+}}}$$ [14.100]

There is a relationship between Eh and pe. Consider the Nernst equation for a generalized reaction, which is written as

$$Eh = Eh^\circ + 2.303 \frac{RT}{nF} \log Q$$ [14.101]

Solving for $\log Q$ yields

$$\log Q = \frac{nF}{2.303RT}(Eh - Eh^\circ)$$ [14.102]

pe for the same reaction is given by

$$pe = pe^\circ - \frac{1}{n} \log Q$$ [14.103]

If we substitute the value of log Q in Equation 14.102 into Equation 14.103, pe becomes

$$pe = \frac{\mathbf{F}\,Eh}{2.303\mathrm{R}T} \qquad\qquad [14.104]$$

and

$$pe^{\circ} = \frac{\mathbf{F}\,Eh^{\circ}}{2.303\mathrm{R}T} \qquad\qquad [14.105]$$

At 25°C and 1 bar,

$$pe = \frac{Eh}{0.0592} \qquad\qquad [14.106]$$

◼ Oxygen Fugacity Diagrams

Eh and *pe* are used to give the extent of reduction and oxidation, that is, redox, in surface and groundwater. Oxygen fugacity, however, is a more convenient unit of measure for the state of oxidation at elevated P and T in the interior of the earth.

Consider the reaction

$$6Fe_2O_3 = 4Fe_3O_4 + O_2 \qquad\qquad [14.107]$$
$$\text{hematite} \quad \text{magnetite}$$

which is often referred to as the HM buffer reaction. It is a buffer because as long as hematite and magnetite are present—the f_{O_2} is buffered to a constant value at constant pressure and temperature. For pure hematite and magnetite, their activities are taken as unity. The equilibrium constant expression for the HM buffer is, therefore,

$$\log \mathbf{K}_{HM} = \log f_{O_2} = \frac{-\Delta G^{\circ}_{HM}}{2.303\mathrm{R}T} \qquad\qquad [14.108]$$

where ΔG°_{HM} is the standard-state Gibbs energy of the reaction at the pressure and temperature of interest.

Using the SUPCRT92 values of Gibbs energy, the log f_{O_2} for this Reaction 14.107 at 25°C and 1 bar is

$$\log f_{O_2} = -\frac{(4 \times -242{,}574) - (6 \times -178{,}155)\ \text{cal mol}^{-1}}{2.303 \times 1.987\ \text{cal mol}^{-1}\text{K}^{-1} \times 298.15\ \text{K}} = -72.3 \qquad [14.109]$$

Note the exceedingly low f_{O_2} compared with that in the surface environment. This implies that hematite is the stable iron oxide phase at surface conditions and that all phases with more reduced Fe are unstable. The calculated fugacity of O_2 at 1 bar from 350°C to 1050°C (labeled HM) is given in **Figure 14-11**.

If phases such as magnetite are not stable in contact with the earth's atmosphere, what is the oxygen fugacity of the mantle? An oxidation-reduction reaction important in the earth's mantle is

$$2Fe_3O_4 + 3SiO_2 = 3Fe_2SiO_4 + O_2 \qquad\qquad\qquad\qquad\qquad [14.110]$$
magnetite quartz   fayalite

referred to as the QFM buffer. In the presence of quartz, magnetite, and a fixed activity of fayalite in an olivine, the oxygen fugacity is fixed by this assemblage.

With all solid phases pure, including fayalite, the equilibrium constant for QFM is

$$\log K_{QFM} = \log \left(\frac{a_{Fa}^3 f_{O_2}}{a_{Mag}^2 a_{Qtz}^3} \right) = \log f_{O_2} \qquad\qquad\qquad\qquad [14.111]$$

which is shown as a dashed line in Figure 14-11. In an olivine where the activity of fayalite is 0.1, a mantle-type olivine, Equation 14.111 is modified to

$$\log K_{QFM} = \log (a_{Fa}^3 f_{O_2}) = \log 0.1^3 f_{O_2} = \log 0.001\ f_{O_2} \qquad\qquad [14.112]$$

In this case, the oxygen fugacity is 3 orders of magnitude greater than QFM.

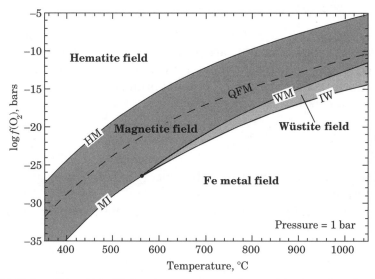

FIGURE 14-11 Base-10 logarithm of O_2 fugacity vs. temperature diagram giving the regions of stability for Fe-containing mineral phases at 1 bar. The fugacity as a function of temperature for equilibrium between quartz + fayalite + magnetite (QFM) is also shown.

Figure 14-11 is calculated at 1 bar. What about the pressure dependence of the equilibria? At 800°C and 1 bar, $K_{HM} = 10^{-9.70}$ and $K_{QFM} = 10^{-14.73}$, whereas at 800°C and 10 kbar, $K_{HM} = 10^{-9.53}$, and $K_{QFM} = 10^{-13.66}$. It is clear that oxygen fugacity for these equilibria shifts upward by somewhat more than an order of magnitude greater as pressure is increased to 10 kbar. These differences should be kept in mind when applying the O_2 buffer equilibrium displayed in Figure 14-11 to conditions of the earth's mantle.

Because the earth's core is thought to contain iron metal, the MI reaction

$$0.5 \ Fe_3O_4 = 1.5 \ Fe + O_2 \qquad\qquad [14.113]$$
$$\text{magnetite} \quad\ \text{iron}$$

where with pure magnetite and iron

$$K_{MI} = f_{O_2} \qquad\qquad [14.114]$$

is also of interest. Magnetite becomes unstable in the presence of iron relative to wüstite at about 562°C at 1 bar. Wüstite is an iron oxide solid solution of composition Fe_xO, where x is between 0.85 and 0.95, depending on f_{O_2} and temperature. Its iron is then mainly Fe^{2+}, but some Fe^{3+} is present.

The equilibrium between wüstite and magnetite is given by

$$\frac{2x}{4x-3} \ Fe_3O_4 = \frac{6x}{4x-3} Fe_xO + O_2 \qquad\qquad [14.115]$$

With pure magnetite, the logarithm of the equilibrium constant for Reaction 14.115 is

$$\log K_{WM} = \left(\frac{6x}{4x-3}\right) \log a_{\text{wüstite}} + \log f_{O_2} \qquad\qquad [14.116]$$

The equilibrium of wüstite with iron is

$$2Fe_xO = 2xFe + O_2 \qquad\qquad [14.117]$$

and with pure iron, the equilibrium constant becomes

$$K_{IW} = \frac{f_{O_2}}{a_{\text{wüstite}}^2} \qquad\qquad [14.118]$$

For most conditions in the earth, the oxygen fugacities are found to be between iron metal in the core (reducing) and hematite found in equilibrium with the O_2-rich atmosphere (oxidizing). The mantle is generally thought to be near equilibrium with quartz, fayalite, and magnetite.

It is clear that f_{O_2} must be related to Eh. Consider the reaction

$$2H_2O = 4H^+ + O_2 + 4e^-$$ [14.119]

The Nernst equation for this reaction with the activity of $H_2O = 1$ is

$$Eh = Eh^\circ + 2.303 \frac{RT}{4F} \log (f_{O_2} a_{H^+}^4)$$ [14.120]

At 25°C and 1 bar, $Eh^\circ = 1.23$ and

$$Eh = 1.23 + 0.0148 \log f_{O_2} - 0.0592 \, pH$$ [14.121]

■ Using O_2 Buffer Reactions to Control H_2

The oxygen buffer reactions discussed above are also used to help control f_{H_2} in phase equilibrium experiments in the presence of H_2O. Consider the reaction

$$H_2O = H_2 + \frac{1}{2}O_2$$ [14.122]

with an equilibrium constant of

$$K_{(14.122)} = \frac{f_{H_2} f_{O_2}^{0.5}}{f_{H_2O}}$$ [14.123]

If the gas phase contains only H_2O, O_2, and H_2, the total pressure is given by

$$P_{total} = \frac{f_{H_2O}}{\chi_{H_2O}} + \frac{f_{O_2}}{\chi_{O_2}} + \frac{f_{H_2}}{\chi_{H_2}}$$ [14.124]

where χ_i is the fugacity coefficient of gas i. Because f_{O_2} is so small relative to χ_{O_2}, the middle term is ignored in this equation.

Equation 14.123 is solved for f_{H_2O} and substituted into this modified Equation 14.124 to give

$$f_{H_2} = \frac{P_{total} \, K_{(14.122)} \, \chi_{H_2O} \, \chi_{H_2}}{f_{O_2}^{0.5} \, \chi_{H_2} + K_{(14.122)} \, \chi_{H_2O}}$$ [14.125]

Knowing f_{O_2}, the fugacity of H_2 at a given pressure is calculated from values of $K_{(14.122)}$, χ_{H_2O}, and χ_{H_2}. χ_{H_2O} and χ_{H_2} are determined by assuming ideal mixing or by using the Redlich-Kwong mixing rules for gases, as outlined in Chapter 4.

■ O$_2$ Fugacities of Magnetite-Ilmenite Pairs

Oxygen fugacities have been determined in igneous rock systems that have crystallized magnetite (Mag) and ilmenite (Ilm) in equilibrium. Because there is Ti in the system along with Fe, the magnetite (Fe_3O_4) phase is typically a solid solution with ulvöspinel (Fe_2TiO_4) and the ilmenite ($FeTiO_3$) phase a solid solution with hematite (Fe_2O_3). The equilibrium between these two phases is written as

$$Fe_3O_4 + FeTiO_3 = Fe_2TiO_4 + Fe_2O_3 \qquad\qquad\qquad [14.126]$$
$$\textit{in}\ \text{Mag} \quad \textit{in}\ \text{Ilm} \qquad \textit{in}\ \text{Mag} \quad \textit{in}\ \text{Ilm}$$

The phase where the component of interest is located is given below the composition of the component in Reaction 14.126.

The solid solution in ilmenite is modeled as random mixing on two sites, Fe^{3+} for Ti^{4+} on one site independent of Fe^{2+} for Fe^{3+} mixing on the other site. Ideal mixing on crystallographic sites in minerals is discussed in Chapter 5. In the case considered here, the activities on the sites are squared. For titanomagnetite, mixing is considered to be independent on one site with a coupled substitution of $2Fe^{3+} \rightarrow Fe^{2+} + Ti^{4+}$ in the structure. In this case, the activities do not need to be squared. Because of the binary nature of the solid solutions, $X_{Fe_2TiO_4}^{Mag} + X_{Fe_3O_4}^{Mag} = 1$ and $X_{Fe_2O_3}^{Ilm} + X_{FeTiO_3}^{Ilm} = 1$. The logarithm of the equilibrium constant for Reaction 14.126 is then

$$\log K_{(14.126)} = \log \frac{X_{Fe_2TiO_4}^{Mag}(1 - X_{FeTiO_3}^{Ilm})^2}{(1 - X_{Fe_2TiO_4}^{Mag})(X_{FeTiO_3}^{Ilm})^2} + \log \frac{\Gamma_{Fe_2TiO_4}^{Mag}(\Gamma_{Fe_2O_3}^{Ilm})^2}{\Gamma_{Fe_3O_4}^{Mag}(\Gamma_{FeTiO_3}^{Ilm})^2} \qquad [14.127]$$

where the second term on the right gives the activity coefficients, based on a Raoult's law standard state.

The reaction for oxidation of Fe between the two phases is

$$6Fe_2O_3 = 4Fe_3O_4 + O_2 \qquad\qquad\qquad\qquad [14.128]$$
$$\text{Ilm} \qquad\ \text{Mag}$$

with an equilibrium constant expression

$$\log K_{(14.128)} = \log \frac{(X_{Fe_3O_4}^{Mag})^4}{((X_{Fe_2O_3}^{Ilm})^2)^6} + \log \frac{(\Gamma_{Fe_3O_4}^{Mag})^4}{((\Gamma_{Fe_2O_3}^{Ilm})^2)^6} + \log f_{O_2} \qquad [14.129]$$

Again, the second term on the right gives the activity coefficients, Γ, based on Raoult's law standard state.

I FIGURE 14-12 Structural formula for pentane, C_5H_{12}.

Doing exchange experiments between magnetite and ilmenite in the presence of an oxygen buffer reaction allowed Spencer and Lindsley (1981) to evaluate the activity expressions as a function of temperature and oxygen fugacity. They used this information to construct a grid in log f_{O_2}-temperature space where the intersection of the compositional line for a magnetite with that for a coexisting ilmenite gave the oxygen fugacity as well as the temperature of equilibrium of the oxide pair. This analysis has confirmed that most igneous rocks from the earth's upper mantle plot near the QFM buffer.

■ Oxidation-Reduction in Organic Compounds

Most reactions in living organisms involve the transfer of electrons. These are then oxidation-reduction reactions. Because organic compounds are covalently bonded, however, assigning valence for electron transfer is more problematic. If hydrogen is considered to have a valence of +1 in an organic species, this implies the valence of carbon in methane, CH_4, is −4. In pentane, C_5H_{12}, however, the valence of carbon would be −2.4 (**Figure 14-12**). This occurs because carbons in pentane appear to have more than one valence type. This is similar to magnetite, Fe_3O_4, in which Fe is in two valence states as outlined above. End carbons in pentane are thought of as having a valence of −3 and the carbon bound between two other carbons a valence of −2.

To consider the oxidation state of a carbon in an organic compound, a −1 is assigned for each element bonded to the C that is less *electronegative* than C (e.g., H) and therefore adds an electron to C. A +1 is assigned to each element that is more electronegative (e.g., N and O), which causes a loss of an electron from C. Another attached carbon counts as 0; therefore, carbon valence in CH_4 is −4, but because of the double C=O bonds, it is +4 in CO_2. The indicated carbon in ethanol in **Figure 14-13** has a valence of −1, and the other carbon in the species has a valence of −3.

I FIGURE 14-13 Structural formula of ethanol, C_2H_5OH.

Summary

Oxidation-reduction (redox) reactions transfer electrons between species. These reactions are characterized by measuring a voltage between an anode and cathode in the reaction. The voltages are related to the Gibbs energy change of the reaction. The half-reactions occurring at the anode and cathode are considered separately by relating them to the standard hydrogen electrode reaction

$$2H^+ + 2e^- = H_{2(g)}$$

which by convention is assigned a value of zero volts at standard conditions. Half-reaction voltages relative to the standard hydrogen electrode designated as Eh are then combined to give the potential for a redox reaction.

Eh of a reaction is related to its activity product through the Nernst equation, given by

$$Eh = Eh^\circ + 2.303 \frac{RT}{n\mathrm{F}} \log Q$$

where Eh° is the standard-state voltage for the reaction, n is the number of electrons transferred in the reaction, F is Faraday's constant, and Q is the activity product for the reaction. At equilibrium, Q is equal to the equilibrium constant of the reaction.

In natural waters, O_2 is the main oxidant and organic matter the main reductant. The redox state of the water is then controlled by the presence of organic matter and the water's ability to exchange with O_2 in air. Equilibrium between redox species does not occur in most waters. Anaerobic waters occur because bacteria feed on (i.e., oxidize) organic matter and reduce available oxidants such as O_2, using them as sinks for electrons. The order in which they are used is O_2, NO_3^-, SO_4^{2-}, $N_{2(aq)}$, HCO_3^-, and finally oxidized organic species.

Redox reactions in water typically also involve the species H^+. Eh–pH diagrams are, therefore, helpful in understanding the stability of Fe aqueous species and iron oxide minerals. These are usually limited by $f_{O_2} = 1$ as the highest oxidation state possible in a 1 bar atmosphere in equilibrium with water, where $Eh = 1.23 - 0.059 \, pH$ and $f_{H_2} = 1$. The most reduced conditions are where $Eh = -0.059 \, pH$.

Redox determinations of the interior of the earth indicate that it is quite reducing, with an oxygen fugacity generally $<10^{-10}$ bar. It appears that much of the mantle is likely near the QFM buffer given by

$$2\,Fe_3O_4 + 3\,SiO_2 = 3\,Fe_2SiO_4 + O_2$$
$$\text{magnetite} \quad \text{quartz} \qquad \text{fayalite}$$

Key Terms Introduced

activity product	Nernst equation
aerobic	oxidation
anaerobic	oxygen fugacity
electronegative	reduction
eutrophic	standard state
Faraday's constant	valence
half-reactions	water electrode
hydrolysis	

Questions

1. What is Eh°? Units?
2. Why are some Eh° values positive, whereas others are negative?
3. How is Gibbs energy related to voltages of oxidation-reduction reactions?
4. How does Eh° differ from Eh?
5. Give the Nernst equation, and explain the terms.
6. How is the fugacity of O_2 related to Eh?
7. Discuss the range of Eh and pH of natural waters.
8. Define pe.
9. What is the oxidation state of the mantle?
10. If the earth's atmosphere was produced by degassing the mantle, why is it so oxidizing?

Problems

1. What is the relation between Eh and oxygen fugacity?
2. An electrochemical cell has a standard potential of 1.5 V. With a steady current of 2.0 ampere for 10 minutes, how much electrical work is done by the cell?
3. What is the oxidation state of carbon in the following species:
   (a) methanol (CH_3OH)
   (b) formate ($HCOO^-$)
   (c) formaldehyde (CH_2O)
4. What is the average oxidation state of carbon for an organism with the Redfield ratio of elements ($C_{106}H_{263}O_{110}N_{16}P_1$)? Note that H, O, N, P are bonded with 1, 2, 3 and 5 bonds, respectively.
5. Balance the following unbalanced redox reactions present in an acidic aqueous solution.
   (a) $Fe^{2+} = Fe_3O_4$ (magnetite)
   (b) $CuS_{(covellite)} + HNO_3^- = CuSO_4 \bullet 5H_2O_{(chalcanthite)} + NO_{(gas)}$
   (c) $K_2Cr_2O_{7(chromic\ acid)} = K^+ + Cr^{3+}$

6. Balance the following unbalanced redox reactions present in an alkaline aqueous solution.
   (a) $H_2O_{2(hydrogen\ peroxide)} + ClO_{2(chlorine\ dioxide)} = ClO_2^- + O_{2(gas)}$
   (b) $Ag_{(metal)} + HS^-_{(aq)} + CrO_4^{2-} = Ag_2S_{(argentite)} + Cr(OH)_{3(Cr-hydroxide)}$
   (c) $KMnO_{4(K-permanganate)} + As_4O_{6(As\ oxide)} = Mn^{2+} + K^+ + H_3AsO_{4(arsenic\ acid)}$

7. A pH of 5 and a redox potential of +0.30 V was measured in a lake at 25°C. What concentration of Cu^{2+} would be in equilibrium with metallic copper?

8. What redox potential must an environment possess for the activities of Fe^{2+} and Fe^{3+} to be equal?

9. Calculate the standard state Gibbs energy of the reaction: $2Fe_{(s)} + 3Pb^{2+} = 2Fe^{3+} + 3Pb_{(s)}$ at 25°C and 1 bar and determine which side is more stable at standard conditions.

10. If a stream of pH = 6.8 is in equilibrium with the atmosphere at 25°C and 1 bar what is its pe?

11. If the pe of stream water at 25°C and 1 bar = 13.80 what is the ratio of $a_{Fe^{3+}}$ to $a_{Fe^{2+}}$ in the stream?

12. The oxidation-reduction couple O_2/H_2O produces a pe of 12. If redox equilibrium is assumed, how would a total molality of Fe in solution of 10^{-6} be distributed between Fe^{3+} and Fe^{2+}? Assume the $a_i = m_i$ for the aqueous species.

13. Calculate the ΔG_R°, and determine the Eh° at 25°C and 1 bar of the half-reaction for the oxidation of ferrous Fe to hematite (Reaction 14.88) from the values given in Appendixes F-3 and F-5. Plot the results on an Eh–pH diagram from pH 1 to 7 if the hematite is pure and the concentration of Fe^{2+} in solution is 10^{-2} m. Compare with Figure 14-10.

14. Automotive (Pb-acid) batteries typically consist of six cells connected in series. The redox reaction

$$Pb_{(s)} + PbO_{2(s)} + 4H^+ + 2SO_4^{2-} \rightarrow 2PbSO_{4(s)} + 2H_2O$$

occurs in each cell. If the battery is operating at standard conditions at 25°C, what is its output voltage?

Problem Making Use of SUPCRT92

15. Determine the standard potential, Eh°, for the following half-cell reactions in water at vapor saturation pressure (liq-vap saturation) and at 25°C, 50°C, 100°C, 150°C, 200°C, 250°C, 300°C, and 350°C. Plot your results.
    (a) $3Fe^{2+} + 4H_2O = Fe_3O_{4(magnetite)} + 8H^+ + 2e^-$
    (b) $Cu_2S_{(chalcocite)} + 2H^+ = 2Cu^{2+} + H_2S_{(aq)} + 2e^-$
    (c) $Cu_2S_{(chalcocite)} = CuS_{(covellite)} + Cu^{2+} + 2e^-$

    At what temperature do reactions A and C have a similar Eh° (give value)?

References

Baas-Becking, L. G. M., Kaplan, I. R. and Moore, D., 1960, Limits of the natural environment in terms of pH and oxidation-reduction potentials. *J. Geol.*, v. 68, pp. 243–284.

Baes, C. F., Jr. and Mesmer, R. E., 1976, *The Hydrolysis of Cations*. Wiley, New York, 489 pp.

Berner, R. A., 1971, *Principles of Chemical Sedimentology*. McGraw-Hill, New York, 240 pp.

Bockris, J. O'M. and Oldfield, L. F., 1954, The oxidation-reduction reactions of hydrogen peroxide at inert metal electrodes. *Tran. Faraday Soc.*, v. 51, pp. 249–259.

Conrad, R., Goodwin, S., and Zeikus, J. G., 1987, Hydrogen metabolism and sulfate-dependent inhibition of methanogenesis in a eutrophic lake sediment (Lake Mendota). *FEMS Microbiol. Ecol.* v. 45, pp. 107–115.

Goodwin, S., Conrad, R., and Zeikus, J. G., 1988, Relation of pH and microbial hydrogen metabolism in diverse sedimentary ecosystems. *Appl. Environ. Microbial.* v. 54, pp. 590–593.

Krumbein, W. C. and Garrels, R. M., 1952, Origin and classification of chemical sediments in terms of pH and oxidation-reduction potentials. *J. Geol.*, v. 60, pp. 1–33.

Lindberg, R. E. and Runnells, D. D., 1984, Ground water redox reactions: An analysis of equilibrium state applied to *Eh* measurements and geochemical modeling. *Science*, v. 225, pp. 925–927.

Lovley, D. R., Dwyer, D. F., and Klug, M. J., 1982, Kinetic analysis of competition between sulfate reducers and methanogens for hydrogen in sediments. *Appl. Environ. Microbial.* v. 43, pp. 1373–1379.

Lovley, D. R. and Goodwin, S., 1988, Hydrogen concentrations as an indicator of the predominant terminal electron-accepting reactions in aquatic sediments. *Geochim. Cosmochim. Acta*, v. 52, pp. 2993–3003.

Lovley, D. R. and Klug, M. J., 1982, Intermediary metabolism of organic matter in the sediments of a eutrophic lake. *Appl. Environ. Microbial.* v. 43, pp. 552–560.

Lovley, D. R. and Phillips, E. J. P., 1987, Competitive mechanisms for the inhibition of sulfate reduction and methane production in the zone of ferric iron reduction in sediments. *Appl. Environ. Microbial.* v. 53, pp. 2636–2641.

Manahan, S. E., 2000, *Environmental Chemistry*, 7th edition. CRC Press, Boca Raton, FL, 898 pp.

Nordstrom, D. K. and Munoz, J. L., 1985, *Geochemical Thermodynamics*. Benjamin/Cummings, Menlo Park, CA, 477 pp.

Spencer, K. J. and Lindsley, D. H., 1981, A model for coexisting iron-titanium oxides. *Am. Mineral.*, v. 66, pp. 1189–1201.

15 | Organic Geochemistry

Organic geochemistry concerns the abundance and behavior of both synthetic and natural organic compounds. It is important to know the makeup, distribution, and fate of organic matter produced on the earth to understand such processes as the production of fossil fuels, development of soils, the consequences of organic pollutants, and the cycling of carbon through the atmosphere and ocean. The major constituent of organic matter is carbon. Carbon exhibits a very diverse chemistry existing in both organic and inorganic forms. Inorganic carbon occurs principally in carbonates as a CO_3^{2-} group. Organic carbon is more reduced and occurs in a wide range of natural and artificial *polymers*. A polymer is a large molecule consisting of many units of one or more monomer molecules (chains) bonded together (e.g., *starch* and *cellulose* are natural polymers). Organic carbon has four bonding electrons that it shares with other elements. This can be in four single bonds or a lesser number of multiple electron covalent bonds. Because a carbon can bond with up to four other carbons, carbon chains of enormous variability in length, bonding character, and element composition occur.

■ Organic Chemistry Nomenclature

In a general sense, organic molecules are regarded as frameworks of connected carbon atoms to which a variety of *functional groups* attach. A functional group is a specifically bonded group of atoms in an organic molecule that often contains one or more elements other than hydrogen and carbon. The nomenclature used to describe organic compounds derives from the number of carbon atoms in the framework plus the attached functional groups. **Table 15-1** lists some International Union of Pure and Applied Chemistry rules of nomenclature. The names for the number of carbons in chains of carbon atoms in a molecule are listed in **Table 15-2**. Some common functional groups are outlined in **Table 15-3**. The names that specify the number of functional groups are given in **Table 15-4**.

Table 15-1

Table 15-1	International union of pure and applied chemistry (IUPAC) rules of organic compound nomenclature

A. The parent name of the compound is based on the longest continuous chain of carbon atoms (l-c-c) whose root name is given in Table 15-2.

B. All other carbons not in the l-c-c are named as functional groups with -yl endings as given in Table 15-2. Other common functional group names are given in Table 15-3.

C. The carbons on the l-c-c are numbered to locate the positions of the functional group.

D. A prefix is used to indicate the number of times a functional group appears, as given in Table 15-4.

Organic carbon compounds are divided into two large groups: *aliphatic* and *aromatic*. Aliphatic compounds are open chains and branches of carbon atoms, whereas aromatics have one or more closed *benzene rings*. A benzene ring is a ring of six carbon atoms that are bonded together with alternating single and double bonds, with the fourth carbon bond bonded to hydrogen in benzene or a substituted functional group in a substituted benzene compound. This arrangement of carbon bonding is particularly stable. **Figure 15-1** shows different ways to represent a benzene ring.

Aliphatic Molecules

Alkanes are the simplest and least reactive aliphatic compounds. They are hydrocarbons, meaning that they are compounds that contain only carbon and hydrogen. Their common structural formula is C_nH_{2n+2}, where *n* is the num-

Table 15-2	Carbon chain terminology

Number of Carbons	Term
1	Methyl or meth
2	Ethyl or eth
3	Propyl or prop
4	Butyl or but
5	Pentyl or pen
6	Hexyl or hex
7	Hepyl or hep
8	Octyl or oct
9	Nonyl or non
10	Decyl or dec
11	Undecyl or undec
12	Dodecyl or dodec

| Table 15-3 | Common functional groups |

Structural formula	Group	Compound types
	Acidic groups	
−C⟨O OH⟩	Carboxyl	Carboxylic acid
−OH	Hydroxyl	Phenols (aromatics)
		Alkenol or enol
	Neutral groups	
C=C	Alkene	Alkenes
C≡C	Alkyne	Alkynes
−OH	Hydroxyl	Alcohols (aliphatic)
−Cl	Chloro	Chlorides
−O−	Oxo	Esters
O ‖ −C−	Carbonyl	Ketones
−C⟨O H⟩	Carbonyl	Aldehydes
−F	Fluoro	Fluorides
	Basic groups	
−N⟨	Amino	Amines

ber of carbons in the chain. Alkanes are said to be *saturated*, meaning that the compounds have the maximum H per C in their structure. They, therefore, contain only single bonds between carbons. The compounds are named for the number of carbons, as given in Table 15-2, with the ending -ane to indicate it is an alkane. The three simplest alkanes are outlined in **Figure 15-2**.

Butane, pentane, hexane, heptane, and octane continue the chain length with four, five, six, seven, and eight carbons, respectively. The structural formula of octane, a constituent of gasoline, is shown in **Figure 15-3**. 2-Methylpropane, a propane with a methyl functional group on the second carbon atom, is shown

| Table 15-4 | Naming in carbon chains |

Repeats of functional groups	Prefix
2	Di
3	Tri
4	Tetra
5	Penta

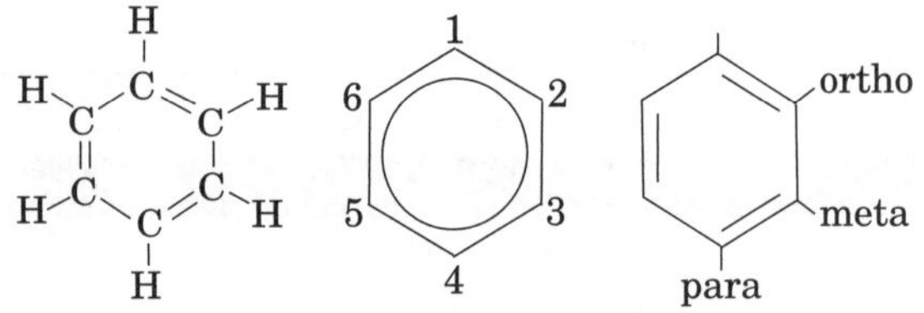

FIGURE 15-1 Three representations of the hexagonal benzene ring. When these are functional groups bonded on other compounds, they are referred to as phenyls. The representation on the left shows the actual carbon atoms present in benzene. The other two show the carbon ring as a schematic representation. Note the alternating single and double carbon–carbon bonds in the structure. The electrons in the double bonds actually resonate between the single- and double-bond positions. The middle representation indicates the position with numbers needed to specify the location of functional groups on the ring. The representation on the right specifies the location of functional groups as names when the ring is a phenyl group.

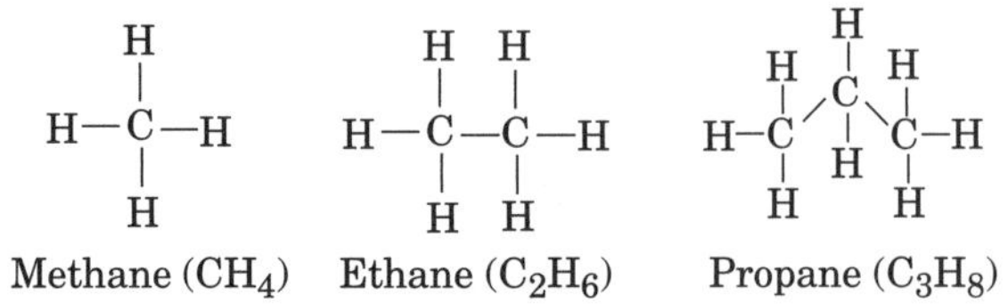

Methane (CH_4)   Ethane (C_2H_6)   Propane (C_3H_8)

FIGURE 15-2 Structural formula of the alkanes with one, two, and three carbons.

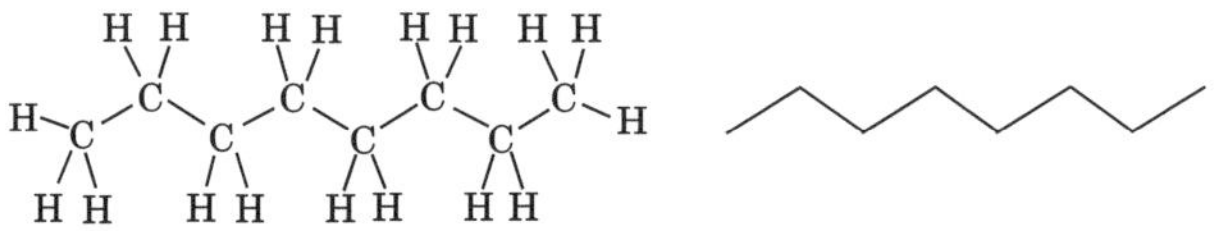

FIGURE 15-3 Octane (C_8H_{18}) showing its structural formula on the left. On the right is a schematic representation of octane where lines represent C–C bonds while C–H bonds as well as H atoms are omitted.

in **Figure 15-4**. The structural formula of 2,3-dimethylbutane, a butane with a methyl group on the second and third carbon atoms, is displayed in **Figure 15-5**. As the length of the carbon chain increases, the vapor pressure of the alkane decreases. As a result, CH_4 (methane) to C_4H_{10} (butane) are gaseous at room temperature and pressure, but with increasing carbons, the liquid alkanes found in gasoline of C_5H_{12} (pentane) to $C_{10}H_{22}$ (decane) are produced. At a chain length of 12, dodecane, $C_{12}H_{26}$, is created, which also goes by the common name kerosene. Increasing the chain length, the compounds become more viscous as heavy oils and finally tars are produced. Alkanes also have cyclic forms that are designated with the prefix cyclo-; **Figure 15-6** shows cyclopentane and cyclohexane.

Alkenes are aliphatic compounds containing a carbon–carbon double-bond functional group. The number of hydrogen–carbon bonds is therefore lower than in the alkanes. Their systematic name is given by the number of carbons in the molecule with the suffix -ene. They are unsaturated compounds because they have an "additional" electron in their double bond that is capable

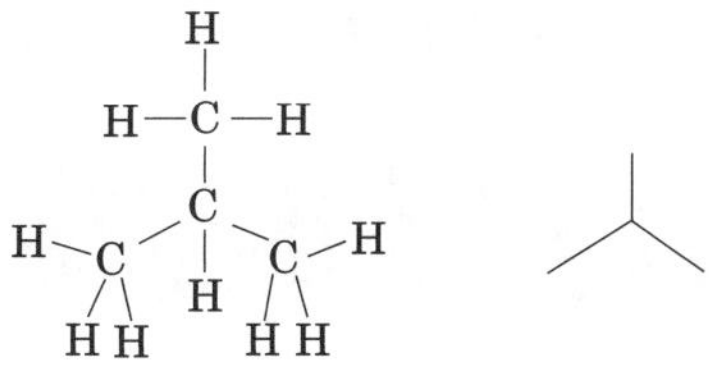

I FIGURE 15-4 2-Methylpropane showing its structural formula and schematic representation.

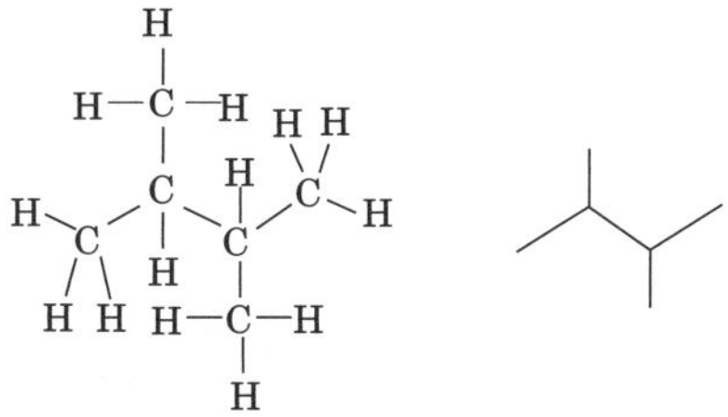

I FIGURE 15-5 2,3-Dimethylbutane showing its structural formula and schematic representation.

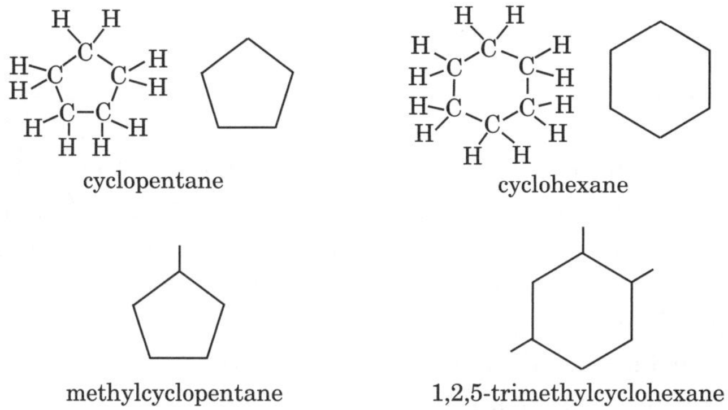

cyclopentane

cyclohexane

methylcyclopentane

1,2,5-trimethylcyclohexane

I FIGURE 15-6 Cyclopentane and cyclohexane with and without functional groups.

of forming an additional bond in the molecule without losing any atoms in the compound. For instance, ethene reacts with H_2 to produce ethane according to the reaction

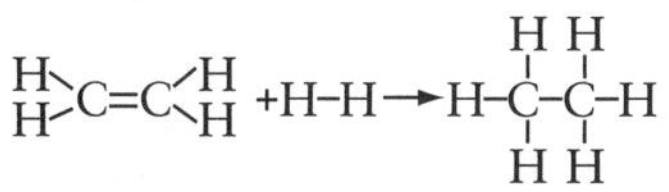

$$[15.1]$$

This reaction is termed a hydrogenation reaction. Other molecules besides H_2 react in a similar way with chemically reactive alkenes. Some simple alkenes are given in **Table 15-5**. The numbers 1 and 2 in front of butene give the location of the carbon–carbon double bond. The numbering convention starts at the end closest to the functional group with the positions numbered sequentially.

Table 15-5	Name and structure of simple alkene molecules

Systematic name	Common name	Structure	Schematic representation
Ethene	Ethylene	$\text{H} \atop \text{H}$ $>$C$=$C$<$ $\text{H} \atop \text{H}$	=
Propene	Propylene	(structure)	/=
1-Butene		(structure)	∧=
2-Butene		(structure)	/=\

Alkynes are aliphatic compounds with a carbon–carbon triple bond. Like alkenes, therefore, they are also unsaturated and capable of undergoing hydrogenation and other electron transfer reactions. Their systematic name ends in -yne. Some simple alkynes are given in **Table 15-6**. Rare in nature, they are found in polluted air.

Alcohols are a group of carbon compounds that have a hydroxyl *radical*, that is, an OH functional group bound to a carbon. They are similar to H_2O, which can be thought of as a hydrogen with a hydroxyl radical. As a result, they dissolve readily in water and are common both in nature and industry. Their systematic name ends in -anol. Methanol, also referred to as methyl or wood alcohol, is derived from the hydrolysis and fermentation of wood (**Table 15-7**). Ethanol, commonly called ethyl or grain alcohol, can be produced by bacteria from the fermentation of sugar by a reaction like

$$C_6H_{12}O_6 \rightarrow 2C_2H_5OH + 2CO_2 \qquad [15.2]$$

Also common are fatty alcohols with a chemical formula $CH_3(CH_2)_yCH_2OH$ where y is an integer from 20 to 30. They are the waxy films on many leaves and fruits and come in normal, branching, saturated, and unsaturated forms. Alcohols can also have additional OH groups.

Carboxylic acids are organic compounds that contain the *carboxyl functional group*. A carbon atom is double bonded to an oxygen and single bonded

Table 15-6   Name and structure of simple alkynes

Systematic name	Structure	Schematic representation
Ethyne	H—C≡C—H	≡
Propyne		
1-Butyne		
2-Butyne		

Table 15-7   Name and structure of common alcohols

Systematic name	Common name	Structure	Schematic representation
Methanol	Methyl alcohol		
Ethanol	Ethyl alcohol		OH
1-Propanol	n-Propyl alcohol		OH
2-Propanol	iso-propyl alcohol		OH

to a hydroxyl group (OH). As weak acids, they are partially dissociated in aqueous solutions as H^+ is lost from the OH group. They add to the acidity of soils when a carboxylic acid–containing compound is present. The anions produced with loss of the H^+ complex with positive species such as aluminum and transition metals. This increases the metal's solubility in aqueous solutions. In addition, the charged nature of the dissociated carboxylic groups increases the stability of these compounds in water over uncharged organic molecules. This occurs because the charges produced by these groups react with the charge distribution in polar water molecules. **Table 15-8** lists some simple carboxylic acids.

Amino Acids and Proteins

Amines are organic derivatives of ammonia (NH_3). They are called primary, secondary, or tertiary, depending on the number of radicals they possess, as shown in **Figure 15-7**, where the R stands for an attached radical, for instance, a carbon group. A radical as used in chemistry is a group of atoms bound together as a single unit and forming part of a larger molecule. Most of the nitrogen in organisms is tied up in the 20 primary amines of common *amino acids*. Amino acids have an amine group and a carboxylic group bonded to a common carbon atom. The structural formulas of amino acids are shown in **Figure 15-8**. Amino functional groups, $-NH_3^+$, are important locations of positive charge

| Table 15-8   Name and structure of some simple carboxylic acids

Systematic name	Common name	Structure
Methanoic acid	Formic acid	$H-\overset{\displaystyle O}{\underset{\displaystyle OH}{C}}$
Ethanoic acid	Acetic acid	$H-\overset{\displaystyle H}{\underset{\displaystyle H}{C}}-\overset{\displaystyle O}{\underset{\displaystyle OH}{C}}$
Ethanedioic acid	Oxalic acid	$\overset{\displaystyle OH}{\underset{\displaystyle O}{C}}-\overset{\displaystyle O}{\underset{\displaystyle OH}{C}}$
Propanoic acid	Propionic acid	$H-\overset{\displaystyle H}{\underset{\displaystyle H}{C}}-\overset{\displaystyle H}{\underset{\displaystyle H}{C}}-\overset{\displaystyle O}{\underset{\displaystyle OH}{C}}$

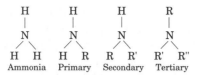

FIGURE 15-7 Structure of ammonia along with primary, secondary, and tertiary amines. R represents an attached radial or functional group.

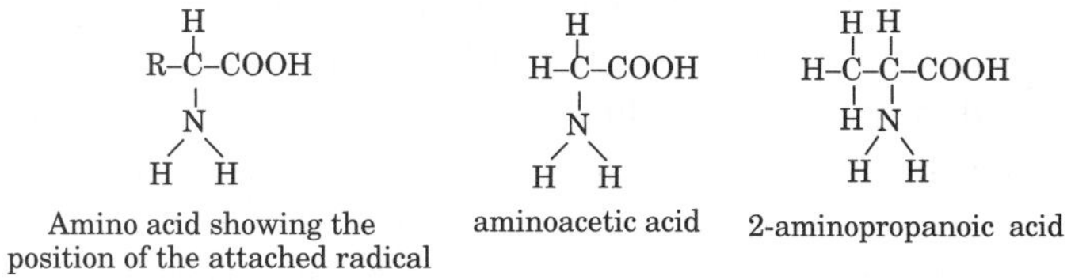

FIGURE 15-8 Structures of some simple amino acids.

on cell walls and in other cellular locations. When these groups deprotonate by the reaction $-NH_3^+ \rightarrow -NH_2 + H^+$ as *pH* increases, they become neutrally charged, similar to the transformation of ammonium, NH_4^+ to ammonia, NH_3 in aqueous solutions.

Long chains of amino acids are known as *peptides*. When chains reach 50 or more carbons in length, they are termed *proteins*. In animals, protein chains are folded to produce fibers that serve as supportive tissue in bone (collagen) and claws and hooves (keratin). This supportive tissue plays the same role as cellulose in plants. Globular proteins include hormones, enzymes, and antibodies. DNA and RNA are proteins responsible for carrying the genetic code for reproduction of life.

Carbohydrates

Carbohydrates are a large class of organic compounds that contain only C, H, and O. As implied by their name, their general formula is $C_m(H_2O)_n$, where *m* and *n* are generally large integers. They are the main constituent of plants and are formed by plants during photosynthesis. Although the details of the reactions that occur are complicated, the overall reaction for oxygenic photosynthesis to form carbohydrates, $C_m(H_2O)_n$, is described by

$$nH_2O + mCO_2 + sunlight \rightarrow C_m(H_2O)_n + mO_2 \qquad [15.3]$$

Different reaction pathways exist for Reaction 15.3 depending on the plant. The 3-carbon C_3 Calvin cycle is used by marine plants and most terrestrial plants. Maize, sugarcane, sorghum, and hot-weather grasses, however, use the 4-carbon C_4 Hatch-Slack cycle to produce carbohydrates during photosynthesis. C_4 plants do not change their rates of growth with changes in atmospheric

CO_2 as readily. The crassulacean acid metabolism (CAM) pathway is more similar to C_4 but tends to produce carbohydrates more enriched in deuterium relative to the water used. This occurs because plants using the CAM pathway include succulents from water-limited environments. One of the distinguishing characteristics is that the C_4 and CAM pathways produce a smaller carbon isotopic fractionation ($\delta^{13}C$ from −9‰ to −16‰, averaging about −13‰ at 25°C) for the incorporation of atmospheric C from CO_2 into carbohydrates than the C_3 pathway ($\delta^{13}C$ from −23‰ to −33‰, averaging about −26‰ at 25°C).

The fractionation difference has implications for using carbon isotopic systematics for determining paleoenvironmental conditions. Mammalian herbivores produce about a +11‰ fractionation to tooth apatite. Analysis of $\delta^{13}C$ in teeth, therefore, is used to help put constraints on their diet of C_3 and C_4 plants (Koch et al., 1994). **Figure 15-9** shows the range of values of $\delta^{13}C$ for various carbon reservoirs on earth. Organic matter and its products are more negative than atmospheric CO_2 ($\delta^{13}C = −7$) and marine carbonates ($\delta^{13}C \sim 0$), irrespective of age.

Carbohydrates have building blocks of either five or six carbon units and are divided into what are called *monosaccharides*, disaccharides, and *polysaccharides*. Monosaccharides have both single chain and ring structures that bind together to form what are termed disaccharides. When three or more monosaccharide units are bonded together, the carbohydrate is termed a polysaccharide. The binding reaction is written as

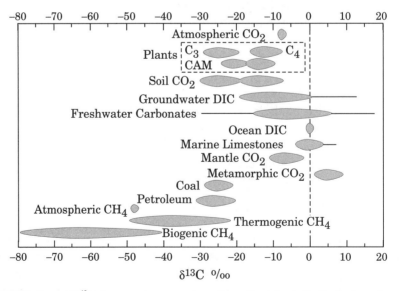

FIGURE 15-9 Range of $\delta^{13}C$ of some carbon reservoirs on the earth relative to the Peedee formation belemnite standard.

H H H OH H H
| | | | | |
H – C – C – C – C – C – C = O
| | | | |
OH OH OH H OH

H H H OH H H
| | | | | |
H – C – C – C – C – C – C – OH
| | |
OH   OH H OH
|_____ O _____|

Glucose ($C_6H_{12}O_6$)   Glucose (monosaccharide ring) ($C_6H_{12}O_6$)

I FIGURE 15-10 Two forms of glucose.

$$mC_6H_{12}O_6 \rightarrow (C_6H_{10}O_5)_m + mH_2O \qquad [15.4]$$
glucose   polysaccharide

Glucose is a simple six-carbon monosaccharide sugar whose structural formula is given in **Figure 15-10**. Sucrose, a simple disaccharide sugar, is two monosaccharide rings (glucose and fructose) bonded together with the removal of an H_2O as indicated in Reaction 15.4. Common polysaccharides are complex sugars, cellulose (units with molecular weights of 300,000–5,000,000), or starch. Chiton, the hard structural material in arthropods and mollusks, is also a polysaccharide.

Cellulose, $(C_6H_{10}O_5)_n$, is the tough insoluble constituent of the cell wall of plants consisting of long, straight, and unbranched units of thousands of glucose ring units in length bonded by Reaction 15.4. Wood is 50% cellulose, whereas cotton is 100% cellulose. Cellulose has glucose units attached by β-linkages, where oxygens bound between units to the same carbon are adjacent to each other. Starches of the chemical formula $(C_6H_{10}O_5)_n$ are branched forms of fused glucose units that are a major storage of carbohydrates as food supplies in plants. The potato is nearly 100% starch. The glucose units are bound through a carbon where oxygens are opposite each other, termed an α-linkage.

Starch is then a *stereoisomer* of cellulose. This stereochemical difference means that different enzymes are required to break down starch as opposed to cellulose. Humans and many animals do not have the cellulase enzymes in their digestive tract necessary to break down cellulose. Both cellulose and starch are broken down by metabolism into glucose units before the heat is released as an energy source during glucose oxidation. Starches break down more rapidly than cellulose in oxygenated environments and, therefore, are generally poorly preserved in soils and in later diagenesis.

Petroleum and Lipids

Lipids are a heterogeneous collection of fats, oils, and waxes. Like carbohydrates they contain only C, H, and O; however, H relative to O in lipids is greater than the 2:1 ratio in carbohydrates. Lipids include fatty acids such as oleic acid ($C_{17}H_{33}COOH$) found in plants and stearic acid ($C_{17}H_{35}COOH$) found in animals. They are substances that have low solubility in water because

---

Major Components of Crude Oil

1. Alkanes from CH_4 (methane) to $C_{70}H_{142}$. "Sweet oil" has many short chain alkanes. Crude oil is "cracked" to break long chains into octane (C_8H_{18}) length chains. Gasoline is made up of chains between C_5H_{12} and $C_{10}H_{22}$. The chains in crude oil are generally separated by boiling. The boiling point increases with length of chain, with octane at 126°C, kerosene, C_{12}, at 210°C, and residue chains of C_{36} and up above 500°C.
2. Cycloalkanes, alkanes in a circular structure; primarily cyclopentane and cyclohexane both of which are rather unreactive.
3. Aromatic hydrocarbons (compounds with benzene rings) are separated on boiling crude oil when gasoline is produced.

---

of the limited charge from disassociation of their carboxyl group but dissolve in organic solvents such as chloroform or acetone.

Petroleum is formed from oceanic organisms that have a larger component of lipids and therefore less oxygen per carbon than the $C_6H_{10}O_5$ of carbohydrates found as the major component of land plants. Organic-rich shales are produced from the slow deposition of oceanic microscopic organisms, generally the remains of phytoplankton. These are deposited with clay minerals in shales and become the main source of the compounds that produce petroleum. With burial and heating, the organisms are broken apart and produce shorter chained, lower O to H, lipid-like liquids.

These liquids do not dissolve in the interstitial H_2O but create a separate fluid phase. This liquid-oil phase is less dense than water and migrates out of the shale. The lipid-rich fluid produced in shale is collected and concentrated in rocks above, which have a significant interconnected pore space, to become crude oil. To become a deposit, the oil also needs to be trapped below an impermeable layer (often a capping shale layer). If it is not trapped, it rises to the top of the water table where its interaction with oxygenated water causes a tar to develop. This is what happened in places like the LaBrea tar pits in California.

■ Organic Carbon in Natural Waters

Organic carbon in water is divided into dissolved organic carbon (DOC), which passes through a 0.45-μm filter, and particulate organic carbon (POC), which does not. Most natural waters have a larger fraction of DOC than POC, although rivers with a high suspended load typically contain a large fraction of POC. Disassociated *humic* and *fulvic acids* (see below) generally dominate DOC and give organic-rich waters their yellow color. Carbohydrates and hydrocarbons are also present in DOC. Zooplankton, phytoplankton, and bacteria are found in POC.

Seawater at the interface with the atmosphere contains a thin microlayer with DOC of 1.4 to 18 mg l^{-1}. With depth, seawater in the layer above the thermocline has a range of DOC between 0.3 and 2.0 mg l^{-1}, with POC ~0.1 mg l^{-1}. This decreases with depth so that below 300 m DOC is between 0.2 and 0.8 mg l^{-1} and POC ~ 0.01 mg l^{-1}. Rainfall above continental areas has an average DOC of 1 mg l^{-1}. Rain that falls through a canopy of trees, however, often has a DOC of 2 to 3 mg l^{-1}. If the rainfall drips from the trees, its DOC increases to 5 to 25 mg l^{-1}. Soil waters have a range of DOC from 2 to 30 mg l^{-1}. The disassociated humic and fulvic acids in these waters complex Al and Fe in solution. The Al and Fe are transported downward in the soil where they reach clay layers. The surfaces of the clays typically adsorb the Al and Fe from solution as well as organic ligands. The DOC of soil water then decreases as soil water is incorporated into groundwater. Groundwater generally has a range of DOC from 0.2 to 15 mg l^{-1}, although those associated with coals and oil shales range from 2 to 10 mg l^{-1}.

River water DOC depends on both the climate and season because values depend on the amount of primary production of plant matter and its decomposition rate. Rivers in cool temperate climates have a DOC from 2 to 8 mg l^{-1}, whereas in warm temperate climates the concentration is generally higher. Rivers draining wetlands have DOCs of up to 60 mg l^{-1}.

■ Diagenesis of Organic Material

Life forms are produced by the input of energy. In the case of photosynthetic organisms, this is energy from the sun. When these organism die, the resulting organic molecules in the environment are high energy and unstable. They will release energy with time as they revert to simpler lower energy compounds. The transformation of organic material during burial and diagenesis is promoted by biologic agents. These organisms use the energy released during decay to perform cellar reactions (see below). Decomposers break down high molecular weight compounds into assailable components. Lipids appear more resistant to degradation than carbohydrates and proteins. Microbial activity, however, reduces the concentrations of all of the types of compounds.

Humic substances are the degradation-resistant residues from plant decay as outlined in **Figure 15-11**. They are the major component of peat deposits and the precursor to coal deposits. Humic substances have a high molecular weight with a highly aromatic character composed dominantly of C (45–55%), O (30–45%), H (3–6%), and N (1–5%). They are commonly characterized based on their solubility in strong base. The insoluble plant material in solution is

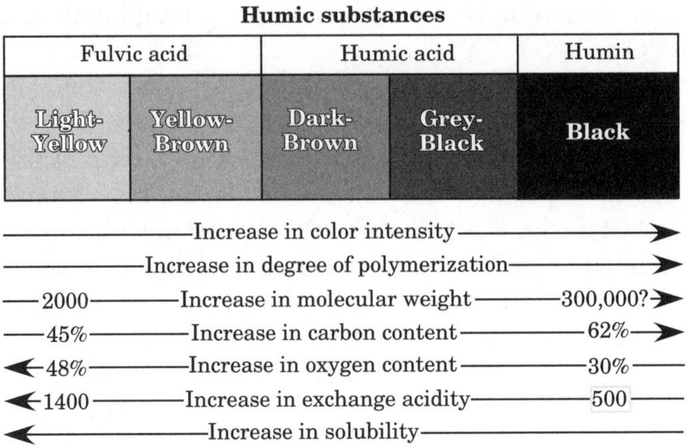

Humic substances

Fulvic acid		Humic acid		Humin
Light-Yellow	Yellow-Brown	Dark-Brown	Grey-Black	Black

---Increase in color intensity----→

---Increase in degree of polymerization----→

--2000---Increase in molecular weight---300,000?-→

--45%---Increase in carbon content---62%--→

←-48%---Increase in oxygen content---30%---

←-1400---Increase in exchange acidity---500---

←---Increase in solubility---

I FIGURE 15-11 Makeup of humic substances. (Adapted from Stevenson, 1994.)

considered *humin*. The dissolved material is then acidified. What precipitates out of solution is termed humic acid, and that which remains in the acid solution is termed fulvic acid. Fulvic acids generally have a lower molecular weight than humic acids, with a greater number of carboxyl groups. It is the charge that occurs from disassociation of H^+ from these groups that makes fulvic acid soluble in H_2O. **Figure 15-12** shows the types of acid groups that occur on natural organic matter. One should also note that bacteria, algae, and fungi all have cell walls that contain functional groups similar to those shown in **Figure 15-13** and are therefore important in binding metals from the aqueous solution with which they are in contact.

Figure 15-13 shows a tabulation of $-\log K$ (pK_a) values of the ionization equilibrium constants for reactions of the acid groups on organic matter such as

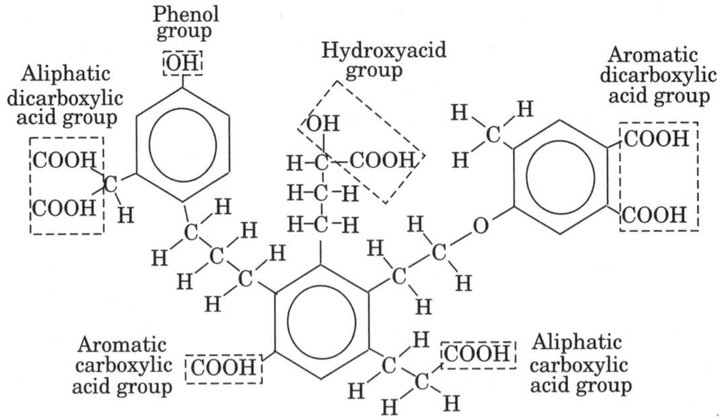

I FIGURE 15-12 Various acid groups on a complex organic polymer.

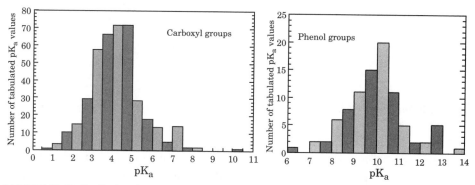

$$R-COOH = R-COO^- + H^+ \qquad \log K_{R-COOH} = \log \frac{a_{R-COO^-}\, a_{H^+}}{a_{R-COOH}} \qquad [15.5]$$

and

$$Ar-OH = Ar-O^- + H^+ \qquad \log K_{Ar-OH} = \log \frac{a_{Ar-O^-}\, a_{H^+}}{a_{Ar-OH}} \qquad [15.6]$$

where R and Ar are aliphatic and aromatic ring frameworks, respectively. The pK$_a$ then would be the *pH* where there are equal activities of charged and uncharged species. The location of the carboxyl and *phenol* groups affects their pK$_a$ values. If they possess dicarboxyl groups, as outlined in Figure 15-12, their pK$_a$ values are lower. If soil waters have a *pH* between 5 and 8, most of the carboxyl groups are disassociated and negatively charged, whereas the phenol groups remain associated and uncharged. The negative charges produced on the organic compounds increase their solubility in water because of the dielectric properties of water, as outlined in Chapter 6, on aqueous solutions. In addition, the H$^+$ contributed to solution lowers the *pH* of the water and increases its ability to weather minerals during diagenesis.

Humin is typically insoluble in most waters, and its effect on water composition is through surface exchange processes. The fulvic acid component of humic substances dissolves as a number of high molecular weight aqueous species in water that have many similar properties. Often they bond +2 and +3 charged metal ions in solution as a *chelation* between two carboxyl groups on the humic substance or as a complex on a single carboxyl group. These include some of the rare metals such as Mo, V, Pb, and Cd as well as Ni, Fe, and Al. **Figure 15-14** shows a chelate produced between two carboxyl groups and a complex with a single carboxyl group on an aromatic compound.

After diagenesis, the small amount of sedimentary organic material in rock that is soluble in organic solvents is termed *bitumen*, whereas the remainder

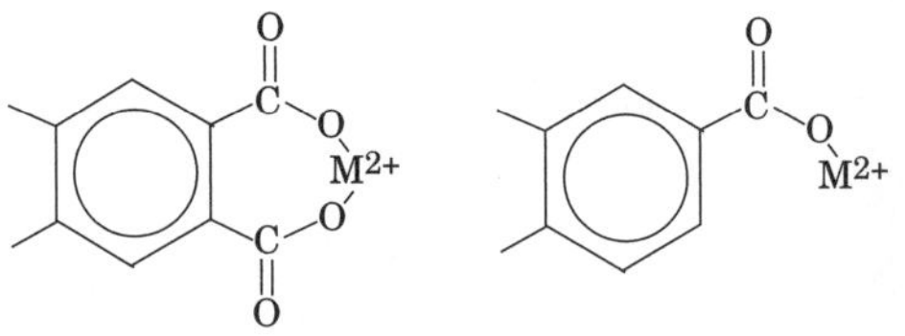

FIGURE 15-14 Binding of a M^{2+} metal ion by chelation between two carboxyl groups and by complexation with one carboxyl group.

is termed *kerogen*. Bitumens have a large lipid component, and the shorter chains become liquid petroleum upon further diagenesis. Kerogen is an aggregation of a variety of organic macromolecules. If it has significant aliphatic compounds from phytoplankton remains, it is a good source for crude oil. If it is dominated by aromatic compounds with identifiable plant remains, it has poor crude oil potential. Kerogen is classified based on the atomic H/C and O/C ratios. Type I kerogen has H/C $\geq$ 1.5 and O/C $\leq$ 0.1. This has high petroleum potential. The Eocene Green River Formation in the western United States has type I kerogen. Type II kerogen has O/C $\leq$ 2 but H/C near 1.25, with intermediate petroleum potential.

Type III kerogen, derived from land plants, has H/C $\leq$ 1.0 and O/C > 2, giving it a poor petroleum potential. It produces some natural gas during thermal maturation, however, if buried to depths of 3 or 4 km where temperatures are between 100°C and 150°C. Typically, type III kerogen is derived from swamps where organic plant matter dominates the depositional environment. It becomes peat and then brown coal on burial with the loss of water during diagenetic reactions. If the swamp is in an estuarian environment, the coal has a high sulfur content. In deposits dominated by plant organic matter with diagenesis, as the temperature increases from 40°C to about 100°C the O/C ratio decreases to less than 0.1. At this point, the organic-rich deposit is termed bituminous coal because it still has significant bitumen content. Above 100°C methane is given off, which decreases the bitumen content, and the H/C ratio falls to less than 0.001. At this point, the coal is termed anthracite. Above about 150°C, the coal structure is destroyed, and the remaining carbon transforms into a graphite type structure.

■ Aquatic Microbial Chemistry

Algae are the primary producers of organic material in water. They are *eukaryotic* microscopic organisms that subsist on inorganic nutrients and through added energy produce organic matter. If the energy is from the sun, they are called photosynthetic. Algae also metabolize organic matter without the sun by using the stored energy in starches and oils. Algae may be single-cell organisms, but most have simple multicellular structures and consist of large colonies, as in seaweed. Fungi are nonphotosynthetic organisms typically composed of filaments that resemble plants. They are *heterotrophic*, growing in and through their food, and are a major decomposer of land plants.

Bacteria are single-cell *prokaryotic* microorganisms with a typical size range of 0.1 to 50 µm. They are divided into *autotrophic* and heterotrophic types. Autotrophic bacteria obtain their nutrients from inorganic sources. Heterotrophic bacteria use organic compounds. A division of bacteria is also made on whether they require oxygen as an electron receptor (see Chapter 14). Aerobic bacteria require oxygen, whereas anaerobic bacteria exist in the absence of molecular oxygen; molecular oxygen is toxic to anaerobic bacteria. Clays, because of their large surface areas, are the sites of many biologic reactions. The soil zone is the most biologically active region on earth. Mineral particles have organic coatings bonded through hydrogen and covalent oxygen bonds. Bacteria also attach as coatings on mineral grains through filaments and mucilage excrement composed of a polysaccharide compound.

The bacteria both feed on the organic matter and obtain nutrients from the mineral *substrate*. After the initial partial decay, most reactions that occur in organic matter do so through a bacterial intermediary. These are concentrated near the soil surface because the number of bacteria typically decreases with depth in the soil; however, they have been reported at depths as great as 23 km.

Viruses are submicroscopic infectious entities without an independent metabolic mechanism. They are about two orders of magnitude smaller than bacteria, being on the order of 100 Å in size. They must have a host to survive and reproduce and are therefore not a living organism. A virus consists of DNA or RNA inside a protein coating. Viruses infect both plants and animals and in fact are commonly found on bacteria. A liter of seawater at times contains 10^{11} viral particles.

■ Hazardous Organic Compounds

Numerous artificial organic compounds have been synthesized. *Chlorofluorocarbons* (CFCs), developed by DuPont in the 1930s under the trade name Freon, are one such set of these compounds. Two simple CFCs are shown in **Figure 15-15**. CFCs are completely substituted methane compounds and were developed as a stable, inert, nontoxic fluid that can be used safely in industry and homes. They are excellent refrigerants. For instance, liquid CFC-12 is sprayed into a chamber in which warm interior air is blown. The droplets of liquid CFC-12 absorb a large amount of heat from air when they boil (i.e., evaporate) at temperatures above −30°C at 1 atm. The heat containing CFC-12 gas that is produced is transported outdoors, where a condensing unit com-

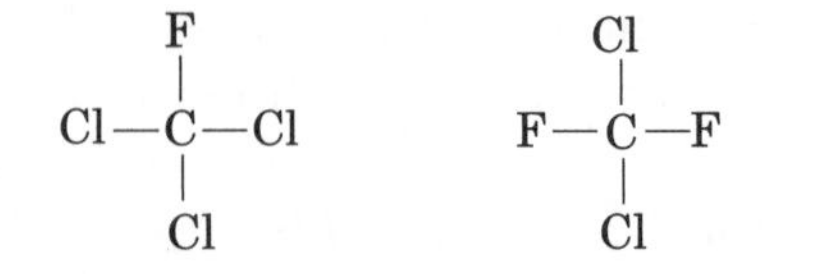

$$
\begin{array}{cc}
\quad\text{F} & \quad\text{Cl} \\
\quad| & \quad| \\
\text{Cl—C—Cl} & \text{F—C—F} \\
\quad| & \quad| \\
\quad\text{Cl} & \quad\text{Cl} \\
\text{CCl}_3\text{F (CFC-11)} & \text{CCl}_2\text{F}_2\text{ (CFC-12)}
\end{array}
$$

I FIGURE 15-15 Two common CFCs.

presses the CFC-12 gas, returning it to a liquid with the release of its heat. The process is repeated, cooling the interior space. CFCs have also been used as a foaming agent. If CFC-12 is vaporized from a liquid state in liquid plastic, it creates bubbles. On cooling of the plastic, rigid plastic foam is produced. This rigid foam has found great use as trays for meat and fast food packaging as well as in home insulation. CFC-11 and CFC-12 have also been used as aerosol spray propellants (e.g., deodorant and paint). At room temperature under pressure, they are liquid that is transformed to a large volume gas when the pressure is released through the spray nozzle.

Their stability gives CFC gas long lifetimes in the atmosphere. The average atmospheric residence time of CFC-11 is 65 years and of CFC-12 is 130 years. Because they are not easily broken down in the lower atmosphere, there is a significant flux into the stratosphere. In the stratosphere, CFCs are disassociated because of the more intense short wavelength solar radiation present that cleaves a Cl atom from the compound. The Cl atom then reacts with ozone, O_3, according to the reaction

$$O_3 + Cl \rightarrow O_2 + ClO \tag{15.7}$$

followed by the reaction

$$ClO + O \rightarrow O_2 + Cl \tag{15.8}$$

The Cl, therefore, destroys the ozone molecule and is reconfigured; thus, it continues to destroy ozone molecules.

Because of the destruction of stratospheric ozone, CFCs are being phased out in the United States and Europe. In their place, hydrogenated CFCs are now being used (**Figure 15-16**). Because of the hydrogen in hydrogenated CFCs, they are thought to break down more easily in the lower atmosphere than CFCs, and, therefore, insignificant amounts reach the stratosphere. One problem with hydrogenated CFCs is that they are significantly more expensive to produce than CFCs and require refitting of existing air-conditioning units that run on CFCs.

Polychlorinated biphenyls (PCBs) are another common type of artificial organic. These are very stable aromatic organic compounds consisting of two linked benzene rings, as shown in **Figure 15-17**. Because benzene rings are referred

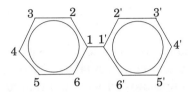

CHF_2Cl (HCFC-22)
Substituted methane

CCl_2FCH_3 (HCFC-141b)
Substituted ethane

I FIGURE 15-16 Two hydrogenated CFCs.

I FIGURE 15-17 PCBs indicating the position from two to six on each phenyl group, of the possible 10 locations of Cl substitution for H in the structure.

to as phenyl groups when they are radicals on another compound, the two linked phenyl groups are termed a biphenyl. PCBs are chemically very stable even at elevated temperatures and have high electrical resistance. PCBs are used in transformers to separate the metal plates and added to lubricating and cutting oils to increase their thermal stability. PCBs have 1 to 10 Cl substituting for H at the indicated locations (Figure 15-17). In the early 1970s, PCBs were found in the fat of animals in the Arctic and Antarctic. This increased concentration in animal fat is called *biomagnification* and occurs when animals existing at higher levels in the food chain consume contaminated animals and concentrate PCBs in their fat. High concentrations of PCB have been shown to decrease the reproductive rate in animals. A concern developed when a high concentration of PCBs was found in human breast milk ($10 \ \mu g \ dm^{-3}$). The manufacture of PCBs was banned by international agreement in the late 1970s. They still exist, however, in some landfills, as well as in sediments at the bottom of the ocean, lakes, and rivers.

Polycyclic aromatic hydrocarbons (PAHs) are a set of potentially harmful organic compounds produced as a byproduct when carbon-containing materials are incompletely burned. They are two or more fused aromatic (benzene)

I FIGURE 15-18 Two different representations of the naphthalene molecule.

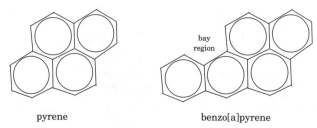

I FIGURE 15-19 Structure of pyrene and benzo[a]pyrene.

rings. Other than naphthalene (two fused rings, **Figure 15-18**), they are not manufactured because they have no commercial use. With four or fewer rings, they remain as gases. When compounds with greater than four rings are produced, they quickly condense and become adsorbed on atmospheric dust particles. PAHs are transported into the lungs by breathing in the dust particles. Soot itself is mainly a graphitic-like PAH carbon compound condensing as a collection of tiny crystals. Each crystal is composed of stacks of planar layers of fused aromatic rings. Graphite is, therefore, the ultimate PAH.

Natural sources of PAHs are from forest and prairie fires and volcanic eruptions. Anthropogenic sources include motor vehicles (particularly diesel engines), fossil fuel furnaces, cigarettes, burning of charcoal briquettes, and asphalt manufacturing plants. There is a relation between PAHs and cancer in that some PAHs are carcinogenic. The most notorious and common is benzo[a]pyrene, shown in **Figure 15-19**. The relative position in space of the fused rings in PAHs determines their level of carcinogenic behavior. Most potent carcinogens possess a bay region. They become carcinogenic because the bay regions are highly biochemically reactive.

Pesticides

Many organic pesticides have been developed to kill or otherwise control (e.g., interfere with reproductive process) a population of unwanted insects or plants. These include insecticides, herbicides, and fungicides. One billion kilograms of pesticides are used annually in North America, with about half used in agriculture. Most insecticides are used in growing corn, whereas most herbicides are used in growing corn and soybeans.

The active ingredients of insecticides are typically organochlorines; organophosphates are also used. These are, however, more toxic to humans and other mammals. One of the most successful insecticides developed was the compound para-*d*ichloro*d*iphenyl*t*richloroethane, DDT. As shown in **Figure 15-20**, DDT is a substituted ethane, with the 3 H on one C replaced by 3 Cl and 2 H on the other C each replaced by phenyl groups. In each phenyl, a Cl replaces H in the para (two opposite) positions.

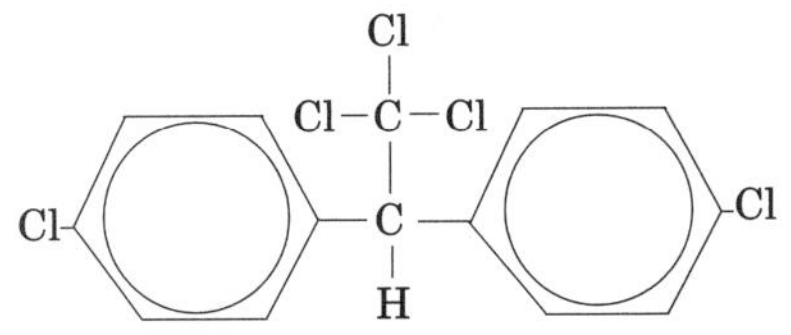

I FIGURE 15-20 DDT, paradichlorodiphenyltrichloroethane.

Insecticide Properties

1. Stability against decomposition in the environment.
2. Very low solubility in water unless oxygen or nitrogen is present in the molecule.
3. High solubility in hydrocarbon-like environments (e.g., fatty material in living matter).
4. Relatively high toxicity to insects but low toxicity to humans.

DDT is an effective insect killer because it is stable in the environment. Insects have highly permeable shells and readily absorb it on contact. The skins of higher animals prevent its intake through absorption. Once in the insect, the chlorine reacts by interfering with the Na balance in nerve membranes. DDT was thought to be the perfect insecticide until 1962. In that year, Rachel Carson published *Silent Spring*, which pointed out a decline in robin populations after DDT was used to spray for bugs carrying Dutch Elm disease. It was later shown to accumulate in the bird's body fat by biomagnification from the eating of affected insects. Its use was banned in 1972 in the United States, but it is still used in some developing countries, especially in malaria-prone regions.

Phenoxy Herbicides and Dioxins

Hexachlorobenzene, whose structure is shown in **Figure 15-21**, was used after World War II as an effective herbicide. It is easy to produce from chlorine and benzene and very soluble in organic media but nearly insoluble in water. Later experiments showed an effective weed killer could be produced by attaching an oxyacetic group ($-O-CH_2COOH$) on a benzene ring where Cl had been substituted for H. Using chlorobenzene, a phenol was first produced. A phenol is a benzene ring (phenyl group) with an OH radical replacing an H. This was synthesized by reacting chlorobenzene with NaOH at high temperature, producing the OH radical as shown in **Figure 15-22**.

The phenol compounds are mildly acidic; therefore, in strong base, the H of OH is lost as H^+ which combines with OH^- to produce H_2O. The overall reaction is

$$C_6H_5OH + NaOH \rightarrow C_6H_5ONa + H_2O \qquad [15.9]$$

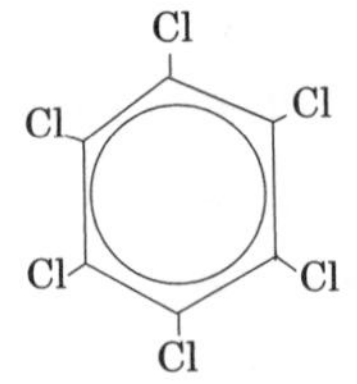

I FIGURE 15-21 Hexachlorobenzene.

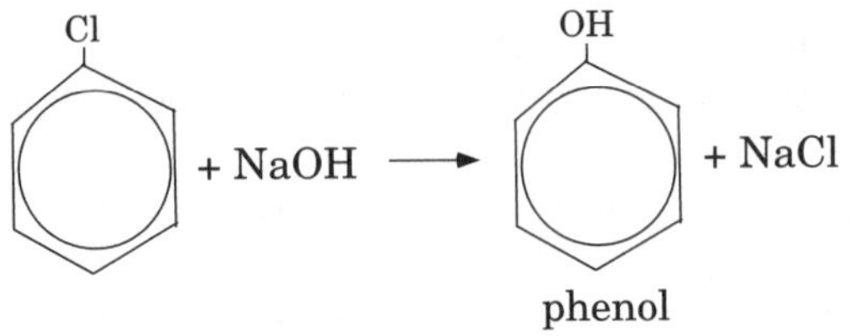

$$C_6H_5Cl + NaOH \rightarrow C_6H_5OH + NaCl$$

FIGURE 15-22 Chlorobenzene reacts with sodium hydroxide at high temperatures to produce a phenol and sodium chloride.

The Na^+ is weakly held on the phenoxy compound so that at high temperatures this is fused to an R-Cl molecule where the R is a radical:

$$C_5H_5ONa + Cl-R \rightarrow NaCl + C_6H_5O-R \qquad [15.10]$$

If the radical is oxyacetic acid and two or three H are replaced by Cl, 2, 4-dichlorophenoxyacetic and 2,4,5-trichlorophenoxyacetic acid, respectively, are produced, whose structures are given in **Figure 15-23**.

2,4-Dichlorophenoxyacetic acid and 2,4,5-trichlorophenoxyacetic are very good herbicides for both agricultural and domestic settings. In the reaction to synthesize 2,4,5-trichlorophenoxyacetic from the trichloropheno, the trichlorophenos are linked to form *dioxin* plus HCl. A dioxin is two pheno groups linked through the oxygens of the pheno radicals, as shown in **Figure 15-24**.

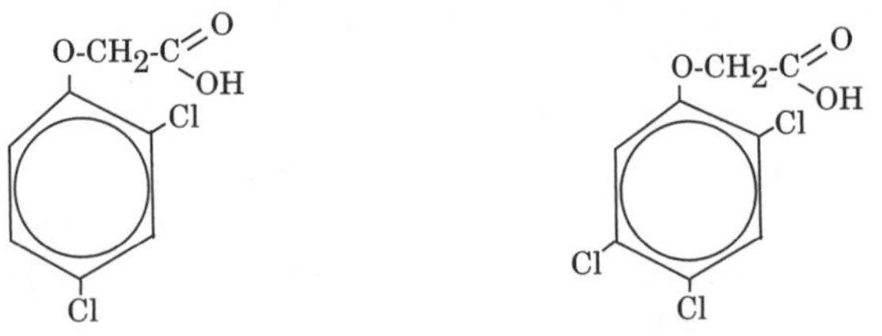

2,4 dichlorophenoxyacetic acid   2,4,5- trichlorophenoxyacetic acid

I FIGURE 15-23 Structural formulas of 2,4-dichlorophenoxyacetic acid and 2,4,5-trichlorophenoxyacetic acid.

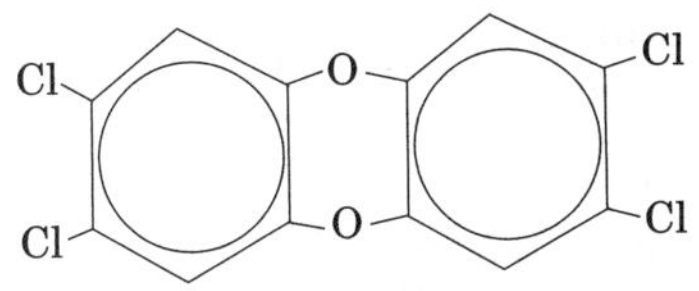

I FIGURE 15-24 Structure of tetrachloride dioxin (2,3,7,8–TCDD).

Environmental contamination by "dioxin" occurred as a result of an explosion at a chemical factory in Italy in 1976. Production of 2,4,5-trichlorophenol by the reaction outlined above was undertaken in a furnace. The reaction was not brought to completion before leaving for the weekend. Uncontrolled, it heated to produce a large amount of dioxin, and the furnace exploded. The wildlife population in the area was decimated, and humans suffered long-term health effects.

In addition to their use as starting materials in the production of herbicides, chlorophenols are used as wood preservatives (fungicides). Trichlorophenol, tetrachlorophenol, and pentachlorophenol (**Figure 15-25**) are sold as wood preservatives. Very large amounts of chlorophenols have been used for this purpose in railroad ties. They are also weak acids as an H^+ is lost from their structure in alkaline solutions. The negatively charged molecule produced complexes with metals in solution at near-neutral and high pH.

If wood with chlorophenols is burned dioxins are produced, as given by the reaction shown in **Figure 15-26**. Chlorophenols are also used in the bleaching of pulp to make white paper. Dioxins are produced when this material is burned. Dioxin production seems unavoidable whenever combustion of organic matter containing Cl, particularly chlorophenols, occurs.

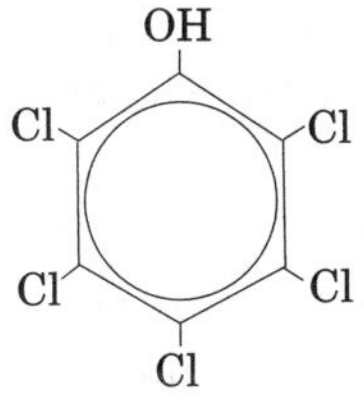

I FIGURE 15-25 PCP, pentachlorophenol.

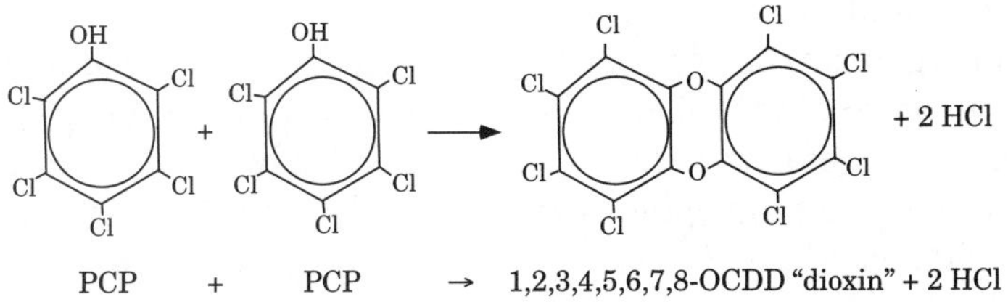

PCP   +   PCP   →   1,2,3,4,5,6,7,8-OCDD "dioxin" + 2 HCl

I FIGURE 15-26 Fusing of two pentachlorophenol to produce dioxin plus two HCl molecules.

Americans employed in industries that produce dioxins have increased levels of cancer. In the early 1970s, around Times Beach, Missouri, oil containing dioxin was used for dust control. In the surrounding area, horses died, and children got ill. The town was evacuated because of problems with skin pigmentation of newborns, retarded growth of children, and abnormal tooth development. Artificial organic compounds have improved human existence; however, an understanding of the long-term consequences is needed to use them most effectively.

■ Cellular Chemical Reactions and Kinetics

A continual supply of energy is required of cells. This is because they are non-equilibrium systems. Cellular reactions then have negative Gibbs energy changes that drive them in the forward direction allowing the synthesis of new molecules and transport of molecules and ions up concentration gradients and across membranes in the cell. In animals, energy is also required to contract proteins to produce cellular movements. Only two sources of energy are available to living organisms: sunlight (used for photosynthesis) and oxidation-reduction (redox) reactions. They both rely on the movement of electrons between species.

Photosynthetic reactions can be divided into two stages. In the first stage, termed the light reactions, red and blue light photons from sunlight promote electrons to higher energy states in chlorophyll atoms so that oxygen in H_2O is oxidized (loses 2 electrons) as it is transformed to O_2. These electrons are then transmitted along an electron transport chain until NADP+ is finally reduced (attaches a H^+ and 2 electrons) to become NADPH and an ADP is transformed to ATP in the process (see below). The NADPH and ATP produced in the light stage is then used in the second stage, the dark (without sunlight) reactions, in the Calvin cycle to reduce, that is add electrons to CO_2 to make glucose.

Glucose stores energy in a cell that can be released by oxidation when it is required. Animals obtain their energy by oxidation of the reduced carbon in a variety of carbohydrates, proteins, and fats. Oxidation reactions produce energy by giving up electrons that can then do electrochemical work. Before oxidation, complex carbohydrates are typically broken down into the simple sugars, glucose ($C_6H_{12}O_6$), and its *isomer*, fructose, by the reverse of Reaction 15.4.

How did organisms early in earth's history obtain energy for cellular functions? The ability to obtain energy from the sun through chlorophyll had not yet evolved, and therefore, the carbohydrates needed by animals were not present. One way early heterotrophic organisms could obtain energy is by the conversion of CO to acetic acid (CH_3COOH) in the redox reaction

$$\overset{+2}{4CO}_{(aq)} + 2H_2O \rightarrow \overset{+4}{2CO}_{2(aq)} + \overset{-3\ +3}{CH_3COOH} + \Delta E \qquad [15.11]$$

The numbers above the carbons in the reaction give its oxidation state in the compound (oxidation-reduction and valence in organic compounds are outlined in Chapter 14). ΔE stands for the energy released. With the standard Gibbs energy of $CO_{(aq)}$ equal to $-29,990$ cal mol^{-1} and using the values for the other species from SUPCRT92, nearly 46 kcal mol^{-1} are available for metabolic reactions at standard conditions. The acetic acid produced can then be stored. When needed, acetic acid can be used to produce 9.6 kcal mol^{-1} at standard conditions for latter metabolic reactions by breakdown to methane gas and aqueous carbon dioxide, as given by

$$\overset{-3\ +3}{CH_3COOH} \rightarrow \overset{-4}{CH}_{4(g)} + \overset{+4}{CO}_{2(aq)} + \Delta E \qquad [15.12]$$

Early organisms could therefore produce their needed energy and have a molecule, acetic acid, which could store energy for later use. In this way, early life could survive as a *chemoautotrophic* organism.

The oxidation (loss of electrons) of simple sugars or other reduced substrate species to obtain energy in a cell is termed *cellular respiration*. Acceptors of substrate electrons can be O_2 (aerobic respiration), NO_3^- (denitrification), SO_4^{2-} (sulphate reduction), Fe^{3+} (iron reduction), Mn^{4+} (manganese reduction), or more oxidized organic molecules (see the section on Eh in sedimentary fluids in Chapter 14). Before discussing the reactions that produce energy in cells, it is helpful to first introduce the reactions and molecules that "store" energy in nearly all organisms that have existed on the earth.

ATP to ADP Reaction

Within cells, most energy is stored in the molecule *adenosine triphosphate (ATP)*. The highly negatively charged ATP molecule ($C_{10}H_{16}N_5O_{13}P_3^{4-}$) consists of a central five-membered sugar ring (ribose) with one side attached to a double ring of nitrogen and carbon atoms (adenine) and the other side to a string of three phosphate groups. The phosphates are where the negative charge centers occur and are the key to the activity of ATP. ATP works by losing the end-most phosphate group when instructed to do so by an enzyme. The reaction products are **adenosine diphosphate (ADP)**, $C_{10}H_{12}N_5O_{10}P_2^{3-}$, and a phosphate group. This loss of an electron from ATP when the oxidized product, ADP, is produced releases energy, which the organism can then use to build proteins, contract muscles, and so forth. The phosphate group ends up primarily as the dihydrogen phosphate ($H_2PO_4^-$) ion with near neutral *pH* conditions inside the cell or attaches itself to another molecule (e.g., an alcohol); however, at 25°C, the *pH* above which monohydrogenphosphate (HPO_4^{2-}) becomes the dominant phosphorous species, is only 7.2.

The energy plus ADP producing reaction can be written as

$$C_{10}H_{12}N_5O_{13}P_3^{4-} + H_2O \rightarrow C_{10}H_{12}N_5O_{10}P_2^{3-} + H_2PO_4^- + \Delta E \qquad [15.13]$$
$$\text{ATP} \qquad\qquad\qquad \text{ADP}$$

where $\Delta E = 7.3$ kcal mol^{-1}. If the cell requires more energy, ADP can convert to *adenosine monophosphate (AMP)* by oxidizing another phosphorous group on the molecule according to the reaction

$$C_{10}H_{12}N_5O_{10}P_2^{3-} + H_2O \rightarrow C_{10}H_{12}N_5O_7P^{2-} + H_2PO_4^- + \Delta E \qquad [15.14]$$
$$\text{ADP} \qquad\qquad\qquad \text{AMP}$$

When the cell does not require energy, the reverse of Reactions 15.13 and 15.14 can take place with $H_2PO_4^-$ being reduced and reattached to the molecule using energy obtained from cellular respiration (see below). Like a rechargeable battery, the ATP molecule can "store" energy, rapidly release energy when the cell requires it, and then "recharge" AMP and ADP to ATP by reattaching phosphate groups. This process typically occurs thousands of times a day on a single ATP molecule in an active cell.

NADH to NAD+ Reaction

Another energy intermediate that stores energy is **h**ydrogenated **n**icotinamide **a**denine **d**inucleotide (NADH). It can release energy by oxidizing to produce nicotinamide adenine dinucleotide (NAD+) by the reaction

$$C_{21}H_{28}N_7O_{14}P_2^{2-} + H^+ + \text{Organic-molecule} \rightarrow$$
$$\text{NADH}$$

$$[15.15]$$

$$\text{Organic-molecule-}H_2 + C_{21}H_{27}N_7O_{14}P_2^- + \Delta E$$
$$\text{NAD+}$$

NADH is oxidized, by giving up two electrons and one proton, in the reaction that produces NAD+. The two electrons and one proton together with an H^+ from solution attach to an organic molecule as a H_2 group. The organic molecule is therefore reduced. The energy stored in NADH is released as NAD+ is produced and then stored in ATP by converting AMP and ADP to ATP by the reverse of Reactions 15.14 and 15.15. In plants, NAD+ and NADH are replaced by nicotinamide adenine dinucleotide phosphate NADP+ ($C_{21}H_{29}$ $N_7O_{17}P_3^-$) and hydrogenated nicotinamide adenine dinucleotide phosphate NADPH ($C_{21}H_{30}N_7O_{17}P_3^{2-}$). They have similar properties and functions but contain an extra phosphate group.

Cellular Respiration

Organisms can obtain energy from breaking down and oxidizing organic molecules through cellular respiration. Much of this energy is first stored in ATP and NADH before it is used to do cellular work. The first stage of cellular respiration is the breakdown of glucose, *glycolysis*. This begins in the *cytosol* of a cell, as shown in **Figure 15-27**. The cytosol is the internal fluid of the cell outside of the *organelles*. Glycolysis starts with the cleaving of glucose with the energy derived from the oxidation of 2 ATP to 2 ADP producing two glyceraldehyde 3-phosphates (G3P) by the reaction:

$$C_6H_{12}O_6 + 2C_{10}H_{12}N_5O_{13}P_3^{4-} + 2H^+ \rightarrow 2C_3H_7O_6P + 2C_{10}H_{12}N_5O_{10}P_2^{3-} + \Delta E \qquad [15.16]$$

Glucose $\qquad$ ATP $\qquad\qquad\qquad\qquad$ G3P $\qquad\qquad$ ADP

Glycolysis continues with the oxidation of G3P to produce pyruvate. This reduces 2 ADP to 2 ATP and NAD+ to NADH in the process by the overall reaction

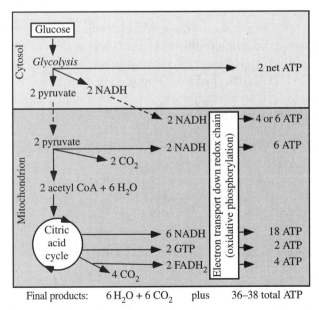

Final products: $6\,H_2O + 6\,CO_2$ plus   36–38 total ATP

FIGURE 15-27 Glucose oxidation reaction occurring both in the cytosol and a mitochondrion in the cell which energize the production of ATP from ADP.

$$C_3H_7O_6P + 2C_{10}H_{12}N_5O_{10}P_2^{3-} + C_{21}H_{27}N_7O_{14}P_2^- + H_2PO_4^- \rightarrow$$

$\quad$ G3P $\qquad\qquad$ ADP $\qquad\qquad$ NAD+

[15.17]

$$CH_3(C=O)_2O^- + 2C_{10}H_{12}N_5O_{13}P_3^{4-} + C_{21}H_{28}N_7O_{14}P_2^{2-} + 3H^+ + H_2O + \Delta E$$

$\quad$ Pyruvate $\qquad\qquad$ ATP $\qquad\qquad$ NADH

Therefore, the conversion of one mole of glucose to two moles of pyruvate is accompanied by the net production of two moles each of the energy storing molecules ATP and NADH.

The pyruvate molecules produced enter *mitochondria* in the cell. Mitochondria are membrane bound organelles that number between a few hundred to thousands in an active cell as they continue to fission and fuse together. With the aid of coenzyme A (CoA), also called pyruvate dehydrogenase, a molecule of CO_2 is removed from pyruvate by the reaction

$$CH_3(C=O)_2O^- + C_{21}H_{36}N_7O_{16}P_3S + C_{21}H_{27}N_7O_{14}P_2^- \rightarrow$$

$\quad$ Pyruvate $\qquad\qquad$ CoA $\qquad\qquad$ NAD+

[15.18]

$$C_{21}H_{35}N_7O_{16}P_3S-C_2H_3O + CO_2 + C_{21}H_{28}N_7O_{14}P_2^{2-} + \Delta E$$

$\quad$ Acetyl CoA $\qquad\qquad\qquad$ NADH

The coenzyme CoA becomes Acetyl CoA, a pyruvate dehydrogenase with an acetyl group attached on its sulfur atom.

Acetyl CoA then enters the *citric acid metabolic cycle*, where it transfers its acetyl group to an acetate compound to produce a six-carbon citrate. The citrate is broken down releasing two more molecules of CO_2 as well as producing ATP, 3 NADH, and an $FADH_2$. The acetate compound is reconstituted in the cycle, allowing it to participate again in the cycle. The net electron exchange reaction in this metabolic cycle is

$$C_{21}H_{35}N_7O_{16}P_3S-C_2H_3O + 2H_2O + 3C_{21}H_{27}N_7O_{14}P_2^- + C_{27}H_{33}N_9O_{15}P_2 +$$

$\quad$ Acetyl CoA $\qquad\qquad\qquad$ NAD+ $\qquad\qquad$ FAD

$$C_{10}H_{15}N_5O_{11}P_2^{3-} + H_2PO_4^- \rightarrow$$

$\qquad$ GDP

$$C_{21}H_{36}N_7O_{16}P_3S + 2CO_2 + 3C_{21}H_{28}N_7O_{14}P_2^{2-} + C_{27}H_{35}N_9O_{15}P_2 +$$

$\quad$ CoA $\qquad\qquad\qquad$ NADH $\qquad\qquad$ FADH$_2$

$$C_{10}H_{15}N_5O_{14}P_3^{4-} + 3H^+ + \Delta E$$

$\qquad$ GTP $\qquad$ [15.19]

where flavin adenine dinucleotide (FAD) is a coenzyme, $C_{27}H_{33}N_9O_{15}P_2$. This is a derivative of riboflavin and is reduced to $FADH_2$, that is, 1,5-dihydro-FAD, $C_{27}H_{35}N_9O_{15}P_2$, in the reaction as two hydrogen atoms are bound to the molecule. The needed energy is supplied by the oxidation of succinate to fumarate with the loss of hydrogen atoms in the subreaction

$$C_4H_4O_4^{2-} + C_{27}H_{33}N_9O_{15}P_2 \rightarrow C_4H_2O_4^{2-} + C_{27}H_{35}N_9O_{15}P_2 + \Delta E \qquad [15.20]$$

Succinate   FAD   Fumarate $FADH_2$

Guanosine diphosphate (GDP), $C_{10}H_{15}N_5O_{11}P_2^{3-}$, in Reaction 15.19 is a nucleotide that can combine with $H_2PO_4^-$ using the energy of the reaction to form guanosine triphosphate GTP, $C_{10}H_{16}N_5O_{14}P_3^{4-}$. This is a nucleotide similar to ATP, composed of guanine, ribose, and three phosphate groups. This subreaction can be written as

$$\Delta E + C_{10}H_{15}N_5O_{11}P_2^{3-} + H_2PO_4^- \rightarrow C_{10}H_{15}N_5O_{14}P_3^{4-} + H_2O \qquad [15.21]$$
$$GDP GTP$$

The energized reduced species FADH2, GTP, and NADH produced in the citric acid metabolic cycle can revert to FAD2, GDP, and NAD+, respectively, and during the process convert ADP to ATP by electron transport along a chain of intermediaries. Also, energy released from NADH produced in Reactions 15.17 to 15.19 can be used to fuel mitochondrial ATP synthesis. During this conversion, either two or three ADPs are transformed to ATP, depending on whether the glycerol phosphate or the malate-aspartate shuttle is used to transport the electrons from the cytosol produced NADH into the mitochondria.

In summary, the total net yield of the oxidation of 1 mole of glucose to 2 moles of pyruvate is either 6 or 8 moles of ATP production. The 2 moles of pyruvate, reacting through the citric acid metabolic cycle, can produce an additional 30 moles of ATP from ADP. The total ATP production, therefore, is either 36 or 38 moles from the complete oxidation of 1 mole of glucose to 6 moles of CO_2 and H_2O. All of the carbon in glucose has been oxidized from an average valence state of 0 in glucose to +4 in CO_2.

Metabolic Pathways for Proteins and Lipids

Cellular energy can also be obtained from breakdown of proteins and lipids (see **Figure 15-28**). With protein, the bonds that connect individual amino acids are cleaved. Those with acetyl groups can be attached to CoA by a reaction similar to Reaction 15.18 and then enter the citric acid cycle. Other amino acids are broken down to pyruvate, whereas still others enter the citric acid cycle directly.

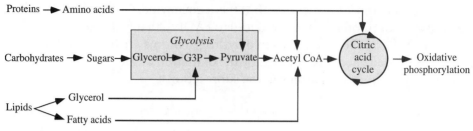

I FIGURE 15-28 Paths of protein, carbohydrate, and fat metabolism.

Lipids are broken down to glycerol and fatty acids as shown for triacylglycerol (often called triglyceride) in **Figure 15-29**. Glycerol ($C_3H_5(OH)_3$) can then be broken down to glyceraldehydes 3-phosphate (G3P). This is done with the aid of ATP to ADP oxidation and with reduction of NAD+ to NADH by the reaction

$$C_3H_5(OH)_3 + C_{10}H_{12}N_5O_{13}P_3^{4-} + C_{21}H_{27}N_7O_{14}P_2^- \rightarrow$$

Glycerol	ATP	NAD+

[15.22]

$$C_{10}H_{12}N_5O_{10}P_2^{3-} + C_{21}H_{28}N_7O_{14}P_2^{2-} + C_3H_7O_6P + \Delta E$$

ADP	NADH	G3P

Fatty acids, $C_nH_{2n+1}COOH$, are organic acids consisting of carbon chains with a carboxyl group at one end. They are broken down to acyl-CoA. For instance, the reaction of the fatty acid, butanoic acid, to acyl-CoA can be written as

$$C_3H_7COOH + C_{10}H_{12}N_5O_{13}P_3^{4-} + 2C_{21}H_{36}N_7O_{16}P_3S + 2H_2O \rightarrow$$

Butanoic acid	ATP	CoA

[15.23]

$$2C_{21}H_{36}N_7O_{16}P_3S-CH_3CO + C_{10}H_{14}N_5O_7P^{2-} + 2H_2PO_4^- + \Delta E$$

Acyl-CoA	AMP

$$
\begin{array}{ccc}
\overset{\displaystyle O}{\overset{\displaystyle \|}{H_2C\text{-}O\text{-}C}}\text{-}R & \overset{\displaystyle O}{\overset{\displaystyle \|}{HO\text{-}C}}\text{-}R & \\
\overset{\displaystyle O}{\overset{\displaystyle \|}{HC\text{-}O\text{-}C}}\text{-}R \longrightarrow & \overset{\displaystyle O}{\overset{\displaystyle \|}{HO\text{-}C}}\text{-}R \ + & H_2C\text{-}OH \\
& & HC\text{-}OH \\
\overset{\displaystyle O}{\overset{\displaystyle \|}{H_2C\text{-}O\text{-}C}}\text{-}R & \overset{\displaystyle O}{\overset{\displaystyle \|}{HO\text{-}C}}\text{-}R & H_2C\text{-}OH
\end{array}
$$

triacylglycerol   3 fatty acids   glycerol

I FIGURE 15-29 The cleaving of lipid triacylglycerol to three fatty acids and glycerol. *R* denotes an alkyl chain of single bonded carbons and hydrogens with the general formula C_nH_{2n+1}.

Acyl-CoA then enters the citric acid metabolic cycle, where it is further oxidized releasing energy to make ATP.

Fermentation Reactions

The fermentation process involves redox reactions between carbons to obtain energy in cells during the breakdown of simple sugar molecules without the aid of O_2 or changing the average oxidization state of the carbons. Two types of fermentation of glucose occur. In both, pyruvate is produced by Reactions 15.16 and 15.17. With fermentation to lactate, these reactions are followed by the reaction

$$CH_3(C{=}O)_2O^- + C_{21}H_{28}N_7O_{14}P_2^{2-} + H^+$$

Pyruvate   NADH   [15.24]

$$\rightarrow CH_3CH(OH)COO^- + C_{21}H_{27}N_7O_{14}P_2^- + \Delta E$$

Lactate   NAD+

In fermentation to ethanol, Reactions 15.16 and 15.17 are followed by the production of acetaldehyde and CO_2 as given by

$$CH_3(C{=}O)_2O^- + C_{21}H_{27}N_7O_{14}P_2^- \rightarrow H_3CH(C{=}O) + C_{21}H_{28}N_7O_{14}P_2^{2-} + CO_2$$

Pyruvate   NAD+   Acetaldehyde   NADH   [15.25]

The acetaldehyde produced breaks down to ethanol. This requires the energy from a NADH to NAD+ oxidation to proceed so the reaction becomes

$$H_3CH(C{=}O) + C_{21}H_{28}N_7O_{14}P_2^{2-} + H^+ \rightarrow C_2H_5OH + C_{21}H_{27}N_7O_{14}P_2^-$$

Acetaldehyde   NADH   Ethanol   HAD+   [15.26]

Therefore, the overall reactions of fermentation of glucose to lactate acid and glucose to ethanol can be written, respectively, as

$$C_6H_{12}O_6 \rightarrow 2CH_3CH(OH)COO^- + 2H^+$$

Glucose   Lactate   [15.27]

and

$$C_6H_{12}O_6 \rightarrow 2C_2H_5OH + 2CO_2$$

Glucose   Ethanol   [15.28]

As shown in **Figure 15-30**, in both fermentation reactions, although some carbons in glucose have been oxidized, others have been reduced. The average oxidation state of the carbons in the reactions is unchanged. Compared with

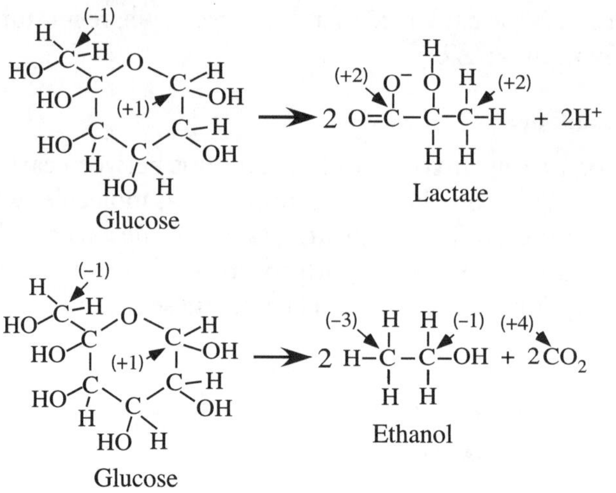

FIGURE 15-30 Fermentation reactions of $C_6H_{12}O_6$ (glucose) to lactic acid and to ethanol (energy released: 118 kJ/mol). The valence of the carbon atoms is indicated. Undesignated carbons have a valence of zero.

oxidative phosphorylation of glucose, in fermentation, the amount of energy stored in ATP production during the fermentation process is only 2 moles of ATP per mole of glucose, as the total oxidation of the organic carbon to CO_2 does not occur.

Molecular Transport Through Cell Membranes

Concentrations of molecules are generally different inside a cell than outside being nonequilibrium systems. Processes that transport molecules across the cell membrane maintain these differences. Cell membranes are lipid bilayers with the interiors having two *hydrophobic* chains of hydrocarbon that exclude charged and polar molecules. These membranes allow the passage of CO_2, O_2, and other uncharged molecules by simple diffusion down concentration gradients; however, most organic molecules in cells are charged (generally with a negative charge) or have polar functional groups on them (*hydrophilic*) and cannot be transported by simple diffusion through cell membranes.

Special transport proteins in the membrane control charged molecule transport. Some proteins can facilitate diffusion by neutralizing the hydrophobic hydrocarbon ends of membrane lipids. Others can transport molecules against their concentration gradient. For instance, the concentration of K^+ is greater inside a cell than outside. The "uphill" transport need to obtain K^+ for the cell is facilitated by added energy, typically from the conversion of ATP to ADP. For macromolecules such as polysaccharides, polypeptides, and polynucleotides, transport is by *vesicle* formation. Secretion of macromolecules from the cell occurs by the vesicle membrane becoming part of the cellular membrane and opening to release their contents outside the cell. Uptake of macromolecules occurs by a cavity formation on the outside of the cell membrane at the location of the macromolecule. It then pinches off creating a vesicle in the cytosol.

Both carrier proteins in the cell membrane and an energy source are required to transport small molecules and ions against a concentration gradient. In the case of outward Na^+ and inward K^+ transport, the same carrier protein is used and is termed the *sodium-potassium pump*. Using ATP to ADP energy, the carrier protein changes shape, and three Na^+ are transported outward for every two potassium transported inward. The cytosol therefore becomes negatively charged and develops a larger K^+ and smaller Na^+ concentration than the environment outside the cell.

Organisms also store energy in some cellular processes by translocating protons from inside the cell to outside through the cell membrane leading to a net negative charge at higher *pH* inside the cell. An exception is alkalophilic bacteria, which exist in alkali environments. They translocate protons into their cells to help maintain a more near neutral *pH*. This electrical potential (voltage) between the cell cytosol and outside environment is used to do cellular work by moving the protons (charge) through the cell membrane. Another observation is that fermentative organisms exhibit a greater range of cytosol *pH* than those that use the citric acid cycle.

Enzyme-Catalyzed Reactions

The reactions described above lower the Gibbs energy of the system as reactants are transformed to products as required by the laws of thermodynamics. The activation energy for reaction by a direct pathway from reactants to products is typically quite large, and an enzyme catalyst is required for reactions to proceed at significant rates. Enzymes are molecules that can manipulate other molecules. Enzyme, E, catalyzing reactant R to produce product P can be written as

$$R + E \underset{k_{-1}}{\overset{k_1}{\Longleftrightarrow}} E_{R/P} \underset{k_{-2}}{\overset{k_2}{\Longleftrightarrow}} E + P \qquad [15.29]$$

E denotes an unoccupied enzyme center, whereas $E_{R/P}$ stands for the enzyme center occupied by the reacted molecule (not exactly either a R or P molecule but with the same molecular formula). k_1, k_{-1}, k_2, and k_{-2} stand for the forward rate constant to produce a reactant/product bound enzyme center, the reverse reaction rate constant for decay to reactant plus unoccupied enzyme, the forward rate constant for producing the product from the reactant/product bound enzyme center and the reverse reaction rate constant for producing the reactant/product occupied enzyme center from the products, respectively.

The change in the number of occupied enzyme centers with time is given by

$$\frac{d[E_{R/P}]}{dt} = k_1[R][E] + k_{-2}[E][P] - k_{-1}[E_{R/P}] - k_2[E_{R/P}]$$

[15.30]

[R], [P], [E], and [$E_{R/P}$] denote the concentration of reactants, products, free catalytic centers, and reactant/product bound catalytic centers, respectively. In most reactions, the intermediate, $E_{R/P}$, is typically observed to equilibrate rapidly and be short-lived relative to other reactants (i.e., R). In this case, a *quasi steady-state approximation* can be made where $d[E_{R/p}]/dt = 0$.

In the simplest case, it is also assumed that $k_{-2}[E][P]$ is insignificant. One argument for this is that typically k_{-2} is very small. That is, P has difficulty reacting with the enzyme to produce $E_{R/P}$. Another argument is that P when produced is further reacted in another reaction keeping [P] small. With $k_{-2}[E][P]$ insignificant and using the quasi steady-state approximation, Equation 15.30 becomes

$$\frac{d[E_{R/P}]}{dt} = 0 = k_1[R][E] - k_{-1}[E_{R/P}] - k_2[E_{R/P}]$$

[15.31]

Solving Equation 15.31 for [$E_{R/P}$] gives

$$[E_{R/P}] = \frac{k_1[R][E]}{k_2 + k_{-1}}$$

[15.32]

Combining the rate constants in Equation 15.32 into a *Michaelis-Menton rate constant*, K_m, results in

$$[E_{R/P}] = \frac{[R][E]}{K_m}$$

[15.33]

The concentration of total enzymes present, [E_{tot}], is equal to the bound plus free enzymes so that

$$[E] = [E_{tot}] - [E_{R/P}]$$

[15.34]

Therefore, from Equation 15.33

$$K_m = \frac{[R][E_{tot}] - [R][E_{R/P}]}{[E_{R/P}]}$$

[15.35]

Solving Equation 15.35 for [$E_{R/P}$] yields

$$[E_{R/P}] = \frac{[E_{tot}][R]}{K_m + [R]}$$

[15.36]

The rate of product formation, $d\mathrm{P}/dt$, is defined as the concentration of reactant bound catalytic centers times their rate of destruction or from Equation 15.36:

$$\frac{d[\mathrm{P}]}{dt} = k_2 \frac{[\mathrm{E_{tot}}][\mathrm{R}]}{\mathrm{K_m} + [\mathrm{R}]} = \frac{v_{max}[\mathrm{R}]}{\mathrm{K_m} + [\mathrm{R}]} \qquad [15.37]$$

where k_2 is the rate constant for destruction of $\mathrm{E_{R/P}}$ with units of time^{-1}. v_{max} stands for the maximum rate (velocity) at which P can form and is equal to $k_2[\mathrm{E_{tot}}]$.

Building on the earlier work of Victor Henri (1903), Equation 15.37 is referred to as the Michaelis and Menten (1913) equation. It is also often referred to as the Monod (1949) equation when used to consider microbial growth. Plotting $d[\mathrm{P}]/dt$ versus [R] in Equation 15.37 produces a rectangular hyperbola so that Equation 15.37 is sometimes referred to as the microbial hyperbolic rate law.

At low [R] relative to $\mathrm{K_m}$ where there are abundant catalytic centers, the reaction equation shows linear rate behavior given by

$$\frac{d[\mathrm{P}]}{dt} = \frac{v_{max}}{\mathrm{K_m}}[\mathrm{R}] \qquad [15.38]$$

At high [R] relative to $\mathrm{K_m}$, the denominator in Equation 15.37 $\mathrm{K_m} + [\mathrm{R}] \approx [\mathrm{R}]$ and Equation 15.37 becomes

$$\frac{d[\mathrm{P}]}{dt} = v_{max} \qquad [15.39]$$

Therefore, at high [R], the number of catalytic centers, $[\mathrm{E_{tot}}]$, controls the reaction rate. A large number of different enzyme catalyzed mechanisms generate behavior that can be described by the Michaelis-Menton rate equation; however, it is difficult to relate the constants k_2 and $\mathrm{K_m}$ to rate constants in elementary reactions of possible mechanism.

Figure 15-31 shows the velocity (reaction rate) as a function of reactant concentration for Michaelis-Menton kinetics with no inhibitors (see below) and where $v_{max} = 1$ and $\mathrm{K_m} = 10$. In a reaction involving more than one reactant (substrate), a multiple substrate Michaelis-Menton equation can be used:

$$v = \frac{d[\mathrm{P}]}{dt} = v_{max} \frac{[\mathrm{A}]}{\left[\mathrm{K_m^A} + [\mathrm{A}]\right]} \times \frac{[\mathrm{B}]}{\mathrm{K_m^B} + [\mathrm{B}]} \times \frac{[\mathrm{C}]}{\mathrm{K_m^C} + [\mathrm{C}]} \qquad [15.40]$$

where [A], [B], and [C] stand for the concentrations of three different reactants (substrates) that combine to form the product, each with their own Michaelis-Menton constants given as $\mathrm{K_m^A}$, $\mathrm{K_m^B}$, and $\mathrm{K_m^C}$, respectively.

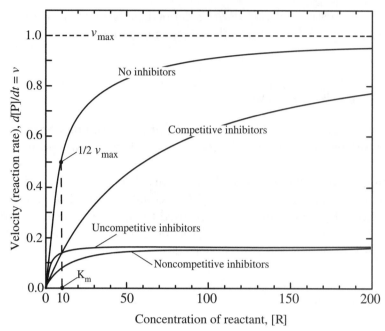

FIGURE 15-31 The rate of a reaction as a function of reactant concentration for a reaction that obeys Michaelis-Menton rate equation behavior ($v_{max} = 1$, $K_m = 10$) with ($[I]/K_I = 5$) and without ($[I]/K_I = 0$) inhibitors. Without inhibitors, K_m is the concentration in which the rate is one half of the maximum rate.

Inhibitors can also be present in enzyme-mediated reactions. Inhibition can be what is termed competitive, noncompetitive, or uncompetitive, depending on their effect on the rate constants, v_{max} and K_m (e.g., Cornish-Bowden, 2004). With competitive inhibition, the inhibitor, I, and reactant, R, compete for the same reaction site on the enzyme. The Michaelis-Menton equation becomes

$$v = \frac{d[P]}{dt} = \frac{v_{max}[R]}{K_m \left(1 + \dfrac{[I]}{K_I}\right) + [R]} \qquad [15.41]$$

where [I] is the inhibitor concentration and K_I stands for the inhibition coefficient. At a particular concentration of inhibitor, the effect on the Michaelis-Menton equation is to increase the constant in the denominator to a value greater than K_m, but v_{max} is unaffected. For instance, under the conditions shown in Figure 15-30 where $[I]/K_I = 5$, the difference between Michaelis-Menton rates with no inhibitors and competitive inhibitors is that it takes a greater concentration of reactant to closely approach the maximum rate of reaction, v_{max}.

With noncompetitive inhibition, the inhibitor binds to both the enzyme and the $E_{R/P}$ complex. The Michaelis-Menton equation becomes

$$v = \frac{d[P]}{dt} = \frac{v_{max}[R]}{(K_m + [R])\left(1 + \dfrac{[I]}{K_I}\right)}$$ [15.42]

In going from Equation 15.37 to Equation 15.42, v_{max} is replaced by $v_{max}/(1 + [I]/K_I)$. With noncompetitive inhibition and $[I]/K_I = 5$, at high reactant concentration, the rate approached a maximum value of 0.2 rather than 1.0 (see Figure 15-30).

With uncompetitive inhibition, the inhibitor binds to the enzyme-reactant/product complex, $E_{R/P}$, but not the free enzyme, E. The Michaelis-Menton equation becomes

$$v = \frac{d[P]}{dt} = \frac{v_{max}[R]}{K_m + [R]\left(1 + \dfrac{[I]}{K_I}\right)}$$ [15.43]

Similar to noncompetitive inhibitors, the maximum rate approached at elevated reactant concentrations is $v_{max}/(1 + [I]/K_I) = 0.2$ for the values of the rate constants outlined in Figure 15-31. With Equations 15.37 and 15.40 to 15.43, most of the effects of enzyme-mediated reactions can be modeled; however, it should be noted that some inhibition is irreversible such that enzyme-activated sites are inactivated and the rate decays exponentially.

■ Life Originating as a Catalyst?

First, what is life? A reasonable definition is a system capable of undergoing Darwinian (random perturbation) style evolution. That is, an entity adapting to the environment through internal changes in succeeding generations. Living organisms have the ability to undertake metabolism. Metabolism is a series of out-of-equilibrium chemical reactions running forward for the entity's maintenance, reproduction, and adaptation. Models of the origin of life must address the problems of the source of energy, materials, and reactions responsible for maintenance, reproduction, and adaptation, as well as the need to transfer information about reproduction and adaptation from one organism to an offspring.

Numerous investigators have demonstrated the ability for natural chemical reactions to produce the essential molecules of life as we know it: amino acids, sugars, and lipids in the *prebiotic* young earth's ocean and its atmosphere (Luisi, 2006). Another source of these molecules has been found in meteorites.

But how did they become the self-replicating systems we call life? For life as we know, it must have started in water. Another important observation is that most amino acids come in left- and right-handed forms, and life employs the left-handed ones almost exclusively. Minerals also occur in left- and right-handed forms, with some crystal faces in water adsorbing and sorting for left-handed molecules (Goldschmidt, 1952). Minerals in water also have the ability to self-assemble lipids and polymerize amino and nucleic acids (Hazen, 2006).

Living systems process energy and possess an organization produced by self-reproduction that is autocatalytic. Far from equilibrium, nonliving systems can also display autocatalytic features (see the discussion of pattern formation in Chapter 13). For instance, hydrophobic carbon chains in aqueous solutions can spontaneously self-aggregate into vesicles that contain water and water-soluble molecules.

Consider the possibility that life started as a result of a far from equilibrium reaction. Far from equilibrium, reactions typically occur where a phase change is involved in the reaction. It is also likely this was an oxidation-reduction reaction as these have large energy changes. A possibility could be the dissolution of the mineral pyrrhotite ($Fe_{1-x}S$) in solution. Given the large number of organic species in the prebiotic ocean an organic catalyst species can form on the pyrrhotite surface to lower the activation energy of reaction by transporting electrons between the mineral and aqueous phase. This organic catalyst may have initiated as a *Fischer-Tropsch process*, a catalytic mediated chemical reaction in which CO and H_2 is transformed into liquid hydrocarbons. The flux of energy in the pyrrhotite surface reaction is now through the catalyst as reactant mineral species are transformed to aqueous products by the movement of electrons. The catalyst species becomes more complex and grows from both solution and mineral species as it adjusts to the changing chemical flux in the environment on the mineral surface. Being involved in a far from equilibrium reaction, it could also have the ability to self-organize.

At some point, its size is no longer stable, and it is cleaved in two parts and detaches from the mineral surface substrate. There are now two catalytic molecules that can be attached to new mineral surfaces. Those with selected advantages to cleave and attach "survive in a Darwinian sense" and are bounded to a new surface substrate. The catalyst continues to grow and increase in complexity, adapting to the far from equilibrium energy of reaction before it is cleaved and detaches again. The self-replication mechanism is initially the catalyst itself with its ability to cleave. Whether simple RNA developed as an independent catalyst or whether it evolved from a pre-existing metabolic catalyst is still a matter of debate. A "cell" wall evolves that allows the charge separation to occur more efficiently. From this humble beginning, a bacteria could evolve that processes energy from reactants to products by moving electrons and develops a genetic code to pass along traits that process this energy more efficiently than by simple cleaving. In many ways, all living organisms, including humans, can be considered catalysts, processing energy from reactants to products more rapidly than is possible through abiotic reactions.

Summary

Organic carbon has four bonding electrons and is more reduced than carbon found in carbonates. The bonding electrons form up to four covalent bonds with other carbons, producing carbon chains of enormous variability in length, bonding character, and element composition. Lesser numbers of double and triple bonds are also stable. Carbon molecules are considered a framework of carbons to which functional groups are attached. The compounds are divided into two main groups: aromatic, which possess a benzene ring, and aliphatic, which do not. Important aliphatic compounds include alkanes, which are chains of singly bonded carbons with hydrogen (e.g., octane, C_8H_{18}).

Carbohydrates, $C_m(H_2O)_n$, where m and n are large integers, are the major constituents of plants produced during photosynthesis and built on glucose ($C_6H_{12}O_6$) units. Two main pathways, the C_3 Calvin cycle and C_4 Hatch-Slack cycle, are used to produce the carbohydrates. There are different $\delta^{13}C$ fractionations for the pathways, which help to determine paleoenvironmental conditions. Lipids are a set of fats, oils, and waxes that, like carbohydrates, contain only C, H, and O, but H relative to O is greater than the ratio found in carbohydrates. They are the basic building block of crude oil.

Organic carbon either dissolves in natural waters typically as disassociated humic and fulvic acids or is present as particulate matter. The dissolved constituents complex Al, Fe, and other metals as chelates. During digenesis, organic matter is transformed into bitumen and kerogen. Depending on its H/C ratio, kerogen produces lipids or becomes peat and finally coal during later digenesis.

There are numerous hazardous anthropogenic organic compounds in the environment. These include CFCs, PCBs, DDT, and dioxins. An understanding of their formation and stability is necessary to control the problems their usage poses.

Reactions in cells require a continuous supply of energy. This is obtained either from sunlight or form the energy stored in oxidation-reduction reactions. Organisms store energy in molecules of NADH (NADPH in plants) and ATP. Cellular respiration involves the breakdown and oxidation of organic matter first by the process of glycolysis before it enters the citric acid metabolic cycle. To transport Na^+ outward and K^+ inward through a cell membrane, the sodium-potassium pump is used.

Most of the reactions in cells occur because of the presence of enzymes. With the quasi steady-state approximation that assumes the number of enzymes is constant, a Michaelis-Menton reaction rate expression is obtained. This rate expression can be modified to account for multiple substrates and the presence of inhibitors. It is argued that life itself may have started as an organic catalyst.

Key Terms Introduced

activation energy
adenosine mono+phosphate (AMP)
adenosine triphos+phate (ATP)
alcohols
aliphatic
alkanes
alkenes
alkynes
amines
amino acids
aromatic
autocatalytic
autotrophic
benzene ring
biomagnification
bitumen
Calvin cycle
carboxyl functional group
carboxylic acid
cellular respiration
cellulose
chelation
citric acid metabolic cycle
cytosol
dioxin
eukaryotic
Fischer-Tropsch process
fulvic acid
functional groups
glycolysis

heterotrophic
humic acid
humic substances
humin
hydrophilic
hydrophobic
kerogen
lipids
Michaelis-Menton rate constant
mitochondria
monosaccharide
organelles
oxidative phosphorylation
peptides
petroleum
phenol
polymer
polysaccharide
prebiotic
prokaryotic
proteins
quasi steady-state approximation
radical
saturated
sodium-potassium pump
starch
stereoisomer
substrate
vesicle

Questions

1. Describe the chemistry of a benzene ring, octane, and kerosene.
2. What is a carboxylic acid? Amino acid?
3. Give the chemical formula of carbohydrates, and describe how they are formed.

4. What is a lipid and how is it related to petroleum?
5. What is chelation, and why can it be important?
6. Describe the role of kerogen in the formation of crude oil.
7. Describe how CFCs destroy stratospheric ozone.
8. What is the structure of a PCB?
9. Describe the process of biomagnification of toxins.
10. What is a dioxin?
11. What is a phenol?
12. What are the three main categories of pesticides?
13. How does POC differ from DOC?

Problems

1. Beer is naturally carbonated. What is the reaction that causes this to occur?
2. A factory produces the sugar glucose by the hydrolysis of cellulose. For each 100 kg of cellulose used, how many kg of water are needed?
3. Fulvic and humic acids isolated from peat (Hertkorn et al., 2002) had average wt % elemental compositions of (C = 49.0, H = 4.2, N = 2.0, O = 44.8) and (C = 55.2, H = 4.3, N = 2.4, O = 38.1), respectively.
   (a) What is the molar ratio of H to C in the acids?
   (b) If put on a 100 carbon basis, what is the approximate chemical formula of each?
   (c) If oxygen in the fulvic acid is all bonded in carboxylic and phenol groups and there are 10 phenol groups in the formula and all the nitrogen is in primary amines, how many H in the formula calculated in (b) are bonded directly to carbons?
4. Write the structural formula of the following compounds:
   (a) Ethanol
   (b) 2,2,4-Trimethylpentane
   (c) 2,4-Dichlorophenol
   (d) 2,3,7,8-TCDD (tetrachlorodibenzo-p-dioxin)
   (e) 2,3,2′,3′-Tetraclorobiphenyl
5. Write out the structural formulas of the following compounds:

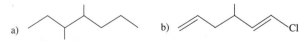

6. What are the structural formulas of the following?
   (a) Triethylamine
   (b) 1-Aminopropane
7. What are the structural formulas of
   (a) methylcyclopentane
   (b) 3-methylhexane?

8. What mass of fish containing 0.40 ppm of Hg would one have to consume to have ingested a total of 80 mg of Hg?

9. Oil is being produced in Argentina from igneous reservoir rocks consisting of thick tuffs of Late Triassic age. Thomas Gold (2001) argues that large volumes of methane are being released by the mantle. It has been argued that this methane complexes to produce the short-chain aliphatic compounds found in petroleum deposits. Outline the reasons why petroleum formation from sedimentary organic matter is the more accepted mechanism.

10. Explain why lipids are the primary source material for petroleum.

Problems Making Use of SUPCRT92

11. Write the reaction of the oxidation of ETHANOL,AQ, N-BUTANE,AQ, and N-OCTANE,AQ to CO_2,aq and H_2O. Using aqueous standard states for the species, calculate the standard-state Gibbs energy change using SUPCRT92. Per $O_{2(aq)}$ consumed, which reaction releases more energy if oxidized at 25°C and 1 bar? Why?

12. What is the pH of a solution of 1 molal ACETIC-ACID,AQ (CH_3COOH) at 20°C? Assume activity coefficients and the activity of H_2O are unity.

References

Cornish-Bowden, A., 2004, *Fundamentals of Enzyme Kinetics,* 3rd edition. Portland Press, London, 422 pp.

Gold, T., 2001, *The Deep Hot Biosphere: The Myth of Fossil Fuels*, Springer, New York, 264 pp.

Goldschmidt, V. M., 1952, Geochemical aspects of the origin of complex organic molecules on the earth, as precursors to organic life. *New Biology*, v. 12, pp. 97–105.

Hazen, R. M., 2006, Mineral surfaces and the prebiotic selection and organization of biomolecules. *Amer. Mineral.*, v. 91, pp. 1715–1729.

Henri, V., 1903, *Lois Générales de l'Action des Diastases*. Librairie Scientifique A. Hermann, Paris, pp. 85–93.

Hertkorn, N., Perminb, A., Perminovab, I., Kovalevskiib, D., Yudovb, M., Petrosyanb, V. and Kettrup, A., 2002, Comparative analysis of partial structures of a peat humic and fulvic acid using one- and two-dimensional nuclear magnetic resonance spectroscopy. *J. Environ. Qual.*, v. 31, pp. 375–387.

Koch, P. L., Fogel, M. L. and Tuross, N., 1994, Tracing the diets of fossil animals using stable isotopes. In Lajtha, K. and Michener, R. H. (Eds.) *Stable Isotopes in Ecology and Environmental Science*. Blackwell Scientific, London, UK, pp. 63–92.

Luisi, P. L., 2006, *The Emergence of Life: From Chemical Origins to Synthetic Biology.* Cambridge University Press, Cambridge, 315 pp.

Michaelis, L. and Menten, M., 1913, Die kinetik der invertinwirkung. *Biochem. Z.,* v. 49, pp. 333–369.

Monod, J., 1949. The growth of bacterial cultures. *Ann. Rev. Microbiol.,* v. 3, pp. 371–393.

Perdue, E. M., 1985, Acidic functional groups on humic substances. In Aiken, G. R., McKnight, D. M., Wershaw, R. L. and MacCarthy, P. (Eds.) *Humic Substances in Soil, Sediment and Water.* John Wiley & Sons, New York, 692 pp.

Stevenson, F. J., 1994, *Humus chemistry: Genesis, Composition, Reactions,* 2nd edition. John Wiley & Sons, New York, 512 pp.

Other Helpful References

Baird, C., 1995, *Environmental Chemistry.* W.H. Freeman, New York, 484 pp.

Killops, S. D. and Killops, V. J., 1993, *An Introduction to Organic Geochemistry.* Longman, Essex, England, 265 pp.

O'Neil, P., 1993, *Environmental Chemistry,* 2nd edition. Chapman & Hall, London, UK, 268 pp.

Thurman, E. M., 1985, *Organic Geochemistry of Natural Waters.* Martinus Nijhoff, Dordrecht, The Netherlands, 497 pp.

16 Atmospheric Chemistry

■ Early Earth Atmosphere and Ocean

The earth's original atmosphere is believed to have been lost when a Mars-sized body collided with the young earth during the formation of the moon. The atmosphere then likely developed from accretion to the solid earth's surface of volatile-rich ice planetesimals during the earth's early history. Material is also believed to be added from thermal outgassing of the earth during the time continental crust was formed. As with magmatic gases released from volcanoes today, the major components of outgassing were H_2O and CO_2 with over an order of magnitude less N_2 and HCl in a mildly reducing oxidation state. A direct comparison cannot be made because of the probable recycling of volatile material back into the mantle over time because of plate tectonic subduction of hydrous basalt and sediments.

With cooling of the earth after its initial formation, atmospheric H_2O condensed to form the oceans about 4.4 billion years (b.y.) ago (Cavosie et al., 2005; Wilde et al., 2001). The ocean produced was likely near its present volume as some of the oldest rocks are shallow marine sedimentary rocks formed on the "*freeboard*" of continental crust. This implies that the sea level was high enough to flood the continental margins. Although sea level relative to continental margins has changed significantly, this can be attributed to changing shapes of the ocean basins rather than changing volume of seawater. This suggests any loss caused by plate subduction is compensated by degassing at mid-oceanic ridges and above subduction zones.

The CO_2 in the early atmosphere reacted with silicate rocks on the surface to produce the abundance of carbonate rocks now present in the crust. A representative simplified reaction responsible for the loss of CO_2 from the atmosphere is from the conversion of the igneous feldspar anorthite to sedimentary calcite and kaolinite by the reaction

$$CO_2 + 2H_2O + \underset{\text{anorthite}}{CaAl_2Si_2O_8} \rightarrow \underset{\text{calcite}}{CaCO_3} + \underset{\text{kaolinite}}{Al_2Si_2O_5(OH)_4} \qquad [16.1]$$

The clays produced, like the kaolinite produced by Reaction 16.1, gave rise to pelitic rocks (shales) present in sedimentary basins. Clay-rich pelites were also produced by the reaction of HCl in the early atmosphere with the Na-rich feldspars present by a reaction like

$$2HCl + H_2O + 2NaAlSi_3O_8 \rightarrow Al_2Si_2O_5(OH)_4 + 4SiO_2 + 2Na^+ + 2Cl^- \qquad [16.2]$$
$$\text{albite} \qquad\qquad \text{kaolinite} \quad \text{quartz}$$

releasing $Na^+ + Cl^-$ to seawater. The extent of the reaction was limited by the HCl present, and the reaction gave rise to a saline ocean.

The salinity of the oceans has varied over time because of the NaCl that was sequestered and released from evaporites and saline groundwater as well as changes in the amount of fresh water tied up in ice during ice ages. Over the *Phanerozoic*, the salinity has likely decreased from more than 50‰ (parts per thousand) at the start of the Cambrian to the present day value of 34.7‰, as shown in **Figure 16-1** as more salt is tied up in evaporties and saline pore waters.

This change in salinity has implications for the evolution of life and climate change. For instance, the higher salinities of the Paleozoic could allow deepwater formation on cooling without the heat exchange from sea ice formation that occurs today (Hay et al., 2001). It has been argued that if the original ocean contained all of the salt in sedimentary rocks, its salinity would be 1.5 to 2 times its present value (Knauth, 1998).

The seawater Mg/Ca ratio is thought by many to have oscillated significantly over time. There have been three periods during Phanerozoic time when aragonite was the dominate *nonskeletal carbonate* deposited in seawater, "aragonite

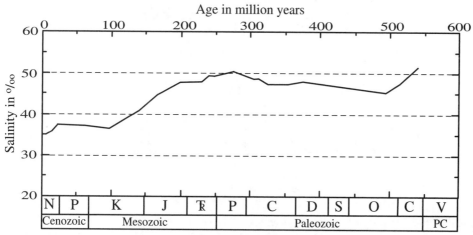

FIGURE 16-1 Salinity in parts per thousand (‰) by weight of seawater as a function of age. (Adapted from Hay et al. 2001.)

seas," which appear synchronistic with MgSO₄ evaporites and high seawater Mg/Ca ratio determined from fluid inclusions. These are separated by two periods in which calcite was the nonskeletal carbonate, "calcite seas," synchronistic with KCl evaporates, low Mg/Ca ratio in fluid inclusions and extensive *epicontinental seas* (**Figure 16-2**). The correlation of Mg/Ca ratio with aragonite/calcite seas likely reflects the observation that high magnesium concentrations in seawater have an inhibiting effect on calcite growth. A high Mg/Ca ratio of seawater can push seawater over an *evaporation divide* (Hardie, 1990) so that the precipitation sequence of mineral out of seawater gives rise to silvite (KCl) with significant carnallite (KMgCl₃•6H₂O) but no or little MgSO₄.

Most investigators believe that the oscillations are produced by changes in Mg uptake from seawater at mid-oceanic ridges (see Chapter 7) with changing spreading rates. High spreading rates causes expansion of the ridges displacing seawater from the deep ocean. The rise in sea level produces epicontinental seas. Increased spreading rates could also lead to greater Mg uptake at the ridges from increased reaction with seawater, which will lower seawater's Mg/Ca ratio (Berner, 2004a; Stanley and Hardie, 1998). It is also possible that the change in Mg/Ca ratio is the result of presence or absence of epicontinental seas, the environment where the calcite seas and KCl evaporates are deposited. The low Mg/Ca ratio produced by the input of low Mg/Ca river waters (see Table 7-5) to the small mass of seawater in epicontinental seas and the subsurface converts calcite to dolomite in these environments.

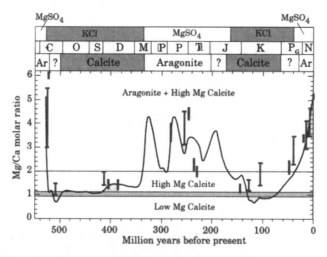

FIGURE 16-2 Molar Mg to Ca ratio of ocean water as a function of time. (Adapted from Hardie, 1996; Stanley et al., 2002.) Given above, the diagram are periods dominated by nonskeletal calcite or aragonite precipitation in seawater and KCl or MgSO₄ dominate evaporites. The vertical hachured bars give Mg/Ca ratios from fluid inclusions from Lowenstein et al. (2001) and Horita et al. (2002).

A modern analogy might be the Caspian Sea, a land-locked, highly saline body of water with input from the Volga and Kura rivers but output only by evaporation. The Mg/Ca ratio, controlled by the river input, is approximately 3 (Millero and Chetirkin, 1980). Because these epicontinental evaporite deposits are better preserved in the geological record, they may be more often sampled. In this case, the deep ocean has likely retained a more nearly uniform high Mg/Ca ratio during most of Phanerozoic time.

■ Present Composition of the Atmosphere

The present average composition of the lower atmosphere is given in **Table 16-1**. The atmosphere is divided into a number of layers based on its temperature (**Figure 16-3**). In contact with the earth's surface is the *troposphere*, the layer where temperature decreases with increasing elevation. Its thickness varies with location and time of year from about 10 km to 16 km, but averages 11 km. Its temperature decreases with increasing elevation because it is being heated from the earth's surface.

Above the troposphere is the *stratosphere*, which extends from about 11 to 50 km from the earth's surface. The troposphere–stratosphere boundary occurs where the temperature starts to increase with increasing elevation. The increase in temperature in the stratosphere is due to the absorption of *ultraviolet* light from the sun by the ozone, O_3, present there. This ultraviolet radiation is reradiated as *infrared* (heat) radiation that warms the stratosphere. At about 50 km, the atmosphere no longer contains O_3; therefore, the atmosphere

|Table 16-1   Average composition of the earth's lower atmosphere to 25 km

Gas	Volume (%)	Current source	Current variability
N_2	78.08	Biologic	Constant
O_2	20.95	Biologic	Constant
Ar	0.93	Radiogenic/interior	Constant
Ne	0.0018	Interior	Constant
He	0.0005	Radiogenic/interior	Escaping
Xe	0.000009	Interior	Constant
H_2O	0 to 4	Interior	Significant
CO_2	0.036	Biologic/industrial	Increasing
CH_4	0.00017	Biologic	Increasing
N_2O	0.00003	Biologic/industrial	Increasing
O_3	0.000004	Photochemical	Variable

The current source for change and the variability of the gas are also given. See Eby (2004).

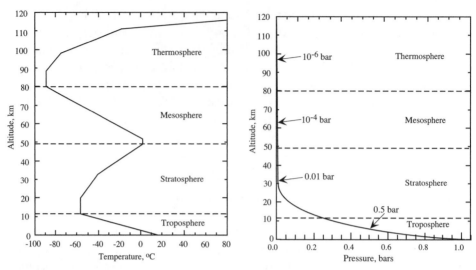

FIGURE 16-3 Temperature and pressure as a function of altitude in the atmosphere according to the U.S. Standard Atmosphere (1976).

is no longer internally heated, and the temperature starts to decrease again with increasing elevation as the *mesosphere* is entered.

At a height above the earth's surface of about 85 km, the temperature starts to increase again as short wavelength solar radiation and bombardment by protons and electrons given off by the sun on the limited number of resident gas molecules generate heat. This layer is termed the *thermosphere* and extends to about 700 km. In this layer, however, the small number of molecules in a given volume means that the amount of heat stored is quite small and subject to wide temperature changes. For instance, its temperature increases during periods of sunspot activity.

Chemical reactions in the atmosphere are dominated by molecular gas-phase collisions, but reactions also occur on the surfaces of solid particles and in water droplets in the lower atmosphere. Atmospheric particles, *aerosols*, range from aggregates of a few molecules to dust particles of greater than 100 μm. Small particles (<0.1 μm) are kept in constant random motion by bombardment of atmospheric gas molecules (Brownian motion). These tend to aggregate rapidly and settle out. When particles are large (>20 μm), they also do not stay aloft in the atmosphere against the force of gravity and have residence times of a day or less before settling out. Thus, most aerosols are in the intermediate size range between 0.2 and 2 μm and stay aloft for months. Aerosols include clays, sea spray, smoke, volcanic emissions, cloud droplets, and pollen particles. The solid particles serve as nuclei for condensation of water vapor to droplets. Rain does not form in the absence of aerosols. Because most aerosols are large enough to cause incoherent scattering of visible light, when their concentrations are high, they produce a haze in the atmosphere.

Figure 16-3 shows the pressure of the atmosphere decreases rapidly as the distance from the earth's surface increases. Over 99.9% of the mass of the atmosphere is contained in the troposphere plus stratosphere. Compared with the average radius of the earth (6371 km), the thickness of these layers is less than 0.8% so that on a planetary scale it is very thin. Above these layers in the upper atmosphere, the density of gas species becomes quite small. For instance, at an altitude of 90 km, the pressure of the atmosphere is about 1.8×10^{-6} bars, and density is about 3.4×10^{-6} kg m^{-3}. With 28.8 g of air ($0.8\, N_2 + 0.2\, O_2$) having about Avogadro's number of molecules (6.02×10^{23}), there are only 7.15×10^{19} molecules m^{-3}. With a molecular radius of about 10^{-10} m, the volume of a molecule is 4.2×10^{-30} m^3; therefore, only 3×10^{-8} percent of the volume is occupied by molecules. Their ability to collide and react is greatly diminished at these concentrations.

To obtain the concentration of particles and gases in the atmosphere, the mass per unit volume is typically expressed as μg m^{-3}. This unit, however, is a function of the temperature and pressure of the air mass. For gases, this problem is avoided by expressing the concentration as a volume *mixing ratio*. This gives the volume of the gas at a particular pressure and temperature to the total volume of the gas mixture that makes up air (i.e., ppm by volume). As the density of air decreases with elevation above the earth's surface, the volume measured by the ideal gas law becomes more accurate, and the mixing ratio is equal to the ratio of number of molecules of the gas to the total number of molecules of gas in an atmospheric volume. This concentration unit is then independent of pressure and temperature.

■ Reactions in the Atmosphere

Many reactions that occur between two gas molecules in a gas mixture require a third molecule to absorb the excess energy of the typical exothermic reaction. In the upper atmosphere, the small number of molecules per unit volume causes the rate of reaction of the three reactants needed to be slow and disequilibrium is maintained for extended periods. For example, the *ionosphere*, a layer in the thermosphere, which extends from 100 to 400 km above the earth's surface, is composed of cations and electrons produced from interaction of molecules with ultraviolet light. They exist for a long time because their collision to form more stable neutral molecules occurs so infrequently.

Reactions in the atmosphere are brought about by the adsorption of light in what are termed *photochemical reactions*. **Figure 16-4** gives the electromagnetic spectrum showing the regions where atomic and molecular excitation

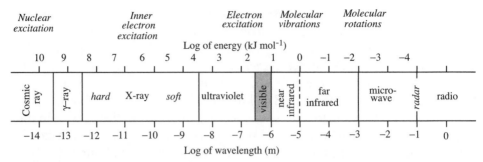

FIGURE 16-4 The electromagnetic spectrum of energy indicating the regions of atomic and molecular excitation.

occur. By excitation, what is meant is that the electrons in species are promoted into higher energy states, allowing reactions between species to occur. The light from the sun consists of a whole spectrum of different frequencies. How much light energy is there? This is given by the *solar constant*, which is the amount of energy in the entire spectrum from the sun that fluxes through a cross-sectional area normal to the sun's rays at the mean earth–sun distance. It is equal to 1360 W m^{-2}. Clearly, the amount at a particular location on the earth depends on its latitude and the time of year. The energy flux at a particular wavelength is given in **Figure 16-5**. Most of the sun's energy is in the ultravi-

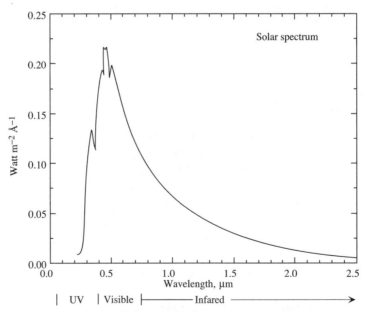

FIGURE 16-5 Solar spectrum of electromagnetic radiation outside the earth's atmosphere that includes ultraviolet (UV), visible, and infrared radiation.

olet and visible light range. The energy, E, of light depends on its frequency as given by

$$E = h\nu \qquad\qquad [16.3]$$

where h is *Planck's constant* (6.626075×10^{-34} J s). Because $\lambda\nu = c$, where c is the speed of light (2.997925×10^8 m s^{-1}) and λ denotes the wavelength of the energy,

$$E = h\nu = \frac{hc}{\lambda} \qquad\qquad [16.4]$$

Short wavelength light, therefore, possesses more energy than long wavelength light.

The solar spectrum is dominantly a continuum as a function of wavelength in the visible and near-ultraviolet region. As shown in **Figure 16-6**, however, below $\lambda = 0.15$ µm a continuum of overlapping lines is not present, and the emission is almost entirely made up of distinct lines at a given wavelength. The strongest individual line is Lyman-α at $\lambda = 0.1216$ µm; however, less than 0.1% of the energy of the spectrum is at wavelengths less than 0.24 µm.

Photochemistry of the upper atmosphere occurs because of *photolysis* of the O_2, N_2, and O_3 that is present. The extreme ultraviolet below about $\lambda = 0.12$ µm is absorbed above 100 km by atomic and molecular oxygen and nitrogen. Radiation with $0.12 < \lambda < 0.30$ µm is absorbed mainly in the mesosphere and stratosphere by molecular oxygen and ozone because N_2 is not an

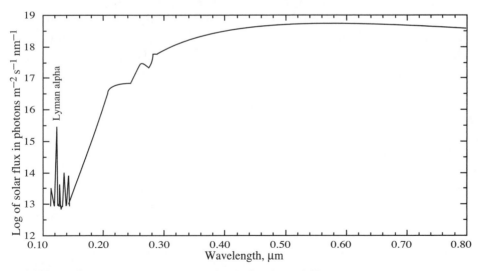

I FIGURE 16-6 Solar emission spectrum at wavelengths from 0.11 to 0.80 µm.

effective absorber at these wavelengths. With $\lambda > 0.30\ \mu m$, the radiation penetrates into the troposphere.

Different reactions require photons of different energies to occur and therefore different frequencies of light (**Figure 16-7**). The photonic energy input in photochemical reactions is usually specified as hv. Consider the energy needed to dissociate O_2. At stratospheric temperature and pressure, the energy is about 495 kJ mol^{-1}. The wavelength of light needed is, therefore

$$\lambda = \frac{hc}{E} = \frac{6.626 \times 10^{-34}\text{J s} \times 2.998 \times 10^{8}\text{m s}^{-1} \times 6.022 \times 10^{23}\text{atom mol}^{-1}}{4.95 \times 10^{5}\text{J mol}^{-1}\text{atom}}$$

$$= 0.242\,\mu m = 242\ nm$$

[16.5]

where the energy is given per mole of atoms so the numerator is multiplied by Avogadro's number (6.022×10^{23} atoms mol^{-1}). The short-wavelength, high-energy photon computed in Equation 16.5 is in the ultraviolet part of the sun's spectrum of light.

Free Radicals

The extra energy from absorption of a photon breaks bonds in molecules to form *free radicals*. A single bond in a molecule is made up of two electrons of opposite spin. When the bond is broken, the electrons either stay together in one of the separate molecular fragments, or they split up with each new species, taking one electron. If they stay together, the species become electrically charged and are referred to as ions. The classic case is NaCl that splits to form Na$^+$ and Cl$^-$. If the electrons split up on separation, however, the species become free radicals. Although free radials have neutral charge, they possess at least one single electron in a highly energetic state and seek out other electrons with which to pair (i.e., bond). Because most electrons exist in a paired state, free radicals often react with species containing only paired electrons. When they do, the former free radical obtains an electron and creates a new free radical

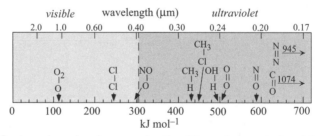

FIGURE 16-7 Bond energies and maximum wavelengths with enough energy to dissociate the indicated bond in the gas molecule. Bond energies are somewhat approximate because they depend on pressure, temperature, and, for the gases shown, the bonding environment.

in the process. When one free radical combines with another or the free radical dissociates to the ground state species with the production of a photon, the free radical is lost.

A photochemical reaction of NO_2 with light to produce the excited free radical NO_2^* is written as

$$NO_2 + h\nu \rightarrow NO_2^* \qquad [16.6]$$

where the superscript * is an energetically unstable electronically excited free radical molecule. Again, these molecules have electrons that are in orbitals of higher energy than in the ground state. This extra energy of NO_2^* is transferred in a number of ways. Often the energy is given to another species by contact in what is called *physical quenching*. The contacted species then develops higher translational energy that is eventually dissipated as heat. Excited molecules use the extra energy to dissociate into two ground state species, as in the reaction of excited O_2 in the upper atmosphere:

$$O_2^* \rightarrow O + O \qquad [16.7]$$

or they ionize, as occurs in the ionosphere with the free radical N_2^* in the reaction

$$N_2^* \rightarrow N_2^+ + e^- \qquad [16.8]$$

where e^- is an electron. A direct contact reaction of the free radical with another species results in

$$O_2^* + O_3 \rightarrow 2O_2 + O \qquad [16.9]$$

Photonic energy is also lost from a species as electrons drop to lower energy orbitals by luminescence during a reaction in what is called *fluorescence* if the radiant energy is lost during the reaction. If the energy continues to radiate after the reaction has stopped, the loss is termed *phosphorescence*.

Free radicals are very important in atmospheric reactions because they are quite energetic and therefore have only limited lifetimes in the atmosphere before reacting. Of particular importance is the hydroxyl free radical, HO^*, a fragment of a water molecule. HO^* forms in the upper atmosphere by photolysis of water:

$$H_2O + h\nu \rightarrow HO^* + H \qquad [16.10]$$

Once formed, the hydroxyl free radical is involved in two types of reactions that occur in less than a second in the troposphere. One is to detach a hydrogen atom from another molecule to create H_2O, and the other is to bond with a molecule to decrease its electron deficiency or increase the bonding numbers. The hydrogen produced typically either combines with HO* to produce water or combines with another H atom to produce H_2.

In the lower atmosphere (i.e., troposphere), the photons that penetrate do not have enough energy for Reaction 16.10 to occur. HO* is produced by the photolysis of ozone instead with photons of wavelengths less than 0.315 μm. Two reactions are required. The first is

$$O_3 + h\nu \rightarrow O^* + O_2 \qquad [16.11]$$

This is followed by the reaction

$$O^* + H_2O \rightarrow 2HO^* \qquad [16.12]$$

Where organic carbon is present, as in polluted air, HO* is also produced by oxidation. In the case of methane, the reaction is

$$CH_4 + O^* \rightarrow H_3C^* + HO^* \qquad [16.13]$$

The methyl free radical, H_3C^*, reacts with O_2 to form a very reactive peroxyl free radical, H_3COO^*. Gases with long residence times in the atmosphere are those that have low reactivity with HO*.

Reaction of HO* occurs with a number of other gas molecules as well. In the case of NO_2, the reaction is

$$NO_2 + HO^* \rightarrow HNO_3 \qquad [16.14]$$

where the excited (warmed) nitric acid molecule undergoes physical quenching and is generally washed out of the atmosphere, contributing to acid rain. HO* reacts with SO_2 by the reaction

$$HO^* + SO_2 \rightarrow HSO_3^* \qquad [16.15]$$

which is followed by the reaction

$$HSO_3^* + O_2 + H_2O \rightarrow H_2SO_4 + HO_2^* \qquad [16.16]$$

Again, the sulfuric acid produced is typically washed out of the atmosphere. Reaction of HO* with carbon monoxide produces carbon dioxide by the reaction

$$HO^* + CO \rightarrow CO_2 + H^* \qquad [16.17]$$

The H* formed then reacts with O_2 by the following reaction

$$H^* + O_2 \rightarrow HO_2^* \qquad [16.18]$$

Table 16-2	Atmospheric trace gases in near ground level dry air

Species	Volume (%)	Major sources[†]	Residence time (removal process)[†]
CH_4	1.6×10^{-4}	Bio	3.6 yr (photo)
CO	~1.2×10^{-5}	Photo, anthro	0.1 yr (photo)
N_2O	3×10^{-5}	Bio	20–30 yr (photo)
NO_x[‡]	10^{-10} to 10^{-6}	Photo, lig, anthro	4 day (photo)
HNO_3	10^{-9}–10^{-7}	Photo	(Wash out)
NH_3	10^{-8}–10^{-7}	Bio	2 day (photo, wash out)
H_2	5×10^{-5}	Bio, photo	(Photo)
H_2O_2	10^{-8}–10^{-6}	Photo	(Wash out)
HO*	10^{-13}–10^{-10}	Photo	(Photo)
HO_2^*	10^{-11}–10^{-9}	Photo	(Photo)
H_2CO	10^{-8}–10^{-7}	Photo	(Photo)
CS_2	10^{-9}–10^{-8}	Anthro, bio	40 day (photo)
OCS	10^{-8}	Anthro, bio, photo	(Photo)
SO_2	~2×10^{-8}	Anthro, photo, volc	3–7 day (photo)
CCl_2F_2	2.8×10^{-5}	Anthro	(Photo)
H_3CCCl_3	~1×10^{-8}	Anthro	(Photo)

[†]photo, photochemical; bio, biologic; anthro, anthropogenic; wash out, wash out by precipitation; lig, lightning; volc, volcanic.

[‡]NO_x = NO + NO_2.

From Manahan (2000); residence times from Brimblecombe (1996).

The hydroxyl radical causes the oxidation of the reduced gas molecules in the atmosphere given in **Table 16-2**.

■ The Carbon Cycle and Atmospheric CO₂

In Chapter 7, the concept of geochemical cycles was discussed by considering the water cycle. The same approach is used to consider the cycle of carbon, that is, the flux of carbon between reservoirs in the earth and its atmosphere. **Figure 16-8** is a representation of the present-day carbon cycle. Because carbon can exist in different oxidation states, the cycle is reasonably complex. The amounts of carbon in the reservoirs are typically determined by estimating the size of the reservoir together with the concentration of reduced and oxidized carbon in it. The fluxes are determined keeping in mind whether the reservoirs are in steady state. That is whether the fluxes in equal the fluxes out.

There are large fluxes of carbon between the a̲tmosphere, t̲errestrial biomass, and o̲cean, the "*ATO*." There are also fluxes from the mantle and sedimentary

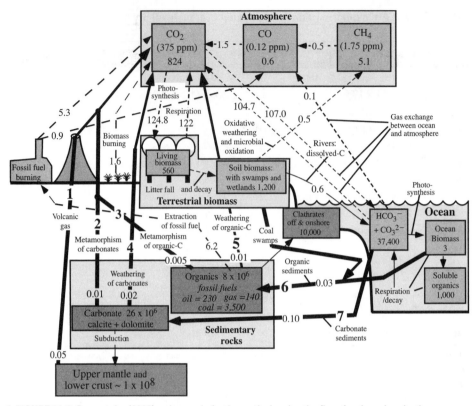

FIGURE 16-8 Present-day (2005) carbon cycle for the earth showing the flux of carbon given by the arrows (units of Pg yr^{-1} = 10^{15} g yr^{-1}). The reservoirs of carbon are shown as boxes (units of Pg). All of the fluxes and reservoirs are best estimates, and there is significant disagreement in detail among investigators as the values are inferred from indirect data except the concentrations in the atmosphere. (Adapted from O'Neil, 1993, and other sources.)

rocks to the atmosphere and from the ocean to sedimentary rocks (as given by the thicker arrows in Figure 16-8), but these are three orders of magnitude or less than the fluxes between the reservoirs in the ATO. The anthropologic fluxes of fossil fuel and biomass burning are intermediate in value. The small yearly fluxes between the ATO and the mantle and sedimentary rock would only be important on long (>10,000 yr) time scales and do not appear to significantly effect short-term changes in atmospheric carbon.

Fluxes in the Present-Day Carbon Cycle

The major carbon species in the atmospheric reservoir are CO_2, CH_4, and CO. With time, the flux of CH_4 into the atmosphere is oxidized to produce CO and finally CO_2 by the reactions

$$2CH_4 + 3O_2 \rightarrow 2CO + 4H_2O \qquad [16.19]$$

and

$$2CO + O_2 \rightarrow 2CO_2 \hspace{4cm} [16.20]$$

The concentration of CH_4 and CO in the atmosphere depends on their rate of supply and their rate of oxidation. With the fluxes in Figure 16-8, the residence times for CH_4 and CO in the atmosphere are 10.2 years and 146 days, respectively (see Equation 7.2). CO_2 and CH_4 are *greenhouse gases*, allowing the shorter wavelength ultraviolet and visible light part of the solar spectrum to pass through without adsorption but adsorbing parts of the longer wavelength infrared radiation emitted from the earth's surface. Global warming occurs if greenhouse gas concentration in the atmosphere is increased and global cooling if their concentration is decreased.

Is the amount of carbon as CO_2 in the atmosphere in steady state? If not, how rapidly is it changing? The major flux of carbon into and out of the atmosphere is through the biomass with nearly as great a flux occurring because of atmosphere-ocean exchange as given in Figure 16-8 and Table 16-3. According to the fluxes given, 3.4×10^{15} g of carbon is accumulating each year in the atmosphere. Other models have been built on somewhat different fluxes, but most would conclude CO_2 in the atmosphere is increasing. This agrees with the direct measurements over time of CO_2 concentrations in the atmosphere.

The measurements of atmospheric CO_2 made at Mauna Loa Observatory for the past 50 years are shown in **Figure 16-9**. These concentrations as a function of time are termed the "Keeling curve," after the late Charles D. Keeling, a professor at Scripps Institution of Oceanography who had been undertaking the measurements to 2005. The yearly cycle occurs because CO_2 declines in the spring when photosynthesis

| Table 16-3 | Present-day carbon fluxes into and out of the atmospheric CO$_2$ reservoir as given by Figure 16-9 in Pg (= 10^{15} g) |

Atmospheric CO$_2$ reservoir (units of Pg carbon/yr)	
Flux in	**Flux out**
122.0 = Respiration	124.8 = Land photosynthesis
104.7 = Ocean degassing	107.0 = Ocean absorption
5.3 = Fossil fuel burning	231.8 = Total
1.5 = Oxidation of CO	
1.6 = Biomass burning	
0.1 = Long-term fluxes labeled 1 + 2 + 3 + 4 + 5	In – Out = 235.2 – 231.8 = 3.4 Pg/yr
235.2 = Total	

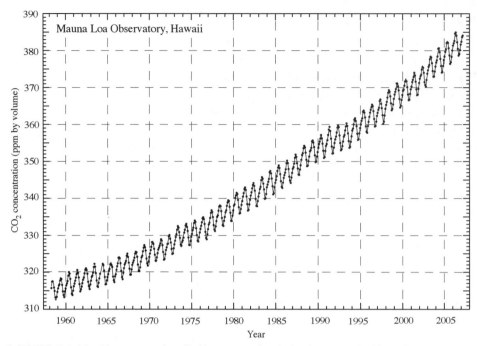

FIGURE 16-9 Monthly average carbon dioxide concentrations in dry air measured at Mauna Loa Observatory, Hawaii for the indicated year. (Modified from C. D. Keeling, T. P. Whorf, and the Carbon Dioxide Research Group, Scripps Institution of Oceanography.)

$$n\text{H}_2\text{O} + m\text{CO}_2 + sunlight \rightarrow \text{C}_m(\text{H}_2\text{O})_n + m\text{O}_2 \qquad [16.21]$$
$$\text{organic matter}$$

is a maximum in the northern hemisphere. CO_2 then increases in the autumn, as photosynthesis decreases and dead vegetation starts to decay, releasing CO_2 and CH_4 to the atmosphere. Increases in burning of fossil fuels containing carbon in the autumn and winter to generate heat also contribute significantly to the yearly cycling of CO_2.

Figure 16-9 also indicates an increase in CO_2 on a year-to-year basis. There is still some debate about how much of the increase is due to anthropogenic contributions as opposed to a long-term natural increase. The anthropogenic contributions are due to the increase in burning/combustion of fossil fuels as well as the clearing of forests, which decreases the photosynthetic CO_2 sequestering effect. The CO_2 increase to the atmosphere is mitigated to some extent by an increased flux by absorption into the ocean when concentrations of CO_2 in the atmosphere increase. A *biologic pump* is at work in the ocean as CO_2 is taken up by phytoplankton in the surface ocean. When the phytoplankton die, they sink to the deep ocean, where they decay and thus transfer CO_2 from shallow to deep water. This then promotes a greater flux of atmospheric CO$_2$ to the shallow ocean.

The carbon cycle is, therefore, not in steady state. Figure 16-8 indicates that anthropogenic fossil fuel and biomass burning fluxes of CO and CO$_2$ contribute 7.8 Pg yr^{-1} of carbon to the atmosphere. Only about half of this shows up in the increase of CO$_2$ in the atmosphere. Besides accumulating in the atmosphere, the model indicates the anthropogenic flux is probably leading to an increase in biomass carbon because of increased photosynthesis, causing more rapid plant growth because of higher atmospheric CO$_2$ concentrations. Some of the increase in CO$_2$ is most likely stored in the ocean reservoir because the flux of CO$_2$ into the ocean is greater than the flux out. As the anthropogenic input to the atmosphere changes, the carbon cycle helps in understanding how the various reservoirs of carbon adjust. The changes observed are about as large as the measurement uncertainties, and thus, it is difficult to make definitive statements.

The non–steady-state fluxes outlined in Figure 16-8 can be compared with the increases measured in the atmosphere outlined in Figure 16-9. From Figure 16-9, the average increase of CO$_2$ in the atmosphere with time is about 1.55 ppm by volume per year for the past few years. With the atmosphere modeled as an ideal gas, this ppm by volume increase is equal to its mole fraction increase per year, 1.55×10^{-6} yr^{-1}. The increase in carbon is given by this mole fraction times the molecular weight of carbon, 12.01 g mol^{-1}, times the mass of the atmosphere in moles. This molar mass equals the mass of the atmosphere in grams, 5.3×10^{-21} g (Campbell, 1977), divided by the grams in a mole of air, 28.97 g mol^{-1} (see the discussion of Table 1-3), or

$$\frac{1.55 \times 10^{-6} \text{yr}^{-1} \times 12.01 \text{ g mol}^{-1} \times 5.3 \times 10^{21} \text{g}}{28.97 \text{ g mol}^{-1}} = 3.4 \text{ Pg yr}^{-1} \qquad [16.22]$$

This is consistent with the value outlined in Table 16-3.

To determine the CO$_2$ concentration in the atmosphere before direct measurements were made, the concentration measured from extraction of air bubbles trapped in annual layers of ice deposited in the Arctic or Antarctica can be used. Shown in **Figure 16-10** are determinations for the last 250 years from an Antarctica ice core. Before 1750, ice cores record a constant CO$_2$ of 280 ± 3 ppm by volume for 1000 years (e.g., Etheridge et al., 1996). The increase to the present concentrations started with humankind's burning of fossil fuels at the start of the 19th century with the advent of the Industrial Revolution. This suggests that historical natural fluxes were in steady state at about 280 ppm by volume until the anthropogenic flux caused an increase starting around 1800.

Because of the link between atmospheric CO$_2$ changes and climate changes, a number of investigators have attempted to estimate past atmospheric CO$_2$ concentrations before humankind's contribution. One concern is increased

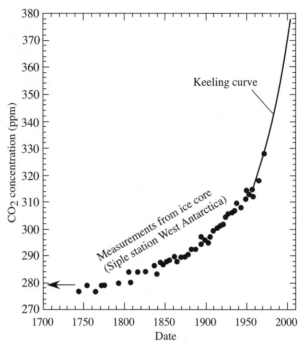

FIGURE 16-10 Yearly average CO_2 concentration in the atmosphere as a function of time given by the Keeling curve and ice core determination. (Adapted from Friedli et al., 1986.)

concentrations of CH_4 derived from organic matter. Its oxidation when released to the atmosphere could increase CO_2 levels dramatically.

Large amounts of organic matter have been deposited in permafrost areas and brought to the ocean by rivers and deposited on continental shelves under seawater at high latitudes. When buried, an anoxic environment can develop (see the discussion on *Eh* in sedimentary fluids in Chapter 14). In anoxic, reducing (*Eh* < 400 mV) environments methane gas is produced by methanogenic Archaeobacteria from the organic matter. If this process takes place at high enough pressures and low enough temperatures, a stable, ice-like methane-*clathrate* forms in the sediments. Methane-clathrates are cage-like structures of cubic ice crystals enclosing a methane gas molecule. Their stability at depth, relative to free $CH_{4(gas)}$ and H_2O on the land surface and below the sediment-water interface, is limited to about 500 m because of increased temperature from the earth's geothermal gradient with depth (**Figure 16-11**).

As given in Figure 16-8, the quantity of carbon in clathrates is larger than all the carbon in fossil fuels; therefore, investigators have considered the possibility of developing methane clathrates as a fossil fuel source. The technical problems of large-scale economic recovery to date, however, have been intractable. Investigators have suggested that large-scale, natural release of CH_4

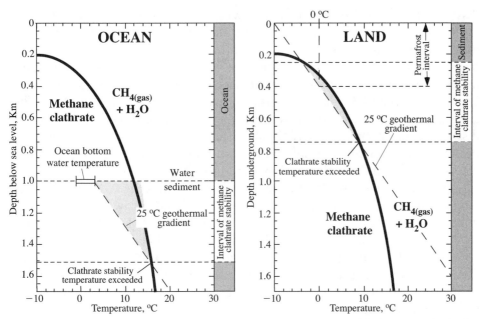

FIGURE 16-11 Methane clathrate stability as a function of temperature and depth in the ocean and on land surfaces given by the gray area.

from clathrates has occurred in the past. In particular, it is argued that the release of methane gas from oceanic clathrates limits the extent of ice ages. As sea level is lowered with accumulation of continental ice, pressure on the clathrates found on continental shelves decreases. This causes the clathrates to become unstable, releasing methane to the atmosphere. The methane and its oxygenated product, CO$_2$, produce an increased greenhouse effect causing global climate warming. It has also been suggested that the methane released could react with O$_2$ in the ocean making a large part of the ocean anoxic for a significant amount of time. This could be the mechanism that leads to the oceanic anoxic events that have occurred in the Phanerozoic (see the discussion for Figure 11-16).

Long-Term Carbon Cycle

Besides possible release of methane from clathrates and their oxidation to CO$_2$, can other natural processes cause significant changes in atmospheric CO$_2$ concentrations and therefore greenhouse warming over Phanerozoic time? When considering the atmosphere over the long term (i.e., millions of years), the major reservoirs of carbon in the mantle and sedimentary rocks as given by dark gray boxes in Figure 16-8 must be considered.

Three fourths of the sedimentary rock carbon is in carbonate minerals, and one fourth is organic carbon. Most of the sedimentary organic carbon is disseminated throughout pelitic rocks, with less than 0.1% accumulated into large enough concentrations to be considered fossil-fuel deposits. Over the long term, the surface reservoirs of the atmosphere, terrestrial biomass, and ocean (ATO) can be considered to be in an internal steady state, changing their internal compositions rapidly as they react to different fluxes to and from sedimentary rocks and the mantle. This is shown schematically in **Figure 16-12**. The same approach was used in the reaction path calculations section for mineral dissolution in Chapter 6. There the slow step was dissolution of minerals in a beaker of water while internal equilibrium with the new species in solution was achieved rapidly. Here the slow step is the flux into and out of the ATO.

The seven important long-term fluxes to and from the ATO are listed in **Table 16-4** and are labeled 1 to 7, given by heavy lined arrows between reservoirs, in Figure 16-8 and solid line arrows in Figure 16-12. Of these fluxes, the burial of carbonate rocks (flux 7) is the largest. What determines this flux? As occurs today, during most of Phanerozoic time, the ocean was likely supersaturated with calcite at some locations precipitating calcium carbonate and at others undersaturated and dissolving calcium carbonate. To determine which is dominant, consider the reaction

$$Ca^{2+} + 2HCO_3^- = CaCO_3 + CO_2 + H_2O \qquad [16.23]$$

where the CO_2 is in equilibrium with the atmosphere. To precipitate calcite requires increased bicarbonate or Ca^{2+} in the ocean or lowering CO_2 in the atmosphere. Changes in the opposite sense will dissolve calcite.

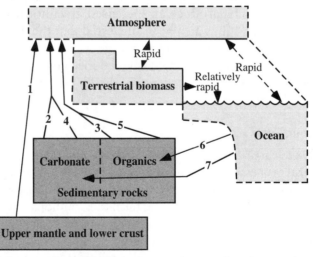

FIGURE 16-12 Long-term carbon cycle indicating the seven important fluxes between the earth's crust and mantle and the short-term ATO cycle.

Table 16-4 **Estimates of important long-term fluxes to and from the surficial environment (ATO). See Berner (2004b).**

Label	Type of flux	Symbol	Value: Pg yr^{-1}
1	Volcanic gas emissions of carbon	F_{volc}	0.05–0.15
2	Diagenesis and metamorphism of carbonates	F_{m-carb}	0.005–0.04
3	Diagenesis and metamorphism of organic carbon	F_{m-org}	0.001–0.02
4	Weathering of carbonates	F_{w-carb}	0.01–0.06
5	Weathering of organic carbon	F_{w-org}	0.005–0.04
6	Burial of organic carbon	F_{b-org}	0.02–0.08
7	Burial of carbonates	F_{b-carb}	0.02–0.3

Calcium silicate minerals can weather by reacting with carbonic acid, releasing Ca^{2+} to the ocean by the reaction:

$$2H_2CO_3 + CaSiO_3 \rightarrow Ca^{2+} + 2HCO_3^- + H_2O + SiO_{2(aq)} \qquad [16.24]$$
$$\text{Ca-silicate}$$

The carbonic acid needed is related to CO_2 in the atmosphere by

$$CO_2 + H_2O \rightarrow 2H_2CO_3 \qquad [16.25]$$

The aqueous silica produced in Reaction 16.24 will eventually precipitate as biogenic silica and revert to quartz on burial which can be represented by the reaction

$$SiO_{2(aq)} \rightarrow SiO_{2(qtz)} \qquad [16.26]$$

Summing Reactions 16.23 to 16.26 results in the Ebelmen (1845)-Urey (1952) reaction:

$$CO_2 + CaSiO_3 \rightarrow CaCO_3 + SiO_{2(qtz)} \qquad [16.27]$$

This represents the important transfer of carbon from the atmosphere by weathering of calcium silicates followed by the precipitation of oxidized carbon in marine carbonates (flux 7). The reverse of Reaction 16.27 denotes the production of CO_2 by volcanism (flux 1), metamorphism, and diagensis of calcite (flux 2) and weathering of carbonates (flux 4). Because of the large abundance of dolomite as well as calcite in sedimentary rocks, weathering of magnesium silicates is also important. By determining the concentration of aqueous species in the reactions as a function of time, Equation 16.23 can be used to determine the amount of CO_2 in the atmosphere.

With M_{ATO} denoting the total mass of carbon in the ATO, mass balance implies

$$\frac{dM_{ATO}}{dt} = F_{volc} + F_{m-carb} + F_{m-org} + F_{w-carb} + F_{w-org} - (F_{b-org} + F_{b-carb}) \qquad [16.28]$$

where F_i stands for the flux of carbon from or to the ATO reservoir by process i. Because the carbon fluxes have different isotopic ratios ($\delta^{13}C$), an isotope mass balance equation is also helpful in determining the fluxes:

$$\frac{d(\delta^{13}C \times M_{ATO})}{dt} = \delta^{13}C_{volc} \times F_{volc} + \delta^{13}C_{m-carb} \times F_{m-carb} + \delta^{13}C_{m-org} \times F_{m-org}$$

$$+ \delta^{13}C_{w-carb} \times F_{w-carb} + \delta^{13}C_{w-org} \times F_{w-org} - (\delta^{13}C_{b-org} \times \qquad [16.29]$$

$$F_{b-org} + \delta^{13}C_{b-carb} \times F_{b-carb})$$

By determining changes in the mass and isotopic composition of the reservoirs as a function of time with various proxies, a model of mass flux to and from the ATO reservoir can be determined. After this change in the mass of carbon in the ATO has been determined, the mass of carbon in the atmosphere as CO_2 can be calculated by distributing the mass between the terrestrial biomass, ocean, and atmosphere in a new steady state. Given in **Figure 16-13** are calculations of Phanerozoic atmospheric CO_2 concentrations as a function of time based on the GEOCARB III model of Berner and Kothavala (2001), the COPSE

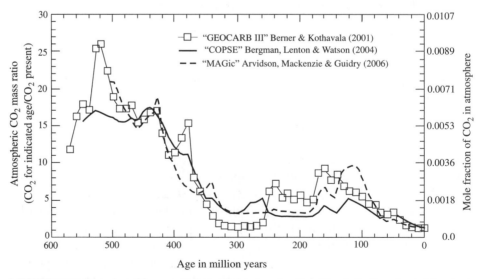

FIGURE 16-13 Atmospheric CO_2 concentrations calculated as a function of time in the Phanerozoic determined from three different models.

model of Bergman, Lenton, and Watson (2004), and the MAGic model of Arvidson, MacKenzie, and Guidry (2006).

Note the higher calculated CO_2 concentration in the geological past. Many investigators argue this is due to higher ocean crust production rates in the past and therefore increased magmatic CO_2 release. Values greater than 15 times the present preindustrial concentrations are produced by the models between 425 and 550 m.y. CO_2 concentrations decrease to more than four times the present preindustrial concentrations around 300 m.y. (Permo-Carboniferous). The decrease in CO_2 is modeled as an effect of the evolution and rise of large vascular land plants. These plants accelerated the weathering of Ca and Mg silicates by producing carbonic and other acids in their root systems. This increases the Ca and Mg flux to the ocean causing sequestering of CO_2 by the precipitation of carbonates. Also, decomposition and burial of the vascular land plants added to a flux of organic carbon to sedimentary rocks.

CO_2 in the atmosphere is modeled as rising again at ~250 m.y. in the GEO-CARB III model but not until ~180 m.y. in the COPSE and MAGic models. CO_2 appears to remain high during the Mesozoic before a slow decline to present pre-industrial concentrations. It is argued (Berner, 2004a) that this decline is due to increased mountain uplift as well as increase solar input with time, both of which increased Ca and Mg silicate weathering rates. With an increased flux of Ca and Mg to the ocean, an increase CO_2 sequestering occurred from carbonate precipitation (flux 7).

Interestingly, the times of lowest CO_2 concentrations shown in Figure 16-13 correspond to the two major Phanerozoic glaciations, the Permo-Carboniferous and the last 30 m.y. Times of high atmospheric CO_2 appear to correspond to periods of increased warmth as deduced from the fossil record between 65 and 250 m.y. These observations support the greenhouse theory of long-term climate change (see Royer et al., 2004).

■ Atmospheric O₂ as a Function of Time

The oxygen that is a major component of our present atmosphere has accumulated over time after the advent of photosynthetic life by Reaction 16.21. Because of the solar energy needed to form the organic matter, $C_m(H_2O)_n$, this organic matter is out of equilibrium with the O_2 produced. If the dead organic matter is not isolated from the oxygen-rich atmosphere by burial, oxidation causes CO_2 and CH_4 to be released to the atmosphere with the consumption of O_2. Early in earth history the initial O_2 produced by Reaction 16.21 is thought to have reacted with atmospheric H_2 to produce water. Later, atmospheric CH_4 was oxidized to CO_2 and H_2O. This was followed in the presence of water by the oxidation of some sulfide, S^{2-}, in surface rocks to sulfate, S^{6+} and ferrous iron,

Fe^{2+} to ferric iron, Fe^{3+}, before O_2 began to accumulate in the atmosphere (see Chapter 11 for a discussion of the isotopic ramifications of this oxidation).

There is still active debate about the timing of the increase in atmospheric O_2 to its present level. Most investigators believe, however, that it was about 1% of its present value of 20.9% 2.0 b.y. ago, the time at which the Fe^{3+} found in *continental red-beds* first appeared in the geologic record. O_2 probably increased to at least 10% of its present value 700 m.y. ago. Model calculations of O_2 concentrations in the atmosphere during Phanerozoic time by Berner and Canfield (1989), Bergman et al. (2004), and Arvidson et al. (2006) are given in **Figure 16-14**. This requires knowing the change in reduced and oxidized species as a function of time, as these changes require a flux of electrons which can be obtain from oxygen as electrons flux between oxygen in O_2 and oxygen in H_2O + silicates. Besides oxygen, the significant redox elements in the crust are iron, carbon, and sulfur.

To oxidize a mole of Fe from Fe^{2+} to Fe^{3+} requires the loss of 1 mole of electrons. This can be accomplished with the reduction of one fourth of a mole of O_2 as it contains two oxygen atoms whose change in valence is from zero in O_2 to −2 in H_2O. In the case of carbon, oxidation of a mole of methane, CH_4, to CO_2 with a valence change from −4 to +4, requires the transfer of 8 moles of electrons or reduction or 2 moles of O_2. Oxidation of organic matter, CH_2O, to CO_2 requires 4 moles of electrons or 1 mole of O_2 reduction. Finally, in the case of sulfur, oxidation of 1 mole of FeS_2 with valence −1 to +6 in sulfates (SO_4^{2-})

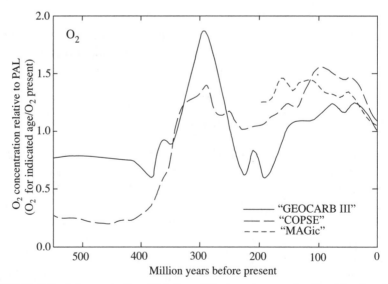

FIGURE 16-14 Estimated concentration of O_2 in the earth's atmosphere as a function of time from the GEO-CARB III model of Berner and Canfield (1989), COPSE model of Bergman, Lenton, and Watson (2004), and MAGic model of Arvidson, Mackenzie and Guidry (2006).

requires seven times the amount of oxygen as need for a mole of iron oxidation or 7/4 of a mole of O_2.

Fe is the most abundant redox element in the crust, and significant changes in the oxidation state of iron in sediments exist; however, Fe plays nearly an insignificant role in controlling oxygen in the atmosphere because it takes so much less oxygen (i.e., electrons) to change its oxidation state than for C or S, and its redox, unlike C and S, is not used to a significant extent by organisms that flux electrons in the earth's surface environment.

The problem of determining O_2 concentration in the atmosphere then becomes that of determining the rate of reactions adding and depleting O_2 in the atmosphere. Burial of organic matter adds O_2 (Reaction 16.21). The burial of pyrite + organic sulfur also adds O_2 to the atmosphere by the reaction

$$2Fe_2O_3 + 8SO_4^{2-} + 16H^+ \rightarrow 15O_2 + 4FeS_2 + 8H_2O \qquad [16.30]$$

where FeS_2 stands for pyrite + organic sulfur buried. O_2 is lost to the atmosphere by oxidation of organic matter (reverse of Reaction 16.21) and reduced carbon gases released from deep in the crust by diagenesis, metamorphism, and volcanism. O_2 is also lost by oxidation of pyrite + organic sulfur (reverse of Reaction 16.30). Similar to the determination of CO_2 in the atmosphere, isotope mass balance of carbon and sulfur can be used as constraints on the magnitude of the fluxes. In this case, $\delta^{13}C$ of calcite and $\delta^{34}S$ of sulfates as a function of time are typically used.

The calculated peak in O_2 near 300 million years ago shown in Figure 16-14 is produced by the rise of vascular land plants in the Carboniferous period, producing O_2 with the burial of organic matter from the plants in vast coal swamps. The models then follow this by decreasing O_2 concentrations because of drying up and oxidation of the coal swamps when the climate changed to more arid conditions. Sediments were then deposited in carbon and sulfur-free continental red-beds. The Bergman, Lenton, and Watson model includes the role of forest fires and phosphorus as linked to changing oceanic productivity of O_2. The MAGic model (Arvidson et al., 2006) includes continental and sea floor weathering of silicates and carbonates, seawater/basalt interactions, precipitation and diagenesis of minerals, decarbonation by subduction, but no isotopic constraints. Problems with the MAGic model have been outlined in Walther (2007).

■ Stratospheric Ozone

In addition to concerns about CO_2 increases in the atmosphere and the effects of global warming, humankind is also worried about changes in the protective stratospheric ozone layer. With photons of wavelengths less than 0.2424 μm, the following reaction occurs in the stratosphere:

$$O_2 + h\nu \rightarrow 2O^* \tag{16.31}$$

This is followed by a reaction to produce ozone, O_3,

$$O^* + O_2 + M \rightarrow O_3 + M \tag{16.32}$$

where M is an energy absorbing N_2 or O_2 molecule. Ozone is destroyed by a photon of wavelength less than 0.325 μm by the reaction

$$O_3 + h\nu \rightarrow O_2 + O^* \tag{16.33}$$

and the single O^* produced reacts by a series of reactions that are summarized as

$$O^* + O_3 \rightarrow 2O_2 \tag{16.34}$$

The concentration of O_3 in the stratosphere is therefore in a steady state, balanced by its production and destruction. The ozone provides an effective shield against the penetration of ultraviolet radiation for wavelengths between 0.23 and 0.32 μm, which is harmful to most organisms. Near the equator where the photon flux density is nearly constant, ozone levels are typically 250 to 300 *Dobson units* (DU) year-round. A DU for a gas is the amount of gas equal to a layer of the pure gas that is 10^{-5} m in thickness at 1 atm and 0°C and is equal to 4.4615×10^{-4} mole m^{-2}. Production of O_3 is greatest in the tropics where the intensity of ultraviolet radiation is the greatest. At higher latitudes, seasonal variations produce large changes in ozone levels. Levels are greater in the spring and summer in the northern hemisphere when increased radiation from the sun causes an increase in Reaction 16.31, producing ozone concentrations as high as 500 DU. Averaged over the year for the entire earth, ozone in the atmosphere is about 300 DU.

The preanthropogenic ozone balance has been upset by humankind's addition to the stratosphere of new species that react with O_3. The most serious are *chlorofluorocarbons* (*CFCs*). These compounds are very stable in the troposphere but undergo photochemical dissociation by short-wavelength ultraviolet light in the stratosphere to produce electronically excited Cl^*. For CFC-12 the reaction is

$$CF_2Cl_2 + h\nu \rightarrow Cl^* + CClF_2^* \tag{16.35}$$

The Cl^* produced reacts with O_3 by the reaction

$$Cl^* + O_3 \rightarrow ClO^* + O_2 \tag{16.36}$$

and is reconstituted by the reaction

$$ClO^* + O^* \rightarrow Cl^* + O_2 \tag{16.37}$$

This results in O_3 destruction and reconstitution of Cl^* to participate in another O_3 destruction reaction. The reactive stratospheric Cl^* is limited by ClO^* combining with NO_2 to create the unreactive species, $ClONO_2$, by the reaction

$$ClO^* + NO_2 \rightarrow ClONO_2 \qquad\qquad\qquad [16.38]$$

The inefficiency of Reaction 16.38 implies ozone destruction by humankind's introduction of CFCs to the stratosphere will last hundreds of years.

NO has been introduced into the lower stratosphere by burning of petroleum to power supersonic transport aircraft in this region. A similar catalytic reaction occurs with NO in the stratosphere. The NO is oxidized by O_3 according to the reaction

$$NO + O_3 \rightarrow NO_2 + O_2 \qquad\qquad\qquad [16.39]$$

The NO_2 produced reacts with a single O^* produced by Reaction 16.31 by the reaction

$$NO_2 + O^* \rightarrow NO + O_2 \qquad\qquad\qquad [16.40]$$

The net reaction is

$$O + O_3 \rightarrow 2O_2 \qquad\qquad\qquad [16.41]$$

with the catalyst, NO, reconstituted; however, with the abandonment of the supersonic transport program because of both economic and environmental factors, there are presently insufficient aircraft to cause much O_3 destruction relative to the amount present.

A large depletion of up to 70% of the normal ozone occurs over Antarctica in early spring, the "Antarctic ozone hole." The size and extent of the region of depletion have been increasing significantly as a function of time, as indicated by the ozone measurements above Halley Bay, Antarctica given in **Figure 16-15**. In the Antarctic winter, strong winds circulate around the Antarctic continent because of the earth's rotation and the greater cooling of the land mass as opposed to the surrounding ocean water, creating the Antarctic polar wind vortex. This isolates the Antarctic stratosphere from the rest of the stratosphere. Temperature decreases dramatically as the Antarctic air mass sinks in the high pressure system produced. When temperatures drop below $-77°C$ in the stratosphere, nitric acid trihydrate (NAT) particles, $HNO_3 \bullet 3H_2O$, start to crystallize.

$ClONO_2$ is removed from the air by the following reactions that occur on the NAT particle surfaces:

$$ClONO_2 + H_2O \rightarrow HOCl + HNO_3 \qquad\qquad\qquad [16.42]$$

and

$$ClONO_2 + HCl \rightarrow Cl_2 + HNO_3 \qquad\qquad\qquad [16.43]$$

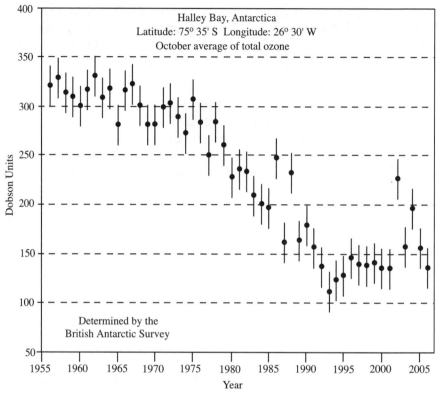

I FIGURE 16-15 Total atmospheric ozone above Halley Bay Station, Antarctica in October of the indicated year.

Therefore, HOCl and Cl_2 accumulate in the Antarctic stratosphere in winter on NAT particles in the absence of sunlight rather than as $ClONO_2$.

When spring arrives with increased ultraviolet light from the sun, the NAT particles are no longer stable and the following photochemical reactions occur:

$$HOCl + h\nu \rightarrow HO^* + Cl^* \qquad\qquad [16.44]$$

and

$$Cl_2 + h\nu \rightarrow 2Cl^* \qquad\qquad [16.45]$$

The Cl* produced reacts to destroy ozone according to Reaction 16.36, creating the Antarctic ozone hole. In the Arctic, stratospheric temperatures are not as low as in the Antarctic, and the NAT clouds form less often. The depletion observed is therefore less. In late spring, the Antarctic polar vortex subsides, and ozone produced in the stratosphere above more equatorial regions is then transported into the Antarctic region. The ozone hole disappears as it is filled in while typically shifting to somewhat more northerly latitudes with time.

■ Photochemical Smog

Photochemical smog pollution is created in the troposphere when hydrocarbons and oxides of nitrogen produced in internal combustion engines interact in intense sunlight in hot dry atmospheres. This is contrasted with the sulfurous smog of London, England in the mid 20th century produced from burning sulfur-rich coal and petroleum in cool humid air masses. Given in **Figure 16-16** are the particulates and SO_2 in smog that killed over 3000 people between December 5th and 9th in 1952. Sulfurous smog often has pH in the range of 2 to 3, whereas acid rain typically has a somewhat higher pH. One reason for the low pH is the lower water content of fog relative to normal cloud droplets for the same amount of sulfate present.

In contrast, the two most important constituents of photochemical smog are ozone and aerosols. In the production of smog, the primary smog-forming pollutant produced by the internal combustion engine is NO. Nitrogen oxide rapidly reacts with O_2 in the atmosphere to form NO_2. Nitrogen dioxide dissociates in sunlight according to the reaction

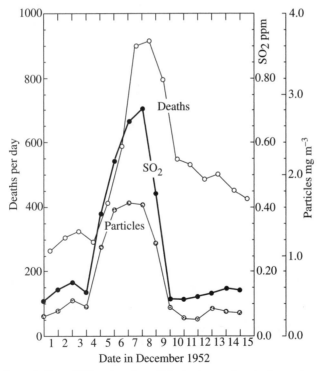

FIGURE 16-16 Concentrations of SO_2 and smoke particles in air as well as deaths per day on the given date in London, England. (Adapted from Wilkins, 1954.)

$$NO_2 + h\nu \rightarrow NO + O^* \qquad \qquad [16.46]$$

The atomic oxygen produced reacts with O_2 by the reaction

$$O^* + O_2 + M \rightarrow O_3 + M \qquad \qquad [16.47]$$

where M is an energy absorbing N_2 or O_2 molecule. The ozone produced then reacts with nitrogen oxide by

$$O_3 + NO \rightarrow NO_2 + O_2 \qquad \qquad [16.48]$$

A near steady-state amount of O_3 molecules and O atoms is, therefore, produced; the concentration depends on the amount of NO_2 and the intensity of sunlight. Because the needed NO_2 is derived primarily from the burning of petroleum products such as gasoline, O_3 concentrations are typically higher in cities than in the surrounding countryside. O_3 is greatest on sunny summer days when the sun's radiation is strong and winds do not dissipate the air mass. Shown in **Figure 16-17** is the diurnal variation of O_3, NO_2, and NO in air from Pasadena, California measured on July 25, 1973. NO reaches a maximum as automobile traffic increases toward its peak in the morning. NO_2 then peaks near the traffic maximum. The O_3 levels are low in the early morning but continue to increase to a maximum as noon is approached with increasing intensity of sunlight. The concentration of O_3 then decreases as the intensity of

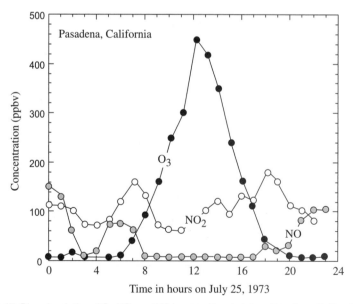

FIGURE 16-17 Diurnal variation of O_3, NO_2, and NO in ground level air from Pasadena, California on July 25, 1973, in parts per billion by volume. (Adapted from Finlayson-Pitts and Pitts, 1977.)

sunlight decreases in the afternoon, and Reaction 16.48 consumes O_3 causing NO_2 to peak again in the late afternoon.

Aerosols are the visible 0.1- to 1-μm airborne particles that scatter light in smog, reducing the visual contrast between objects and leading to the phenomena called haze. These particles are primarily ammonium sulfate, $(NH_4)_2SO_4$, and ammonium nitrate, NH_4NO_3. They are formed from the neutralization of sulfuric and nitric acid in polluted air with ammonia, NH_3, by the reactions

$$H_2SO_4 + 2NH_3 \rightarrow (NH_4)_2SO_4 \qquad\qquad [16.49]$$

and

$$HNO_3 + NH_3 \rightarrow NH_4NO_3 \qquad\qquad [16.50]$$

The atmospheric NH_3 is generally obtained from the decomposition of animal (principally livestock) urine and ammonia fertilizers from agricultural regions.

Summary

The original atmosphere was lost when a Mars-sized body collided with the earth in the formation of the moon. The major components similar to outgassing from the mantle today were then H_2O and CO_2 with over an order of magnitude less N_2 and HCl. With cooling, atmospheric H_2O condensed to form the oceans 4.4 billion years ago. The CO_2 reacted with silicate rocks to produce carbonates and clays. Seawater obtained its salinity early when HCl in the atmosphere reacted with Na-rich feldspars and water producing Na^+ and Cl^-.

The present atmosphere is a thin shell of gases, primarily N_2 (~78%) and O_2 (~21%), held to the surface of the earth by gravity whose pressure decreases rapidly with height. It is divided into four regions based on its temperature structure: troposphere, stratosphere, mesosphere, and thermosphere.

In the atmosphere because of the low concentration of species, equilibrium is the exception rather than the rule. Concentrations of a gas are typically reported as mixing ratios in ppmv (parts per million by volume) that is equal to its mole fraction because at the low pressures considered the ideal gas law is appropriate.

Aerosols in the atmosphere can be solids (e.g., clays) or liquids (sea spray). They range in size from about 100 Å to 100 μm. There is also a significant flux of anthropogenic aerosols into the atmosphere as a result of burning of fossil fuels.

Residence times typically control the concentrations of species observed in the atmosphere rather than equilibrium. High-energy photons from the sun cause photochemical reactions to occur. These are particularly important in the upper atmosphere. Free radicals are produced, and these cause species to react toward equilibrium. A free radical is an atom or molecule with a high-energy electron orbital occupied by a singe electron. These are quite reactive. Of particular importance is the hydroxyl free radial, OH*, a fragment of a water molecule.

Changes in CO_2 in the atmosphere are controlled by the carbon cycle. There is a yearly cycle in atmospheric concentrations of the greenhouse gas, CO_2, that declines in the spring when photosynthesis is a maximum and increases in the autumn as photosynthesis decreases and dead vegetation starts to decay. Increases in burning of fossil fuels in the autumn and winter also contribute to the yearly cycling of CO_2. As documented by the Keeling curve, concentrations of CO_2 in the atmosphere have been increasing on a year-to-year basis.

To determine the CO_2 in the atmosphere before direct measurement were made, the CO_2 in air bubbles trap in ice as a function of its age can be used. Interest has developed in the large amount of methane-clathrates found in sediments. Both in the future and for the past release of the methane from these clathrates can cause a very large atmospheric greenhouse effect.

Models of CO_2 levels in the atmosphere have generally been greater in the geological past than at present. Over these time scales, volcanic emissions, weathering, diagenesis, and metamorphism of carbonates and organic matter as well as rates of burial of organic carbon and carbonates become important in the models. O_2 levels in the past have been calculated to be both greater and less than current concentrations. They can be determined by estimating changes in the abundance of reduced and oxidized species in the earth's surface environment as a function of time.

Ozone, O_3, has the ability to absorb harmful ultraviolet radiation in the stratosphere. Photochemical reactions there cause the formation and destruction of ozone, producing a steady-state concentration. Anthropogenic additions of CFCs to the atmosphere have introduced Cl to the stratosphere, where it undergoes photochemical reactions that lead to additional O_3 destruction. During winter, the wind pattern around Antarctica isolates it from the rest of the earth's atmosphere. NAT particles in the stratosphere absorb the Cl during winter. With spring, the Cl is released and then undergoes photochemical reactions that cause O_3 destruction, leading to the formation of an Antarctic ozone hole.

Ozone is also created in the troposphere from excited O produced from a photochemical reaction involving NO_2 derived from the burning of petroleum. The O_3 produced reacts with NO in the atmosphere to produce O_2. It is, therefore, in steady state, with a concentration that depends on the intensity of sunlight and the concentration of nitrous compounds in the atmosphere. This ozone is a major constituent of photochemical smog and is detrimental because breathing it causes respiratory problems in sensitive people.

Key Terms Introduced

aerosols

ATO

biologic pump

Brownian motion

chlorofluorocarbons (CFCs)

clathrate

continental red-beds

Dobson unit (DU)

epicontinental sea

evaporation divide

fluorescence

free radicals

freeboard

infrared

ionosphere

mesosphere

mixing ratio

nonskeletal carbonate

Phanerozoic

phosphorescence

photochemical reactions

photolysis

physical quenching

Planck's constant

solar constant

stratosphere

thermosphere

troposphere

ultraviolet

Questions

1. What is a free radical?
2. Why does temperature decease with distance from the earth's surface in the troposphere but increase with distance in the stratosphere?
3. Explain why a hole develops in the ozone layer above Antarctica during early spring each year.
4. What causes the yearly cycle in CO_2 concentrations in the atmosphere?
5. Why is CO_2 increasing in the atmosphere on a yearly basis? What are the inputs and outputs?
6. What is the effect of aerosols in the earth's atmosphere?
7. How is the wavelength of light related to its energy?
8. Describe how ozone is produced in polluted air.
9. What are the main chemical constituents of photochemical smog?

Problems

1. Using Figure 16-8, calculate the average residence time of CO in the atmosphere. How does this compare with that of CH_4?
2. If fossil fuel burning was stopped tomorrow but other fluxes into and out of the atmosphere stayed the same, how long before atmospheric CO_2 would return to preindustrial levels of 270 ppm?
3. Give an example of a photochemical reaction.
4. If air pollutants rise in the earth's atmosphere, why do they not continue into outer space?

5. Because of their reactive surfaces, breathing in ultrafine particles present in the atmosphere causes inflammation of the lungs. With particles of density 2.7 g cm^{-3}, what is the number of particles and the total surface area of 1 g of ultrafine particles of 0.2 μm in diameter as opposed to fine particles of 20 μm in diameter if they are modeled as spheres?

6. Give the concentration of O_3 in number of molecules per cm^3 and μg m^{-3} for 1 ppbv (parts per billion by volume) of O_3 at 1 atm pressure and 25°C.

7. Show that at 1 atm and 25°C, μg m^{-3} = ppmv × 40.87 × (MW), where (MW) gives the molecular weight of the gas (g/mol) and ppmv gives the ratios in volume.

8. The concentration of NO_2 measured in polluted air at 22°C and 1.003 bar is 82 ppbv (ppm by volume). Considering ideal gas behavior, what is this concentration in μg m^{-3} and molec (molecules) per cm^3?

9. Calculate the half-life of NO_2 in the atmosphere in hours if the average, daytime, steady-state [OH*] = 1.4×10^6 radicals cm^{-3} and the decay of NO_2 is by the reaction

$$NO_2 + OH^* \rightarrow HNO_3$$

with a rate constant of k = 2.0×10^{-11} cm^3 molecule^{-1} s^{-1}.

10. The lifetime of a reactant species in the atmosphere, τ, is generally defined as the time it takes for its concentration to decrease to 1/e of its original value, where e is the base of the natural logarithm. For a second-order reaction as given in Problem 9,

$$\tau (NO_2) = \frac{1}{k[OH^*]}$$

How many hours is the lifetime of NO_2 in the atmosphere?

11. NO is an air pollutant that reacts readily with O_2 in air by the following reaction: $2NO + O_2 = 2NO_2$. If 1 mole of NO and 2 moles of O_2 are added to a 4-liter container and allowed to come to equilibrium, producing a final concentration of NO_2 of 0.20 M:
    (a) What is the equilibrium constant of the reaction?
    (b) If the volume of the container is reduced, which way will the reaction go?
    Assume the gases behave ideally.

12. If the rate constant for Reaction 16.31 is $k_{(16.31)}$, that for Reaction 16.32 is $k_{(16.32)}$, and that for Reaction 16.33 is $k_{(16.33)}$.
    (a) Give the rate expression for the change in the concentration of O_3, i.e., $[O_3]$, as a function of time.
    (b) Give a similar expression for O*.

(c) If Reaction 16.32 is so fast that excited oxygen atoms react as soon as they are formed by Reaction 16.31, what is the steady-state ozone concentration given by?

13. Consider that the following elementary reactions operate in the stratosphere:

1. $O_3 \rightarrow O^* + O_2$
2. $O^* + O_3 \rightarrow 2O_2$

(a) What is the overall chemical reaction?

(b) What are the rate laws for these reactions?

Problem Making Use of SUPCRT92

14. To what temperature need a 10 bar primordial atmosphere, with the mole fraction of CO_2 and H_2O gas both equal to 0.4, cool before the net reaction rate for Reaction 16.1 is zero?

References

Arvidson, R. S., Mackenzie, F. T. and Guidry, M., 2006, MAGic: A Phanerozoic model for the geochemical cycling of major rock-forming components. *Amer. Jour. Sci.*, v. 306, pp. 135–190.

Bergman, N. M., Lenton, T. M. and Watson, A. J., 2004, COPSE: A new model of biogeochemical cycling over Phanerozoic time. *Amer. Jour. Sci.*, v. 304, pp. 397–437.

Berner, R. A., 2004a, A model for calcium, magnesium and sulfate in seawater over Phanerozoic time. *Amer. Jour. Sci.*, v. 304, pp. 438–459.

Berner, R. A., 2004b, *The Phanerozoic Carbon Cycle: CO2 and O2*. Oxford University Press, Oxford, England, 150 pp.

Berner, R. A. and Canfield, D. E., 1989, A new model for atmospheric oxygen over Phanerozoic time. *Amer. Jour. Sci.*, v. 289, pp. 333–361.

Berner, R. A. and Kothavala, Z., 2001, GEOCARB III: A revised model of atmospheric CO2 over Phanerozoic time. *Amer. Jour. Sci.*, v. 301, pp. 182–204.

Brimblecombe, P., 1996, *Air Composition and Chemistry*, 2nd edition. Cambridge University Press, Cambridge, UK, 253 pp.

Campbell, I. M. *Energy and the Atmosphere*. New York: Wiley, 1977, p. 350.

Cavosie, A. J., Valley, J. W., Wilde, S. A. and E.I.M.F., 2005, Magmatic $\delta^{18}O$ in 4400-3900 Ma detrial zircons: A record of the alteration and recycling of crust in the Early Archean. *Earth Planetary Sci. Let.*, v. 235, pp. 663–681.

Ebelmen, J. J., 1845, Sur les produits de la decomposition des especes minérals de la famile des silicates. *Annu. Rev. Moines*, v. 12, pp. 627–654.

Eby, G. N., 2004, *Principles of Environmental Geochemistry*. Brooks/Cole-Thomson Learning, Pacific Grove, CA, 514 pp.

Etheridge, D. M., Steele, L. P., Langenfelds, R. L., Francey, R. J., Barnola, J.-M. and Morgan, V. I., 1996, Natural and anthropogenic changes in atmospheric CO_2 over the last 1000 years from air in Antarctic ice and firn. *Jour. Geophysical Res.*, v. 101, pp. 4415–4128.

Finlayson-Pitts, B. J. and Pitts, J. N., Jr., 1977, The chemical basis of air quality: Kinetics and mechanisms of photochemical air pollution and application to control strategies. *Adv. Environ. Sci. Technol.*, v. 7, pp. 76–162.

Friedli, H., Lötscher, H., Oeschger, H., Siegenthaler, U. and Stauffer, B., 1986, Ice core record of 13C/12C ratio of atmospheric CO_2 in the past two centuries. *Nature* v. 324, pp. 237–238.

Hardie, L. A., 1990, The roles of rifting and hydrothermal $CaCl_2$ brines in the origin of potash evaporites: An hypothesis. *Amer. Jour. Sci.,* v. 290, pp. 43–106.

Hardie, L. A., 1996, Secular variations in seawater chemistry: An explanation for coupled secular variation in the mineralogies of marine limestones and potash evaporites over the past 600 m.y. *Geology*, v. 24, pp. 279–283.

Hay, W. W., Wold, C. N., Söding, E. and Floegel, S., 2001, Evolution of sediment fuxes and ocean salinity. In Merriam, D. F. and Davis, J. C. (Eds.) *Geological Modeling and Simulation: Sedimentary Systems*. Kluwer Academic/Plenum Publishers, pp. 153–167.

Horita, J., Zimmermann, H. and Holland, H. D., 2002, Chemical evolution of seawater during the Phanerozoic: Implications from the record of marine evaporates. *Geochim. Cosmoch. Acta*, v. 66, pp. 3733–3756.

Keeling, C. D. and Whorf, T. P., 2003, Atmospheric CO_2 records from sites in the SIO air sampling network. In *Trends: A Compendium of Data on Global Change*. Carbon Dioxide Information Analysis Center, Oak Ridge National Laboratory, U.S. Department of Energy, Oak Ridge, TN.

Knauth, L. P., 1998, Salinity history of the Earth's early ocean. *Nature*, v. 395, pp. 554–555.

Lowenstein, T. K., Timofeeff, M. N., Brennan, S. T., Hardie, L. A. and Demicco, R. V., 2001, Oscillations in Phanerozoic seawater chemistry: Evidence from fluid inclusions. *Science*, v. 294, pp. 1086–1088.

Manahan, S. E., 2000, *Environmental Chemistry*, 7th edition. Lewis Publishers, Boca Raton, FL, 898 pp.

Millero, F. J. and Chetirkin, P. V., 1980, The density of Caspian waters. *Deep-Sea Research*, v. 27A, pp. 265–271.

Royer, D. L., Berner, R. A., Montañez, I. P., Tabor, N. J. and Beering, D. J., 2004, CO_2 as a primary driver of Phanerozoic climate. *GSA Today*, v. 14, No. 3, pp. 4–10.

Stanley, S. M. and L. A. Hardie, 1998, Secular oscillations in the carbonate mineralogy or reef-building and sediment producing organisms driven by tectonically forced shifts in seawater chemistry. *Palaeogeography. Palaeoclimatology Palaeoecology*, v. 144, pp. 3–19.

Stanley, S. M., Ries, J. B. and Hardie, L. A., 2002, Low-magnesium calcite produced by coralline algae in seawater of Late Cretaceous composition. *Proceedings of the National Academy*, v. 99, no. 24, pp. 15323–15326.

Urey, H. C., 1952, *The Planets: Their Origin and Development*. Yale University Press, New Haven, Conn., 245 pp.

U.S. Standard Atmosphere, 1976, National Oceanic and Atmospheric Administration, Washington, DC.

Walther, J. V., 2007, Comment: MAGic: A Phanerozoic model for the geochemical cycling of major rock-forming components: *Amer. Jour. Sci.*, v. 307, pp. 856–857.

Wilde, S. A., Valley, J. W., Peck, W. H. and Graham, C. M., 2001, Evidence from detrital zircons for the existence of continental crust and oceans on the earth 4.4 Gyr ago. *Nature*, v. 409, pp. 175–178.

Wilkins, E. T., 1954, Air pollution and the London fog of December, 1952. *Jour. Royal Society for the Promotion of Health*, v. 74, no. 1, pp. 1–21.

Other Helpful References

Brasseur, G. P., Orlando, J. J. and Tyndall, G. S., 1999, *Atmospheric Chemistry and Global Change*. Oxford University Press, New York, 645 pp.

McEwan, M. J. and Phillips, L. F., 1975, *Chemistry of the Atmosphere*. John Wiley & Sons, New York, 301 pp.

Warneck, P., 2000, Chemistry of the Natural Atmosphere, 2nd edition. *International Geophysics Series*, v. 71, Academic Press, San Diego, CA, 927 pp.

Symbols and Abbrevations

a = acceleration

a_i = thermodynamic activity of $i = f_i/f_i°$

$a_\pm$ = mean ionic activity of a species in solution

a_{RK} = Redlich–Kwong attraction constant

a_w = Van der Waals attraction constant

$å$ = electrolyte's mean ionic size or distance of closest approach of the ions in solution

A = chemical affinity of a reaction

A = Helmholtz free energy

A = area

$Å$ = Ångstrom = 10^{-10} m

ADP = adenosine diphosphate

AMP = adenosine monophosphate

ATO = Atmosphere-Terrestrial biosphere-Ocean reservoir

ATP = adenosine triphosphate

b_{RK} = Redlich–Kwong constant for volume occupied by gas molecules

b_w = van der Waals constant for volume occupied by gas molecules

bar = pressure unit = 10^5 pascals

c = speed of light

cal = thermochemical calorie, a unit of energy equal to exactly 4.184 J

°C = degrees Celsius. A temperature scale in which °C = K + 273.15

C = coulomb, the SI unit of electric charge = 6.2415×10^{18} electron charges

CAM = crassulacean acid metabolism

CCD = carbonate compensation depth

CDM = Canyon Diablo iron meteorite

CFC = chlorofluorocarbon

CHUR = chondritic uniform reservoir

CoA = coenzyme A

C_3 = type of plant that uses a 3-carbon cycle to produce carbohydrates during photosynthesis

C_4 = type of plant that uses a 4-carbon cycle to produce carbohydrates during photosynthesis

C_P = heat capacity at constant pressure = quantity of heat that increases the temperature of a mole of the substance by 1°C at constant pressure

C_V = heat capacity at constant volume = quantity of heat that increases the temperature of a mole of the substance by 1°C at constant volume

d = width

D = daughter isotope produced

DOC = dissolved organic carbon

Da = dalton

D_i = Nernst distribution coefficient of i

D_i = diffusion coefficient of i in units of $cm^2\ s^{-1}$

e = base of the natural log = 2.7182818 . . .

e^- = electron

E = enzyme

Eh = voltage of a reaction relative to the standard hydrogen electrode

$Eh°$ = standard state voltage of reaction relative to the hydrogen electrode

E_a = activation energy of a chemical reaction

E_k = kinetic energy

E_m = energy of mass

f_a = fraction remaining of a mass of atmospheric H_2O

f_i = thermodynamic fugacity of i

F = Faraday constant, a unit of electric charge equal to Avogadro's number of electron charges = 96,485 coulombs per mole of charges

F = force

F = fraction of melt

F = carbon flux

g = mean gravitational acceleration = 9.80665 m s^{-2} (average of 9.780 m s^{-2} at earth's equator to 9.832 m s^{-2} at poles)

g = gram

g.a.o. = gram atoms of oxygen

$G \equiv$ Gibbs "free" energy $= U + PV - TS$

G3P = glyceraldehyde 3-phosphate

G_i^o = standard-state Gibbs energy of the ith species

Ga = giga-years

h = Planck's constant $= 6.626075 \times 10^{-34}$ J s

h = hydraulic head

H = enthalpy $= U + PV$

H_i^o = standard-state enthalpy of the ith species

HREE = heavy rare earth element

HFS = high field strength

I = ionic strength

J = joule, SI unit of energy $= 1$ v C $= 1$ W s

J_i = mass flux of i

J_i^{dif} = diffusional flux of i

J_q = heat flux

k = Boltzmann's constant $= 1.38066 \times 10^{-23}$ J/K

k_S = Setchénow coefficient

k_H = Henry's law constant

k_r^+ = forward reaction rate constant

k_{mes}^+ = macroscopically measured forward rate constant

K = kelvins, the thermodynamic temperature scale in which K $= 0$ at a temperature of absolute zero. 0 K $= -273.15°$C

K = hydraulic conductivity

K_f = Freundlich sorption coefficient

K = thermal conductivity constant (J K^{-1} m^{-1} s^{-1})

K_m = Michaelis-Menton rate constant

K = equilibrium constant of a reaction

$\mathbf{K}_D$ = distribution coefficient

$\mathbf{K}_{iap}$ = ion activity product

$\mathbf{K}_{sp}$ = molal solubility product

$\mathbf{K}_{act}^+$ = equilibrium constant of the reaction to produce an activated complex in the forward direction

K–T = Cretaceous Period (K) – Tertiary Period (T) boundary

l = length

$\ln x$ = natural logarithm of $x = e^x = 2.302585 \log x$

$\log x$ = base-10 logarithm of $x = 10^x$

L_i = phenomenologic coefficient for i in units of mol^2 $joule^{-1}$ cm^{-1} s^{-1}

LIL = large ion lithophile

LREE = light rare earth element

m = meter

m_i = mass or molality of species i = mol per 1000 g H_2O

$m_{\pm}$ = mean ionic molality

mol = mole, that is, Avogadro's number of particles

m.y. = million years

M_i = molarity of species i = mol per 1000 cm^3 of solution

MOR = mid-oceanic ridge

MORB = mid-oceanic ridge basalt

n = moles of electrons transferred per mole of reaction

N = Newton

N = number of isotopes

= gram atoms of oxygen per cm^3

$N°$ = Avogadro's number = 6.02214×10^{23} mol^{-1}

NAD+ = oxidized form of nicotinamide adenine dinucleotide

NADH = hydrogenated nicotinamide adenine dinucleotide

NADPH = hydrogenated nicotinamide adenine dinucleotide phosphate

NADP+ = oxidized form of nicotinamide adenine dinucleotide phosphate

OAE = oceanic anoxic event

p = momentum of a particle

ppb = parts per billion either by weight or by volume

ppm = parts per million either by weight or by volume

pe = minus logarithm of the activity of the electron, e^-, in solution

$pe°$ = minus logarithm of the activity of the electron in solution under standard-state conditions

P = pressure either in pascals (Pa) = 1 N m^{-1}, bars = 10^5 Pa, or atmospheres (atm) = 1.013×10^5 Pa

P = parent atom that can decay

PAH = polycyclic aromatic hydrocarbon

PCB = polychlorinated biphenyl

PDB = Peedee formation belemnite from South Carolina

POC = particulate organic carbon

Pa = pascal = SI unit of pressure = 1 newton m^{-2}

P_C = critical pressure

P_o = permittivity constant (8.8542×10^{-12} C^2 N^{-1} mP)

pK_a = negative logarithm of the acid dissociation constant

q = flux of either heat or fluid

q_{rev} = heat adsorbed from the surroundings during a reversible process

Q = activity product of a reaction

QFM = quartz + fayalite + magnetite oxygen buffer

r = rate of reaction

r = radius

R = molar gas constant that is equal to the Boltzmann constant times Avogadro's number and has a value of 8.31424 J mol^{-1} K^{-1} = 83.1424 cm^3 bar mol^{-1} K^{-1} = 1.98726 cal mol^{-1} K^{-1}

R_+ = forward rate of reaction

R_{net} = net rate of forward minus backward reaction

REE = rare earth element

s = second

s = surface area

S = entropy

$SMOW$ = standard mean ocean water, a standard for isotopic analysis

S_{abs} = absolute entropy of a substance

t = time, often in seconds

t_{ft} = fission track age

t_R = average amount of time a substance spends in a reservoir

T = temperature either in K or °C

T_{ALK} = total alkalinity of a solution in molality

T_C = critical temperature

TDS = total dissolved solids

T_{fus} = fusion temperature in K

U = Internal Energy

v_i = velocity of i

v_i = voltage of reaction i, the SI unit of electrical potential difference

v_{max} = maximum rate (velocity) at which product can form

V_i = volume of i, which in SI units is m^3

$\overline{V}$ = molar volume

w = work

w_d = width of diffuse layer of charge in solution at a solid surface

watt = SI unit of power = 1 joule second^{-1}

$\overline{W}$ = molar work

X_i = mole fraction of species i

Z_i = compressibility factor of pure gas i

Z_j = charge on the jth species

ZPC = zero point of charge

α = coefficient of isobaric thermal expansion

α^{+2} = alpha particle

$\alpha_{H_2O}^{CO_2}$ = oxygen isotope fractionation factor between CO_2 and H_2O

β = isothermal compression

β^- = beta particle (electron ejected from nucleus)

β^+ = positron particle

β_1 = constant related to the binding energy for adsorption

β_2 = adsorption maximum for a sample

γ_s = Gibbs energy per unit surface area

γ_A = molal activity coefficient of A

$\gamma_\pm$ = mean ionic molal activity coefficient

Γ_{LJ} = Lennard-Jones 6-12 potential energy function of interaction between molecules

Γ_A = Raoult's law activity coefficient of A

δx = path dependent variable of x

δS_{heat} = entropy added to a system by transfer of heat

δS_{irr} = irreversible entropy produced due to increased randomness

$\delta^{(heavy)}A$ = ratio of heavy to light isotope of element A in a sample in per mil (‰)

ΔC_{P_R} = heat capacity change at constant pressure of reaction

ΔE = energy change

ΔE = voltage change

ΔE° = voltage of the reaction under standard-state conditions

ΔE_{grav} = gravitation energy

$\Delta G_{act}^{o\ddagger}$ = standard-state Gibbs energy of reaction to form the activated complex in the forward direction

ΔG_f = Gibbs energy of formation of a substance from its elements in their most stable state at 25°C and 1 bar.

ΔG_R = Gibbs energy of reaction

ΔG_R^o = standard-state Gibbs energy change of a reaction

ΔH_f = enthalpy of formation of a substance from its elements in their most stable state at 25°C and 1 bar.

ΔH_R = enthalpy of reaction

ΔH_{xtal} = heat of crystallization

$\Delta^{18}O_Y^X = \delta^{18}O_X - \delta^{18}O_Y$ where X and Y are different phases or species

ΔP_{vis} = viscous pressure gradient

ΔS_{fus} = entropy change on fusion

ΔS_R = entropy of reaction

ΔV_{fus} = volume change on fusion

ΔV_R = volume of reaction

ε = dielectric constant

ε = negative of the Lennard-Jones 6-12 energy at the equilibrium distance between gas molecules

ε_o = permittivity in a vacuum (8.854×10^{-14} C V^{-1} cm^{-1})

ε_{Nd} = epsilon isotopic value of Nd

η = viscosity of a fluid

κ = thermal diffusivity = ratio of the thermal conductivity to heat capacity at constant volume

κ = transmission coefficient, the probability of breakdown of the rate-limiting activated complex to products

λ = wavelength

$\mu = {}^{238}U/{}^{204}Pb$ in a reservoir as it exists today

μ_i = chemical potential of component i = change in Gibbs energy with change in i when holding everything else constant

μ_i^o = standard chemical potential of component i = change in Gibbs energy with change in i in the chosen standard state

μ_i^* = non-standard chemical potential of component i = change in Gibbs energy from the standard state to the state of interest for 1 mole of i.

ν_i = stoichiometric coefficient of species i in a reaction

ξ = extent of chemical reaction with time variable

π = pi, which has a value of 3.14159265 . . .

ρ = density

ρ_f = density of fluid

ρ_R = density of rock

σ = distance of molecular separation where Lennard-Jones 6-12 potential is zero

σ = entropy production per unit volume of a system

σ_c = surface charge produced by loss of cations to solution at a solid surface

σ_{cr} = solid surface charge contributed by coordinative reactions with the aqueous solution

σ_d = diffuse layer of charge in solution near solid surface

σ_p = permanent solid surface charge produced by imperfections

σ_o = total net charge on a solid surface

σ_β = surface charge in solution from a Stern layer

$\sum CO_2$ = total molality of inorganic carbon in solution

τ = tortuosity

χ = Langmuir adsorption isotherm equation variable

χ_A = fugacity coefficient of gas A = f_A/P_A

Ψ = electrostatic potential (V)

Ψ_d = electrostatic potential at the start of the diffuse layer

Ψ^o = electrostatic potential at the solid surface

ω = regular solution or Margules interaction parameter

$\equiv$ = species on a solid surface that is bonded into the solid

$^o/_{oo}$ = stable isotope delta value in parts per thousand

[A] = concentration of A

Prefix multipliers for metric quantities

Factor	Prefix	Symbol	Factor	Prefix	Symbol
10^{-1}	deci-	d	10^{1}	deca-	da
10^{-2}	centi-	c	10^{2}	hecto-	h
10^{-3}	milli-	m	10^{3}	kilo-	k
10^{-6}	micro-	μ	10^{6}	mega-	M
10^{-9}	nano-	n	10^{9}	giga-	G
10^{-12}	pico-	p	10^{12}	tera-	T
10^{-15}	femto-	f	10^{15}	peta-	P
10^{-18}	atto-	a	10^{18}	exa-	E

B

Common Conversion Factors

The most widely used systems of measurements are SI and CGS units. Given below are conversion factors between the two systems and some other common units.

Force = mass × acceleration

System	Force unit
CGS	dyne = force to accelerate a mass of 1 g by 1 cm s^{-2}
SI (MKS)	newton = force to accelerate a mass of 1 kg by 1 m s^{-2}

Pressure = force × area^{-1} or energy × volume^{-1} or force × distance × volume^{-1}

Pressure unit	Bar	Atmosphere	psi	mm Hg	Pascal
CGS: 1 µbar = dyne cm^{-2} =	10^{-6}	0.98692×10^{-6}	14.504×10^{-6}	7.50×10^{-4}	0.1
SI: 1 pascal = newton m^{-2} =	10^{-5}	0.98692×10^{-5}	14.504×10^{-5}	7.50×10^{-3}	1
1 bar =	1	0.98692	14.504	750.1	10^{5}
1 atmosphere =	1.01325	1	14.696	760	1.01325×10^{5}

Energy = force × distance or pressure × volume

Energy unit	Erg	Joule	Thermochemical calorie	cm^{3} × bar
CGS: 1 erg = dyne cm =	1	10^{-7}	2.38901×10^{-8}	10^{-6}
SI: 1 joule = newton meter =	10^{7}	1	0.23901	10.00
1 calorie =	4.184×10^{7}	4.1840	1	41.84

Other helpful conversions

Entropy =	Energy $\times$ K^{-1}
Volume =	Distance^3 or energy $\times$ pressure^{-1}
Dynamic viscosity =	CGS unit = poise = 1 dyne s cm^{-2}
	SI unit = poiseuille = 1 newton s m^{-2} = pascal s

APPENDIX C1
Table of Elements with Atomic Number, Atomic Weight, and Valence States

Element	Symbol	Atom. num.	Atomic weight	Valence states
Actinium	Ac	89	227.03	+3
Aluminum	Al	13	26.98	+3
Americium	Am	95	(243)	+3,+4,+5,+6
Antimony	Sb	51	121.75	+3,+5,−3
Argon	Ar	18	39.95	0
Arsenic	As	33	74.92	+3,+5,−3
Astatine	At	85	(210)	?
Barium	Ba	56	137.34	+2
Berkelium	Bk	97	(247)	+3,+4
Beryllium	Be	4	9.01	+2
Bismuth	Bi	83	208.98	+3,+5
Boron	B	5	10.81	+3
Bromine	Br	35	79.91	+1,+5,−1
Cadmium	Cd	48	112.30	+2
Calcium	Ca	20	40.08	+2
Californium	Cf	98	(251)	+3
Carbon	C	6	12.01	+2,+4,−4
Cerium	Ce	58	140.12	+3,+4
Cesium	Cs	55	132.90	+1
Chlorine	Cl	17	35.45	+1,+5,+7,−1
Chromium	Cr	24	52.00	+2,+3,+6
Cobalt	Co	27	58.93	+2,+3
Copper	Cu	29	63.54	+1,+2
Curium	Cm	96	(247)	+3
Dysprosium	Dy	66	162.50	+3
Einsteinium	Es	99	(254)	+3

Continued

Appendix C1   continued

Element	Symbol	Atom. num.	Atomic weight	Valence states
Erbium	Er	68	167.26	+3
Europium	Eu	63	151.96	+2,+3
Fermium	Fm	100	(257)	+3
Fluorine	F	9	19.00	−1
Francium	Fr	87	(223)	+1
Gadolinium	Gd	64	157.25	+3
Gallium	Ga	31	69.72	+3
Germanium	Ge	32	72.59	+2,+4
Gold	Au	79	196.97	+1,+3
Hafnium	Hf	72	178.49	+4
Helium	He	2	4.00	0
Holmium	Ho	67	164.93	+3
Hydrogen	H	1	1.01	+1,−1
Indium	In	49	114.82	+3
Iodine	I	53	126.90	+1,+5,+7,−1
Iridium	Ir	77	192.20	+3,+4
Iron	Fe	26	55.85	+2,+3
Krypton	Kr	36	83.80	0
Lanthanum	La	71	174.97	+3
Lead	Pb	82	207.19	+2,+4
Lithium	Li	3	6.94	+1
Lutetium	Lu	71	174.97	+3
Magnesium	Mg	12	24.31	+2
Manganese	Mn	25	54.94	+2,+3,+4,+7
Mendelevium	Md	101	(258)	+2,+3
Mercury	Hg	80	200.59	+1,+2
Molybdenum	Mo	42	95.95	+6
Neodymium	Nd	60	144.24	+3
Neon	Ne	10	20.18	0
Neptunium	Np	93	237.05	+3,+4,+5,+6
Nickel	Ni	28	58.71	+2,+3
Niobium	Nb	41	92.91	+3,+5
Nitrogen	N	7	14.01	+1,+2,+3,+4, +5,0,−1,−2,−3
Nobelium	No	102	(259)	+2,+3
Osmium	Os	76	190.20	+3,+4
Oxygen	O	8	16.00	−2
Palladium	Pd	46	106.40	+2,+4
Phosphorus	P	15	30.97	+3,+5,−3
Platinum	Pt	78	195.09	+2,+4
Plutonium	Pu	94	(244)	+3,+4,+5,+6

Element	Symbol	Atom. num.	Atomic weight	Valence states
Polonium	Po	84	(209)	+2,+4
Potassium	K	19	39.10	+1
Praseodymium	Pr	59	140.91	+3
Promethium	Pm	61	(145)	+3
Protactinium	Pa	91	231.04	+4,+5
Radium	Ra	88	226.03	+2
Radon	Rn	86	(222)	0
Rhenium	Re	75	186.20	+4,+6,+7
Rhodium	Rh	45	102.90	+3
Rubidium	Rb	37	85.47	+1
Ruthenium	Ru	44	101.07	+3
Samarium	Sm	62	150.35	+2,+3
Scandium	Sc	21	44.96	+3
Selenium	Se	34	78.96	+4,+6,–2
Silicon	Si	14	28.09	+2,+4,–4
Silver	Ag	47	107.87	+1
Sodium	Na	11	22.99	+1
Strontium	Sr	38	87.62	+2
Sulfur	S	16	32.06	+4,+6,–2
Tantalum	Ta	73	180.95	+5
Technetium	Tc	43	98.91	+4,+6,+7
Tellurium	Te	52	127.60	+4,+6,–2
Terbium	Tb	65	158.92	+3
Thallium	Tl	81	204.37	+1,+3
Thorium	Th	90	232.04	+4
Thulium	Tm	69	168.93	+3
Tin	Sn	50	118.69	+2,+4
Titanium	Ti	22	47.90	+2,+3,+4
Tungsten	W	74	183.85	+6
Uranium	U	92	238.03	+3,+4,+5,+6
Vanadium	V	23	50.94	+2,+3,+4,+5
Xenon	Xe	54	131.30	0
Ytterbium	Yb	70	173.04	+2,+3
Yttrium	Y	39	88.90	+3
Zinc	Zn	30	65.37	+2
Zirconium	Zr	40	91.22	+4

Values in parentheses are the atomic weight of the longest lived isotope.

C2 Electron Configuration of Neutral Elements

Filling of the Indicated Orbitals with Electrons in Their Ground State

Atomic no.	Element	1s	2s	2p	3s	3p	3d	4s	4p	4d	4f	5s	5p	5d	5f	6s	6p	6d	7s
1	H	1																	
2	He	2																	
3	Li	2	1																
4	Be	2	2																
5	B	2	2	1															
6	C	2	2	2															
7	N	2	2	3															
8	O	2	2	4															
9	F	2	2	5															
10	Ne	2	2	6															
11	Na	2	2	6	1														
12	Mg	2	2	6	2														
13	Al	2	2	6	2	1													
14	Si	2	2	6	2	2													
15	P	2	2	6	2	3													
16	S	2	2	6	2	4													
17	Cl	2	2	6	2	5													
18	Ar	2	2	6	2	6													
19	K	2	2	6	2	6		1											
20	Ca	2	2	6	2	6		2											
21	Sc	2	2	6	2	6	1	2											
22	Ti	2	2	6	2	6	2	2											
23	V	2	2	6	2	6	3	2											
24	Cr	2	2	6	2	6	5	1											

Atomic no.	Element	1s	2s	2p	3s	3p	3d	4s	4p	4d	4f	5s	5p	5d	5f	6s	6p	6d	7s
25	Mn	2	2	6	2	6	5	2											
26	Fe	2	2	6	2	6	6	2											
27	Co	2	2	6	2	6	7	2											
28	Ni	2	2	6	2	6	8	2											
29	Cu	2	2	6	2	6	10	1											
30	Zn	2	2	6	2	6	10	2											
31	Ga	2	2	6	2	6	10	2	1										
32	Ge	2	2	6	2	6	10	2	2										
33	As	2	2	6	2	6	10	2	3										
34	Se	2	2	6	2	6	10	2	4										
35	Br	2	2	6	2	6	10	2	5										
36	Kr	2	2	6	2	6	10	2	6										
37	Rb	2	2	6	2	6	10	2	6			1							
38	Sr	2	2	6	2	6	10	2	6			2							
39	Y	2	2	6	2	6	10	2	6	1		2							
40	Zr	2	2	6	2	6	10	2	6	2		2							
41	Nb	2	2	6	2	6	10	2	6	4		1							
42	Mo	2	2	6	2	6	10	2	6	5		1							
43	Tc	2	2	6	2	6	10	2	6	6		1							
44	Ru	2	2	6	2	6	10	2	6	7		1							
45	Rh	2	2	6	2	6	10	2	6	8		1							
46	Pd	2	2	6	2	6	19	2	6	10									
47	Ag	2	2	6	2	6	10	2	6	10		1							
48	Cd	2	2	6	2	6	10	2	6	10		2							
49	In	2	2	6	2	6	10	2	6	10		2	1						
50	Sn	2	2	6	2	6	10	2	6	10		2	2						
51	Sb	2	2	6	2	6	10	2	6	10		2	3						
52	Te	2	2	6	2	6	10	2	6	10		2	4						
53	I	2	2	6	2	6	10	2	6	10		2	5						
54	Xe	2	2	6	2	6	10	2	6	10		2	6						
55	Cs	2	2	6	2	6	10	2	6	10		2	6			1			
56	Ba	2	2	6	2	6	10	2	6	10		2	6			2			
57	La	2	2	6	2	6	10	2	6	10		2	6	1		2			
58	Ce	2	2	6	2	6	10	2	6	10	2	2	6			2			
59	Pr	2	2	6	2	6	10	2	6	10	3	2	6			2			
60	Nd	2	2	6	2	6	10	2	6	10	4	2	6			2			

Continued

Appendix C2   continued

Atomic no.	Element	1s	2s	2p	3s	3p	3d	4s	4p	4d	4f	5s	5p	5d	5f	6s	6p	6d	7s
61	Pm	2	2	6	2	6	10	2	6	10	5	2	6			2			
62	Sm	2	2	6	2	6	10	2	6	10	6	2	6			2			
63	Eu	2	2	6	2	6	10	2	6	10	7	2	6			2			
64	Gd	2	2	6	2	6	10	2	6	10	7	2	6	1		2			
65	Tb	2	2	6	2	6	10	2	6	10	9	2	6			2			
66	Dy	2	2	6	2	6	10	2	6	10	10	2	6			2			
67	Ho	2	2	6	2	6	10	2	6	10	11	2	6			2			
68	Er	2	2	6	2	6	10	2	6	10	12	2	6			2			
69	Tm	2	2	6	2	6	10	2	6	10	13	2	6			2			
70	Yb	2	2	6	2	6	10	2	6	10	14	2	6			2			
71	Lu	2	2	6	2	6	10	2	6	10	14	2	6	1		2			
72	Hf	2	2	6	2	6	10	2	6	10	14	2	6	2		2			
73	Ta	2	2	6	2	6	10	2	6	10	14	2	6	3		2			
74	W	2	2	6	2	6	10	2	6	10	14	2	6	4		2			
75	Re	2	2	6	2	6	10	2	6	10	14	2	6	5		2			
76	Os	2	2	6	2	6	10	2	6	10	14	2	6	6		2			
77	Ir	2	2	6	2	6	10	2	6	10	14	2	6	7		2			
78	Pt	2	2	6	2	6	10	2	6	10	14	2	6	9		1			
79	Au	2	2	6	2	6	10	2	6	10	14	2	6	10		1			
80	Hg	2	2	6	2	6	10	2	6	10	14	2	6	10		2			
81	Tl	2	2	6	2	6	10	2	6	10	14	2	6	10		2	1		
82	Pb	2	2	6	2	6	10	2	6	10	14	2	6	10		2	2		
83	Bi	2	2	6	2	6	10	2	6	10	14	2	6	10		2	3		
84	Po	2	2	6	2	6	10	2	6	10	14	2	6	10		2	4		
85	At	2	2	6	2	6	10	2	6	10	14	2	6	10		2	5		
86	Rn	2	2	6	2	6	10	2	6	10	14	2	6	10		2	6		
87	Fr	2	2	6	2	6	10	2	6	10	14	2	6	10		2	6		1
88	Ra	2	2	6	2	6	10	2	6	10	14	2	6	10		2	6		2
89	Ac	2	2	6	2	6	10	2	6	10	14	2	6	10		2	6	1	2
90	Th	2	2	6	2	6	10	2	6	10	14	2	6	10		2	6	2	2
91	Pa	2	2	6	2	6	10	2	6	10	14	2	6	10	2	2	6	1	2
92	U	2	2	6	2	6	10	2	6	10	14	2	6	10	3	2	6	1	2
93	Np	2	2	6	2	6	10	2	6	10	14	2	6	10	5	2	6		2
94	Pu	2	2	6	2	6	10	2	6	10	14	2	6	10	6	2	6		2
95	Am	2	2	6	2	6	10	2	6	10	14	2	6	10	7	2	6		2
96	Cm	2	2	6	2	6	10	2	6	10	14	2	6	10	7	2	6	1	2
97	Bk	2	2	6	2	6	10	2	6	10	14	2	6	10	8	2	6	1	2

Atomic no.	Element	1s	2s	2p	3s	3p	3d	4s	4p	4d	4f	5s	5p	5d	5f	6s	6p	6d	7s
98	Cf	2	2	6	2	6	10	2	6	10	14	2	6	10	10	2	6		2
99	Es	2	2	6	2	6	10	2	6	10	14	2	6	10	11	2	6		2
100	Fm	2	2	6	2	6	10	2	6	10	14	2	6	10	12	2	6		2
101	Md	2	2	6	2	6	10	2	6	10	14	2	6	10	13	2	6		2
102	No	2	2	6	2	6	10	2	6	10	14	2	6	10	14	2	6		2
103	Lr	2	2	6	2	6	10	2	6	10	14	2	6	10	14	2	6	1	2

APPENDIX D

Atomic Abundance of Elements for the Entire Earth

Abundances Relative to Si = 10⁶

Element	Atomic abundance	Element	Atomic abundance	Element	Atomic abundance
H	15,200	Fe	148,000	Te	1.43
He	—	Co	227	I	0.026
Li	76.1	Ni	68,100	Xe	—
Be	1.20	Cu	176	Cs	0.087
B	8.30	Zn	279	Ba	7.2
C	5710	Ga	15.4	La	0.675
N	127	Ge	37.2	Ce	1.78
O	3,490,000	As	9.4	Pr	0.226
F	546	Se	15.1	Nd	1.18
Ne	—	Br	0.33	Sm	0.338
Na	1350	Kr	—	Eu	0.129
Mg	1,060,000	Rb	1.3	Gd	0.461
Al	128,000	Sr	40.6	Tb	0.083
Si	1,000,000	Y	7.2	Dy	0.542
P	13,600	Zr	42.2	Ho	0.12
S	112,000	Nb	2.1	Er	0.339
Cl	138	Mo	6.0	Tm	0.051
Ar	—	Ru	2.8	Yb	0.328
K	851	Rh	0.60	Lu	0.055
Ca	94,300	Pd	1.8	Hf	0.318
Sc	52.6	Ag	0.15	Ta	0.031
Ti	4210	Cd	0.037	W	0.266
V	396	In	0.0046	Re	0.080
Cr	18,000	Sn	1.16	Os	1.13
Mn	2100	Sb	0.103	Ir	1.08

Element	Atomic abundance	Element	Atomic abundance	Element	Atomic abundance
Pt	2.11	Tl	0.0047	Th	0.0548
Au	0.288	Pb	0.123	U	0.0148
Hg	0.0097	Bi	0.0045		

Values from Ganapathy, R. and Anders, E., 1974, Bulk compositions of the moon and earth, estimated from meteorites. *Proceedings of the 5th Lunar Science Conference*. Lunar and Planetary Institute, Houston, TX.

E

Gibbs Energy and Volume of Liquid H₂O

Gibbs energy of liquid H₂O in cal mol⁻¹ as a function of pressure and temperature calculated from SUPCRT92

Temperature (°C)	Psat*	0.5 kbar	1.0 kbar	2.0 kbar	3.0 kbar	5.0 kbar	10.0 kbar
0	−56,289	−56,077	−55,869	−55,464	−55,071		
25	−56,688	−56,475	−56,265	−55,857	−55,461	−54,695	
50	−57,124	−56,909	−56,697	−56,285	−55,884	−55,111	−53,291
100	−58,098	−57,877	−57,659	−57,236	−56,825	−56,033	−54,172
150	−59,193	−58,964	−58,738	−58,301	−57,878	−57,065	−55,165
200	−60,395	−60,158	−59,921	−59,466	−59,029	−58,192	−56,251
250	−61,690	−61,449	−61,199	−60,722	−60,268	−59,406	−57,420
300	−63,071	−62,833	−62,564	−62,061	−61,587	−60,696	−58,664
350	−64,529	−64,306	−64,011	−63,476	−62,980	−62,056	−59,975
400		−65,874	−65,539	−64,963	−64,440	−63,482	−61,347
500		−69,364	−68,829	−68,135	−67,548	−66,507	−64,259
1000		−90,479	−88,775	−87,047	−85,950	−84,334	−81,370

*Psat is the pressure at which liquid H₂O is saturated with vapor. Values labeled Psat below 100°C are at 1 bar, with the Psat value at 0°C really at 0.01°C.

Molar volumes of H_2O in cm^3 mol^{-1} as a function of pressure and temperature calculated from SUPCRT92

Temperature (°C)	Psat*	0.5 kbar	1.0 kbar	2.0 kbar	3.0 kbar	5.0 kbar	10.0 kbar
0	18.0	17.6	17.2	16.7	16.2		
25	18.1	17.7	17.4	16.8	16.4	15.7	
50	18.2	17.9	17.5	17.0	16.5	15.9	14.7
100	18.8	18.4	18.0	17.4	17.0	16.2	15.0
150	19.6	19.1	18.7	18.0	17.4	16.6	15.3
200	20.8	20.1	19.5	18.6	18.0	17.1	15.6
250	22.5	21.4	20.5	19.4	18.6	17.5	15.9
300	25.3	23.2	21.9	20.3	19.4	18.0	16.2
350	31.3	26.0	23.6	21.4	20.2	18.6	16.5
400		31.2	26.0	22.7	21.1	19.2	16.8
500		70.1	34.1	26.1	23.3	20.6	17.5
1000		206.8	102.6	54.1	39.9	29.6	21.7

*Psat is the pressure at which liquid H_2O is saturated with vapor. Values labeled Psat below 100°C are at 1 bar, with the Psat value at 0°C really at 0.01°C.

F1

Interactive Session with SUPCRT92

■ Starting a Session

Go to the website for the book: http://www.jbpub.com/catalog/9780763726423/

Open "Student Resources"

Download the appropriate SUPCRT92 application file to your computer and open it.

SUPCRT92 will be used to determine the thermodynamic properties of the reaction

$$2H^+ + CaCO_3 + 2NaAlSi_3O_8 = CaAl_2Si_2O_8 + 4SiO_2 + 2Na^+ + CO_2 + H_2O$$

   aq   calcite   albite   anorthite   quartz   aq   gas   liquid

The prompts in this program are given in **bold,** whereas the replies you input for this calculation are given in normal type. Wording in *italics* gives explanations.

would you like to use the default thermodynamic database? (y/n)

    y   *(Short for "yes"; a "no" allows one to use a modified set of parameters)*

choose file option for specifying reaction-independent parameters:

        1 = select one of three default files

        2 = select an existing nondefault file

        3 = build a new file

    1   *(A file that tells the application which pressures and temperatures to use)*

input solvent phase region

        1 = one phase

        2 = liq-vap saturation curve

        3 = EQ3/6 one-phase/sat curve

    1   *(These are the three possible default files that contain an array of pressures and temperatures to use. In this calculation pressures and temperatures in the one-phase liquid region will be used.)*

choose file option for specifying a reaction

> 1 = use an existing reaction file
>
> 2 = build a new reaction file

2   (*Because this is the first time the calculation is to be performed on your computer, a new file with the information for the reaction needs to be specified.*)

specify number of reactions to be processed:

1   (*Only the reaction indicated above will be input.*)

input title for reaction 1 of 1: (*Any identifying title of up to 80 characters can be input*)

2H+ + CaCO3 + 2NaAlSi3O8 = CaAl2Si2O8 + 4SiO2 + 2Na+ + CO2 + H2O

Reactions are input with one species per line with the coefficient in the reaction coming first, and after a space, the species identifier is entered.

The names of minerals are spelled out completely in capital letters (e.g., WOLLAS-TONITE); gases are entered as the formula of the gas, with each element written starting with a capital and followed by a lowercase letter if needed. This is followed by a comma and lowercase g (e.g., He,g or CH4,g). Aqueous species are input as the species formula, with each element written starting with a capital and followed by a lowercase letter if needed followed by its charge (e.g., Fe3+ or HS–). If the species is uncharged, this is followed by a comma and lowercase aq (e.g., SiO2,aq or CaCl2,aq). Liquid water is entered as H2O, whereas steam as H2O,g. The species available and their designation are given in Appendices F3–F5.

enter [coeff species] pairs, separated by

blanks, one pair per line, for reaction i

(conclude with [0])

1.0 ANORTHITE	(*The reaction is written with the product coefficients*
4.0 QUARTZ	*in the reaction input as positive numbers and the*
2.0 Na+	*reactants as negative numbers*)
1.0 CO2,g	
1.0 H2O	
−2.0 H+	
−1.0 CALCITE	
−2.0 ALBITE	
0	(*The last line input contains only a "zero." This tells the program you are done*)

is this correct? (y/n)

y   (*y for "yes." If you make a mistake inputting a line, you will be prompted for the correct values*)

would you like to save these reactions to a file? (y/n)

n   (*n for "no." If you want to save the inputted reaction, enter y and specify file name when prompted in a format like xxxxx.rea*)

specify name of tabulated output file:

reaction1.txt   (*Any identifying file name can be used. To help identify it and open it in WORD, a format like xxxxxx.txt is helpful if using a PC*)

specify options for *x-y* plot files:

1 do not generate plot files

2 generate plot files in generic format

3 generate plot files in KaleidaGraph format

1   (*If you want separate files that contain calculated values of each of the thermodynamic variables, enter "2" or "3"*)

execution in progress...

... execution completed

Close "Sup92pc" or "Output from SUPCRT92"

(*A "zero.dat" file may appear that can be deleted or ignored.*)

Double click on the file "reaction1.txt" to open it.

Scroll to see thermodynamic properties of the reaction from 500 bar to 5 kbar

First pages of output from <u>reaction1.txt</u>

```
***** SUPCRT92: input/output specifications for this run

        USER-SPECIFIED CON FILE containing
        T-P-D grid & option switches: file not saved

        USER-SPECIFIED RXN FILE containing
        chemical reactions: file not saved

        THERMODYNAMIC DATABASE: dprons92.dat

        SUPCRT-GENERATED TAB FILE containing
        tabulated reaction properties (this file): reaction1.txt

***** summary of option switches

        isat, lopt, iplot, univar, noninc:    0   2   1   0   0

***** summary of state conditions

        ISOBARS(bars)   : min, max, increment:    500.0000    5000.0000    500.0000
        TEMP(degC) range: min, max, increment:      0.0000    1000.0000    100.0000
```

`****************************************** REACTION   1   ******************************************`

REACTION TITLE:

2H+ + CaCO3 + 2NaAlSi3O8 = CaAl2Si2O8 + 4SiO2 + 2Na+ + CO2 +H2O

REACTION STOICHIOMETRY:

COEFF.	NAME	FORMULA
-1.000	CALCITE	CaCO3
-2.000	ALBITE	Na(AlSi3)O8
-2.000	H+	H(+)
1.000	ANORTHITE	Ca(Al2Si2)O8
4.000	QUARTZ	SiO2
1.000	CARBON-DIOXIDE	CO2,g
2.000	Na+	Na(+)
1.000	H2O	H2O

STANDARD STATE PROPERTIES OF THE SPECIES AT 25 DEG C AND 1 BAR

...... MINERALS ......

NAME	DELTA G (cal/mol)	DELTA H (cal/mol)	S (cal/mol/K)	V (cc/mol)	Cp (cal/mol/K)
ANORTHITE	-954078.	-1007552.	49.100	100.790	50.4
QUARTZ	-204646.	-217650.	9.880	22.688	10.6
CALCITE	-269880.	-288552.	22.150	36.934	19.6
ALBITE	-886308.	-939680.	49.510	100.250	49.0

NAME	MAIER-KELLY COEFFICIENTS					PHASE TRANSITION DATA		
	a(10**0)	b(10**3)	c(10**-5)	T limit (C)	Htr (cal/mol)	Vtr (cc/mol)	dPdTtr (bar/K)	
ANORTHITE	63.311	14.794	-15.440	1426.85				
QUARTZ	11.220	8.200	-2.700	574.85	290.	0.372	38.500	
post-transition 1	14.410	1.940	0.000	1726.85				
CALCITE	24.980	5.240	-6.200	926.85				
ALBITE	61.700	13.900	-15.010	199.85				
post-transition 1	81.880	3.554	-50.154	926.85				

...... GASES ......

NAME	DELTA G (cal/mol)	DELTA H (cal/mol)	S (cal/mol/K)	V (cc/mol)	Cp (cal/mol/K)
CO2,g	-94254.	-94051.	51.085	0.	8.9

NAME	MAIER-KELLY COEFFICIENTS			T limit (C)
	a(10**0)	b(10**3)	c(10**-5)	
CO2,g	10.570	2.100	-2.060	2226.85

...... AQUEOUS SPECIES ......

NAME	DELTA G (cal/mol)	DELTA H (cal/mol)	S (cal/mol/K)	V (cc/mol)	Cp (cal/mol/K)
Na+	-62591.	-57433.	13.960	-1.2	9.1
H+	0.	0.	0.000	0.0	0.0
H2O	-56688.	-68317.	16.712	18.1	18.0

EQUATION-OF-STATE COEFFICIENTS

NAME	a1(10**1)	a2(10**-2)	a3(10**0)	a4(10**-4)	c1(10**0)	c2(10**-4)	omega(10**-5)
Na+	1.8390	-2.2850	3.2560	-2.7260	18.1800	-2.9810	0.3306
H+	0.0000	0.0000	0.0000	0.0000	0.0000	0.0000	0.0000

STANDARD STATE PROPERTIES OF THE REACTION AT 25 DEG C AND 1 BAR

PRES(bars)	TEMP(degC)	DH2O(g/cc)	LOG K	DELTA G (cal)	DELTA H (cal)	DELTA S (cal/K)	DELTA V (cc)	DELTA Cp (cal/K)
1.000	25.000	0.997	4.610	-6290.	12526.	63.2	-30.2	20.5

STANDARD STATE PROPERTIES OF THE REACTION AT ELEVATED TEMPERATURES AND PRESSURES

PRES(bars)	TEMP(degC)	DH2O(g/cc)	LOG K	DELTA G (cal)	DELTA H (cal)	DELTA S (cal/K)	DELTA V (cc)	DELTA Cp (cal/K)
500.000	0.000	1.024	4.098	-5121.	10910.	58.8	-31.0	42.5
500.000	100.000	0.980	6.753	-11530.	13971.	68.4	-23.8	27.2
500.000	199.850	0.897	PHASE TRANSITION #1 for ALBITE					
500.000	200.000	0.897	8.629	-18682.	16392.	74.2	-24.4	TRANSITION
500.000	300.000	0.777	10.011	-26255.	17644.	76.6	-35.6	1.8
500.000	400.000	0.578	11.117	-34241.	*** DELTA G ONLY (CHARGED AQUEOUS SPECIES) ***			

500.000	500.000	0.257	*** BEYOND RANGE OF APPLICABILITY OF AQUEOUS SPECIES EQNS ***					
500.000	587.811	0.170	PHASE TRANSITION #1 for QUARTZ					
500.000	600.000	0.164	*** BEYOND RANGE OF APPLICABILITY OF AQUEOUS SPECIES EQNS ***					
500.000	700.000	0.130	*** BEYOND RANGE OF APPLICABILITY OF AQUEOUS SPECIES EQNS ***					
500.000	800.000	0.110	*** BEYOND RANGE OF APPLICABILITY OF AQUEOUS SPECIES EQNS ***					
500.000	900.000	0.097	*** BEYOND RANGE OF APPLICABILITY OF AQUEOUS SPECIES EQNS ***					
500.000	1000.000	0.087	*** BEYOND RANGE OF APPLICABILITY OF AQUEOUS SPECIES EQNS ***					
1000.000	0.000	1.045	4.380	-5474.	9883.	56.3	-28.1	71.9
1000.000	100.000	1.000	6.914	-11805.	13620.	68.2	-22.3	29.0
1000.000	199.850	0.924	PHASE TRANSITION #1 for ALBITE					
1000.000	200.000	0.924	8.757	-18960.	16250.	74.5	-22.3	TRANSITION
1000.000	300.000	0.823	10.146	-26608.	18153.	78.1	-25.7	15.9
1000.000	400.000	0.693	11.217	-34549.	19707.	80.6	-29.5	16.0
1000.000	500.000	0.528	12.083	-42745.	21794.	83.5	-20.3	27.7
1000.000	600.000	0.374	12.831	-51265.	24461.	86.7	0.1	TRANSITION
1000.000	600.798	0.373	PHASE TRANSITION #1 for QUARTZ					
1000.000	700.000	0.282	*** BEYOND RANGE OF APPLICABILITY OF AQUEOUS SPECIES EQNS ***					
1000.000	800.000	0.231	*** BEYOND RANGE OF APPLICABILITY OF AQUEOUS SPECIES EQNS ***					
1000.000	900.000	0.198	*** BEYOND RANGE OF APPLICABILITY OF AQUEOUS SPECIES EQNS ***					
1000.000	1000.000	0.176	*** BEYOND RANGE OF APPLICABILITY OF AQUEOUS SPECIES EQNS ***					
1500.000	0.000	1.064	4.638	-5796.	9032.	54.3	-26.0	94.6

1500.000	100.000	1.017	7.066	-12065.	13290.	68.0	-21.2	30.4
1500.000	199.850	0.947		PHASE TRANSITION #1 for ALBITE				TRANSITION
1500.000	200.000	0.946	8.877	-19218.	16069.	74.6	-21.0	16.9
1500.000	300.000	0.858	10.257	-26900.	18128.	78.6	-23.3	13.3
1500.000	400.000	0.751	11.327	-34889.	19639.	81.0	-27.2	13.3
1500.000	500.000	0.630	12.177	-43080.	20853.	82.7	-30.7	12.5

Revised HKF Model for the Thermodynamic Properties of Aqueous Species

Helgeson and Kirkham (1976) and Helgeson et al. (1981) developed a model to calculate the thermodynamic properties of aqueous species from 0°C to 1000°C and 1 bar to 5.0 kbar, the HKF equation of state. The model was later revised and extended (Oelkers and Helgeson 1988a, 1988b, 1988c; Shock, 1995; Shock and Helgeson, 1988, 1989, 1990; Shock et al., 1989, 1992, 1997; Tanger and Helgeson, 1988) and is generally identified as the revised HKF model. The model considers that a species has a set of intrinsic properties, and when it is placed in water, the species will cause local disruption of the water around it and thus a change in the energy of the solution. In addition, if it is charged, an energy change will occur by solvation of water to the aqueous species from the disrupted water structure produced. The HKF model divides the molal infinitely dilute properties of aqueous species into separate nonsolvation and solvation contributions. The nonsolvation contribution of a species is the sum of its intrinsic properties together with the solvent collapse it causes when placed in solution (or in the case of neutral species, local solvent expansion caused by cavity formation). The solvation contribution then accounts for the change in energy due to electrostatic attraction of the aqueous species for the solvent after the collapse/expansion.

In the HKF model, therefore, the standard infinitely dilute molal volume of the jth aqueous species is given by

$$\bar{V}_j^\circ = \Delta\bar{V}_{n,j}^\circ + \Delta\bar{V}_{s,j}^\circ \qquad \text{[F2.1]}$$

where $\Delta\bar{V}_{n,j}^\circ$ is the change in volume due to nonsolvation contributions and $\Delta\bar{V}_{s,j}^\circ$ adds the contributions due to solvation at standard-state conditions. The standard nonsolvated volume in the HKF model is characterized by

$$\Delta\bar{V}_{n,j}^\circ = \sigma + \left(\frac{\xi}{T - \theta}\right) \qquad \text{[F2.2]}$$

where T gives the temperature in K, and θ represents a solvent temperature parameter with a value of 228 K. The σ and ξ parameters are given by

$$\sigma = a_1 + \left(\frac{a_2}{\Psi + P}\right) \qquad \text{[F2.3]}$$

and

$$\xi = a_3 + \left(\frac{a_4}{\Psi + P}\right) \qquad \text{[F2.4]}$$

where Ψ represents a solvent pressure parameter and has a value of 2600 bars. The terms a_1, a_2, a_3, and a_4 are pressure- and temperature-independent parameters unique for each aqueous species that are fit to the determined standard-state properties. They are given in Appendix F5. The nonsolvated standard volume of the jth aqueous species is therefore calculated from

$$\Delta \overline{V}^{\circ}_{n,j} = a_1 + \left(\frac{a_2}{\Psi + P}\right) + \left[a_3 + \left(\frac{a_4}{\Psi + P}\right)\left(\frac{1}{T - \theta}\right)\right] \qquad \text{[F2.5]}$$

The volume (and heat capacity) change on solvation is modeled with the Born (1920) equation. This gives the energy of removing a species from a vacuum and placing it in a solvent, such as water with a dielectric constant of ε,

$$\Delta \overline{G}^{\circ}_{s,j} = \frac{N^{\circ}(Z_j e)^2}{2r_{e,j}}\left(\frac{1}{\varepsilon} - 1\right) \qquad \text{[F2.6]}$$

where N°, Z_j, e, $r_{e,j}$, and ε are Avogadro's number, the species charge (except for neutral species), the absolute electronic charge (4.80298×10^{-10} esu), the electrostatic radius of the species, and the dielectric constant of water, respectively. Conventional properties of single ions are obtained by subtracting the properties of H^+ from the solution so that the conventional Born coefficient of the jth aqueous species is given by

$$\omega_j = \frac{N^{\circ}(Z_j e)^2}{2r_{e,j}} - \frac{N^{\circ}(Z_{H^+} e)^2}{2r_{e,H^+}} Z_j \qquad \text{[F2.7]}$$

The Born energy to account for the solvation energy of the jth aqueous species then becomes

$$\Delta \overline{G}^{\circ}_{s,j} = \omega_j\left(\frac{1}{\varepsilon} - 1\right) \qquad \text{[F2.8]}$$

and the additivity convention of ions to obtain the properties of electrolytes is maintained. Values of ω_j for aqueous species are given in Appendix F5.

In the revised HKF model, the electrostatic radius of the species for the jth cation, $r_{e,j}$, is obtained from

$$r_{e,j} = r_{x,j} + |Z_j|\,(0.94 + g) \qquad \text{[F2.9]}$$

and for the jth anion from

$$r_{e,j} = r_{x,j} + |Z_j|\,g \qquad \text{[F2.10]}$$

where $r_{x,j}$ is the crystallographic radius of the jth aqueous species and g is a universal function of T and P derived by regression of $\overline{V}^{\circ}$ and $\overline{C}^{\circ}_P$ measurements for NaCl in water but for use with all charged aqueous species. The value of g is given by

$$g = 0.5 \left(-b + \sqrt{b^2 - 4c} \right) \qquad \text{[F2.11]}$$

where

$$b = 3.72 - 2\eta \left(\sum_{i=-1}^{4} \sum_{j=0}^{4-i} a_{ij} T^i \rho^j \right)^{-1} \qquad \text{[F2.12]}$$

and

$$c = 3.4571 - 3.72\eta \left(\sum_{i=-1}^{4} \sum_{j=0}^{4-i} a_{ij} T^i \rho^j \right)^{-1} \qquad \text{[F2.13]}$$

In Equations F2.11 to F2.13, η is the Born (1920) constant $= 1.66027 \times 10^5$ Å cal mol^{-1}, and ρ represents the density of H_2O in g cm^{-3} at the pressure and temperature of interest. The coefficients $a_{i,j}$ are compiled in **Table F(2)-1** as determined by Tanger and Helgeson (1988); therefore, ω_j is a function of pressure and temperature and depends on the nature of the aqueous species. Despite the fact that neutral aqueous species have no formal charge, they do have finite values of ω_j in the HKF model because of their polar nature. These ω_j are taken to be independent of pressure and temperature in Equation F2.8.

The pressure derivative of Equation F2.8 gives the aqueous species' standard solvation volume, $\Delta \overline{V}_{s,j}^{\circ}$, or

$$\left(\frac{\partial \Delta \overline{G}_{s,j}^{\circ}}{\partial P} \right)_T = \Delta \overline{V}_{s,j}^{\circ} = \omega_j \left(\frac{\partial \left(\frac{1}{\varepsilon} - 1 \right)}{\partial P} \right)_T + \left(\frac{\partial \omega_j}{\partial P} \right)_T \left(\frac{1}{\varepsilon} - 1 \right) \qquad \text{[F2.14]}$$

The $\Delta \overline{V}_{s,j}^{\circ}$ can be rewritten as

$$\Delta \overline{V}_{s,j}^{\circ} = \omega_j Q + \left(\frac{\partial \omega_j}{\partial P} \right)_T \left(\frac{1}{\varepsilon} - 1 \right) \qquad \text{[F2.15]}$$

|**Table F2-1**   **Values of coefficients in equations (F2.12) and (F2.13)**

a(i,j)	Values	a(i,j)	Values
a(−1,0)	−3,447,160,509	a(1,0)	−32,603.00052
a(−1,1)	6,814,704,812	a(1,1)	56,066.15908
a(−1,2)	−1,761,045,616	a(1,2)	−28,682.03067
a(−1,3)	−4,230,074,117	a(1,3)	3,794.138977
a(−1,4)	3,256,835,062	a(2,0)	27.16227245
a(−1,5)	−658,333,086.1	a(2,1)	−33.68176788
a(0,0)	18,062,334.78	a(2,2)	9.804290400
a(0,1)	−36,060,976.34	a(3,0)	−1.060481048 × 10^{-2}
a(0,2)	21,028,348.14	a(3,1)	6.934581461 × 10^{-3}
a(0,3)	−1,133,277.153	a(4,0)	1.576972120 × 10^{-6}
a(0,4)	−1,416,255.412		

where

$$Q = \frac{1}{\varepsilon}\left(\frac{\partial \ln \varepsilon}{\partial T}\right)_T \qquad \text{[F2.16]}$$

Combining Equations F2.1, F2.5, and F2.15 produces an expression for the conventional standard molal volume of species j:

$$\overline{V}_j^o = a_1 + \left(\frac{a_2}{\Psi + P}\right) + \left[a_3 + \left(\frac{a_4}{\Psi + P}\right)\right]\left(\frac{1}{T - \theta}\right) - \omega Q + \left(\frac{\partial \omega_j}{\partial P}\right)_T\left(\frac{1}{\varepsilon} - 1\right) \qquad \text{[F2.17]}$$
$$\underbrace{\hspace{5cm}}_{\text{nonsolvation terms}} \qquad \underbrace{\hspace{5cm}}_{\text{solvation terms}}$$

To obtain the conventional standard heat capacity of an aqueous species, consider the temperature derivative of Equation F2.8:

$$-\left(\frac{\partial \Delta \overline{G}_s^o}{\partial T}\right)_P = \Delta \overline{S}_s^o = \omega_j\left(\frac{\partial\left(\frac{1}{\varepsilon} - 1\right)}{\partial T}\right)_P - \left(\frac{\partial \omega_j}{\partial T}\right)_P\left(\frac{1}{\varepsilon} - 1\right) \qquad \text{[F2.18]}$$

$\Delta \overline{S}_s^o$ denotes the conventional standard entropy of the aqueous species written as

$$\Delta \overline{S}_s^o = \omega_j Y\left(\frac{\partial \omega_j}{\partial T}\right)_P\left(\frac{1}{\varepsilon} - 1\right) \qquad \text{[F2.19]}$$

where

$$Y = \frac{1}{\varepsilon}\left(\frac{\partial \ln \varepsilon}{\partial T}\right)_P \qquad \text{[F2.20]}$$

The thermodynamic model requires

$$\left(\frac{\partial \Delta \overline{S}_s^o}{\partial T}\right)_P = \left(\frac{\Delta \overline{C}_{P_s}^o}{T}\right) \qquad \text{[F2.21]}$$

where $\Delta \overline{C}_{P_s}^o$ is the conventional standard heat capacity. Recasting Equation F2.21 and substituting the temperature derivative of Equation F2.19, the conventional standard partial molal heat capacity of solvation of the jth aqueous species is given by

$$\Delta \overline{C}_{P_{s,j}}^o = T\left(\frac{\partial \Delta \overline{S}_s^o}{\partial T}\right)_P = \omega TX + 2TY\left(\frac{\partial \omega_j}{\partial T}\right)_P - T\left(\frac{\partial^2 \omega_j}{\partial T^2}\right)_P\left(\frac{1}{\varepsilon} - 1\right) \qquad \text{[F2.22]}$$

where X is defined as

$$X = \frac{1}{\varepsilon}\left[\left(\frac{\partial^2 \ln \varepsilon}{\partial T^2}\right)_P - \frac{1}{\varepsilon}\left(\frac{\partial \ln \varepsilon}{\partial T}\right)_P^2\right] \qquad \text{[F2.23]}$$

Similar to the volume, the conventional standard partial molar heat capacity in the HKF model is given by a nonsolvation and solvation term

$$\Delta \overline{C}_{P_j}^o = \Delta \overline{C}_{P_{n,j}}^o + \Delta \overline{C}_{P_{s,j}}^o \qquad \text{[F2.24]}$$

The conventional standard nonsolvated heat capacity in the revised HKF model is calculated from

$$\Delta \overline{C}^{\circ}_{P_{n,j}} = c_1 + \frac{c_2}{(T - \theta)^2} - \left(\frac{2T}{(T - \theta)^3} \right) \left[a_3(P - P_r) + a_4 \ln \left(\frac{\Psi + P}{\Psi + P_r} \right) \right] \qquad \text{[F2.25]}$$

where c_1 and c_2 are pressure and temperature-independent parameters unique for each aqueous species given in Appendix F5. The term in square brackets in Equation F2.25 determines the pressure behavior of $\Delta \overline{C}^{\circ}_{P_{n,j}}$. Substituting Equations F2.22 and F2.25 into Equation F2.24 gives

$$\Delta \overline{C}^{\circ}_{P_j} = c_1 + \frac{c_2}{(T - \theta)^2} - \left(\frac{2T}{(T - \theta)^3} \right) \left[a_3(P - P_r) + a_4 \ln \left(\frac{\Psi + P}{\Psi + P_r} \right) \right] \qquad \text{[F2.26]}$$

non-solvation terms

$$+ \omega TX + 2TY \left(\frac{\partial \omega_j}{\partial T} \right)_P - T \left(\frac{\partial^2 \omega_j}{\partial T^2} \right)_P \left(\frac{1}{\varepsilon} - 1 \right)$$

solvation terms

Knowing the standard partial molal volume and heat capacity of an aqueous species, the standard partial molal Gibbs energy of the species is calculated by integration from the reference pressure, P_r, and temperature, T_r, to the pressure and temperature of interest with the equation

$$\Delta \overline{G}^{\circ}_{f,P,T} - \Delta \overline{G}^{\circ}_{f,P_r,T_r} - \overline{S}^{\circ}_{P_r,T_r}(T - T_r) + \int_{T_r}^{T} \overline{C}^{\circ}_P dT - \int_{T_r}^{T} \frac{\overline{C}^{\circ}_P}{T} dT + \int_{P_r}^{P} \overline{V}^{\circ} dP \qquad \text{[F2.27]}$$

as outlined in Chapter 3. For an aqueous species, therefore, the apparent standard partial molal Gibbs energy of formation at the pressure and temperature of interest is given by

$$\begin{aligned}
\Delta \overline{G}^{\circ}_{f,P,T} = {} & \Delta \overline{G}^{\circ}_{f,P_r,T_r} - \overline{S}^{\circ}_{P_r,T_r}(T - T_r) - c_1 \left[T \ln \left(\frac{T}{T_r} \right) - T + T_r \right] \\
& + a_1(P - P_r) + a_2 \ln \left(\frac{\Psi + P}{\Psi + P_r} \right) \\
& - c_2 \left\{ \left[\left(\frac{1}{T - \theta} \right) - \left(\frac{1}{T_r - \theta} \right) \right] \left(\frac{\theta - T}{\theta} \right) - \frac{T}{\theta^2} \ln \left[\frac{T_r(T - \theta)}{T(T_r - \theta)} \right] \right\} \\
& + \left(\frac{1}{T - \theta} \right) \left[a_3(P - P_r) + a_4 \ln \left(\frac{\Psi + P}{\Psi + P_r} \right) \right] \\
& + \omega_{P,T} \left(\frac{1}{\varepsilon_{P,T}} - 1 \right) - \omega_{P_r,T_r} \left(\frac{1}{\varepsilon_{P_r,T_r}} - 1 \right) + \omega_{P_r,T_r} Y_{P_r,T_r}(T - T_r)
\end{aligned} \qquad \text{[F2.28]}$$

ω is taken to be independent of pressure and temperature for neutral species so that $\omega_{P,T} = \omega_{P_r,T_r}$ equals a constant.

Inspection of Equation F2.28 indicates to obtain a value of the apparent standard partial molal Gibbs energy of formation of an aqueous species at pressure and temperature, $\Delta \overline{G}^{\circ}_{f,P,T}$, requires a number of values. These include the apparent standard partial molal Gibbs energy of formation of the species at the reference conditions of 1 bar and 25°C, $\Delta \overline{G}^{\circ}_{f,P_r,T_r}$ in cal mol^{-1} and its molal entropy at 25°C and 1 bar, $\Delta S^{\circ}_{P_r,T_r}$ in cal mol^{-1} K^{-1}, together with a_1 and a_3 in cal mol^{-1} bar^{-1}, a_2 and a_4 in cal mol^{-1} bar^{-2}, c_1 in cal mole^{-1} K^{-1}, c_2 in cal K mole^{-1}, and ω_{P_r,T_r} in cal mol^{-1}. These are given in Appendix F5. The value of θ, the solvent temperature parameter, is 228 K, whereas that of Ψ, the solvent pressure parameter, is 2600 bars. The dielectric constant of water at 25°C and 1 bar, ε_{P_r,T_r}, is 78.47 and the Born derivative equation Y at 25°C and 1 bar, Y_{P_r,T_r}, is -5.81×10^{-5} K^{-1}. Given these parameters, ω_{PT} and ε_{PT} must be independently calculated. ω_{PT} requires the ion's crystallographic radius from Ahrens (1952) or estimated (Tanger and Helgeson, 1988) and the density of H_2O, ρ, at the pressure and temperature of interest. ρ can be calculated from the equations given in Helgeson and Kirkham (1974). ε_{PT}, the dielectric constant of water at the pressure and temperature of interest, can also be determined from the equations given in Helgeson and Kirkham (1974). These equations and coefficients are used in the SUPCRT92 computer program (Johnson et al., 1992), which is available at the publisher's website at http://www.jbpub.com/catalog/9780763726423/ under Student Resources. SUPCRT92 can be used to calculate the Gibbs energy of aqueous species at elevated temperature and pressure given by Equation F2.28 along with other standard-state thermodynamic properties of aqueous species from the parameters given in Appendix F5.

The software to run SUPCRT92 and packages to modify the database are available from the online research portal (http://affinity.berkeley.edu/predcent) of the Prediction Central laboratory at the University of California, Berkeley.

References

Ahrens, L. H., 1952, The use of ionization potentials, 1: Ionic radii of the elements. *Geochim. Cosmochim. Acta,* v. 2, pp. 155–169.

Born, M., 1920, Volumen und Hydrationswärme der Ionen. *Zeitschr. Physik,* v. 1, pp. 45–48.

Helgeson, H. C. and Kirkham, D. H., 1974, Theoretical prediction of the thermodynamic properties of aqueous electrolytes at high pressures and temperatures. I. Summary of the thermodynamic/electrostatic properties of the solvent. *Am. J. Sci.,* v. 274, pp. 1089–1198.

Helgeson, H. C. and Kirkham, D. H., 1976, Theoretical prediction of the thermodynamic properties of aqueous electrolytes at high pressures and temperatures. III. Equation of state for aqueous species at infinite dilution. *Am. J. Sci.,* v. 276, pp. 97–240.

Helgeson, H. C., Kirkham, D. H. and Flowers, G. C., 1981, Theoretical prediction of the thermodynamic properties of aqueous electrolytes at high pressures and temperatures. IV. Calculation of activity coefficients, osmotic coefficients, and apparent molal and standard and relative partial molal properties to 600°C and 5 kb. *Am. J. Sci.,* v. 281, pp. 1249–1516.

Johnson, J. W., Oelkers, E. H. and Helgeson, H. C., 1992, SUPCRT92: A software package for calculating the standard molal thermodynamic properties of minerals, gases, aqueous species, and reactions from 1 to 5000 bar and 0 to 1000°C. *Comput. Geosci.,* v. 18, pp. 899–947.

Oelkers, E. H. and Helgeson, H. C., 1988a, Calculation of the thermodynamic properties of aqueous species at high pressures and temperatures: aqueous tracer diffusion coefficients of ions to 1000°C and 5 kb. *Geochim. Cosmochim. Acta,* v. 52, pp. 63–85.

Oelkers, E. H. and Helgeson, H. C., 1988b, Calculation of the thermodynamic properties of aqueous species at high pressures and temperatures: Dissociation constants for supercritical alkali metal halides at temperatures from 400° to 800°C and pressures from 500 to 4000 bars. *J. Phys. Chem.*, v. 92, pp. 1631–1639.

Oelkers, E. H. and Helgeson, H. C., 1988c, Calculation of the thermodynamic properties of aqueous species at high pressures and temperatures: summary of the limiting equivalent conductances, Stokes law radii, apparent solvation numbers, and Walden products of ionic species, and the anomalous mobility of hydrogen and hydroxide ions to 1000°C and 5 kb. *J. Solution Chem.*, v. 52, pp. 63–85.

Shock, E. L., 1995, Organic acids in hydrothermal solutions: standard molal thermodynamic properties of carboxylic acids and estimates of dissociation constants at high temperatures and pressures. *Am. J. Sci.*, v. 295, pp. 496–580.

Shock, E. L. and Helgeson, H. C., 1988, Calculation of the thermodynamic properties of aqueous species at high pressures and temperatures: correlation algorithms for ionic species and equation of state predictions to 5 kb and 1000°C. *Geochim. Cosmochim. Acta*, v. 52, pp. 2009–2036.

Shock, E. L. and Helgeson, H. C., 1989, Correction to Shock and Helgeson (1988). *Geochim. Cosmochim. Acta*, v. 53, p. 215.

Shock, E. L. and Helgeson, H. C., 1990, Calculation of the thermodynamic properties of aqueous species at high pressures and temperatures: Standard partial molal properties of organic species. *Geochim. Cosmochim. Acta*, v. 54, pp. 915–943.

Shock, E. L., Helgeson, H. C. and Sverjensky, D. A., 1989, Calculation of the thermodynamic and transport properties of aqueous species at high pressures and temperatures: Standard partial molal properties of inorganic neutral species. *Geochim. Cosmochim. Acta*, v. 53, pp. 2157–2183.

Shock, E. L., Oelkers, E. H., Johnson, J. W., Sverjensky, D. A. and Helgeson, H. C., 1992, Calculation of the thermodynamic and transport properties of aqueous species at high pressures and temperatures: Effective electrostatic radii, dissociation constants and standard partial molal properties to 1000°C and 5 kbar. *J. Chem. Soc. Faraday Trans.*, v. 88, pp. 803–826.

Shock, E. L., Sassani, D. C., Willis, M. and Sverjensky, D. A. 1997, Inorganic species in geological fluids: Correlations among standard molal thermodynamic properties of aqueous ions and hydroxide complexes. *Geochim. Cosmochim. Acta*, v. 61, pp. 907–950.

Sverjensky, D. A., Shock, E. L. and Helgeson, H. C., 1997, Prediction of the thermodynamic properties of aqueous metal complexes to 1000°C and 5 kb. *Geochem. Cosmochim. Acta*, v. 61, pp. 907–950.

Tanger, J. C. IV and Helgeson, H. C., 1988, Calculation of the thermodynamic and transport properties of aqueous species at high pressures and temperatures: revised equations of state for the standard partial molal properties of ions and electrolytes. *Am. J. Sci.*, v. 288, pp. 19–98.

APPENDIX

F3 | Thermodynamic Data for Minerals[a]

[a]Data from Johnson, J. W., Oelkers, E. H., and Helgeson, H. C., 1992, SUPCRT: A software package for calculating the standard molal thermodynamic properties of minerals, gases, aqueous species, and reactions from 1 to 5000 bar and 0 to 1000°C. *Comput. Geosci.*, v. 18, pp. 899–947.

Table F3-1 Minerals described with one heat capacity equation

Name (SUPCRT92)	Molar formula	Abbreviation	ΔG°_f	ΔH°_f	S°	V°	Maier-Kelly a	C_P b × 10³	Coefficients c × 10⁻⁵	C_P lim °C	Ref.
AKERMANITE	$Ca_2MgSi_2O_7$	Ak	-879362.0	-926497.0	50.03	92.81	60.09	11.40	11.40	1700	1,19
ALABANDITE	MnS	Alb	-52178.0	-51000.0	19.20	21.46	11.40	1.80	0.00	1803	1
ALBITE,LOW	$Na(AlSi_3)O_8$	Lo-Ab	-886308.0	-939680.0	49.51	100.07	61.70	13.90	15.01	1400	1
ALUNITE	$KAl_3(OH)_6(SO_4)_2$	Alu	-1113600.0	-1235600.0	78.40	293.60	153.45	0.00	54.95	650	1
AMESITE,7A	$Mg_2Al(AlSi)O_5(OH)_4$	Ams-7A	—	—	52.00	103.00	81.03	24.738	20.23	1000	1
AMORPH-SILICA	SiO_2	Amor-Si	-202892.0	-214568.0	14.34	29.00	5.930	47.20	22.78	622	1
ANALCIME	$NaAlSi_2O_6*H_2O$	Anl	-738098.0	-790193.0	56.00	97.10	53.49	24.14	8.88	1000	1
ANALCIME, DEHYDR	$NaAlSi_2O_6$	Dhyd-Anl	-674989.0	-714678.0	41.90	89.10	42.09	24.14	8.88	1000	1
ANDALUSITE	Al_2SiO_5	And	-580587.0	-615866.0	22.20	51.53	41.310833	6.292556	12.392063	1043	1
ANDRADITE	$Ca_3Fe_2Si_3O_{12}$	Adr	-1296619.0	-1380345.0	70.13	131.85	113.532	15.636	30.889	1100	1,19
ANGLESITE	$PbSO_4$	Ang	-194353.0	-219870.0	35.51	47.95	10.96	31.00	-4.20	1100	1
ANHYDRITE	$CaSO_4$	Anh	-315925.0	-342760.0	25.50	45.94	16.78	23.60	0.00	1453	1
ANNITE	$KFe_3(AlSi_3)O_{10}(OH)_2$	Ann	-1147156.0	-1232195.0	5.20	154.32	106.43	29.77	19.31	1000	1
ANORTHITE	$Ca(Al_2Si_2)O_8$	An	-954078.0	-1007552.0	49.10	100.79	63.311	14.794	15.44	1700	1,19
ARAGONITE	$CaCO_3$	Arg	-269683.0	-288531.0	21.56	34.15	20.13	10.24	-3.34	600	1,18
ARTINITE	$Mg_2(OH)_2(CO_3)*3H_2O$	Art	-613915.0	-698043.0	55.67	96.90	70.87	27.66	7.43	1000	1
AZURITE	$Cu_3(OH)_2(CO_3)_2$	Azr	-334417.0	-390100.0	66.97	91.01	36.88	77.44	0.92	780	1
BARITE	$BaSO_4$	Brt	-325563.0	-352100.0	31.60	52.10	33.80	0.00	8.43	1422	1
BERNDTITE	SnS_2	Brn	-34750.0	-36700.0	20.90	40.96	15.51	4.20	0.00	1000	12
BOEHMITE	$AlO(OH)$	Bhm	-217250.0	-235078.0	11.58	19.535	14.435	4.20	0.00	500	1
BROMELLITE	BeO	Brm	—	—	3.38	8.309	8.45	4.00	3.17	1200	1
BRUCITE	$Mg(OH)_2$	Brc	-199646.0	-221390.0	15.09	24.63	24.147	4.0124	6.11	900	1
CA-AL-PYROXENE	$CaAl(AlSi)O_6$	Ca-Al-Px	-742067.0	-783793.0	35.00	63.50	54.13	6.42	14.90	1400	1,19
CALCITE	$CaCO_3$	Cal	-269880.0	-288552.0	22.15	36.934	24.98	5.24	6.20	1200	1,18

Name (SUPCRT92)	Molar formula	Abbreviation	ΔG_f°	ΔH_f°	S°	V°	Maier-Kelly a	$b \times 10^3$	Coefficients $c \times 10^{-5}$	C_P lim °C	Ref.
CASSITERITE	SnO_2	Cst	−124260.0	−138800.0	12.50	21.55	17.246	2.8026	4.9001	1500	12
CELADONITE	$K(MgAl)Si_4O_{10}(OH)_2$	Cln	—	—	74.90	157.10	80.25	25.30	18.54	1000	1
CELESTITE	$SrSO_4$	Cls	−320435.0	−347300.0	28.00	46.25	21.80	13.30	0.00	1500	1
CERUSSITE	$PbCO_3$	Cer	−150370.0	−168000.0	31.30	40.59	12.39	28.60	0.00	800	1
CHABAZITE	$Ca(Al_2Si_4)O_{12}*6H_2O$	Cbz	—	—	152.90	247.76	146.00	44.47	16.43	1000	1
CHALCEDONY	SiO_2	Cha	−204276.0	−217282.0	9.88	22.688	11.22	8.20	2.70	848	1
CHAMOSITE,7A	$Fe_2Al(AlSi)O_5(OH)_4$	Chm-7A	—	—	64.70	106.20	84.91	25.40	18.77	1000	1
CHLORARGYRITE	$AgCl$	Crg	−26247.0	−30370.0	23.00	25.727	14.33	1.821	2.43	728	7,15
CHLORITOID	$FeAl_2SiO_5(OH)_2$	Cld	—	—	42.10	69.80	60.63	17.13	13.63	1000	1
CHRYSOTILE	$Mg_3Si_2O_5(OH)_4$	Ctl	−964871.0	−1043123.0	52.90	108.50	75.82	31.60	17.58	1000	1
CINNABAR	HgS	Cin	−10940.0	−12750.0	19.70	28.416	10.46	3.72	−0.0618		1
CLINOCHLORE,14A	$Mg_5Al(AlSi_3)O_{10}(OH)_8$	14A-Cnc	−1961703.0	−2116964.0	111.20	207.11	166.50	42.10	37.47	900	1
CLINOZOISITE	$Ca_2Al_3Si_3O_{12}(OH)$	Czo	−1549240.0	−1643781.0	70.64	136.20	106.118	25.214	27.145	700	1,19
COPPER,NATIVE	Cu	Cu	0.0	0.0	7.923	7.113	5.41	1.50	0.00	1357	1
CORDIERITE	$Mg_2Al_3(AlSi_5)O_{18}$	Crd	−2061279.0	−2183199.0	97.33	233.22	143.83	25.80	38.60	1700	1
CORDIERITE,HYDRO	$Mg_2Al_3(AlSi_5)O_{18}*H_2O$	Hyd-Crd	−2121350.0	−2255676.0	111.43	241.22	155.23	25.80	38.60	17	1
CORUNDUM	Al_2O_3	Cm	−378167.0	−400500.0	12.17	25.575	23.197	9.326	6.30	500	x
COVELLITE	CuS	Cv	−12612.0	−12500.0	15.90	20.42	10.60	2.64	0.00	1273	1
CRISTOBALITE,ALPHA	SiO_2	α-Crs	−203895.0	−216755.0	10.372	25.74	13.98	3.34	3.81	1000	1
CRISTOBALITE,BETA	SiO_2	β-Crs	−203290.0	−215675.0	11.963	27.38	17.39	0.31	9.89	2000	1
CUMMINGTONITE	$Mg_7Si_8O_{22}(OH)_2$	Cum	—	—	127.50	271.90	185.24	185.24	44.66	800	1
CUPRITE	Cu_2O	Cpr	−35384.0	−40830.0	22.08	23.437	14.08	5.88	0.76	1515	1
DAPHNITE,14A	$Fe_5Al(AlSi_3)O_{10}(OH)_8$	Dph-14A	—	—	142.50	213.42	176.21	43.76	33.82	1000	1
DIASPORE	$AlO(OH)$	Dsp	−218402.0	−237170.0	8.43	17.76	14.435	4.20	0.00	500	1
DICKITE	$Al_2Si_2O_5(OH)_4$	Dck	—	—	47.10	99.300	72.77	29.20	21.52	1000	1,7

Table F3-1 continued

Name (SUPCRT92)	Molar formula	Abbreviation	ΔH_f°	ΔG_f°	S°	V°	Maier-Kelly a	C_P b×10³	Coefficients c×10⁻⁵	C_P lim °C	Ref.
DIOPSIDE	$CaMg(SiO_3)_2$	Di	-765378.0	-723780.0	34.20	66.090	52.87	7.84	15.74	1600	1,19
DOLOMITE	$CaMg(CO_3)_2$	Dol	-556631.0	-517760.0	37.09	64.365	41.557	23.952	9.884	1000	1,19
DOLOMITE,DISORDERED	$CaMg(CO_3)_2$	Dis-Dol	-553704.0	-515653.0	39.84	64.390	44.711	17.779	10.948	1000	1,19
DOLOMITE,ORDERED	$CaMg(CO_3)_2$	Ord-Dol	-556631.0	-517760.0	37.09	64.340	44.711	17.779	10.948	1000	1,19
EPIDOTE	$Ca_2FeAl_2Si_3O_{12}(OH)$	Ep	-1543992.0	-1450906.0	75.28	139.200	117.622	12.816	31.864	1100	1,19
EPIDOTE,ORDERED	$Ca_2FeAl_2Si_3O_{12}(OH)$	Ord-Ep	-1544016.0	-1450906.0	75.20	139.2	113.78	14.69	28.92	1100	1,19
FAYALITE	Fe_2SiO_4	Fa	-354119.0	-330233.0	35.45	46.390	36.51	9.36	6.70	1490	1
FE-PARGASITE	$Na(Ca_2Fe_4Al)(Al_2Si_6)O_{22}(OH)_2$	Fe-Prg	—	—	185.50	279.89	213.57	42.99	47.29	1000	1
FE-TREMOLITE	$(Ca_2Fe_5)Si_8O_{22}(OH)_2$	Fe-Tr	—	—	163.50	282.8	197.93	58.95	41.17	800	1
FERROUS-OXIDE	FeO	Frs-Ox	-65020.0	-60097.0	14.52	12.0	12.122	2.072	0.75	1600	1
FLUORITE	CaF_2	Fl	-293000.0	-280493.0	16.39	24.542	14.30	7.28	-0.47	1424	1
F-PHLOGOPITE	$KMg_3(AlSi_3)O_{10}(F)_2$	F-Phl	—	—	—	269.70	190.26	59.36	50.01	800	1
F-TREMOLITE	$(Ca_2Mg_5)Si_8O_{22}(F)_2$	F-Tr	—	—	129.40	270.45	183.74	55.25	43.73	800	1
FORSTERITE	Mg_2SiO_4	Fo	-520000.0	-491564.0	22.75	43.79	35.81	6.54	8.52	1800	1
GALENA	PbS	Gn	-23500.0	-23115.0	21.80	31.49	11.17	2.20	0.00	1388	1
GEHLENITE	$Ca_2(Al_2Si)O_7$	Gh	-951225.0	-903148.0	48.10	90.24	63.74	8.00	15.12	1800	1,19
GIBBSITE	$Al(OH)_3$	Gbs	-309065.0	-276168.0	16.75	31.956	8.65	45.60	0.00	425	1
GLAUCOPHANE	$Na_2(Mg_3Al_2)Si_8O_{22}(OH)_2$	Gln	—	—	130.00	269.70	190.26	59.36	50.01	800	1
GOLD,NATIVE	Au	Au	0.0	0.0	11.33	10.215	5.66	1.24	0.00	1336	1
GRAPHITE	C	Gr	0.0	0.0	1.372	5.298	4.03	1.14	2.04	2500	1
GREENALITE	$Fe_3Si_2O_5(OH)_4$	Grn	—	—	72.60	115.00	81.65	32.6	15.39	1000	1
GROSSULAR	$Ca_3Al_2Si_3O_{12}$	Grs	-1582737.0	-1496307.0	60.87	125.30	104.017	17.013	27.318	1000	1,19
GRUNERITE	$Fe_7Si_8O_{22}(OH)_2$	Gru	—	—	163.30	256.67	198.83	60.93	39.55	800	1
HALITE	$NaCl$	Hl	-98260.0	-91807.0	17.24	27.015	10.98	3.90	0.00	1073	1
HALLOYSITE	$Al_2Si_2O_5(OH)_4$	Hal	—	—	48.52	99.30	72.77	29.20	21.52	1000	1,7
HEDENBERGITE	$CaFe(SiO_3)_2$	Hd	-678276.0	-638998.0	40.70	68.27	54.81	8.17	15.01	1600	1,19
HUNTITE	$CaMg_3(CO_3)_4$	Hun	-1082600.0	-1004710.0	71.59	122.90	84.17	42.86	20.44	1000	1

Name (SUPCRT92)	Molar formula	Abbreviation	ΔH°_f	ΔG°_f	S°	V°	Maier-Kelly a	C_P $b \times 10^3$	Coefficients $c \times 10^{-5}$	C_P lim °C	Ref.
HYDROMAGNESITE	$Mg_5(OH)_2(CO_3)_4*4H_2O$	Hydro-Mgs	-1557090.0	-1401687.0	129.38	208.80	141.46	65.28	21.67	1000	1
JADEITE	$NaAl(SiO_3)_2$	Jd	-722116.0	-679445.0	31.90	60.40	48.16	11.42	11.87	1400	1
K-FELDSPAR	$K(AlSi_3)O_8$	K-Fs	-949188.0	-895374.0	51.13	108.87	76.617	4.311	29.945	1400	1
KAOLINITE	$Al_2Si_2O_5(OH)_4$	Kln	-982221.0	-905614.0	48.53	99.52	72.77	29.20	21.52	1000	1
KYANITE	Al_2SiO_5	Ky	-616897.0	-580956.0	20.00	44.09	41.393132	6.816463	12.882104	466	1
LAUMONTITE	$Ca(Al_2Si_4)O_{12}*4H_2O$	Lmt	-1728664.0	-1596823.0	116.10	207.55	123.20	44.47	16.43	10.0	1,19
LEONHARDITE	$Ca_2(Al_4Si_8)O_{24}*7H_2O$	Lnr	—	—	220.40	407.86	235.00	88.94	32.86	1000	1
LIME	CaO	Lim	-151790.0	-144366.0	9.50	16.764	11.67	1.08	1.56	2000	1
MAGNESITE	$MgCO_3$	Mgs	-265630.0	-245658.0	15.70	28.018	19.731	12.539	4.748	1000	1
MALACHITE	$Cu_2(OH)_2(CO_3)$	Mal	-251900.0	-214204.0	44.50	54.86	27.76	43.78	1.34	80	1
MANGANOSITE	MnO	Mng	-92070.0	-86735.0	14.27	13.221	11.11	1.94	0.88	1800	1,6
MERWINITE	$Ca_3Mg(SiO_4)_2$	Mer	-1090796.0	-1036526.0	60.50	104.40	72.97	11.96	14.44	1700	1,19
METACINNABAR	HgS	Met-Cin	-11800.0	-10437.0	21.20	30.169	10.52	3.63	0.00	1000	1
MICROCLINE,MAX	$K(AlSi_3)O_8$	Max-Mc	-949188.0	-895374.0	51.13	108.741	63.83	12.90	17.05	1400	1
MINNESOTAITE	$Fe_3Si_4O_{10}(OH)_2$	Mnn	—	—	83.50	147.86	88.31	42.61	11.15	800	1
MONTICELLITE	$CaMgSiO_4$	Mtc	-540800.0	-512829.0	26.40	51.47	36.82	5.34	8.00	1400	1
MUSCOVITE	$KAl_2(AlSi_3)O_{10}(OH)_2$	Ms	-1427408.0	-1336301.0	68.80	140.71	97.56	26.38	25.44	1000	1
NEPHELINE	$Na(AlSi)O_4$	Ne	-500241.0	-472872.0	29.72	54.16	35.908062	6.457918	7.327965	1500	1
PARAGONITE	$NaAl_2(AlSi_3)O_{10}(OH)_2$	Pg	-1416963.0	-1326012.0	66.40	132.53	97.43	24.50	26.440	1000	1
PARGASITE	$Na(Ca_2Mg_4Al)(Al_2Si_6)O_{22}(OH)_2$	Prg	-3016624.0	-2846728.0	160.00	273.50	205.80	41.66	50.21	1000	1,19
PERICLASE	MgO	Per	-143800.0	-136086.0	6.44	11.248	10.18	1.74	1.48	2100	1
PHLOGOPITE	$KMg_3(AlSi_3)O_{10}(OH)_2$	Phl	-1488067.0	-1396187.0	76.10	149.66	100.61	28.78	21.50	1000	1
POTASSIUM-OXIDE	K_2O	K-Ox	-86800.0	-77056.0	22.50	40.38	18.510	8.65	0.88	1100	1
PYRITE	FeS_2	Py	-41000.0	-38293.0	12.65	23.94	17.88	1.32	3.05	1000	1
PYROPHYLLITE	$Al_2Si_4O_{10}(OH)_2$	Prl	-212521.0	-1255997.0	23.90	31.075	21.99	9.30	4.69	700	1

Table F3-1 continued

Name (SUPCRT92)	Molar formula	Abbreviation	ΔG_i°	ΔH_i°	S°	V°	Maier-Kelly a	C_P b×10³	Coefficients c×10⁻⁵	C_P lim °C	Ref.
RHODOCHROSITE	$MnCO_3$	Rds	-195045.0	-212521.0	23.90	31.075	21.99	9.30	4.69	700	1
RICHTERITE	$Na_2(CaMg_6)Si_8O_{22}(OH)_2$	Rct		—	137.60	272.80	194.8	61.1	46.15	800	1
ROMARCHITE	SnO	Sn-Ox	-61459.0	-68340.0	13.66	20.895	9.55	3.50	0.00	1237	12
RUTILE	TiO_2	Ru	-212883.0	-226101.0	12.02	18.82	15.014420	2.720501	2.365364	1800	7,9
SANIDINE,HIGH	$K(AlSi_3)O_8$	Hi-Sa	-893738.0	-946538.0	54.53	109.008	63.83	12.90	17.05	1400	1
SEPIOLITE	$Mg_4Si_6O_{15}(OH)_2(H_2O)_2*4H_2O$	Sep	-2211192.0	-2418000.0	146.60	285.60	157.92	104.30	18.68	800	1
SIDERITE	$FeCO_3$	Sd	-162414.0	-179173.0	25.10	29.378	11.63	26.80	0.00	885	1,13
SILLIMANITE	Al_2SiO_5	Sil	-580091.0	-615099.0	23.13	49.90	40.023988	7.390511	11.674053	1200	1
SILVER,NATIVE	Ag	Ag	0.0	0.0	10.17	10.272	5.09	2.04	-0.36	1234	1
SMITHSONITE	$ZnCO_3$	Smt	-174850.0	-194260.0	19.70	28.275	9.30	33.00	0.00	780	1
SODIUM-OXIDE	Na_2O	Na-Ox	-89883.0	-99140.0	17.935	25.00	18.25	4.89	2.89	1000	1
SPHALERITE	ZnS	Sp	-47947.0	-49000.0	14.019	23.83	11.77	1.26	1.16	1300	1
SPINEL	$MgAl_2O_4$	Spn	-517006.0	-546847.0	19.27	39.71	36.772808	6.415456	9.708792	2000	1
STAUROLITE	$Fe_2Al_9Si_4O_{23}(OH)$	St		—	117.10	224.40	207.12	36.94	57.22	1000	1
STRONTIANITE	$SrCO_3$	Str	-275470.0	-294600.0	23.20	39.01	23.52	6.32	5.08	1197	1
SYLVITE	KCl	Sy	-97735.0	-104370.0	19.73	37.524	9.89	5.20	-0.77	1043	1
TALC	$Mg_3Si_4O_{10}(OH)_2$	Tcl	-1320188.0	-1410920.0	62.34	136.25	82.482	41.614	13.342	800	1
TENORITE	CuO	Tn	-30568.0	-37200.0	10.18	12.22	11.53	1.88	1.76	1600	1
TITANITE	$CaTiSiO_5$	Ti	-587129.0	-620960.0	30.88	55.65	42.234220	5.703951	9.525038	1670	7,9,19
TREMOLITE	$(Ca_2Mg_5)Si_8O_{22}(OH)_2$	Tr	-2770245.0	-2944038.0	131.19	272.92	188.222	57.294	44.822	800	1,19
WAIRAKITE	$Ca(Al_2Si_4)O_{12}*2H_2O$	Wai	-1477432.0	-1579333.0	105.10	186.87	100.40	44.47	16.43	1000	1,19
WITHERITE	$BaCO_3$	Wth	-278400.0	-297500.0	26.80	45.81	21.50	11.06	3.91	1079	1
WOLLASTONITE	$CaSiO_3$	Wo	-369225.0	-389590.0	19.60	39.93	26.64	3.60	6.52	1400	1,19
WURTZITE	ZnS	Wur	-44810.0	-45850.0	14.064	23.846	11.82	1.160	1.04	1300	1
ZINCITE	ZnO	Znc		—	10.43	14.338	11.71	1.220	2.18	2000	1
ZOISITE	$Ca_2Al_3Si_3O_{12}(OH)$	Zo	-1549179.0	-1643691.0	70.74	135.90	106.118	25.214	27.145	700	1,19

Table F3-2   Minerals described with two heat capacity equations

Name (SUPCRT) abbreviation	Formula	G_i°	H_i°	S°	V°	Maier-Kelly a	C_p b×10³	Function c×10⁻⁵	Tran T	ΔH_{tr}	ΔV_{tr}	Maier-Kelly a	C_p b×10³	function c×10⁻⁵	C_p Lim	Ref.
ALBITE	$Na(AlSi_3)O_8$															
Ab		−886308	−939680	49.51	100.25	61.70	13.90	15.01	473	—	—	81.88	3.554	50.154	1200	1
ALBITE,HIGH	$NaAlSi_3O_8$															
Hi-Ab		−884509	−937050	52.30	100.43	61.70	13.90	15.01	623	—	—	64.17	13.90	15.01	1400	1
ALMANDINE	$Fe_3Al_2Si_3O_{12}$															
Al		—	—	75.60	115.28	97.52	33.64	18.73	848	—	—	107.09	14.86	10.63	1600	1
AMESITE,14A	$Mg_4Al_2(Al_2Si_2)O_{10}(OH)_8$															
14Å-Am		—	—	108.9	205.40	172.59	34.98	41.67	848	—	—	169.40	41.24	−44.37	1000	1
ANTIGORITE	$Mg_{48}Si_{34}O_{85}(OH)_{62}$															
Atg		−15808020	−17070891	861.36	1749.13	1228.45	513.76	286.68	848	—	—	1234.83	501.24	281.28	1000	1
CLINOCHLORE,7A	$Mg_5Al(AlSi_3)O_{10}(OH)_8$															
7A-Cnc		−1957101	−2113197	106.50	211.50	162.82	50.62	40.88	848	—	—	166.01	44.36	38.18	900	1
COESITE	SiO_2															
Cos		−203541	−216614	9.65	20.641	11.00	8.20	2.70	848	—	—	14.19	1.94	0.0	2000	1
CRISTOBALITE	SiO_2															
Crs		−203895	−216755	10.372	25.74	13.98	3.34	3.81	543	321	—	17.39	0.31	9.89	2000	1
DAPHNITE,7A	$Fe_5Al(AlSi_3)O_{10}(OH)_8$															
7A-Dph		—	—	138.90	221.20	172.53	52.28	37.23	848	—	—	175.72	46.02	34.63	1000	1
EDENITE	$Na(Ca_2Mg_5)(AlSi_7)O_{22}(OH)_2$															
Ed		—	—	161.0	271.0	199.71	48.78	46.01	848	—	—	202.90	42.52	43.31	1000	1
EPISTILBITE	$CaAl_2Si_6O_{16}*5H_2O$															
Eps		—	—	152.9	267.56	157.04	60.87	21.83	848	—	—	163.42	48.35	16.43	1000	1

Table F3-2 continued

Name (SUPCRT) abbreviation	Formula G_f°	H_f°	S°	V°	Maier-Kelly C_P a	$b \times 10^3$	Function $c \times 10^{-5}$	Tran T	ΔH_{tr}	ΔV_{tr}	Maier-Kelly C_P a	$b \times 10^3$	function $c \times 10^{-5}$	C_P.Lim	Ref.
FERROEDENITE	$Na(Ca_2Fe_5)(AlSi_7)O_{22}(OH)_2$														
Fe-Ed	—	—	193.7	280.90	209.42	50.44	42.36	848	—	—	212.61	44.18	39.66	1000	1
FERROSILITE	$FeSiO_3$														
Fs	-267588.0	-285658.0	21.63	32.952	26.49	5.07	5.55	413	37	0.056	23.865	8.78	4.70	1400	1,13
FLUOREDENITE	$Na(Ca_2Mg_5)(AlSi_7)O_{22}(F)_2$														
Fl-Ed	—	—	146.7	272.50	195.23	46.74	44.92	848	—	—	198.42	40.48	42.22	1000	1
HERZENBERGITE	SnS														
Hrz	-25023.0	-25464.0	18.36	29.01	8.53	7.48	0.90	875	160	—	9.78	3.74	0.0	1153	12
HEULANDITE	$Ca(Al_2Si_7)O_{18}*6H_2O$														
Hul	—	—	182.4	316.37	179.66	69.07	24.53	848	—	—	189.23	50.29	16.43	1000	1
KALSILITE	$K(AlSi)O_4$														
Kls	-481750.0	-509408.0	31.85	59.89	29.43	17.36	5.32	810	160	—	42.50	0.0	0.0	1800	1
LAWSONITE	$CaAl_2Si_2O_7(OH)_2*H_2O$														
Lws	-1073408.0	-1158104.0	55.80	101.32	81.80	23.36	16.26	848	—	—	84.99	17.10	13.56	1000	1,19
MAGNETITE	Fe_3O_4														
Mag	-242574.0	-267250.0	34.83	44.524	21.88	48.20	0.00	900	—	—	48.0	0.0	0.0	1800	1
MARGARITE	$CaAl_2(Al_2Si_2)O_{10}(OH)_2$														
Mrg	-1394150.0	-1485803.0	63.80	129.40	102.50	16.35	-28.05	848	—	—	99.31	22.61	0.0	1000	1,19
NATROLITE	$Na_2(Al_2Si_3)O_{10}*2H_2O$														
Ntr	—	—	101.6	169.72	95.76	40.08	15.06	848	—	—	92.57	46.34	17.76	1000	1
NESQUEHONITE	$MgCO_3*3H_2O$														
Nsh	-412035.0	-472576.0	46.76	74.79	-1574.804	3899.1	417.325	-306.5	184	—	25.246	91.289	4.222	340	1
NICKEL	Ni														
Ni	0.0	0.0	7.14	6.588	4.06	7.04	0.00	633	0.0	—	6.00	1.80	0.0	1725	1
PHILLIPSITE,Ca	$Ca(Al_2Si_5)O_{14}*5H_2O$														
Ca-Php	—	—	166.6	265.0	145.82	52.67	19.13	848	—	—	149.01	-46.41	16.43	1000	1

Name (SUPCRT) abbreviation	Formula	G_f°	H_f°	S°	V°	Maier-Kelly C_P Function a	$b \times 10^3$	Function $c \times 10^{-5}$	Tran T	ΔH_{tr}	ΔV_{tr}	Maier-Kelly C_P function a	$b \times 10^3$	function $c \times 10^{-5}$	C_P Lim	Ref.
PHILLIPSITE,K	$K_2(Al_2Si_5)O_{14} *5H_2O$															
K-Php		—	—	172.1	265.0	152.66	60.24	18.45	848	—	—	155.85	53.98	15.75	1000	1
PHILLIPSITE,Na	$Na_2(Al_2Si_5)O_{14} *5H_2O$															
Na-Php		—	—	172.4	265.0	152.40	56.48	20.46	848	—	—	155.59	50.22	17.76	1000	1
PREHNITE	$Ca_2Al_2Si_3O_{10}(OH)_2$															
Prh		-1390097.0	-1481649.0	65.00	140.33	91.60	37.82	19.60	848	—	—	101.17	19.04	-11.50	1000	1,19
PYROPE	$Mg_3Al_2Si_3O_{12}$															
Prp		—	—	62.30	113.27	91.69	32.64	20.92	848	—	—	101.26	13.86	12.82	1800	1,7
QUARTZ	SiO_2															
Qtz		-204646.0	-217650.0	9.88	22.688	11.22	8.20	-2.70	848	290	0.372	14.41	1.94	0.0	2000	1
QUICKSILVER	Hg															
Hg		0.0	0.0	18.17	14.822	6.44	0.00	-0.19	629	1414	—	4.97	0.0	0.0	3000	1
SPESSARTINE	$Mn_3Al_2Si_3O_{12}$															
Sps		—	—	74.50	117.90	94.48	33.24	19.12	848	—	—	104.05	14.46	11.02	1800	1
STILBITE	$NaCa_2(Al_5Si_{13})O_{36} *14H_2O$															
Stb		—	—	399.3	649.91	390.55	137.68	49.84	848	—	—	400.12	118.90	-41.74	1000	1
TIN,NATIVE	Sn															
Sn		0.0	0.0	12.24	16.289	4.42	6.30	0.00	505.0	1680	—	7.30	0.0	0.0	3000	12

Table F3-3 Minerals described with three heat capacity equations

Name (SUPCRT)	Formula	Abbrev.	ΔG°_f	ΔH°_f	S°	V°	a	$b \times 10^3$	$c \times 10^{-5}$	Tran T	ΔH_{tr}
ΔV_{tr}	a	$b \times 10^3$	$c \times 10^{-5}$	2nd Tran T	2nd ΔH_{tr}	2nd ΔV_{tr}	a	$b \times 10^3$	$c \times 10^{-5}$	C_P lim	Ref.
AEGERINE	$NaFe(SiO_3)_2$	Agt	—	—	36.70	64.37	46.16	19.31	-9.46	950	—
—	52.42	10.01	-7.68	1050	0.0	—	50.27	10.89	-7.68	1400	—
ANTHOPHYLLITE	$Mg_7Si_8O_{22}(OH)_2$	Ath	-2715430.0	-2888749.0	128.6	264.4	180.6826	0.574	-38.462	903	—
—	197.542	41.614	-13.342	1258	0.0	—	199.522	41.614	-13.342	1800	1
BORNITE	Cu_5FeS_4	Bn	-86704.0	-79922.0	99.29	98.60	49.76	35.08	-1.35	485	1430
—	-34.31	247.0	0.0	540	0.0	—	80.33	-2.04	0.0	1200	1,16
BUNSENITE	NiO	Bsn	-50573.0	-57300.0	9.08	10.97	-4.99	37.58	3.89	525	0.0
—	13.88	0.0	0.0	565	0.0	—	11.18	2.02	0.0	2000	0.0
CHALCOCITE	Cu_2S	Cc	-20626.0	-19000.0	28.90	27.48	12.63	18.82	0.0	376	920
—	26.78	-7.35	0.0	717	0.0	—	20.32	0.0	0.0	1400	1
CHALCOPYRITE	$CuFeS_2$	Ccp	-44900.0	-44453.0	31.15	42.83	20.79	12.80	-1.34	830	2405
—	-141.4	210.0	0.0	930	0.0	—	41.22	0.0	0.0	1200	1,16
CRONSTEDTITE,7A	$Fe_2FeSiO_5(OH)_4$	7Å-Csd	—	—	73.50	110.9	84.79	41.84	-12.476	950	—
—	97.30	23.24	-8.93	1050	0.0	—	93.01	25.0	-8.93	1500	—
ENSTATITE	$MgSiO_3$	En	-348930.0	-369686.0	16.20	31.276	24.55	4.74	-6.28	903	166
0.02	28.765	0.0	0.0	1050	390	1.09	29.26	0.0	0.0	1800	1
FERROGEDRITE	$(Fe_4Al_2)(Al_2Si_6)O_{22}(OH)_2$	Fe-Ged	—	—	153.5	265.9	196.27	58.39	-39.01	903	—
—	204.70	48.91	-26.45	1258	0.0	—	205.69	48.91	-26.45	1500	—
HASTINGSITE	$Na(Ca_2Fe_4Fe)(Al,Si)_8O_{22}(OH)_2$	Hs	—	—	189.2	280.3	211.57	50.88	-44.88	950	—
—	217.82	41.58	-43.10	1258	0.0	—	215.68	42.46	-43.10	1500	—
HEMATITE	Fe_2O_3	Hem	-178155.0	-197720.0	20.94	30.274	23.49	18.60	-3.55	950	160
—	36.00	0.0	0.0	1050	0.0	—	31.71	1.76	0.0	1800	1

Name (SUPCRT) / ΔV_{tr}	Formula / a	Abbrev. / $b \times 10^3$	ΔG°_f / $c \times 10^{-5}$	ΔH°_f / 2nd Tran T	S° / 2nd ΔH_{tr}	V° / 2nd ΔV_{tr}	a / a	$b \times 10^3$ / $b \times 10^3$	$c \times 10^{-5}$ / $c \times 10^{-5}$	Tran T / C_P lim	ΔH_{tr} / Ref.
LARNITE	Ca_2SiO_4	Lrn	—	—	30.50	51.60	34.87	9.74	-6.26	970	440
—	32.16	11.02	0.0	1710	3390.0	—	49.00	0.0	0.0	2000	1,8
MG-HASTINGSITE	$Na(Ca_2Mg_4Fe)Al_2Si_6O_{22}(OH)_2$	Mg-Hs	—	1050	163.8	273.80	203.8	49.55	-47.8	950	—
—	210.06	40.25	-46.02	—	—	—	207.91	41.13	-46.02	1500	1
MG-RIEBECKITE	$Na_2(Mg_3Fe_2)Si_8O_{22}(OH)_2$	Mg-Rbk	-41.63	1050	138.0	271.3	186.26	75.14	-45.18	950	—
—	198.77	56.54	-41.63	—	—	—	194.48	58.30	-41.63	1500	1
PYRRHOTITE	FeS	Po	-24084.0	-24000.0	14.41	18.20	5.19	26.40	0.0	411	570
—	17.40	0.0	0.0	598	120.0	—	12.20	2.38	0.0	1468	1
RIEBECKITE	$Na_2(Fe_3Fl)Si_8O_{22}(OH)_2$	Rbk	—	1050	156.6	274.9	192.09	76.14	-42.99	950	—
—	204.6	57.54	-39.44	—	—	—	200.31	59.30	-39.44	1500	1
IRON	Fe	Fe	0.0	0.0	6.52	7.092	3.04	7.58	0.60	1033	326
—	11.13	0.0	0.0	1183	215.0	—	5.80	1.98	0.0	1673	7,8
PD-OXYANNITE	$KFe_3(AlSi_3)O_{10}(OH)O^-$	Pd-Oxn	—	950	68.50	143.2	88.34	54.12	-18.06	848	—
—	97.91	35.34	-9.96	—	—	—	-116.68	7.44	-4.63	1050	1

Name (SUPCRT92) gives name needed for use of the SUPCRT92 program. ΔG°_f and ΔH°_f are 25 °C and 1 bar Gibbs energy and enthalpy of formation from the elements, respectively, in cal mol⁻¹. S° gives the entropy in cal mol⁻¹ K⁻¹. V° is in cm³ per mole. These values are for the pure phase at 25 °C and 1 bar. Note that the first set of Maier-Kelly heat capacity equation coefficients is to be used up to the indicated transition temperature given in °C. If there is an enthalpy, ΔH_{tr}, or volume of transition, ΔV_{tr}, these are added at the transition (given in cal mol⁻¹ and cm³ mol⁻¹, respectively). The second set of Maier-Kelly coefficients is to be used from the first transition temperature (°C) to the second transition temperature (°C). The third set of Maier-Kelly coefficients is to be used from the second transition temperature to the C_P limit temperature (°C). Mineral abbreviations are consistent with those recommended by the American Mineralogist and published by Kretz, R., 1983, Symbols for rock-forming minerals. *Am. Mineral.*, v. 68, pp. 277–279.

— indicates no value reported.

References

1. Helgeson, H. C., Delany, J. M, Nesbitt, H. W. and Bird, D. K., 1978, Summary and critique of the thermodynamic properties of rock-forming minerals. *Am. J. Sci.*, v. 278A, 229 pp.

2. Shock, E. L. and Helgeson, H. C., 1988, Calculation of the thermodynamic and transport properties of aqueous species at high pressures and temperatures: correlation algorithms for ionic species and equation of state predictions to 5 kb and 1000°C. *Geochim. Cosmochim. Acta*, v. 52, pp. 2009–2036.

3. Shock, E. L., Helgeson, H. C. and Sverjensky, D. A., 1989, Calculation of the thermodynamic and transport properties of aqueous species at high pressures and temperatures: standard partial molal properties of inorganic neutral species. *Geochim. Cosmochim. Acta*, v. 53, pp. 2157–2183.

4. Shock, E. L. and Helgeson, H. C., 1990, Calculation of the thermodynamic and transport properties of aqueous species at high pressures and temperatures: standard partial molal properties of organic species. *Geochim. Cosmochim. Acta*, v. 54, pp. 915–943.

5. Sverjensky, D. A., Shock, E. L. and Helgeson, H. C., 1997, Prediction of the thermodynamic properties of aqueous metal complexes to 1000°C and 5 kb. *Geochem. Cosmochim. Acta*, v. 61, pp. 907–950.

6. Wagman, D. D., Evans, W. H., Parker, V. B., Schumm, R. H., Halow, I., Bailey, S. M., Churney, K. L., and Nuttall, R. L., 1982, The NBS tables of chemical and thermodynamic properties. *J. Phys. Chem. Ref. Data*, v. 11, supplement no. 2, 392 pp.

7. Robie, R. A., Hemingway, B. S. and Fisher, J. S., 1979, Thermodynamic properties of minerals and related substances at 298.15 K and 1 bar (10**5 Pascals) pressure and at higher temperatures. *U.S. Geol. Surv. Bull.* 1452, 456 pp.

8. Kelley, K. K., 1960, Contributions to the data in theoretical metallurgy. XIII: high temperature heat content, heat capacities and entropy data for the elements and inorganic compounds. *U.S. Bur. Mines Bull.* 584, 232 pp.

9. Bowers, T. S. and Helgeson, H. C., 1983, Calculation of the thermodynamic and geochemical consequences of nonideal mixing in the system $H_2O–CO_2–NaCl$ on phase relations in geologic systems: equation of state for $H_2O–CO_2–NaCl$ fluids at high pressures and temperatures. *Geochim. Cosmochim. Acta*, v. 47, pp. 1247–1275.

10. Sverjensky, D. A., 1987, Calculations of the thermodynamic properties of aqueous species and the solubilities of minerals in supercritical electrolyte solutions. *Rev. Mineral.*, v. 17, pp. 177–209.

11. Shock, E. L., *Personal Calculation in 1992.* Laboratory of Theoretical Geochemistry, Dept. Geol. Geophys., Univ. CA, Berkeley, CA.

12. Jackson, K. J. and Helgeson, H. C., 1985, Chemical and thermodynamic constraints on the hydrothermal transport and deposition of tin. II. Interpretation of phase relations in the Southeast Asian tin belt. *Econ. Geol.*, v. 80, no. 5, pp. 1365–1378.

13. Helgeson, H. C., 1985, Errata II. Thermodynamics of minerals, reactions, and aqueous solutions at high pressures and temperatures. *Am. J. Sci.*, v. 285, pp. 845–855.

14. Shock, E. L. and Helgeson, H. C., 1989, Corrections to Shock and Helgeson (1988). *Geochim. Cosmochim. Acta*, v. 53, p. 215.

15. Pankratz, L. B., 1970, *Thermodynamic Data for Silver Chloride and Silver Bromide.* U.S. Bureau of Mines Report of Investigations #7430, 12 pp.

16. Pankratz, L. B. and King, E. G., 1970, *High-Temperature Enthalpies and Entropies of Calcopyrite and Bornite.* U.S. Bureau of Mines Report of Investigations #7435, 10 pp.

17. Johnson, J. W., *Personal Calculation in 1992*, Earth Sciences Dept., Lawrence Livermore National Lab., Livermore, CA. Parameters given provide smooth metastable extrapolation of one-bar steam properties predicted by the Haar et al. (1984) equation of state to temperatures < the saturation temperature (99.632°C).

18. Gibbs free energies calculated from solubility data reported by Plummer, L. N. and Busenberg, E., 1982, The solubilities of calcite, aragonite, and vaterite in CO_2–H_2O solutions between 0 and 90°C and an evaluation of the aqueous model of the system $CaCO_3$–CO_2–H_2O. *Geochim. Cosmochim. Acta*, v. 46, pp. 1011–1040. Enthalpies adjusted for the correction in the Gibbs free energy.

19. Gibbs free energies and enthalpies were corrected to be consistent with updated values of Gibbs free energies of Ca^{2+} and CO_3^{2-} (Shock and Helgeson, 1988) together with the solubilities of calcite and aragonite reported by Busenberg and Plummer (1982).

APPENDIX F4

Thermodynamic Data for Gases[a]

---

[a]Data from Johnson, J. W., Oelkers, E. H., and Helgeson, H. C., 1992, SUPCRT: A software package for calculating the standard molal thermodynamic properties of minerals, gases, aqueous species, and reactions from 1 to 5000 bar and 0 to 1000°C. *Comput. Geosci.*, v. 18, pp. 899–947.

Name	SUPCRT92	ΔG_f°	ΔH_f°	S°	V°	Maier-Kelly a	C_P $b \times 10^3$	Coefficients $c \times 10^{-5}$	C_P lim. (°C)	T_{crit} (K)	P_{crit} (bar)	Reference
Argon	Ar,g	0.0	0.0	37.008	0.0	4.968	0.0	0.0	8000	150.8	48.6	1,2
Methane	CH4,g	−12122.4	−17880.0	44.518	0.0	0.0565	11.44	0.46	1500	191.0	46.4	1,2
Carbon dioxide	CO2,g	−94254.0	−94051.0	51.085	0.0	10.57	2.10	2.06	2500	301.1	73.9	1,2
Hydrogen	H2,g	0.0	0.0	31.234	0.0	6.52	0.78	−0.12	3000	33.2	13.0	1,2
Steam	H2O,g	−54524.8	−57935.1	44.763	0.0	12.6647	−10.4041	−0.013078	2523	647.3	221.2	3
Hydrogen sulfide	H2S,g	−8021.0	−4931.0	49.185	0.0	7.81	2.96	0.46	2300	373.5	90.1	1,2
Helium	He,g	0.0	0.0	30.151	0.0	4.968	0.0	0.0	8000	5.2	2.29	1,2
Krypton	Kr,g	0.0	0.0	39.217	0.0	4.968	0.0	0.0	8000	209.3	55.0	1,2
Nitrogen	N2,g	0.0	0.0	45.796	0.0	6.83	0.90	0.12	3000	126.1	33.9	1,2
Ammonia	NH3,g	−3932.0	−11021.0	46.000	0.0	7.11	6.00	0.37	1800	405.6	114.0	1,2
Neon	Ne,g	0.0	0.0	34.973	0.0	4.968	0.0	0.0	8000	44.4	27.3	1,2
Oxygen	O2,g	0.0	0.0	49.029	0.0	7.16	1.00	0.40	3000	154.7	50.8	1,2
Radon	Rn,g	0.0	0.0	42.120	0.0	4.968	0.0	0.0	3000	377.2	62.8	1,2
Sulfur	S2,g	18953.0	30681.0	54.540	0.0	8.72	0.16	0.90	3000	—	—	1,2
Sulfur dioxide	SO2,g	−71748.0	−70944.0	59.330	0.0	11.04	1.88	1.84	2000	430.9	78.7	1,2
Xenon	Xe,g	0.0	0.0	40.555	0.0	4.968	0.0	0.0	8000	289.7	59.0	1,2

SUPCRT92 gives the name of the gas needed for use of the SUPCRT92 program. ΔG_f° and ΔH_f° are pure gases at 25 °C and 1 bar Gibbs energy and enthalpy of formation from the elements, respectively, in cal mol^{-1}. S° gives the pure gas entropy at 25 °C and 1 bar in cal mol^{-1} K^{-1}. The 1 bar, cal mol^{-1} K^{-1} Maier-Kelly C_P coefficients a, b, and c are given, multiplied by the indicated powers of 10, along with the equation's temperature limit in °C. Critical temperature (K) and critical pressure (bar) are given along with the indicated references.

References

1. Wagman, D. D., Evans, W. H., Parker, V. B., Schumm, R. H., Halow, I., Bailey, S. M., Churney, K. L. and Nuttall, R. L., 1982, The NBS tables of chemical and thermodynamic properties. *J. Phys. Chem. Ref. Data*, v. 11, supplement no. 2, 392 pp.

2. Kelley, K. K., 1960, *Contributions to the Data in Theoretical Metallurgy XIII: High Temperature Heat Content, Heat Capacities and Entropy Data for the Elements and Inorganic Compounds*. U.S. Bureau of Mines Bulletin, #584, Washington, D.C., 232 pp.

3. Johnson, J. W., *Personal Calculation in 1922*. Earth Sciences Dept., Lawrence Livermore National Lab., Livermore, CA. Parameters given provide smooth metastable extrapolation of 1-bar steam properties predicted by the Haar et al. (1984) equation of state to temperatures < the saturation temperature (99.632°C).

F5

Thermodynamic Data for Aqueous Species[a]

[a] Data from Johnson, J. W., Oelkers, E. H. and Helgeson, H. C., 1992, SUPCRT: A software package for calculating the standard molal thermodynamic properties of minerals, gases, aqueous species, and reactions from 1 to 5000 bar and 0 to 1000°C. *Comput. Geosci.*, v. 18, pp. 899–947.

Name / (Formula)	$\Delta \overline{G}^{\circ}_{f,P_r,T_r}$	$\Delta \overline{H}^{\circ}_{f,P_r,T_r}$	$\overline{S}^{\circ}_{P_r,T_r}$	$a1 \times 10$	$a2 \times 10^{-2}$	$a3$	$a4 \times 10^{-4}$	$c1$	$c2 \times 10^{-4}$	$\omega_{P_r,T_r} \times 10^{-5}$	Reference
ACETATE,AQ (CH₃COO⁻)	−88270.	−116180.	20.60	7.7525	8.6996	7.5825	−3.1385	26.3000	−3.8600	1.3182	3
ACETIC-ACID,AQ (CH₃COOH)	−94760.	−116100.	42.70	8.8031	12.4572	3.5477	−3.2939	40.8037	−0.9218	−0.2337	3
Al+3	−116510.	–	−81.203	−3.3984	−16.0789	12.0699	−2.1143	14.4295	−8.8523	2.7403	9
Al(OH)+2	−166425.	–	−43.290	−0.4532	−8.8878	9.2434	−2.4116	15.4131	−4.8618	1.5897	9
Al(OH)2+	−214987.	–	−6.584	2.4944	−1.6909	6.4146	−2.7091	16.7439	−1.0465	0.5324	9
Al(OH)3,aq	−263321.	–	14.185	5.4624	5.5560	3.5662	−3.0087	20.0270	1.7829	0.0	9
Al(OH)4–	−312087.	–	24.748	8.4938	12.9576	0.6570	−3.3147	55.7265	−11.4047	1.0403	9
CO2,aq	−92250.	−98900.	28.10	6.2466	7.4711	2.8136	−3.0879	40.0325	8.8004	−0.0200	2
CO3-2	−126191.	−161385.	−11.95	2.8524	−3.9844	6.4142	−2.6143	−3.3206	−17.1917	3.3914	1
CaCO3,aq	−262850.	−287390.	2.50	−0.2430	−8.3748	9.0417	−2.4328	9.9510	−8.5406	−0.0300	4
Ca(HCO3)+	−273830.	−294350.	16.00	3.1911	0.0104	5.7459	−2.7794	42.3545	8.5559	0.3084	4
Ca+2	−132120.	−129800.	−13.50	−0.1947	−7.2520	5.2966	−2.4792	9.0000	−2.5220	1.2366	1
CaCl+	−162550.	−169250.	−1.50	2.6670	−1.2694	6.2489	−2.7265	23.6652	1.2044	0.5756	4
CaCl2,aq	−194000.	−211060.	6.00	6.1187	7.1588	2.9363	−3.0749	26.6309	4.1743	−0.0300	4
CaSO4,aq	−31293.	−345900.	5.00	2.7910	−0.9666	6.1300	−2.7390	−8.4941	−8.1271	−0.0010	4
Cl–	−31379.	−39933.	13.56	4.0320	4.8010	5.5630	−2.8470	−4.4000	−5.7140	1.4560	1
Cu+	11950.	17132.	9.70	0.7835	−5.8682	8.0565	−2.5364	17.2831	−0.2439	0.3351	1
Cu+2	15675.	15700.	−23.20	−1.1021	10.4726	9.8662	−2.3461	20.3000	−4.3900	1.4769	1
ETHANOL,AQ (C2H5OH)	−43330.	−68650.	35.900	9.2333	9.9581	12.1445	−3.1906	60.0175	0.1507	−0.2037	3
F–	−67340.	−80150.	−3.15	0.6870	1.3588	7.6033	−2.8352	4.4600	−7.4880	1.7870	1
FORMATE,AQ (HCOO⁻)	−83862.	−101688.	21.70	5.7842	6.3405	3.2606	−3.0410	17.0000	−12.4000	1.3003	3
FORMIC-ACID,AQ (HCOOH)	−88982.	−101680.	38.90	6.3957	4.6630	10.7209	−2.9717	22.1924	−3.1196	−0.3442	3
Fe+2	−21870.	−22050.	−25.30	−0.7803	−9.6867	9.5573	−2.3786	14.9632	−4.6438	1.4574	1,7,8
Fe+3	−4120.	−11850.	−66.30	−3.1784	−15.5422	11.8588	−2.1365	11.0798	−9.9808	2.7025	1,8
FeCl+	−53030.	−61260.	−10.06	2.0756	−2.7134	6.8165	−2.6668	24.6737	1.1555	0.7003	4

Name / (Formula)	$\overline{\Delta G}^{\circ}_{f,P_r,T_r}$	$\overline{\Delta H}^{\circ}_{f,P_r,T_r}$	$\overline{S}^{\circ}_{P_r,T_r}$	$a_1 \times 10$	$a_2 \times 10^{-2}$	a_3	$a_4 \times 10^{-4}$	c_1	$c_2 \times 10^{-4}$	$\omega_{P_r,T_r} \times 10^{-5}$	Reference
FeCl2,aq	-81280.	-100370.	-4.22	5.4085	5.4247	3.6179	-3.0033	25.1189	3.6488	-0.0300	4
H+	0.0	0.0	0.0	0.0	0.0	0.0	0.0	0.0	0.0	0.0	1
H2,aq	4236.	-1000.	13.80	5.1427	4.7758	3.8729	-2.9764	27.6251	5.0930	-0.2090	2
H2S,aq	-6673.	-9001.	30.00	6.5097	6.7724	5.9646	-3.0590	32.3000	4.7300	-0.1000	2
HCO-3	-140282.	-164898.	23.53	7.5621	1.1505	1.2346	-2.8266	12.9395	-4.7579	1.2733	1
HF,aq	-71662.	-76835.	22.50	3.4753	0.7042	5.4732	-2.8081	14.3647	-0.1828	-0.0007	2
HNO3,aq	-24730.	-45410.	42.70	999.00	999.00	999.000	999.00	15.2159	1.2635	-0.3600	2
HS-	2860.	-3850.	16.30	5.0119	4.9799	3.4765	-2.9849	3.4200	-6.2700	1.4410	1
HSO-3	-126130.	-149670.	33.40	6.7014	8.5816	2.3771	-3.1338	15.6949	-3.3198	1.1233	1
HSO-4	-180630.	-212500.	30.00	6.9788	9.2590	2.1108	-3.1618	20.0961	-1.9550	1.1748	1
HSiO-3	-242300.	-271880.	10.00	3.6325	1.0881	5.3224	-2.8240	0.4312	-9.7567	1.4767	4
K+	-67510.	-60270.	24.15	3.5590	-1.4730	5.4350	-2.7120	7.4000	-1.7910	0.1927	1
KCl,aq	-96850.	-96810.	42.25	6.8112	8.8497	2.2717	-3.1448	-2.4989	-5.8864	-0.0500	4
KHSO4,aq	-246550.	-270540.	56.31	9.0596	14.3397	0.1139	-3.3718	28.4277	4.7060	-0.0010	4
KSO-4	-246640.	-276980.	35.00	6.0776	7.0585	2.9757	-3.0708	9.8504	-5.2753	1.0996	4
Mg(CO3),aq	238760.	-270570.	-24.00	-0.5837	-9.2067	9.3687	-2.3984	-8.6676	-8.0945	-0.0300	4
Mg(HCO3)+	-250200.	-275750.	3.00	2.7171	-1.1469	6.2008	-2.7316	49.0065	9.9391	0.5985	4
Mg+2	-108505.	-111367.	-33.00	-0.8217	-8.5990	8.3900	-2.3900	20.8000	-5.8920	1.5372	1
MgCl+	-139700.	-151440.	-20.50	2.2269	-2.3440	6.6713	-2.6821	39.0235	5.6430	0.8565	4
N2,aq	4347.	-2495.	22.900	6.2046	7.3685	2.8539	-3.0836	35.7911	8.3726	-0.3468	3
N-BUTANE,AQ (C4H10)	36.	-36230.	40.020	12.8905	23.6960	-3.5683	-3.7585	79.0569	24.2408	-0.6061	3
NH3,aq	-6383.	-19440.	25.77	5.0911	2.7970	8.6248	-2.8946	20.3000	-1.1700	-0.0500	2
NH+4	-18990.	-31850.	26.57	3.8763	2.3448	8.5605	-2.8759	17.4500	-0.0210	0.1502	1
NO-2	-7700.	-25000.	29.40	5.5864	5.8590	3.4472	-3.0212	3.4260	-7.7808	1.1847	1
NO-3	-26507.	-49429.	35.12	7.3161	6.7824	-4.6838	-3.0594	7.7000	-6.7250	1.0977	1

Appendix F5   continued

Name / (Formula)	$\Delta \bar{G}^{\circ}_{f,P_r,T_r}$	$\Delta \bar{H}^{\circ}_{f,P_r,T_r}$	$\bar{S}^{\circ}_{P_r,T_r}$	$a1 \times 10$	$a2 \times 10^{-2}$	$a3$	$a4 \times 10^{-4}$	$c1$	$c2 \times 10^{-4}$	$\omega_{P_r,T_r} \times 10^{-5}$	Reference
N-OCTANE,AQ (C_8H_{18})	8580.	−59410.	63.80	21.4171	44.5126	−11.7437	−4.6191	124.7924	41.2905	−0.9662	3
Na+	−62591.	−57433.	13.96	1.8390	−2.2850	3.2560	−2.7260	18.1800	−2.9810	0.3306	1
NaCl,aq	−92910.	−96120.	28.00	5.0363	4.7365	3.4154	−2.9748	10.8000	−1.3000	−0.0380	5,6
O2,aq	3954.	−2900.	26.04	5.7889	6.3536	3.2528	−3.0417	35.3530	8.3726	−0.3943	2
OH-	−37595.	−54977.	−2.56	1.2527	0.0738	1.8423	−2.7821	4.1500	−10.3460	1.7246	1
PO-4	−243500.	−305300.	−53.00	−0.5259	−9.0654	9.3131	−2.4042	−9.4750	−26.4397	5.6114	1
SO2,aq	−71980.	−77194.	38.70	6.9502	9.1890	2.1383	−3.1589	31.2101	6.4578	−0.2461	3
SO-3	−116300.	−151900.	−7.00	3.6537	0.3191	7.3853	−2.7922	−7.8368	−18.5362	3.3210	1
SO-4	−177930.	−217400.	4.50	8.3014	−1.9846	−6.2122	−2.6970	1.6400	−17.9980	3.1463	1
SiO2,aq	−199190.	−209775.	18.00	1.9000	1.7000	20.0000	−2.7000	29.1000	−51.2000	0.1291	2
Zn+2	−35200.	−36660.	−26.20	−1.0677	−10.3884	9.8331	−2.3495	15.9009	−4.3179	1.4574	1
ZnCl+	−66850.	−66240.	23.0	1.5844	−3.9128	7.2879	−2.6172	22.4668	1.9825	0.2025	4
ZnCl2,aq	−98300.	−109080.	27.03	5.0570	4.5665	3.9552	−2.9678	32.0555	5.9669	−0.0010	4
ZnCl-3	−129310.	−151060.	25.00	9.5417	15.5168	−0.3487	−3.4205	42.3675	5.5412	1.2513	4
ZnCl4-2	−161890.	−195200.	36.00	14.6628	28.0213	−5.2636	−3.9374	56.1061	5.7856	2.6662	4
ZnCl-3	−129310.	−151060.	25.00	9.5417	15.5168	−0.3487	−3.4205	42.3675	5.5412	1.2513	4

Name gives the name of the aqueous species needed for use in the SUPCRT92 program. $\Delta \bar{G}^{\circ}_{f,P_r,T_r}$ and $\Delta \bar{H}^{\circ}_{f,P_r,T_r}$ are the standard apparent Gibbs energy and enthalpy of formation from the elements, respectively, in cal mol⁻¹ for a hypothetical 1 molal standard state at 25 °C and 1 bar where activity coefficients are taken as unity at infinite dilution. $\bar{S}^{\circ}_{P_r,T_r}$ gives the molal entropy at 25 °C and 1 bar in cal mol⁻¹ K⁻¹. a1 and a3 are in cal mol⁻¹ bar⁻¹, whereas a2 and a4 are in cal mol⁻¹ bar⁻², c1 is in cal mole⁻¹ K⁻¹ and c2 is in cal K mole⁻¹. ω is given in cal mol⁻¹.

References

1. Shock, E. L. and Helgeson, H. C., 1988, Calculation of the thermodynamic and transport properties of aqueous species at high pressures and temperatures: correlation algorithms for ionic species and equation of state predictions to 5 kb and 1000°C. *Geochim. Cosmochim. Acta*, v. 52, pp. 2009–2036.

2. Shock, E. L., Helgeson, H. C. and Sverjensky, D. A., 1989, Calculation of the thermodynamic and transport properties of aqueous species at high pressures and temperatures: standard partial molal properties of inorganic neutral species. *Geochim. Cosmochim. Acta*, v. 53, pp. 2157–2183.

3. Shock, E. L. and Helgeson, H. C., 1990, Calculation of the thermodynamic and transport properties of aqueous species at high pressures and temperatures: standard partial molal properties of organic species. *Geochim. Cosmochim. Acta*, v. 54.

4. Sverjensky, D. A., Shock, E. L. and Helgeson, H. C., 1997, Prediction of the thermodynamic properties of aqueous metal complexes to 1000°C and 5 kb. *Geochim. Cosmochim. Acta*, v. 61, pp. 907–950.

5. Sverjensky, D. A., 1987, Calculations of the thermodynamic properties of aqueous species and the solubilities of minerals in supercritical electrolyte solutions. *Rev. Mineral.*, v. 17, pp. 177–209.

6. Shock, E. L., *Personal calculation in 1992.* Lab. of Theoretical Geochemistry, Dept. Geol. Geophys., University of CA, Berkeley, CA.

7. Helgeson, H. C., 1985, Errata II. Thermodynamics of minerals, reactions, and aqueous solutions at high pressures and temperatures. *Am. J. Sci.*, v. 285, pp. 845–855.

8. Shock, E. L. and Helgeson, H. C., 1989, Corrections to Shock and Helgeson (1988). *Geochim. Cosmochim. Acta*, v. 53, p. 215.

9. Al properties from Tagirov, B. and Schott, J., 2001, Aluminum speciation in crustal fluids revisited. *Geom. Cosmochim. Acta*, v. 65, pp. 3965–3992.

Debye-Hückel A Parameters (in kg$^{1/2}$ mole$^{-1/2}$)

Temperature (°C)	Vapor sat	0.5 kbar	1.0 kbar	1.5 kbar	2.0 kbar	3.0 kbar	4.0 kbar	5.0 kbar
25	0.5092	0.4980	0.4882	0.4807	0.4730	0.4606	0.4505	0.4417
50	0.5336	0.5208	0.5096	0.5013	0.4925	0.4784	0.4670	0.4570
75	0.5639	0.5486	0.5353	0.5251	0.5149	0.4987	0.4860	0.4748
100	0.5998	0.5809	0.5649	0.5522	0.5402	0.5214	0.5070	0.4947
125	0.6416	0.6177	0.5983	0.5828	0.5686	0.5463	0.5299	0.5161
150	0.6898	0.6592	0.6352	0.6168	0.5998	0.5732	0.5541	0.5387
175	0.7454	0.7057	0.6756	0.6538	0.6334	0.6016	0.5795	0.5621
200	0.8099	0.7576	0.7196	0.6934	0.6692	0.6315	0.6057	0.5863
225	0.8860	0.8159	0.7673	0.7353	0.7068	0.6625	0.6327	0.6111
250	0.9785	0.8822	0.8192	0.7795	0.7461	0.6946	0.6605	0.6366
275	1.0960	0.9595	0.8762	0.8263	0.7873	0.7280	0.6894	0.6629
300	1.2555	0.9595	0.9398	0.8766	0.8308	0.7630	0.7195	0.6905
325	1.4943	1.1705	1.0126	0.9317	0.8774	0.8001	0.7515	0.7198
350	1.9252	1.3267	1.0981	0.9934	0.9282	0.8399	0.7857	0.7513
375		1.5464	1.2007	1.0639	0.9845	0.8832	0.8228	0.7854
400		1.8789	1.3262	1.1453	1.0476	0.9305	0.8632	0.8225
425		2.4301	1.4811	1.2402	1.1188	0.9823	0.9069	0.8625
450		3.3553	1.6723	1.3504	1.1988	1.0384	0.9536	0.9049
475		4.5059	1.9065	1.4768	1.2873	1.0977	1.0017	0.9480
500		5.5075	2.1872	1.6181	1.3822	1.1575	1.0485	0.9889
525			2.5092	1.7685	1.4782	1.2127	1.0890	1.0228
550			2.8476	1.9151	1.5648	1.2551	1.1158	1.0428
575			3.1486	2.0366	1.6259	1.2735	1.1199	1.0406
600			3.3281	2.0960	1.6400	1.2556	1.0919	1.0081

From Helgeson, H. C. and Kirkham, D. H.,1974, Theoretical prediction of the thermodynamic behavior of aqueous electrolytes at high pressures and temperatures: II. Debye-Hückel parameters for activity coefficients and relative partial molal properties, *Am. J. Sci.*, v. 274, pp. 1199–1261.

APPENDIX G2

Debye-Hückel B Parameters (in kg$^{1/2}$ mole$^{-1/2}$ Å)

Temperature (°C)	Vapor sat	0.5 kbar	1.0 kbar	1.5 kbar	2.0 kbar	3.0 kbar	4.0 kbar	5.0 kbar
25	0.3283	0.3282	0.3281	0.3281	0.3280	0.3297	0.3297	0.3280
50	0.3325	0.3321	0.3317	0.3315	0.3312	0.3308	0.3306	0.3304
75	0.3371	0.3364	0.3358	0.3353	0.3349	0.3342	0.3337	0.3334
100	0.3422	0.3411	0.3402	0.3395	0.3388	0.3378	0.3372	0.3366
125	0.3476	0.3461	0.3448	0.3439	0.3430	0.3416	0.3407	0.3401
150	0.3533	0.3512	0.3496	0.3483	0.3472	0.3455	0.3443	0.3435
175	0.3592	0.3565	0.3544	0.3528	0.3514	0.3493	0.3479	0.3469
200	0.3655	0.3618	0.3592	0.3573	0.3556	0.3530	0.3514	0.3503
225	0.3721	0.3673	0.3639	0.3616	0.3596	0.3566	0.3547	0.3535
250	0.3792	0.3729	0.3686	0.3658	0.3635	0.3601	0.3580	0.3567
275	0.3871	0.3787	0.3733	0.3699	0.3673	0.3634	0.3611	0.3598
300	0.3965	0.3850	0.3780	0.3739	0.3710	0.3667	0.3643	0.3629
325	0.4085	0.3921	0.3829	0.3780	0.3747	0.3700	0.3674	0.3661
350	0.4256	0.4004	0.3882	0.3822	0.3784	0.3734	0.3707	0.3694
375		0.4104	0.3940	0.3867	0.3823	0.3769	0.3741	0.3729
400		0.4230	0.4004	0.3915	0.3865	0.3806	0.3777	0.3766
425		0.4386	0.4076	0.3968	0.3909	0.3845	0.3815	0.3804
450		0.4548	0.4154	0.4024	0.3956	0.3885	0.3853	0.3843
475		0.4625	0.4237	0.4083	0.4004	0.3924	0.3890	0.3880
500		0.4620	0.4321	0.4141	0.4050	0.3960	0.3923	0.3912
525			0.4397	0.4193	0.4089	0.3986	0.3944	0.3932
550			0.4454	0.4231	0.4113	0.3995	0.3948	0.3934
575			0.4472	0.4240	0.4109	0.3978	0.3924	0.3907
600			0.4428	0.4203	0.4064	0.3922	0.3863	0.3842

From Helgeson, H. C. and Kirkham, D. H.,1974, Theoretical prediction of the thermodynamic behavior of aqueous electrolytes at high pressures and temperatures: II. Debye-Hückel parameters for activity coefficients and relative partial molal properties, *Am. J. Sci.*, v. 274, pp. 1199–1261.

APPENDIX H

Decay Chain of Isotopes ^{238}U, ^{232}Th, and ^{235}U Indicating the Half-Life of Each Isotope

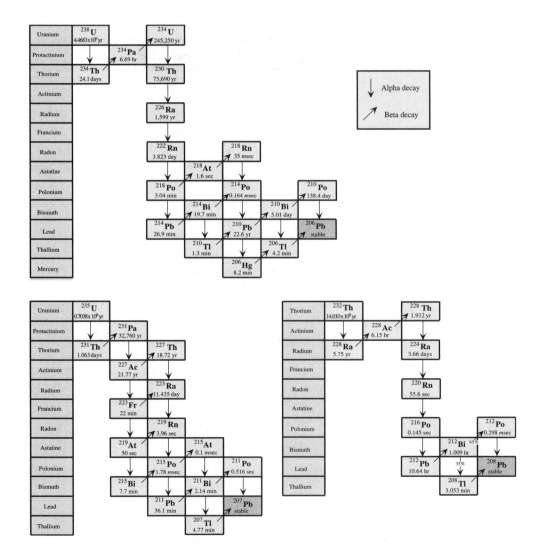

Balancing Oxidation-Reduction Reactions

Oxidation-reduction (redox) reactions involve the transfer of electrons from one species to another or the change in the valence state of an element. Consider balancing the unbalanced redox reaction

$$Mn^{2+}_{(aq)} + Cl_{2(aq)} \rightarrow MnO^-_{4(aq)} + Cl^-_{(aq)}$$

Step 1. Determine which species are being oxidized and which are being reduced and write the half-reactions:

Mn oxidation: $Mn^{2+}_{(aq)} \rightarrow MnO^-_{4(aq)}$

Cl reduction: $Cl_{2(aq)} \rightarrow Cl^-_{(aq)}$

Step 2. Balance the atoms in each half-reaction except hydrogen and oxygen:

Mn oxidation: $Mn^{2+}_{(aq)} \rightarrow MnO^-_{4(aq)}$

Cl reduction: $Cl_{2(aq)} \rightarrow 2Cl^-_{(aq)}$

Step 3. Add water to balance the oxygen and H^+ to balance the hydrogen:

Mn oxidation: $Mn^{2+}_{(aq)} + 4H_2O \rightarrow MnO^-_{4(aq)} + 8H^+$

Cl reduction: $Cl_{2(aq} \rightarrow 2Cl^-_{(aq)}$

Step 4. Add electrons to balance the charges:

Mn oxidation: $Mn^{2+}_{(aq)} + 4H_2O \rightarrow MnO^-_{4(aq)} + 8H^+ + 5e^-$

Cl reduction: $Cl_{2(aq)} + 2e^- \rightarrow 2Cl^-_{(aq)}$

Step 5. Multiply each half-reaction by some integer to equalize the electrons in each reaction:

Mn oxidation: $2(Mn^{2+}_{(aq)} + 4H_2O \rightarrow MnO^-_{4(aq)} + 8H^+ + 5e^-)$

Cl reduction: $5(Cl_{2(aq)} + 2e^- \rightarrow 2Cl^-_{(aq)})$

Step 6. Add the two half-reactions with the electrons, canceling

Mn oxidation: $2Mn^{2+}_{(aq)} + 8H_2O \rightarrow MnO^-_{4(aq)} + 16H^+ + 10e^-$

Cl reduction: $5Cl_{2(aq)} + 10e^- \rightarrow 10Cl^-_{(aq)}$

$2Mn^{2+}_{(aq)} + 8H_2O + 5Cl_{2(aq)} = 2MnO^-_{4(aq)} + 16H^+ + 10Cl^-_{(aq)}$

Step 7. If the reaction is in an alkaline solution where OH^- is dominate rather than H^+, add the same number of OH^- to each side as H^+ present:

$16OH^- + 2Mn^{2+}_{(aq)} + 8H_2O + 5Cl_{2(aq)} = 2MnO^-_{4(aq)} + 16H^+ + 10Cl^-_{(aq)} + 16OH^-$

Step 8. Combine any OH^- and H^+ on the same side of the reaction to produce H_2O. Cancel any H_2O that are on both sides of the reaction to give

$16OH^- + 2Mn^{2+}_{(aq)} + 5Cl_{2(aq)} = 2MnO^-_{4(aq)} + 8H_2O + 10Cl^-_{(aq)}$

Step 9. The reaction should always be checked to make sure the equation contains the same number of atoms of each type and the same total charge on each side of the reaction.

APPENDIX J

Conversion Factors Between Solution Concentration Scales for a Single Solute

Needed Concentration of Solute i	Known Concentration of Solute i			
	Mole fraction X_i	Molality m_i	Molarity M_i	Weight percent (= ppm by weight/10^4) wt%
X_i	1	$\dfrac{m_i Mw_s}{1000 + m_i Mw_i}$	$\dfrac{M_i Mw_s}{M_i(Mw_s - Mw_i) + 1000\rho}$	$\dfrac{wt\%/Mw_i}{wt\%/Mw_i + (100 - wt\%/Mw_s)}$
m_i	$\dfrac{1000 X_i}{Mw_s - X_i Mw_s}$	1	$\dfrac{1000 M_i}{1000\rho - M_i Mw_i}$	$\dfrac{100\,wt\%}{Mw_i(100 - wt\%)}$
M_i	$\dfrac{1000\rho X_i}{X_i Mw_i + (1 - X_i)Mw_s}$	$\dfrac{1000\rho\, m_i}{1000 + m_i Mw_i}$	1	$\dfrac{10\rho\, wt\%}{Mw_i}$
wt%	$\dfrac{100 X_i Mw_i}{X_i Mw_i + (1 - X_i)Mw_s}$	$\dfrac{100\, m_i Mw_i}{1000 + m_i Mw_i}$	$\dfrac{M_i Mw_i}{10\rho}$	1

Mw_s = molecular weight of solvent, Mw_i = molecular weight of solute i, ρ = density of solution (g/cm^3).

Glossary

absorption: Incorporation of chemical species from a fluid or gas into a solid or a gas into a liquid.

acceleration: The differential of velocity with respect to time or the second derivative of position with respect to time.

acceleration due to gravity: The acceleration of a mass toward another mass because of attractive forces between them. The mass of the earth produces an acceleration, g, equal to 9.7803 m s^{-2} at the equator and increases to 9.8322 at the poles. A standard value of 9.8066 m s^{-2} is used in most calculations.

accelerator mass spectrometry: A technique for measuring abundances of atoms. Atoms of an element in a sample are converted into ions. These ions are accelerated by an electric field in a particle accelerator. Large magnets and detectors are used to separate and count the isotopes present according to their mass-to-charge ratio. Atoms of single isotopes are detected in the presence of 1×10^{12} to 1×10^{15} other isotopes of the same element.

accretionary prism: A wedge of material that is scraped from the top of a subducting oceanic lithospheric slab and produces a fore arc on the edge of the overlying continental crust.

achondrites: Stony meteorites that do not contain rounded grains called chondrules.

acid: A chemical substance that dissociates in water to give free protons (H$^+$).

acid rain: Precipitation containing nitric and/or sulfuric acids with a pH less than 5.6. It can be either wet (rain, snow, or fog) or dry precipitation (absorbed gas on particulate matter, aerosol particles, or dust).

actinides: A series of 15 chemical elements of the Periodic Table with atomic numbers 89 to 103.

activated complex: A high-energy configuration of atoms through which reactants in a chemical reaction must pass before the atoms in the reactants decay to the stable product species.

activation energy: The energy above the ground state energy necessary to produce an activated complex to start a chemical reaction.

activity: In thermodynamics, the ratio of the fugacity of a substance to that in a specified standard state.

activity–activity diagram: A phase diagram at constant temperature and pressure that gives the phase relations between the logarithm of the activities of two components while balancing on a third and fixing as constant the values of any additional components.

activity coefficient: A coefficient that when multiplied by the concentration of a substance gives its thermodynamic activity.

activity product (Q): The product of the thermodynamic activities of all of the species in a reaction raised to their coefficient in the reaction with product coefficients positive and reactant coefficients negative. If the reaction is at equilibrium, the activity product equals the equilibrium constant.

adiabatic: Change or process that occurs in a mass with no transfer of heat across its boundaries.

adiabatic gradient: The change in temperature as a function of distance in a pressure field for a mass closed to heat exchange with the surroundings. With decreased pressure on raising toward the earth's surface, most mantle rocks cool as they expand. It is likewise for an air mass ascending in the earth's atmosphere.

ADP (adenosine diphosphate): A necleotide containing two charged phosphate groups that with the aid of extra energy can accept a third phosphate group and become ATP or give off energy by detaching a phosphate group and becoming AMP.

adsorption: Attachment of chemical species from a fluid or gas on the surface of a solid or from a gas onto the surface of a liquid.

adsorption isotherm: The amount of adherence of ions or molecules to a solid or liquid surface as a function of its concentration in the phase adjoining the surface at a constant temperature and pressure.

advection: The transport of a volume or mass from one place to another by moving the medium that contains the volume or mass.

aerobic: An environment where free oxygen molecules are available or an organism that requires free oxygen is present. It is generally considered to be at a pressure of O_2 greater than 10^{-4} bars, although some investigators use 10^{-6} bars.

aerosols: Small solid particles or liquid droplets of about 0.01 to 10 μm dispersed in a gas such as the earth's atmosphere.

albite: A milky white to colorless feldspar mineral whose endmember composition is $NaAlSi_3O_8$. In plagioclase, albite solid solution is from pure albite to a 90% albite endmember.

alcohols: Organic compounds that contains an OH functional group.

aliphatic: Organic compounds that consist of open chains and branches of carbon atoms with no ring structures.

alkali: Referring to the first column of the Periodic Table of elements that includes Li, Na, K, Rb, Cs, and Fr.

alkaline basalt: A low-silica basalt with normative nepheline and/or acmite. It is silica undersaturated as it contains no quartz or orthopyroxene, but olivine is common as phenocrysts and in the groundmass.

alkaline earths: Elements in group IIA of the Periodic Table of the elements that include Be, Mg, Ca, Sr, Ba, and Ra.

alkaline solutions: A solution where $m_{OH^-} > m_{H^+}$. At 25°C and 1 bar, an alkaline solution has $pH > 7$.

alkalinity: Gives the moles of protons that must be added to a water to achieve neutral *pH*. It is determined from the molality of charges on nonconservative anions minus that on nonconservative cations in solution given as $m_{HCO_3^-} + 2m_{CO_3^{2-}} + m_{OH^-} - m_{H^+} + m_{H_3SiO_4^-} + m_{HS^-} + m_{organic\ anions} \cdots$ and is equal to Σ conservative cations $- \Sigma$ conservative anions.

alkanes: Organic molecules that contain only carbon and hydrogen in which all linkages are single bonds. Their chemical formula is C_nH_{2n+2}, where n is a positive integer.

alkynes: Organic molecules that contain only carbon and hydrogen in which one or more pairs of carbon atoms are joined by triple bonds.

alpha decay: A transformation of the nucleus of an atom with the ejection of an alpha particle consisting of two protons plus two neutrons.

amines: An organic compound with an amine group, that is, a nitrogen atom in threefold coordination, bound to a carbon atom or chain.

amino acids: An organic compound with both an amine group and a carboxylic acid group.

ammonification: Heterotrophic bacteria breaking down amino acids from animal wastes into $NH_{3(gas)}$.

AMP (adenosine monophosphate): A necleotide consisting of a charged phosphate group, the pentose sugar ribose, and the nucleobase adenine.

amphibolite: A metamorphic rock composed mainly of amphibole and plagioclase.

anaerobic: An environment that lacks free oxygen molecules or an organism that does not require free oxygen. Generally put at a pressure of O_2 less than 10^{-4} bars, although some investigators use 10^{-6} bars.

andesite: An extrusive igneous rock with 53–63 wt% SiO_2 that contains phenocrysts composed primarily of chemically zoned andesine (An^{70}-An^{85}) and one or more mafic minerals. The groundmass is similar, but quartz is generally also present.

anion: A negatively charged atom or bonded group of atoms.

anorthite: A grayish to white feldspar mineral of endmember composition $CaAl_2Si_2O_8$ that in plagioclase can be a solid solution from pure anorthite to 90% anorthite endmember.

anoxic: Refers to an environment where measurable O_2 is absent ($O_2 < 10^{-6}$ molal).

antineutrino: The antiparticle of the neutrino. A stable subatomic particle of spin one half with kinetic energy but with near zero mass and zero electric charge given off with an electron during β^- decay.

antiparticle: A subatomic particle that has the same mass as another particle but has opposite values for its other properties, including charge, baryon number, and strangeness. When an antiparticle and normal particle combine, they annihilate.

apparent values: A set of values used in thermodynamic determinations of the Gibbs energy, enthalpy, and entropy of species and phases across a reaction. Rather than using formation from the elements at elevated pressures and temperatures, for apparent values, the change in the thermodynamic property for the elements as a function of pressure and temperature is not included when the values for species and phases are determined as the values of the elements cancel across a reaction.

aromatic: Organic compounds that contain one or more closed benzene rings.

Arrhenius plot: A plot of ln k versus $1/T$, where k is a rate constant. It is based on application of the Arrhenius equation, ln k = ln A $- E_a/RT$, where A is the pre-exponential factor and E_a is the activation energy of the process.

assimilation: In biological systems, refers the process of transforming external substances into substances internal to an organism.

asthenosphere: A layer in the upper mantle where seismic wave velocities are lower than elsewhere in the mantle. It is thought to occur because of the presence of a small amount (<1%) of melt softening the layer. It typically lies between a depth of 65 and 165 km below old stable oceanic crust and between a depth of 120 to 220 km below stable continental crust. At active ocean ridges, it approaches the surface and can be present at the shallow crust–mantle boundary under extended continental crust.

atomic absorption: A type of chemical analysis where an element's concentration is determined by measuring the extent of absorption by the element of a wavelength of light that is shined through the sample.

atomic mass number: The number of protons plus neutrons in the nucleus of an isotope of an element.

atomic number: The number of protons in the nucleus of an atom of an element.

atomic weight: The average weight of an atom relative to 1/12th that of the neutral atom of the ^{12}C isotope of carbon.

ATO: As used in this book is a short hand for referring to the Atmosphere, Terrestrial biomass, and Ocean.

ATP (adenosine triphosphate): A necleotide containing three charged phosphate groups that releases energy when it breaks down to ADP plus a phosphate ion. The energy is generally used in cell processes.

autocatalytic process: A reaction in which one of the products of the reaction increases the rate of reaction.

autotrophic: An organism that produces its own food, needing only CO_2 or carbonate for its carbon. It is therefore capable of living on inorganic chemicals.

Avogadro's number, N°: A constant equal to the number of atoms in 12 g of neutral ^{12}C that has a value of 6.022136×10^{23}. It defines the mole, the quantity unit in the SI system of units. One mole has Avogadro's number of objects in it.

back-arc basins: Sedimentary basins that occur on the landward side of volcanic arcs and are formed by extension of the earth's crust.

baryon number: A number equal to the difference between the number of baryons and the number of antibaryons. A baryon is a subatomic particle that interacts via the strong nuclear force and has a half-integral spin, like the proton and neutron. The baryon number, similar to the charge, energy, and angular momentum, must be conserved in any reaction between elementary particles. It keeps a proton from decaying to a positron and gamma ray even though the other conservation laws are satisfied.

basalt: An extrusive mafic (dark-colored) igneous rock that contains about 50 wt% SiO_2 composed chiefly of calcic plagioclase and clinopyroxene with or without olivine.

bauxite deposits: Concentrations of hydrated alumina phases (Al hydroxides) formed by the reaction of large quantities of rainwater with soil layers leaching most of the metals except Al from the soil.

beachrock: Rock formed at beaches from cementation of quartz grains by precipitation of calcium carbonate from seawater after some evaporation.

bed load: The solid debris carried in a flowing river by rolling, sliding, or saltation (jumping) along the river bed rather than by being supported by the water as with the suspended load.

Benioff zone: A zone within the earth defined by earthquakes that is considered to mark the upper surface of a subducting oceanic lithospheric plate.

benzene ring: A six-member ring of carbon atoms that are bound by alternating single and double bonds.

beta decay: Negative beta decay involves the transformation of a neutron to a proton with the ejection of an electron, whereas positive beta decay involves the transformation of a proton to a neutron with the ejection of a positron.

bicarbonate: An aqueous species that can be denoted as HCO_3^-.

bidentate: With regard to surface species, a species bonded to two different active surface sites.

biologic pump: When the earth's atmosphere is discussed, this refers to the transfer of CO_2 from shallow to deep ocean water by the uptake of CO_2 when phytoplankton are produced in the ocean's photic zone and sinking of the dead phytoplankton to the deep ocean where the CO_2 is released by phytoplankton decay.

biomagnification: The process by which an increase of a substance in an organism at higher levels on the food chain occurs by concentration from eating animals containing the substance at lower levels on the food chain.

bioturbation: Digging, shuffling, ingestion, and shifting of material by animals in sediments at the seawater–sediment interface.

bitumen: Organic material found in rocks that is soluble in organic solvents.

black shales: Shales that accumulated and retained enough organic material during their formation to be black rather than the gray color of normal shales (>1% total organic carbon).

black smoker: A plume of fine-grained black sulfide minerals precipitated rapidly after seawater is first heated and altered from contact with rocks emplaced at mid-oceanic ridges and then vented from a fracture into the colder ocean water.

blackbody: A perfect emitter/absorber of radiation. A blackbody radiates energy in a spectral distribution and intensity that depends on its temperature.

Boltzmann distribution: The distribution specifying the number, n, of particles of energy, E, in a system of particles in thermal equilibrium at temperature, T, in Kelvin. $n = 1/e^{(E-\mu)kT}$ where k denotes the Boltzmann constant, and μ stands for the chemical potential of the system.

Boltzmann's constant (k): A constant that relates temperature to energy that is equal to two thirds of the energy required to raise the temperature of a particle in an ideal gas by 1 K—equal to 1.38066×10^{-23} J K^{-1}.

boundaries: In a thermodynamic system, these separate the mass in the system from the surroundings. Boundaries can be specified as not movable, not allowing the transfer of heat, mass, and so on.

Bowen's reaction series: Two series for the crystallization of minerals from a magma with decreasing temperature. Early formed minerals can react with the magma to produce new minerals further down in the series at lower temperatures. The discontinuous series is olivine $\rightarrow$ pyroxene $\rightarrow$ amphibole $\rightarrow$ biotite $\rightarrow$ K-feldspar + muscovite + quartz. The parallel continuous series is for plagioclase, being calcic at high temperature to alkalic at lower temperature.

brecciated zones: Rocks consisting of fragmented angular pieces as in fault, talus, or eruption breccias.

bridging oxygen: An oxygen bound between two tetrahedral cations in different silicon–oxygen tetrahedrons.

Brownian motion: The motion of small particles in a fluid where the particles are kept suspended against the force of gravity by the momentum transferred by the random collisions that occur with molecules in the fluid.

buoyancy forces: The ability of material to rise in a gravitational field because of the decrease in energy occurring because of lower density material rising from the center of gravity and being replaced by more dense material.

Calvin cycle: The second or dark (light independent) stage of photosynthesis in which a series of biochemical enzyme-mediated reactions reduces atmospheric carbon dioxide, producing simple sugar molecules.

carbonaceous chondrite: A type of stony meteorite that has chondrules in a matrix of low-temperature phyllosilicates with up to 3% carbon and 9% water.

carbonate: When considering an aqueous solution, it refers to the sum of the oxidized carbon containing species in solution of H_2CO_3, HCO_3^-, and CO_3^{2-}.

carbonate alkalinity: The molality of HCO_3^- plus two times the molality of CO_3^{2-} in solution.

carbonate compensation depth (CCD): The depth in the ocean where the rate of dissolution of carbonates balances its rate of accumulation; therefore, it marks the depth at which carbonates can no longer accumulate in sediments. It varies from 5 to 2.5 km, being shallower in warmer water.

carbonate ion: An aqueous species that can be denoted as CO_3^{2-}.

carbonic acid: The uncharged carbonate aqueous species that can be denoted as $CO_{2(aq)}$ or H_2CO_3 that acts like an acid in H_2O by producing an H^+ and the bicarbonate ion.

carboxyl functional group: A carbon atom double bonded to an oxygen and single bonded to a hydroxyl group (OH).

carboxylic acids: Organic compounds that contain a carboxylic functional group, COOH, where the H partially dissociates in aqueous solutions.

catalytic reaction: An additional reaction path that speeds the conversion of reactants to products by reducing the activation energy along the catalytic reaction path.

cation exchange: When considering mineral surfaces the replacement of a cation at the surface of a mineral by a different cation from solution.

cations: A positively charged atom or bonded group of atoms.

cellular respiration: A metabolic oxidation/reduction processes where energy is obtained from organic molecules (glucose, amino acids, and fatty acids) being oxidized. In aerobic respiration, O_2 is reduced (gains electrons), and the organic carbon is oxidized (loses electrons) to produce CO_2. With anaerobic respiration, other organic molecules are reduced to oxidize glucose and produce CO_2. In some cases, NO_3^-, SO_4^{2-}, or CO_2 are reduced producing N_2 (denitrification), reduced S^{2-} (sulfate reduction), or CH_4 (methanogenesis), respectively.

cellulose: A polysaccharide that forms the basic structural unit of plants with a formula of $(C_6H_{10}O_5)_n$, where n is generally greater than 10,000.

chelates: Organic ring structures that bond metal ions to two or more of their atoms.

chelation: A process by which a metal aqueous species is bonded between two or more functional groups on an organic aqueous species.

chemical affinity (A): The change in the heat produced by a reaction in a system divided by the extent of reaction. A is positive if a reaction is proceeding in a system but zero at equilibrium. It gives the extent a reaction is out of equilibrium in terms of the Gibbs energy per mole of the reaction.

chemical potential (μ_i): The infinitesimal change in Gibbs energy with an infinitesimal change in the number of moles of a component at constant temperature, pressure, and moles of all of the other components in the system.

chemical potential gradient: A gradient in the Gibbs energy (chemical potential) of a component with distance $(d\mu_i/dx)$.

chemoautotrophic: Obtaining metabolic energy from chemical reactions rather than from light or by incorporating preexisting organic material.

chlorofluorocarbons (CFCs): Short-chain alkanes that have Cl and F substituted for H.

chondritic meteorites or **chondrites:** A type of stony meteorite that has chondrules, 0.5- to 1-mm diameter spherical volumes consisting of olivine, pyroxene, or plagioclase crystals.

chondrules: Spherical volumes of 0.5- to 1-mm diameter consisting of olivine, pyroxene, or plagioclase crystals that are found in chondritic meteorites.

chromatographic: A process that separates different chemical constituents of a fluid plus solid mixture by passing the fluid through a reactive solid where constituents in the fluid can be absorbed or react with the solids.

citric acid metabolic cycle: Also known as the **tricarboxylic acid (TCA) cycle.** This is a series of enzyme-catalyzed chemical reactions that occur in cellular respiration. It is part of a metabolic pathway involved in the chemical conversion of carbohydrates, fats, and proteins into carbon dioxide and water with the generation of ATP from ADP. During reactions in the cycle, an acetyl group is oxidized to two CO_2 molecules. It is a cycle because the reacted molecules are regenerated so that they can react again.

Clapeyron equation: An equation from thermodynamics that relates the change in pressure and temperature of a phase transition in a closed system at equilibrium to its change in entropy, ΔS_R, over the change in volume, ΔV_R, given by $dP/dT = \Delta S_R/\Delta V_R$.

clathrate: A solid where a cage-like arrangement of H_2O molecules is stabilized by a central low molecular weight gas molecule (e.g., CH_4).

clinopyroxene: A calcium-rich monoclinic pyroxene that includes the minerals diopside ($CaMgSi_2O_6$), hedenbergite ($CaFeSi_2O_6$), and augite (Ca,Na,Mg,Fe^{2+}, $Mn,Fe^{+3},Al,Ti)_2(Si,Al)_2O_6$. Monoclinic structures have a single, two-fold axis of symmetry and a plane of symmetry normal to it.

closed system: A thermodynamic system in which processes taking place do not involve the transfer of mass across the system's boundaries.

closure temperature: With regard to radioactive isotopes, the temperature below which the parent and daughter isotopes behave as a closed system.

coefficient of isothermal compression (β): A dimensionless quantity that gives the relative amount a phase compresses with increasing pressure at constant temperature. $\beta = -1/V(\partial V/\partial P)_T$ where V is the molar volume of the phase.

coefficient of thermal expansion (α): A dimensionless quantity that gives the relative amount a phase expands on heating at constant pressure, $\alpha = 1/V(\partial V/\partial T)_P$, where V is the molar volume of the phase.

colloids: Small particles in the range of 10 to 10^{-3} μm in diameter.

common ion effect: The phenomena that the solubility of a mineral will decrease relative to its value in pure H_2O in a solution in which any ions that are common to ions in the solubility reaction for the mineral are present.

complexes: Aqueous complexes refer to species in solution that are formed by combining together simple ions. These include aqueous species like $MgCl^+$, $NaSO_4^-$, and $MgSO_4^o$.

component: A chemical formula used to specify the compositions of all of the phases in a chemical system in terms of endmembers. In thermodynamic models, the number of these endmember components must be the minimum necessary to describe the system.

compressibility: The change in volume of a substance with pressure. It is usually measured at constant temperature.

concordia: A curve on a plot of the ratios $^{206}Pb*/^{238}U$ versus $^{207}Pb*/^{235}U$, giving the development of these ratios as a function of time, where no Pb or U was been lost or added to the sample since it crystallized. The superscript * gives the radiogenic portions of the Pb isotope produced.

conductive heat flux: The rate of heat transfer per unit area because of vibrating atoms transferring kinetic energy to other atoms that they contact.

conservative elements: In seawater, elements that retain their relative proportion during evaporation and precipitation and show little gradient with depth in the ocean. They typically have residence times of greater than 10^6 years in the ocean.

conservative species: When referring to aqueous species, identifies those species whose concentration is unaffected by changes in P, T, or pH.

contact metamorphism: Metamorphism caused by the heat provided through contact with an intruding or extruding magma. The country rock will reach its highest temperature and therefore greatest degree of metamorphism at the magma contact.

continental crust: The earth's outer solid layer underlying the continental surface as opposed to the oceans. It ranges in thickness from 20 to 80 km and averages 35 km. It is less dense than the mantle below and thus "floats" on top of it.

continental red-beds: Sedimentary rocks, typically sandstones, deposited on the land surface that contain ferric minerals such as hematite or ferric anhydrite, giving them a red color. The ferric iron is stable relative to ferrous iron in the presence of atmospheric O_2.

convective heat flux: The rate of heat transfer caused by the movement of material under the force of gravity produced by density differences from differential heating.

coordinative reactions: Surface reactions that occur at the site of oxygens that are in coordination with cations within a solid.

cosmic microwave background: Radiation produced a few hundred thousand years after the Big Bang that created the universe. Soon after its formation, the universe was extremely hot and dense, made up of a plasma of photons, protons, and neutrons, as well as a number of unstable, heavier particles. The photons decoupled from the matter in the universe, creating a radiation in all directions that has since cooled to 2.7 K, the majority of which is presently in the microwave radiation band.

cotectic: Refers to the minimum melting temperature and its composition in a three or greater component system.

coulombic: Relating to the interaction of electrical charges.

coulometric titration: A measurement of the amount of matter transformed from one oxidation sate to another during a redox reaction by measuring the amount of electricity (in coulombs) consumed or produced. Seawater samples are acidified to transform the three inorganic carbon species to CO_2 gas. The gas produced can be reduced where the number of moles of inorganic carbon is proportional to the coulombs consumed.

crackle: Rock covered with a network of fine surface cracks.

crassulacean acid metabolism (CAM): An elaborate photosynthetic carbon fixation pathway that conserves water and is used by plants found in arid conditions, including cacti and pineapples.

critical point: The pressure and temperature where the liquid and gas of a substance lose their identities. For an upper critical point, pressures and temperatures greater than this point, in which only one phase exists, are sometimes referred to as supercritical.

crystal field stabilization: Describes the effect of the electrical field of neighboring ions on the energies of the valence orbitals of an ion in a crystal needed to obtain an accurate description of its bond energies. It is particularly important for transition metal compounds because of the different shapes and occupation of their d-orbitals.

cytosol: The internal fluid of the cell outside the organelles.

daltons: A unit of atomic mass equal to 1/12th of the mass of the neutral atom of ^{12}C or 1.660540×10^{-24} g.

Darcy's law: A relationship used to calculate the flux of fluid through a porous media given by $q = -K(dh/dx)$, where q is the fluid volume flux, K gives the hydraulic conductivity, and (dh/dx) denotes the change in hydraulic head with distance.

Debye-Hückel extended term: A term of the form bI that is sometimes added to an equation for the logarithm of the activity coefficient, γ, of an ion or electrolyte to account for the behavior of γ at high I, where I is the solution's ionic strength and b is a constant.

Debye-Hückel limiting law: A law that calculates the value of the activity coefficient of an electrolyte or charged species in the limit of infinite dilution in water.

decay equation: The equation $n = n^o e^{-\lambda t}$, which gives the decrease of n from a starting value of n^o, where λ is the decay constant and t denotes time.

delta notation: In isotopic analysis, the presentation of isotopic ratios in phase A as $\delta^{(heavy)}A = (R_A - R_{standard})/R_{standard}) \times 1000$, where the ratios, R, are the heavy to light isotope values of the subscripted phase.

denitrification: A process in which anaerobic bacteria break down nitrates, producing N_2.

deterministic chaos: Describes the behavior of certain nonlinear dynamical systems that under certain conditions exhibit behavior that is sensitive to initial conditions. An arbitrarily small perturbation of the trajectory may lead to significantly different future behavior.

Desulphovibrio: A sulfate-reducing bacteria.

devolatilization reaction: A reaction that releases a volatile (i.e., readily vaporized) species such as H_2O or CO_2.

diagenesis: The chemical and physical alterations that occur during burial of sediments to produce a solid rock.

diapir: A domal, mushroom-shaped body produced when a less dense plastic body intrudes into a greater density plastic body under the force of gravity.

dielectric constant: The permittivity of a substance relative to a vacuum denoted as ε, where $\varepsilon = \text{permittivity}_{sub}/\text{permittivity}_{vac}$.

differential: The infinitesimal change in a function with respect to a corresponding infinitesimal change in a variable.

differentiation: In geochemistry, a process by which a solution or mixture is separated into two or more chemically distinct parts. In mathematics, it is a process of obtaining the derivative of a function.

diffusion: The process by which material is transported as a function of time in a stationary medium because of a gradient in chemical potential. This chemical potential gradient is often also a concentration gradient so that diffusion is often characterized in terms of concentration gradients.

dimensional analysis: A procedure in which the units of quantities are used to guide the mathematical operations—that is, the units must be balanced across an equation. It is used to obtain conversions and to check work done with formulas and in simple problems.

dimorphism: In a solid, the property of crystallizing in two different crystal structures such as with $CaCO_3$ as calcite or aragonite.

dioxin: A chemical compound in which two phenol radials are linked through the oxygens of the phenols with the loss of the hydrogen atoms.

dipole: A species or system with oppositely charged ends (electric) or magnetic ends (magnetic).

discordia: A linear array of values that subtend a concordia curve, indicating that samples have lost Pb or gained U since they were formed.

distribution coefficient: A partition coefficient between two entities. For adsorption, the change in the mass of material adsorbed with change in solution concentration.

Dobson unit (DU): A unit of measure of the amount of a gas present through a thickness of the earth's atmosphere equal to 4.4615×10^{-4} mole m^{-2} that is equivalent to a 0.01-mm thickness at 0°C and 1 atm.

dosimeter: A device for measuring doses of x-rays or radioactivity.

double layer: On solids in solution, this refers to two electric layers of opposite charge, a surface layer that may contain coordinated ions and in particular hydroxyl groups. This layer is balanced by a diffuse layer of oppositely charged ions so that the double layer remains neutrally charged.

DSDP core: Deep Sea Drilling Program core obtained between 1968 and 1983 when 624 sites on the seafloor were drilled to obtain solid cores of the sediments and the rocks below. Results were published in volumes entitled "Initial Reports of the Deep Sea Drilling Project."

eclogite: A granular rock consisting of garnet and pyroxene with some jadeitic component, which forms during high temperature and pressure metamorphism.

edenite exchange: A coupled atomic substitution of an empty cubo-octahedral site plus Si^{4+} in a tetrahedral site exchanged for Na^+ in the cubo-octahedral site and Al^{3+} in a tetrahedral site in a crystal structure.

electric dipole: A neutrally charged entity but with oppositely charged ends.

electrical moment: The charge times the distance of separation in a dipole.

electronegative: The state in which an atom or radical attracts electrons from another atom or radical in the formation of an ionic bond.

electronegativity: The ability of an atom to attract electrons.

electrophoresis: The migration of ions or charged particles under the influence of an electric field. Particles with a positive charge move toward the cathode and negative to the anode.

electrostatic: In bonding, refers to a bond in which one element gives up an electron that then resides in another atom's electron orbital.

electrostatic potential: A scalar property that depends on what charges exist and how they are distributed. It therefore depends on location (x, y, z). It gives the energy available to do electrical work (potential) on a unit charge. At a point in which a +1 charge is attracted, the potential is negative. At a point in which the +1 charge is repelled, the potential is positive. The units of potential of electromotive force are given in CGS units as statvolts (statvolt = 299.792 volts).

element: A particular element is all of the atoms that have the same number of protons in their nuclei.

elementary reaction: A description of one of the events that occurs during a reaction at the molecular level.

elevation head: A measure of fluid energy given by its height in a gravitational field.

emulsion: An intimate mixture of two immiscible liquids.

endothermic: A process or reaction that absorbs heat.

energy: A measure of the quantity of work that a system is capable of doing. The total energy can be taken as that caused by mass + motion + potential in a force field.

enthalpy (H): Defined as the internal energy + pressure times volume; $H = U + PV$. dH is equivalent to the amount of heat added at equilibrium during a constant pressure process.

entropy (S): A measure of the randomness of the energy states in a system. It has a statistical basis where the entropy at equilibrium is the state of greatest probability. The change in entropy for a system in going from one equilibrium state to another is the heat absorbed from the surrounding for a reversible process divided by the absolute temperature at which the heat was absorbed (dq_{rev}/T).

epicontinental seas: Inland seas, bodies of seawater covering part of the interior of a continent; compare the Mediterranean Sea.

epilimnion: The top wind-mixed layer of a lake that can experience large seasonal temperature changes.

epithermal deposits: Ore deposits formed along fractures in rocks at shallow depths in the earth deposited from solutions at elevated temperatures.

epsilon value (ε_{Nd}): The ratio of $^{143}Nd/^{144}Nd$ in a rock at the time it was crystallized relative to the value in a model chondritic uniform reservoir (CHUR) given by a chondrite meteorite at the same time as calculated from:

$$\varepsilon_{Nd} = [\frac{(^{143}Nd_o/^{144}Nd)_{Rock}}{(^{143}Nd/^{144}Nd)_{CHUR}} - 1] \times 10^4$$

equation of state: An equation that describes the volume of a substance in terms of its temperature, pressure, and composition.

equilibrium: A state of a system where none of its properties changes with time and if a force operating on the system is changed slightly and then returned to its original value, the system will change but will return to its original state.

equilibrium constant: A constant equal to the product of all the nonstandard state Gibbs energy terms of the products raised to the power of their coefficients in a reaction divided by all of the nonstandard state Gibbs energy terms of the reactants raised to the power of their coefficients in the reaction at equilibrium. The natural logarithm of the equilibrium constant is equal to the negative of the standard state Gibbs energy of reaction divided by the gas constant and absolute temperature.

equilibrium isotope effects: Isotopic exchange that occurs when pressure, temperature, or composition conditions change while equilibrium is maintained between reactants and products in a reaction.

ethanol: Ethyl alcohol with the chemical formula C_2H_5OH.

eukaryotic: An organism containing membrane bound intracellular structures for specialized functions, such as a nucleus.

eutectic: The minimum melting temperature and its composition in a binary system in which complete solid solution does not occur.

eutrophic: A body of water with a high nutrient content that, therefore, produces abundant organic matter. This organic matter tends to deplete the water of oxygen as it decays.

euxinic: Restricted circulation. In euxinic basins, the water typically becomes anoxic at depth.

evaporation divide: A solution composition boundary encountered during the evaporation process. For solutions on one side of the composition boundary, the minerals crystallization sequence as evaporation proceeds will be different than solutions on the other side of the boundary.

evaporite mineral: A mineral precipitated when seawater is evaporated.

exchange reaction: With regard to minerals, a reaction written to indicate how different atoms can be substituted for each other in a mineral at particular atomic sites.

exothermic: A process or reaction that produces heat.

extensive variable: A variable that describes a thermodynamic system that depends on the total mass in the system such as the volume.

extent of reaction (ξ): A measure of the progress of a reaction where ξ is given by the final mass minus the original mass divided by the stoichiometric coefficient of the species in the reaction. ξ will be the same for all species in the reaction.

extrusive: A rock produced from magma that has either erupted at or flowed out on the earth's surface.

Faraday's constant (F): A unit of electrical charge equal to a mole (Avogadro's number) of charges whose value is 96,485.3 coulombs.

feldspar: Any of a group of minerals consisting of silica and aluminum oxide in a three-dimensional framework structure of the general formula, $NAl(Al,Si)_3O_8$ where the N site most commonly contains K, Na, and Ca, but Ba, Rb, Sr, Cs, or Li can also be present. They make up nearly 60% of the crust of the earth.

felsic: An adjective pertaining to rocks indicating a predominance of feldspars and a high concentration of Si (from feldspar + silica).

fermentation: Process in which a living cell is able to obtain energy through the breakdown of glucose and other simple sugar molecules in organic matter without any net oxidation of the organic molecule.

first-order transformation: In solids, a transformation in which there is a finite change in heat and volume at the transition pressure and temperature.

Fischer-Tropsch process: A catalytic mediated chemical reaction of $(2n + 1)\,H_2 + nCO \rightarrow C_nH_{(2n+2)} + nH_2O$) in which CO and H_2 are transformed into liquid hydrocarbons.

flocculate: To combine a number of minute suspended (colloidal) particles together in a clot-like mass due to electrostatic interactions of their surfaces that is then capable of settling out of solution.

fluorescence: The emission of light from atoms of a substance during a reaction.

flux: The amount of energy or mass that flows from one place to another in unit time. It is often given as a mass per unit cross-sectional area per unit time.

force: The rate of change of momentum with time. A vector quantity equal to mass times acceleration.

forces: Variables that describe how energy is transferred into a system. One speaks of a pressure force transferring pressure-volume work, a force of temperature transferring heat energy, and the force of gravity transferring gravitational potential energy.

Fourier's law: The law of heat conduction that states that the rate of heat transfer, dq/dt through a substance is proportional to the gradient of temperature, dT/dx and the area, A, perpendicular to that gradient, through which the heat is fluxing. It is given by $dq/dt = k\,A\,dT/dx$ where k is a constant for a given substance.

fractionation: A chemical or physical process other than radioactive decay that changes the amount of an element or isotope in a phase. In fractional crystallization, the magma phase changes composition as minerals crystallize. In isotope fractionation, certain mass-dependent reactions can cause slight variations in the concentration of stable isotopes in a phase as reaction occurs.

fractionation factor (α): A factor that relates the change in isotopic ratio for a molecule or phase during a chemical reaction or a physical change in state from A to B. $\alpha = R_A/R_B$, where R is the ratio of the heavy-to-light isotopes in the subscripted state.

free radicals: An atom or molecule with an unpaired electron in an outer more energetic orbital.

freeboard: When referring to continents, this is the area of the continental crust that is flooded with seawater.

Freundlich sorption: Sorption that obeys the equation $C_i' = K_f C_i^m$, where C_i' is the mass of i sorbed per mass of sample. C_i is the concentration of i in solution, and m is an exponent on C_i. K_f is termed the Freundlich adsorption coefficient, which is typically a constant as a function of concentration.

fugacity (*f*): A thermodynamic variable, for which the logarithm of its change is directly related to the change in Gibbs energy of a species or phase as given by $dG = RTd \ln f$. It is sometimes referred to as the "thermodynamic" pressure of a gas as $f = P$ for an ideal gas, where P is pressure.

fugacity coefficient (χ): A coefficient that when multiplied by the pressure gives the fugacity of a substance at a specified pressure, temperature, and composition.

fulvic acid: The organic material that dissolves and remains in an aqueous solution when a sample is first added to strong base and then is acidified. It is soluble at all *pH* values and is light yellow to yellow brown in color in aqueous solutions.

functional groups: In organic chemistry, a configuration of specifically bonded atoms in organic molecules that typically contains one or more elements other than hydrogen and carbon.

gamma rays: Electromagnetic energy consisting of photons between 1000 and 100,000 electron volts.

geocline: The packet of sediments and lava deposited on a subsiding continental margin.

GEOSECS: Geochemical Ocean Sections Study, a global survey in the Atlantic from July 1972 to May 1973, the Pacific from August 1973 to June 1974, and the Indian Ocean from December 1977 to March 1978. It made measurements of the chemical and isotopic composition of the ocean.

geothermal gradient: The change in temperature as a function of depth in the earth due to local production of heat and its transfer from below, usually given as °C per km.

Gibbs-Dühem equation: An equation, $0 = SdT - VdP + \sum_{i=1}^{k} n_i d\mu_i$, that states at equilibrium how, pressure, P, temperature, T, and the chemical potential of each component i, μ_i, must changes to retain equilibrium in the system.

Gibbs energy (*G*): A relative measure of a system's capacity to do work at constant pressure and temperature. $G \equiv U + PV - TS$, where U is internal energy, P is pressure, V is volume, T is temperature, and S is entropy.

global meteoric water line (GMWL): Because the evaporation of water from the oceans and the reverse process of condensation of atmospheric vapor fractionate oxygen and hydrogen isotopes based almost solely on temperature, there is a relation between $\delta^{18}O$ and δD of precipitation given by the line $\delta D = 8 \, \delta^{18}O + 10$.

glycolysis: The conversion of glucose to pyruvate in the cytosol of cells with the production of two molecules of ATP.

gram atoms of oxygen (g.a.o.): The number of moles of oxygen in a mole of a particular species or phase.

gram negative: Bacteria that do not retain crystal violet dye in the gram-staining technique. Staining is controlled by the structure of the cell wall surrounding the bacteria and influences which antibiotics can kill the bacteria.

granite: A plutonic igneous rock that is generally composed of alkali feldspar and has quartz (10% to 40%) with some plagioclase.

granulites: Rocks that have undergone metamorphism at very high temperatures (>650°C), destroying any OH-bearing minerals and producing an anhydrous granular textured rock consisting of quartz, feldspar, and the ferromagnesium minerals pyroxene and garnet. These high temperatures are encountered in the lower continental crust.

gravity: The force of attraction between two masses.

greenhouse gas: A gas in the atmosphere that traps the infrared background radiation from a planet's surface. On earth, H_2O and CO_2 are the two most important greenhouse gases.

groundmass: The fine material between conspicuous crystals in an igneous rock.

H_2O steam saturation pressure: A pressure defined for liquid H_2O where at the temperature specified the solution would be in equilibrium with its vapor. This is the lowest pressure of H_2O liquid stability below which H_2O liquid is transformed into vapor.

half-life: For a radioactive isotope, the length of time it takes for it to decay to one half its concentration.

half-reactions: When considering oxidation-reduction reactions, reactions that are written to produce free electrons. These must be combined with reactions that consumes these electron to produce an oxidation-reduction reaction.

halogens: Elements in group VIIA of the Periodic Table of elements that include F, Cl, Br, I, and At.

harzburgite: An ultramafic igneous rock that is dominantly olivine and pyroxene, making it a type of peridotite.

Hatch-Slack cycle: A metabolic cycle which temporarily fixes CO_2 into a four-carbon organic acid. This acid is next broken down to three-carbon organic acids plus CO_2. The CO_2 is then fixed into carbohydrates in a normal Calvin type cycle.

heat capacity: The amount of heat energy needed to raise the temperature of a specified amount of matter by 1°C. This heat can be determined at constant volume, C_v, or at constant pressure, C_P.

heat energy: The energy that flows from place to place because of a difference in temperature.

heat flow unit: A measure of the flux of heat through a cross-sectional area equal to $1.0 \ \mu cal \ cm^{-2} \ s^{-1}$.

heat flux: The amount of heat energy that is transferred across a cross-sectional area in a given amount of time. For a conductive heat flux, the heat is transferred because of a temperature gradient. For the earth, this is typically given in heat flow units (HFU), which are $10^{-6} \ cal \ cm^{-2} \ s^{-1}$.

Helmholtz free energy (A): A type of energy potential defined as $A \equiv U - TS$ where U is the internal energy, T is temperature in K, and S is the entropy. It gives the potential for work in isothermal processes. A decreases in any spontaneous reaction at fixed V and T.

Henry's law: A dilute solution law that states that the pressure of a gas above a solution is directly proportional to its concentration in the solution, which implies that at low concentrations the fugacity of a solute increases linearly with increased solute concentration in the solvent from an infinitely dilute state.

heterogeneous reaction: A reaction occurring between two or more different phases.

heterotrophic: Refers to an organism that must obtain organic molecules and energy from other organisms.

high field strength elements: Elements that form ions of valency greater than +2 that are not easily incorporated into the atomic sites present in common rock-forming silicates but are rather partitioned into a magma phase. These include Zr^{4+}, Hf^{4+}, Ta^{4+}, Nb^{5+}, Th^{4+}, U^{4+}, Mo^{6+}, W^{6+}, and the rare earth elements of +3 valence. They are typically found concentrated in pegmatite or in minerals precipitated from hydrothermal solutions.

homogeneous reactions: A reaction occurring in a single phase.

hot spots: Locations in the mantle where hot upwelling molten rocks are produced. Hot spots appear stationary with respect to the moving overlying lithospheric plate. The magma produced therefore forms chains of volcanoes on the lithospheric plate as a function of time.

humic acid: The organic material that goes into solution in a strong base but precipitates when the solution is acidified ($pH < 2$).

humic substances: The degradation-resistant residual material from plant decay. They are of high molecular weight and yellow to black in color.

humin: The insoluble black organic material remaining when a soil or rock sample is put in a strong base. It remains insoluble at all pH values.

hydraulic conductivity: A coefficient of proportionality relating the rate at which water can move through a permeable medium because of a gradient in the hydraulic head.

hydraulic head: A measure of fluid energy for a fluid volume in the crust given in units of length. It is equal to the height relative to a given depth that the fluid would rise in a pipe open to the earth's atmosphere.

hydrofracturing: The fracturing of rock by a fluid under pressure.

hydrologic cycle: Gives the movement as a function of time of H_2O as liquid, vapor, and ice on the earth between the various reservoirs where it resides.

hydrolysis: With regard to aqueous species, the incorporation of H_2O into the species. That is, the first hydrolysis reaction of Fe^{2+} is $Fe^{2+} + H_2O = Fe(OH)^+ + H^+$. The second is $Fe^{2+} + 2H_2O = Fe(OH)_2 + 2H^+$.

hydrophilic: Molecules that bond with and are therefore stabilized by polar water molecules.

hydrophobic: Uncharged nonpolar molecules that do not bond with polar water molecules.

hydrostatic equilibrium: A volume in which pressure on any surface through the volume has the same magnitude. A static fluid is in hydrostatic equilibrium as gravity and pressure are in balance.

hydrostatic pressure: The pressure exerted by a fluid. In a static fluid, this pressure is exerted by the overlying column of fluid and equals $\rho_f g h$, where ρ_f is the density of the fluid, g is the acceleration of gravity, and h is the height of interconnected fluid above the surface of interest.

hypolimnion: The cold bottom layer in a stratified lake that exists below the thermocline.

ideal gas: A state of real gases reached only at very low pressures and/or very high temperatures. This "idealized" gas is modeled as point masses that have perfectly elastic collisions with the walls of its confining container. Its volume (V) is related to temperature (T) and pressure (P) by $PV = nRT$, where n is the number of moles of gas and R is the gas constant.

ideal mixing: Mixing of two or more components if the species involved are completely indistinguishable. This implies there will be no heat or volume of mixing but the entropy increases and therefore the Gibbs energy decreases on mixing.

idioblastic: Refers to a crystal in a rock that is bounded by well-developed crystal faces (euhedral).

incongruently: For a mineral dissolving in a solution such as a magma, this refers to the situation where the elements added to the solution from the mineral are not in the same ratios as found in the mineral.

infinitely dilute: A state of a solute in H_2O where each solute species is surrounded by only H_2O molecules and therefore does not have interactions with other solute species in solution.

infrared: Electromagnetic radiation at wavelengths from 0.78 to about 300 μm. These wavelengths are greater, and therefore, lower energy than visible red light but at lower wavelength and therefore greater energy than microwaves. One feels this energy as heat.

inner sphere complex: For mineral surfaces, a molecular unit that attaches through bonding electrons directly to atoms on the mineral surface.

intensive variable: A variable that describes a thermodynamic system that does not depend on the mass of the system, such as pressure or temperature.

internal energy (U): The total energy of a body or phase. It is the sum of energy caused by mass, motion, and the presence of potential fields.

intrusive rock: A rock formed from magma that has solidified beneath the earth's surface.

invariant point: A point fixed in pressure, temperature, and composition space.

ion activity product, K_{iap}: The product of the measured molalities of each ion in solution raised to their stoichiometric coefficients for a mineral solubility reaction. If the solution is saturated with the mineral, K_{iap} will equal K_{sp}, the solubility product of the reaction.

ionic strength: A measure of the concentration of electrical forces in an aqueous solution given by $I = 0.5 \; \Sigma_i m_i z_i^2$, where m_i and z_i represent the molality and charge on the ith aqueous species with the summation over all i aqueous species in the solution.

ionosphere: A layer in the thermosphere of the earth's atmosphere beginning at 400 km in altitude and extending outward where ionization of species is prevalent.

irons: A term for meteorites made up predominantly of Fe–Ni alloys.

Irving-Williams series: The order of stability of the first d-block transition metals arranged in terms of bond strength.

isobaric: A process or reaction occurring at constant pressure.

isochron: A line connecting points of equal age as derived by a dating method. Often a straight line on a plot of the ratio of the radioactive isotope to a nonradioactive isotope (e.g., $^{87}Rb/^{86}Sr$) versus the ratio of the daughter isotope to the same nonradioactive isotope (e.g., $^{87}Sr/^{86}Sr$).

isolated system: A thermodynamic system whose boundaries are rigid walls, insulated from heat exchange with the surroundings, that does not allow the transfer of mass in or out and there are no changes in the potential fields acting on the system.

isomer: A molecule with the same composition but with a different arrangement of its atoms.

isostatic equilibrium: A state of matter in a gravitational field where the matter is neutrally buoyant.

isotope: One of a set of atoms having the same number of protons (i.e., the same element) but a different number of neutrons.

I/S transformation: A mineralogic transformation in mixed layer clays where smectite is transformed to illite with increasing degree of diagenesis. $Al^{3+} + K^+$ is substituted for Si^{4+} in the clay layers.

jadeite exchange: A substitution of Na^+ for Ca^{2+} in the M2 octahedral site coupled with Al^{+3} exchange for Mg^{2+} and/or Fe^{2+} in the M1 octahedral site in a crystal structure.

kerogen: The organic bituminous material in a sedimentary rock that is not soluble in organic solvents and is produced during diagenesis. Kerogen is a precursor to oil formation.

kimberlite: An alkalic peridotite porphyry rock containing abundant olivine and phlogopite phenocrysts that sometimes contains diamonds. It is found in "pipes," vertical columns of rock that have risen from a deep mantle source.

kinetic energy: The energy of a mass, m, due to its velocity, v, given by $1/2\ mv^2$.

kinetic isotopic effects: Isotopic exchange caused by concentration or temperature gradients where diffusional processes are important, such as in slow rates of chemical reactions. Under these conditions there are measurably different rates of reaction for heavy vs. light isotope species.

Kopp's rule: The heat capacity of a solid is the sum of contributions from each element or oxide in the compound.

lambda transformation: In minerals, a phase transformation that occurs over a region of pressure and temperature. It is characterized by having a heat capacity of transformation that is a continuous function of temperature across the transformation region.

Langmuir adsorption: Adsorption that obeys the equation $C_i/C_i' = C_i/\beta_2 + 1/\beta_1\beta_2$. C_i' is the mass of i adsorbed per mass of sample, C_i is the concentration of i in solution, and β_1 and β_2 are constants.

lanthanides: The 15 consecutive elements in the first row of the inner transition elements of the Periodic Table starting with lanthanum through lutetium (atomic numbers 57–71). Except for La, Gd, and Lu, they have identical 5 and 6 shells but different electron occupancy in the 4f subshell. Sometimes lanthanum is not included.

large ion lithophile (LIL) elements: Elements with large ionic radii and +1 or +2 valency that prefer a silicate phase rather than a metallic or sulfide phase. They do not fit regularly in the atomic sites found in common rock-forming silicates and include the elements Rb, Cs, Sr, and Ba.

law of mass action: The ratio of molar concentrations of the reactants to those of the products raised to the power equal to their coefficients in a reversible reaction at equilibrium at a given temperature and pressure remains constant.

Lennard-Jones 6-12 potential: A potential energy function, Γ_{LJ}, of interaction between molecules where the intermolecular forces are modeled with an attractive term and a repulsive term given by $\Gamma_{LJ} = 4\varepsilon\ [(\frac{\sigma}{r})^{12} - (\frac{\sigma}{r})^6]$ where σ is the distance of separation at zero potential energy, ε, is the negative of the energy at the equilibrium distance, and r is the distance of separation of the molecules. The Lennard-Jones potential is mildly attractive as two uncharged molecules or atoms approach from a distance (sixth power with distance) but is strongly repulsive when they approach too close (twelve power with distance).

lever rule: A rule for plotting a value along a line between two endmembers where the more of the endmember, the closer the value is to the endmember composition. This is much like a lever, where the position of a mass along a lever toward a particular end increases the force in that endmember's direction. It is calculated by taking difference between the two endmembers, A – B, and between the value, V, and one of the endmembers, A – V. The percentage of A relative to B in V is given by $(1 - [A - V]/[A - B]) \times 100$.

lherzolite: A plutonic ultramafic igneous rock with mafic minerals equal to or greater than 90 wt% with olivine, between 40 and 90 wt%, and having both orthopyroxene and clinopyroxene.

ligands: Molecules that donate electrons through two or more of their atoms.

lignin: A polysaccharide found in cell walls that provides rigidity to plants. It is resistant to biological decomposition.

lipids: A group of compounds, including fats, waxes, and related substances made up of C, H, and O that are soluble in organic solvents but not in water.

liquidus: The locus of temperature-composition points that represent the maximum solubility of a solid phase in the liquid phase.

lithosphere: The outer rigid layer of the earth above the seismic low-velocity zone, that is, asthenosphere. It contains the crust and some mantle with a thickness generally between 65 and 220 km. It is, however, much thinner at mid-oceanic ridges and under extensional continental areas.

lithostatic pressure: The pressure exerted by an overlying column of rock. It equals $\rho_R g h$, where ρ_R is the average density of the rock, g is the acceleration of gravity, and h is the height of rock above the surface of interest.

log normal: A statistical distribution that displays a normal distribution when plotted on a natural logarithmic scale. Examples of log normal distributions include the income of employed persons and species abundance in ecology.

lysocline: The depth in the ocean where calcium carbonate begins to dissolve, marking the depth at which the ocean becomes undersaturated with calcium carbonate.

mafic: A rock that contains a large concentration of ferromagnesian minerals. A ferromagnesian mineral is one that contains abundant Fe and/or Mg.

majorite: A Mg-rich garnet of endmember composition $Mg_3(Fe,Al,Si)_2(SiO_4)_3$.

mantle: A region in the earth extending from the base of the crust at a depth of between 6 and 80 km to the top of the core at 2885 km below the surface.

mantle array: Typically, the general inverse correlation between $^{87}Sr/^{86}Sr$ and $^{143}Nd/^{144}Nd$ for basalts derived from the mantle.

mantle wedge: A wedge-shaped part of the mantle between the crust and a downgoing oceanic lithosphere plate.

Margules formulation: A characterization of the natural logarithm of the activity coefficient of component 1, γ_1, in a binary solution in terms of powers of the mole fraction of component 2, X_2, given by $\ln \gamma_1 = \sum_k a_k X_2^k$ where a_k are constants.

Markov process: A simple stochastic (random) process in which the distribution of future states depends only on the present state and not on how it arrived in the present state.

mean electrostatic site potential: A measure of bond strengths to a particular site (e.g., O) in a solid by assuming each bonding atom is a point with a valence charge and summing all of the charges to separation distances to the site.

mean ionic activity ($a_\pm$): The average activity contributed by each cation and anion in a strong electrolyte given as $a_\pm = a_i^{1/v}$, where a_i is the activity of the electrolyte and v is the sum of cations plus anions in the electrolyte.

mechanical mixture: When mixing two or more pure components, a combination of the pure phases is the correct ratio for the mixture to be considered but before the random mixing that produces a single phase of constant composition occurs.

mesosphere: A layer in the earth's atmosphere between the stratosphere and thermosphere, extending from about 50 to 80 km in altitude.

metamorphic field gradient: The gradient in pressure and temperature of metamorphism determined from a suite of metamorphic rock samples in the field.

metastable: A state that is not at its lowest energy level but is kept from reacting to the lower energy state because of a large activation energy for possible reactions to the lower energy state.

meteorites: Masses of silicates or metals that fall on the earth's surface from outer space. Meteorites are divided into stones, irons, and stony irons. Stony meteorites include a class termed chondrites. Carbon-containing chondrites are termed carbonaceous chondrites.

Michaelis-Menton rate constant, K_m: A constant in an enzyme catalyzed reaction equation given by $v = (v_{max}[R])/(K_m + [R])$ where v denotes the velocity (rate) of the reaction and v_{max} stands for the maximum rate, a value which the reaction approaches at high reactant concentrations, $[R]$.

minerals: Naturally occurring inorganic solids that are crystalline with repeating arrangement of atoms. Natural amorphous solids with defined composition such as in opal are often included, although they are best referred to as mineraloids.

miscibility: The ability of various ratios of components to be homogeneously mixed in a single stable phase.

mitochondria: A remnant of ancient parasitic bacteria, which now form organelle found in cells. It is where most of the cell's ATP is produced.

mixing ratio: When considering concentrations of gases in the earth's atmosphere, this is the ratio of the volume of the gas to the volume of the atmosphere at the same temperature and pressure. Because the ideal gas law is assumed to hold, this is also the ratio of molecules of the gas of interest to total number of gas molecules present in the volume. The concentration given by the mixing ratio is, therefore, independent of pressure and temperature.

mode: A particular independent manifestation of a phenomena.

Moho: Short for the Mohorovicic seismic discontinuity that marks a seismic wave velocity discontinuity that separates the crust from the mantle of the earth at about 10 km below oceans but about 35 km below continents.

molality: The concentration of a solute in a solution in moles per kg of solvent.

molarity: The concentration of a solute in a solution in moles per liter of solution.

mole: Avogadro's number (6.02214×10^{23}) of entities such as atoms, molecules, or species and equals the mass in a molecular weight of a substance in grams.

mole-atom: A quantity based on the number of atoms in moles present in an amount of a species. For instance, 1 mole of H_2O would contain 3 mole-atoms.

mole fraction: The fraction of entities such as species given as moles relative to the total moles of entities considered.

monoclinic: Crystals where the cell dimensions are of unequal length with two of the axes at $90°$ to each other, and the third is not.

monodentate: With regard to surface species, a species bonded to one active surface site.

monosaccharide: A sugar that cannot be decomposed to a simpler sugar. Any of the simple sugars that serve as building blocks for carbohydrates. They are named for their number of carbon atoms (3 or more) ending in -ose (pentose). The carbon atoms are bonded to hydrogen atoms (–H), hydroxyl groups (–OH), and carbonyl groups (–C=O). Each carbon atom bonds a hydroxyl group except for the first and last. These can contain either a –CHO (aldehyde) or C=O (ketone) group. Their molecular configuration at carbon 2 creates two stereoisomers, the D and L configuration. All of these classifications can be combined, resulting in names like D-aldohexose or ketotriose. Most monosaccharides can form cyclic structures, which predominate in aqueous solution.

MORB: An abbreviation for mid-oceanic ridge basalt.

Nernst distribution coefficient: A constant as a function of concentration at constant pressure and temperature that gives the ratio of the molar concentration of a substance dissolved in two different phases.

Nernst equation: An equation that relates voltages of reactions to their activity products given by $Eh = Eh° + (RT/nF) \ln Q$ where Eh is the voltage difference of products to reactants, $Eh°$ denotes the standard state voltage difference, R stands for the gas constant, T represents temperature in Kelvin, F is Faraday's constant, n stands for the number of moles of electrons per mole of reaction, and Q is the activity product for the reaction.

neutrino: A stable subatomic particle of spin 1/2 with kinetic energy but with near-zero mass and zero electric charge given off with a positron during $β^+$ decay.

nitrification: The conversion of ammonia (NH_3) into oxidized nitrogen. NH_3 is first transformed to nitrite and then nitrate.

nitrogen-fixation: The conversion by bacteria or industrial processes of atmospheric N_2 to forms that can be used by plants, typically ammonia.

nonlinear systems: A group of interrelated, or interdependent, elements or components whose mathematical representation is not expressible as a linear function of its descriptors but has a more complex representation such as quadratic.

nonskeletal carbonate: Calcite, aragonite and dolomite that has precipitated directly out of seawater, and therefore, has not formed by growth of a living organism.

non–steady state: A system in which the rate or amount changes or can change with time.

normality: The number of moles of a particular species produced (generally H^+ or OH^-) per liter of solution.

normative hypersthene: The amount of hypersthene ($MgFeSi_2O_6$) that may not be observed in an igneous rock but would form if the melt that produced the rock cooled slowly enough so as to allow it to completely crystallize without groundmass. The normative minerals can be calculated from the bulk chemical composition using the **CIPW** procedure, which proportions the simple oxides of a chemical analysis of a rock's composition into mineral amounts that theoretically would have crystallized if the rock had cooled slowly (given by Cross, C. W., Iddings, J. P., Pirsson, L. V., and Washington, H. S., 1902, A quantitative chemico-mineralogical classification and nomenclature of igneous rocks. *Jour. of Geology*, v. 10, pp. 555–690.).

nuclear fission: The splitting of an atomic nucleus into two nearly equal parts. During the process, free neutrons, photons (usually gamma rays), antineutrinos, and beta particles can be released.

nucleation: The process by which a mass of a given concentration and structure appears and begins to gather or accrete new material.

nuclide: An atom distinguished by a specific number of both protons and neutrons; a single isotope.

oceanic anoxic events (OAEs): Time periods in the geologic past where the bottom waters of the ocean had very low oxygen content and were anoxic. This allowed organic-rich material to be preserved in sediments. OAEs are particularly common in the mid-Cretaceous period.

oceanic crust: The earth's outer solid layer generally underlying the ocean surface as opposed to the continents. Its thickness is about 6 km, and it is made up of basalt and its intrusive equivalent, gabbro. Generally, the sediments/sedimentary rocks on top of the basalt of 0 to 5 km thickness are also included.

Oddo-Harkins rule: An observation that with four exceptions the cosmic abundances of elements of even atomic number exceed those of adjacent elements of odd atomic number.

oil-field brine: Basinal formation waters containing a high concentration of calcium and sodium salts. They are typically encountered during deep drilling for oil and gas recovery.

oligotrophic: A body of water that has a deficiency of plant nutrients.

olivine: A silicate mineral consisting of SiO_4 tetrahedra linked by divalent atoms; commonly Mg and Fe^{2+} to produce forsterite (Mg_2SiO_4) and fayalite (Fe_2SiO_4) or a solid solution between the two. It is a common mineral in mafic and ultramafic rocks.

open system: A thermodynamic system where processes taking place can involve the transfer of mass across the system's boundaries to its surroundings.

orbital: For an atom or molecule, this gives the volume of space within which an electron would have a particular probability of occurring.

orbital subshell: The volume around an atom's nucleus that gives the most probable location of electrons.

organelle: A distinct subcellular structure with a specialized function (e.g., nucleus or mitochondrion).

orographic: Due to the effects of a mountain or mountain range.

orthopyroxene: A Mg plus Fe^{2+} pyroxene that has crystallized in an orthorombic structure. An orthorhombic structure has a unit cell with three mutually perpendicular axes of different lengths. It includes the minerals enstatite ($Mg_2Si_2O_6$) and ortho-ferrosilite ($Fe_2Si_2O_6$) and the solid solutions between them.

outer sphere complex: For mineral surfaces in aqueous solutions, a molecular unit that attaches through an intervening water molecule to atoms on the mineral surface.

oxidation: The loss of electrons from an atom of an element, as in Fe^{2+} being transformed to Fe^{3+}.

oxidative phosphorylation: The final metabolic pathway of cellular respiration after the citric acid cycle. Electrons are transferred from electron donors to electron acceptors down a chain of reactions that lead to the production of ATP.

oxygen fugacity: The fugacity of O_2 of a system. At low fugacities, this is equated with the partial pressure of O_2 in the system.

oxygen minimum zone: A zone in the ocean where oxygen levels are lower because of reactions with sinking organic matter that has been produced near the ocean surface.

Paleozoic: A geologic era from ~543 to 248 m.y. that followed the Proterozoic and preceded the Mesozoic.

partial differential: The instantaneous rate of change of a function with respect to one variable while holding all of the other variables that describe the function constant.

partial equilibrium: A situation in which some of the state variables that describe a system are in equilibrium, whereas others are not.

partial melt: A melt produced from melting only a part of the parent rock. Typically, its composition is not the same as the parent rock from which it was derived.

partial molar quantity: The amount a quantity changes in a solution with the addition of one mole of the substance to an infinite amount of the solution. For instance, the partial molar volume of substance A would be the volume change when 1 mole of A was added to an infinite amount of the specified solution.

partial pressure: A partial pressure of gas A, P_A, is defined as its mole fraction, X_A, times the total pressure of the system, P_{total}.

partition function: A function that indicates how the energies of molecules in a system are partitioned. The probability of seeing a molecule in an energy state, E_j, is $e^{-E_j/kT}/\sum(e^{-E_i/kT})$, where k represents Boltzmann's constant, T denotes temperature in K, and the summation in i is taken over all of the allowed energy states.

pelite or **pelitic rocks:** A clay-rich sedimentary rock, that is shale, composed mainly of aluminum rich phyllosilicates or their metamorphic equivalent.

peptides: Chains of amino acids up to about 50 units in length.

peridotite: A coarse-grained ultramafic plutonic rock consisting of at least 90% of olivine and pyroxene that can also include other mafic minerals such as amphiboles, and micas. It contains little or no feldspar, but can include spinels or garnets as Al-rich phases.

permeability: When considering fluids in the earth the capacity of a rock to allow fluid to be transported through it.

permittivity: The inverse of the constant needed to reduce the force of attraction between unit electrical charges at a unit distance due to having a substance between them. For a vacuum, this is designated as ε_o.

perovskite: A pseudocubic mineral that near the earth's surface has a chemical formula of $CaTiO_3$. At the conditions of the earth's lower mantle, below the 650-km seismic discontinuity, Mg-perovskite with complete substitution of Mg for Ca and Si for Ti to produce $MgSiO_3$ occurs. Ca-perovskite with complete substitution of Si for Ti ($CaSiO_3$), together with Mg-perovskite and wüstite, are considered to be the dominant minerals present in the lower mantle.

petroleum: Crude oil consisting of alkanes, cycloalkanes, and aromatic hydrocarbons.

pH: The negative of the base-10 logarithm of the hydrogen ion activity (H^+) in solution.

pH buffer: A solution whose pH changes little with the addition of H^+ or OH^-.

phase: A homogeneous portion of a system with uniform physical and chemical characteristics. A gas, liquid, and solid of a substance would be three distinct phases. A granite made up of quartz, orthoclase, oligioclase, and biotite would contain four phases.

phase rule: Gibbs' phase rule states f = c − p + 2, where f is the number of degrees of freedom, c is the number of components, and p is the number of phases in the system.

phenocrysts: Conspicuous crystals in a groundmass of smaller crystals or glass.

phenol: An aromatic carbon ring with an attached OH radical.

phenomenological approach: Uses individual observation and measurements as they are perceived to discern how nature operates.

phenomenological coefficients: In irreversible thermodynamics, the coefficients that when multiplied by their thermodynamic force give the irreversible flows of energy.

phosphorescence: Radiant energy loss by an atom after the energy source is removed, produced from energy release as electrons in excited states return to their ground state.

phosphorylation: The process in which phosphate is added to a molecule. For instance, phosphorylation adds inorganic phosphate to AMP and ADP to form the molecule ATP.

photic zone: A zone on the top of the ocean or a lake where there is enough sunlight for photosynthesis to occur, about 100 meters for calm clear water.

photochemical reactions: Reactions that require the energy of light to occur. Typically, the light energy allows reactions to occur at much lower temperatures than in light-free situations.

photolysis: The breaking of bonds in molecules by photons of the appropriate energy.

photosynthesis: The production of carbohydrate material from sunlight, CO_2, and H_2O, with the aid of chlorophyll. O_2 is produced as a by-product.

physical quenching: Rapid physical cooling.

placer deposits: Deposits of sand or gravel that contain nuggets or flakes of metals like Au and Pt or other valuable minerals.

plagioclase exchange: A substitution of Na^+ for Ca^{2+} in the feldspar N site coupled with Al^{3+} for Si^{4+} exchange in a tetrahedral site in a crystal structure.

Planck's constant (h): A constant equal to the energy, E, over its frequency, v, having a value of 6.626075×10^{-34} J s. It gives the quantification of energy into distinct levels, where $E = hv$.

planetesimals: A group of small bodies consisting of silicate minerals and metals embedded in ices of less than 100 km across that are thought to exist in the early solar system and to have developed into the present planets.

Poisson distribution: A frequency distribution, $f(n)$, for random events, n, of increasing unlikelihood per unit measurement, where $f(n) = a^n e^{-a}/n!$ and a is the average number in the distribution.

Poisson's equation: States $\nabla^2 \Psi = -\frac{\rho}{\varepsilon_o}$, where $\nabla^2 \Psi$ is shorthand for $(d^2\Psi/dx^2) + (d^2\Psi/dy^2) + (d^2\Psi/dz^2)$. Ψ, ε_o, and ρ denote the electrostatic potential, the fundamental constant that gives the permittivity of a vacuum and the charge density, respectively. The electrostatic potential is the energy necessary to bring a unit charge to a particular location (x,y,z) in an electric field from infinite distance.

polar bonds: Bonds between elements that show both some ionic and covalent character. They are sometimes referred to as intermediate bonds.

polarizability: The displacement of positive and negative charges in a neutral species to create oppositely charged centers.

polymer: A large molecule consisting of many units of carbon-containing compounds bonded together.

polymorphs: Solids of the same composition that have crystallized in two or more different structures.

polysaccharide: A compound that can be decomposed into many molecules of monosaccharides.

porphyry: A rock that contains conspicuous phenocrysts in a fine-grain microcrystalline groundmass.

positive feedback: The response of a system where a perturbation of the system amplifies the perturbation rather than coming to a steady state.

potential energy: The energy of a mass because of its position in a potential field such as height in a gravitational field.

ppm: Parts per million typically either by mass or volume.

ppb: Parts per billion typically either by mass or volume.

prebiotic: Time and state of the environment before the appearance of life on earth.

pressure: Force per unit area. In the earth, it is given by the average density of the material above (ρ) times the acceleration of gravity (g) times the depth (h), $\rho g h$.

pressure head: A measure of fluid energy for a fluid volume in the crust due to the pressure it is under given in units of length.

pressure solution: The increase in solubility of a solid surface due to a higher pressure on the surface than the average pressure on the solid.

primary magma: A liquid produced by melting of rocks that has not been changed by contamination or differentiation processes following its separation from its crystalline source. Typically, it is more felsic than its source.

processes: In a thermodynamic system the way one goes from one state to another. These can be constrained to be adiabatic, isobaric, or constant volume.

products: The results of a chemical reaction typically given by the right side of a written reaction.

prograde reactions: Metamorphic reactions that occur during the heating of a rock volume to its highest temperature.

prokaryotic: Bacteria without a nucleus or other membrane-bound intracellular structures for specialized functions.

proteins: Chains of amino acids of greater than about 50 units in length.

pycnocline: The zone where there is a pronounced density gradient with depth in stratified water.

pyrolite: A name given by A. E. Ringwood (1966) to a rock of a particular pyroxene-olivine mixture thought to represent the undifferentiated mantle rock before crustal material was extracted.

quadrupole: A neutrally charged entity with internal charge that can be thought of as resulting from two sets of opposite charged centers.

quantum mechanical tunneling: The event of small probability that a particle in a reaction can overcome (tunnel through) an energy barrier by borrowing the energy needed for a short time.

quasi steady-state approximation (QSSA): A way of simplifying the differential equations that describe kinetic systems. As used by biochemists, it assumes a steady-state concentration of reaction intermediates so that the change in their concentration with time can be taken as zero.

Quaternary: A geologic epoch that covers about the last 1.8 million years.

radical: A group of atoms bound together as a single unit and forming part of a larger molecule.

radioactive isotope: An isotope that can undergo spontaneous decay to another isotope.

Raman spectroscopy: Use of the Raman effect where a shift in wavelength of monochromatic light occurs in scattered light due to changes in the rotational and vibrational energy of molecules. The spectra produced can be used to characterize the material.

Raoult's law: A law based on a pure species at the pressure and temperature of interest standard state. The law states the thermodynamic activity of the species is equal to its mole fraction.

rare earth elements (REEs): A group of 17 chemical elements made up of scandium, yttrium, and the 15 consecutive elements in the first row of the inner transition elements of the Periodic Table starting with lanthanum and including lutetium (atomic numbers 57–71).

Rayleigh distillation: A separation process in which the products are immediately separated from the original mass, such as removal of raindrops from a cloud or crystals from magma. The composition of the remaining mass is a function of the fractionation factor of the separation and the amount of mass removed.

reactants: The elements in a reaction that are used to produce the products. They are typically given by the left side of a written reaction.

reaction intermediary: When considering elementary molecular scale reaction mechanisms, an intermediate reaction step between the initial reactant and final product.

reaction order: The order of a reaction is the sum of all of the exponents of the concentration terms in the rate equation. With rate = $[A]^2[B]$, the reaction order is 3.

recycled elements: Elements incorporated into organic matter in the ocean surface zone, usually with residence times of less than 10^6 years. Their concentration in seawater generally decreases as the surface of the ocean is approached.

red giant: A final state of stellar evolution in which a star of a few solar masses or less distends and glows red because of its cooler surface temperature. Our sun is predicted to go through this evolution in the future.

Redfield ratio: The ratio (106:16:1) of C to N to P found in phytoplankton. Most organisms contain nearly this ratio of these major nutrients.

reduction: The gain of electrons by an atom of an element, as in Fe^{3+} being transformed to Fe^{2+}.

reference state: In thermodynamics, the pressure, temperature, and state of matter where Gibbs energy, enthalpy, and entropy are taken as zero. By convention, this is 25°C and 1 bar for elements in their most stable state.

regional metamorphism: Metamorphism of rocks over a wide area caused by increasing temperatures with increasing burial of strata typically caused by regional tectonic forces.

regular solution: A solution in which the components disperse randomly so that ΔS of mixing is $-R \sum_i X_i \ln X_i$ where X_i stands for the mole fraction of component i and with ΔV of mixing zero, but energetics of interaction occur so that ΔH of mixing is finite.

respiration: A metabolic process that uses the energy from the oxidation of organic matter.

retrograde reactions: Metamorphic reactions that occur during the cooling of a rock volume from the highest temperature it has experienced. These reactions are facilitated by the introduction of H_2O into the volume.

retrograde solubility: A saturation reaction in which the saturated phase solubility increases with decreasing temperature rather than typical increase in solubility with increasing temperature.

reversible process or reaction: A process or reaction where no energy is dissipated, only transferred.

Reynolds number (Re): The ratio of inertia to viscous force in a fluid. The value of this dimensionless number determines whether flow is laminar or turbulent. For flow in a tabular fracture, $Re = 2v_{av}d/\eta$, where v_{av} is the average velocity, d is the width of the opening of the fracture, and η represents the viscosity of the fluid. If Re exceeds 2,300, tabular-fracture flow will be turbulent.

rheologic divisions: With reference to the earth, layers with different abilities to flow. A rigid layer termed the lithosphere extends to a depth of 70 to 150 km, depending on location. Below this is the asthenosphere, which is plastic. The mantle below the asthenosphere is rigid but on a long time scale can flow. The outer core convects, producing the magnetic field of the earth whereas the inner core is rigid solid material.

rhyolite: An extrusive igneous rock with phenocrysts of quartz and alkali feldspar that is chemically equivalent to granite.

sabkha: An arid coastal environment that can flood with seawater, producing evaporites and saline brines.

saturated: With respect to organic carbon-hydrogen compounds, refers to a chain of carbons with only single C—C bonds so the species is saturated with hydrogen.

saturation curve: The array of pressure-temperature points along which both a liquid and its vapor coexist. For H_2O saturation, these exist from the triple point at 0.01°C and 0.006 bar to the critical point at 374°C and 220 bars.

scavenged elements: When referring to elements in the ocean, those elements that are adsorbed onto small particles that are then incorporated into large particles and carried downward toward the ocean floor. These elements tend to have decreasing concentrations in seawater as a function of depth.

Schrödinger equation: A fundamental equation of quantum mechanics and wave mechanics that states $H\Psi = E\Psi$. That is the application of the hamiltonian operator, H, to the wavefunction, Ψ, of a system will give the energy of the system. It relates the wavefunction to the allowed energies of the wavefunction.

secondary porosity: Porosity developed in a rock after it first forms. This can be by mineral or glass dissolution in pore solutions, fracturing, or in the case of carbonates by dolomitization.

secular equilibrium: In kinetics, it refers to the state where for all the reaction intermediaries their rates of reaction are the same.

shear strain rate: The rate at which a set of parallel lines in a body have been displaced past one another by deformation.

shear stress: A component of stress that acts tangential to the stress surface.

shield: When referring to continents, an extensive area of old exposed basement rocks generally of Precambrian age.

skarn: Rocks that were originally nearly pure limestones or dolomites that have been altered by contact metamorphism to Ca-Mg-Fe silicates and aluminosilicates.

SMOW (standard mean ocean water): An isotopic standard used for oxygen and hydrogen analysis created in 1967 by Harmon Craig and other researchers from Scripps Institution of Oceanography by mixing distilled ocean waters collected from various locations around the world.

sodium-potassium pump: A transport protein in the plasma membrane that transports three Na^+ out of the cell and two K^+ into the cell against their concentration gradients. This causes the interior of the cell to become negatively charged.

soil air: The gas phase that is an integral part of most soils.

soil fabric: The arrangement, size, shape, and frequency of the individual soil components within the soil, generally considered from a genetic or functional viewpoint.

soil solution: The liquid phase found in soils.

solar constant: The mean amount of radiation from the sun per unit surface area normal to the rays at the mean earth–sun distance equal to 1,360 W m^{-2}.

solar mass: A mass equal to that of our Sun = 1.9891×10^{30} kg or 333,000 times the mass of the earth.

solar nebula: The cloud of gas and dust from which our sun and the planets in our solar system condensed 4.6 billion years ago.

solid solution: A homogeneous mixture of components in a mineral.

solidus: In a temperature versus composition diagram in a two-component system, the line that at any temperature gives the composition of the solid that is in equilibrium with the melt. In a three-component system, it is a surface and in a four-component system, a volume.

solubility product: For a solid, the product of the concentrations of the ions raised to the power of the number of ions in the solubility reaction of the solid.

solutes: Substances that are present in small quantities in a solution relative to the solvent. In ocean water substances like $CO_{2(aq)}$ and Na^+ are solutes in an H_2O solvent.

solution: A homogeneous mixture of two or more substances that has a continuously variable composition over some range of the compositions. Solutions have uniform properties throughout their volume.

solvent: A substance that is the dominant entity in a solution. In an aqueous solution, H_2O is the solvent.

solvus: In a temperature versus composition diagram, a line in a two-component system, a two-dimensional surface in a three-component system, or an $n-1$ surface in an n component system that separates a single solution domain from one having two or more distinct phases.

sorbate: A species that is incorporated into a substance through its surface or attached to that surface.

sorbent: A substance into which a species is incorporated through its surface or attached to its surface.

sorption isotherm: The equilibrium relationship at constant pressure, temperature, and pH between the concentration of a substance in a fluid and that incorporated on or into a solid surface displayed graphically by a curve, the so-called sorption isotherm.

spallation: An atomic nucleus splitting into three or more fragments due to interactions with energetic particles.

species: A chemical entity in a solution, for example, molecule, ion.

stable isotope: An isotope of an element that does not undergo spontaneous radioactive decay.

standard state: The thermodynamic state of a system for a species or compound given by the pressure, temperature, concentration, and state of matter. The standard state is a state between the reference state and the state of interest that is helpful for analysis. A common standard state for gases is the pure gas at 1 bar but at the temperature of interest. For minerals, a state of the pure phase at the pressure and temperature of interest is often used. For aqueous species, a hypothetical state of a 1 molal solution with the species energetics acting as though they were in an infinite volume of H_2O at the pressure and temperature of interest is often used.

starch: A polysaccharide that stores energy in plants with a chemical formula of $(C_6H_{10}O_5)_n$, where n is a positive integer.

state of a system: A set of variables that completely defines the system.

state variables: Independent variables that define the state of a system, for example, moles of each phase, temperature, pressure. The values of these variables depend on the state of the system and not how the system achieved the state.

statistical mechanics: A discipline involving the study of the physical properties and energy levels of a large number of particles based on probability distributions.

steady state: A state in which material fluxing into a system is balanced by a flux out so that the amount stays constant. This state is not necessarily an equilibrium state.

steam saturation pressure: In aqueous solutions, the pressure at which both steam and liquid water can coexist. This increases with temperature along the H_2O pressure-temperature saturation curve.

stereoisomer: A chemical compound that is identical in composition and atomic bonding but has a different arrangement of atoms than another compound.

Stern layer: In models of the solid–water interface, an additional compact surface layer present in triple-layer models, containing ions other than adsorbed H^+ and OH^- that are considered to be bound at the surface of the solid with intervening water molecules.

strangeness: A hadronic quantum number equal to hypercharge minus baryon number, indicating the possible transformations of an elementary particle on strong interaction with another elementary particle. Hadrons are elementary particles that are subject to both the strong and the weak force. They are composed of even smaller entities, termed quarks.

stratosphere: The layer around the earth at greater elevations than the troposphere where temperature increases with height. It extends from 10–16 to 50 km, depending on location and time of year.

stochastic approach: A model of a phenomena that is based on probability.

stock: A plutonic igneous intrusion with steep vertical contacts that covers less than 40 square miles in area.

stockwork: Ore deposit consisting of a mass of veinlets where the deposit is nearly equal in length, width, and height as opposed to having a tabular or sheet form.

stones: A term for a group of meteorites consisting primarily of Mg–Fe silicate minerals and plagioclase.

stony-irons: A term for a group of meteorites consisting of subequal amounts of silicate minerals and Fe–Ni alloys.

strange attractor: A set of values to which a system evolves, that is attracted to, after a long enough time but never reaches. Approach to the attractor produces a kind of regular pattern that never exactly recurs. For instance, with a normal pendulum, it does not matter where you release it from, it will always come to rest in the same position. This rest state is the attractor for the system. A strange attractor occurs in a dissipative (nonequilibrium) dynamical system. This attractor has a noninteger dimension.

strong acid: A substance that when added to water disassociates nearly completely, giving almost its total H^+ to solution at any pH.

strong electrolytes: Substances that dissolve by dissociating almost completely to ions in H_2O.

subduction zones: Planar dipping zones in the earth where lithospheric oceanic plates are sinking into the mantle below an overriding lithospheric plate.

sublimation: The process by which a substance passes directly from a solid to a gaseous state or vice versa.

substrate: A reactant in an enzyme catalyzed reaction. Substrates bind on an enzyme's active site. The enzyme then catalyzes a chemical change in the substrate.

supercritical fluid: A substance above its critical pressure and temperature that has low viscosity.

supergene: Applies to ores that have been precipitated from descending meteoric water at temperature less than approximately 50°C.

supernovae: The explosive death of a particularly massive star, which disperses its energy, both light and mass into the universe.

supported concentration: In isotopic studies, it refers to the steady-state concentration of an isotope in a decay-chain produced by continued decay of a parent isotope minus the amount lost by decay of the isotope produced.

surface energy: The energy of a surface due to the discontinuity in the environments between the two phases present. This energy is positive, making particles with large surface area unstable relative to particles with low surface area.

surface sorption: For a solid, the adherence of species to its surface when placed in a gas or liquid.

suspended load: The material carried by a body of water that is not dissolved in the water or rolled along the bottom but rather is carried as particulate matter kept in suspension.

system: Any region of the universe large or small that is considered for analysis. It is defined by its chemical composition, the physical state of the phases in it, and the forces and fields operating across its boundaries.

Taylor's series: A series expansion of a function $f(x)$ given as a power series in $(x - a)$ of $f(x) = f(a) + (f'(a)/1!)(x - a) + (f''(a)/2!)(x - a)^2 + (f'''(a)/3!)(x - a)^3 \ldots$, where $f^n(a)$ is the nth derivative of the function evaluated at a.

temperature: A scale that indicates the relative amount of internal kinetic energy a mass possesses. Differences in temperature allow energy as heat to be transferred from the higher temperature to the lower temperature mass.

tetrahedron: A polyhedron that has four triangular faces.

thermal conductivity: The ability of a substance to conduct heat. It is the amount of heat transferred per unit area in a unit amount of time per gradient in temperature. In SI units, it is given as $W\ m^{-1}\ K^{-1}$.

thermochronology: Considers the rates and therefore timing responsible for cooling of rocks during exhumation by determining differences in noble gas and fission track closure temperatures in minerals.

thermocline: The zone where there is a pronounced temperature gradient with depth in a stratified lake or seawater.

thermodynamics: A field of study that uses models based on the transfer of energy as heat, work, mass, and changes in position in energy fields to understand under what conditions equilibrium is achieved for a reaction and determining the direction of reactions that are not at equilibrium.

thermosphere: The outer layer in the earth's atmosphere where temperature increases with altitude due to the heat produced from the sun's energy splitting molecules into ions. It extends from about 80 to 700 km in altitude.

tholeiitic basalt: A silica-rich basalt containing mostly clinopyroxene, and calcium-rich plagioclase with minor iron-titanium oxide. It can also contain low-calcium pyroxene. Tholeiite is quartz saturated (normative quartz) with quartz at times present in the groundmass. Xenocrysts of olivine can be present but nepheline is absent.

tortuosity: A nondimensional parameter accounting for the orientation of a fracture relative to the flow path. For single-fracture flow, it is 1, whereas for grain boundary flow, it is between 0.5 and 0.8 (see Bear, 1975, referenced in Chapter 13).

total alkalinity: A measure of the concentration of bases in solution and therefore the power of the solution to neutralize hydrogen ions (H^+).

total differential: The sum of the infinitesimal change of a function with respect to each variable that describes the function while holding all the other variables constant times the infinitesimal change of the variable. The total differential of $y(x,z,t)$, $dy = (\partial y/\partial x)_{z,t}\, dx + (\partial y/\partial z)_{x,t}\, dz + (\partial y/\partial t)_{x,z}\, dt$.

total dissolved solids (TDS): The total mass of material (mg) dissolved in a liter of water. It is typically determined by measuring the evaporated residue of a water sample.

trace elements: Elements that are not essential constituents of common rock-forming minerals but are found in small quantities in its structure, typically less than 1 wt% of the mineral.

transition elements/metals: Elements that have partially filled d subshells that include all the elements in the Periodic Table of the elements in groups IIIB, IVB, VB, VIB, VIIB, VIIIB, IB, and IIB.

transitory state: In kinetic theory, a molecular state of elevated energy that a reactant must pass through to proceed to the product species.

triple-layer model: When considering mineral surfaces in solution, a three-layered model of surface charge where a layer of charge develops as a result of solvated ions between the charged mineral surface layer and the diffuse layer of charge of opposite sign that develops in the solution.

trondhjemite: A white granitic rock whose type locality is from the Trondheim area of Norway. It is composed of sodic plagioclase, typically oligoclase, with quartz, sparse biotite, and little or no alkali feldspar.

trophic effect: The concentration of an element or isotope produced as one organism eats another organisms as one travels up the food chain.

troposphere: The layer in the atmosphere in contact with the earth's surface where temperature decreases with height. It extends from 0 to 10–16 km in elevation depending on location and time of year. It is thickest at the equator and thinnest at the poles.

tschermaks exchange: A coupled substitution of both octahedral and tetrahedral Al for an octahedral Mg and tetrahedral Si in a crystal structure.

ultramafic: A type of rock that contains more than 90% of ferromagnesium, that is, Fe + Mg containing minerals.

ultraviolet: A type of electromagnetic radiation between 0.01 and 0.39 μm in wavelength. It is shorter in wavelength than the violet part of the visible spectrum and longer than x-rays.

valence: The number of electrons an atom will give up or share to form bonds with other atoms. A negative valence indicates that the atom wants to obtain rather than give up electrons. An atom can have more than one valence depending on its chemical environment.

vector: A mathematical quantity with both magnitude and direction.

velocity: The differential of position with respect to time.

vesicle: Small, membrane-bounded, spherical organelle in the cytosol of a cell.

virial equation: An equation of the state of a substance which characterizes the property in increasing powers of a compositional variable of the substance, for example, mole fraction. The squared term accounts for interactions between pairs of molecules, the third power to triplet interactions, et cetera. Accuracy can be increased by considering continuing higher order terms.

viscosity: The tangential force, F, required to move a planar area along its plane when separated from a similar surface by a substance. Reported in poise (CGS units) or poiseuille (SI units). It is given by $F = \eta A(dv_x/dz)$, where η is the coefficient of viscosity, A is the area of the plates, v_x is the velocity of one plate relative to the other in the direction of the force, and z is the distance of separation of the plates.

viscous heating: Heat produced by tangential forces deforming a body. This is what causes the heating at the bend in a metal coat hanger that has been deformed by a back and forth motion.

viscous pressure: A tangential pressure exerted by a viscous fluid perpendicular to its flow direction.

volcanic arcs: A curved chain of volcanic mountains in map view produced when subducted lithosphere releases fluids that cause the production of magma above as they ascend into the hotter mantle.

water electrode: An electrical conductor where H_2O is oxidized to produce O_2, H^+, and electrons.

water table: A surface in the earth above which air can exist in pores and below which all the pores are filled with water.

weak acid: A substance that when added to water only partially disassociates to give H^+ to the solution with a significant amount of the H^+ retained in an undisassociated species.

work: The change in energy of a mass by displacing it with a force. If the force is pressure, the work is termed pressure-volume work. With a gravitational force it is displacement in a gravitational field.

wüstite: A mineral of composition FeO with a cubic rock-salt structure. Magnesiowüstite (Fe,Mg)O is thought to be common in the earth's lower mantle.

xenoblastic: A metamorphic rock texture in which crystals display poor facial development against other crystal types in the rock.

x-ray fluorescence: A type of chemical analysis for elements heavier than Na where a sample is irradiated with a beam of x-rays (electromagnetic energy ranging in wave length from 3×10^{-10} to 3×10^{-13} m). The wavelengths given off are separated and their intensities measured. The concentration of an element present is related to the intensity of a particular wavelength.

x-rays: Electromagnetic energy ranging in wavelength from 3×10^{-10} to 3×10^{-13} m.

Index

Note: An *f* following a page number indicates a figure, and a *t* following a page number indicates tabular material.